PARTICLES, FIELDS, AND GRAVITATION

Ryszard Rączka (1931–1996)

This volume is dedicated to the memory of our great friend Ryszard Rączka who died on 26 August 1996. Ryszard Rączka was a great theoretical physicist and one of the most prominent Polish scientists. His research was devoted to high energy particle theory and mathematical physics. He became an expert in group theory and its physical applications. His papers concerning this subject were widely recognized as landmarks. His intuition, deep insight in physics and mathematical talent allowed him to make also significant contributions to the theory of elementary particles. Due to the typical for him independence of views Ryszard was always ready to subject to the penetrating analysis any commonly accepted opinion concerning basic questions of physics. His criticism was always very valuable and allowed many of us to gain a deeper understanding of the advanced problems of particle physics.

Ryszard Rączka was good, deeply honest and modest man. His warm, friendly attitude towards other people gained him many friends all over the world; some of them participated in this conference. He was always sensitive to people's needs, ready to support anybody and to find time for him despite many engagements. In this respect we, physicists from Łódź, owe him very much. We will miss Ryszard's optimism and enthusiasm which were a constant source of encouragement for many of us. We will all remember him.

In the name of Ryszard's friends
Piotr Kosiński and Jakub Rembieliński

PARTICLES, FIELDS, AND GRAVITATION

Lodz, Poland   April 1998

EDITOR
Jakub Rembieliński
University of Lodz, Poland

American Institute of Physics

AIP CONFERENCE
PROCEEDINGS 453

Woodbury, New York

Editor:

Jakub Rembieliński
Department of Theoretical Physics
University of Lodz
ul. Pomorska 149/153
90-236 Lodz
POLAND

E-mail: jaremb@mvii.uni.lodz.pl

Authorization to photocopy items for internal or personal use, beyond the free copying permitted under the 1978 U.S. Copyright Law (see statement below), is granted by the American Institute of Physics for users registered with the Copyright Clearance Center (CCC) Transactional Reporting Service, provided that the base fee of $15.00 per copy is paid directly to CCC, 222 Rosewood Drive, Danvers, MA 01923. For those organizations that have been granted a photocopy license by CCC, a separate system of payment has been arranged. The fee code for users of the Transactional Reporting Service is: 1-56396-837-1/ 98 /$15.00.

© 1998 American Institute of Physics

Individual readers of this volume and nonprofit libraries, acting for them, are permitted to make fair use of the material in it, such as copying an article for use in teaching or research. Permission is granted to quote from this volume in scientific work with the customary acknowledgment of the source. To reprint a figure, table, or other excerpt requires the consent of one of the original authors and notification to AIP. Republication or systematic or multiple reproduction of any material in this volume is permitted only under license from AIP. Address inquiries to Office of Rights and Permissions, 500 Sunnyside Boulevard, Woodbury, NY 11797-2999; phone: 516-576-2268; fax: 516-576-2499; e-mail: rights@aip.org.

L.C. Catalog Card No. 98-88573
ISBN 1-56396-837-1
ISSN 0094-243X
DOE CONF- 980409

Printed in the United States of America

CONTENTS

Preface ix
Committees xi
In Memory of Ryszard Rączka xiii
    J. Werle

QUANTUM DEFORMATIONS AND NON-COMMUTATIVE GEOMETRY

Line Bundles on Quantum Spheres 3
    T. Brzeziński and S. Majid
Action of a Finite Quantum Group on the Algebra of
Complex $N \times N$ Matrices 9
    R. Coquereaux and G. E. Schieber
Solutions of q-Deformed Equations with Quantum Conformal Symmetry 24
    V. K. Dobrev, B. S. Kostadinov, and S. T. Petrov
Reality Conditions on Quantum Euclidean Planes 39
    G. Fiore and J. Madore
Two Disjoint Aspects of the Deformation Programme: Quantizing
Nambu Mechanics; Singleton Physics 49
    M. Flato
Quantum Deformations of Conformal Algebras with Mass-Like
Deformation Parameters 53
    A. Frydryszak, J. Lukierski, P. Minnaert, and M. Mozrzymas
Differential Structures on $E_\kappa(2)$ from Contraction 67
    P. Kosiński and P. Maślanka
Twisting of Quantum Groups and Integrable Models 75
    P. P. Kulish
Remarks on Quantum Statistics 86
    W. Marcinek
Quantum Minkowski Spaces 97
    P. Podleś
Deformation Quantization: Twenty Years After 107
    D. Sternheimer
q-Derivatives, Quantization Methods, and q-Algebras 146
    R. Twarock
Non-Commutative Space-Time and Quantum Groups 155
    J. Wess

QUANTUM MECHANICS, SOLVABLE AND QUASI-SOLVABLE PROBLEMS

The Complex Pendulum 167
    C. M. Bender
Quasi Exactly Solvable Operators and Abstract Associative Algebras 177
    Y. Brihaye and P. Kosinski

On Conformal Reflections in Compactified Phase Space 186
    P. Budinich
Localization Problem in Quantum Mechanics and Preferred Frame 199
    P. Caban and J. Rembieliński
Solvable Potentials, Non-Linear Algebras, and Associated Coherent States 209
    F. Cannata, G. Junker, and J. Trost
On the Inverse Variational Problem in Classical Mechanics 219
    J. Cisło, J. T. Łopuszański, and P. C. Stichel
Change of Variables, Fundamental Solutions, and Borel Resummation 226
    S. Giller and P. Milczarski
Einstein–Podolsky–Rosen Experiment from Noncommutative Quantum Gravity 234
    M. Heller and W. Sasin
Wigner Quantization Problem for External Forces 242
    E. Kapuścik and A. Horzela
Coherent States for a Quantum Spin 1/2 Particle on a Sphere 249
    K. Kowalski, J. Rembieliński, and L. C. Papaloucas
On Time-Dependent Quasi-Exactly Solvable Models 257
    D. Mayer, A. Ushveridze, and Z. Walczak
Discrete Symmetries and Supersymmetries of Quantum-Mechanical Systems 268
    J. Niederle
Foundations of Quantum Theory and Thermodynamics 276
    V. Olkhov
Integrability of $N=3$ Supersymmetric KdV Equation 285
    Z. Popowicz

DEVELOPMENTS IN QUANTUM AND TOPOLOGICAL FIELD THEORIES

Curved Domain Wall/Vortex Solutions in Relativistic Field Theories 293
    H. Arodź
Three-Dimensional Topological Quantum Field Theory of Witten Type 303
    M. Bakalarska and B. Broda
On Notivarg Propagator 312
    M. Bakalarska and W. Tybor
Lorentz Covariance, Higher-Spin Superspaces and Self-Duality 317
    C. Devchand and J. Nuyts
Deformation Stability of BRST-Quantization 324
    M. Dütsch and K. Fredenhagen
Dynamical Symmetry Breaking in the Nambu–Jona–Lasinio Model under the Influence of External Electromagnetic and Gravitational Fields 334
    E. Elizade and Y. I. Shil'nov
On the Interplay between Perturbative and Nonperturbative QCD 344
    J. Fischer and I. Vrkoč

Chiral Symmetry Breaking in Nambu–Jona–Lasinio Model in
External Constant Electromagnetic Field................................. 349
    E. V. Gorbar
On Quantization of Field Theories in Polymomentum Variables............. 356
    I. V. Kanatchikov
Witten's Integral and the Kontsevich Integral........................... 368
    L. H. Kauffman
Gauge Invariants and Bosonization....................................... 382
    J. Kijowski, G. Rudolph, and M. Rudolph
Conformal Symmetry and Unification...................................... 394
    M. Pawłowski
The Zwanziger Action for Electromagnetodynamics Revisited............... 404
    D. Sorokin

GRAVITATION AND GEOMETRICAL METHODS IN PHYSICS

Non-Generic Symmetries on Extended Taub–NUT Metric...................... 413
    D. Baleanu
Some Properties of Light Propagation in Relativity...................... 421
    S. L. Bażański
On Some Class of Gravitational Lagrangians.............................. 431
    A. Borowiec
Remarks on Symplectic Connections....................................... 437
    M. Cahen
Two-Dimensional Dilaton Gravity... 442
    M. Cavaglià
Integrable Hierarchies in Twistor Theory................................ 449
    M. Dunajski
Nonsingular Cosmological Black Hole..................................... 460
    I. Dymnikova and B. Sołtysek
Relativistic Particle in Singular Gravitational Field................... 472
    G. Jorjadze and W. Piechocki
Black Holes with Yang–Mills Hair.. 478
    B. Kleinhaus, J. Kunz, A. Sood, and M. Wirschins
Gravitational Instantons with Source.................................... 488
    K. B. Marathe
High Current in General Relativity...................................... 498
    B. E. Meierovich
Reflections and Spinors on Manifolds.................................... 518
    A. Trautman
Exact Solutions in Locally Anisotropic Gravity and Strings.............. 528
    S. I. Vacaru
Statistical Mechanics of Black Holes in Induced Gravity................. 538
    A. Zelnikov

List of Participants.. 549
Author Index.. 559

Preface

The conference "Particles, Fields and Gravitation" was held from 15 to 19 April of 1998 in Łódź, the second biggest city in Poland and one of the most important academic centers in the country. It was dedicated to the memory of Ryszard Rączka (1931-1996), the prominent Polish physicist. Lectures took place in lecture halls of the Faculty of Physics and Chemistry of the Łódź University. The conference was attended by more than one hundred scientists from all over the world, namely from: Algeria, Belarus, Belgium, Bulgaria, Canada, Czech Republic, Estonia, France, Georgia, Germany, Great Britain, Italy, Japan, Mexico, Poland, Romania, Russia, Spain, Turkey, Ukraine, USA, and Vatican.

During five working days of the conference, participants presented 80 talks. Plenary lectures as well as contributed talks provided a comprehensive coverage of the main aspects of contemporary theoretical and mathematical physics, such as: quantum deformations and noncommutative geometry, quantum mechanics, quantum and topological field theory, solvable and quasisolvable models, modern gravitation theory and geometrical methods in physics. During the conference many informal meetings and discussions took place, reflecting the interest and excitement related to this field of research.

The conference volume contains 54 contributed papers, and the memorial talk by the late Professor Józef Werle from Warsaw University, teacher and friend of Ryszard Rączka, is included in the front of the book.

Organizers are very grateful to all the Sponsors of the conference for their financial support. Particular thanks are due to the Vice-Rector of the University of Łódź, Prof. Wanda Krajewska for her help in organizing this conference. We are obliged to the family of Ryszard Rączka for attending the conference. Last but not least, we thank all the Members of the Scientific Committee and participants for their collaboration.

By a tragic coincidence two participants, Józef Werle and Stanisław Zakrzewski, have died shortly after the conference.

Professor **Józef Werle** (1923-1998), an eminent scientist, teacher and animator of the scientific life was a senior of theoretical physics in Poland.

Professor **Stanisław Zakrzewski** (1951-1998), our good friend, was a widely recognized specialist in quantum group and Poisson structures theory.

We miss them.

Jakub Rembielinski
Łódź, August, 1998

INTERNATIONAL CONFERENCE
"PARTICLES, FIELDS AND GRAVIATION"

Dedicated to the Memory of Ryszard Rączka (1931-1990)

Łódź, Poland, April 15–19, 1998

Scientific Committee

A. Ashtekar (Penn State Univ.)
P. Budinich (SISSA, Trieste)
H.-D. Doebner (ASI, Clausthal)
M. Flato (Univ. Bourgogne, Dijon)
K. Gawędzki (IHES, Bures-sur-Yvette)
E. Kapuścik (IFJ, Cracow)
J. Kijowski (Warsaw Univ.)
P. Kosiński (Łódź Univ.)
J. Lukierski (Wrocław Univ.)
Y. Ne'eman (Tel-Aviv Univ.)

J. Niederle (AS Czech Rep., Prague)
J. Nuyts (Mons Univ.)
R. Penrose (Oxford Univ.)
L. O'Raifeartaigh (IAS, Dublin)
J. Rembieliński (Łódź Univ.)
D. V. Shirkov (JINR, Dubna)
A. Trautman (Warsaw Univ.)
J. Wess (Munich Univ.)
S. L. Woronowicz (Warsaw Univ.)

Organizing Committee

M. Kibler (Univ. Lyon I)
J. Kijowski (Warsaw Univ.)
P. Kosiński (Łódź Univ.)

M. Pawłowski (IPJ., Warsaw)
J. Rembieliński (Łódź Univ.)

Scientific Secretary

Kordian A. Smoliński (Łódź Univ.)

Supporting Organizations

European Commission Directorate General XII
Stiftung für deutsch-polnische Zusammenarbeit
State Committee for Scientific Research and Ministry of National Education
City of Łódź
University of Łódź

In Memory of Ryszard Rączka

Ryszard Rączka was born in October 1931 in Warsaw to a poor craftsman family of devout Catholics. He was not yet eight years old when the Second World War broke out. During the war the living conditions of the family were gradually deteriorating. The food rations were insufficient so the members of the family were forced to undertake expeditions to sometimes quite remote villages to buy extra food from peasants or to barter it for some attractive goods brought from Warsaw. This was of course a dangerous procedure strictly forbidden by the occupants who wanted to have a monopoly for food supply and distribution. In the beginning Ryszard took part in these expeditions as a companion of adults of his family but at the age of roughly 11-12 years he was undertaking such ventures by himself. This must have developed in him the ability of taking independent and sometimes even risky decisions. He preserved these precious abilities to the end of his life.

In his youth, in the secondary school, Ryszard was a very enthusiastic football player and was dreaming about a career in this sport. However, a contusion of his leg prevented him from realizing this dream, and so by a misfortune which turned only to the good he became a physicist. However, for all his life he was very fond of swimming. To have a good swim in the Mediterranean was the main extra-scientific attraction of his visits to Trieste. He also very much liked sailing and walking. The high discipline of physical exercises kept him fit for very concentrated and heavy intellectual work in physics.

In fall 1952 he started to study physics at the University of Warsaw. He was from the start a very good student. However, at that time also the students of physics had to learn Marxism-Leninism as the only just, allegedly scientific ideology providing the true picture of the world. Apart from lectures there were also some obligatory practical exercises in Marxism. They consisted in the proper Marxist interpretation of the historical or current events, which were provided by the lecturer, and then followed by a discussion of the matter with the students. It happened that one of the well known Polish bishops has been arrested by the communists, put on trial and accused of having committed some imaginary crimes against the party, Polish state and nation. As a devout catholic Ryszard could not stand these slanders. He and his future wife Aleksandra Reklewska and openly the lecturer that these were all libels, that the state media were lying etc. This happened at the very beginning of the second semester of their studies. The reaction of the party to this protest was furious. They were both immediately forbidden to continue the studies and after almost half a year of disciplinary procedure they were in August 1953 officially relegated from the university. Ryszard made frantic attempts to move to another university. He was even accepted to the Wrocław University but the good did not last long. The information about his misdeameneour in Warsaw soon reached Wrocław so he was dismissed again.

FIGURE 1. 20 years old.

In order not to lose the time completely Ryszard decided to start the studies at the evening higher school of technology. This was a school of lower level than the university and was intended only for the working people. So, in order to be accepted to this school Ryszard had to find employment in some official enterprise. He took several jobs not only to satisfy these requirements but also to earn some money. The most scientific of these jobs was that at the institute of basic technological problems of the Polish Academy of Sciences. In the beginning of 1955 when applying to reenter the university he obtained from this place a very high appreciation of his technical abilities. Almost exactly two years after the clash with the Marxist lecturer, in the beginning of march 1955, Ryszard and Ola were readmitted to the University of Warsaw and could continue the study of physics.

In may 1957 I came back to Warsaw after a longer stay in England and a shorter visit to the United States to attend the Rochester conference on elementary particles. It was the time when the discovery of violation of the parity conservation was firmly established. It followed from the proposed form of the Fermi coupling that the electrons and neutrinos emitted in beta decays should be strongly polarized. This fact aroused much interest in the theoretical and experimental studies of various reactions with the participation of polarized electrons. After coming back home I started myself to investigate various polarization phenomena. I also gave some problems of this kind to the students looking for interesting subjects of their diploma works. I proposed to Ryszard and Ola the task of calculating the cross section for the Moeller scattering of arbitrarily polarized electrons. In 1957 they were just married so the dean of the faculty agreed that they make the calculations jointly. I also expected that in this way the paper would appear sooner, which might be of some advantage should some other people do the same. When they finished the calculations I found out that indeed some American physicist has just published in the physical review a paper on the same subject. Fortunately, his paper contained some serious mistakes. So Ryszard and Ola could send a short version of their paper, however containing the correct form of all the relevant formulae. So, their first paper written when they were still students was published in the physical review and awarded by the Polish State Council for Nuclear Energy. After passing very well the last examinations, and on the basis of this paper, in June 1958 they obtained the magister degree in theoretical physics. The dean of the faculty succeeded in concealing in the final documents any data, which could reveal the two years of suspension. Otherwise, the knowledge of suspension might be harmful for Ryszard when applying for employment. So, in his student's book he has the highest mark for Marxism-Leninism taken from the evening school of technology. This was the only examination passed in this school and acknowledged by the university. Sending him for examination to the same people that caused his relegation would be too risky. Also the date of the beginning of his study at the university put in the diploma has been changed so that one cannot guess the gap of two years.

After having obtained the magister degree Ryszard was employed as an assistant in the Institute of Physics of the Polish Academy of Sciences. Simultaneously he was a postgraduate student at the university. In this capacity he had some exercises with the students to various lectures of theoretical physics for which he obtained some modest stipend. This material support was quite important because in 1958 Ola has born the first son Piotr who is also a physicist and is today with us. In the period of twelve years Ola has born 5 children and the problem of earning enough money for maintaining the rapidly growing family was for a long period very essential. In 1960 Ryszard got the job at the institute of nuclear research where he was working all the time to his death in 1996.

After 1958 Ryszard did not continue the investigation of the polarization phenomena and I did not press him. He came back to these problems later. Around 1959 or 1960 he tried a cooperation with the Japanese nuclear physicist Ziro Koba who was for several years working at our institute in Warsaw. Koba told us that he settled in Poland because he appreciated very much the Polish climate that allows for intensive work all the year round. Having a weak constitution he could not stand the Japanese hot and humid summers. After spending a couple of years in Poland he was surprised that the Poles do not take the advantages of their climate. Ryszard published two common papers with Ziro Koba and some Polish young physicists on the elastic scattering of nucleons and pions at very high energies. However, at that time a new vast area of research emerged: the multiparticle reactions. I got myself more and more interested in this field and suggested Ryszard that the calculations of angular and energy distributions in some interesting multiparticle reactions would a good subject for his doctor thesis. We published only one joint paper on the nucleon-antinucleon annihilation. Ryszard concentrated on the statistical approach to the multiparticle production with angular momentum conservation. In 1962 he went to work at the joint institute of nuclear research in Dubna where he stayed almost two years. He continued there the line started in Warsaw. In Dubna Ryszard finished in 1963 a fine paper: "Statistical theory of multiparticle production at very high energy collisions" which he submitted as his doctor thesis which he defended in the very beginning of 1964 at the University of Warsaw.

The results obtained by Ryszard in his doctor thesis were very interesting for experimentalists. For example a group in Krakow observed since more than a decade strange phenomena in very high-energy collisions caused by cosmic rays. In some percentage of cases the final state consisted of a large number of particles which in the overall center of mass frame looked like two separate sources of multiparticle emission moving in opposite directions. Since no definite mass could be attributed to these sources they were called fireballs. Their nature remained mysterious for quite a time. Ryszard has shown that such an effect appears when the particles collide with very high energies and a large value of the impact parameter that means with high angular momentum. Thus the fireballs turned out to be ordinary kinematical and statistical effects of high-energy collisions with no new physics behind.

I have followed a different path, investigating the rigorous S-matrix formulation and applying group theoretical methods to the relativistic description of reactions, including multiparticle processes. In 1964 we both received a common prize of the highest degree from the state council for nuclear research for our contributions to the theory of multiparticle reactions. About 1962 I started to work on a book on model independent methods in the relativistic theory of reactions. By this I meant mainly general methods of quantum theory and the applications of various groups and symmetry principles. In 1964 Abdus Salam received Italian and international support to create the international center for theoretical physics in Trieste. Salam—whom I knew since several years—invited me to join the center for the first year of its existence, i.e. for the academic year 1964–65. I recommended Ryszard as a very brilliant and hard working young physicist with very good training in mathematics and particle physics. Salam accepted my suggestion and so in the fall of this year we both started to work at ICTP in the provisional building at Piazza Oberdan.

In fall 1964 my book was already well advanced but because of many teaching and administrative duties in Warsaw I had some difficulties with finding enough time to finish it. In Trieste I could concentrate on this task but it took more time than I thought in the beginning, so I hardly did any other research at the center for several months of my stay at the center. I sent it to the editor for publication in spring

FIGURE 2. At Piazza Oberdan 1964.

1965. I believe that during numerous discussions with Ryszard still in Warsaw about problems I had when writing my book I inseminated him with interest in the theory of groups. It happened that the investigation of various groups and their representations was one of the main subjects of research of many physicists who were working in the center in this first year. Ryszard joined this main stream with his typical eagerness. His abilities and attained results were soon duly appreciated and Salam extended his invitation for Ryszard for the second year. During his two years stay at ICTP he was investigating the problem of higher symmetries in the theories of elementary particles as well as representations of locally compact groups. Another field of his research consisted in investigations of relativistic field equations for two body systems. In 1965–66 he was heading a seminar at ICTP on the theory of elementary particles and had a series of lectures on representations of groups. When working at the ICTP he was also tutoring two Ph.D. students in spites of the fact the he himself received this degree only two years earlier. For the work entitled "The properties of group representations of higher symmetries" the Warsaw University gave him in 1967 the habilitatus degree and the title of docent. This work contained essentially the results of papers written during his first two years stay in Trieste. In May 1975 he was nominated extraordinary professor at the Institute for Nuclear Research.

Apart from the stay in Dubna and frequent longer and shorter visits to ICTP Ryszard spent one year in Boulder with Barut. I am not able to list his numerous scientific visits to other places; visits that lasted sometimes only one week, sometimes several months. He was taking part, very often as invited speaker, in many international conferences, seminars, workshops, winter and summer schools. He

FIGURE 3. Seattle 1969; Mathematics–Physics.

FIGURE 4. Göteborg 1968; Nobel Symposium.

FIGURE 5. Schladming 1969.

was often invited by several scientific institutions as a visiting professor. Especially frequent visits he paid to Paris and Dijon where he had established tight scientific and personal contacts. I know that he recognized me as his teacher and master, however, I must honestly admit that in many respects he soon surpassed his master. It certainly refers to physical insight and mathematical depth. It was one of the happy cases when the pupil surpassed his teacher and the teacher was happy with that.

The close scientific collaboration with people he met at the center lasted very long, in some cases up to the nineties, i.e. Almost to his death. I may mention; Barut, Budinich, Fischer, Flato, Fronsdal, Furlan, Niederle, Rashid, Wess and others. Some of these people are here and will be talking at this session. The most spectacular and highly appreciated result of this collaboration was the voluminous (over 700 pages) book on the "Theory of group representations and applications", written with Asim Barut. It was published in 1977 but it had two subsequent English editions and one edition in Russian. Unfortunately Barut died several years ago and cannot be with us. The book written with Barut was and still is highly valued by particle theorists. I should like to quote the other main fields of research of Ryszard apart from those mentioned earlier. A large number of his papers is devoted to various nonlinear field equations. The integral representations proposed by Ryszard allowed for a reduction of quantum field equations to some classical field equations. Much interest he devoted to field equations with polynomial nonlinearities, in particular those with fourth power that are playing important role in the Standard Model of the strong interactions. With different co-authors he was investigating the P, C, and time reversal symmetries in nonlinear classical and quantum field equations, nonlinear spinor representations and non-linear

FIGURE 6. With Barut in Paris 1978.

spinor field equations. Another group of papers is dealing with various properties of QCD, for example renormalization scheme and gauge dependence of perturbative QCD four loop order (and beyond) approximations and their physical meaning, etc. He questioned the physical meaning of the perturbative expansions in QCD. Several papers are dealing with Weyl and conformal covariant field theories and their possible implications, new approach to unified field theory, geometrical origin of some physical properties, cosmological properties and their implications, etc. In the last few years of his life Ryszard was trying to prove that one can eliminate the Higgs field from the Standard Model of strong interactions, transforming it out to the gravitational sector. The origin of the idea of the gravitational origin of the Higgs field may be found in an earlier paper published in 1988 with M. Flato. However, the physical meaning of the transformation applied by Ryszard in this proof aroused many doubts that he could not remove up to his death. In spite of this shortcoming (deficiency) Ryszard was widely propagating his conviction that the Higgs field is a spurious structure and one should not spend such a lot of effort and money for its experimental discovery. This is only one example of Ryszard's strong scientific convictions that sometimes were well justified and very fruitful, sometimes

FIGURE 7. ~ 50 y.

rather harmful as bordering to unfounded obstinacy. He had some troubles with accepting critical remarks.

I am not able to present here in a more detailed form all his scientific fields of interest, or list the papers and the names of all collaborators. It would not make sense in such a lecture devoted to my sketchy, necessarily incomplete and personal memory of Ryszard. In general, I may say that he was extremely active in theoretical physics. He was very good not only in phenomenological evaluation of the experimental data but also in very profound mathematical analysis of the current theoretical models, as well as in proposing some interesting new models. It is rather natural that not all his proposals and views have been accepted by the scientific community of physicists. We know that even Einstein and Heisenberg were not infallible in their proposals. He was cofounder and co-editor of "Letters in Mathematical Pphysics", "Mathematical Physics and Applied Physics—International Monography Series", "Mathematical Physics Studies". He was co-editor of proceedings of several international symposia, workshops and conferences: "Symposium on Mathematical Physics", Warsaw 1974; "Workshop of Stefan Banach Centre", Warsaw 1988; "Polarization Dynamics in Nuclear and Particle Physics", Trieste 1992; "Particles and Fields", Rydzyna Castle 1994; "New Ideas in the Theory of Fundamental Interactions", Zakopane 1995.

Ryszard was a good diplomat in the sense that having initiative and courage to act as well as good persuasive powers he was able to convince people to some common actions of organizational type. This referred not only to his fellow physicists but also to sometimes very gloomy administrative officials. In 1967 he convinced Salam and—with the help of the latter—the International Atomic Energy Agency in Vienna to create a net of selected theoretical institutes in eastern Europe affiliated with ICTP. The idea was accepted also by the local authorities in those countries. In effect several institutes from this area signed cooperation agreements with ICTP, which provided full coverage of living costs for young physicists coming to ICTP—for visits not extending few weeks—to attend seminars, workshops or conferences organized in Trieste. The sending institutions had to pay only for the travel expenses. In the years 1967–72 about 80 Polish physicists took advantage of these federation agreements to visit ICTP.

He also organized successful German–Polish cooperation that included DESY in Hamburg, the universities in Bielefeld and in Clausthal. A couple of international symposia have been organized within this cooperation (Rydzyna, Zakopane). He was also one of the organizers of a successful series of lectures on pure mathematics and mathematical physics known as "the Banach Symposium" held in Warsaw in 1988. Together with I. Birula-Białynicki he lead a seminar in Warsaw in which also some people from Łódź participated.

He was holding several partly administrative positions in the Institute for Nuclear Research. So—in 1967–70 he was charged with care for the theory of elementary particles. In 1969–71 he was the head of postgraduate studies. In 1971–75 he was the head of the Division of the Theory of Nuclei and Elementary Particles. In 1995 he was elected chairman of the scientific council of the whole Institute of Nuclear Studies including not only the mentioned—rather small—theoretical division but also several very much larger experimental divisions. He was very happy with this election that was a sign of acknowledgement and high appreciation by his colleagues. I remember how delighted and proud he was when he ran into my office with this information. However, very soon it turned out that it was not a proper position for him. In fact it required adherence to strict laws, rules, regulations that could be replaced by charity, compassion, mercy and love. In other words this was a position for a person who knows when it is necessary to say: "not". And Ryszard did not like to say this word even when the legal situation was forcing him to do it.

Ryszard Rączka was a deeply religious man and devoted catholic in the strange English meaning of this word. in English the word catholic means: sympathizing with all people, tolerant, liberal. of course it also refers to the members of the roman catholic religion, that claims to be universal, i.e. good and acceptable for all human beings. In the Polish language the words tolerance and liberal can be never replaceable with the adjective catholic. One may distinguish two groups of attributes of the Christian God. The first group comprises justice, law, judgement, verdict, heavenly reward for the virtuous, hell for the sinners, etc. In the second

FIGURE 8. Trieste 1986.

FIGURE 9. A walk along the Vistula.

FIGURE 10. Close to 60.

group of God's attributes we have: charity, mercy, tolerant love, forgiveness, unbounded goodness, graciousness, helpfulness, etc. For the human mind these two groups of God's attributes are contradictory, even opposite. Only God knows how they can be reconciled but it remains His mystery. The human beings have to make a choice between these two. A choice that taken rigidly may lead to tragic results. Another way out of this dilemma consists in working out some discomfortable but bearable compromises. 60 years ago I read the drama "Brand" by Ibsen. It describes a devout pastor who is strictly professing the first group of the mentioned attributes. Thus brand is a just, law-abiding judge passing merciless sentences, a man who forgot mercy, charity, forgiveness and tenderness. His high and rigid principles finally bring disaster in contact with real human beings. In the tragic end of the drama a reminding voice from heaven says: "God is also Deus Caritatis".

Ryszard rejected the first choice and took decidedly the second. Was it really much better than the choice of Brand? I shall refrain from giving a ready answer to this difficult question. It requires evaluation and judgement and I may not qualify for a good judge. You may try to find the answer yourself, and for yourself, if you like.

At the end of my talk I'd like to pay my homage to Ola. Without her support and screening Ryszard could not become what he was. May I present her in my—but not only in my name—this bouquet of roses?

Józef Werle

Łódź, April 17, 1998

QUANTUM DEFORMATIONS AND

NON-COMMUTATIVE GEOMETRY

Line Bundles on Quantum Spheres

Tomasz Brzeziński[*][1] and Shahn Majid[†][2]

[*]*Department of Mathematics, University of York, Heslington, York YO10 5DD, England*
[†] *DAMTP, University of Cambridge, Cambridge CB3 9EW, England*

Abstract. The (left coalgebra) line bundle associated to the quantum Hopf fibration of any quantum two-sphere is shown to be a finitely generated projective module. The corresponding projector is constructed and its monopole charge is computed. It is shown that the Dirac q-monopole connection on any quantum two-sphere induces the Grassmannian connection built with this projector.

INTRODUCTION

In the standard approach to non-commutative geometry [7] (see [11] for an accessible introduction) the notion of a vector bundle is identified with that of a finitely generated projective module E of an algebra B. Recall that E is a finitely generated projective left B-module if E is isomorphic to $B^n e := \{(b_1, \ldots, b_n)e \mid b_1, \ldots, b_n \in B\}$, where e is an $n \times n$ matrix with entries from B such that $e^2 = e$ (e is called a *projector* of E). The algebra B plays the role of an algebra of functions on a manifold and E is thought of as a module of sections (matter fields) on a vector bundle over this manifold. A (universal) connection or a covariant derivative on E is a map $\nabla : E \to \Omega^1 B \otimes_B E$, where $\Omega^1 B$ denotes the space of (universal) differential 1-forms on B, such that for all $\rho \in \Omega^1 B$, $x \in E$,

$$\nabla(\rho \otimes x) = \rho \nabla(x) + d\rho \otimes x. \qquad (1)$$

A different approach to non-commutative gauge theories, based on quantum groups, was introduced in [4]. In this approach one begins with a quantum principal bundle P constructed algebraically as a Hopf-Galois extension of B by a Hopf algebra (quantum group) H (see [13] for a review of Hopf-Galois extensions). One then constructs a vector bundle as a module E associated to B and an H-comodule (corepresentation) V. Any strong connection (gauge field) in P induces then a covariant derivative in E. An example of a quantum principal bundle is the quantum

---

[1] Lloyd's Tercentenary Fellow. On leave from: Department of Theoretical Physics, University of Łódź, Pomorska 149/153, 90–236 Łódź, Poland. E-MAIL: `tb10@york.ac.uk`
[2] E-MAIL: `majid@damtp.cam.ac.uk`

Hopf fibration over the standard quantum two-sphere in [4]. The Dirac q-monopole is the canonical connection on this bundle. In a recent paper [10] the line bundle associated to the q-monopole bundle has been constructed. It has been shown that it is a finitely generated projective module. The covariant derivative induced by the q-monopole connection is the *Grassmannian connection* (see eq. (6) below).

On the other hand, the standard quantum two-sphere is only one of the infinite family of quantum two-spheres constructed in [15]. To describe gauge theory on all such spheres one needs the notions of a *coalgebra principal bundle*, introduced in [5], and an *associated module* as an algebraic version of a coalgebra vector bundle, introduced in [2]. In this note we announce some of the results of a forthcoming paper [6], in which the theory of [5] is generalised and combined with [2] to give the theory of connections on coalgebra principal and associated bundles, illustrated by the example of monopole connection for all quantum two-spheres. The new results of this note then consist of an explicit projective module description of line bundles over all quantum two-spheres, following the method used for the standard quantum sphere in [10].

We work over the field k of real or complex numbers, and we use the standard coalgebra notation. Thus, for a coalgebra C, Δ is the coproduct and ϵ is the counit. In a Hopf algebra S denotes the antipode. We use Sweedler's notation for the coproduct, $\Delta(c) = c_{(1)} \otimes c_{(2)}$ (summation understood). If P is a right C-comodule then $\Delta_P : P \to P \otimes C$ denotes the right coaction. On elements we write $\Delta_P(p) = p_{(0)} \otimes p_{(1)}$ (summation understood). If V is a left C-comodule then $_V\Delta : V \to C \otimes V$ denotes the left coaction. The reader not familiar with coalgebras is referred to [16] or to any modern textbook on quantum groups.

COALGEBRA GAUGE THEORY

We begin with the following definition from [3] (which generalises the earlier definition of a coalgebra principal bundle in [5]).

DEFINITION 1 *Let C be a coalgebra, P an algebra and a right C-comodule, and B a subalgebra of P, $B := \{b \in P \mid \forall p \in P\ \Delta_P(bp) = b\Delta_P(p)\}$. We say that P is a* coalgebra-Galois extension *(or a* coalgebra principal bundle*) of B iff the canonical left P-module right C-comodule map can $: P \otimes_B P \to P \otimes C$, $can(p \otimes p') = pp'_{(0)} \otimes p'_{(1)}$ is bijective. Such a coalgebra-Galois extension is denoted by $P(B)^C$.*

The definition of a coalgebra principal bundle implies [3] that there exists a unique *entwining map* $\psi : C \otimes P \to P \otimes C$ (cf. [5, Definition 2.1] for a definition of an entwining map). Explicitly, $\psi(c \otimes p) = can \circ (can^{-1}(1 \otimes c)p)$. The properties of ψ [5, Proposition 2.2] imply that the map $\psi^2 = (\mathrm{id}_P \otimes \psi) \circ (\psi \otimes \mathrm{id}_P) : C \otimes P \otimes P \to P \otimes P \otimes C$, restricts to $\psi^2 : C \otimes \Omega^1 P \to \Omega^1 P \otimes C$, where $\Omega^1 P = \{\sum_i p_i \otimes p'_i \in P \otimes P \mid \sum_i p_i p'_i = 0\}$ is the bimodule of universal one-forms on P. Also $\psi^2 \circ (\mathrm{id}_C \otimes d) = (d \otimes \mathrm{id}_C) \circ \psi$, where $d : P \to \Omega^1 P$, $d(p) = 1 \otimes p - p \otimes 1$ is the universal differential. In other words, the universal differential calculus on P is covariant with respect to ψ. This allows one to introduce the following definition in [6], which extends the earlier one in [5].

DEFINITION 2 *Let $P(B)^C$ be a coalgebra principal bundle with the entwining map ψ. A connection one-form in $P(B)^C$ is a linear map $\omega : C \to \Omega^1 P$ such that:*
  (i) $1_{(0)}\omega(1_{(1)}) = 0$, where $1_{(0)} \otimes 1_{(1)} = \Delta_P(1)$.
  (ii) For all $c \in C$, $\chi \circ \omega(c) = 1 \otimes c - \epsilon(c)\Delta_P(1)$, where $\chi : P \otimes P \to P \otimes C$, $\chi(p \otimes p') = p\Delta_P(p')$.
  (iii) For all $c \in C$, $\psi^2(c_{(1)} \otimes \omega(c_{(2)})) = \omega(c_{(1)}) \otimes c_{(2)}$.

Conditions (i)-(iii) have the following geometric meaning. By (i), ω can be thought of as acting on $\ker \epsilon$, which can be interpreted as a dual of the "Lie algebra" of C. Condition (ii) means that an application of ω to a vector field obtained by lifting an element of a Lie algebra gives back this element of a Lie algebra. Finally, (iii) is the covariance of ω under the adjoint coaction. As shown in [5] [6], connection 1-forms are in bijective correspondence with *connections*, i.e., left P-linear projections Π in $\Omega^1 P$ with the kernel $P(\Omega^1 B)P$ and such that $\Pi \circ d$ is right C-covariant. A connection is said to be a *strong connection*, if for all $p \in P$, $dp - p_{(0)}\omega(p_{(1)}) \in (\Omega^1 B)P$ [9].

Given a coalgebra principal bundle $P(B)^C$, and a left C-comodule V with the coaction $_V\Delta : V \to C \otimes V$, one constructs the left B-module

$$E := \{\sum_i p^i \otimes v^i \in P \otimes V \mid \sum_i \Delta_P(p_i) \otimes v^i = \sum_i p_i \otimes {_V\Delta}(v^i)\}.$$

The module E, introduced in [2], plays the role of a fibre bundle associated to P with fibre V. The coalgebra gauge theory in [6] ensures that if there exists $c \in C$ such that $\Delta_P(1) = 1 \otimes c$, then any strong connection ω in P induces a covariant derivative $\nabla : E \to \Omega^1 B \otimes_B E$ on E (in the sense of equation (1)) given by

$$\nabla(\sum_i p^i \otimes v^i) = \sum_i dp^i \otimes v^i - \sum_i {p^i}_{(0)}\omega({p^i}_{(1)}) \otimes v^i. \qquad (2)$$

Since the existence of a connection in a module E is equivalent to E being projective [8, Corollary 8.2], we conclude that if there is a strong connection in $P(B)^C$, then every associated left B-module E is projective.

LEFT COALGEBRA LINE BUNDLES ON QUANTUM SPHERES

Recall that the quantum group $SU_q(2)$ is a free algebra generated by 1 and the matrix of generators $\mathbf{t} = \begin{pmatrix} \alpha & \beta \\ \gamma & \delta \end{pmatrix}$, subject to the relations:

$$\alpha\beta = q\beta\alpha, \quad \alpha\gamma = q\gamma\alpha, \quad \beta\gamma = \gamma\beta, \quad \beta\delta = q\delta\beta,$$

$$\gamma\delta = q\delta\gamma, \quad \alpha\delta = \delta\alpha + (q - q^{-1})\beta\gamma, \quad \alpha\delta - q\beta\gamma = 1.$$

$SU_q(2)$ is a Hopf algebra of a matrix type, i.e.

$$\Delta(t_{ik}) = \sum_{j=1}^{2} t_{ij} \otimes t_{jk}, \quad \epsilon(\mathbf{t}) = 1, \quad S\begin{pmatrix} \alpha & \beta \\ \gamma & \delta \end{pmatrix} = \begin{pmatrix} \delta & -q^{-1}\beta \\ -q\gamma & \alpha \end{pmatrix}.$$

The quantum two-spheres S_{qs}^2 [15] are homogeneous spaces of $SU_q(2)$. For a given q they can be viewed as subalgebras of $SU_q(2)$ generated by 1 and

$$\xi = s(\alpha^2 - q^{-1}\beta^2) + (s^2 - 1)q^{-1}\alpha\beta, \quad \eta = s(q\gamma^2 - \delta^2) + (s^2 - 1)\gamma\delta,$$
$$\zeta = s(q\alpha\gamma - \beta\delta) + (s^2 - 1)q\beta\gamma,$$

where $s \in [0, 1]$. The standard quantum sphere corresponds to $s = 0^3$.

It is shown in [1] that for any quantum sphere there is a coalgebra principal bundle with the base $B = S_{qs}^2$ and the total space $P = SU_q(2)$. The structure coalgebra is a quotient $C_s = SU_q(2)/J_s$ where $J_s = \{\xi - s, \eta + s, \zeta\}SU_q(2)$. We denote by π_s the canonical projection $SU_q(2) \to C_s$. The coproduct on C_s is obtained by projecting down the coproduct in $SU_q(2)$ by π_s. The coaction of C_s on $SU_q(2)$ is given by $\Delta_{SU_q(2)} = (\mathrm{id}_{SU_q(2)} \otimes \pi_s) \circ \Delta$. It has been recently found in [14], that the coalgebras C_s are spanned by group-like elements ($c \in C_s$ is group-like if $\Delta(c) = c \otimes c$). These are computed explicitly in [6] and are given by

$$\pi_s(1), \quad g_n^+ = \pi_s(\prod_{k=0}^{n-1}(\alpha + q^k s\beta)), \quad g_n^- = \pi_s(\prod_{k=0}^{n-1}(\delta - q^{-k}s\gamma)), \quad n = 1, 2, \ldots$$

(all products increase from left to right).

Similarly to [10], consider a one-dimensional left corepresentation $V = k$ of C_s given by the coaction $_k\Delta : k \to C_s \otimes k$, $_k\Delta(1) = g_1^+ \otimes 1$. The left S_{qs}^2-module E_s associated to $SU_q(2)$ and this corepresentation comes out as

$$E_s = \{x(\alpha + s\beta) + y(\gamma + s\delta) \mid x, y \in S_{qs}^2\}.$$

The module E_s is the *line bundle* over S_{qs}^2. Our first goal is to construct the projector for E_s.

Consider the following matrix with entries from S_{qs}^2,

$$e_s = \frac{1}{1+s^2}\begin{pmatrix} 1-\zeta & \xi \\ -\eta & s^2 + q^{-2}\zeta \end{pmatrix}.$$

Notice that e_s can be also written in the following form

$$e_s = \frac{1}{1+s^2}\begin{pmatrix} (\alpha + s\beta)(\delta - qs\gamma) & (\alpha + s\beta)(s\alpha - q^{-1}\beta) \\ (\gamma + s\delta)(\delta - qs\gamma) & (\gamma + s\delta)(s\alpha - q^{-1}\beta) \end{pmatrix}$$
$$= \frac{1}{1+s^2}\begin{pmatrix} \alpha + s\beta \\ \gamma + s\delta \end{pmatrix}(\delta - qs\gamma, s\alpha - q^{-1}\beta).$$

---

[3] It is perhaps more customary to parametrise quantum spheres by the parameter $c \geq 0$ [15]. The parameter s which is used here is related to p in [1] via $s - s^{-1} = p^{-1}$, and p^2 should be identified with c in [15].

The use of the relation
$$(\delta - qs\gamma)(\alpha + s\beta) + (s\alpha - q^{-1}\beta)(\gamma + s\delta) = 1 + s^2, \tag{3}$$
makes it clear that $e_s^2 = e_s$, i.e., e_s is a projector.

Now consider the left S_{qs}^2-module $(S_{qs}^2)^2 e_s := \{(x,y)e_s \mid x, y \in S_{qs}^2\}$, and the linear map $\Theta_s : (S_{qs}^2)^2 e_s \to E_s$ given by $\Theta_s : (x,y)e_s \mapsto x(\alpha + s\beta) + y(\gamma + s\delta)$. This map is well-defined because $(x,y)e_s = 0$ if and only if
$$\begin{cases} (x(\alpha + s\beta) + y(\gamma + s\delta))(\delta - qs\gamma) = 0 \\ (x(\alpha + s\beta) + y(\gamma + s\delta))(s\alpha - q^{-1}\beta) = 0 \end{cases} \tag{4}$$
Multiplying the first equation by $\alpha + s\beta$ and the second one by $\gamma + s\delta$ and then using (3) one deduces that $x(\alpha + s\beta) + y(\gamma + s\delta) = 0$. The map Θ_s is clearly surjective. Also, if $x(\alpha + s\beta) + y(\gamma + s\delta) = 0$, then (4) are satisfied, so that $(x,y)e_s = 0$. This implies that Θ_s is an isomorphism of left S_{qs}^2-modules. Therefore, E_s is isomorphic to $(S_{qs}^2)^2 e_s$ and hence it is a finitely generated projective module.

Next we compute the Chern number or the monopole charge of E_s. The formula (4.4) in [12] for the Chern character of S_{qs}^2 gives $\tau^1(1) = 0$, $\tau^1(\zeta) = \frac{1+s^2}{1-q^{-2}}$. Hence the Chern number of E_s is $\mathrm{ch}(E_s) := \tau^1(\mathrm{tr}\, e_s) = \tau^1(1 - (\zeta - q^{-2}\zeta)/(1 + s^2)) = -1$, and is independent of s. By the same argument as in [10] we conclude that E_s is not isomorphic to $S_{qs}^2 \otimes V$. This implies [2] that $SU_q(2)$ is not isomorphic to $S_{qs}^2 \otimes C_s$ as a left S_{qs}^2-module and a right C_s-comodule (i.e. $SU_q(2)(S_{qs}^2)^{C_s}$ is not *cleft*). This means that the coalgebra Hopf bundle of S_{qs}^2 is not trivial.

Finally we show that the *monopole connection* in $SU_q(S_{qs}^2)^{C_s}$, constructed in [6], is the Grassmannian connection in E_s. Its connection one-form is given by
$$\omega(g_n^{\pm}) = Si(g_n^{\pm})_{(1)} di(g_n^{\pm})_{(2)}, \tag{5}$$
where
$$i(g_n^+) = \prod_{k=0}^{n-1} \frac{\alpha + q^k s(\beta + \gamma) + q^{2k} s^2 \delta}{1 + q^{2k} s^2}, \quad i(g_n^-) = \prod_{k=0}^{n-1} \frac{\delta - q^{-k} s(\beta + \gamma) + q^{-2k} s^2 \alpha}{1 + q^{-2k} s^2}.$$
The connection (5) is strong and, since $\Delta_{SU_q(2)}(1) = 1 \otimes \pi_s(1)$, there is the covariant derivative ∇ on E_s given by (2). Explicitly, $\nabla(u) = du - u\omega(g_1^+)$, for all $u \in E_s$. Let $u = x(\alpha + s\beta) + y(\gamma + s\delta)$ for some $x, y \in S_{qs}^2$. The explicit form of $\omega(g_1^+)$ yields
$$\nabla(u) = d(x(\alpha + s\beta)) + d(y(\gamma + s\delta))$$
$$-\frac{1}{1+s^2}(x(\alpha + s\beta) + y(\gamma + s\delta))((\delta - qs\gamma)d(\alpha + s\beta) + (s\alpha - q^{-1}\beta)d(\gamma + s\delta)).$$
With the help of the Leibniz rule and (3) this can be gathered in the following form
$$\nabla(u) = (d(x,y) + (x,y)\frac{1}{2}d\begin{pmatrix} (\alpha + s\beta)(\delta - qs\gamma) & (\alpha + s\beta)(s\alpha - q^{-1}\beta) \\ (\gamma + s\delta)(\delta - qs\gamma) & (\gamma + s\delta)(s\alpha - q^{-1}\beta) \end{pmatrix})\begin{pmatrix} \alpha + s\beta \\ \gamma + s\delta \end{pmatrix}$$
$$= ((dx, dy) + (x,y)de_s)\begin{pmatrix} \alpha + s\beta \\ \gamma + s\delta \end{pmatrix}.$$

Now viewing u in $(S_{qs}^2)^2 e_s$ via Θ_s^{-1} and ∇ as a map $(S_{qs}^2)^2 e_s \to \Omega^1 S_{qs}^2 \otimes_{S_{qs}^2} (S_{qs}^2)^2 e_s$ via Θ_s one obtains

$$\nabla((x,y)e_s) = (d(x,y) + (x,y)de_s)e_s. \tag{6}$$

The covariant derivative of the form (6) is called the *Grassmannian connection* on a projective module. Therefore, similarly to the standard quantum sphere case discussed in [10], the covariant derivative corresponding to the q-monopole connection on any quantum sphere S_{qs}^2 is the Grassmannian connection on E_s.

REFERENCES

1. T. Brzeziński. Quantum homogeneous spaces as quantum quotient spaces. *J. Math. Phys.*, 37:2388–2399, 1996. (q-alg/9509015)
2. T. Brzeziński. On modules associated to coalgebra Galois extensions. *Preprint*, q-alg/9712023
3. T. Brzeziński and P.M. Hajac. Coalgebra extensions and algebra coextensions of Galois type. *Preprint*, q-alg/9708010, 1997 (to appear in *Commun. Algebra*).
4. T. Brzeziński and S. Majid. Quantum group gauge theory on quantum spaces. *Commun. Math. Phys.*, 157:591–638, 1993. Erratum 167:235, 1995.
5. T. Brzeziński and S. Majid. Coalgebra bundles. *Commun. Math. Phys.*, 191:467–492, 1998.
6. T. Brzeziński and S. Majid. Quantum geometry of algebra factorisations and coalgebra bundles. *Preprint* DAMTP/97-139 (in preparation).
7. A. Connes. *Non-Commutative Geometry*. Academic Press, 1994.
8. J. Cuntz and D. Quillen. Algebra extensions and nonsingularity. *J. Amer. Math. Soc.*, 8:251–289, 1995.
9. P. M. Hajac. Strong connections on quantum principal bundles. *Commun. Math. Phys.* 182:579–617, 1996.
10. P.M. Hajac and S. Majid. Projective module description of the q-monopole. *Preprint* math.QA/9803003
11. G. Landi. *An Introduction to Noncommutative Spaces and their Geometries*. Springer-Verlag, 1997.
12. T. Masuda, Y. Nakagami and J. Watanabe. Noncommutative differential geometry on the quantum two sphere of Podleś. I: An algebraic viewpoint. *K-Theory*, 5:151–175, 1991.
13. S. Montgomery. *Hopf Algebras and Their Actions on Rings*. CBMS Lectures vol. 82, AMS, 1993.
14. E.F. Müller and H.-J. Schneider. Quantum homogeneous spaces with faithfully flat module structure. *University of Munich preprint*, 1998.
15. P. Podleś. Quantum spheres. *Lett. Math. Phys.*, 14:193–202, 1987.
16. M. E. Sweedler. *Hopf Algebras* W.A.Benjamin, Inc., New York, 1969.

Action of a Finite Quantum Group on the Algebra of Complex $N \times N$ Matrices

R. Coquereaux[1], G. E. Schieber[2]

Centre de Physique Théorique - CNRS - Luminy, Case 907
F-13288 Marseille Cedex 9 - France

Abstract. Using the fact that the algebra $\mathcal{M} \doteq M_N(\mathbb{C})$ of $N \times N$ complex matrices can be considered as a reduced quantum plane, and that it is a module algebra for a finite dimensional Hopf algebra quotient $\mathcal{H}$ of $U_q sl(2)$ when q is a root of unity, we reduce this algebra $\mathcal{M}$ of matrices (assuming N odd) into indecomposable modules for $\mathcal{H}$. We also show how the same finite dimensional quantum group acts on the space of generalized differential forms defined as the reduced Wess Zumino complex associated with the algebra $\mathcal{M}$.

INTRODUCTION

When q is a root of unity ($q^N = 1$), the quantized enveloping algebra $U_q sl(2, \mathbb{C})$ posesses interesting quotients that are finite dimensional Hopf algebras. The structure of the left regular representation of such an algebra was investigated in [3] and the pairing with its dual in [2]. We call $\mathcal{H}$ the Hopf algebra quotient of $U_q sl(2, \mathbb{C})$ defined by the relations $K^N = 1, X_\pm^N = 0$ (we shall define the generators $K, X_\pm$ in a later section), and $\mathcal{F}$ its dual. It was shown[3] in [3] that the non semi-simple algebras $\mathcal{H}$ is isomorphic with the direct sum of a complex matrix algebra and of several copies of suitably defined matrix algebras with coefficients in the ring $Gr(2)$ of Grassmann numbers with two generators. The explicit structure (for all values of N) of those algebras, including the expression of generators themselves, in terms of matrices with coefficients in $\mathbb{C}$ or $Gr(2)$, was obtained by [7]. Using these results, the representation theory of $\mathcal{H}$, for the case $N = 3$, was presented in [4]. Following this work, the authors of [1], studied the action of $\mathcal{H}$ (case $N = 3$) on the algebra of complex matrices $M_3(\mathbb{C})$. In the letter [8], a reduced Wess-Zumino complex $\Omega_{WZ}(\mathcal{M})$ was introduced, thus providing a differential calculus bicovariant with respect to the action of the quantum group $\mathcal{H}$ on the algebra $M_3(\mathbb{C})$ of complex

---

[1] Email: coquecpt.univ-mrs.fr
[2] Email: schiebercpt.univ-mrs.fr
[3] Warning: the authors of [3] actually consider a Hopf algebra quotient defined by $K^{2N} = 1, X_\pm^N = 0$, so that their algebra is, in a sense, twice bigger than ours.

matrices. This differential algebra (that could be used to generalize gauge field theory models on an auxiliary smooth manifold) was also analysed in terms of representation theory of $\mathcal{H}$ in the same letter. In particular, it was shown that $M_3(\mathbb{C})$ itself can be reduced into the direct sum of three indecomposable representations of $\mathcal{H}$. A general discussion of several other properties of the dually paired Hopf algebras $\mathcal{F}$ and $\mathcal{H}$ (scalar products, star structures, twisted derivations etc.) can also be found there, as well as in the article [9]. Other properties of $SL_q(2,\mathbb{C})$ at third (or fourth) root of unity should also be discussed in [11].

In the present paper, after recalling several basic definitions and properties, we show that the algebra of usual $N \times N$ complex matrices (assuming N odd) decomposes, under the action of the quantum group $\mathcal{H}$, into a direct sum of N indecomposable representations of dimension N that we call N_p. Every *indecomposable* module of this type contains an invariant *irreducible* subspace of dimension p. In particular, N_N itself is irreducible. We also show how these representations appear as particular subrepresentations of the projective indecomposable modules of $\mathcal{H}$.

Finally we shall give the action of generators of $\mathcal{H}$ on the elements of the reduced Wess-Zumino complex.

A short discussion about the use of those structures in physics is given at the end.

THE DUALLY PAIRED FINITE DIMENSIONAL QUANTUM GROUPS $\mathcal{F}$ AND $\mathcal{H}$

$\mathcal{M} \doteq M_N(\mathbb{C})$ as a finite dimensional quantum plane

The algebra of $N \times N$ matrices can be generated by two elements x and y with relations :

$$xy = qyx \quad \text{and} \quad x^N = y^N = \mathbb{1} , \qquad (1)$$

where q denotes a N-th root of unity ($q \neq 1$) and $\mathbb{1}$ denotes the unit matrix. Explicitly, x and y can be taken as the following matrices:

$$x = \begin{pmatrix} 1 & & & \\ & q^{-1} & & \\ & & q^{-2} & \\ & & & \ddots \\ & & & & q^{-(N-1)} \end{pmatrix} \qquad y = \begin{pmatrix} 0 & & & & \\ \vdots & & & & \\ \vdots & & \mathbb{1}_{N-1} & & \\ \vdots & & & & \\ 1 & 0 & \cdots & \cdots & 0 \end{pmatrix} \qquad (2)$$

This result can be found in [6].

Warning : for technical reasons, we shall assume in all this paper that N is odd.

The dually paired quantum groups $\mathcal{F}$ and $\mathcal{H}$

The quantum group $\mathcal{F}$ coacts on $\mathcal{M}$

We now consider a free associative algebra generated by four *a priori* non commuting symbols a, b, c, d and define the following "change of variables" (notice that the symbol $\otimes$ does not denote the usual tensor product since it involves also a matrix multiplication):

$$\begin{pmatrix} x' \\ y' \end{pmatrix} = \begin{pmatrix} a & b \\ c & d \end{pmatrix} \otimes \begin{pmatrix} x \\ y \end{pmatrix} , \tag{3}$$

and

$$\begin{pmatrix} \tilde{x} & \tilde{y} \end{pmatrix} = \begin{pmatrix} x & y \end{pmatrix} \otimes \begin{pmatrix} a & b \\ c & d \end{pmatrix} . \tag{4}$$

One then imposes that quantities x', y' (and $\tilde{x}, \tilde{y}$) obtained by the previous matrix equalities should satisfy the same relations as x and y (the multiplication of these elements make perfect sense within the tensor product of the corresponding algebras). One obtains in this way the relations:

$$\begin{array}{ll} qca = ac & qdb = bd \\ qba = ab & qdc = cd \\ cb = bc & ad - da = (q - q^{-1})bc \end{array} \tag{5}$$

together with

$$\begin{array}{ll} a^N = \mathbb{1} , & b^N = 0 \\ c^N = 0 , & d^N = \mathbb{1} \end{array} . \tag{6}$$

One also take the central element $\mathcal{D} \doteq da - q^{-1}bc = ad - qbc$ to be equal to $\mathbb{1}$. The algebra defined by a, b, c, d and the above set of relation will be called $\mathcal{F}$. It is clearly a quotient of $Fun(SL_q(2, \mathbb{C}))$, the algebra of polynomial functions on the quantum group $SL_q(2, \mathbb{C})$. Since $a^N = \mathbb{1}$, multiplying the relation $ad = \mathbb{1} + qbc$ from the left by a^{N-1} leads to

$$d = a^{N-1}(\mathbb{1} + qbc) \tag{7}$$

so that d can be eliminated. The algebra $\mathcal{F}$ can therefore be *linearly* generated —as a vector space— by the elements $a^\alpha b^\beta c^\gamma$ where indices α, β, γ run in the set $\{0, 1, \cdots, N-1\}$. We see that $\mathcal{F}$ is a *finite dimensional* associative algebra, whose dimension is

$$\dim(\mathcal{F}) = N^3 .$$

$\mathcal{F}$ is not only an associative algebra but a Hopf algebra (cf. previously given references). The coproduct of generators can be read directly from the above 2×2 matrix with entries a, b, c, d, for instance, $\Delta a = a \otimes a + b \otimes c$, etc. .

This quantum group, by construction, *coacts* on $\mathcal{M}$.

The quantum group $\mathcal{H}$ acts on $\mathcal{M}$

Multiplication and comultiplication being interchanged by duality, it is clear that the dual $\mathcal{H}$ of $\mathcal{F}$ is also a quantum group (of the same dimensionality). For compatibility reasons with previous references, we choose the linear basis $K^\alpha X_+^\beta X_-^\gamma$ in $\mathcal{H}$, where K, X_+ and X_- are defined by duality as follows:

$$\begin{array}{llll}
<K,a>=q & <K,b>=0 & <K,c>=0 & <K,d>=q^{-1} \\
<X_+,a>=0 & <X_+,b>=1 & <X_+,c>=0 & <X_+,d>=0 \\
<X_-,a>=0 & <X_-,b>=0 & <X_-,c>=1 & <X_-,d>=0
\end{array} \quad (8)$$

From multiplication and comultiplication in $\mathcal{F}$, one gets :

Multiplication:

$$KX_\pm = q^{\pm 2} X_\pm K$$
$$[X_+, X_-] = \frac{1}{(q-q^{-1})}(K - K^{-1}) \quad (9)$$
$$K^N = \mathbb{1}$$
$$X_+^N = X_-^N = 0 \ .$$

Comultiplication: The comultiplication is an algebra morphism, i.e., $\Delta(XY) = \Delta X \, \Delta Y$. It is given by

$$\begin{aligned}
\Delta X_+ &\doteq X_+ \otimes \mathbb{1} + K \otimes X_+ \\
\Delta X_- &\doteq X_- \otimes K^{-1} + \mathbb{1} \otimes X_- \\
\Delta K &\doteq K \otimes K \\
\Delta K^{-1} &\doteq K^{-1} \otimes K^{-1} \ .
\end{aligned} \quad (10)$$

There is also an antipode and a counit but we shall not need them in the sequel.

The quantum group $\mathcal{H}$ acts on itself (by left or right multiplication), it acts also on its dual $\mathcal{F}$ from the left or from the right, as follows:

$$\langle y, h^L[f] \rangle = \langle yh, f \rangle \qquad \langle y, h^R[f] \rangle = \langle hy, f \rangle , \quad (11)$$

where $y, h \in \mathcal{H}, \{ \in \mathcal{F}$. These actions can also be expressed as :

$$h^L[f] = \langle id \otimes h , \Delta f \rangle_{(2)} \qquad h^R[f] = \langle id \otimes h , \Delta f \rangle_{(1)} , \quad (12)$$

where the notation $\langle , \rangle_{(1)}$ means that we only pair the first term in the tensor product (resp. the second).

Finally, $\mathcal{H}$ also *acts* on the reduced quantum plane $\mathcal{M}$ (the algebra of $N \times N$-

matrices) since its dual $\mathcal{F}$ coacts on it. There are again two possibilities, left or right, but we shall use the left action.

The left action of $\mathcal{H}$ on $\mathcal{M}$ is generally defined as follows. If we denote the right coaction of $\mathcal{F}$ on $\mathcal{M}$ as :

$$\delta_R[m] = m_{(1)} \otimes f_{(2)} \qquad \text{for} \quad m \in \mathcal{M}, \{ \in \mathcal{F}, \tag{13}$$

then :

$$h^L[m] = m_{(1)} \langle h , f_{(2)} \rangle \qquad h \in \mathcal{H} . \tag{14}$$

The action of generators of $\mathcal{H}$ on generators of $\mathcal{M}$ is given by the following table.

Left	K	X_+	X_-
x	qx	0	y
y	$q^{-1}y$	x	0

(15)

Using the coproduct, one finds the expression of these generators on an arbitrary element of $\mathcal{M}$

$$K^L[x^r y^s] = q^{(r-s)} x^r y^s$$
$$X_+^L[x^r y^s] = q^r (1 + q^{-2} + \cdots + q^{-2(s-1)}) x^{r+1} y^{s-1}$$
$$= q^r \left(\frac{1-q^{-2s}}{1-q^{-2}}\right) x^{r+1} y^{s-1} \tag{16}$$
$$X_-^L[x^r y^s] = q^s (1 + q^{-2} + \cdots + q^{-2(r-1)}) x^{r-1} y^{s+1}$$
$$= q^s \left(\frac{1-q^{-2r}}{1-q^{-2}}\right) x^{r-1} y^{s+1}$$

with $1 < r, s < N$.

Remember that $\mathcal{M}$ itself is *not* a quantum group, but a module (a representation space) for the quantum group $\mathcal{H}$. However, taking $h \in \mathcal{H}$, $m_1, m_2 \in \mathcal{M}$ and writing $\Delta h = \sum h_1 \otimes h_2$, one can check the following supplementary compatibility condition $h[m_1 m_2] = \sum h_1[m_1] h_2[m_2]$ between the module structure of $\mathcal{M}$, the algebra structure of $\mathcal{M}$ and the coproduct in $\mathcal{H}$. This makes $\mathcal{M}$ a module algebra over $\mathcal{H}$.

In order to reduce this module into indecomposable modules, it is necessary to know at least part of the representation theory of $\mathcal{H}$. This is recalled in the next subsection.

Representation theory of $\mathcal{H}$

As already stated in the introduction, using a result by [3], the explicit structure (for all values of N) of those algebras, including the expression of generators $X_\pm, K$

themselves, in terms of matrices with coefficients in $\mathbb{C}$ or in the Grassmann[4] algebra $Gr(2)$ with two generators θ_1, θ_2, was obtained by [7]. We shall not need the general theory but only the following fact: when N is odd, $\mathcal{H}$ is isomorphic with the direct sum

$$\mathcal{H} = \mathcal{M}_N \oplus (M_{N-1|1}(\Lambda^2))_0 \oplus (M_{N-2|2}(\Lambda^2))_0 \oplus \cdots\cdots \oplus (M_{\frac{N+1}{2}|\frac{N-1}{2}}(\Lambda^2))_0 \qquad (17)$$

where:

- M_N is a $N \times N$ complex matrix
- An element of the $(M_{N-1|1}(\Lambda^2))_0$ block (space that we shall just call $M_{N-1|1}$) is of the following form :

$$\begin{pmatrix} \bullet & \bullet & \cdots & \bullet & \bullet & \circ \\ \bullet & \bullet & \cdots & \bullet & \bullet & \circ \\ \vdots & \vdots & & \vdots & \vdots & \\ \bullet & \bullet & \cdots & \bullet & \bullet & \circ \\ \circ & \circ & \cdots & \circ & \circ & \bullet \end{pmatrix} \qquad (18)$$

We have introduced the following notation:
• is an even element of the ring $Gr(2)$ of Grassmann numbers with two generators, i.e., of the kind :

$$\bullet = \alpha + \beta\theta_1\theta_2, \qquad \alpha, \beta \in \mathbb{C}.$$

◦ is an odd element of the ring $Gr(2)$ of Grassmann numbers with two generators, i.e., of the kind :

$$\circ = \gamma\theta_1 + \delta\theta_2 \qquad \gamma, \delta \in \mathbb{C}$$

- An element of the $M_{N-2|2}$ block is of the kind :

$$\begin{pmatrix} \bullet & \bullet & \cdots & \bullet & \circ & \circ \\ \bullet & \bullet & \cdots & \bullet & \circ & \circ \\ \vdots & \vdots & & \vdots & \vdots & \\ \bullet & \bullet & \cdots & \bullet & \circ & \circ \\ \circ & \circ & \cdots & \circ & \bullet & \bullet \\ \circ & \circ & \cdots & \circ & \bullet & \bullet \end{pmatrix} \qquad (19)$$

- etc.

Notice that $\mathcal{H}$ is **not** a semi-simple algebra : its Jacobson radical $\mathcal{J}$ is obtained by selecting in equation 17 the matrices with elements proportionnal to Grassmann variables. The quotient $\mathcal{H}/\mathcal{J}$ is then semi-simple ... but no longer Hopf!

Projective indecomposable modules (PIM's, also called principal modules) for $\mathcal{H}$ are directly given by the columns of the previous matrices.

---

[4] Remember that $\theta_1^2 = \theta_2^2 = 0$ and that $\theta_1\theta_2 = -\theta_2\theta_1$

- From the M_N block, one obtains N equivalent irreducible representations of dimension N that we shall denote N_{irr}.
- From the $M_{N-p|p}$ block (<u>assume</u> $p < N - p$), one obtains
  - $(N - p)$ equivalent indecomposable projective modules of dimension $2N$ that we shall denote P_{N-p} with elements of the kind

$$(\underbrace{\bullet \bullet \cdots \bullet}_{N-p} \underbrace{\circ \circ \cdots \circ}_{p}) \tag{20}$$

  - p equivalent indecomposable projective modules (also of dimension $2N$) that we shall denote P_p with elements of the kind

$$(\underbrace{\circ \circ \cdots \circ}_{N-p} \underbrace{\bullet \bullet \cdots \bullet}_{p}) \tag{21}$$

Other submodules can be found by restricting the range of parameters appearing in the columns defining the PIM's and imposing stability under multiplication by elements of $\mathcal{H}$. In this way, one can determine, for each PIM the lattice of its submodules. For a given PIM of dimension $2N$ (with the exception of N_{irr}), one finds totally ordered sublattices (displayed below) with exactly three non trivial terms : the radical (here, it is the biggest non trivial submodule of a given PIM), the socle (here it is the smallest non trivial submodule), and one "intermediate" submodule of dimension exactly equal to N. However the definition of this last submodule (up to equivalence) depends on the choice of an arbitrary complex parameter λ, so that we have a chain of inclusions for every such parameter. The collection of all these sublattices fully determines the lattice structure of submodules of a given principal module. Since we have two types of principal modules (besides the irreducible ones, N_{irr}), we obtain explicitly the following two types of chains of inclusions [5]:

- First type : submodules of P_{N-p}.

$$0 \hookrightarrow \underline{N-p} \hookrightarrow \underline{N}_{N-p} \hookrightarrow \underline{N+p} \hookrightarrow \underline{2N} = P_{N-p}$$

where $\hookrightarrow$ represent inclusion.

An element of the submodule $\underline{N}_{N-p}$ (which has dimension N) is of the kind :

$$\underline{N}_{N-p} = \begin{pmatrix} \beta_1\theta_1\theta_2 & \beta_2\theta_1\theta_2 & \cdots & \beta_{N-p}\theta_1\theta_2 & \gamma_1\theta_\lambda & \cdots & \gamma_p\theta_\lambda \end{pmatrix} \tag{22}$$

where :

$$\theta_\lambda = \lambda_1\theta_1 + \lambda_2\theta_2 \qquad \lambda = \lambda_1/\lambda_2 \in \mathbb{C}P^1$$

Notice that the submodule $\underline{N}_{N-p}$ itself is the direct sum of an invariant submodule of dimension $(N-p)$, and a vector subspace of dimension p. We shall denote this as follows :

$$\underline{N}_{N-p} = \underline{N-p} \oplus p$$

---

[5] Here we label the submodules by underlining their dimension.

with
$$\underline{N-p} = \begin{pmatrix} \beta_1\theta_1\theta_2 & \beta_2\theta_1\theta_2 & \cdots & \beta_{N-p}\theta_1\theta_2 & 0 & \cdots & 0 \end{pmatrix} \quad (23)$$

- Second type : submodules of P_p.

$$0 \hookrightarrow \underline{p} \hookrightarrow \underline{N}_p \hookrightarrow \underline{2N-p} \hookrightarrow \underline{2N'} = P_p$$

An element of the submodule $\underline{N}_p$ (which has dimension N) is of the kind :

$$\underline{N}_p = \begin{pmatrix} \gamma_1\theta_{\lambda_1} & \gamma_2\theta_{\lambda_2} & \cdots & \gamma_{N-p}\theta_{\lambda_{N-p}} & \beta_1\theta_1\theta_2 & \beta_2\theta_1\theta_2 & \cdots & \beta_p\theta_1\theta_2 \end{pmatrix} \quad (24)$$

Notice that the submodule $\underline{N}_p$ itself is the direct sum of an invariant submodule of dimension p, and a vector subspace of dimension $N-p$

$$\underline{N}_p = \underline{p} \not\in (N-p)$$

with
$$\underline{p} = \begin{pmatrix} 0 & 0 & \cdots & 0 & \beta_1\theta_1\theta_2 & \beta_2\theta_1\theta_2 & \cdots & \beta_p\theta_1\theta_2 \end{pmatrix} \quad (25)$$

Notice that the quotient of a PIM by its own radical defines an irreducible representation for the quantum group $\mathcal{H}$. The irreducible representations for $\mathcal{H}$ are therefore of dimensions $1, 2, 3 \ldots N$. Warning : the projective cover of a given indecomposable module appearing in one of the above sublattices is not necessarily equal to the principal module that appears as the maximum element of the same sublattice.

We have already noticed that each PIM contains indecomposable submodules N_p of dimension exactly equal to N, each such submodule containing itself one invariant irreducible subspace of dimension p (the precise definition involves the choice of a parameter λ). As we shall see, these submodules of dimension N are exactly those that appear in the decomposition of the algebra of complex $N \times N$ matrices into representations of $\mathcal{H}$.

As an example, let us explicitly describe the case $N = 5$. Then, $dim\mathcal{H} = 5^3 = 125$. We can write $5 = 5 + 0 = 4 + 1 = 3 + 2$, so that $\mathcal{H} = M_5 \oplus (M_{4|1}(\Lambda^2))_0 \oplus (M_{3|2}(\Lambda^2))_0$ and we have five principal modules, one is irreducible ($\underline{5}_{irr}$, of dimension 5), the others (P_4, P_1, P_3, P_2) are projective indecomposable and have the same dimension 10. One can also write $\mathcal{H} = 5\,\underline{5}_{irr} \oplus 4\,P_4 + 1\,P_1 + 3\,P_3 + 2\,P_2$.

The lattices of submodules read:

$$0 \hookrightarrow \underline{4} \hookrightarrow \underline{5}_4 \hookrightarrow \underline{6} \hookrightarrow \underline{10'} = P_4$$
$$0 \hookrightarrow \underline{1} \hookrightarrow \underline{5}_1 \hookrightarrow \underline{9} \hookrightarrow \underline{10''} = P_1$$
$$0 \hookrightarrow \underline{3} \hookrightarrow \underline{5}_3 \hookrightarrow \underline{7} \hookrightarrow \underline{10'''} = P_3$$
$$0 \hookrightarrow \underline{2} \hookrightarrow \underline{5}_2 \hookrightarrow \underline{8} \hookrightarrow \underline{10''''} = P_2$$

Besides the five-dimensional irreducible representation which is itself a PIM, there are four others : $\underline{4}_{irr} = P_4/\underline{6}$, $\underline{1}_{irr} = P_1/\underline{9}$, $\underline{3}_{irr} = P_3/\underline{7}$ and $\underline{2}_{irr} = P_2/\underline{8}$.

REDUCTION OF THE ALGEBRA $\mathcal{M} \doteq M_N(\mathbb{C})$, N ODD, INTO INDECOMPOSABLE REPRESENTATIONS OF $\mathcal{H}$

We shall now focus our attention on the case of $N \times N$ complex matrices, in the case where N is odd, and show how this familiar algebra can be reconstructed as a sum of representations for the finite dimensional quantum group $\mathcal{H}$.

A vectorial basis of this algebra is given by matrices

$$\{x^r y^s\} \qquad r, s \in \{0, 1, \cdots, N-1\}$$

where x and y are the particular generators already defined in section 2. The left action of generators of $\mathcal{H}$ on an arbitrary element of $\mathcal{M}$ is given by:

$$K[x^r y^s] = q^{(r-s)} x^r y^s$$

$$X_+[x^r y^s] = q^r \frac{1-q^{-2s}}{1-q^{-2}} x^{r+1} y^{s-1}$$

$$X_-[x^r y^s] = q^s \frac{1-q^{-2r}}{1-q^{-2}} x^{r-1} y^{s+1}$$

The generator K always acts as an automorphism, for this reason, in order to study the invariant subspaces of $\mathcal{M}$ under the left action of $\mathcal{H}$, we shall only have to consider the action of X_+ et X_-.

Forgetting numerical factors, the action de X_+ and of X_- on a given element of $\mathcal{M}$ can be written as follows:

$$x^{r+1} y^{s-1} \underset{X_-}{\overset{X_+}{\longleftrightarrow}} x^r y^s \underset{X_-}{\overset{X_+}{\longleftrightarrow}} x^{r-1} y^{s+1}$$

It is then easy to decompose the space of $N \times N$ matrices into invariant subspaces for this action.

• Starting from the element x^{N-1}, we have :

$$0 \overset{X_+}{\uparrow} x^{N-1} \underset{X_-}{\overset{X_+}{\longleftrightarrow}} x^{N-2} y \cdots xy^{N-2} \underset{X_-}{\overset{X_+}{\longleftrightarrow}} y^{N-1} \overset{X_-}{\downarrow} 0$$

We obtain in this way an **irreducible** subspace of dimension N, denoted N_{irr}. A base of this subspace is given by:

$$\{x^r y^s\} \qquad \text{avec} \qquad (r+s) = N-1$$

- Starting from the element x^{N-2}, we obtain the following diagram:

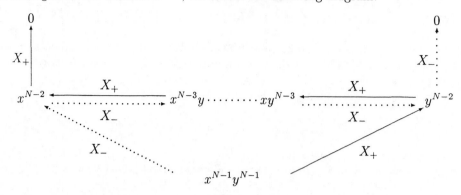

We obtain again an invariant subspace of the same dimension N, denoted N_{N-1}. A base of this subspace is given by:

$$\{x^r y^s\} \quad \text{avec} \quad (r+s) = N-2 \quad [\text{modulo } N]$$

This space is the direct sum of an invariant subspace of dimension $N-1$ (hence the notation), a basis of which being given by

$$\{x^r y^s\} \quad \text{avec} \quad (r+s) = N-2,$$

and a non-invariant vector subspace of dimension 1 generated by $\{x^{N-1} y^{N-1}\}$.
- The process can be repeated, up to:

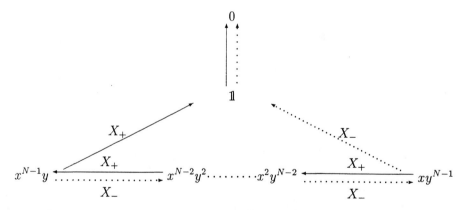

We obtain in this last case a subspace of still the same dimension N, denoted N_1 with a base given by:

$$\{x^r y^s\} \quad \text{avec} \quad (r+s) = 0 \quad [\text{modulo } N]$$

It is the direct sum of an invariant vector subspace of dimension 1, generated by $\mathbb{1}$, and a non-invariant vector subspace of dimension $N-1$.

To conclude, we see that, under the left action of $\mathcal{H}$, the algebra of $N \times N$ matrices can be decomposed into a direct sum of invariant subspaces of dimension N, according to

- $N_N = N_{irr}$: irreducible

- N_{N-1} : reducible indecomposable, with an invariant subspace of dimension $N-1$.

- N_{N-2} : reducible indecomposable, with an invariant subspace of dimension $N-2$.

- $\vdots$

- N_1 : reducible indecomposable, with an invariant subspace of dimension 1.

These representations of dimension N coincide exactly with those also called N_p (or N_{N-p}) in the previous section. Using these notations, the algebra of matrices N×N can be written

$$\mathcal{M} = \mathcal{N}_N \oplus \mathcal{N}_\infty \oplus \mathcal{N}_\epsilon \oplus \cdots \oplus \mathcal{N}_{N-\infty} \qquad (26)$$

with:

$$\begin{aligned} N_N &= N_{irr} \\ N_{N-1} &= (N-1) \not\in 1 \\ N_{N-2} &= (N-2) \not\in 2 \\ &\vdots \\ N_1 &= 1 \not\in (N-1) \end{aligned}$$

Continuing the example given at the end of section 2 we see that

$$M(5,\mathbb{C}) = \underline{5} \oplus \underline{5}_4 \oplus \underline{5}_3 \oplus \underline{5}_2 \oplus \underline{5}_1$$

ACTION OF $\mathcal{H}$ ON THE REDUCED WESS-ZUMINO COMPLEX $\Omega_{WZ}(\mathcal{M})$

The reduced Wess-Zumino complex $\Omega_{WZ}(\mathcal{M})$

The Wess-Zumino complex was constructed in [5] as the unique (up to a redefinition of $q \to q^{-1}$) quadratic differential algebra on the quantum plane, bicovariant with respect to the action of the quantum group $SL_q(2,\mathbb{C})$. The commutation relations between the $N \times N$ matrices x, y and their differentials dx, dy, and between the differentials themselves are given by:

$$xy = qyx$$
$$x\,dx = q^2 dx\,x \qquad x\,dy = q\,dy\,x + (q^2-1)dx\,y$$
$$y\,dx = q\,dx\,y \qquad y\,dy = q^2 dy\,y \qquad\qquad (27)$$
$$dx^2 = 0 \qquad dy^2 = 0$$
$$dx\,dy + q^2 dy\,dx = 0$$

The reduced quantum plane itself is not a quadratic algebra, since it contains relations like $x^N = y^N = 1$, but Leibniz rule, together with the fact that N is a root of unity imply that such compatibility relations like $dx^N = dy^N = 0$ are automatically satisfied [8] [9], indeed

$$d(x^N) = d(x^{N-1})x + x^{N-1}dx = \ldots = (1 + q + q^2 + \ldots + q^{N-1})x^{N-1}dx = 0$$
$$d(y^N) = d(y^{N-1})y + y^{N-1}dy = \ldots = (1 + q + q^2 + \ldots + q^{N-1})x^{N-1}dy = 0$$

We call "reduced Wess-Zumino complex" $\Omega_{WZ}(\mathcal{M})$ the quotient of the differential algebra of Wess-Zumino by the corresponding differential ideal. Note that $\dim(\Omega^0_{WZ}(\mathcal{M})) = N^2$, $\dim(\Omega^1_{WZ}(\mathcal{M})) = 2N^2$ and $\dim(\Omega^2_{WZ}(\mathcal{M})) = N^2$. Therefore, $\dim(\Omega_{WZ}(\mathcal{M}) = 4N^2$.

Remarks

Here is a list of questions that can be asked about $\Omega_{WZ}(\mathcal{M})$.

- Study the action of $\mathcal{H}$.
- Decompose this differential algebra into representations of $\mathcal{H}$
- Study the cohomology of d
- Extend the star operation(s) of $\mathcal{M}$ to star (or graded star) operations of the differential algebra
- Study algebraic connections in $\Omega_{WZ}(\mathcal{M})$ and their curvature
- etc.

These questions are studied in [8] [9], mostly with the particular choice $N = 3$. Here we shall only give explicitly the action of $\mathcal{H}$ on this differential algebra, for an arbitrary N.

Action of $\mathcal{H}$

We already gave in ?? the left action of generators of $\mathcal{H}$ on an arbitrary element of $\mathcal{M}$. The left action of the same elements on generators dx, dy of the Manin dual

$\mathcal{M}'$ of $\mathcal{M}$ (although the later is not quadratic, we can use this terminology !) is given by :

$$K^L[dx] = qdx \qquad X_+^L[dx] = 0 \qquad X_-^L[dx] = dy$$
$$K^L[dy] = q^{-1}dy \qquad X_+^L[dy] = dx \qquad X_-^L[dy] = 0$$

Using these two pieces of information, we now find the action of generators of $\mathcal{H}$ on arbitrary elements of $\Omega^1_{WZ}(\mathcal{M})$ thanks to the coproduct :
Let us write : $\Delta(h) = h_{(1)} \otimes h_{(2)}$ whenever $h \in \mathcal{H}$, then, for $m_1 \in \mathcal{M}$ and $m_2 \in \mathcal{M}'$, we have:

$$h^L[m_1 m_2] = h^L_{(1)}[m_1] \otimes h^L_{(2)}[m_2]$$

The left action of generators of $\mathcal{H}$ on arbitrary elements of $\Omega^1_{WZ}(\mathcal{M})$ is then given by :

$$K[x^r y^s dx] = q^{r+1-s} x^r y^s dx$$
$$K[x^r y^s dy] = q^{r-s-1} x^r y^s dy$$
$$X_+[x^r y^s dx] = q^r \frac{1-q^{-2s}}{1-q^{-2}} x^{r+1} y^{s-1} dx$$
$$X_+[x^r y^s dy] = q^r \frac{1-q^{-2s}}{1-q^{-2}} x^{r+1} y^{s-1} dy + q^{r-s} x^r y^s dx$$
$$X_-[x^r y^s dx] = q^{s-1} \frac{1-q^{-2r}}{1-q^{-2}} x^{r-1} y^{s+1} dx + x^r y^s dy$$
$$X_-[x^r y^s dy] = q^{s+1} \frac{1-q^{-2r}}{1-q^{-2}} x^{r-1} y^{s+1} dx$$

COMMENTS

The usual algebra $\mathcal{M}$ of $N \times N$ complex matrices (or the group of its unitary elements) is very often used in various fields of physics. The observation that it is a module algebra for a finite dimensional non commutative, non cocommutative (and non semi-simple) quantum group, and that it can be correspondingly decomposed into the sum of N indecomposable representations (of dimension N) came up — for us — as a surprise. One could then be tempted to speak of "hidden symmetry", whenever $\mathcal{M}$ plays a role in the description of some physical process, but we do not know yet the physical meaning of such a "symmetry", and its interpretation should certainly depend upon the particular situation at hand. The above properties nevertheless suggest that, in several branches of physics, it may be worth to study the appearance and meaning of "symmetries" associated with this quantum group $\mathcal{H}$ (which is a kind of "fat" version of the algebra of the group $\mathbb{Z}_N$).

Notice that representations of $\mathcal{H}$ are also, *a priori*, particular representations of $U_q sl(2, \mathbb{C})$ when q is a root of unity (representations for which $K^N = 1$, and $X_\pm^N = 0$). Such representations appear in several examples of conformal field theories, and also in the study of abelian anyons (particles obeying a one-dimensional statistics associated with the braid group, or with a Hecke algebra). It may be simpler and conceptually more appropriate to discuss such problems in terms of $\mathcal{H}$ than with the infinite dimensional Hopf algebra $U_q sl(2, \mathbb{C})$.

One of the purposes of the letter [8] was to construct generalized differential forms as elements of the tensor product Ξ of the space of usual (De Rham) forms on a (usual) space-time and of the differential algebra defined by the reduced Wess-Zumino complex. One obtains naturally on Ξ an action of a Hopf algebra which is the product of $\mathcal{H}$, times the envelopping algebra of the Lie algebra of the Lorentz group. $\mathcal{H}$ appears therefore as a discrete (but neither commutative nor cocommutative) analogue of the Lorentz group.

Finally, the group of unitary elements of the quotient of $\mathcal{H}$ by its radical, when N is odd, is isomorphic with

$$U(N) \times (U(N-1) \times U(1)) \times (U(N-2) \times U(2)) \times (U(N-3) \times U(3)) \times \ldots$$

For $N = 3$, one obtains $U(3) \times U(2) \times U(1)$. After appropriate identification of several $U(1)$ factors, one recognizes the gauge group describing of the standard model of elementary particles. This last observation was made in [10] where it was suggested that $\mathcal{H}$ could play the role of a Lorentz group for the "internal space" and that it would be tempting to devise a generalized gauge theory using this finite quantum group as a basic ingredient. This last comment can also be related to the construction described in [8].

REFERENCES

1. L. Dąbrowski, F. Nesti and P. Siniscalco, *A finite quantum symmetry of $M(3, \mathbb{C})$*, SISSA 63/97/FM, `hep-th/9705204`.
2. D. V. Gluschenkov and A. V. Lyakhovskaya, *Regular representation of the quantum Heisenberg double (q is a root of unity)*, Zapiski LOMI 215 (1994).
3. A. Alekseev, D. Gluschenkov and A. Lyakhovskaya, *Regular representation of the quantum group $SL_q(2)$ (q is a root of unity)*, St. Petersburg Math. J., Vol **6**, N5, 88 (1994).
4. R. Coquereaux, *On the finite dimensional quantum group $M_3 \oplus (M_{2|1}(\Lambda^2))_0$*, CPT-96/P.3388, `hep-th/9610114`, Lett. Math. Phys. **42**, 309 (1997).
5. J. Wess and B. Zumino, *Covariant differential calculus on the quantum hyperplane*, Nucl. Phys. B (Proc. Suppl.) **18B**, 302 (1990).
6. H. Weyl, *The theory of groups and quantum mechanics*, Dover Publications (1931).
7. O. Ogievetsky, *Matrix structure of $SL_q(2)$ when q is a root of unity*, CPT-96/P.3390, to appear.

8. R. Coquereaux, A. O. García and R. Trinchero, *Finite dimensional quantum group covariant differential calculus on a complex matrix algebra*, CPT-98/P.3630, math.QA/9804021.
9. R. Coquereaux, A. O. García and R. Trinchero, *Differential calculus and connections on a quantum plane at a cubic root of unity*, CPT-98/P.3632, math.QA/***
10. A. Connes, *NonCommutative Geometry and Reality*, IHES/M/95/52.
11. D. Kastler, Valavane, Proceedings of the conference of Palermo (1997), to appear.

Solutions of q-Deformed Equations with Quantum Conformal Symmetry

V.K. Dobrev, B.S. Kostadinov and S.T. Petrov

Bulgarian Academy of Sciences
Institute of Nuclear Research and Nuclear Energy
72 Tsarigradsko Chaussee, 1784 Sofia, Bulgaria
dobrev@bgearn.acad.bg

Abstract. We consider the construction of explicit solutions of a hierarchy of q-deformed equations which are (conditionally) quantum conformal invariant. We give two types of solutions - polynomial solutions and solutions in terms of q-deformation of the plane wave. We give a q-deformation of the plane wave as a formal power series in the noncommutative coordinates of q-Minkowski space-time and four-momenta. This q-plane wave has analogous properties to the classical one, in particular, it has the properties of q-Lorentz covariance, and it satisfies the q-d'Alembert equation on the q-Lorentz covariant momentum cone. On the other hand, this q-plane wave is not an exponent or q-exponent. Thus, it differs conceptually from the classical plane wave and may serve as a its regularization. Using this we give also solutions of the massless Dirac equation involving two conjugated q-plane waves - one for the neutrino, the other for the antineutrino. It is also interesting that the neutrino solutions are deformed only through the q-pane wave, while the prefactor is classical. Thus, we can speak of a definite left-right asymmetry of the quantum conformal deformation of the neutrino-antineutrino system.

INTRODUCTION

One of the purposes of quantum deformations is to provide an alternative of the regularization procedures of quantum field theory. Applied to Minkowski space-time the quantum deformations approach is also an alternative to Connes' noncommutative geometry [1]. The first step in such an approach is to construct a noncommutative quantum deformation of Minkowski space-time. There are several possible such deformations, cf. [2-8]. We shall follow the deformation of [5] which is different from the others, the most important aspect being that it is related to a deformation of the conformal group.

The first problem to tackle in a noncommutative deformed setting is to analyze the behavior of the wave equations analogues. Thus, we start here with the study of the solutions of the deformed d'Alembert equation derived in [9] with the use of quantum conformal symmetry. The most interesting question of course is the deformation of the usual plane wave. We give now a description of our setting.

It is well known that the d'Alembert equation

$$\Box\, \varphi(x) \;=\; 0\,, \qquad \Box \;=\; \partial^\mu \partial_\mu \;=\; (\partial_0)^2 - (\vec\partial)^2\,, \qquad (1)$$

is conformally invariant, cf., e.g., [10]. Here φ is a scalar field of fixed conformal weight, $x = (x_0, x_1, x_2, x_3)$ denotes the Minkowski space-time coordinates. Not known was the fact that (1) may be interpreted as conditionally conformally invariant equation and thus may be rederived from a subsingular vector of a Verma module of the algebra $sl(4)$, the complexification of the conformal algebra $su(2,2)$ [9].

The same idea was used in [9] to derive a q-d'Alembert equation, namely, as arising from a subsingular vector of a Verma module of the quantum algebra $U_q(sl(4))$. The resulting equation is a q-difference equation and the solution spaces are built on the noncommutative q-Minkowski space-time of [5].

Besides the q-d'Alembert equation in [9] were derived a whole hierarchy of equations parametrized by a natural number a. In fact, the case $a=1$ corresponds to the q-d'Alembert equation, while for each $a > 1$ there are two couples of equations involving fields of conjugated Lorentz representations of dimension a. For instance, the case $a=2$ coresponds to the massless Dirac equation, one couple of equations describing the neutrino, the other couple of equations describing the antineutrino, while the case $a=3$ corresponds to the Maxwell equations.

In [9] the solution spaces were given only via their transformation properties under $U_q(sl(4))$ and quantum conformal symmetry. However, no explicit solutions, which are important for the applications, were given. This is what we review in the present paper following mostly [11] and [12]. First we give polynomial solutions for all equations of the hierarchy. Further, we give a q-deformation of the usual plane wave. It is a formal power series in the noncommutative coordinates of q-Minkowski space-time and four-momenta. This q-plane wave has analogous properties to the classical one. In particular, it has the properties of q-Lorentz covariance, and it satisfies the q-d'Alembert equation on the q-Lorentz covariant momentum cone.

On the other hand, this q-plane wave is not an exponent or q-exponent, cf. [13]. Thus, it differs conceptually from the classical plane wave and may serve as a regularization of the latter. In the same sense it differs from the q-plane wave in the paper [14], which is not surprising, since there is used different q-Minkowski space-time (from [2-4] and different q-d'Alembert equation both based only on a (different) q-Lorentz algebra, and not on q-conformal (or $U_q(sl(4))$) symmetry as in our case. In fact, it is not clear whether the q-Lorentz algebra of [2-4] used in [14] is extendable to a q-conformal algebra.

We give also solutions of the massless Dirac equation involving two conjugated q-plane waves - one for the neutrino, the other for the antineutrino. It is also interesting that the neutrino solutions are deformed only through the q-pane wave, while the prefactor is classical. Thus, we can speak of a definite left-right asymmetry of the quantum conformal deformation of the neutrino-antineutrino system.

SOLUTIONS OF D'ALEMBERT EQUATION

In this short section we write down the solutions of (1) which we shall deform below. First we introduce new Minkowski variables:

$$x_\pm \equiv x_0 \pm x_3, \quad v \equiv x_1 - ix_2, \quad \bar{v} \equiv x_1 + ix_2, \qquad (2)$$

which, (unlike the x_μ), have definite group-theoretical interpretation as part of a coset of the conformal group $SU(2,2)$ (or of $SL(4)$ with the appropriate conjugation) [5]. The d'Alembert equation in terms of these variables is:

$$\Box \, \varphi = (\partial_- \partial_+ - \partial_v \partial_{\bar{v}}) \, \varphi = 0, \qquad (3)$$

while the Minkowski length is: $\mathcal{L} = x_- x_+ - v\bar{v} = x_0^2 - \vec{x}^2$.

In view of the generalizations to the q-deformed case we first recall the polynomial solutions of (3). There are two series of such solutions. One series is parametrized by four constants $a, b, c, d \in \mathbb{Z}_+$ and (up to a nonzero multiplicative constant) the solutions are given by:

$$\varphi = \sum_{n=0}^{n_{a,b}} \frac{(-a)_n \, (-b)_n}{(c+1)_n \, (d+1)_n} \, v^{n+d} \, x_-^{a-n} \, x_+^{b-n} \, \bar{v}^{c+n}, \qquad n_{a,b} = \min(a,b), \qquad (4)$$

where $(\alpha)_n = \Gamma(\alpha+n)/\Gamma(\alpha)$ is the Pochhammer symbol (in terms of the standard gamma function Γ).) If $cd = 0$ then (4) is given in terms of the standard hypergeometric function, e.g., for $d = 0$:

$$\varphi_{a,b,c} = x_-^a \, x_+^b \, \bar{v}^c \, {}_2F_1(-a, -b, c+1; \tfrac{v\bar{v}}{x_- x_+}), \qquad (5a)$$

$$\,_2F_1(\alpha, \beta, \gamma; \zeta) = \sum_{n=0}^\infty \frac{(\alpha)_n \, (\beta)_n}{(\gamma)_n \, n!} \zeta^n. \qquad (5b)$$

The other general solution is obtained from (4) by interchanging $(x_+, x_-) \longleftrightarrow (v, \bar{v})$.

These $sl(4)$ symmetric solutions are well known [15]. (Note that (4) may be extended to arbitrary complex a, b, c, d, except that $c, d \notin -\mathbb{N}$.) Not so well known is that one can construct the standard plane wave from such solutions. We consider $\exp(k \cdot x)$, where

$$(k \cdot x) = k^\mu x_\mu = \tfrac{1}{2}(k_- x_+ + k_+ x_- - k_v \bar{v} - k_{\bar{v}} v), \qquad (6)$$

and $(k_v, k_-, k_+, k_{\bar{v}})$ are related to the components k_μ of the four-momentum as the variables $(v, x_-, x_+, \bar{v})$ are related to x_μ. Clearly, $\exp(k \cdot x)$ satisfies (3) iff the four-momentum is on the light-cone:

$$\mathcal{L}^k = k_- k_+ - k_v k_{\bar{v}} = 0, \qquad (7)$$

since one has:

$$\square \, (k \cdot x)^s \;=\; \tfrac{1}{4}\, s(s-1)\, (k_- k_+ - k_v k_{\bar v})\, (k \cdot x)^{s-2} \,. \tag{8}$$

The seemingly unknown fact which we would like record is that $\exp(k \cdot x)$ may be obtained as a formal linear combination of all solutions (5a). [1] Explicitly, we have:

$$\exp(k \cdot x) \;=\; \sum_{s \in \mathbb{Z}_+} \frac{1}{s!}\, (k \cdot x)^s \;=\; \tag{9a}$$

$$= \sum_{a,b,c \in \mathbb{Z}_+} \frac{(-1)^c\, 2^{-(a+b+c)}}{a!\, b!\, c!}\, k_v^c\, k_-^b\, k_+^a\, \varphi_{a,b,c}(v, x_-, x_+, \bar v) \,. \tag{9b}$$

To obtain (9b) we substitute in (9a) the general binomial expansion rewritten as follows:

$$(k \cdot x)^s \;=\; \tfrac{s!}{2^s} \sum_{a,b,n \in \mathbb{Z}_+} \frac{(-1)^{s-a-b}}{\Gamma(a-n+1)\,\Gamma(b-n+1)\,\Gamma(s-a-b+n+1)\, n!} \times$$
$$\times\, (k_{\bar v} v)^n\, (k_+ x_-)^{a-n}\, (k_- x_+)^{b-n}\, (k_v \bar v)^{s-a-b+n} \;=\; \tag{10a}$$

$$= \tfrac{s!}{2^s} \sum_{a,b \in \mathbb{Z}_+} \frac{(-1)^{s-a-b}}{a!\, b!\, \Gamma(s-a-b+1)}\, k_v^{s-a-b}\, k_-^b\, k_+^a\, \varphi_{a,b,s-a-b} \,, \tag{10b}$$

where summation is truncated as necessary by the property: $1/\Gamma(-m) = 0$ if $m \in \mathbb{Z}_+$, the property: $(-a)_n = (-1)^n (a-n+1)_n$ is used, and passing from (10a) to (10b) is used (7). Formulae (10) are important since they are generalizable to the q-deformed setting as we show below.

SOLUTIONS OF THE Q-D'ALEMBERT EQUATION

Polynomial Solutions

This section is based mostly on [11]. Here we give the solutions of the q-d'Alembert equation of [9] formulated on the noncommutative q-Minkowski space-time of [5]. The latter is given by the following commutation relations:

$$x_\pm v = q^{\pm 1} v x_\pm \,, \quad x_\pm \bar v = q^{\pm 1} \bar v x_\pm \,,$$
$$x_+ x_- - x_- x_+ = \lambda v \bar v \,, \quad \bar v v = v \bar v \,, \quad (\lambda \equiv q - q^{-1}) \,, \tag{11}$$

with the deformation parameter being a phase: $|q| = 1$. The q-Minkowski length is:

$$\mathcal{L}_q \;=\; x_- x_+ - q^{-1} v \bar v \,. \tag{12}$$

It commutes with the q-Minkowski coordinates and has the correct classical limit $\mathcal{L}_{q=1} = \mathcal{L}$. Relations (11) are preserved by the anti-linear anti-involution ω acting as :

---

[1] This fact was noticed by the first two authors and P.A. Terziev.

$$\omega(x_\pm) = x_\pm, \quad \omega(v) = \bar{v}, \quad \omega(q) = \bar{q} = q^{-1}, \quad (\omega(\lambda) = -\lambda), \tag{13}$$

from which follows also that $\omega(\mathcal{L}_q) = \mathcal{L}_q$.

The solution space consists of formal power series in the q-Minkowski coordinates:

$$\hat{\varphi} = \sum_{j,n,\ell,m \in \mathbb{Z}_+} \mu_{jn\ell m}\, \hat{\varphi}_{jn\ell m}, \qquad \hat{\varphi}_{jn\ell m} = v^j\, x_-^n\, x_+^\ell\, \bar{v}^m. \tag{14}$$

The solution space (14) is a representation space of the quantum algebra $U_q(sl(4))$. The latter is defined, (cf. [16]), as the associative algebra over $\mathcal{C}$ with unit element $1_\mathcal{U}$, generators $k_i^\pm$, $X_i^\pm$, $i = 1,2,3$, and nontrivial relations ($[x]_q \equiv (q^x - q^{-x})/\lambda$):

$$k_i k_j = k_j k_i, \quad k_i k_i^{-1} = k_i^{-1} k_i = 1_\mathcal{U}, \quad k_i X_j^\pm = q^{\pm c_{ij}} X_j^\pm k_i, \tag{15a}$$

$$[X_i^+, X_j^-] = \delta_{ij}\left(k_i^2 - k_i^{-2}\right)/\lambda, \tag{15b}$$

$$\left(X_i^\pm\right)^2 X_j^\pm - [2]_q X_i^\pm X_j^\pm X_i^\pm + X_j^\pm \left(X_i^\pm\right)^2 = 0, \quad |i-j| = 1. \tag{15c}$$

Further we suppose that q is not a nontrivial root of unity. The action of $U_q(sl(4))$ on $\hat{\varphi}$ is given explicitly by (cf. [17]):

$$\pi(k_1)\, \hat{\varphi}_{jn\ell m} = q^{(j-n+\ell-m)/2}\, \hat{\varphi}_{jn\ell m}, \tag{16a}$$

$$\pi(k_2)\, \hat{\varphi}_{jn\ell m} = q^{n+(j+m+1)/2}\, \hat{\varphi}_{jn\ell m}, \tag{16b}$$

$$\pi(k_3)\, \hat{\varphi}_{jn\ell m} = q^{(-j-n+\ell+m)/2}\, \hat{\varphi}_{jn\ell m}, \tag{16c}$$

$$\pi(X_1^+)\, \hat{\varphi}_{jn\ell m} = q^{-1+(j-n-\ell+m)/2}\, [n]_q\, \hat{\varphi}_{j+1,n-1,\ell m} + $$
$$+ q^{-1+(j-n+\ell-m)/2}\, [m]_q\, \hat{\varphi}_{jn,\ell+1,m-1}, \tag{17a}$$

$$\pi(X_2^+)\, \hat{\varphi}_{jn\ell m} = q^{(-j+m)/2}\, [j+n+m+1]_q\, \hat{\varphi}_{j,n+1,\ell m} + $$
$$+ q^{1+(j+n+3m)/2}\, [\ell]_q\, \hat{\varphi}_{j+1,n,\ell-1,m+1}, \tag{17b}$$

$$\pi(X_3^+)\, \hat{\varphi}_{jn\ell m} = -\, q^{-1+(j+n-\ell-m)/2}\, [j]_q\, \hat{\varphi}_{j-1,n,\ell+1,m} - $$
$$-\, q^{-1+(3j+n-3\ell-m)/2}\, [n]_q\, \hat{\varphi}_{j,n-1,\ell,m+1}, \tag{17c}$$

$$\pi(X_1^-)\, \hat{\varphi}_{jn\ell m} = q^{2+(-j+n-\ell+m)/2}\, [j]_q\, \hat{\varphi}_{j-1,n+1,\ell m} + $$
$$+ q^{2+(j-n-\ell+m)/2}\, [\ell]_q\, \hat{\varphi}_{jn,\ell-1,m+1}, \tag{18a}$$

$$\pi(X_2^-)\, \hat{\varphi}_{jn\ell m} = -\, q^{(j-m)/2}\, [n]_q\, \hat{\varphi}_{j,n-1,\ell m}, \tag{18b}$$

$$\pi(X_3^-)\, \hat{\varphi}_{jn\ell m} = -\, q^{(-j-3n+\ell+3m)/2}\, [\ell]_q\, \hat{\varphi}_{j+1,n,\ell-1,m} - $$
$$-\, q^{(-j-n+\ell+m)/2}\, [m]_q\, \hat{\varphi}_{j,n+1,\ell,m-1}. \tag{18c}$$

(Note that the representation formulae in [17] are for general holomorphic representations of $U_q(sl(4))$ characterized by three integers (r_1, r_2, r_3), (which here are $(0, -1, 0)$), and the functions depend in general on two additional variables $z, \bar{z}$.)

Next we recall the q-d'Alembert equation from [9]:

$$\Box_q \hat{\varphi} = \sum_{j,n,\ell,m \in \mathbb{Z}_+} \mu_{jn\ell m} \Box_q \hat{\varphi}_{jn\ell m} = 0 , \tag{19a}$$

$$\Box_q \hat{\varphi}_{jn\ell m} = q^{1+n+2m+2j+\ell}[n]_q[\ell]_q \hat{\varphi}_{j,n-1,\ell-1,m} - q^{n+j+\ell+m}[j]_q[m]_q \hat{\varphi}_{j-1,n,\ell,m-1} \tag{19b}$$

First we write down the analogue of (4), (with $a, b, c, d \in \mathbb{Z}_+$):

$$\hat{\varphi} = \sum_{n=0}^{n_{a,b}} q^{n(c+d+n)} \frac{(-a)_n^q (-b)_n^q}{(c+1)_n^q (d+1)_n^q} v^{n+d} x_-^{a-n} x_+^{b-n} \bar{v}^{c+n} , \tag{20}$$

where $(\alpha)_n^q = \Gamma_q(\alpha + n)/\Gamma_q(\alpha)$ is the q-Pochhammer symbol (in terms of the q-gamma function Γ_q), and then the analogue of (5):

$$\hat{\varphi}_{a,b,c} = \sum_{n=0}^{n_{a,b}} q^{n(c+n)} \frac{(-a)_n^q (-b)_n^q}{(c+1)_n^q [n]_q!} v^n x_-^{a-n} x_+^{b-n} \bar{v}^{c+n} , \tag{21}$$

where $[n]_q! = \Gamma_q(n+1) = [n]_q[n-1]_q \cdots [1]_q$, $[0]_q! = 1$, and the property $1/\Gamma_q(-m) = 0$ if $m \in \mathbb{Z}_+$ is used.

Remark: Equation (19) may be rewritten in a form closer to the $q = 1$ in (3) by introducing q-difference operators. For this we first define the operators:

$$\hat{M}_\kappa^\pm \hat{\varphi} = \sum_{j,n,\ell,m \in \mathbb{Z}_+} \mu_{jn\ell m} \hat{M}_\kappa^\pm \hat{\varphi}_{jn\ell m} , \quad \kappa = \pm, v, \bar{v} , \tag{22a}$$

$$T_\kappa^\pm \hat{\varphi} = \sum_{j,n,\ell,m \in \mathbb{Z}_+} \mu_{jn\ell m} T_\kappa^\pm \hat{\varphi}_{jn\ell m} , \quad \kappa = \pm, v, \bar{v} , \tag{22b}$$

and $\hat{M}_+^\pm, \hat{M}_-^\pm, \hat{M}_v^\pm, \hat{M}_{\bar{v}}^\pm$, resp., acts on $\hat{\varphi}_{jn\ell m}$ by changing by ± 1 the value of j, n, ℓ, m, resp., while $T_+^\pm, T_-^\pm, T_v^\pm, T_{\bar{v}}^\pm$, resp., acts on $\hat{\varphi}_{jn\ell m}$ by multiplication by $q^{\pm j}, q^{\pm n}, q^{\pm \ell}, q^{\pm m}$, resp. Using the above we define the q-difference operators as follows:

$$\hat{\mathcal{D}}_\kappa \hat{\varphi} = \frac{1}{\lambda} \hat{M}_\kappa^{-1} \left(T_\kappa - T_\kappa^{-1} \right) \hat{\varphi} . \tag{23}$$

Using (22) and (23) then (19) may be rewritten as:

$$\left(q \hat{\mathcal{D}}_- \hat{\mathcal{D}}_+ T_v T_{\bar{v}} - \hat{\mathcal{D}}_v \hat{\mathcal{D}}_{\bar{v}} \right) T_v T_- T_+ T_{\bar{v}} \hat{\varphi} = 0 . \tag{24}$$

Note that the operators in (22), (23), (24) for different variables commute, i.e., using these one is technically passing to commuting variables. Note that keeping the normal ordering it is straightforward to interchange commuting and noncommuting variables. ◇

q-Plane Wave

We want to q-deform the plane wave. Clearly, the most general q-deformation is:

$$(\exp(k \cdot x))_q = \sum_{s=0}^{\infty} \frac{1}{[s]_q!} f_s(v, x_-, x_+, \bar{v}) , \qquad (25)$$

where f_s is a homogeneous polynomial of degree s in both sets of variables $(k_v, k_-, k_+, k_{\bar{v}})$ and $(v, x_-, x_+, \bar{v})$, such that $(f_s)|_{q=1} = (k \cdot x)^s$. Thus, we set $f_0 = 1$. One may expect that f_s for $s > 1$ would be equal or at least proportional to $(f_1)^s$, but the outcome would be that this is not the case. In order to proceed systematically we have to impose the conditions of q-Lorentz covariance and the q-d'Alembert equation.

The complexification of the q-Lorentz subalgebra of the q-conformal algebra is generated by $k_j^{\pm}$, $X_j^{\pm}$, $j = 1, 3$. Using (17a,c), (18a,c) it is easy to check that:

$$\pi(X_j^{\pm}) \mathcal{L}_q = 0 , \implies \pi(X_j^{\pm}) (\mathcal{L}_q)^s = 0 , \qquad j = 1, 3. \qquad (26)$$

Since $(k \cdot x)^s$ is a scalar as $(\mathcal{L}_q)^s$, then also the q-deformations f_s should be scalars, and thus also should obey (26). In order to implement this we suppose that the momentum components are also non-commutative obeying the same rules (11) as the q-Minkowski coordinates, and that they commute with the coordinates. Also the ordering of the momentum basis will be the same for the coordinates. Taking all this into account we can see that a natural expression for f_s is:

$$f_s = \sum_{a,b,n \in \mathbb{Z}_+} \beta_{a,b,n}^s \frac{(-1)^{s-a-b}}{\Gamma_q(a-n+1) \Gamma_q(b-n+1) \Gamma_q(s-a-b+n+1) [n]_q!} \times$$
$$\times k_v^{s-a-b+n} k_-^{b-n} k_+^{a-n} k_{\bar{v}}^n v^n x_-^{a-n} x_+^{b-n} \bar{v}^{s-a-b+n} , \qquad (27)$$

where we have introduced some factors that are obvious from the correspondence with (10a) for $q = 1$. (The expression in (27) does not involve terms that would vanish for $q = 1$. Actually, we shall see that such expressions would lead to non-covariant momenta light-cone.) In order to implement q-Lorentz covariance we impose the conditions:

$$\pi(X_j^{\pm}) f_s = 0, \qquad j = 1, 3 . \qquad (28)$$

For this calculation we suppose that the q-Lorentz action on the noncommutative momenta is given by (16a,c), (17a,c), (18a,c). We also have to use the twisted derivation rule [17]:

$$\pi(X_j^{\pm}) \psi \cdot \psi' = \pi(X_j^{\pm}) \psi \cdot \pi(k_j^{-1}) \psi' + \pi(k_j) \psi \cdot \pi(X_j^{\pm}) \psi' , \qquad (29)$$
$$\psi = k_v^{s-a-b+n} k_-^{b-n} k_+^{a-n} k_{\bar{v}}^n , \qquad \psi' = v^n x_-^{a-n} x_+^{b-n} \bar{v}^{s-a-b+n} .$$

The four conditions (28) bring eight relations between the coefficients β, however only three are independent, namely, the relations:

$$\beta^s_{a,b,n} = q^{-s-2n+a+2b}\,\beta^s_{a,b-1,n}\,, \tag{30a}$$

$$\beta^s_{a,b,n} = q^{s-2n-2a+b}\,\beta^s_{a-1,b,n}\,, \tag{30b}$$

$$\beta^s_{a,b,n} = q^{s+4n-2a-2b-2}\,\beta^s_{a,b,n-1}\,, \tag{30c}$$

solving which we find the following solution:

$$\beta^s_{a,b,n} = q^{n(s-2a-2b+2n)\,+\,a(s-a-1)\,+\,b(-s+a+b+1)}\,\beta^s_{0,0,0}\,, \tag{31}$$

i.e., for each $s \geq 1$ only one constant remains to be fixed.

Next we impose the q-d'Alembert equation:

$$\Box_q\,f_s = 0\,, \tag{32}$$

which holds trivially for $s = 0, 1$. For $s \geq 2$ we substitute (27) to obtain:

$$\Box_q\,f_s = \sum_{a,b,n \in \mathbb{Z}_+} \frac{\beta^s_{a,b,n}\,(-1)^{s-a-b}}{\Gamma_q(a-n+1)\,\Gamma_q(b-n+1)\,\Gamma_q(s-a-b+n+1)\,[n]_q!} \times$$
$$\times\,k_v^{s-a-b+n}\,k_-^{b-n}\,k_+^{a-n}\,k_{\bar v}^n\,\Box_q\,\hat\varphi_{n,a-n,b-n,s-a-b+n} = \tag{33a}$$

$$= \sum_{a,b,n \in \mathbb{Z}_+} \beta^s_{a,b,n}\,\frac{(-1)^{s-a-b}\,q^{2s+1+2n-a-b}}{\Gamma_q(a-n)\,\Gamma_q(b-n)\,\Gamma_q(s-a-b+n+1)\,[n]_q!} \times$$
$$\times\,k_v^{s-a-b+n}\,k_-^{b-n}\,k_+^{a-n}\,k_{\bar v}^n\,\hat\varphi_{n,a-n-1,b-n-1,s-a-b+n}\, -$$

$$-\sum_{a,b,n \in \mathbb{Z}_+} \beta^s_{a,b,n+1}\,\frac{(-1)^{s-a-b}\,q^s}{\Gamma_q(a-n)\,\Gamma_q(b-n)\,\Gamma_q(s-a-b+n+1)\,[n]_q!} \times$$
$$\times\,k_v^{s-a-b+n+1}\,k_-^{b-n-1}\,k_+^{a-n-1}\,k_{\bar v}^{n+1}\,\hat\varphi_{n,a-n-1,b-n-1,s-a-b+n} = \tag{33b}$$

$$= (q\,k_-\,k_+ \,-\, k_v\,k_{\bar v})\sum_{a,b,n \in \mathbb{Z}_+} \beta^s_{a,b,n}\,\frac{(-1)^{s-a-b}\,q^{2s+2n-a-b}}{\Gamma_q(a-n)\,\Gamma_q(b-n)\,[n]_q!} \times$$
$$\times\,\frac{1}{\Gamma_q(s-a-b+n+1)}\,k_v^{s-a-b+n}\,k_-^{b-n-1}\,k_+^{a-n-1}\,k_{\bar v}^n \times$$
$$\times\,\hat\varphi_{n,a-n-1,b-n-1,s-a-b+n} = \tag{33c}$$

$$= (k_-\,k_+ \,-\, q^{-1}k_v\,k_{\bar v})\,\frac{q^{2s}\,\beta^s_{0,0,0}}{\beta^{s-2}_{0,0,0}}\,f_{s-2}\,, \tag{33d}$$

where to pass from (33b) to (33c) we have used the commutation rules of the momenta and (30c). If (32) holds then for every $s \geq 2$ we obtain (as for q=1) the condition that the momentum operators are on the q-Lorentz covariant q-light cone (cf. (7), (12)):

$$\mathcal{L}^k_q = k_-k_+ \,-\, q^{-1}\,k_v k_{\bar v} = 0\,. \tag{34}$$

31

By (33d) we have also the analogue of the classical property (8).

We have also the analogue of (10b):

$$f_s = \beta^s_{0,0,0} \sum_{a,b \in \mathbb{Z}_+} (-1)^{s-a-b} q^{a(s-a-1) + b(-s+a+b+1)} \times$$

$$\times \sum_n \frac{q^{n(s-2a-2b+2n)}}{\Gamma_q(a-n+1)\,\Gamma_q(b-n+1)\,\Gamma_q(s-a-b+n+1)\,[n]_q!} \times$$

$$\times k_v^{s-a-b+n}\, k_-^{b-n}\, k_+^{a-n}\, k_{\bar v}^n\, v^n\, x_-^{a-n}\, x_+^{b-n}\, \bar v^{s-a-b+n} = \quad (35a)$$

$$= \beta^s_{0,0,0} \sum_{a,b \in \mathbb{Z}_+} (-1)^{s-a-b} q^{a(s-a-1) + b(-s+a+b+1)} \times$$

$$\times \sum_n \frac{q^{n(s-a-b+n)}}{\Gamma_q(a-n+1)\,\Gamma_q(b-n+1)\,\Gamma_q(s-a-b+n+1)\,[n]_q!} \times$$

$$\times k_v^{s-a-b}\, k_-^{b}\, k_+^{a}\, v^n\, x_-^{a-n}\, x_+^{b-n}\, \bar v^{s-a-b+n} = \quad (35b)$$

$$= \beta^s_{0,0,0} \sum_{a,b \in \mathbb{Z}_+} \frac{(-1)^{s-a-b}\, q^{a(s-a-1)+b(-s+a+b+1)}}{[a]_q!\,[b]_q!\,\Gamma_q(s-a-b+1)} \times$$

$$\times k_v^{s-a-b}\, k_-^b\, k_+^a\, \hat\varphi_{a,b,s-a-b}, \quad (35c)$$

where in passing from (35a) to (35b) we have used the commutation relations of the momenta and (34).

Now it remains only to fix the coefficient $\beta^s_{0,0,0}$. We note that for q=1 it holds:

$$(k \cdot x)|_{k \to x} = (x \cdot x) = \mathcal{L}, \quad (36)$$

and thus we shall impose the conditions:

$$(f_s)|_{k \to x} = (\mathcal{L}_q)^s. \quad (37)$$

Next we note that:

$$(\mathcal{L}_q)^s = \sum_{n=0}^{s} (-1)^n \binom{s}{n}_q q^{n(n-s-1)}\, v^n\, x_-^{s-n}\, x_+^{s-n}\, \bar v^n. \quad (38)$$

A tedious calculation shows that:

$$(f_s)|_{k \to x} = \beta^s_{0,0,0}\, (\mathcal{L}_q)^s \sum_{p=0}^{s} \frac{q^{(s-p)(p-1)+p}}{[p]_q!\,[s-p]_q!}, \quad (39)$$

and comparing (39) with (37) we finally obtain:

$$\left(\beta^s_{0,0,0}\right)^{-1} = \sum_{p=0}^{s} \frac{q^{(s-p)(p-1)+p}}{[p]_q!\,[s-p]_q!}. \quad (40)$$

Note that $\left(\beta^s_{0,0,0}\right)^{-1}\big|_{q=1} = 2^s/s!$, as expected.

It remains to explain why we have not used $(f_1)^s$ for $s > 1$ instead of f_s, since $(f_1)^s$ satisfies q-Lorentz covariance (28) if f_1 does. First, let us stress that this would have been a different solution since f_s for $s > 1$ is not equal, and not even proportional, to $(f_1)^s$. This is seen already by comparing the cases $s = 1, 2$:

$$f_1 = \beta^1_{0,0,0} (qk_-x_+ + q^{-1}k_+x_- - k_v\bar{v} - k_{\bar{v}}v), \tag{41}$$

$$\begin{aligned}f_2 = \beta^2_{0,0,0} (&q^2k_-^2x_+^2 + q^{-2}k_+^2x_-^2 + k_v^2\bar{v}^2 + k_{\bar{v}}^2v^2 + \\&+ q[2]_q k_-k_+x_-x_+ - [2]_q k_v k_+ x_- \bar{v} - q^{-2}[2]_q k_+ k_{\bar{v}} vx_- - \\&- [2]_q k_v k_- x_+ \bar{v} - q^2[2]_q k_- k_{\bar{v}} vx_+ + q[2]_q k_v k_{\bar{v}} v\bar{v}).\end{aligned} \tag{42}$$

Clearly, $(f_1)^2$ contains terms proportional to λ that are not present in f_2. The problem with $(f_1)^s$ is that imposing on it the q-d'Alembert equation will bring a s-dependent relation between the momenta, which is not q-Lorentz covariant. For instance, for $s = 2$ imposing: $\Box_q (f_1)^2 = 0$ results in the following condition on the momenta: $[2]_q k_-k_+ = (3 - q^2) k_v k_{\bar{v}}$ instead of (34).

POLYNOMIAL SOLUTIONS FOR NONZERO SPIN

This section is based mostly on [12]. As we mentioned for $a \in I\!N + 1$ there are two couples of equations involving fields of conjugated Lorentz representations of dimension a. One of the couples is [9]:

$$\{q^a[a - 1 - N_z]_q \hat{\mathcal{D}}_v T_- - \hat{\mathcal{D}}_- \hat{\mathcal{D}}_z\} T_z T_v T_+ \hat{\varphi} = 0 \tag{43}$$

$$\{q^a[a - 1 - N_z]_q \hat{\mathcal{D}}_+ T_- T_{\bar{v}} T_v - \hat{\mathcal{D}}_{\bar{v}} \hat{\mathcal{D}}_z\} T_z T_v T_+ \hat{\varphi} = 0 \tag{44}$$

The solutions are polynomials in z of degree $a - 1$ and formal power series in the q-Minkowski coordinates:

$$\hat{\varphi} = \sum_{i,j,n,l,m \in \mathbb{Z}_+} p_{ijnlm} z^i v^j x_-^n x_+^l \bar{v}^m \tag{45}$$

Substituting (45) in (43) and (44) we obtain two recurrence relations:

$$p_{i+1,j,n+1,lm} = q^{a+n} \frac{[a - i - 1]_q [j + 1]_q}{[i + 1]_q [n + 1]_q} p_{i,j+1,nlm} \tag{46}$$

$$p_{i+1,jnl,m+1} = q^{a+j+n+m} \frac{[a - i - 1]_q [l + 1]_q}{[i + 1]_q [m + 1]_q} p_{ijn,l+1,m} \tag{47}$$

Solving (46) and (47) we obtain, respectively:

$$p_{ijnlm} = q^{i(a+n-1)-\frac{i(i-1)}{2}} \frac{[a-1]_q! [j+i]_q! [n-i]_q!}{[a-i-1]_q! [i]_q! [j]_q! [n]_q!} p_{0,j+i,n-i,lm} \qquad (48)$$

$$p_{ijnlm} = q^{i(a+j+n+m-1)} \frac{[a-1]_q! [l+i]_q! [m-i]_q!}{[a-i-1]_q! [i]_q! [l]_q! [m]_q!} p_{0jn,l+i,m-i} \qquad (49)$$

Combining (48) and (49) we obtain:

$$p_{nmdbc} = q^{n(a+d+c+\frac{n}{2}-\frac{1}{2})+m(c+m+n)} \frac{[a-1]_q! [b]_q! [d]_q! [c]_q!}{[a-n-1]_q! [n]_q! [m]_q! [d-m]_q!} \times$$

$$\times \frac{1}{[b-n-m]_q! [c+n+m]_q!} p_{00dbc} \qquad (50)$$

Accordingly, the solution of (43) and (44) is given using (45) as follows (since p_{00dbc} are constants):

$$\hat{\varphi}_{dbc} = \sum_{m=0}^{min(d,b)} \sum_{n=0}^{a-1} q^{n(a+d+c+\frac{n}{2}-\frac{1}{2})+m(c+m+n)} \frac{[a-1]_q! [b]_q! [d]_q! [c]_q!}{[a-n-1]_q! [n]_q! [m]_q! [d-m]_q!} \times$$

$$\times \frac{1}{[b-n-m]_q! [c+n+m]_q!} z^n v^m x_-^{d-m} x_+^{b-n-m} \bar{v}^{c+n+m} \qquad (51)$$

Formula (51) is valid also for noncommutative coordinates. For commuting coordinates it may be written in more compact form as follows:

$$\hat{\varphi}_{dbc} = x_-^d x_+^b \bar{v}^c F_D^q \left(-b, 1-a, -d, c+1; q^{a+d+c-\frac{1}{2}} \frac{z\bar{v}}{x_+}, q^c \frac{v\bar{v}}{x_+ x_-} \right) \qquad (52)$$

where F_D^q is a q-deformation of a double hypergeometric function:

$$F_D^q[a, b_1, b_2, c, z_1, z_2] = \sum_{n=0}^{\infty} \sum_{m=0}^{\infty} \frac{q^{\frac{n^2}{2}+m^2+nm} (a)_{n+m}^q (b_1)_n^q (b_2)_m^q}{(c)_{n+m}^q [n]_q! [m]_q!} z_1^n z_2^m \qquad (53)$$

which for $q = 1$ is given by [18], (f-la 5.7.1.6). For $a = 1$ these solutions coincide with (5).

We pass now to the other couple of equations [9]:

$$(q^{-a}[a - N_{\bar{z}} - 1]_q \hat{\mathcal{D}}_+ T_{\bar{v}} - \hat{\mathcal{D}}_v \hat{\mathcal{D}}_{\bar{z}} T_-) T_- T_{\bar{v}} \hat{\varphi}' = 0 \qquad (54)$$

$$(q^{-a}[a - N_{\bar{z}} - 1]_q \hat{\mathcal{D}}_{\bar{v}} - \hat{\mathcal{D}}_- \hat{\mathcal{D}}_{\bar{z}} T_v^2 T_-) T_{\bar{v}} \hat{\varphi}' = 0 \qquad (55)$$

The solutions are polynomials in $\bar{z}$ of degree $a-1$ and formal power series in the q-Minkowski coordinates:

$$\hat{\varphi}' = \sum_{i,j,n,l,m \in \mathbb{Z}_+} p_{ijnlm}\, v^m x_-^n x_+^l \bar{v}^j \bar{z}^i \tag{56}$$

Substituting (56) in (54) and (55) we obtain two recurrence relations:

$$p_{i+1,jnl,m+1} = q^{-a+j-n-m}\frac{[a-i-1]_q[l+1]_q}{[i+1]_q[m+1]_q}\, p_{ijn,l+1,m} \tag{57}$$

$$p_{i+1,j,n+1,l,m} = q^{-a-n-2m}\frac{[a-i-1]_q[j+1]_q}{[i+1]_q[n+1]_q}\, p_{i,j+1,nlm} \tag{58}$$

The solution of (57) and (58) is:

$$p_{nmd'b'c'} = q^{-n(a+d'+c'+\frac{n}{2}-\frac{1}{2})+m(c'+m+n)}\frac{[a-1]_q![b']_q![d']_q![c']_q!}{[a-n-1]_q![n]_q![m]_q![d'-m]_q!} \times$$
$$\times \frac{1}{[b'-n-m]_q![c'+n+m]_q!}\, p_{00d'b'c'} \tag{59}$$

Accordingly, the solution of (54) and (55) is given by:

$$\hat{\varphi}'_{d'b'c'} = \sum_{m=0}^{\min(d',b')}\sum_{n=0}^{a-1} q^{-n(a+d'+c'+\frac{n}{2}-\frac{1}{2})+m(c'+m+n)}\frac{[a-1]_q![b']_q![d']_q![c']_q!}{[a-n-1]_q![n]_q![m]_q![d'-m]_q!} \times$$
$$\times \frac{1}{[b'-n-m]_q![c'+n+m]_q!}\, v^{c'+m+n} x_-^{d'-m} x_+^{b'-n-m} \bar{v}^m \bar{z}^n =$$
$$= x_-^{d'} x_+^{b'} \bar{v}^{c'}\, F_D''^{q}\left(-b', 1-a, -d', c'+1; q^{-(a+d+c-\frac{1}{2})}\frac{z\bar{v}}{x_+}, q^{c'}\frac{v\bar{v}}{x_+ x_-}\right) \tag{60}$$

where another deformation of the same double hypergeometric function as above is used:

$$F_D''^{q}[a', b'_1, b'_2, c', z_1, z_2] = \sum_{n=0}^{\infty}\sum_{m=0}^{\infty} \frac{q^{\frac{-n^2}{2}+m^2+nm}(a')^q_{n+m}(b'_1)^q_n(b'_2)^q_m}{(c')^q_{n+m}[n]_q![m]_q!}\, z_1^n z_2^m \tag{61}$$

The first formula in (60) is valid also for noncommutative coordinates. For $a=1$ these solutions coincide with (5) and (51), (52).

SOLUTIONS OF THE MASSLESS Q-DIRAC EQUATION IN TERMS OF Q-PLANE WAVES

This section is based mostly on [12]. As it was shown in [9] if a function satisfies (43) and (44) or (54) and (55) then it satisfies also the q-d'Alembert equation. Thus, it is justified to look for solutions in terms of q-plane waves.

Here we shall restrict to the case $a = 2$. In this case the equations (43), (44) and (54), (55) take the following form:

$$(q^2[1 - N_z]_q \hat{\mathcal{D}}_v T_- - \hat{\mathcal{D}}_- \hat{\mathcal{D}}_z) T_z T_v T_+ \, \hat{\varphi} = 0 \tag{62}$$

$$(q^2[1 - N_z]_q \hat{\mathcal{D}}_+ T_- T_{\bar{v}} T_v - \hat{\mathcal{D}}_{\bar{v}} \hat{\mathcal{D}}_z) T_z T_v T_+ \, \hat{\varphi} = 0 \tag{63}$$

$$(q^{-2}[1 - N_{\bar{z}}]_q \hat{\mathcal{D}}_+ T_{\bar{v}} T_v^{-1} - \hat{\mathcal{D}}_v \hat{\mathcal{D}}_{\bar{z}} T_-) \, T_- T_{\bar{v}} \, \hat{\varphi}' = 0 \tag{64}$$

$$(q^{-2}[1 - N_{\bar{z}}]_q \hat{\mathcal{D}}_{\bar{v}} - \hat{\mathcal{D}}_- \hat{\mathcal{D}}_{\bar{z}} T_v^2) \, T_{\bar{v}} \, \hat{\varphi}' = 0 \tag{65}$$

where the functions $\hat{\varphi}$, $\hat{\varphi}'$ are polynomials of first degree in z, $\bar{z}$, respectively, and can be wriiten as:

$$\hat{\varphi} = \hat{\varphi}_0 + z\hat{\varphi}_1, \qquad \hat{\varphi}' = \hat{\varphi}'_0 + \bar{z}\hat{\varphi}'_1 \tag{66}$$

We note that the above equations are a q-deformation of the massless Dirac equation. Indeed, for $q = 1$ they can be rewritten in the two-component form of the massless Dirac equation. It is well known that the latter splits into independent equations for the neutrino $\Phi_{(-)}$ and the antineutrino $\Phi_{(+)}$:

$$(\partial_{x_0} \pm (\sigma_1 \partial_{x_1} + \sigma_2 \partial_{x_2} + \sigma_3 \partial_{x_3})) \, \Phi_{(\pm)}(x) = 0 \tag{67}$$

where σ_k are the Pauli matrices:

$$\sigma_1 = \begin{pmatrix} 0 & 1 \\ 1 & 0 \end{pmatrix}, \quad \sigma_2 = \begin{pmatrix} 0 & -i \\ i & 0 \end{pmatrix}, \quad \sigma_3 = \begin{pmatrix} 1 & 0 \\ 0 & -1 \end{pmatrix} \tag{68}$$

It is easy to see that $\Phi_{(\pm)}$ are expressed through our functions (for $q = 1$) as:

$$\Phi_{(+)} = \tfrac{1}{2} \begin{pmatrix} \hat{\varphi}_0 \\ -\hat{\varphi}_1 \end{pmatrix}, \qquad \Phi_{(-)} = \tfrac{1}{2} \begin{pmatrix} \hat{\varphi}'_1 \\ \hat{\varphi}'_0 \end{pmatrix}. \tag{69}$$

Thus our field $\hat{\varphi}$ corresponds to the antineutrino, while $\hat{\varphi}'$ corresponds to the neutrino.

We start first with the q-deformation of the neutrino equations (64) and (65). We shall look for solutions in terms of the q-plane wave (27):

$$\hat{\varphi}' = \sum_{s=0}^{\infty} \frac{1}{[s]_q!} \, \psi'_s \tag{70}$$

where ψ'_s are the analogs of f_s, so we shall solve:

$$(q^{-2}[1 - N_{\bar{z}}]_q \hat{\mathcal{D}}_+ T_{\bar{v}} T_v^{-1} - \hat{\mathcal{D}}_v \hat{\mathcal{D}}_{\bar{z}} T_-) \, T_- T_{\bar{v}} \, \psi'_s = 0 \tag{71}$$

$$(q^{-2}[1 - N_{\bar{z}}]_q \hat{\mathcal{D}}_{\bar{v}} - \hat{\mathcal{D}}_- \hat{\mathcal{D}}_{\bar{z}} T_v^2) \, T_{\bar{v}} \, \psi'_s = 0 \qquad (72)$$

Furthermore we shall make the following Ansatz:

$$\psi'_s = (\alpha k_+ + \beta k_- + \gamma k_v + \delta k_{\bar{v}} + \bar{z}(\alpha' k_+ + \beta' k_- + \gamma' k_v + \delta' k_{\bar{v}})) \, f_s \qquad (73)$$

where $\alpha, \beta, \gamma, \delta, \alpha', \beta', \gamma', \delta'$ are constants to be determined. We substitute (73) in (71) and (72) for commutative Minkowski coordinates and noncommutative momenta on the q-light cone. Solving we find that:

$$\beta = 0; \quad \gamma = 0; \quad \alpha' = 0; \quad \delta' = 0; \quad \beta' = -\delta; \quad \gamma' = -\alpha \qquad (74)$$

Thus, the general solution of (71) and (72) is:

$$\psi'^{\alpha,\delta}_s = (\alpha k_+ + \delta k_{\bar{v}} - (\delta k_- + \alpha k_v)\bar{z}) \, f_s \qquad (75)$$

and so the two independent solutions are given in terms of the q-plane wave:

$$\psi'^{(1)} = (k_+ - k_v \bar{z}) \, \exp_q(k \cdot x) \qquad (76a)$$
$$\psi'^{(2)} = (k_{\bar{v}} - k_- \bar{z}) \, \exp_q(k \cdot x) \qquad (76b)$$

Let us stress that the prefactors do not depend on q, i.e., they coincide with the classical ones (which, of course, are obtained by a much shorter calculation).

Now we pass to the antineutrino field:

$$\hat{\varphi} = \sum_{s=0}^{\infty} \frac{1}{[s]_q!} \, \psi_s \qquad (77)$$

where we shall solve the q-deformed equations:

$$(q^2[1 - N_z]_q \hat{\mathcal{D}}_v T_- - \hat{\mathcal{D}}_- \hat{\mathcal{D}}_z) T_z T_v T_+ \, \psi_s = 0 \qquad (78)$$

$$(q^2[1 - N_z]_q \hat{\mathcal{D}}_+ T_- T_{\bar{v}} T_v - \hat{\mathcal{D}}_{\bar{v}} \hat{\mathcal{D}}_z) T_z T_v T_+ \, \psi_s = 0 \qquad (79)$$

Analogously to the above we shall write:

$$\psi_s = (\alpha k_+ + \beta k_- + \gamma k_v + \delta k_{\bar{v}} + z(\alpha' k_+ + \beta' k_- + \gamma' k_v + \delta' k_{\bar{v}})) \, \tilde{f}_s \qquad (80)$$

where $\tilde{f}_s$ is from the conjugated q-plane wave, namely, it is written in the conjugated basis (w.r.t. ω, cf. (13)):

$$\widetilde{\exp}_q(k \cdot x) = \sum_{s=0}^{\infty} \frac{1}{[s]_q!} \, \tilde{f}_s$$
$$\tilde{f}_s = \sum_{a,b,n} P(n,s,a,b) \, k_{\bar{v}}^n k_+^{a-n} k_-^{b-n} k_v^{s-a-b+n} \, \bar{v}^{s-a-b+n} x_+^{b-n} x_-^{a-n} v^n \qquad (81)$$

[If we use the standard basis the prefactors will depend on s and we would not be able to express the solutions in terms of a q-plane wave.] Now the general solution of (78) and (79) is:

$$\psi_s^{\alpha,\gamma} = (\alpha k_+ + \gamma k_v - q^4 z(\alpha k_{\bar{v}} + \gamma k_-)) \, \tilde{f}_s \qquad (82)$$

and so the two independent solutions of (62) and (63) are:

$$\psi^{(1)} = (k_+ - q^4 k_{\bar{v}} z) \, \widetilde{\exp}_q(k \cdot x) \qquad (83a)$$
$$\psi^{(2)} = (k_v - q^4 k_- z) \, \widetilde{\exp}_q(k \cdot x) \qquad (83b)$$

Note that unlike the neutrino case, the antineutrino prefactors are q-deformed.

Acknowledgments. V.K.D. is thankful to the organizers of the conference for the financial support.

REFERENCES

1. A. Connes, *Noncommutative Geometry*, (Academic Press, 1994).
2. U. Carow-Watamura et al, Zeit. f. Physik **C48** (1990) 159.
3. W.B. Schmidke, J. Wess and B. Zumino, Zeit. f. Physik **C52** (1991) 471.
4. S. Majid, J. Math. Phys. **32** (1991) 3246.
5. V.K. Dobrev, Phys. Lett. **341B** (1994) 133 & **346B** (1995) 427.
6. J.A. de Azcarraga, P.P. Kulish, F. Rodenas, Lett. Math. Phys. **32** (1994) 173.
7. S. Majid, J. Math. Phys. **35** (1994) 5025.
8. J.A. de Azcarraga, F. Rodenas, J. Phys. A: Math. Gen. **29** (1996) 1215.
9. V.K. Dobrev, J. Phys. A: Math. Gen. **28** (1995) 7135.
10. A.O. Barut and R. Rączka, *Theory of Group Representations and Applications*, II edition, (Polish Sci. Publ., Warsaw, 1980).
11. V.K. Dobrev and B.S. Kostadinov, Solutions of deformed d'Alembert equation with quantum conformal symmetry, ICTP preprint IC/97/164 (1997).
12. V.K. Dobrev and S.T. Petrov, Solutions of q-deformed equations with quantum conformal symmetry, ICTP preprint in preparation (1998).
13. G. Gaspar and M. Rahman, *Basic Hypergeometric Series*, (Cambridge U. Press, 1990).
14. U. Meyer, Comm. Math. Phys. **174** (1995) 447.
15. W. Miller, *Symmetry and Separation of Variables*, (Addison-Wesley, Reading, 1977).
16. M. Jimbo, Lett. Math. Phys. **10** (1985) 63-69; Lett. Math. Phys. **11** (1986) 247-252.
17. V.K. Dobrev, J. Phys. **A27** (1994) 4841 & 6633.
18. H. Bateman and A. Erdelyi, *Higher Transcendental Functions*, Vol. 1 (New-York, McGraw-Hill, 1953).

Reality Conditions on Quantum Euclidean Planes

G. Fiore*, and J. Madore[†]

*Dip. di Matematica e Applicazioni, Fac. di Ingegneria
Università di Napoli, V. Claudio 21, 80125 Napoli
[†] Max-Planck-Institut für Physik
Föhringer Ring 6, D-80805 München

Abstract. A general discussion is given of reality conditions within the context of noncommutative geometry. In particular a real differential calculus is constructed over the quantum euclidean planes.

INTRODUCTION AND MOTIVATION

We recall that noncommutative geometry is geometry in which the algebra of functions of a smooth manifold has been replaced by a noncommutative algebra $\mathcal{A}$. Although the expression was introduced by von Neumann it was Connes [2] who stressed the fact that geometry is more than just an algebra just as a manifold is more than just a set of points. He introduced the notion of 'noncommutative differential geometry' to refer in general to a noncommutative geometry with an associated differential calculus $\Omega^*(\mathcal{A})$. Just as it is possible to give many differential structures to a given topological space it is possible to define many differential calculi over a given algebra. We shall study here one particular family of noncommutative geometries known as quantum euclidean spaces [8] each of which is described by an associative algebra $\mathbb{R}_q^n$ whose structure is determined by the action of a quantum group [17,18]. Here n is an integer, the 'dimension' of the space and $q \in [1, \infty)$. The quantum euclidean spaces have [1] two covariant differential calculi which are in a sense complex conjugate one to another. There have been several attempts [14] to construct from them a real differential calculus in a natural way. We shall propose a solution to this problem in the last section.

The differential calculi which we shall consider are those for which the space of 1-forms $\Omega^1(\mathcal{A})$ is free as a left or right module over the algebra. We shall impose a slightly stronger condition and require the existence of a 'frame' or 'Stehbein', a set of basis elements θ^a which commutes with all the elements of the algebra. This condition is satisfied by a large class of examples of noncommutative geometries which can be considered as the noncommutative analogues of parallelizable manifolds. The 'frame' is a natural generalization to noncommutative geometry of the 'moving frame' of E. Cartan but one cannot speak of 'movement' since the geometry is pointless. We shall in the second section give a general description of reality conditions for these calculi. We shall show that the reality condition plays a special role in the definition of a linear connection and that in fact it places severe restrictions on the existence of such connections. To a certain extent each differential calculus has a unique torsion-free, metric-compatible linear connection. In noncommutative geometry [7] a linear connection is a couple (D, σ) consisting of a covariant derivative D and a 'generalized flip' σ, which we shall see, in the cases which interest us, is uniquely determined by the reality condition. Both can be defined in terms of the frame as the maps

$$D\theta^a = -\omega^a{}_{bc}\theta^b \otimes \theta^c, \qquad \sigma(\theta^a \otimes \theta^b) = S^{ab}{}_{cd}\theta^c \otimes \theta^d. \tag{1}$$

The covariant derivative must satisfy both a left and a right Leibniz rule

$$D(f\theta^a) = df \otimes \theta^a + fD\theta^a, \qquad D(\theta^a f) = \sigma(\theta \otimes df) + D\theta^a f. \tag{2}$$

We shall see that the condition that the curvature be real leads to a braid condition on the flip σ.

REALITY CONDITIONS

In noncommutative geometry (or algebra), reality conditions are not as natural as they can be in the commutative case; the product of two hermitian elements is no longer necessarily hermitian. The product of two hermitian differential forms is also not necessarily hermitian. If the reality condition is to be extended to a covariant derivative then there is [6,9] a unique correspondence between its existence and the existence of a left and right Leibniz rule. One can show also [9] that the matrix which determines the reality condition must satisfy the Yang-Baxter condition if the extension of the covariant derivative to tensor products is to be well-defined. This is equivalent to the braid condition for the matrix which determines the right Leibniz rule. It is necessary in discussing the reality of the curvature form.

Suppose now that $\mathcal{A}$ is a $*$-algebra. We would like to choose the differential calculus such that the reality condition $(df)^* = df^*$ holds. This can at times be difficult as we shall see below. If the calculus is based on a set of derivations [5] then we must require that a derivation X satisfy the reality condition $X^* = X$ where $X^*f = (Xf^*)^*$. We shall suppose that the frame is hermitian:

$$(\theta^a)^* = \theta^a. \tag{3}$$

This will be the case if it is dual to a set of real derivations. One finds that for general $f \in \mathcal{A}$ and $\xi \in \Omega^1(\mathcal{A})$ one has

$$(f\xi)^* = \xi^* f^*, \qquad (\xi f)^* = f^* \xi^*. \tag{4}$$

There are elements $I^{ab}{}_{cd}, J^{ab}{}_{cd} \in \mathcal{Z}(\mathcal{A})$ such that

$$(\theta^a \theta^b)^* = \imath_2(\theta^a \theta^b) = I^{ab}{}_{cd} \theta^c \theta^d, \qquad (\theta^a \otimes \theta^b)^* = \jmath_2(\theta^a \otimes \theta^b) = J^{ab}{}_{cd} \theta^c \otimes \theta^d. \tag{5}$$

The compatibility condition with the product π of the algebra $\Omega^*(\mathcal{A})$

$$\pi \circ \jmath_2 = \imath_2 \circ \pi \tag{6}$$

becomes

$$(P^{ab}{}_{cd})^* J^{cd}{}_{ef} = I^{ab}{}_{cd} P^{cd}{}_{ef} = I^{ab}{}_{ef}. \tag{7}$$

At least if the frame is associated to derivations one can show that

$$I^{ab}{}_{cd} = -P^{ba}{}_{cd}. \tag{8}$$

The compatibility condition with the product implies then that

$$(P^{ab}{}_{cd})^* P^{dc}{}_{ef} = P^{ba}{}_{ef}. \tag{9}$$

For general $\xi, \eta \in \Omega^1(\mathcal{A})$ it follows from (8) that

$$(\xi\eta)^* = -\eta^* \xi^*. \tag{10}$$

One has necessarily also the relations

$$(f\xi\eta)^* = (\xi\eta)^* f^*, \qquad (f\xi \otimes \eta)^* = (\xi \otimes \eta)^* f^* \tag{11}$$

for arbitrary $f \in \mathcal{A}$. An involution can be introduced on the algebra of forms even if they do not have a frame [3].

We shall also require the reality condition

$$D\xi^* = (D\xi)^* \tag{12}$$

on the connection. This must be consistent with the Leibniz rules. There is little one can conclude in general but if the differential is based on real derivations then from the equalities

$$(D(f\xi))^* = D((f\xi)^*) = D(\xi^* f^*) \tag{13}$$

one finds the conditions

$$(fD\xi)^* = (D\xi^*)f^* \tag{14}$$

and [7,10,9]

$$(\xi \otimes \eta)^* = \sigma(\eta^* \otimes \xi^*). \tag{15}$$

A change in σ therefore implies a change in the definition of an hermitian tensor. It is clear that there is an intimate connection between the reality condition and the right-Leibniz rule. The expression (15) for the involution on tensor products becomes the identity

$$J^{ab}{}_{cd} = S^{ba}{}_{cd}. \tag{16}$$

Equation (16) can be also read from right to left as a definition of the right-Leibniz rule in terms of the hermitian structure. The condition that the connection (1) be real can be written as

$$(\omega^a{}_{bc})^* = \omega^a{}_{de}(J^{de}{}_{bc})^*. \tag{17}$$

In order for the curvature to be real we must require that the extension D_2 [7] of the involution to the tensor product of two elements of $\Omega^1(\mathcal{A})$ be such that

$$\pi_{12} \circ D_2(\xi \otimes \eta)^* = \left(\pi_{12} \circ D_2(\xi \otimes \eta)\right)^*. \tag{18}$$

We shall impose a stronger condition. We shall require that D_2 itself be real:

$$D_2(\xi \otimes \eta)^* = (D_2(\xi \otimes \eta))^*. \tag{19}$$

This condition can be made more explicit when a frame exists. In this case the map D_2 is given by

$$D_2(\theta^a \otimes \theta^b) = -(\omega^a{}_{pq}\delta^b_r + S^{ac}{}_{pq}\omega^b{}_{cr})\theta^p \otimes \theta^q \otimes \theta^r. \tag{20}$$

To solve the reality condition (19) we introduce elements $J^{abc}{}_{def} \in \mathcal{Z}(\mathcal{A})$ such that

$$(\theta^a \otimes \theta^b \otimes \theta^c)^* = \jmath_3(\theta^a \otimes \theta^b \otimes \theta^c) = J^{abc}{}_{def}\theta^d \otimes \theta^e \otimes \theta^f. \tag{21}$$

Using (16) one finds then that the equality

$$D_2 \circ \jmath_2 = \jmath_3 \circ D_2 \tag{22}$$

can be solved [9] for $J^{abc}{}_{def}$:

$$J^{abc}{}_{def} = J^{ab}{}_{pq}J^{pc}{}_{dr}J^{qr}{}_{ef} = J^{bc}{}_{pq}J^{aq}{}_{rf}J^{rp}{}_{de}. \tag{23}$$

The second equality is the Yang-Baxter Equation written out with indices. Using this equation it follows that (21) is indeed an involution:

$$(J^{abc}{}_{pqr})^* J^{pqr}{}_{def} = \delta^a_d \delta^b_e \delta^c_f. \tag{24}$$

It is reasonable to suppose that even in the absence of a frame the Yang-Baxter condition holds. The map $\jmath_3$ can be written as

$$(\xi \otimes \eta \otimes \zeta)^* \equiv \jmath_3(\xi \otimes \eta \otimes \zeta) = \sigma_{12}\sigma_{23}\sigma_{12}(\zeta^* \otimes \eta^* \otimes \xi^*). \tag{25}$$

Because of (16) the Yang-Baxter condition for $\jmath_2$ becomes the braid equation

$$\sigma_{12}\sigma_{23}\sigma_{12} = \sigma_{23}\sigma_{12}\sigma_{23} \tag{26}$$

for the map σ.

QUANTUM EUCLIDEAN SPACES

The q-deformed euclidean 'spaces' $\mathbb{C}_q^n$ and $\mathbb{R}_q^n$ are associative algebras which are covariant under the coaction of the quantum groups $SO_q(n)$. To describe them it is convenient to introduce the projector decomposition of the corresponding braid matrix

$$\hat{R} = qP_s - q^{-1}P_a + q^{1-n}P_t \qquad (27)$$

where the P_s, P_a, P_t are $SO_q(n)$-covariant q-deformations of respectively the symmetric trace-free, antisymmetric and trace projectors. They are mutually orthogonal and their sum is equal to the identity:

$$P_s + P_a + P_t = 1. \qquad (28)$$

The trace projector is 1-dimensional and its matrix elements can be written in the form

$$P_t{}^{ij}{}_{kl} = (g^{mn}g_{mn})^{-1}g^{ij}g_{kl}, \qquad (29)$$

where g_{ij} is the q-deformed euclidean metric. The quantum euclidean space is the formal associative algebra $\mathbb{C}_q^n$ with generators x^i and relations

$$P_a{}^{ij}{}_{kl}x^k x^l = 0$$

for all i, j. One obtains the real quantum euclidean space $\mathbb{R}_q^n$ by choosing $q \in \mathbb{R}^+$ and by giving the algebra an involution defined by

$$x_i^* = x^j g_{ji}.$$

This condition is an $SO_q(n, \mathbb{R})$-covariant condition and n linearly independent, real coordinates can be obtained as combinations of the x^i. The 'length' squared

$$r^2 := g_{ij}x^i x^j = x_i^* x^i \qquad (30)$$

is $SO_q(n, \mathbb{R})$-invariant, real and generates the center $\mathcal{Z}(\mathbb{R}_q^n)$ of $\mathbb{R}_q^n$. We can extend $\mathbb{R}_q^n$ by adding to it the square root r of r^2 and the inverse r^{-1}. For various reasons having to do with the extensions of homogeneous quantum groups $SL_q(n)$, $SO_q(n)$ and q-Lorentz [15,16,13] one adds also an extra generator Λ called the dilatator and its inverse Λ^{-1} chosen such that

$$x^i \Lambda = q\Lambda x^i. \qquad (31)$$

We shall choose Λ to be unitary. Since r and Λ do not commute the center of the new extension is trivial.

There are two [1] $SO_q(n)$-covariant differential calculi $\Omega^*(\mathbb{R}_q^n)$ and $\bar{\Omega}^*(\mathbb{R}_q^n)$ over the algebra $\mathbb{R}_q^n$. Let d and $\bar{d}$ be the respective differentials and set $\xi^i = dx^i$ and $\bar{\xi}^i = \bar{d}x^i$. The calculi are determined respectively by the commutation relations

$$x^i \xi^j = q\, \hat{R}^{ij}{}_{kl} \xi^k x^l. \tag{32}$$

for $\Omega^1(\mathbb{R}_q^n)$ and

$$x^i \bar{\xi}^j = q^{-1}\, \hat{R}^{-1\,ij}{}_{kl} \bar{\xi}^k x^l \tag{33}$$

for $\bar{\Omega}^1(\mathbb{R}_q^n)$. For neither calculus is it possible to extend the involution. To construct a real calculus one must mix two conjugate ones.

We have two modules of 1-forms $\Omega^1(\mathbb{R}_q^n)$ and $\bar{\Omega}^1(\mathbb{R}_q^n)$ and we can define an involution on the direct sum $\Omega^1(\mathbb{R}_q^n) \oplus \bar{\Omega}^1(\mathbb{R}_q^n)$ by the equations

$$(f_i \xi^i)^* = \bar{\xi}^i f_i^*, \qquad (\xi^i f_i)^* = f_i^* \bar{\xi}^i. \tag{34}$$

We define the submodule

$$\Omega^1_R(\mathbb{R}_q^n) \subset \Omega^1(\mathbb{R}_q^n) \oplus \bar{\Omega}^1(\mathbb{R}_q^n) \tag{35}$$

to be the set of real 1-forms, invariant under the involution. It can be considered as the module of 1-forms of a real differential calculus. One can also define on the direct sum an involution given by

$$(\theta^a)^* = \bar{\theta}^a, \qquad (\bar{\theta}^a)^* = \theta^a.$$

We define then

$$\theta^a_R = \theta^a + \bar{\theta}^a$$

and within the direct sum a submodule $\Omega^1_R(\mathbb{R}_q^n)$ to be the set of all elements ξ of the right-hand side of (35) of the form $\xi = f_a \theta^a_R$ with $f_a = f_a^*$.

The algebra $\mathbb{R}_q^1$ has only two generators x and Λ which satisfy the commutation relation $x\Lambda = q\Lambda x$. This is a modified version of the Weyl algebra with q real instead of with unit modulus. We can represent the algebra on a Hilbert space $\mathcal{R}_q$ with basis $|k\rangle$ by

$$x|k\rangle = q^k |k\rangle, \qquad \Lambda|k\rangle = |k+1\rangle. \tag{36}$$

The Formula (32) becomes

$$x\,dx = q\,dx\,x, \qquad dx\,\Lambda = q\Lambda\,dx \tag{37}$$

for the 1-forms. If we choose

$$\lambda_1 = -\frac{q}{q-1}\Lambda \tag{38}$$

and introduce the derivation $e_1 = \text{ad}\,\lambda_1$ then

$$e_1 x = q\Lambda x, \qquad e_1 \Lambda = 0 \tag{39}$$

and the calculus (37) is defined by the condition $df(e_1) = e_1 f$ for arbitrary $f \in \mathbb{R}_q^1$.

Since Λ is unitary we have $(\lambda_1)^* \neq -\lambda_1$ and e_1 is not a real derivation. We use therefore the second differential calculus $\bar{\Omega}^*(\mathbb{R}_q^1)$ defined by the relations

$$x\bar{d}x = q^{-1}\bar{d}xx, \qquad \bar{d}x\Lambda = q\Lambda\bar{d}x \tag{40}$$

and based on the derivation $\bar{e}_1$ formed using $\bar{\lambda}_1 = -\lambda_1^*$. This calculus is defined by the condition $\bar{d}f(\bar{e}_1) = \bar{e}_1 f$ for arbitrary $f \in \mathbb{R}_q^1$. The derivation $\bar{e}_1$ is also not real. It is easy to see however that

$$e_1^* = \bar{e}_1. \tag{41}$$

considered as an equality between derivations. The frame elements θ^1 and $\bar{\theta}^1$ dual to the derivations e_1 and $\bar{e}_1$ are given by

$$\begin{array}{ll}\theta^1 = \theta_1^1 dx, & \theta_1^1 = \Lambda^{-1}x^{-1}, \\ \bar{\theta}^1 = \bar{\theta}_1^1 \bar{d}x, & \bar{\theta}_1^1 = q^{-1}\Lambda x^{-1}.\end{array} \tag{42}$$

We have now two differential calculi $\Omega^*(\mathbb{R}_q^1)$ and $\bar{\Omega}^*(\mathbb{R}_q^1)$ and we can define an involution on the direct sum $\Omega^1(\mathbb{R}_q^1) \oplus \bar{\Omega}^1(\mathbb{R}_q^1)$ by the Equations (34)

To define a real differential d_R we must embed the algebra $\mathbb{R}_q^1$ in the larger algebra $M_2(\mathbb{R}_q^1)$, the algebra of 2×2 matrices with values in $\mathbb{R}_q^1$. It will suffice to consider $\mathbb{R}_q^1 \times \mathbb{R}_q^1 \subset M_2(\mathbb{R}_q^1)$. Let $a^\dagger$ be the hermitian adjoint of the matrix a. Consider the element

$$\lambda_R = \begin{pmatrix} \lambda_1 & 0 \\ 0 & \bar{\lambda}_1 \end{pmatrix} = \frac{q}{q-1}\begin{pmatrix} -\Lambda & 0 \\ 0 & \Lambda^{-1} \end{pmatrix} \tag{43}$$

of $\mathbb{R}_q^1 \times \mathbb{R}_q^1$. By the definition of the adjoint we have

$$\lambda_R^* = \epsilon \lambda_R^\dagger \epsilon = -\lambda_R, \qquad \epsilon = \begin{pmatrix} 0 & 1 \\ 1 & 0 \end{pmatrix}. \tag{44}$$

and the associated derivation $e_R = \text{ad}\,\lambda_R$ is real. If for arbitrary $f \in \mathbb{R}_q^1$ we define

$$d_R f(e_R) = e_R f \tag{45}$$

then by definition

$$\mathbb{R}_q^1 \xrightarrow{d_R} \Omega_R^1(\mathbb{R}_q^1). \tag{46}$$

A second definition of $\Omega_R^1(\mathbb{R}_q^1)$ can be given directly in terms of d and $\bar{d}$. We have introduced a vector space of derivations $\text{Der}(\mathbb{R}_q^1)$ spanned by e_1 and $\bar{e}_1$ which is a subspace of the infinite-dimensional space of all derivations of $\mathbb{R}_q^1$. We can extend the differentials d and $\bar{d}$ to all of $\text{Der}(\mathbb{R}_q^1)$ by setting

$$df(\bar{e}_1) = 0, \qquad \bar{d}f(e_1) = 0. \tag{47}$$

The differentials could in fact be extended to act on the entire set of all derivations. From the definition (45) we find then d_R is given by

$$d_R = d + \bar{d} \tag{48}$$

and that it satisfies

$$d_R f(e_1) = df(e_1) = e_1 f, \qquad d_R f(\bar{e}_1) = \bar{d}f(\bar{e}_1) = \bar{e}_1 f. \tag{49}$$

By construction then we have

$$\begin{aligned}\theta^1(e_1) &= 1, & \theta^1(\bar{e}_1) &= 0, \\ \bar{\theta}^1(e_1) &= 0, & \bar{\theta}^1(\bar{e}_1) &= 1.\end{aligned} \tag{50}$$

From the these relations follows that the frame dual to the derivation e_R is

$$\theta_R^1 = \theta^1 + \bar{\theta}^1, \qquad \theta_R^1(e_R) = 1. \tag{51}$$

This means that the differential of $f \in \mathbb{R}_q^1$ is given by

$$d_R f = e_R f \theta_R^1 = \begin{pmatrix} e_1 f & 0 \\ 0 & \bar{e}_1 f \end{pmatrix} \theta_R^1. \tag{52}$$

The module $\Omega_R^1(\mathbb{R}_q^1)$ can be defined to be the set of all elements ξ of the right-hand side of (35) of the form $\xi = f_1 \theta_R^1$ with

$$f_1^* = \epsilon f_1^\dagger \epsilon = f_1. \tag{53}$$

This condition reduces effectively $\mathbb{R}_q^1 \times \mathbb{R}_q^1$ to $\mathbb{R}_q^1$. In particular we have

$$(d_R f)^* = \epsilon \begin{pmatrix} e_1 f & 0 \\ 0 & \bar{e}_1 f \end{pmatrix}^\dagger \epsilon \, \theta_R^1 = d_R f^*. \tag{54}$$

Within the calculus $\Omega_R^1(\mathbb{R}_q^1)$ we have

$$e_R x = \begin{pmatrix} q\Lambda & 0 \\ 0 & \Lambda^{-1} \end{pmatrix} x. \tag{55}$$

From this we conclude that

$$x d_R x = \begin{pmatrix} q & 0 \\ 0 & q^{-1} \end{pmatrix} d_R x x, \qquad d_R x \Lambda = q \Lambda d_R x \tag{56}$$

which are the real-calculus equivalent of the relations (37) and (40).

If d_R is to be a differential then the extension to higher order forms much be such that $d_R^2 = 0$, which implies that the d and $\bar{d}$ must be extended so that

$$d\bar{d} + \bar{d}d = 0. \tag{57}$$

From (37) and (40) one finds the relations

$$(dx)^2 = 0, \qquad (\bar{d}x)^2 = 0. \tag{58}$$

Consider now the differential d of (40) and the differential $\bar{d}$ of (37). Using the relations (57) one finds the relation

$$dx\bar{d}x + q^{-1}\bar{d}xdx = 0. \tag{59}$$

This is a relation within the calculus $\Omega^*(\mathbb{R}_q^1) \otimes \bar{\Omega}^*(\mathbb{R}_q^1)$. Within the calculus $\Omega_R^*(\mathbb{R}_q^1)$ we have

$$(d_R x)^2 = 0 \tag{60}$$

which is a different relation from (59). The algebraic structure of the calculus $\Omega^*(\mathbb{R}_q^1) \otimes \bar{\Omega}^*(\mathbb{R}_q^1)$ is given by the relations (37), (40) and (57). The algebraic structure of

$$\Omega_R^*(\mathbb{R}_q^1) \subset \Omega^*(\mathbb{R}_q^1) \otimes \bar{\Omega}^*(\mathbb{R}_q^1) \tag{61}$$

is given by the relations (56) and the condition (60). Within the calculus $\Omega_R^*(\mathbb{R}_q^1)$ we impose also

$$d_R \theta_R^1 = 0, \qquad (\theta_R^1)^2 = 0. \tag{62}$$

The extension of the involution to the 2-forms in the calculus $\Omega^*(\mathbb{R}_q^1) \otimes \bar{\Omega}^*(\mathbb{R}_q^1)$ is given by

$$(dx\bar{d}x)^* = -dx\bar{d}x \qquad (\theta^1\bar{\theta}^1)^* = -\theta^1\bar{\theta}^1. \tag{63}$$

These elements are antihermitian. The forms θ^1, $\bar{\theta}^1$ and θ_R^1 are closed. They are also exact.

ACKNOWLEDGMENT

One of the authors (JM) would like to thank J. Wess for his hospitality and the Max-Planck-Institut für Physik in München for financial support.

REFERENCES

1. Carow-Watamura U., Schlieker M. and Watamura S., $SO_q(N)$ *covariant Differential Calculus on Quantum Space and Quantum Deformation of Schrödinger Equation*, Z. Phys. C **49** (1991) 439.
2. Connes A., *Noncommutative Geometry*, Academic Press, (1994).
3. Connes A., *Noncommutative Geometry and Reality*, J. Math. Phys. **36** 6194 (1995).
4. Cuntz A., and Quillen D., *Algebra extensions and nonsingularity*, J. Amer. Math. Soc. **8** 251 (1995).
5. Dimakis A. and Madore J., *Differential Calculi and Linear Connections*, J. Math. Phys. **37** 4647 (1996).
6. Dubois-Violette M., Madore J., Masson T. and Mourad J., *Linear Connections on the Quantum Plane*, Lett. Math. Phys. **35** 351 (1995).
7. Dubois-Violette M., Madore J., Masson T. and Mourad J., *On Curvature in Noncommutative Geometry*, J. Math. Phys. **37** 4089 (1996).
8. Faddeev L.D., Reshetikhin N.Y. and Takhtajan L.A., *Quantization of Lie Groups and Lie Algebras*, Algebra i Analysis, **1** (1989), 178; translation: Leningrad Math. J. **1** 193 (1990).
9. Fiore G. and Madore J., *Leibniz Rules and Reality Conditions*, Naples Preprint 98-13, (1998).
10. Kastler, D., Testard D. and Madore J., *Connections of bimodules in noncommutative geometry*, Contemp. Math. **203** 159 (1997).
11. Madore J., *An Introduction to Noncommutative Differential Geometry and its Physical Applications*, Cambridge University Press (1995).
12. Madore J. and Mourad J., *Quantum Space-Time and Classical Gravity*, J. Math. Phys. **39** 423 (1998).
13. Majid S., *Braided Momentum in the q-Poincaré group*, J. Math. Phys. **34** 2045 (1993).
14. Ogievetsky O.and Zumino B., *Reality in the Differential Calculus on q-euclidean Spaces*, Lett. Math. Phys. **25** 121 (1992).
15. O. Ogievetsky, W. B. Schmidke, J. Wess and B. Zumino, *q-deformed Poincaré algebra*, Commun. Math. Phys. **150** 495 (1992).
16. M. Schlieker, W. Weich and R. Weixler, *Inhomogeneous Quantum Groups*, Z. Phys. C **53** 79 (1992).
17. Woronowicz S.L., *Twisted SU(2) Group. An example of a Non-Commutative Differential Calculus*, Publ. RIMS, Kyoto Univ. **23** 117 (1987).
18. Woronowicz S.L., *Compact Matrix Pseudogroups*, Commun. Math. Phys. **111** 613 (1987).

Two Disjoint Aspects of the Deformation Programme: Quantizing Nambu Mechanics; Singleton Physics

Moshé Flato

Laboratoire Gevrey de Mathématique Physique, CNRS ESA 5029
Département de Mathématiques, Université de Bourgogne
BP 400, F-21011 Dijon Cedex France
e-mail: flato@u-bourgogne.fr

Abstract. We present briefly the deformation philosophy and indicate, with references, how it was applied to the quantization of Nambu mechanics and to particle physics in anti De Sitter space.

Deformation theory of algebraic structures has proved itself extremely efficient and very much so in the last two decades. In this short contribution (in fact, a long abstract with references) we shall indicate two recent examples of new applications of the deformation philosophy. But before doing so let us present the essence of our deformation philosophy [1].

Physical theories have their domain of applicability mainly depending on the velocities and distances concerned. But the passage from one domain (of velocities and distances) to another one does not appear in an uncontrolled way. Rather, a new fundamental constant enters the modified formalism and the attached structures (symmetries, observables, states, etc.) *deform* the initial structure; namely, we have a new structure which in the limit when the new parameter goes to zero coïncides with the old formalism. In other words, to *detect* new formalisms we have to study deformations of the algebraic structures attached to a given formalism.

The only question is in which category we perform this research of deformations. Usually physics is rather conservative and if we start e.g. with the category of associative or Lie algebras, we tend to deform in this category. However recently examples were given in the literature of generalizations of this principle. For instance quantum groups [2] are in fact deformations of Hopf algebras. Other examples include more general deformations like in the quantization of Nambu mechanics or in non Abelian deformations [3]. Here we shall point out two recent developments based on our deformation philosophy.

The first example has to do with the generalization of classical mechanics suggested by Nambu in 1973 [4] and the recent attempts to quantize it. The question there, which has to do with generalized deformations, is of general interest. Apparently Nambu mechanics presents a new alternative to classical mechanics (see however [5]). Does it possess a quantum version attached to it and what price do we have to pay to have such a version? The question is partially answered in [6,7] and a detailed presentation can be found in [8].

The second example has to do with recent developments in field theories based on supergravity, conformal field theories, compactification of higher dimensional field theories, string theory, M-theory, p-branes, etc. for which people rediscovered the efficiency and advantages of anti De Sitter theories (which are stable deformations of Poincaré field theories in the category of Lie groups; see however [9] for quantum groups, at roots of unity in that case). There are many reasons for the advantages of anti De Sitter (often abbreviated as AdS) theories among which we can mention that AdS field theory admits an invariant natural infrared regularization of the theories in question and that the kinematical spectra (angular momentum and energy) are naturally discrete. But in addition AdS theories have a great bonus: the existence of *singleton* representations discovered by Dirac [10] for $SO(3,2)$, corresponding to a "square root" of AdS massless representations. We discovered that fact around 20 years ago [11] and developed rather extensively its physical consequences in the following years [9,12–25].

Singleton theories are topological in the sense that the corresponding singleton field theories live naturally on the boundary at infinity of the De Sitter bulk (boundary which has one dimension less than the bulk). They are new types of gauge theories which in addition permit to consider massless particles, e.g. the photon, as *dynamically* composite AdS particles [17,19,21]. Some of the beautiful properties of singleton theories can be extended to higher dimensions, and this is the main point of the recent huge interest in these AdS theories, which touched a large variety of aspects of AdS physics. More explicitly, in several of the recent articles among which we can mention [26–29,31,30,32,33], the new picture permits to study duality between CFT on the boundary at infinity and the corresponding AdS theory in the bulk. That duality, which has also interesting dynamical aspects in it, utilizes among other things the great notational simplifications permitted by singleton physics.

REFERENCES

1. Flato, M. "Deformation view of physical theories", Czechoslovak J. Phys. **B32** (1982), 472-475.
2. Drinfeld V.G. "Quantum Groups", in: *Proc. ICM86, Berkeley*, **1**, 101-110, Amer. Math. Soc., Providence (1987).
3. Pinczon G. "Non commutative deformation theory", Lett. Math. Phys. **41** (1997), 101-117.

4. Nambu Y. "Generalized Hamilton dynamics", Phys. Rev **D7** (1973), 2405-2412.
5. Bayen F. and Flato M. "Remarks concerning Nambu's generalized mechanics", Phys. Rev. **D11** (1975), 3049-3053.
6. Dito G., Flato M., Sternheimer D. and Takhtajan L. "Deformation Quantization and Nambu Mechanics", Commun. Math. Phys. **183** (1997), 1-22.
7. Dito G. and Flato M. "Generalized Abelian Deformations: Application to Nambu Mechanics", Lett. Math. Phys. **39** (1997), 107-125.
8. Flato M., Dito G. and Sternheimer D. "Nambu mechanics, n-ary operations and their quantization" in *Deformation theory and symplectic geometry, Proceedings of Ascona meeting, June 1996* (D. Sternheimer, J. Rawnsley and S. Gutt, Eds.), Math. Physics Studies **20**, 43-66, Kluwer Acad. Publ., Dordrecht (1997).
9. Flato M., Hadjiivanov L.K. and Todorov I.T. "Quantum Deformations of Singletons and of Free Zero-Mass Fields", Foundations of Physics **23** (1993), 571-586.
10. Dirac, P. A. M. "A remarkable representation of the $3+2$ de Sitter group", J. Math. Phys. **4** (1963), 901-909.
11. Flato M. and Fronsdal C. "One massless particle equals two Dirac singletons", Lett. Math. Phys. **2** (1978), 421-426.
12. Angeloupoulos E. and Flato M. "On unitary implementability of conformal transformations", Lett. Math. Phys. **2** (1978), 405-412.
13. Angeloupoulos E., Flato M., Fronsdal C. and Sternheimer D. "Massless particles, conformal group and De Sitter universe", Phys. Rev. **D23** (1981), 1278-1289.
14. Bayen F., Flato M., Fronsdal C. and Haidari A. "Conformal invariance and gauge fixing in QED", Phys. Rev. **D 32** (1985), 2673-2682.
15. Binegar B., Flato M., Fronsdal C. and Salamo S. "De Sitter conformal field theories", Czechoslovak J. Phys. **B32** (1982), 439-471.
16. Binegar B., Fronsdal C. and Heidenreich W. "de Sitter QED", Ann. Physics **149** (1983), 254-272; "Conformal QED", J. Math. Phys. **24** (1983), 2828-2846; "Linear conformal quantum gravity", Phys. Rev. D **27** (1983), 2249-2261.
17. Flato M. and Fronsdal C. "Composite Electrodynamics". J. Geom. Phys. **5** (1988), 37-61.
18. Flato M. and Fronsdal C. "On Dis and Racs", Physics Letters B **97** (1980), 236-240; "Quantum field theory of singletons. The Rac", J. Math. Phys. **22** (1981), 1100-1105; "Representations of conformal supersymmetry", Lett. Math. Phys. **8** (1984), 159-162; "Quarks or Singletons?", Physics Letters B, **172** (1986), 412-416; "Spontaneously generated field theories, zero-center modules, colored singletons and the virtues of $N=6$ supergravity" in *Essays in Supersymmetry* (C. Fronsdal Ed.) 123-162, Mathematical Physics Studies **8**, D. Reidel Publ., Dordrecht (1986); "Singletons: Fundamental Gauge Theory" in: *Topological and geometrical methods in field theory (Espoo, 1986)*, 273-290, World Sci. Publ., Teaneck NJ (1986); "The singleton dipole", Commun. Math. Phys. **108** (1987), 469-482; "How BRST invariance of QED is induced from the underlying field theory", Physics Letters B **189** (1987), 145-148; "Parastatistics, highest weight $\mathfrak{osp}(n,\infty)$ modules, singleton statistics and confinement", J. Geom. Phys. **6**, (1989), 293-309; "Three-dimensional singletons", Lett. Math. Phys. **20** (1989), 65-74; "Non abelian singletons", J. Math. Phys. **32** (1991), 524-531.

19. Flato M. and Frønsdal C. "Interacting Singletons", Lett. Math. Phys. **44** (1998), 249-259 (`hep-th/9803013`).
20. Flato M., Fronsdal C. and Gazeau J.P. "Masslessness and light-cone propagation in 3+2 De Sitter and 2+1 Minkowski spaces", Phys. Rev. **D 33** (1986), 415-420.
21. Flato M., Fronsdal C. and Sternheimer D. "Singletons as a basis for composite conformal quantum electrodynamics" in *Quantum Theories and Geometry*, Mathematical Physics Studies **10**, 65-76, Kluwer Acad. Publ. Dordrecht (1988).
22. Fronsdal C. "Singletons and massless, integral-spin fields on de Sitter space", Phys. Rev. D **20** (1979), 848-856.
    Fang J. and Fronsdal C. "Massless, half-integer-spin fields in de Sitter space", Phys. Rev. D **22** (1980), 1361-1367.
23. Fronsdal C. "The supersingleton I, free dipole and interactions at infinity; II, composite electrodynamics" Lett. Math. Phys. **16** (1988), 163-172; 173-177;
    "Dirac supermultiplet", Phys. Rev. D **26** (1982), 1988-1995.
24. Fronsdal C. and Heidenreich W.F. "Linear de Sitter gravity", J. Math. Phys. **28** (1987), 215-220.
25. Harun Ar Rashid A.M., Flato M. and Fronsdal C. "Three D Singletons and 2-D C.F.T.", Int. J. Modern Physics **A 7** (1992), 2193-2206.
26. Maldacena J.M. "The Large N Limit of Superconformal Field Theories and Supergravity", `hep-th/9711200`.
27. Witten E. "Anti De Sitter Space And Holography", `hep-th/9802150`;
    "Anti-de Sitter Space, Thermal Phase Transition, And Confinement In Gauge Theories", `hep-th/9803131`;
    "Baryons And Branes In Anti de Sitter Space", `hep-th/9805112`.
28. Ferrara S. and Frønsdal C. "Conformal Maxwell theory as a singleton field theory on AdS_5, IIB three-branes and duality", `hep-th/9712239`;
    "Gauge fields as composite boundary excitations", `hep-th/9802126`;
    "Gauge Fields and Singletons of AdS_{2p+1}", `hep-th/9806072`.
29. Ferrara S., Frønsdal C. and Zaffaroni A. "On N=8 Supergravity on AdS_5 and N=4 Superconformal Yang-Mills theory", `hep-th/9802203`.
30. Ferrara S., Kehagias A., Partouche H. and Zaffaroni A. "Membranes and Fivebranes with Lower Supersymmetry and their AdS Supergravity Duals", `hep-th/9803109`.
31. Ferrara S. and Zaffaroni A. "N=1,2 4D Superconformal Field Theories and Supergravity in AdS_5", `hep-th/9803060`;
    "Bulk Gauge Fields in AdS Supergravity and Supersingletons `hep-th/9807090`.
32. Freedman D.Z., Mathur S.D., Matusis A. and Rastelli L. "Correlation functions in the CFT(d)/AdS(d+1) correspondence", `hep-th/9804058`.
33. Sezgin E. and Sundell P. "Higher Spin N=8 Supergravity", `hep-th/9805125`.

Quantum Deformations of Conformal Algebras with Mass-Like Deformation Parameters

Andrzej Frydryszak*, Jerzy Lukierski*, Pierre Minnaert[†] and Marek Mozrzymas*

*Institute for Theoretical Physics, University of Wrocław, 50-204 Wrocław, Poland
[†]Laboratoire de Physique Théorique, Université Bordeaux 1,
33-175 Gradignan, France

Abstract. We recall the mathematical apparatus necessary for the quantum deformation of Lie algebras, namely the notions of coboundary Lie algebras, classical r-matrices, classical Yang-Baxter equations (CYBE), Fröbenius algebras and parabolic subalgebras. Then we construct the quantum deformation of $D = 1$, $D = 2$ and $D = 3$ conformal algebras, showing that this quantization introduce fundamental mass parameters. Finally we consider with more details the quantization of $D = 4$ conformal algebra. We build three classes of $sl(4,\mathbb{C})$ classical r-matrices, satisfying CYBE and depending respectively on 8, 10 and 12 generators of parabolic subalgebras. We show that only the 8-dimensional r-matrices allow to impose the $D = 4$ conformal $o(4,2) \simeq su(2,2)$ reality conditions. Weyl reflections and Dynkin diagram automorphisms for $o(4,2)$ define the class of admissible bases for given classical r-matrices.

INTRODUCTION

There are three classes of relativistic $D = 4$ symmetries occuring in present theory of fundamental interactions:
i) Poincaré symmetries, described by commuting fourmomentum generators P_μ, and the Lorentz algebra described by six generators $M_{\mu\nu}$.
ii) de-Sitter and anti-de-Sitter symmetries described by noncommuting "curved" fourmomentum generators P_μ

$$[P_\mu, P_\nu] = \pm \frac{i}{R^2} M_{\mu\nu} \qquad (1)$$

and the Lorentz generators $M_{\mu\nu}$.
iii) Conformal symmetries, extending Poincaré symmetries by dilatations (generator D) and four conformal accelerations (generators K_μ).

It appears that $D = 4$ conformal algebra is the same irrespectively of the scale in which we measure the distances; in other words if we rescale the generators

$$P_\mu \to \frac{1}{\lambda} P_\mu, \quad K_\mu \to \lambda K_\mu \qquad (2)$$

the conformal algebra remains the same. If we consider quantum deformations of conformal algebra one can introduce two types of quantum deformations:
a) with dimensionless deformation parameter q, which is invariant under the rescaling (2).
b) with dimensionfull deformation parameter κ, which under the rescaling (2) undergoes the following change

$$\kappa \to \frac{1}{\lambda} \kappa \qquad (3)$$

These so-called κ-deformations are of special physical interest, because they introduce same fundamental mass parameter κ as new geometric constant in the theory. In the present lecture we shall decribe such deformations using the formalism of classical r-matrices. Such a formalism describes infinitesimal deformations. In order to introduce full "finite" quantum deformation in the form of noncommutative and nococommutative Hopf algebra we should integrate these infinitesimal forms (see eg. [1]).

THE MATHEMATICAL APPARATUS

In this section we recall the definitions of Lie bialgebras and coboundary Lie bialgebras and their relations with Yang-Baxter equations. We also show how the solution of the classical Yang-Baxter equation is related to the construction of Fröbenius subalgebras. For more information and detailed proofs the reader is referred to the book by Chari and Presley [2].

Lie bialgebra

Definition. Let $\mathcal{G}$ be a Lie algebra. A Lie bialgebra structure on $\mathcal{G}$ is defined by a linear, skew-symmetric mapping $\delta : \mathcal{G} \to \mathcal{G} \otimes \mathcal{G}$, called the cocommutator, with the two conditions :
D1: the dual mapping $\delta^* : \mathcal{G}^* \to \mathcal{G}^* \otimes \mathcal{G}^*$ is a Lie bracket,
D2: δ is a 1-cocycle of $\mathcal{G}$ with values in $\mathcal{G} \otimes \mathcal{G}$.

Comments. Lie bialgebras are the infinitesimal structures associated to Poisson-Lie groups. In order to see how one can build such a structure, let us first make some comments on the definition.

1. Lie bracket. Let $Z \in \mathcal{G}$ and $X^* \otimes Y^* \in \mathcal{G}^* \otimes \mathcal{G}^*$. The dual map δ^* is defined by

$$\langle X^* \otimes Y^*, \delta(Z) \rangle = \langle \delta^*(X^*, Y^*), Z \rangle \qquad (4)$$

Condition D1 implies that

$$\delta^*(X^*, Y^*) = [X^*, Y^*]^* \qquad (5)$$

and that this bracket is skewsymmetric and satisfies Jacobi identity.

2. The adjoint action $ad_X^\otimes$. In order to explain the meaning of condition D2, we first recall the definition of the adjoint action. To each $X \in \mathcal{G}$ one associates a mapping $ad_X : \mathcal{G} \to \mathcal{G}$ defined by $ad_X(A) = [X, A]$. Because of the Jacobi identity for the Lie bracket, this mapping satisfies the identity

$$ad_{[X,Y]} = ad_X ad_Y - ad_Y ad_X . \qquad (6)$$

Similarly one defines the adjoint action $ad_X^\otimes$ in the tensor product $\mathcal{G} \otimes \mathcal{G}$, $ad_X^\otimes : \mathcal{G} \otimes \mathcal{G} \to \mathcal{G} \otimes \mathcal{G}$ by

$$ad_X^\otimes(A \otimes B) = [X, A] \otimes B + A \otimes [X, B], \qquad (7)$$

and the adjoint action in the triple tensor product $\mathcal{G} \otimes \mathcal{G} \otimes \mathcal{G}$ is defined by

$$ad_X^\otimes(A \otimes B \otimes C) = [X, A] \otimes B \otimes C + A \otimes [X, B] \otimes C + A \otimes B \otimes [X, C]. \quad (8)$$

These mappings satisfy the identity

$$ad_{[X,Y]}^\otimes = ad_X^\otimes ad_Y^\otimes - ad_Y^\otimes ad_X^\otimes . \qquad (9)$$

3. Condition D2. The mapping δ of the definition is said to be a 1-cocycle in $\mathcal{G}$ if it satisfies the relation

$$\delta([X, Y]) = ad_X^\otimes \delta(Y) - ad_Y^\otimes \delta(X). \qquad (10)$$

Coboundary Lie bialgebra

In order to built bialgebras explicitly, one must find mappings δ that satisfy conditions D1 and D2 of the definition.

1. Condition D2. Comparing the structure of equations (9) and (10) one sees that the 1-cocycle condition (10) can be satisfied by choosing the mapping δ in the following way:

$$\delta(X) = ad_X^\otimes(r) \qquad (11)$$

where r is any tensor in $\mathcal{G} \otimes \mathcal{G}$. In this case δ is said to be a coboundary.

2. Condition D1. The tensor $r \in \mathcal{G} \otimes \mathcal{G}$ can be written $r = \sum_i a_i \otimes b_i$. One associates to r

- the transpose $\tau(r)$ defined by

$$\tau(r) = \sum_i b_i \otimes a_i. \tag{12}$$

- the imbeddings of r in $\mathcal{G} \otimes \mathcal{G} \otimes \mathcal{G}$:

$$r_{12} = \sum_i a_i \otimes b_i \otimes 1,$$
$$r_{13} = \sum_i a_i \otimes 1 \otimes b_i, \tag{13}$$
$$r_{23} = \sum_i 1 \otimes a_i \otimes b_i.$$

- the Schouten bracket $[[r,r]] \in \mathcal{G} \otimes \mathcal{G} \otimes \mathcal{G}$ defined by

$$[[r,r]] = [r_{12}, r_{13}] + [r_{12}, r_{23}] + [r_{13}, r_{23}], \tag{14}$$

where, for instance

$$[r_{12}, r_{23}] = \sum_{i,j} a_i \otimes [b_i, a_j] \otimes b_j. \tag{15}$$

3. Then, condition D1 is equivalent to the following theorem [2]

Theorem. The mapping δ defined by $\delta(X) = ad_X^\otimes(r)$ is a cocommutator if, for any $X \in \mathcal{G}$, the following conditions are satisfied :
T1 : $\quad ad_X^\otimes(r + \tau(r)) = 0,$
T2 : $\quad ad_X^\otimes([[r,r]]) = 0.$

Lie bialgebras built in this way are called coboundary Lie bialgebras.

Yang-Baxter equations

We shall now show that the construction of coboundary Lie algebras is related to the solution of Yang-Baxter equations. Indeed, because of condition T1 of the preceding theorem, one may consider only skew symmetric tensors : $r \in \mathcal{G} \wedge \mathcal{G}$. Then, the Schouten bracket $[[r,r]]$ is in $\mathcal{G} \wedge \mathcal{G} \wedge \mathcal{G}$ and because of condition T2, one has $[[r,r]] = k\omega$, where ω is the unique ad-invariant trivector in $\mathcal{G} \wedge \mathcal{G} \wedge \mathcal{G}$, i.e., $ad_X^\otimes \omega = 0$.

- For $k \neq 0$, the equation $[[r,r]] = k\omega$, is the Modified Classical Yang-Baxter equation (MCYBE).

- For $k = 0$, the equation $[[r, r]] = 0$, is the Classical Yang-Baxter equation (CYBE).

A coboundary Lie bialgebra that arises from a skew symmetric tensor r, solution of the CYBE is called a triangular bialgebra. If the tensor r is not skew symmetric, the algebra is called quasi triangular.

Borel, parabolic and Fröbenius subalgebras

In the Cartan-Weyl basis, a Lie algebra $\mathcal{G}$ is defined by the elements $\{h_i, e_{+a}, e_{-a}\}$ where the abelian generators h_i ($i = 1...r =$ rank of $\mathcal{G}$) form the Cartan subalgebra $\mathcal{H}$ and e_{+a} (e_{-a}) ($a = 1...N = \frac{1}{2}(\dim \mathcal{G} - r)$) are the positive (negative) roots which form the sets $\boldsymbol{b}_{\pm}$. The algebra $\mathcal{G}$ is the direct sum $\mathcal{G} = \boldsymbol{b}_{-} \oplus \mathcal{H} \oplus \boldsymbol{b}_{+}$. The Borel subalgebras are the sets $\mathcal{B}_{\pm} = \mathcal{H} \oplus \boldsymbol{b}_{\pm} = \{h_i, e_{\pm a}\}$.

In the Cartan-Chevalley basis, the Lie algebra $\mathcal{G}$ is defined by the abelian generators $\{h_i\}$ and by the set of simple roots which is a subset of the set $\{e_{+a}, e_{-a}\}$ of positive and negative roots. A parabolic subalgebra of $\mathcal{G}$ is a subalgebra obtained by removing from the Cartan-Chevalley basis some simple negative (or positive) roots. Otherwise stated, a parabolic subalgebra can be built by addition of some simple negative (or positive) roots to the Borel subalgebra $\mathcal{B}_{+}$ (or $\mathcal{B}_{-}$).

Definition. A Lie algebra is called a *quasi* Fröbenius algebra if it exists a skew-symmetric bilinear form $B : \mathcal{G} \wedge \mathcal{G} \to \mathbb{C}$, such that for any $X, Y, Z \in \mathcal{G}$ one has

$$B([X, Y], Z) + B([Y, Z], X) + B([Z, X], Y) = 0. \tag{16}$$

If the bilinear form B is generated by an element $g_B^* \in \mathcal{G}^*$ in such a way that $B(X, Y) = \langle g_B^*, [X, Y] \rangle$ the Lie algebra is called a Fröbenius algebra.

Fröbenius algebras are related to Yang Baxter equations by the following

Theorem. Let $B_{ij} = \langle e_i^* \otimes e_j^*, B \rangle$ be the skew symmetric matrix that describes the bilinear form B in a basis $\{e_i\}$ of $\mathcal{G}$. Then, the inverse matrix r^{ij} such that $r^{ij} B_{jk} = \delta_k^i$ is skew symmetric and it defines a 2-tensor $r = r^{ij} e_i \otimes e_j$ that satisfies Classical Yang-Baxter Equation (CYBE).

The classification of the classical r-matrices of a Lie algebra $\mathcal{G}$ is therefore reduced to the classification of the *quasi* Fröbenius subalgebras of $\mathcal{G}$. These subalgebras are even dimensional and can be identified with a set of parabolic subalgebras defined above. Solutions of the Classical Yang-Baxter Equation for simple Lie algebra have been classified by Belavin and Drinfel'd [3] and more solutions have been obtained by Gerstenhaber and Giaquinto [4], Alexeevsky and Perelomov [5] and Stolin [6]

QUANTUM DEFORMATION OF CONFORMAL ALGEBRA WITH FUNDAMENTAL MASS PARAMETERS

$D = 1$ and $D = 2$ conformal algebras

The $D = 1$ conformal algebra contains three generators : the time translations (energy) P, the scaling transformations D and the conformal accelerations K. They satisfy the commutation relations

$$[D, P] = P, \qquad [D, K] = -K, \qquad [P, K] = 2D, \qquad (17)$$

After the identification

$$e_+ = P, \qquad e_- = K, \qquad h = 2D, \qquad (18)$$

these relations can be written as the $sl(2, \mathbb{R}) \simeq o(2,1)$ algebra in Cartan-Chevalley basis

$$[h, e_\pm] = \pm 2 e_\pm, \qquad [e_+, e_-] = h. \qquad (19)$$

For $sl(2; \mathbb{R})$ there exist only two inequivalent deformations, described by the following antisymmetric classical r-matrices:

1. The standard deformation generated by the r-matrix

$$r_s = c_s e_+ \wedge e_-. \qquad (20)$$

It is easy to see that the invariance of (20) under the rescalings $P \to \lambda^{-1} P$, $K \to \lambda K$ imply that the deformation parameter c_s is dimensionless.

2. The nonstandard deformations generated by the r-matrices

$$r_\pm = c_\pm h \wedge e_\pm. \qquad (21)$$

Using (18) one obtains

$$r_+ = 2c_+ D \wedge P, \qquad r_- = 2c_- D \wedge K. \qquad (22)$$

We see that the quantum deformations generated by the classical r-matrices $r_\pm$ provide deformation parameters that transform as the inverse of a mass ($c_+ = \frac{1}{M}$ or as a mass $c_- = \tilde{M}$, where M, $\tilde{M}$ are fundamental masses.

The quantization of the Lie algebra $sl(2;\mathbb{R})$ generated by r_+ has been given firstly by Ohn [7]. The commutation relations (17) are deformed as follows:

$$[D,P] = M \sinh \frac{P}{M}, \qquad [P,K] = 2D,$$
$$[D,K] = \frac{1}{2}\left(-K \cosh \frac{P}{M} - (\cosh \frac{P}{M})K\right). \qquad (23)$$

The coproduct and the antipode take the form:

$$\Delta(P) = P \otimes 1 + 1 \otimes P,$$
$$\Delta(D) = e^{-\frac{P}{M}} \otimes D + D \otimes e^{\frac{P}{M}}, \qquad (24)$$
$$\Delta(K) = e^{-\frac{P}{M}} \otimes K + K \otimes e^{\frac{P}{M}},$$

$$S(P) = -P, \qquad S(K) = -K - \frac{1}{M}(D - \sinh \frac{P}{M})$$
$$S(D) = -D - 2 \sinh \frac{P}{M}. \qquad (25)$$

Because the $D=2$ conformal algebra $o(2,2)$ is isomorphic to the direct sum $o(2,1) \oplus o(2,1)$, the deformation of $D=2$ conformal algebra is determined by the (standard or nonstandard) deformations of $o(2,1)$. In fact the decomposition of the conformal algebra into one-dimensional conformal algebras of right-movers and left-movers permits to introduce different M and $\tilde{M}$ parameters in these two sectors.

The description of quantum $D=2$ conformal algebra by a pair of nonstandard deformations was proposed recently in [8].

$D=3$ conformal algebra

The $D=3$ complex conformal algebra is the algebra $sp(4,\mathbb{C}) \simeq so(5,\mathbb{C})$, with 10 generators in the Cartan-Weyl basis. The Borel subalgebras are $\mathcal{H} \oplus b_\pm = \{h_1, h_2, e_{\pm 1}, e_{\pm 2}, e_{\pm 3}, e_{\pm 4}\}$, where $e_{\pm 1}, e_{\pm 2}$ are the simple roots and $e_{\pm 3}, e_{\pm 4}$ are composite roots

$$e_{\pm 3} = [e_{\pm 1}, e_{\pm 2}], \qquad e_{\pm 4} = [e_{\pm 1}, e_{\pm 3}]. \qquad (26)$$

The maximal nonstandard r-matrix takes the form

$$r_+ = c_3 (h_1 \wedge e_4 - e_1 \wedge e_3) + c'_3 h_2 \wedge e_4. \qquad (27)$$

The real form $sp(4,\mathbb{R}) \simeq so(3,2)$ is defined by introduction of a †-involution [9] such that

$$h_i^\dagger = -h_i,$$
$$e_1^\dagger = \lambda e_1, \qquad e_2^\dagger = \varepsilon e_2, \tag{28}$$
$$e_3^\dagger = -\lambda \varepsilon e_3, \qquad e_4^\dagger = \varepsilon e_4,$$

where $\lambda^2 = \varepsilon^2 = 1$. The invariance of (27) under the †-involution (28) implies $\lambda = \pm, \varepsilon = -1$. In the following we make the choice $\lambda = -\varepsilon = 1$.

The $so(3,2)$ algebra can be defined by the set of 10 generators M_{AB}, $(A, B = 0, 1, 2, 3, 4)$, $M_{AB} = -M_{BA}$ that satisfy the commutation relations:

$$[M_{AB}, M_{CD}] = \eta_{BC} M_{AD} + \eta_{AD} M_{BC} - \eta_{AC} M_{BD} - \eta_{BD} M_{AC}, \tag{29}$$

with $\eta_{AB} = \text{diag}(-1, 1, -1, 1, 1)$. The Cartan-Weyl basis is related to the M_{AB} generators by [9], [10]

$$h_1 = M_{12}, \qquad h_2 = M_{04} - M_{12},$$
$$e_1 = \frac{1}{\sqrt{2}}(M_{23} + M_{31}), \qquad e_3 = \frac{1}{\sqrt{2}}(M_{03} + M_{34}),$$
$$e_2 = -\frac{1}{\sqrt{2}}(M_{14} + M_{24} + M_{01} + M_{02}), \tag{30}$$
$$e_4 = \frac{1}{\sqrt{2}}(M_{14} - M_{24} + M_{01} - M_{02}).$$

The †-involution (28) implies $M_{AB}^\dagger = -M_{BA}$.

The physical basis of the $D = 3$ conformal algebra contains the 3 generators J, L_1, L_2 of the $so(2,1)$ subalgebra, the 3 translation generators P_0, P_1, P_2, the 3 generators of conformal acceleration K_0, K_1, K_2 and the scaling generator D. These generators are related to the M_{AB} generators by

$$J = M_{31}, \qquad L_1 = M_{12}, \qquad L_2 = M_{23},$$
$$P_0 = \frac{1}{\sqrt{2}}(M_{02} + M_{42}), \qquad K_0 = \frac{1}{\sqrt{2}}(M_{02} - M_{42}),$$
$$P_1 = \frac{1}{\sqrt{2}}(M_{01} + M_{41}), \qquad K_1 = \frac{1}{\sqrt{2}}(M_{01} - M_{41}), \tag{31}$$
$$P_2 = \frac{1}{\sqrt{2}}(M_{03} + M_{43}), \qquad K_2 = \frac{1}{\sqrt{2}}(M_{03} - M_{43}),$$
$$D = M_{04}.$$

Substituting the physical basis in the expression (27) of the r-matrix, one obtains

$$r_+ = \frac{1}{M_1}(L_1 \wedge P_1 - (L_2 + J) \wedge P_2) + \frac{1}{M_2}(D - L_1) \wedge P_-, \tag{32}$$

where $P_\pm = P_0 \pm P_1$, and $M_1 = \frac{2}{c_3}, M_2 = \frac{2}{c_3'}$ describe fundamental mass parameters.

QUANTUM DEFORMATION OF THE $D=4$ CONFORMAL ALGEBRA

Cartan-Weyl basis and its real forms.

One can write the complex $sl(4,\mathcal{C})$ algebra in Cartan Weyl basis e_{AB} ($A, B = 1,\ldots,4$), where the choice of indices (A, B) is taken from the position of non-vanishing entry in the 4×4 fundamental matrix representation. In particular, the diagonal elements $h_1 = e_{11} - e_{22}$, $h_2 = e_{22} - e_{33}$, $h_3 = e_{33} - e_{44}$ describe three commuting Cartan generators, and the simple root generators $e_1 = e_{12}$, $e_2 = e_{23}$ and $e_3 = e_{34}$ describe Cartan-Chevalley basis. The composite roots extending Cartan-Chevalley basis to Cartan-Weyl basis are defined by the expressions:

$$e_4 \equiv e_{13} = [e_{12}, e_{23}], \quad e_5 \equiv e_{24} = [e_{23}, e_{34}], \quad e_6 \equiv e_{14} = [e_{12}, e_{24}] = [e_{13}, e_{34}]. \quad (33)$$

The classical $sl(4)$ algebra is generated by the relations satisfied by the Cartan-Weyl basis generators $e_{\pm A}, h_A$ ($A = 1, 2, \ldots 6$; $h_4 = h_1 + h_2$, $h_5 = h_2 + h_3$, $h_6 = h_1 + h_2 + h_3$):

$$\begin{aligned}
[h_A, e_{\pm B}] &= \pm \alpha_{AB} e_{\pm B}, \\
[e_i, e_{-j}] &= \delta_{ij} h_j, \quad i = 1, 2, 3, \\
[e_a, e_{-a}] &= h_a, \quad a = 4, 5, 6,
\end{aligned} \quad (34)$$

where the Cartan matrix α_{AB} is

$$(\alpha_{AB}) = \begin{pmatrix} 2 & -1 & 0 & 1 & -1 & 1 \\ -1 & 2 & -1 & 1 & 1 & 0 \\ 0 & -1 & 2 & -1 & 1 & 1 \\ 1 & 1 & -1 & 2 & 0 & 1 \\ -1 & 1 & 1 & 0 & 2 & 1 \\ 1 & 0 & 1 & 1 & 1 & 2 \end{pmatrix} \quad (35)$$

The remaining generators of Cartan-Weyl basis of $U(sl(4,\mathcal{C}))$ are defined by Serre relations and the above definitions. They have the following form:

$$\begin{aligned}
[e_{\pm 1}, e_{\pm 3}] &= [e_{\pm 1}, e_{\pm 4}] = [e_{\pm 1}, e_{\pm 6}] = 0, \\
[e_{\pm 2}, e_{\pm 4}] &= [e_{\pm 2}, e_{\pm 5}] = [e_{\pm 2}, e_{\pm 6}] = 0, \\
[e_{\pm 3}, e_{\pm 5}] &= [e_{\pm 3}, e_{\pm 6}] = 0, \\
[e_{\pm 4}, e_{\pm 5}] &= [e_{\pm 4}, e_{\pm 6}] = 0, \\
[e_{\pm 5}, e_{\pm 6}] &= 0,
\end{aligned} \quad (36)$$

and

$$\begin{aligned}
{[e_{\pm 1}, e_{\mp 4}]} &= e_{\mp 2}, & [e_{\pm 1}, e_{\mp 5}] &= 0, & [e_{\pm 1}, e_{\mp 6}] &= 0, \\
[e_{\pm 2}, e_{\mp 4}] &= e_{\mp 1}, & [e_{\pm 2}, e_{\mp 5}] &= e_{\mp 3}, & [e_{\pm 2}, e_{\mp 6}] &= 0, \\
[e_{\pm 3}, e_{\mp 4}] &= 0, & [e_{\pm 3}, e_{\mp 5}] &= e_{\mp 2}, & [e_{\pm 3}, e_{\mp 6}] &= e_{\mp 4}, \\
[e_{\pm 4}, e_{\mp 5}] &= 0, & [e_{\pm 4}, e_{\mp 6}] &= e_{\mp 3}, & [e_{\pm 5}, e_{\mp 6}] &= e_{\mp 1}.
\end{aligned} \quad (37)$$

In order to describe the real forms of the complex Lie algebra $sl(4,\mathbb{C})$ we introduce an involutive anti-automorphism $X \to X^*$ such that for any $X, Y \in U(sl(4,\mathbb{C}))$

$$(XY)^* = Y^* X^*, \qquad (\mu X + \lambda Y)^* = \mu^* X^* + \lambda^* Y^*, \qquad \mu, \lambda \in \mathbb{C} \tag{38}$$

The real $o(4,2)$ algebra can be defined by the following inequivalent reality conditions [11] ($j = 1, 2, 3$)

$$\begin{align}
\text{(i)} \quad & h_j^* = -h_{4-j}, & e_{\pm j}^* &= e_{\pm(4-j)}, \tag{39}\\
\text{(ii)} \quad & h_j^* = h_j, & e_{\pm j}^* &= \epsilon_j e_{\mp j}, \tag{40}
\end{align}$$

with three nonequivalent choices of $(\epsilon_1, \epsilon_2, \epsilon_3)$, namely:

$$(1, -1, 1), \qquad (-1, 1, -1), \qquad (-1, -1, -1). \tag{41}$$

The algebra $sl(4,\mathbb{C})$ is isomorphic to the complex conformal algebra $o(4,2,\mathbb{C})$ and the latter can be described by the 15 generators $M_{PQ} = -M_{QP}$ ($P, Q = 0, 1, 2, 3, 4, 5$) that satisfy the commutation relations

$$[M_{PQ}, M_{RS}] = \eta_{QR} M_{PS} + \eta_{PS} M_{QR} - \eta_{PR} M_{QS} - \eta_{QS} M_{PR} \tag{42}$$

where $\eta_{AB} = \mathrm{diag}(-1, 1, 1, 1, 1, -1)$. The physical basis of the $D = 4$ conformal algebra is related to the generators M_{PQ} in the following way: the Lorentz generators are $M_{\mu\nu} = (M_i = \frac{1}{2}\epsilon_{ijk} M_{jk}, L_i = M_{i0})$; $\mu, \nu = 0, 1, 2, 3$, and the translation generators P_μ, the conformal acceleration generators K_μ and the scaling operator D are

$$P_\mu = (M_{4\mu} + M_{5\mu}), \qquad K_\mu = (M_{5\mu} - M_{4\mu}), \qquad D = M_{45}. \tag{43}$$

We recall that $\mathcal{G} = \mathbf{b}_- \oplus \mathcal{H} \oplus \mathbf{b}_+$ where $\mathcal{H} = (h_1, h_2, h_3)$ is the Cartan subalgebra and that $\mathcal{B}_\pm = \mathcal{H} \oplus \mathbf{b}_\pm$ are the two Borel subalgebras. The reality conditions lead to the following two ways of defining the real $D = 4$ conformal algebra generators in terms of Cartan-Weyl basis of $sl(4,\mathbb{C})$.

(i) For the reality condition $(\mathbf{b}_\pm)^* \subset \mathbf{b}_\pm$ one has

$$\begin{array}{llll}
M_+ = e_1 + e_{-3}, & M_- = -(e_3 + e_{-1}), & M_3 = \frac{i}{2}(h_1 - h_3), & \\
L_+ = i(e_{-3} - e_1), & L_- = -i(e_3 - e_{-1}), & L_3 = \frac{1}{2}(h_1 + h_3), & \\
P_1 = -(e_4 + e_5), & P_2 = i(e_4 - e_5), & P_3 = i(e_2 - e_6), & (44) \\
K_1 = e_{-4} - e_{-5}, & K_2 = i(e_{-4} + e_{-5}), & K_3 = i(e_{-2} - e_{-6}), & \\
P_0 = -i(e_2 + e_6), & K_0 = i(e_{-2} + e_{-6}), & D = \frac{1}{2}(h_1 + 2h_2 + h_3), &
\end{array}$$

where $M_\pm = M_1 \pm iM_2$, $L_\pm = L_1 \pm iL_2$. We see that the Cartan subalgebra $\mathcal{H}$ is described by the non compact algebra (M_3, L_3, D), and under the $*$-operation the operators are real.

(ii) The reality condition $(\mathbf{b}_\pm)^* \subset \mathbf{b}_\mp$ cannot be applied to the solutions of the

CYBE for $sl(4)$ with a dimension of solution (number of generators) less than eight. Hence the assignment of the conformal generators for this case will not be needed in further considerations and it is omitted here.

It appears that the Cartan subalgebra $\mathcal{H}$ is described by the compact Abelian subgroup $(M_{12} = M, M_{34} = \frac{1}{2}(P_3 - K_3), M_{50} = \frac{1}{2}(P_0 + K_0))$. The choices of the generators can be modified if we take into consideration the discrete group of Weyl reflections, which preserve the Lie-algebra relations. There are three basic Weyl reflections $\sigma_1, \sigma_2, \sigma_3$ describing the automorphism of $sl(4,\mathcal{C})$ Lie algebra. For instance, explicit relations defining σ_1 are as follows:

$$\begin{array}{lll} \sigma_1(e_{\pm 1}) = (a_1)^{\mp 1} e_{\mp 1}, & \sigma_1(e_{\pm 2}) = (a_4)^{\pm 1} e_{\pm 4}, & \sigma_1(e_{\pm 3}) = (a_3)^{\pm 1} e_{\pm 3}, \\ \sigma_1(e_{\pm 4}) = (a_2)^{\pm 1} e_{\pm 2}, & \sigma_1(e_{\pm 5}) = (a_6)^{\pm 1} e_{\pm 6}, & \sigma_1(e_{\pm 6}) = (a_5)^{\pm 1} e_{\pm 5}, \end{array} \quad (45)$$

where the a_i coefficients satisfy the relations: $a_4 = a_1 a_2 \quad a_5 = a_2 a_3 \quad a_6 = a_1 a_2 a_3$. There exists also an isomorphism of Dynkin diagram $(\alpha_1 \leftrightarrow \alpha_3)$ which implies the following isomorphism of $sl(4;\mathcal{C})$ Lie algebra:

$$\beta(e_{\pm 1}) = e_{\pm 3}, \quad \beta(e_{\pm 2}) = e_{\pm 2}, \quad \beta(e_{\pm 4}) = e_{\pm 5}, \quad \beta(e_{\pm 6}) = e_{\pm 6}. \quad (46)$$

Any product of Weyl reflections is again an isomorphism of $sl(4,\mathcal{C})$, but not all these isomorphisms commute with the $*$-operations defining real forms. The condition

$$\sigma_{i_1 \ldots i_k} \circ * = * \circ \sigma_{i_1 \ldots i_k} \quad (47)$$

is necessary for defining the restriction of Weyl reflections to $o(4,2)$ algebra. We obtain

i) for the $*$-operation $(\boldsymbol{b}_\pm)^* \subset \boldsymbol{b}_\pm$ the involutions σ_2, $\sigma_1 \sigma_3 = \sigma_3 \sigma_1$, β are also isomorphisms of the real algebra $o(4,2)$, provided that $b_1^* = b_3$, $b_2^* = b_2$.

ii) for the $*$-operation $(\boldsymbol{b}_\pm)^* \subset \boldsymbol{b}_\mp$ we obtain the following isomorphisms of $o(4,2)$: σ_2, β, provided that $b_i^* b_i = 1$.

Classical r-matrices for $sl(4)$ and $o(4,2)$ reality conditions

For $\mathcal{G} = sl(4,\mathcal{C})$, we can distinguish three relevant families of parabolic subalgebras spanning corresponding classical r-matrices. Let $\mathcal{B}_+ = (h_i, e_A); i = 1, 2, 3; A = 1, \ldots 6$ denote a Borel subalgebra. Then, according to the number d of elements in the parabolic subalgebra, we have the following classification of classical r-matrices:

d=12. Parabolic subalgebra $P_{(-2,-3)} = (\mathcal{B}_+, e_{-2}, e_{-3})$. In this case one obtains the one-parameter generalization of the solution given by Gerstenhaber and Giaquinto [4]

$$r_{(12)} = \frac{1}{4}(3h_1 + 2h_2 + h_3) \wedge e_1 + \frac{1}{4}(h_1 + 2h_2 + 3h_3) \wedge e_3 + e_4 \wedge e_{-2} +$$
$$-e_6 \wedge e_{-5} + \lambda(\frac{1}{2}(h_1 + 2h_2 + h_3) \wedge e_2 + (e_4 + e_5) \wedge e_{-3}) \quad (48)$$

This solution of the CYBE has the following properties: the parameter λ is arbitrary (it has inverse of mass dimension), each part of r satisfies CYBE separately, it does not permit the restriction of $sl(4,\mathcal{C})$ to real $o(4,2)$.

$d=10$. Parabolic subalgebras $P_{(j)} = (\mathcal{B}_+, e_{-j})$, $j = 1, 2, 3$. In order to define Fröbenius algebras, we have to consider three different sets of nonsingular bilinear forms generated by elements $g^* \in \mathcal{G}^*$:

i) Parabolic subalgebra P_1.

$$\begin{array}{ll} g_{1a}^* = e_5^* + e_4^* + e_1^*, & g_{1d}^* = e_6^* + e_2^* + e_{-1}^*, \\ g_{1b}^* = e_5^* + e_4^* + e_3^*, & g_{1e}^* = e_6^* + e_2^* + e_1^* + e_3^*, \\ g_{1c}^* = e_6^* + e_2^* + e_3^*. & \end{array} \quad (49)$$

They yield the following r-matrices:

$$\begin{aligned} r_{1a}^{(10)} &= -e_2 \wedge e_3 + e6-1 + \tfrac{1}{2}(e_1 + e_3) \wedge (h_1 + h_3) \\ &\quad + \tfrac{1}{4}e_4 \wedge (h_1 + 2h_2 - h_3) + \tfrac{1}{4}e_5 \wedge (h_1 + 2h_2 + 3h_3), \\ r_{1b}^{(10)} &= -e_1 \wedge e_2 + e_6 \wedge e_{-1} + \tfrac{1}{2}(e_1 + e_3) \wedge (h_1 + h_3) \\ &\quad + \tfrac{1}{4}e_4 \wedge (3h_1 + 2h_2 + h_3) + \tfrac{1}{4}e_5 \wedge (-h_1 + 2h_2 + 3h_3), \\ r_{1c}^{(10)} &= -e_1 \wedge e_5 + e_4 \wedge e_{-1} + \tfrac{1}{2}(e_3 + e_{-1}) \wedge (-h_1 + h_3) \\ &\quad + \tfrac{1}{4}e_2 \wedge (-h_1 + 2h_2 + h_3) + \tfrac{1}{4}e_6 \wedge (3h_1 + 2h_2 + h_3), \\ r_{1d}^{(10)} &= -e_1 \wedge e_5 + e_3 \wedge e_4 + \tfrac{1}{2}(e_3 + e_{-1}) \wedge (-h_1 + h_3) \\ &\quad + \tfrac{1}{4}e_2 \wedge (h_1 + 2h_2 - h_3) + \tfrac{1}{4}e_6 \wedge (h_1 + 2h_2 + 3h_3), \\ r_{1e}^{(10)} &= -\tfrac{1}{2}e_1 \wedge e_2 + e_6 \wedge e_{-1} + \tfrac{1}{4}(e_1 + e_3) \wedge (h_1 + h_3) \\ &\quad - \tfrac{1}{2}e_2 \wedge (e_3 + e_4 - e_5) + \tfrac{1}{4}e_4 \wedge (h_1 + h_2) + \tfrac{1}{4}e_5 \wedge (h_2 + h_3). \end{aligned} \quad (50)$$

ii) Parabolic subalgebra $P_{(2)}$. We have shown explicitly, by considering the most general ansatz, that there does not exist a Fröbenius algebra structure on $P_{(2)}^*$.

iii) Parabolic subalgebra $P_{(3)}$. We consider the nonsingular bilinear forms generated by the elements:

$$\begin{array}{ll} g_{3a} = e_5^* + e_4^* + e_1^* & g_{3d} = e_6^* + e_2^* + e_{-3}^*, \\ g_{3b} = e_5^* + e_4^* + e_3^* & g_{3e} = e_5^* + e_4^* + e_1^* + e_3^*, \\ g_{3c} = e_6^* + e_2^* + e_{+1}^*. & \end{array} \quad (51)$$

They yield the following r-matrices:

$$\begin{aligned}
r_{3a}^{(10)} &= -e_2 \wedge e_3 - e_6 \wedge e_{-3} + \tfrac{1}{2}(e_1 + e_3) \wedge (h_1 + h_3) \\
&\quad + \tfrac{1}{4} e_4 \wedge (h_1 + 2h_2 - h_3) + \tfrac{1}{4} e_5 \wedge (h_1 + 2h_2 + 3h_3), \\
r_{3b}^{(10)} &= -e_1 \wedge e_2 - e_6 \wedge e_{-3} + \tfrac{1}{2}(e_1 + e_3) \wedge (h_1 + h_3) \\
&\quad + \tfrac{1}{4} e_4 \wedge (3h_1 + 2h_2 + h_3) + \tfrac{1}{4} e_5 \wedge (-h_1 + 2h_2 + h_3), \\
r_{3c}^{(10)} &= e_3 \wedge e_4 - e_5 \wedge e_{-3} - \tfrac{1}{2}(e_1 + e_3) \wedge (-h_1 + h_3) \\
&\quad + \tfrac{1}{4} e_2 \wedge (h_1 + 2h_2 - h_3) + \tfrac{1}{4} e_6 \wedge (h_1 + 2h_2 + 3h_3), \\
r_{3d}^{(10)} &= -e_1 \wedge e_5 + e_3 \wedge e_4 - \tfrac{1}{2}(e_1 + e_{-3}) \wedge (-h_1 + h_3) \\
&\quad + \tfrac{1}{4} e_2 \wedge (-h_1 + 2h_2 + h_3) + \tfrac{1}{4} e_6 \wedge (3h_1 + 2h_2 + h_3), \\
r_{3e}^{(10)} &= -\tfrac{1}{2} e_1 \wedge e_5 - e_6 \wedge e_{-3} + \tfrac{1}{4}(e_1 + e_3) \wedge (-h_1 + h_3) \\
&\quad - \tfrac{1}{2} e_2 \wedge (e_3 + e_4 - e_5) + \tfrac{1}{2} e_4 \wedge (h_1 + h_2) + \tfrac{1}{2} e_5 \wedge (h_2 + h_3).
\end{aligned} \quad (52)$$

It can be shown that all 10-dimensional classical r-matrices generated in this way are not compatible with the $o(4,2)$ reality conditions. The reason is that σ_2 commutes with the $*$ operation and $(\sigma_2 \otimes \sigma_2) r_3^{(10)} = r_1^{(10)}$, but these r-matrices are not compatible, i.e., $[[r_1^{(10)}, r_3^{(10)}]] \neq 0$.

$d=8$. Here the subalgebra is only the Borel subalgebra $\mathcal{B}_+$. The corresponding r-matrix is [12]

$$r_1^{(8)} = e_4 \wedge e_3 - e_5 \wedge e_1 + a h_2 \wedge e_6 + h_6 \wedge e_6 \quad (53)$$

Taking into account that this classical r-matrix is real under the $*$-operation (39) and using the Weyl automorphism that commutes with this $*$-involution, we obtain another form of the d=8 solution:

$$\begin{aligned}
r_2^{(8)} &= (\sigma_2 \otimes \sigma_2) \circ r_1^{(8)} = [(\sigma_1 \sigma_3) \otimes (\sigma_1 \sigma_3)] \circ r_1^{(8)} \\
&= e_5 \wedge e_{-3} - e_4 \wedge e_{-1} + h_2 \wedge e_2 + a h_6 \wedge e_2
\end{aligned} \quad (54)$$

Let us note that in the physical basis, the r-matrices above are spanned by the generators of the $d=4$ Weyl subalgebra (M_i, L_i, P_μ, D).

CONCLUSIONS

Our main result here is the following: the quantum deformation of $D=4$ conformal symmetry which introduces fundamental mass as deformation parameter is span by generators of $D=4$ Weyl algebra. We see therefore that deformations of conformal algebra are induced by deformations of its subalgebra, obtained by supplementing $D=4$ Poincaré algebra by eleventh dilatation generator. The main task now is to describe these deformations in the integrated form. Following general results of Drinfeld [13] in the case when classical r-matrix satisfies CYBE, "finite" deformation formulae can be obtained by introducing the twist function modifying primitive coproduct. The work in this direction is under considerations.

REFERENCES

1. P. Etingof and D. Kazhdan, Quantization of Lie Bialgebras, I, Harvard University preprint (1996)
2. V. Chari and A. Presley, A guide to quantum groups, Cambridge University Press (1994)
3. A.A. Belavin and V.G. Drinfel'd, Funct. Anal. Appl. **16**, No 3 (1982) 1-29
4. M. Gerstenhaber and A. Giaquinto, Lett. Math. Phys. **40** (1997) 337.
5. D.V. Alexeevsky and A.M. Perelomov, Poisson brackets on simple Lie algebras and symplectic Lie algebras, Cern preprint, CERN-TH.6984/93 (unpublished)
6. A. Stolin, On rational solutions of the classical Yang-Baxter equations, Ph. D. Thesis, Stockholm, (1991) (unpublished)
7. C. Ohn, Lett. Math. Phys. **22** (1991) 287
8. A. Ballesteros, F.J. Harranz, M.A. del Olmo and M. Santander, Jour. Phys. **A 28** (1995) 941
9. C. Juszczak, Jour. Phys. **A 27** (1994) 385
10. J. Lukierski, A. Nowicki and H. Ruegg, Phys. Lett. **B 271** (1991) 321
11. J. Lukierski, A. Nowicki and J. Sobczyk, Jour. Phys. **A 26** (1993) 4047
12. J. Lukierski, P. Minnaert and M. Mozrzymas, Phys. Lett. **B 371** (1996) 215
13. V. G. Drinfeld, Sov. Math. Dokl. **28**, 667 (1983)

Differential Structures on $E_\kappa(2)$ from Contraction

Piotr Kosiński*, Paweł Maślanka[1]

Theoretical Physics Department II
University of Lodz
ul. Pomorska 149/153, 90 236 Lodz, Poland

Abstract. The deformed double covering of $E(2)$ group, denoted by $\tilde{E}_\kappa(2)$, is obtained by contraction from the $SU_\mu(2)$. The contraction procedure is then used for producing a new examples of differential calculi: 3D–left covariant calculus on both $\tilde{E}_\kappa(2)$ and the deformed Euclidean group $E_\kappa(2)$ and two different 4D–bicovariant calculi on $\tilde{E}_\kappa(2)$ which correspond to the one 4D–bicovariant calculi on $E_\kappa(2)$ described in Ref. [7].

INTRODUCTION

We present here the result of our recent paper [1] concerning the contraction of quantum group $SU_\mu(2)$ and various differential calculi on it. In the following section, the contraction procedure from $SU_\mu(2)$ to $\tilde{E}_\kappa(2)$, the double covering of deformed euclidean $E_\kappa(2)$ group, is outlined. Then, in the next section, we discuss the differential calculi on $\tilde{E}_\kappa(2)$ and $E_\kappa(2)$ resulting, by contraction, from the ones on $SU_\mu(2)$ group. Some explicite formulae are given in Appendix.

THE CONTRACTION OF $SU_\mu(2)$

Let us recall that $SU_\mu(2)$ is a matrix quantum group, [2]:

$$g = \begin{pmatrix} \sigma & -\mu\rho^* \\ \rho & \sigma^* \end{pmatrix}$$

$$\Delta(g) = g \otimes g$$

which matrix elements satisfy the following relations:

$$\begin{aligned} \rho\rho^* &= \rho^*\rho & \mu(\rho - \rho^*)\sigma &= \sigma(\rho - \rho^*) \\ \sigma\sigma^* + \mu^2\rho\rho^* &= I & \mu(\rho + \rho^*)\sigma &= \sigma(\rho + \rho^*) \\ & & \sigma^*\sigma + \rho\rho^* &= I \end{aligned} \quad (1)$$

---

[1] Supported by KBN grant 2P03B 130 12

The contraction (R) and the deformation (κ) parameters are introduced by the relations:

$$\mu = e^{\frac{1}{\kappa R}}$$

$$\sigma = \frac{1}{1+\mu}\left(\mu a + a^* + \frac{1}{R}(w^* - \mu w)\right)$$

$$\sigma^* = \frac{1}{1+\mu}\left(\mu a^* + a + \frac{1}{R}(w - \mu w^*)\right) \qquad (2)$$

$$\rho = \frac{1}{1+\mu}\left(a - a^* - \frac{1}{R}(w + w^*)\right)$$

$$\rho^* = -\frac{1}{1+\mu}\left(a - a^* + \frac{1}{R}(w + w^*)\right)$$

We assume that $a(R)$ and $w(R)$ have well–defined limits as $R \to \infty$:

$$\lim_{R\to\infty} a(R) = a_0 \quad \lim_{R\to\infty} w(R) = w_0 \qquad (3)$$

We are now ready to perform the contraction of $SU_\mu(2)$ by taking the limit $R \to \infty$ in eq.(1). As a result we obtain the quantum counterpart of the double covering of $E(2)$, denoted below by $\tilde{E}_\kappa(2)$. The structure of $\tilde{E}_\kappa(2)$ is described by the following [1]:

Theorem 1
$\tilde{E}_\kappa(2)$ is Hopf algebra generated by the elements: a_0, a_0^*, w_0 and w_0^* subject to the relations:

$$[a_0^*, w_0] = -[a_0^*, w_0^*] = \frac{1}{2\kappa}(a_0^{*2} - I)$$

$$[a_0, w_0] = -[a_0, w_0^*] = \frac{1}{2\kappa}(a_0^2 - I)$$

$$[w_0, w_0^*] = -\frac{1}{2\kappa}(a_0 + a_0^*)(w_0 + w_0^*)$$

$$[a_0, a_0^*] = 0 \quad a_0^* a_0 = a_0 a_0^* = I$$

$$\Delta a_0 = a_0 \otimes a_0 \quad \Delta a_0^* = a_0^* \otimes a_0^* \qquad (4)$$

$$\Delta w_0 = w_0 \otimes a_0^* + a_0 \otimes w_0$$

$$\Delta w_0^* = w_0^* \otimes a_0 + a_0^* \otimes w_0^*$$

$$S(a_0) = a_0^* \quad S(a_0^*) = a_0$$

$$S(w_0) = -a_0^* w_0 a_0 \quad S(W_0^*) = -a_0 w_0^* a_0^*$$

$$\epsilon(a_0) = \epsilon(a_0^*) = 1 \quad \epsilon(w_0) = \epsilon(w_0) = 0$$

For proof see Ref.[1].
In order to get the deformed twodimensional Euclidean group $E_\kappa(2)$ we put:

$$A = a_0^2, \quad A^* = a_0^{*2}, \quad v_+ = -ia_0 w_0, \quad v_- = iw_0^* a_0 \qquad (5)$$

As a result the following Hopf algebra $E_\kappa(2)$ (cf. [5], [6], [7]) is obtained from the eqs.(4):

$$[A, v_-] = \frac{i}{\kappa}(I - A)[A^*, v_-] = \frac{i}{\kappa}(A^* - A^{*2})$$
$$[A, v_+] = \frac{i}{\kappa}(A - A^2)[A^*, v_+] = \frac{i}{\kappa}(I - A^*)$$
$$[v_+, v_-] = \frac{i}{\kappa}(v_- - v_+)[A, A^*] = 0$$
$$AA^* = A^*A = I$$
$$\Delta A = A \otimes A \Delta A^* = A^* \otimes A^* \qquad (6)$$
$$\Delta v_+ = A \otimes v_+ + v_+ \otimes I \Delta v_- = A^* \otimes v_- + v_- \otimes I$$
$$S(A) = A^*S(A^*) = A$$
$$S(v_+) = -A^*v_+ S(v_-) = -Av_-$$
$$\epsilon(A) = \epsilon(A^*) = 1 \epsilon(v_+) = \epsilon(v_-) = 0$$

THE DIFFERENTIAL CALCULI

The contraction procedure described in sec.II can be applied to the variety of differential calculi on $SU_\mu(2)$ constructed by Woronowicz [2], [3] and Stachura [4]. As a result we obtain the differential calculi on $\tilde{E}_\kappa(2)$ and, consequently, also on $E_\kappa(2)$. Technically, one can look for the right ideal R (the main object is Woronowicz theory [3]) appearing from the contraction of group algebra or appply the contraction scheme directly to the relevant differential calculi on $SU_\mu(2)$. Actually we have used both techniques and obtained for $\tilde{E}_\kappa(2)$ one threedimensional (3D) leftcovariant calculus and two bicovariant fourdimensional calculi ($4D_\pm$ –they are described explicitly in Appendix).

Having obtained the calculi on $\tilde{E}_\kappa(2)$ one can ask for their counterparts on $E_\kappa(2)$. This problem can be solved in a quite straightforward way for 3D calculus. The relevant calculus on $E_\kappa(2)$ is described by the following explicit formulae (φ_0, φ_1, φ_2 span the basis in the space of left-invariant 1–forms):

$$[A, \varphi_0] = 0 \qquad [A^*, \varphi_0] = 0$$
$$[A, \varphi_1] = 0 \qquad [A^*, \varphi_1] = 0$$
$$[A, \varphi_2] = 0 \qquad [A^*, \varphi_2] = 0$$
$$[v_-, \varphi_0] = \frac{i}{\kappa}A^*\varphi_0$$
$$[v_-, \varphi_1] = \frac{i}{2\kappa}A^*(3\varphi_1 + \varphi_2) - \frac{1}{4\kappa^2}A^*\varphi_0$$
$$[v_-, \varphi_2] = \frac{i}{2\kappa}A^*(\varphi_1 + 3\varphi_2) - \frac{1}{4\kappa^2}A^*\varphi_0 \qquad (7)$$

$$[v_+, \varphi_0] = \frac{i}{\kappa}A\varphi_0$$
$$[v_+, \varphi_1] = \frac{i}{2\kappa}A(3\varphi_1 + \varphi_2) - \frac{1}{4\kappa^2}A\varphi_0$$
$$[v_+, \varphi_2] = \frac{i}{2\kappa}A(\varphi_1 + 3\varphi_2) - \frac{1}{4\kappa^2}A\varphi_0$$

To complete the description of the first order differential calculus we must introduce the $*$–operator. It is easy to see that the involution acts as follow:

$$\varphi_0^* = -\varphi_0, \quad \varphi_1^* = \varphi_2, \quad \varphi_2^* = \varphi_1 \tag{8}$$

$$\varphi_0 \wedge \varphi_0 = 0$$
$$\varphi_0 \wedge \varphi_1 + \varphi_1 \wedge \varphi_0 = 0$$
$$\varphi_0 \wedge \varphi_0 + \varphi_2 \wedge \varphi_0 = 0 \tag{9}$$
$$\varphi_2 \wedge \varphi_1 + \varphi_1 \wedge \varphi_2 + \frac{i}{4\kappa}\varphi_2 \wedge \varphi_0 - \frac{i}{4\kappa}\varphi_1 \wedge \varphi_0 = 0$$
$$\varphi_2 \wedge \varphi_2 - \frac{3i}{8\kappa}\varphi_1 \wedge \varphi_0 - \frac{5i}{8\kappa}\varphi_2 \wedge \varphi_0 = 0$$
$$\varphi_1 \wedge \varphi_1 + \frac{5i}{8\kappa}\varphi_1 \wedge \varphi_0 + \frac{3i}{8\kappa}\varphi_2 \wedge \varphi_0 = 0$$

The following Cartan–Maurer equations complete the description of the calculus:

$$d\varphi_0 = 0$$
$$d\varphi_1 = -\frac{1}{2}\varphi_0 \wedge \varphi_1 \tag{10}$$
$$d\varphi_2 = \frac{1}{2}\varphi_0 \wedge \varphi_2$$

Following the general framework we introduce the counterparts of left invariant vector fields

$$dx = (\chi_0 * x)\varphi_0 + (\chi_1 * x)\varphi_1 + (\chi_2 * x)\varphi_2 \tag{11}$$

where $x \in E_\kappa(2)$ and $\chi * x = (id \otimes \chi)\Delta x$.
The resulting quantum Lie algebra reads

$$[\chi_1, \chi_0] = \frac{5i}{8\kappa}\chi_1^2 - \frac{1}{2}\chi_1 - \frac{i}{2\kappa}\chi_1\chi_2 - \frac{3i}{8\kappa}\chi_2^2$$
$$[\chi_2, \chi_0] = \frac{3i}{8\kappa}\chi_1^2 + \frac{i}{4\kappa}\chi_1\chi_2 + \frac{1}{2}\chi_2 - \frac{5i}{8\kappa}\chi_2^2 \tag{12}$$
$$[\chi_1, \chi_2] = 0$$

It is easy to check that the involution acts as follow:

$$\chi_0^* = \chi_0, \qquad \chi_1^* = -\chi_2, \qquad \chi_2^* = -\chi_1 \qquad (13)$$

The case of fourdimensional calculi is more interesting. The bicovariant fourdimensional calculus on $E_\kappa(2)$ was constructed in Ref.[7]. From the results contained there it follows that this calculus is unique; i.e. there is only one fourdimensional bicovariant calculus on $E_\kappa(2)$. In fact, a careful analysis of the $4D_\pm$ calculi on $\tilde{E}_\kappa(2)$ shows that they both lead to the same calculus on $E_\kappa(2)$ described in [7]. This can be roughly explained as follows in the classical case the differential calculus is obtained with the choice $R = (\ker \epsilon)^2$, i.e. it is determined by the ideal consisting of functions that vanish, up to second order, at the group identity. Therefore, local diffeomorphism gives unique relation between differential calculi. However, in the quantum case situation looks differently. Two different calculi on $\tilde{E}_\kappa(2)$ reduce to the single one on $E_\kappa(2)$. In order to get some insight let us consider Hopf subalgebra of $\tilde{E}_\kappa(2)$ generated by a_0, a_0^*. It is commutative Hopf algebra so we can speak in terms of algebra of functions on $U(1)$. In the D_+ case we obtain the standard calculus on $U(1)$. Indeed, denoting $a_0 = a^{i\Theta}$ we see that the corresponding ideal is generated by $\cos\Theta - 1 \approx \Theta^2$. On the other hand in the D_- case the ideal is generated by $\cos 2\Theta - 1$ and $\sin\Theta(\cos\Theta + 1)$. Therefore, it consists of functions vanishing not only at $\Theta = 0$ but also at $\Theta = \Pi$. However, under the mapping $a_0 \to A = a_0^2$, which is double covering, it procedures the ideal of functions vanishing at the group identity (in a special way). We see that in the quantum case, generically, the relation between calculi depends on global properties of the mapping.

APPENDIX

We write out here explicitly the relations defining $4D_\pm$ calculi.
A) D_+ ($\psi_1, \psi_2, \psi_3, \psi_4$ span the space of left invariant 1 forms)

$$\begin{aligned}
&[a_0, \psi_1] = 0 \qquad\qquad [w_0, \psi_1] = 0 \\
&[a_0^*, \psi_1] = 0 \qquad\qquad [w_0^*, \psi_1] = 0 \\
&[a_0, \psi_2] = \frac{1}{2\kappa} a_0 \psi_1 \\
&[a_0^*, \psi_2] = -\frac{1}{2\kappa} a_0^* \psi_1 \\
&[w_0, \psi_2] = -\frac{1}{2\kappa} w_0 \psi_1 + \frac{1}{\kappa} a_0 \psi_2 \\
&[w_0^*, \psi_2] = \frac{1}{2\kappa} w_0^* \psi_1 - a_0^* \psi_4 \\
&[a_0, \psi_3] = -\frac{1}{2\kappa} a_0 \psi_1 \\
&[a_0^*, \psi_3] = \frac{1}{2\kappa} a_0^* \psi_1
\end{aligned} \qquad (14)$$

$$[w_0, \psi_3] = \frac{1}{2\kappa} w_0 \psi_1 - a_0 \psi_4 - \frac{1}{\kappa} a_0 (\psi_1 + \psi_3)$$
$$[w_0^*, \psi_3] = -\frac{1}{2\kappa} w_0^* \psi_1 - \frac{1}{\kappa} a_0^* \psi_3$$
$$[a_0, \psi_4] = \frac{1}{2\kappa^2} a_0 \psi_1$$
$$[a_0^*, \psi_4] = -\frac{1}{2\kappa^2} a_0^* \psi_1$$
$$[w_0, \psi_4] = -\frac{1}{2\kappa^2} w_0 \psi_1 + \frac{1}{\kappa^2} a_0 \psi_2$$
$$[w_0^*, \psi_4] = \frac{1}{2\kappa^2} w_0^* \psi_1 + \frac{1}{\kappa} a_0^* \psi_4 + \frac{2}{\kappa^2} a_0^* \psi_3$$

The external product identies read:

$$\psi_1 \wedge \psi_1 = 0$$
$$\psi_1 \wedge \psi_2 + \psi_2 \wedge \psi_1 = 0$$
$$\psi_1 \wedge \psi_3 + \psi_3 \wedge \psi_1 = 0$$
$$\psi_1 \wedge \psi_4 + \psi_4 \wedge \psi_1 = 0$$
$$\psi_2 \wedge \psi_2 - \frac{1}{\kappa} \psi_1 \wedge \psi_2 = 0 \quad (15)$$
$$\psi_2 \wedge \psi_3 + \psi_3 \wedge \psi_2 + \frac{1}{\kappa} \psi_1 \wedge (\psi_3 + \psi_2) = 0$$
$$\psi_2 \wedge \psi_4 + \psi_4 \wedge \psi_2 = 0$$
$$\psi_3 \wedge \psi_3 - \frac{1}{\kappa} \psi_1 \wedge \psi_3 = 0$$
$$\psi_3 \wedge \psi_4 + \psi_4 \wedge \psi_3 + \frac{1}{\kappa^2} \psi_1 \wedge (\psi_3 - \psi_2) = 0$$
$$\psi_4 \wedge \psi_4 + \frac{1}{\kappa^3} \psi_1 \wedge \psi_2 = 0$$

while Cartan–Maurer equations are given by:

$$d\psi_1 = 0$$
$$d\psi_2 = -\psi_1 \wedge \psi_2$$
$$d\psi_3 = \psi_1 \wedge \psi_3 \quad (16)$$
$$d\psi_4 = \frac{1}{\kappa} \psi_1 \wedge \psi_2$$

B) $4D_-$ ($\Phi_1, \Phi_2, \Phi_3, \Phi_4$ span the space of left invariant 1–forms)

$$\Phi_1 a_0 = a_0 (\Phi_2 - 2\Phi_1)$$
$$\Phi_1 a_0^* = -a_0^* \Phi_2$$
$$\Phi_2 a_0 = -a_0 \Phi_1$$

$$\Phi_2 a_0^* = a_0^*(\Phi_1 - 2\Phi_2)$$
$$\Phi_3 a_0 = -a_0\Phi_3 + \frac{1}{2\kappa}a_0(\Phi_1 - \Phi_2)$$
$$\Phi_3 a_0^* = -a_0^*\Phi_3 + \frac{1}{2\kappa}a_0^*(\Phi_2 - \Phi_1)$$
$$\Phi_4 a_0 = -a_0\Phi_4 + \frac{1}{2\kappa}a_0(\Phi_2 - \Phi_1)$$
$$\Phi_4 a_0^* = -a_0^*\Phi_4 + \frac{1}{2\kappa}a_0^*(\Phi_1 - \Phi_2) \tag{17}$$
$$\Phi_1 w_0 = -w_0\Phi_2 - 2a_0\Phi_3$$
$$\Phi_1 w_0^* = w_0^*\Phi_2 - 2w_0^*\Phi_1 - 2a_0^*\Phi_4$$
$$\Phi_2 w_0 = w_0\Phi_1 - 2w_0\Phi_2 - 2a_0\Phi_3$$
$$\Phi_2 w_0^* = -w_0^*\Phi_1 - 2a_0^*\Phi_4$$
$$\Phi_3 w_0 = -w_0\Phi_3 - \frac{1}{2\kappa}w_0(\Phi_1 - \Phi_2) + \frac{1}{\kappa}a_0\Phi_3$$
$$\Phi_3 w_0^* = -w_0^*\Phi_3 + \frac{1}{2\kappa}w_0^*(\Phi_1 - \Phi_2) + \frac{1}{\kappa}a_0^*\Phi_3 + \frac{1}{\kappa^2}a_0^*\Phi_2$$
$$\Phi_4 w_0 = -w_0\Phi_4 + \frac{1}{2\kappa}w_0(\Phi_1 - \Phi_2) - \frac{1}{\kappa}a_0\Phi_4 + \frac{1}{\kappa^2}a_0\Phi_2$$
$$\Phi_4 w_0^* = -w_0^*\Phi_4 + \frac{1}{2\kappa}w_0^*(\Phi_2 - \Phi_1) - \frac{1}{\kappa}a_0^*\Phi_4$$

The external product identies read:

$$\Phi_1 \wedge \Phi_1 = 0$$
$$\Phi_2 \wedge \Phi_2 = 0$$
$$\Phi_1 \wedge \Phi_2 + \Phi_2 \wedge \Phi_1 = 0$$
$$\Phi_3 \wedge \Phi_1 + 3\Phi_1 \wedge \Phi_3 - 1\Phi_2 \wedge \Phi_3 = 0$$
$$\Phi_3 \wedge \Phi_2 - \Phi_2 \wedge \Phi_3 + 2\Phi_1 \wedge \Phi_3 \tag{18}$$
$$\Phi_3 \wedge \Phi_3 - \frac{1}{\kappa}\Phi_1 \wedge \Phi_3 + \frac{1}{\kappa}\Phi_2 \wedge \Phi + 3 = 0$$
$$\Phi_4 \wedge \Phi_1 - \Phi_1 \wedge \Phi_4 + 2\Phi_2 \wedge \Phi_4 = 0$$
$$\Phi_4 \wedge \Phi2 + 3\Phi_2 \wedge \Phi_4 - 2\Phi_1 \wedge \Phi_4 = 0$$
$$\Phi_4 \wedge \Phi_3 + \Phi_3 \wedge \Phi_4 + \frac{1}{\kappa}(\Phi_1 - \Phi_2) \wedge \Phi_3 + \frac{1}{\kappa}(\Phi_1 - \Phi_2) \wedge \Phi_4 = 0$$
$$\Phi_4 \wedge \Phi_4 - \frac{1}{\kappa}(\Phi_1 - \Phi_2) \wedge \Phi_4 = 0$$

The following Cartan–Maurer formulas complete the description of the second order calculus:

$$d\Phi_1 = 0$$
$$d\Phi_2 = 0 \tag{19}$$

$$d\Phi_3 = (\Phi_1 - \Phi_2) \wedge \Phi_3$$
$$d\Phi_4 = (\Phi_2 - \Phi_1) \wedge \Phi_4$$

REFERENCES

1. P.Kosiński, P.Maślanka *The contraction of $SU_\mu(2)$ and its differential structures to $E_\kappa(2)$*,
2. Woronowicz S.L., RIMS **29**, 117 (1987)
3. Woronowicz S.L., Commun. Math. Phys. **122**, 125 (1989)
4. Stachura P., Lett. Math. Phys. **25**, 175 (1992)
5. Maślanka P., J. Math. Phys. **35**, 1976 (1994)
6. Ballesteros A., Celeghini E., Giachetti R., Sorace E., Tarlini M., J. Phys. **A 26**, 7495 (1993)
7. Giller S., Gonera C., Kosiński P., Maślanka P., Acta Phys. Pol. **B28**, 1121 (1997)

Twisting of Quantum Groups and Integrable Models

P. P. Kulish[1]

St.Petersburg Department of the Steklov Mathematical Institute,
Fontanka 27, St.Petersburg, 191011, Russia.

Abstract. Few new twisting elements for extented jordanian quantization of $sl(N)$, for the Drinfeld-Jimbo quantum algebra $\mathcal{U}_q(sl(3))$, and for the Lie superalgebra $osp(1|2)$ are given. Applications of these twisting elements to the spin chain models integrable by the quantum inverse scattering method are discussed.

Quantum groups as a solid mathematical object in the framework of the theory of Hopf algebras, were extracted from the quantum inverse scattering method. The detailed formulation of the theory of quantum groups [1] does not contain description of their possible transformations. The popular FRT-formalism [2] also refers to a particular form of the underlying R-matrix. A deformation quantization approach to quantum groups [3] gave rise to the notion of twist transformation and a twisting element $\mathcal{F}$ [4], which is a similarity transformation of the coproduct Δ. In this contribution few new exact twisting elements are given and some applications of the twist transformations to models integrable by the quantum inverse scattering method (QISM) [5,6] are given. Three twisting elements are presented: extension of the jordanian deformation of $sl(2)$ [8,9] to $sl(N)$ case, a twist from the Drinfeld-Jimbo quantum algebra $\mathcal{U}_q(sl(3))$ to the Cremmer-Gervais one, and a twist deformation of the Lie superalgebra $osp(1|2)$. It is worthwhile to mention recent papers, where the twist transformation and corresponding elements are intensively discussed [7]. Some part of the content of this contribution is partially covered in the papers [10,12,11,13–15].

Recall the twisting of Hopf algebras. A Hopf algebra $\mathcal{A}(m, \Delta, \epsilon, S)$ with multiplication $m: \mathcal{A} \otimes \mathcal{A} \to \mathcal{A}$, coproduct $\Delta: \mathcal{A} \to \mathcal{A} \otimes \mathcal{A}$, counit $\epsilon: \mathcal{A} \to C$, and antipode $S: \mathcal{A} \to \mathcal{A}$ (see definitions in Refs. [1,2]) can be transformed with an invertible element $\mathcal{F} \in \mathcal{A} \otimes \mathcal{A}$, $\mathcal{F} = \sum f_i^{(1)} \otimes f_i^{(2)}$ into a twisted one $\mathcal{A}_t(m, \Delta_t, \epsilon, S_t)$ [4]. This Hopf algebra $\mathcal{A}_t$ has the same multiplication and counit maps but the twisted coproduct and antipode

$$\Delta_t(a) = \mathcal{F}\Delta(a)\mathcal{F}^{-1}, \quad S_t(a) = vS(a)v^{-1}, \quad v = \sum f_i^{(1)} S(f_i^{(2)}), \quad a \in \mathcal{A}.$$

---

[1] Partially supported by the RFFI grants N 96-01-00851 and N 98-01-00310.

The twisting element has to satisfy the equations

$$(\epsilon \otimes id)(\mathcal{F}) = (id \otimes \epsilon)(\mathcal{F}) = 1, \tag{1}$$

$$\mathcal{F}_{12}(\Delta \otimes id)(\mathcal{F}) = \mathcal{F}_{23}(id \otimes \Delta)(\mathcal{F}). \tag{2}$$

A quasitriangular Hopf algebra $\mathcal{A}(m, \Delta, \epsilon, S, \mathcal{R})$ has additionally an element $\mathcal{R} \in \mathcal{A} \otimes \mathcal{A}$ (a universal R-matrix) [1], which relates the coproduct Δ and its opposite coproduct Δ^{op} by the similarity transformation

$$\Delta^{op}(a) = \mathcal{R}\Delta(a)\mathcal{R}^{-1}, \quad a \in \mathcal{A}.$$

A twisted quasitriangular quantum algebra $\mathcal{A}_t(m, \Delta_t, \epsilon, S_t, \mathcal{R}_t)$ has the twisted universal R-matrix

$$\mathcal{R}_t = \tau(\mathcal{F}) \, \mathcal{R} \, \mathcal{F}^{-1}, \tag{3}$$

where τ means the permutation of the tensor factors: $\tau(f \otimes g) = (g \otimes f), \tau(\mathcal{F}) = \mathcal{F}_{21}$.

Note that the composition of appropriate twists can be defined $\mathcal{F} = \mathcal{F}_2\mathcal{F}_1$. The element $\mathcal{F}_1$ has to satisfy the twist equation with the coproduct of the original Hopf algebra, while $\mathcal{F}_2$ must be its solution for Δ_{t_1} of the intermediate Hopf algebra twisted by $\mathcal{F}_1$. In particular, if $\mathcal{F}$ is a solution to the twist equation (2) then $\mathcal{F}^{-1}$ satisfies this equation with $\Delta \to \Delta_t$.

An important subclass of factorizable twists consists of elements satisfying the following equations

$$(\Delta \otimes id)(\mathcal{F}) = \mathcal{F}_{13}\mathcal{F}_{23}, \tag{4}$$

$$(id \otimes \Delta_t)(\mathcal{F}) = \mathcal{F}_{12}\mathcal{F}_{13}. \tag{5}$$

It is easy to see that the universal R-matrix $\mathcal{R}$ satisfies these equations for $\Delta_t = \Delta^{op}$. Another well developed case is the jordanian twist of $sl(2)$ with $\mathcal{F}^{(j)}$ described by [8]

$$\mathcal{F}^{(j)} = \exp\left(\frac{1}{2}h \otimes \ln(1 + 2\xi\, X)\right) = e^{h \otimes \sigma}. \tag{6}$$

Due to the fact that the Cartan element h is primitive in $sl(2) : \Delta(h) = h \otimes 1 + 1 \otimes h$, and σ is primitive in the jordanian $\mathcal{U}_\xi(sl(2)) : \Delta_t(\sigma) = \sigma \otimes 1 + 1 \otimes \sigma$, one gets

$$(\Delta \otimes id)e^{h \otimes \sigma} = e^{h \otimes 1 \otimes \sigma} e^{1 \otimes h \otimes \sigma},$$

$$(id \otimes \Delta_t)e^{h \otimes \sigma} = e^{h \otimes \sigma \otimes 1} e^{h \otimes 1 \otimes \sigma}.$$

It will be shown that the element $\mathcal{F}^{(j)}$ (6) can be extended by an extra factor $\mathcal{F}^{(ej)} = \mathcal{F}^{(e)}\mathcal{F}^{(j)}$ to twist the universal enveloping algebra of $sl(N)$.

Extended jordanian twist for $sl(N)$. For the case of $\mathcal{U}(sl(N))$ the following form of twisting element $\mathcal{F}^{(ej)}$ can be chosen

$$\mathcal{F}^{(ej)} = \mathcal{F}^{(e)}\mathcal{F}^{(j)} = \prod_{j=2}^{N-1} \exp\left(2\xi E_{1j} \otimes E_{jN} e^{-\sigma}\right) \exp\left(H_{1N} \otimes \sigma\right),$$

where $H_{1N} = E_{11} - E_{NN}, \sigma = \frac{1}{2}\ln(1 + 2\xi E_{1N})$. This twist of $\mathcal{U}(sl(N))$ is generated by the twist of $\mathcal{U}(\mathbf{B}^\vee)$ (here $\mathbf{B}^\vee$ is the restricted Borel subalgebra of $sl(N)$ with the basic elements $\{H_{1N}, E_{1N}, E_{1j}, E_{jN}\}_{j=2,\ldots,N-1}$) leading to the Hopf algebra $\mathcal{U}_\xi(\mathbf{B}^\vee)$ with the initial commutation relations, the twisted coproducts

$$\begin{aligned}
\Delta_\mathcal{F} H_{1N} &= H_{1N} \otimes e^{-2\sigma} + 1 \otimes H_{1N} - 4\xi \sum_{j=2}^{N-1} E_{1j} \otimes E_{jN} e^{-3\sigma}, \\
\Delta_\mathcal{F} E_{1i} &= E_{1i} \otimes e^{-\sigma} + 1 \otimes E_{1i}, \\
\Delta_\mathcal{F} E_{iN} &= E_{iN} \otimes e^\sigma + e^{2\sigma} \otimes E_{iN}, \\
\Delta_\mathcal{F} E_{1N} &= E_{1N} \otimes e^{2\sigma} + 1 \otimes E_{1N},
\end{aligned} \quad (7)$$

antipodes

$$\begin{aligned}
S_\mathcal{F}(\sigma) &= -\sigma, & S_\mathcal{F}(E_{1i}) &= -E_{1i}e^\sigma, \\
S_\mathcal{F}(E_{iN}) &= -E_{iN}e^{-3\sigma}, & S_\mathcal{F}(E_{1N}) &= -E_{1N}e^{-2\sigma}, \\
S_\mathcal{F}(H_{1N}) &= -H_{1N}e^{2\sigma} - 4\xi \sum_{j=2}^{N-1} E_{1j} E_{jN}
\end{aligned} \quad (8)$$

and the universal R-matrix of the form

$$\mathcal{R} = \mathcal{F}_{21}\mathcal{F}^{-1}$$
$$= \prod_j \exp\left(2\xi E_{jN}e^{-\sigma} \otimes E_{1j}\right) \exp\left(\sigma \otimes H_{1N}\right) \exp\left(-H_{1N} \otimes \sigma\right) \prod_j \exp\left(-2\xi E_{1j} \otimes E_{jN}e^{-\sigma}\right). \quad (9)$$

The coproducts and antipodes for other elements of $\mathcal{U}_\xi(sl(N))$ can be calculated using the standard formulas. The obtained expressions are rather cumbersome. Thus, for example, in the case of $\mathcal{U}_\xi(sl(3))$ the coproduct of E_{32} looks like

$$\begin{aligned}
\Delta_\mathcal{F} E_{32} &= E_{32} \otimes e^{-\sigma} + 1 \otimes E_{32} \\
&+ \xi H_{13} \otimes E_{12}e^{-2\sigma} + 2\xi E_{12} \otimes H_{23}e^{-\sigma} - \xi H_{13}E_{12} \otimes (e^{-\sigma} - e^{-3\sigma}) \\
&- 4\xi^2 E_{12} \otimes E_{23}E_{12}e^{-3\sigma} - 4\xi^2 E_{12}^2 \otimes E_{23}e^{-4\sigma}.
\end{aligned}$$

Twisting the coproducts is acting by the exponential of the adjoint operator defined on the tesor product $\mathcal{U}(sl(N)) \otimes \mathcal{U}(sl(N))$. This operator is nilpotent and all the twisted coproducts can be expressed through the finite number of its powers.

The Cremmer-Gervais universal R-matrix for $gl(3)$. The study of the quantum $sl(N)$ Toda field theory gave rise to the Cremmer-Gervais solution of the Yang-Baxter equation R_{CG}, which is different from the standard R-matrix of the Drinfeld-Jimbo quantum algebra $\mathcal{U}_q(sl(N))$

$$R_{DJ} = q\sum_i e_{ii} \otimes e_{ii} + \sum_{i\neq j} e_{ii} \otimes e_{jj} + \omega \sum_{i<j} e_{ij} \otimes e_{ji}, \quad q=e^\gamma, \quad \omega = q - q^{-1}. \tag{10}$$

The R-matrix R_{CG} for $N=3$ is

$$\begin{aligned}R_{CG} = & R_S + (p-1)\left(e_{11} \otimes e_{22} + e_{22} \otimes e_{33}\right) \\ & + (p^{-1}-1)\left(e_{22} \otimes e_{11} + e_{33} \otimes e_{22}\right) + (p^2/q - 1)e_{11} \otimes e_{33} \\ & + (q/p^2 - 1)e_{33} \otimes e_{11} + q\nu\left(e_{32} \otimes e_{12} - p^2/q^2 e_{12} \otimes e_{32}\right).\end{aligned} \tag{11}$$

Here p and ν are two additional independent deformation parameters.

The quasitriangular Hopf algebra $\mathcal{U}_q(sl(3))$ can be defined by the two triples of generators $\{h_i, e_i, f_i\}$, $i=1,2$, subjected to the relations [1,2]

$$q^{h_i}e_j = q^{a_{ij}}e_j q^{h_i}, \quad q^{h_i}f_j = q^{-a_{ij}}f_j q^{h_i}, \quad [e_i, f_j] = \delta_{ij}\frac{q^{h_i} - q^{-h_i}}{q - q^{-1}},$$
$$e_k^2 e_l - (q+q^{-1})e_k e_l e_k + e_l e_k^2, \quad f_k^2 f_l - (q+q^{-1})f_k f_l f_k + f_l f_k^2, \quad k \neq l, \tag{12}$$

where the Cartan matrix elements are $a_{ii} = 2$, $a_{ii+1} = a_{i+1i} = -1$. The coproduct on these generators reads

$$\Delta(h_i) = h_i \otimes 1 + 1 \otimes h_i, \quad \Delta(e_i) = e_i \otimes q^{h_i} + 1 \otimes e_i, \quad \Delta(f_i) = f_i \otimes 1 + q^{-h_i} \otimes f_i. \tag{13}$$

Having introduced elements corresponding to the composite root

$$e_{13} = e_1 e_2 - q e_2 e_1, \quad f_{13} = f_2 f_1 - q^{-1} f_1 f_2,$$

one gets the universal R-matrix in the factorized form:

$$\mathcal{R}_{DJ} = q^{t_0} \exp_{q^{-2}}(\omega e_2 \otimes f_2) \exp_{q^{-2}}(\omega e_{13} \otimes f_{13}) \exp_{q^{-2}}(\omega e_1 \otimes f_1), \tag{14}$$

where $t_0 = \sum_{ij}(a^{-1})_{ij} h_i \otimes h_j$ is the canonical element of the Cartan subalgebra $\mathcal{H} \otimes \mathcal{H}$ and the q-exponential is

$$\exp_q(x) = \sum_{n=0}^\infty \frac{x^n}{[n;q]!} = \{\prod_{k=0}^\infty (1 - (1-q)xq^k)\}^{-1}, \quad [n;q]! = \frac{q^n - 1}{q-1}. \tag{15}$$

The new matrix elements of R_{CG} in (11) correspond to contributions of $e_1 \otimes f_2$ and $f_2 \otimes e_1$ in the fundamental representation. Although commuting with each other, the elements e_1 and f_2 do not generate independent Hopf subalgebras because h_1 does not commute with f_2 nor does h_2 with e_1. To overcome this obstacle let us extend $\mathcal{U}_q(sl(3))$ with the central element $C = e_{11} + e_{22} + e_{33}$, $h_1 = e_{11} - e_{22}$, $h_2 = e_{22} - e_{33}$, and perform a diagonal twist to separate the above mentioned Hopf subalgebras

$$\mathcal{F}^{(1)} = \exp(\frac{\gamma}{2}[e_{11} \wedge e_{22} + e_{11} \wedge e_{33} + e_{22} \wedge e_{33}]), \quad q = e^\gamma.$$

This is a particular case of the Reshetikhin twist with fixed parameters to separate the Hopf subalgebras we are interested in. The twisted coproducts $\Delta_t = \mathcal{F}\Delta\mathcal{F}^{-1}$ of the generators $\tilde{e}_1 = e_1 q^{\frac{1}{2}(e_{11}+e_{22})}$ and $\tilde{f}_2 = f_1 q^{-\frac{1}{2}(e_{22}+e_{33})}$ do not contain common elements:

$$\Delta_t(\tilde{e}_1) = \tilde{e}_1 \otimes q^{2e_{11}} + 1 \otimes \tilde{e}_1, \quad \Delta_t(\tilde{f}_2) = \tilde{f}_2 \otimes 1 + q^{2e_{33}} \otimes \tilde{f}_2.$$

The Hopf subalgebra $\mathcal{B}(1)_-$ generated by $\{e_{33}, \tilde{f}_2\}$ appears to be dual but with opposite product $\mathcal{B}(1)_{op}^*$ to the Hopf subalgebra $\mathcal{B}(1)$ spanned by $\{e_{11}, \tilde{e}_1\}$. So, the corresponding canonical element is

$$\mathcal{F}^{(2)} = \exp_{q^2}(\mu\, \tilde{e}_1 \otimes \tilde{f}_2) q^{2e_{11}\otimes e_{33}} \tag{16}$$

with independent parameter μ which does not change the factorization property (4), (5). This element can be used for further twisting already twisted $\mathcal{U}_q(gl(3))$. With extra diagonal twist $\mathcal{F}^{(3)}$ depending on $(e_{11} - e_{33}) \otimes C$ we get three parameter universal R-matrix, reduced to (11) in the fundamental representation.

An interesting feature of the constructed universal R-matrix

$$\mathcal{R}_{CG} = (\mathcal{F}^{(3)}\mathcal{F}^{(2)}\mathcal{F}^{(1)})_{21} \mathcal{R}_{DJ} (\mathcal{F}^{(3)}\mathcal{F}^{(2)}\mathcal{F}^{(1)})^{-1}$$

is that due to the fixed non-trivial q dependence of the first twist $\mathcal{F}^{(1)}$ there is no values of the parameters p, ν for which $\mathcal{R}_{CG}$ coincides with $\mathcal{R}_{DJ}$. At the same time, for $q = 1$ and fixed p, ν one gets a triangular Hopf algebra $\sigma(\mathcal{R})\mathcal{R} = 1$.

Let us briefly discuss the relations among the FRT-generators of the Drifeld-Jimbo (standard) quantum algebra and the twisted one. Taking the first factor of $\mathcal{A} \otimes \mathcal{A}$ in the fundamental representation, we get three 3×3 matrices

$$F_{21} = (\rho \otimes id)\sigma(\mathcal{F}), \quad L_{DJ}^{(+)} = (\rho \otimes id)\mathcal{R}_{DJ}, \quad F_{12} = (\rho \otimes id)\mathcal{F},$$

entries of which are expressed in terms of the standard generators (12). Multiplying these matrices one gets the L-matrix of the FRT-approach $L_{CG}^{(+)} = F_{21} L_{DJ}^{(+)} F_{12}^{-1}$ entries of which are generators of the esoteric quantum algebra, adding the same formulas for

$$L_{CG}^{(-)} = (\rho \otimes id)\,\sigma(\mathcal{R}_{CG}^{-1}) = F_{21} L_{DJ}^{(-)} F_{12}^{-1}.$$

Twisting of $osp(1|2)$. The quasitriangular Hopf superalgebra $\mathcal{U}_q(osp(1|2))$ [17] is generated by three elements $\{h, v_-, v_+\}$, analogously to the universal enveloping of $osp(1|2)$ or $sl(2)$, subject to the relations

$$[h, v_\pm] = \pm v_\pm, \qquad [v_+, v_-] = -\frac{1}{4}(q^h - q^{-h})/(q - q^{-1}), \tag{17}$$

where the commutator $[\,,\,]$ is understood as the Z_2-graded one: $[a, b] = ab - (-1)^{p(a)p(b)} ba$, with $p(a) = 0, 1$ being the parity (even or odd) of the element. The coproduct is

$$\Delta_q(h) = h \otimes 1 + 1 \otimes h \,,$$
$$\Delta_q(v_\pm) = v_\pm \otimes q^{h/2} + q^{-h/2} \otimes v_\pm \,. \tag{18}$$

The $osp(1|2)$ commutation relations follow from (17) in the limit $q \to 1$ and adding $X_\pm = \pm 4(v_\pm)^2$ as the Lie superalgebra generators. It is worthy to note that, while $sl(2)$ is embedded into $osp(1|2)$, such embedding does not exist for $sl_q(2)$ into $osp_q(1|2)$.

The jordanian twist $\mathcal{F}^{(j)}$ (6) preserving the algebraic relations among the generators of $\mathcal{U}(osp(1|2))$, results in the twisted coproduct Δ_j [12]

$$\begin{aligned}
\Delta_j(h) &= h \otimes e^{-2\sigma} + 1 \otimes h \,, \\
\Delta_j(v_+) &= v_+ \otimes e^\sigma + 1 \otimes v_+ \,, \\
\Delta_j(v_-) &= v_- \otimes e^{-\sigma} + 1 \otimes v_- + \xi h \otimes v_+ e^{-2\sigma} \,.
\end{aligned} \tag{19}$$

A contraction procedure of [8] applying to $osp_q(1|2)$ R-matrix, results in a twisting element $\mathcal{F}^{sj}$ with an extra contribution due to the odd generator v_+ [15]. This element is factorised $\mathcal{F}^{(sj)} = \mathcal{F}^{(s)} \mathcal{F}^{(j)}$. The form of the super-twist part $\mathcal{F}^{(s)}$ is

$$\mathcal{F}^{(s)} = \exp\left(-2\xi \left(v \otimes v\right) \varphi(\sigma \otimes 1, 1 \otimes \sigma)\right) \,,$$

where $\varphi(\sigma_1, \sigma_2)$ is a symmetric function of its arguments. The universal R-matrix of the twisted $\mathcal{U}(osp(1|2))$ is

$$\mathcal{R}^{(sj)} = \mathcal{F}^{(s)}_{21} \mathcal{F}^{(j)}_{21} (\mathcal{F}^{(j)})^{-1} (\mathcal{F}^{(s)})^{-1} \,.$$

Fixing the structure of $\mathcal{F}^{(s)}$ and considering the fundamental representation ρ only for the first factor in $\mathcal{A} \otimes \mathcal{A}$, one can take the upper triangular $L^{(+)}$ matrix of the FRT-formalism [6]

$$L^{(+)} = \begin{pmatrix} E^{-1} & V & H \\ 0 & 1 & W \\ 0 & 0 & E \end{pmatrix} = (\rho \otimes id) \mathcal{R}^{(sj)} \,, \tag{20}$$

and define the commutation relations of the generators H, E, V, W from the FRT-relation

$$R^{(sj)}(\xi) L^{(+)}_1 L^{(+)}_2 = L^{(+)}_2 L^{(+)}_1 R^{(sj)}(\xi) \,,$$

where the Z_2-graded tensor product is used defining $L^{(+)}_1 = L^{(+)} \otimes I$ and $L^{(+)}_2 = I \otimes L^{(+)}$ [5]. These commutation relations are (V and W are odd)

$$[E, V] = 0 \,, \quad [E, W] = 0 \,,$$
$$[H, E] = \xi(E^2 - 1) \,, \quad [H, V] = \xi(V(E^{-1} - E) - W) \,,$$
$$[V, V] = \xi(1 - E^{-2}) \,, \quad [W, W] = \xi(E^2 - 1) \,,$$

$$VW + WV = -\xi(E - E^{-1}), \quad V^2 + W^2 = \frac{1}{2}\xi(E^2 - E^{-2}).$$

It is easy to see that one can take H and W as two independent generators, while $V = -WE^{-1}$ and $E^2 = 1 + 2W^2/\xi$.

Taking into account the property of the universal R-matrix $\mathcal{R}$ [1], and the definition $L^{(+)} = (\rho \otimes id)\mathcal{R}^{(sj)}$, one gets the coproducts of the $L^{(+)}$ entries

$$\begin{aligned}
\Delta(E) &= E \otimes E, \\
\Delta(V) &= V \otimes E^{-1} + 1 \otimes V, \\
\Delta(W) &= W \otimes 1 + E \otimes W, \\
\Delta(H) &= H \otimes E^{-1} + E \otimes H - W \otimes V.
\end{aligned}$$

We have to find the expressions of the generators H, E, V, W in terms of h, v, and define the super-twist $\mathcal{F}^{(s)}$ from the intertwining relation. The known form of this element $\mathcal{F}^{(s)}$ in question and restriction of the first factor to the fundamental representation ρ only give the following expressions of the generators H, E, V, W

$$H = \xi h e^\sigma - 2(\xi v)^2 e^{-\sigma}, \; E = \exp(\frac{1}{2}\ln(1 + 2\xi X)) = e^\sigma, \; V = -2\xi v e^{-\sigma}, \; W = 2\xi v. \tag{21}$$

Hence the coproducts of the generators h, v are the following

$$\begin{aligned}
\Delta_{sj}(h) &= h \otimes E^{-2} + 1 \otimes h + 4\xi\, vE^{-1} \otimes vE^{-2}, \\
\Delta_{sj}(v) &= v \otimes 1 + e^\sigma \otimes v.
\end{aligned} \tag{22}$$

Thus the super-jordanian twist of the Borel subalgebra sB_+ of $osp(1|2)$ is defined.

To define the corresponding deformation of the $\mathcal{U}(osp(1|2))$ we have to find $\mathcal{F}^{(sj)}$ and the coproduct of the generator v_-

$$\mathcal{F}^{(sj)}\Delta(v_-)(\mathcal{F}^{(sj)})^{-1} = \mathcal{F}^{(s)}(v_- \otimes E^{-1} + 1 \otimes v_- + \xi h \otimes v_+ E^{-2})(\mathcal{F}^{(s)})^{-1}. \tag{23}$$

The knowledge of the exact form of the super-part of the twisting element seems necessary to get a final expression for $\Delta_{sj}(v_-)$. Although one can prove using (2) (with appropriate modifications due to the Z_2-grading) the existence of $\mathcal{F}^{(s)}$ order by order in ξ we do not have a closed form. The intertwining relation

$$\mathcal{F}^{(s)}\Delta_j(v_+)(\mathcal{F}^{(s)})^{-1} = (v_+ \otimes E + 1 \otimes v_+)(\mathcal{F}^{(s)})^{-2} = v_+ \otimes 1 + E \otimes v_+$$

also can be used to find $\mathcal{F}^{(s)}$. The conjectured form of $\varphi(\sigma_1, \sigma_2)$ is the following

$$\varphi(\sigma \otimes 1, 1 \otimes \sigma) = \sum_{k=1}^{\infty} f_k(\sigma) \otimes f_k(\sigma).$$

Each $f_k(\sigma)$ of this expression is characterized by its non zero contribution starting from the irreducible representation of spin $s = k/2$, $dimV_s = 4s + 1$ and by the first term $(\xi X_+)^{k-1}$ with a coefficient to be defined. In particular, one gets $f_1(\sigma) = 2/(e^\sigma + 1)$.

Integrable models of the QISM. According to the QISM [5,6] one can construct a variety of integrable models, related to a given solution of the YBE (an R-matrix). The importance of the corresponding spin-chain models is connected with possibilities to get different field-theoretical integrable models using different limiting procedures [?]. The extra parameters adding to the R-matrix by a twist $\mathcal{F}$ are quite useful in this way. It was pointed out [10], that the twist of an R-matrix

$$R(u) \to F_{21} R_{12}(u) F^{-1} \equiv R^{(F)}(u),$$

preserves the regularity property [5]: the existence of the spectral parameter value $u = u_0$, where $R(u_0)$ is proportional to the permutation operator $\mathcal{P}$. It is obvious, that from $R(u_0) \simeq \mathcal{P}$ it follows

$$R^{(F)}(u_0) = F_{21} R_{12}(u_0) F^{-1} \simeq \mathcal{P}.$$

This regularity property is important to get local integrals of motion.

The simplest integarble XXX-model is related to the Yangian $\mathcal{Y}(gl(2))$. The twist of the Lie algebra $sl(2)$: $\Delta_\xi = \mathcal{F}^{(j)} \Delta (\mathcal{F}^{(j)})^{-1}$ can be extended to the Yangian due to the embedding $sl(2) \subset \mathcal{Y}(gl(2))$.

Let us consider the Yang–Baxter algebra

$$R(u-v) T_1(u) T_2(v) = T_2(v) T_1(u) R(u-v),$$

for the entries of the 2×2 matrix

$$T_t(u) = \begin{pmatrix} A(u) & B(u) \\ C(u) & D(u) \end{pmatrix},$$

given by twisting the Yang solution,

$$R(u-v) = F_{21} \left(I - \frac{\eta}{u-v} \mathcal{P} \right) F_{12}^{-1} = R_\xi - \frac{\eta}{u-v} \mathcal{P}, \tag{24}$$

with $\mathcal{P}$ as the permutation matrix and $F_{12} = (\rho \otimes \rho) e^{h \otimes \sigma}$, where ρ is the fundamental representation of $sl(2)$. From the transfer matrix $t(u) = tr T_t(u) = A(u) + D(u)$ a deformed Hamiltonian for the Heisenberg chain of length N (XXX_ξ-model) follows

$$H = \sum_n (\sigma_n^x \sigma_{n+1}^x + \sigma_n^y \sigma_n^y + \sigma_n^z \sigma_{n+1}^z + \xi^2 \sigma_n^- \sigma_{n+1}^- + \xi(\sigma_n^- - \sigma_{n+1}^-)),$$

where ξ is a deformation parameter, σ_n^x, σ_n^y, and σ_n^z are Pauli sigma-matrices acting in C_n^2 related to the nth site of the chain and $\sigma_n^- = \frac{1}{2}(\sigma_n^x - i\sigma_n^y)$. According to the general scheme of the QISM one can construct integrable models for other values of spin $s = 1, \frac{3}{2}, 2, \ldots$ as well. However, it is easy to see that this Hamiltonian is not Hermitian, and thus creates extra difficulties in constructing the algebraic Bethe

Ansatz for this model. Due to the triangular structure of the twist, the spectrum of the transfer matrix and the Bethe equations of the XXX_ξ-model coincide with the usual ones.

The local structure of the monodromy matrix $T_t(u)$ leads to the following asymptotic expansion,

$$(T_t)_N(u) = L_N(u)\ldots L_1(u) = \prod_{k=1}^{N} R_{ak}(\xi) + \frac{1}{u}\sum_{k=1}^{N} M_k P_{ak} M_{N-k-1} + O\left(\frac{1}{u^2}\right),$$

where $M_k = \prod_{m=1}^{k-1} R_{am}(\xi) = (id \otimes \Delta^{(k-1)})R(\xi)$. Using for the constant term the notation

$$T_0 = \begin{pmatrix} E^{-1} & 0 \\ G & E \end{pmatrix} = \prod_{k=1}^{N} R_{ak}(\xi) = (id \otimes \Delta^{(N-1)})R(\xi),$$

one gets the symmetry algebra $sl_\xi(2)$ of the quantum scattering data. The element E is the group-like one, and it commutes with the transfer matrix $t(u)$, while G has the following simple coproduct: $\Delta(G) = G \otimes E^{-1} + E \otimes G$.

The Hecke condition preserving by twisiting, leads to the spectral parameter dependent solution of the YBE through the Yang-Baxterization procedure

$$\check{R}(u) = u\check{R} - \frac{1}{u}\check{R}^{-1} = (u - u^{-1})\check{R}_{CG} + \omega u^{-1} I. \tag{25}$$

The L-operator of the integrable spin chain coincides with the R-matrix $R(u)$ and the density of the Hamiltonian is $H = \sum_n h_{n,n+1}$, $(u_0 = \pm 1)$

$$h_{1,2} \simeq \mathcal{P}\tfrac{d}{du}R(u)|_{u=u_0} = 2\check{R}_{CG} + const.$$

$$= \sum_{i=1}^{3} q\, e_{ii} \otimes e_{ii} + (p\, e_{21} \otimes e_{12} + p^{-1} e_{12} \otimes e_{21}) + q\nu\left(p\, e_{12} \otimes e_{32} - (p/q)^2 e_{32} \otimes e_{12}\right)$$
$$+ (p^2/q\, e_{31} \otimes e_{13} + q/p^2 e_{13} \otimes e_{31}) + \omega \sum_{a<b} e_{bb} \otimes e_{aa}.$$

Due to the fact that the twist transformation leads to the explicit expression for the new R-matrix in terms of the old one and the twist element $\mathcal{F}$, one can get connection between quantum scattering data (the transition matrix) of initial and deformed integrable models

$$T_t(u) = (\rho \otimes \pi)(id \otimes \Delta_t^{(N)})\mathcal{F}_{21} \mathcal{R} \mathcal{F}^{-1}, \tag{26}$$

where ρ and π are representations of the quantum algebra in the auxiliary and quantum space respectively [10]. However, the general explicit expression for $T_t(u)$ looks rather cumbersome. The formulae (26) includes $(N-1)$ iterations of the twisted coproduct $\Delta_t = \mathcal{F}\Delta\mathcal{F}^{-1}$.

Although the iterated coproduct is a similarity transformation of the original one, the transition matrices are related by a more complicated transformation, since the factors are not the inverse of each other. The Yang-Baxter algebra of the quantum scattering data $(T_t(u))_{ij}$ for the twisted model is more complicated than the $SL(3)$-spin chain due to the extra non-zero elements of $R_{CG}(u)$ proportional to ν. However, as in the case of the XXX_ξ-spin chain [10], the spectrum of the transfer matrix $t(u) = trT_t(u)$ is similar to the $SL(3)$-spin chain case, taking into account the changed eigenvalues of the diagonal elements of $T_t(u)$ on the reference (vacuum) state $(0, 0, 1)^t$. The Bethe equations defining the two sets of quasimomenta have similar structure with obvious changes due to the parameter p. The second parameter ν enters into the eigenvectors and adjoint vectors.

Acknowledgements The author is grateful to Professor J. Rembielinski for the invitation to this representative conference. Discussions with Professor J. Lukierski and the hospitality of the Institute for Theoretical Physics of the Wroclaw University are appreciated.

REFERENCES

1. Drinfeld, V.G., "Quantum groups", in: *Proc. Int. Cong. Math. Berkeley, 1986*, **1**, ed. Gleason, A.V. (AMS, Providence, 1987) pp. 798-820.
2. Faddeev, L.D., Reshetikhin, N.Yu. and Takhtajan, L.A., *Algebra i Analiz* **1** 178 (1989), English transl. *Leningrad Math. J.* **1** 193 (1990).
3. Bayen, F., Flato, M., Lichnerowicz, A., Stenheimer, D., *Ann. Phys.* **111**, 61 (1978).; Flato, M. and Sternheimer, D., *Lett. Math. Phys.* **22**, 155 (1991).
4. Drinfeld, V.G., *DAN USSR*, **273**, (3) 531 (1983); *Leningrad Math. J.* **1**, 1419 (1990).
5. Kulish, P.P. and Sklyanin, E.K., *Lect. Notes Phys.* **153**, 69 (1982).
6. Faddeev, L.D., *How algebraic Bethe Ansatz works for integrable modles*, in: Les Houches Lectures, 1996; hep-th/9605187.
7. Maillet, J.-M. and Sanchez de Santos, J., *Drinfeld twists and algebraic Bethe Ansatz*, (1996); q-alg/9612012;
   Fiore, G., *Drinfeld twist and q-deforming maps for Lie group covariant Heisenberg algebras*, (1996); q-alg/9708017;
   Fronsdal, C., *Publ. RIMS, Kyoto Univ.* **33**, 91 (1997);
   Jimbo, M., Konno, H., Odake, S. and Shiraishi, J., Quasi-Hopf twistors for elliptic quantum groups, (1997), q-alg/9712029;
   Arnaudon, D., Buffenoire, E., Ragoucy, E. and Roche, Ph.: Universal solutions of quantum dynamical Yang-Baxter equations, (1997), q-alg/9712037.
8. Gerstenhaber, M., Giaquinto, A. and Schack, S.D., in *Quantum groups. Proc. EIMI, 1990*, ed. Kulish, P.P., *Lect. Notes Math.* **1510**, (Springer-Verlag, Berlin, 1992) pp. 9-46;
   Ogievetsky, O.V., in *Proc. Winter School Geometry and Physics, Zidkov, Suppl. Rendiconti cir. Math. Palermo, Serie II* **N 37**, 185 (1993); Preprint MPI-Ph/92-99, Munich, (1992) 14p.
9. Zakrzewski, S., *Lett. Math. Phys.* **22**, 287 (1991).
10. Kulish, P. P. and Stolin, A. A., *Czech. J. Phys.* **47** (12), 1207–1212 (1997).
11. Chaichian, M., Kulish, P.P. and Damaskinsky, E.V.: *Dynamical systems related to*

the Cremmer-Gervais R-matrix, Teor. Mat. Fiz. **116**, N 3 (1998), (to be published) q-alg/9712016.
12. Celeghini, E. and Kulish, P.P., *J. Phys.A:* **31**, L79 (1998); preprint q-alg/9712024.
13. Kulish, P.P. and Mudrov, A.I., *Lett.Math. Phys.* **41**, 111 (1998); math.QA/9804006.
14. Kulish, P.P., Lyakhovsky, V.D. and Mudrov, A.I., *Extended jordanian twists for Lie algebras*, preprint math.QA/9806014 (1998).
15. Kulish, P.P., *Super-jordanian deformation of the orthosymplectic Lie superalgebras*, preprint DFF 315/6/98 (1998); math.QA/9806104.
16. Kulish, P.P. and Mudrov, A. I., *Twist related geometries of deformed Minkowski spaces*, (under preparation) (1998).
17. Kulish, P.P., and Reshetikhin, N.Yu., *Lett. Math. Phys.* **18**, 143 (1989).

Remarks on Quantum Statistics

Władysław Marcinek[1]

Institute of Theoretical Physics, Uniwersity of Wrocław, Poland

Abstract. Some problems related to an algebraic approach to quantum statistics are discussed. Generalized quantum statistics is described as a result of interactions. The Fock space representation is discussed. The problem of existence of well-defined scalar product is considered. An example of physical effect in system with generalized statistics is also given.

INTRODUCTION

In the last years a few different approaches to quantum statistics which generalize the usual boson or fermion statistics had been intensively developed by several authors. The so-called q-statistics and corresponding q-relations have been studied by Greenberg [1,2], Mohapatra [3], Fivel [4] and many other, see [5–7] for example. The deformation of commutation relations for bosons and fermions corresponding to quantum groups $SU_q(2)$ has been given by Pusz and Woronowicz [8,9]. The q-relations corresponding to superparticles has been considered by Chaichian, Kulish and Lukierski [10]. Quantum deformations have been also studied by Vokos [11], Fairle and Zachos [12] and many others.

Note that there is also and approach to particle systems with some nonstandard statistics in low dimensional spaces based on the notation of the braid group B_n [13,14]. In this approach the configuration space for the system of n-identical particles moving on a manifold $\mathcal{M}$ is given by the formula

$$Q_n(\mathcal{M}) = (\mathcal{M}^{\times n} - D)/S_n,$$

where D is the subcomplex of the Cartesian product $\mathcal{M}^{\times n}$ on which two or more particles occupy the same position and S_n is the symmetric group. The group $\pi_1(Q_n(\mathcal{M})) \equiv B_n(\mathcal{M})$ is known as the n-string braid group on $\mathcal{M}$. Note that there is a group $\Sigma_n(\mathcal{M})$ which is a subgroup of $B_n(\mathcal{M})$ and is an extension of the symmetric group S_n describing the interchange process of two arbitrary indistinguishable particles. It is obvious that the statistics of the given system od particles by the

---

[1] The work is partially sponsored by Polish Committee for Scientific Research (KBN) under Grant 2P03B130.12

group Σ_n [13,14]. The mathematical formalism related to the braid group statistics has been developed intensively by Majid, see [15–21] for example. It is interesting that all commutation relations for particles equipped with arbitrary statistics can be described as representations of the so-called quantum or Wick algebra W. Such algebraic formalism has been considered by Jorgensen, Schmith and Werner [22] and further developed by the author in series of papers [23–31] and also by Rałowski [32–34]. Similar approach has been also considered by other authors, see [35–37] and [38,39]. An interesting approach to quantum statistics has been also given in [40,41]. A proposal for the general algebraic formalism for description of particle systems equipped with an arbitrary generalized statistics based on the concept of monoidal categories with duality has been given by the author in [42]. The physical interpretation for this formalism was shortly indicated. A few examples of applications for this formalism are considered in [43,44].

The generalization of concept of quantum statistics is motivated by many different applications in quantum field theory and statistical physics. Some problems in condensed matter physics, magnetism or quantum optics lead to study of particle systems obeying nonstandard statistics. It is interesting that in the last years new and highly organized structures of matter have been discovered. For example in fractional quantum Hall effect a system with well defined internal order has appeared [45]. Another interesting structures appear in the so-called $\frac{1}{2}$ electronic magnetotransport anomaly [46,47], high temperature superconductors or laser excitations of electrons. In these cases certain anomalous behaviour of electron has appeared. Another exaple is given by the so-called Lutinger liquid [48]. The concept of statistical–spin liquids has been studied by Byczuk and Spalek [49]. It is interesting that half of the available single-particle states are removed by the statistical interaction between the particles with opposite spins. The study of highly organized structures leads to the investigation of correlated systems of interacting particles. The essential problem is such study is to transform the system of interacting particles into an effective model convenient for the description of ordered structures. A system with generalized statistics seems to be one of the best candidates for such model. Hence there is an interest on the developement of formalism related to the particle systems with unusual statistics and the possible physical applications. In this paper we would like to discuss some problems of possibility of application of systems with generalized statistics for the further description of ordered structures. All our considerations are based on the assumptions that the quantum statistics of charged particles is determined by some specific interactions.

FUNDAMENTAL ASSUMPTIONS

The starting point for our discussion is a system of charged particles interacting with certain quantum field. The proper physical nature of the system is not essential for our consideration. The fundamental assumption is that the problem of interacting particles can be reduced to the study of a system consisting n charged particles and N-species of quanta of the field. In this way we can restrict our attention to study of such system. It is natural to expect that some new excited states of the system have appeared as a result of certain specific interaction. The existence of new ordered structures depends on the existence of such additional excitations. Hence we can restrict our attention to the study of possibility of appearance for these excitations. For the description of such possible excited states we use concept of dressed particles. We assume that every charged particle is equipped with ability to absorb quanta of the external field. A system which contains a particle and certain number of quanta as a result of interaction with the external field is said to be dressed particle. A particle without quantum is called undressed or a quasihole. The particle dressed with two quanta of certain species is understand as a system od two new objects called quasiparticles. A quasiparticle is in fact the charged particle dressed with a single quantum. Two quasiparticles are said to be identical if they are dressed with quanta of the same species. In the opposite case when the particle is equipped with two different species of quanta then we have different quasiparticles. We describe excited states as composition of quasiparticles and quasiholes. It is interesting that quasiparticles and quasiholes have also their own statistics. We give the follwing assumption for the algebraic description of excitation spectrum of single dressed particle.

Assumption 0. The ground state. *There is a state $|0\rangle = \mathbf{1}$ called the ground one. There is also the conjugate ground state $\langle 0| \equiv \mathbf{1}^*$. This is the state of the system before intersection.*

Assumption 1. Elementary states. *There is an ordered (finite) set of single quasiparticle states*

$$S := \{x^i \,:\, i = 1, \ldots, N < \infty\}. \tag{1}$$

These states are said to be elementary (simple). They represent elementary excitations of the system. We assume that the set S of elementary states forms a basis for a finite linear space E over a field of complex numbers $\mathbb{C}$.

Assumption 2. Elementary conjugate states. *There is also a corresponding set of single quasihole states*

$$S^* := \{x^{*i} \,:\, i = N, N-1, \ldots, 1\}. \tag{2}$$

These states are said to be conjugated. The set S^* of conjugate states forms a basis for the complex conjugate space E^*. The pairing $(.|.) : E^* \otimes E \to \mathbb{C}$ is given by

$$(x^{*i}|x^j) := \delta^{ij}. \tag{3}$$

Assumption 3. Composite states. *There is a set of projectors*

$$\Pi_n \,:\, E^{\otimes n} \to E^{\otimes n} \tag{4}$$

such that we have a n-multinilear mapping

$$\odot_n : E^{\times n} \to E^{\otimes n}. \tag{5}$$

defined by the following formula

$$x^{i_1} \odot \ldots \odot x^{i_n} := \Pi_n(x^{i_1} \otimes \ldots \otimes x^{i_n}). \tag{6}$$

The set of n-multiquasiparticle states is denoted by $P^n(S)$. All such states are result of composition (or clustering) of elementary ones. These states are also called composite states of order n. They represent additional excitations charged particles under interaction. In this way for multiquasiparticle states we have the follwing set of states

$$P^n(S) := \{x^\sigma \equiv x^{i_1} \odot \ldots \odot x^{i_n} \; : \; \sigma = (i_1, \ldots, i_n) \in I\}, \tag{7}$$

here I is set of sequences of indices such that the above set of states forms a basis for a linear space $\mathcal{A}^n$. We have

$$\mathcal{A}^n = \text{Im}(\Pi_n). \tag{8}$$

Obviously we have here $\mathcal{A}^0 \equiv 1\mathbb{C}$, $\mathcal{A}^1 \equiv E$ and $\mathcal{A}^n \subset E^{\otimes n}$.

Assumption 4. Composite conjugated states. *We also have a set of projectors*

$$\Pi_n^* \; : \; E^{*\otimes n} \to E^{*\otimes n} \tag{9}$$

and the corresponding set of composite conjugated states of length n

$$P^n(S^*) := \{x^{*\sigma} \equiv x^{*i_1} \odot \ldots \odot x^{*i_n} \; : \; \sigma = (i_1, \ldots, i_n) \in I\}. \tag{10}$$

The set $P^n(S^*)$ of composite conjugated states of length n forms a basis for a linear space $\mathcal{A}^{*n}$.

Assumption 5. Algebra of states. *The set of all composite states of arbitraty length is denoted by $P(S)$. For this set of states we have the following linear space*

$$\mathcal{A} := \bigoplus_n \mathcal{A}^n. \tag{11}$$

If the formula

$$m(s \otimes t) \equiv s \odot t := \Pi_{m+n}(\tilde{s} \otimes \tilde{t}) \tag{12}$$

for $s = \Pi_m(\tilde{s})$, $t = \Pi_n(\tilde{t})$, $\tilde{s} \in E^{\otimes n}$, $\tilde{t} \in E^{\otimes m}$, defines an associative multiplication in $\mathcal{A}$, then we say that we have an algebra of states. This algebra represents excitation spectrum for single dressed particle.

Assumption 6. Algebra of conjugated states. *The set of composite conjugated states od arbitrary length is denoted by $P(S^*)$. We have a linear space*

$$\mathcal{A}^* := \bigoplus_n \mathcal{A}^{*n}. \tag{13}$$

If m is the multiplication in $\mathcal{A}$, then the multiplication in $\mathcal{A}^*$ corresponds to the opposite multiplication in $\mathcal{A}$

$$m^{\mathrm{op}}(t^* \otimes s^*) = (m(s \otimes t))^*. \tag{14}$$

CREATION AND ANNIHILATION OPERATORS

We define creation operators for our model as multiplication in the algebra $\mathcal{A}$

$$a_s^+ t := s \odot t, \quad \text{for } s,t \in \mathcal{A}, \tag{15}$$

where the multiplication is given by (12). For the ground state and annihilation operators we assume that

$$\langle 0|0 \rangle = 0, \quad a_{s^*}|0\rangle = 0 \quad \text{for } s^* \in \mathcal{A}^*. \tag{16}$$

The proper definition of action of annihilation operators on the whole algebra $\mathcal{A}$ is a problem. For the pairing $\langle -|-\rangle^n : \mathcal{A}^{*n} \otimes \mathcal{A} \to \mathbb{C}$ we assume in addition that we have the following formulae

$$\langle 0|0 \rangle^0 := 0, \quad \langle i|j \rangle^1 := \left(x^{*i} | x^j \right) = \delta^{ij}, \quad \langle s|t \rangle^n := \langle \tilde{s} | P_n \tilde{t} \rangle_0^n \quad \text{for } n \geq 2 \tag{17}$$

where $\tilde{s}, \tilde{t} \in E^{\otimes n}$, $P_n : E^{\otimes n} \to E^{\otimes n}$ is an additional linear operator and

$$\langle i_1 \ldots i_n | j_1 \ldots j_n \rangle_0^n := \langle i_1|j_1 \rangle^1 \ldots \langle i_n|j_n \rangle^1. \tag{18}$$

Observe that we need two sets $\Pi := \{\Pi_n\}$ and $P := \{P_n\}$ of operators and the action

$$a : s^* \otimes t \in \mathcal{A}^{*k} \otimes \mathcal{A}^n \to a_{s^*}t \in \mathcal{A}^{n-k} \tag{19}$$

of annihilation operators for the algebraic description of our system. In this way the triple $\{\Pi, P, a\}$, where Π and P are sets of linear operators and a is the action of annihilation operators, is the initial data for our model. The problem is to find and classify all triples of initial data which lead to the well-defined models. The general solution for this problem is not known for us. Hence we must restrict our attention for some examples.

Definition. If operators P and Π and the action a of annihilation operators are given in such a way that there is unique, nondegenerate, positive definite scalar product, creation operators are adjoint to annihilation ones and vice versa, then we say that we have a well-defined system with generalized states.

Example 1. We assume here that $\Pi_n \equiv P_n \equiv \mathrm{id}_{E^{\otimes n}}$. This means that the algebra of states $\mathcal{A}$ is identical with the full tensor algebra TE over the space E, and the second algebra $\mathcal{A}^*$ is identical with the tensor algebra TE^*. The action (19) of annihilation operators is given by the formula

$$a_{x^{*i_k} \otimes \ldots \otimes x^{*i_1}}(x^{j_1} \otimes \ldots \otimes x^{j_n}) := \delta_{i_1}^{j_1} \ldots \delta_{i_k}^{j_k} x^{j_{n-k+1}} \otimes \ldots \otimes x^{j_n}. \tag{20}$$

For the scalar product we have the equation

$$\langle i_n \ldots i_1 | j_1 \ldots j_n \rangle^n := \delta^{i_1 j_1} \ldots \delta^{i_n j_n}. \tag{21}$$

It is easy to see that we have the relation and

$$a_{x^{*i}} a_{x_j} := \delta_i^j \mathbf{1}. \tag{22}$$

In this way we obtain the most simple example of well-defined system with generalized statistics. The corresponding statistics is the so-called infinite (Boltzmann) statistics [1,2].

Example 2. For this example we assume that $\Pi_n \equiv \mathrm{id}_{E^{\otimes n}}$. This means that $\mathcal{A} \equiv TE$ nad TE^*. For the scalar product and for the action of annihilation operators we assume that there is a linear and invertible operator $T : E^* \otimes E \to E \otimes E^*$ defined by its matrix elements

$$T(x^{*i} \otimes x^j) = \sum_{k, *l} T^{*ij}_{k*l} x^k \otimes x^{*l}, \tag{23}$$

such that we have

$$(T^{*ij}_{k*l})^* = \bar{T}^{*ji}_{l*k}, \quad \text{i.e. } T^* = \bar{T}^t, \tag{24}$$

and $(T^t)^{*ij}_{k*l} = T^{*ji}_{l*k}$. Note that this operator need not to be linear, one can also consider the case of nonlinear one. We also assume that the operator T^* act to the left, i.e. we have the relation

$$(x^{*j} \otimes x^i) T^* = \sum_{l, *k} (x^l \otimes x^{*k}) \bar{T}^{*ji}_{l*k}, \tag{25}$$

and

$$(T(x^{*i} \otimes x^j))^* \equiv (x^{*j} \otimes x^i) T^*. \tag{26}$$

The operator T given by formula (23) is said to be a *twist* or a *cross* operator. The operator T describes the cross statistics of quasiparticles and quasiholes. The set P of projectors is defined by induction

$$P_{n+1} := (\mathrm{id} \otimes P_n) \circ R_{n+1}, \qquad (27)$$

where $P_1 \equiv \mathrm{id}$ and the operator R_n is given by the formula

$$R_n := \mathrm{id} + \tilde{T}^{(1)} + \tilde{T}^{(1)}\tilde{T}^{(2)} + \ldots + \tilde{T}^{(1)} \ldots \tilde{T}^{(n-1)}, \qquad (28)$$

where $\tilde{T}^{(i)} := \mathrm{id}_E \otimes \ldots \otimes \tilde{T} \otimes \ldots \otimes \mathrm{id}_E$, $\tilde{T}$ on i-th place, and

$$(\tilde{T})^{ij}_{kl} = T^{*ki}_{l*j}. \qquad (29)$$

If the operator $\tilde{T}$ is a bounded operator acting on some Hilbert space such that we have following Yang–Baxter equation on $E \otimes E \otimes E$

$$(\tilde{T} \otimes \mathrm{id}_E) \circ (\mathrm{id}_E \otimes \tilde{T}) \circ (\tilde{T} \otimes \mathrm{id}_E) = (\mathrm{id}_E) \otimes \tilde{T}) \circ (\tilde{T} \otimes \mathrm{id}_E) \circ (\mathrm{id}_E \otimes \tilde{T}), \qquad (30)$$

and $\|\tilde{T}\| \leq 1$, then according to Bożejko and Speicher [7] there is a positive definite scalar product

$$\langle s|t\rangle^n_T := \langle s|P_n t\rangle^n_0 \qquad (31)$$

for $s, t \in \mathcal{A}^n \equiv E^{\otimes n}$. Note that the existence of nontrivial kernel of operator $P_2 \equiv R_1 \equiv \mathrm{id}_{E \otimes E} + \tilde{T}$ is essential for the nondegeneracy of the scalar product [22]. One can see that if this kernel is trivial, then we obtain well-defined system with generalized statistics [33,34].

Example 3. If the kernel of P_2 is nontrivial, then the scalar product (31) is degenerate. Hence we must remove this degeneracy by factoring the mentioned above scalar product by the kernel. We assume that there is an ideal $I \subset TE$ generated by a subspace $I_2 \subset \ker P_2 \subset E \otimes E$ such that

$$a_{s^*} I \subset I \qquad (32)$$

for every $s^* \in \mathcal{A}^*$, and for the corresponding ideal $I^* \subset E^* \otimes E^*$ we have

$$a_{s^*} t = 0 \qquad (33)$$

for every $t \in TE$ and $s^* \in I^*$. The above ideal I is said to be Wick ideal [22]. We have here the following formulae

$$\mathcal{A} := TE/I, \quad \mathcal{A}^* := TE^* \qquad (34)$$

for our algebras. The projection Π is the quotient map

$$\Pi : \tilde{s} \in TE \to s \in TE/I \equiv \mathcal{A}. \qquad (35)$$

For the scalar product we have here the following relation

$$\langle s|t\rangle_{B,T} := \langle \tilde{s}|\tilde{t}\rangle_T \tag{36}$$

for $s = P_m(\tilde{s})$ and $t = P_n(\tilde{t})$. One can define here the action of annihilation operators in such a way that we obtain well-defined system with generalized statistics [34].

Example 4. If a linear and invertible operator $B : E \otimes E \to E \otimes E$ defined by its matrix elements

$$B(x^i \otimes x^j) := B^{ij}_{kl}(x^k \otimes x^l) \tag{37}$$

is given such that we have the following conditions

$$B^{(1)}B^{(2)}B^{(1)} = B^{(2)}B^{(1)}B^{(2)}, \tag{38}$$
$$B^{(1)}T^{(2)}T^{(1)} = T^{(2)}T^{(1)}B^{(2)}, \tag{39}$$
$$(\mathrm{id}_{E\otimes E} + \tilde{T})(\mathrm{id}_{E\otimes E} - B) = 0, \tag{40}$$

then one can prove that there is well defined action of annihilation operators and scalar product. In this case we need two operators T and B satisfying the above consistency conditions for the model with generalized statistics [32–34].

Example 5. If $B = \frac{1}{\mu}\tilde{T}$, where μ is parameter, then the third condition (40) is equivalent to the well known Hecke condition for $\tilde{T}$ and we obtain the well-known relations for Hecke symmetry and quantum groups [8,9,50].

PHYSICAL EFFECT

Let us consider the system equipped with generalized statistics and described by two operators T and B like in Example 4. We assume here in addition that a linear and Hermitian operator $S : E \otimes E \to E \otimes E$ such that

$$S^{(1)}S^{(2)}S^{(1)} = S^{(2)}S^{(1)}S^{(2)}, \quad \text{and} \quad S^2 = \mathrm{id}_{E\otimes E} \tag{41}$$

is given. If we have the following relation

$$\tilde{T} \equiv B \equiv S, \tag{42}$$

then it is easy to see that the conditions (40) are satisfied and we have well-defined system with generalized statistics. Let us assume for simplicity that the operator S is diagonal and is given by the following equation

$$S(x^i \otimes x^j) = \epsilon^{ij} x^j \otimes x^i, \tag{43}$$

for $i, j = 1, \ldots, N$, where $\epsilon^{ij} \in \mathbb{C}$, and $\epsilon^{ij}\epsilon^{ji} = 1$. In general we have

$$c_{ij} = -(-1)^{\Sigma_{ij}}(-1)^{\Omega_{ij}}, \qquad (44)$$

where $\Sigma := (\Sigma_{ij})$ and $\Omega = (\Omega_{ij})$ are integer-valued matrices such that $\Sigma_{ij} = \Sigma_{ji}$ and $\Omega_{ij} = -\Omega_{ji}$. The algebra $\mathcal{A}$ is here a quadratic algebra generated by relations

$$x^i \odot x^j = \epsilon^{ij} x^j \odot x^i, \quad \text{and} \quad (x^i)^2 = 0 \text{ if } \epsilon^{ii} = -1. \qquad (45)$$

We also assume that $\epsilon^{ii} = -1$ for every $i = 1, dots, N$. In this case the algebra $\mathcal{A}$ is denoted by $\Lambda_\epsilon(N)$. Now let us study the algebra $\Lambda_\epsilon(2)$, where $\epsilon^{ii} = -1$ for $i = 1, 2$, and $\epsilon^{ij} = 1$ for $i \neq j$, in more details. In this case our algebra is generated by x^1 and x^2 such that we have

$$x^1 \odot x^2 = x^2 \odot x^1, \quad (x^1)^2 = (x^2)^2 = 0. \qquad (46)$$

Note that the algebra $\Lambda_\epsilon(2)$ is an example of the so-called $Z_2 \oplus Z_2$-graded commutative colour Lie superalgebra [51]. Such algebra can be transformed into the usual Grassmann algebra Λ_2 generated by Θ^1 and Θ^2 such that we have the anticommutation relation

$$\Theta^1 \Theta^2 = -\Theta^2 \Theta^1, \qquad (47)$$

and $(\Theta^1)^2 = (\Theta^2)^2 = 0$. In order to do such transformation we use the Clifford algebra C_2 generated by e^1, e^2 such that we have the relations

$$e^i e^j + e^j e^i = 2\delta^{ij} \quad \text{for } i, j = 1, 2. \qquad (48)$$

For generators x^1 and x^2 of the algebra $\Lambda_\epsilon(2)$ the transformation is given by

$$\Theta^1 := x^1 \otimes e^1, \quad \text{and} \quad \Theta^2 := x^2 \otimes e^2. \qquad (49)$$

It is interesting that the algebra $\Lambda_\epsilon(2)$ can be represented by one Grassmann variable Θ, $\Theta^2 = 0$

$$x^1 = (\Theta, 1), \quad x^2 = (1, \Theta). \qquad (50)$$

For the product $x^1 \odot x^2$ we obtain

$$x^1 \odot x^2 = (\Theta, \Theta). \qquad (51)$$

In physical interpretation of generators Θ^1 and Θ^2 of the algebra Λ_2 represents two fermions. They anticommuta and according to Pauli exclusion principle we cannot put them into one energy level. Observe that the corresponding generators x^1 and x^2 of the algebra $\Lambda_\epsilon(2)$ commute, their squares disappear and they describe two different quasiparticles. This means that these qausiparticles behave partially like bosons, we can put them simultaneously into one energy level. This also means that single fermion can be transformed under certain interactions into a system of two different quasiparticles.

REFERENCES

1. Greenberg, O. W., *Phys. Rev. Lett.* **64**, 705 (1990).
2. Greenberg, O. W., *Phys. Rev.* **D43**, 4111 (1991).
3. Mohapatra, R. N., *Phys. Lett.* **B242**, 407 (1990).
4. Fivel, D. I., *Phys. Rev. Lett.* **65**, 3361 (1990).
5. Zagier, D., *Commun. Math. Phys.* **147**, 199 (1992).
6. Meljanac, S., and Perica, A., *Mod. Phys. Lett.* **A9**, 3239 (1994).
7. Bożejko, M., and Speicher, R., *Math. Ann.* **300**, 97 (1994).
8. Pusz, W., *Rep. Math. Phys.* **27**, 394 (1989).
9. Pusz, W., and Woronowicz, S. L., *Rep. Math. Phys.* **27**, 231 (1989).
10. Chaichian, M., Kulish, P., and Lukierski, J., *Phys. Lett.* **B262**, 43 (1991).
11. Vokos, S. P., *J. Math. Phys.* **32**, 2979 (1991).
12. Fairle, D. B., and Zachos, C. K., *Phys. Lett.* **B256**, 43 (1991).
13. Wu, Y. S., *J. Math. Phys.* **52**, 2103 (1984).
14. Imbo, T. D., and March-Russel, J., *Phys. Lett.* **B252**, 84 (1990).
15. Majid, S., *Int. J. Mod. Phys.* **A5**, 1 (1990).
16. Majid, S., *J. Math. Phys.* **34**, 1176 (1993).
17. Majid, S., *J. Math. Phys.* **34**, 4843 (1993).
18. Majid, S., *J. Math. Phys.* **34**, 2045 (1993).
19. Majid, S., "Algebras and Hopf algebras in braided categories", in *Advances in Hopf Algebras*, New York: Plenum Press, 1993.
20. Majid, S., *J. Geom. Phys.* **13**, 169 (1994).
21. Majid, S., *AMS Cont. Math.* **134**, 219 (1992).
22. Jorgensen, P. E. T., Schmith, L. M., and Werner, R. F., *J. Funct. Anal.* **134**, 33 (1995).
23. Marcinek, W., *J. Math. Phys.* **33**, 1631 (1992).
24. Marcinek, W., *Rep. Math. Phys.* **34**, 325 (1994).
25. Marcinek, W., *Rep. Math. Phys.* **33**, 117 (1993).
26. Marcinek, W., *J. Maht. Phys.* **35**, 2633 (1994).
27. Marcinek, W., *Int. J. Math. Phys.* **A10**, 1465 (1995).
28. Marcinek, W., "On the deformation of commutation relations", in *Proceedings of the XIII Workshop in Geometric Methods in Physics*, Białowieża (Poland) 1994; New York: Plenum Press, 1995.
29. Marcinek, W., "On algebraic model of composite fermions and bosons", in *Proceedings in the IXth Max Born Symposium*, Karpacz (Poland), 1996.
30. Marcinek, W., "On quantum Weyl algebras and generalized quons", in *Proceedings of the Symposium: Quantum Groups and Quantum Spaces*, Warsaw, 1995; Warsaw: Banach Center Publ., 1997.
31. Marcinek, W., *Rep. Math. Phys.* **41**, 155 (1998).
32. Marcinek, W., and Rałowski, R., "Particle operators from braided geometry", in *Quantum Groups: Formalism and Applications*, Warsaw: PWN Polish Sci. Publ, 1995, pp. 149–154.
33. Marcinek, W., and Rałowski, R., "On Wick algebras with braid relations", preprint IFT UWr 876/9 (1994) and *J. Math. Phys.* **36**, 2803 (1995).

34. Rałowski, R., *J. Phys.* **A30**, 2633 (1997).
35. Scipioni, R., *Phys. Lett.* **B327**, 56 (1994).
36. Ting, Y., Wu, Z. Y., *Science in China* **A37**, 1472 (1994).
37. Meljanac, S., and Perica, A., *Mod. Phys. Lett.* **A9**, 3293 (1994).
38. Pillin, M., *Commun. Math. Phys.* **180**, 23 (1996).
39. Mishira, A. K., and Rajasekaran, G., *J. Math. Phys.* **38**, 23 (1997).
40. Fiore, G., and Schupp, P., "Statistics and quantum group symmetries", in *Proceedings of the Symposium: Quantum Groups and Quantum Spaces*, Warsaw 1995; Warsaw: Banach Center Publ., 1997.
41. Meljanac, S., and Molekovic, M., *Int. J. Mod. Phys.* **A11**, 193 (1996).
42. Marcinek, W., "Categories and quantum statistics", in *Proceedings of the Symposium: Quantum Groups and Their Application in Physics*, Poznań, 1995; *Rep. Math. Phys.* **38**, 149 (1996).
43. Marcinek, W., "Topology and quantization", in *Proceedings of the IVth International School on Theoretical Physics: Symmetry and Structural Properties*, Zajączkowo k. Poznania (Poland), 1996.
44. Marcinek, W., *J. Math. Phys.* **39**, 818 (1998).
45. Zee, A., "Quantum Hall fluids in field theory, topology and condensed matter physics", in *Lecture Notes in Physics*, Heidelberg: Springer, 1995.
46. Jain, J. K., *Phys. Rev. Lett.* **63**, 199 (1989); *Phys. Rev.* **B40**, 8079 (1989); **B41**, 7653 (1990).
47. Du, R. R., Stormer, H. L., Tsui, D. C., Yeh, A. S., Pfeiffer, L. N., and West K. W., *Phys. Rev. Lett.* **73**, 3274 (1994).
48. Haldane, F. D. M., *J. Phys.* **C14**, 2585 (1981).
49. Byczuk, K., and Spalek, J., *Lett. Math. Phys.* **26**, 11 (1995).
50. Kempf, A., *Lett. Math. Phys.* **26**, 11 (1992).
51. Lukierski, J., and Rittenberg, V., *Phys. Rev.* **D18**, 385 (1978).

Quantum Minkowski Spaces

Piotr Podleś

Department of Mathematical Methods in Physics
Faculty of Physics, Warsaw University

One of the main problems of theoretical physics is to find a satisfactory theory which would generalize both the quantum field theory and the general theory of relativity. It is widely recognized that in such a theory the geometry of the space–time should drastically change at very small distances, comparable with the Planck's length. One of possibilities is to replace the space–time by so called quantum space. In such an approach the role of the commutative algebra of functions on the space–time (generated in the simplest case by the coordinates) is played by some noncommutative algebra. At the present stage we are only able to test this idea in particular examples, which are important in physics. Many classical objects were already deformed in the above sense (cf. e.g. [17], [8], [15], [9], [1], [2], [18]). Here we deal with the most interesting case, namely that of Minkowski space M endowed with the action of Poincaré group P. Examples of quantum Poincaré groups and their actions on quantum Minkowski spaces appeared e.g. in [5], [3], [19] (the case of quantum Poincaré groups and algebras extended by dilatations was considered e.g. in [4], [16], [7], [6]). The aim is to find the classification of quantum Poincaré groups and quantum Minkowski spaces as well as mathematical and physical properties of those objects. We sketch the results of four papers [11], [12], [13], [14].

The (connected component of) vectorial Poincaré group

$$\tilde{P} = SO_0(1,3) \ltimes \mathbf{R}^4 = \{(M,a) : M \in SO_0(1,3), a \in \mathbf{R}^4\}$$

has the multiplication $(M,a) \cdot (M',a') = (MM', a + Ma')$. By the Poincaré group we mean spinorial Poincaré group (which is more important in quantum field theory then $\tilde{P}$)

$$P = SL(2, \mathbf{C}) \ltimes \mathbf{R}^4 = \{(g,a) : g \in SL(2, \mathbf{C}), a \in \mathbf{R}^4\}$$

with multiplication $(g,a) \cdot (g',a') = (gg', a + \lambda_g(a'))$ where the double covering $SL(2, \mathbf{C}) \ni g \longrightarrow \lambda_g \in SO_0(1,3)$ is the standard one. The group homomorphism $\pi : P \ni (g,a) \longrightarrow (\lambda_g, a) \in \tilde{P}$ is also a double covering. In particular, $(-\mathbf{1}_2, 0) \in P$ can be treated as rotation about 2π which is trivial in $\tilde{P}$ but nontrivial in P (it changes the sign of wave functions for fermions). Both P and $\tilde{P}$ act on Minkowski space $M = \mathbf{R}^4$ as follows $(g,a)x = (\lambda_g, a)x = \lambda_g x + a$, $g \in SL(2, \mathbf{C})$, $a, x \in \mathbf{R}^4$.

Let us consider continuous functions w_{AB}, y_i on P defined by

$$w_{AB}(g,a) = g_{AB}, \qquad y_i(g,a) = a_i.$$

We introduce Hopf *-algebra Poly$(P) = (\mathcal{B}, \Delta)$ of polynomials on the Poincaré group P as the *-algebra $\mathcal{B}$ with identity I, generated by w_{AB} and y_i, $A, B = 1, 2$, $i \in \mathcal{S} = \{0, 1, 2, 3\}$ endowed with the comultiplication Δ given by $(\Delta f)(x, y) = f(x \cdot y)$, $f \in \mathcal{B}$, $x, y \in P$ ($f^*(x) = \overline{f(x)}$). In particular,

$$\Delta w_{CD} = w_{CF} \otimes w_{FD}, \qquad (1)$$

$$\Delta y_i = y_i \otimes I + \Lambda_{ij} \otimes y_j, \qquad (2)$$

$y_i^* = y_i$, where

$$\Lambda = V^{-1}(w \otimes \bar{w})V, \qquad V = \begin{pmatrix} 1 & 0 & 0 & 1 \\ 0 & 1 & -i & 0 \\ 0 & 1 & i & 0 \\ 1 & 0 & 0 & -1 \end{pmatrix} \qquad (3)$$

(we sum over repeated indices, one has $V_{CD,i} = (\sigma_i)_{CD}$ where σ_i are the Pauli matrices). Moreover, w is a representation, i.e. it is invertible (as 2×2 matrix with elements in $\mathcal{B}$) and satisfies (1). Equivalences of representations are defined as usual (by means of invertible matrices with complex entries), e.g. $\Lambda \simeq w \otimes \bar{w}$. We put $y = (y_i)_{i \in \mathcal{S}}$. One can also treat w_{CD} as continuous functions on the Lorentz group $L = SL(2, \mathbf{C})$ ($w_{CD}(g) = g_{CD}$, $g \in L$). We define Hopf *-algebra Poly$(L) = (\mathcal{A}, \Delta)$ of polynomials on L as *-algebra with I, generated by all w_{CD} endowed with Δ obtained by restriction of Δ for $\mathcal{B}$ to $\mathcal{A}$. Clearly w and Λ are representations of L. It is easy to check that

1. $\mathcal{B}$ is generated as algebra by $\mathcal{A}$ and the elements y_i, $i \in \mathcal{S}$.
2. $\mathcal{A}$ is a Hopf *-subalgebra of $\mathcal{B}$.
3. $\mathcal{P} = \begin{pmatrix} \Lambda & y \\ 0 & I \end{pmatrix}$ is a representation of $\mathcal{B}$ where Λ is given by (3).
4. There exists $i \in \mathcal{S}$ such that $y_i \notin \mathcal{A}$.
5. $\Gamma \mathcal{A} \subset \Gamma$ where $\Gamma = \mathcal{A}X + \mathcal{A}$, $X = \text{span}\{y_i : i \in \mathcal{S}\}$.
6. The left $\mathcal{A}$-module $\mathcal{A} \cdot \text{span}\{y_i y_j, y_i, I : i, j \in \mathcal{S}\}$ has a free basis consisting of $10 + 4 + 1$ elements.

(5. and 6. follow from the relations $y_i a = a y_i$, $y_i y_j = y_j y_i$, $a \in \mathcal{A}$, and elementary computations, a free basis is given by $\{y_i y_j, y_i, I : i \leq j, i, j \in \mathcal{S}\}$). According to [18], Poly(L) satisfies:

i. $(\mathcal{A}, \Delta)$ is a Hopf *-algebra such that $\mathcal{A}$ is generated (as *-algebra) by matrix elements of a two–dimensional representation w

ii. $w \otimes w \simeq I \oplus w^1$ where w^1 is a representation

iii. the representation $w \otimes \bar{w} \simeq \bar{w} \otimes w$ is irreducible

iv. if $\mathcal{A}', \Delta', w'$ satisfy i.–iii. and there exists a Hopf *-algebra epimorphism $\rho : \mathcal{A}' \longrightarrow \mathcal{A}$ such that $\rho(w') = w$ then ρ is an isomorphism (the universality condition).

We define quantum Lorentz groups and quantum Poincaré groups as objects having the same properties as the classical Lorentz and Poincaré groups:

Definition 1 *We say [18] that H is a quantum Lorentz group if* $\mathrm{Poly}(H) = (\mathcal{A}, \Delta)$ *satisfies i.-iv. We say [11] that G is a quantum Poincaré group if Hopf *-algebra* $\mathrm{Poly}(G) = (\mathcal{B}, \Delta)$ *satisfies the conditions 1.–6. for some quantum Lorentz group H with* $\mathrm{Poly}(H) = (\mathcal{A}, \Delta)$ *and a representation w of H.*

Remark. The condition 5. follows from $\mathcal{P} \otimes w \simeq w \otimes \mathcal{P}$, $\mathcal{P} \otimes \bar{w} \simeq \bar{w} \otimes \mathcal{P}$, while 6. is suggested by the requirement $W(\mathcal{P} \otimes \mathcal{P}) = (\mathcal{P} \otimes \mathcal{P})W$ for a "twist-like" matrix W. Moreover, the condition 4. is superfluous (it follows from the condition 6.).

Theorem 2 *Let G be a quantum Poincaré group, $\mathrm{Poly}(G) = (\mathcal{B}, \Delta)$. Then $\mathcal{A}$ is linearly generated by matrix elements of irreducible representations of G, so $\mathcal{A}$ is uniquely determined. Moreover, we can choose w in such a way that $\mathcal{A}$ is the universal *-algebra generated by* w_{AB}, $A, B = 1, 2$, *satisfying*

$$(w \otimes w)E = E,$$

$$E'(w \otimes w) = E',$$

$$X(w \otimes \bar{w}) = (\bar{w} \otimes w)X,$$

where the triples $E \in M_{4\times 1}(\mathbf{C}), E' \in M_{1\times 4}(\mathbf{C}), X \in M_{4\times 4}(\mathbf{C})$ are listed in Theorem 1.4 of [11]. We can (and will) choose y_i in such a way that $y_i^* = y_i$.

In particular, it turns out that only quantum Lorentz groups with the parameter $q = \pm 1$ are admissible. Nevertheless, there are many families of admissible quantum Lorentz groups (numbered by some other parameters). In the following we assume that G is a quantum Poincaré group, $\mathrm{Poly}(G) = (\mathcal{B}, \Delta)$ and w, y are as in Theorem 2. Using the general theory of inhomogeneous quantum groups [10], we find the full system of commutation relations for $\mathcal{B}$:

Theorem 3 *$\mathcal{B}$ is the universal *-algebra with I, generated by $\mathcal{A}$ and y_i with relations*

$$y \otimes v = G_v(v \otimes y) + H_v v - (\Lambda \otimes v)H_v,$$

$$(R - 1)(y \otimes y - Zy + T - (\Lambda \otimes \Lambda)T) = 0,$$

$$y_i^* = y_i, \quad i \in \mathcal{S},$$

for any $v \in \mathrm{Rep}\, G$ (the set of all representations of G), where $(G_v)_{iC,Dj} = f_{ij}(v_{CD})$, $(H_v)_{iC,D} = \eta_i(v_{CD})$, $R = G_\Lambda$, $Z = H_\Lambda$ and $T_{ij} \in \mathbf{C}$, $f_{ij}, \eta_i \in \mathcal{A}'$ $(i, j \in \mathcal{S})$ satisfy the conditions of Theorem 1.5 of [11]. Δ is given by (1),(2). Moreover, $(\mathcal{B}, \Delta)$ gives a quantum Poincaré group if and only if a system of linear and quadratic equations listed in the proof of Theorem 1.6 of [11] is fulfilled.

It turns out that there are two choices for f determined by a number $s = \pm 1$ - the calculations are made for each s separately and the results are given in terms of $H_{EFCD} = V_{EF,i}\eta_i(w_{CD})$ and $T_{EFCD} = V_{EF,i}V_{CD,j}T_{ij}$. We solve [11] the above system of equations for almost all quantum Lorentz groups (except two cases, including the classical Lorentz group for which a large class of solutions is known). Moreover, we single out unisomorphic objects. The classification is presented in Theorem 1.6 of [11]. We also identify few examples which were known earlier (cf. [5], [3], [19]). We prove that $\mathcal{B}$ has exactly the sime "size" as in the undeformed case. Namely,

$$\mathcal{B}^N = \mathcal{A} \cdot \mathrm{span}\{y_{i_1} \cdot \ldots \cdot y_{i_n} : i_1, \ldots, i_n \in \mathcal{S}, \quad n = 0, 1, \ldots, N\}$$

is a free left $\mathcal{A}$-module and

$$\mathrm{dim}_{\mathcal{A}} \mathcal{B}^N = \sum_{n=0}^{N} d_n$$

where d_n is the number of classical monomials of nth degree in 4 variables.

We denote by $l : P \times M \longrightarrow M$ the action of Poincaré group on Minkowski space, $\mathcal{C} = \mathrm{Poly}(M)$ denotes the unital algebra generated by coordinates x_i ($i \in \mathcal{S}$) of the Minkowski space $M = \mathbf{R}^4$. The coaction $\Psi : \mathcal{C} \longrightarrow \mathcal{B} \otimes \mathcal{C}$ and $*$ in $\mathcal{C}$ are given by $(\Psi f)(x, y) = f(l(x, y))$, $f^*(y) = \overline{f(y)}$, $x \in P$, $y \in M$. One gets

$$\Psi x_i = \Lambda_{ij} \otimes x_j + y_i \otimes I. \tag{4}$$

We define a quantum Minkowski space as object having the same properties as the classical Minkowski space:

Definition 4 *We say that $(\mathcal{C}, \Psi)$ describes a quantum Minkowski space associated with a quantum Poincaré group G, $\mathrm{Poly}(G) = (\mathcal{B}, \Delta)$, if $\mathcal{C}$ is a unital $*$-algebra generated by x_i, $i \in \mathcal{S}$, $\Psi : \mathcal{C} \longrightarrow \mathcal{B} \otimes \mathcal{C}$ is a unital $*$-homomorphism, (4) holds and other conditions of Definition 1.10 of [11] are satisfied.*

Theorem 5 *Let G be a quantum Poincaré group with w, y as in Theorem 2. Then there exists a unique (up to a $*$-isomorphism) pair $(\mathcal{C}, \Psi)$ describing associated Minkowski space:*

$\mathcal{C}$ is the universal unital $*$-algebra generated by x_i, $i = 0, 1, 2, 3$, satisfying $x_i^* = x_i$ and

$$(R - \mathbf{1})(x \otimes x - Zx + T) = 0,$$

and Ψ is given by (4). Moreover,

$$\mathrm{dim}\,\mathcal{C}^N = \sum_{n=0}^{N} d_n,$$

where $\mathcal{C}^N = \mathrm{span}\{x_{i_1} \cdot \ldots \cdot x_{i_n} : i_1, \ldots, i_n \in \mathcal{S}, \quad n = 0, 1, \ldots, N\}$.

The next our goal is to find the differential structure on quantum Minkowski spaces. It turns out [12] that there exists a unique 4-dimensional covariant first order differential calculus on a quantum Minkowski space provided $\tilde{F} = 0$ where

$$\tilde{F} = [(R-1) \otimes \mathbf{1}]\{(\mathbf{1} \otimes Z)Z - (Z \otimes \mathbf{1})Z + T \otimes \mathbf{1} - (\mathbf{1} \otimes R)(R \otimes \mathbf{1})(\mathbf{1} \otimes T)\}$$

(otherwise there is no such a calculus). This condition singles out a large class of quantum Minkowski spaces [12]. From now on we assume that this condition is fulfilled. In particular, there exists a $\mathcal{C}$-bimodule $\Gamma^{\wedge 1}$ (of differential forms of the first order) and a linear mapping $d : \mathcal{C} \longrightarrow \Gamma^{\wedge 1}$ such that $d(ab) = a(db) + (da)b$, $a, b \in \mathcal{C}$, and dx_i, $i \in \mathcal{S}$, form a basis of $\Gamma^{\wedge 1}$ (as right $\mathcal{C}$-module). We prove

$$x_i dx_j = R_{ij,kl} dx_k x_l + Z_{ij,k} dx_k, \ i,j \in \mathcal{S}.$$

This calculus prolongates to a unique exterior algebra of differential forms, with the same properties as in the undeformed case. In particular it possesses $*$ such that $(dx_i)^* = dx_i$.

The partial derivatives $\partial_i : \mathcal{C} \longrightarrow \mathcal{C}$ are uniquely defined by

$$da = dx_i \partial_i(a), \quad a \in \mathcal{S}.$$

They can be also obtained as

$$\partial_i = (Y_i \otimes id)\Psi$$

where $Y_i \in \mathcal{A}'$ are introduced in the proof of Proposition 3.1.2 of [12].

The metric tensor $g = (g_{ij})_{i,j \in \mathcal{S}}$ (its entries are called in [12] by g^{ij}) is defined by the equations $(\Lambda \otimes \Lambda)g = g$ (or $\Lambda g \Lambda^T = g$) and $\overline{g_{ij}} = g_{ji}$. After the choice of a real factor we fix it as

$$g = -2q^{1/2}(V^{-1} \otimes V^{-1})(\mathbf{1} \otimes X \otimes \mathbf{1})(E \otimes \tau E),$$

where τ is always a standard twist. Then the Laplacian is defined by $\Box = g_{ij}\partial_j\partial_i$. It commutes with the partial derivatives and therefore the momenta $P_l = i\partial_l$ are well defined in the spaces of solutions of the Klein–Gordon equation $(\Box + m^2)\varphi = 0$. One proves that the momenta[1] $P^k = g_{kl}P_l$ and Laplacian are hermitian and have good transformation properties. The commutation relations among partial derivatives and with the coordinates are as follows:

$$\partial_l \partial_k = R_{ij,kl} \partial_j \partial_i,$$

$$\partial_i x_k = \delta_{ki} + (R_{kl,in} x_n + Z_{kl,i})\partial_l.$$

Let us now pass to the particles of spin 1/2 [14]. First we define the space of bispinors as $\mathbf{C}^4$ endowed with a representation $\mathcal{G} \simeq w \oplus \bar{w}$ of a quantum Poincaré

---

[1] in this case we relax our rule of writing all indices in the subscript position

group. We choose $\mathcal{G} = {}^c w \oplus \bar{w}$ where ${}^c w = (w^T)^{-1} \simeq w$. We are going to find the gamma matrices $\gamma_i \in M_{4\times 4}(\mathbf{C})$, $i \in \mathcal{S}$. At the moment they are not determined yet. The Dirac operator has form $\partial\!\!\!/ = \gamma_i \otimes \partial_i$. It acts on the bispinor functions $\phi \in \tilde{\mathcal{C}} \equiv \mathbf{C}^4 \otimes \mathcal{C}$ (in a more advanced approach we should consider square integrable functions ϕ). They can be written as $\phi = \varepsilon_a \otimes \phi_a$ where ε_a, $a = 1,2,3,4$, form the standard basis of $\mathbf{C}^4$. We define the action $\tilde{\Psi} : \tilde{\mathcal{C}} \longrightarrow \mathcal{B} \otimes \tilde{\mathcal{C}}$ of quantum Poincaré group on $\tilde{\mathcal{C}}$ by

$$\tilde{\Psi}(\varepsilon_a \otimes \phi^a) = \mathcal{G}_{al}\phi_a^{(1)} \otimes \varepsilon_l \otimes \phi_a^{(2)}$$

where $\Psi(\phi^a) = \phi_a^{(1)} \otimes \phi_a^{(2)}$ (Sweedler's notation, exception of the summation convention). Then the classical condition of invariance of the Dirac operator is generalized to

$$\tilde{\Psi}(\partial\!\!\!/ \phi) = (\mathrm{id} \otimes \partial\!\!\!/)[\tilde{\Psi}(\phi)], \quad \phi \in \tilde{\mathcal{C}}.$$

According to Theorem III.1 of [14], the above condition is equivalent to

$$\gamma_i = \begin{pmatrix} 0 & bA_i \\ a\sigma_i & 0 \end{pmatrix}, \quad i \in \mathcal{S},$$

where

$$A_i = q^{-1/2} E^T (\sigma_i \circ D) E, \qquad (5)$$

$(\sigma_i \circ D)_{KL} = (\sigma_i)_{AB} D_{AB,KL}$, $D = \tau X^{-1} \tau$, $a, b \in \mathbf{C}$ (E is regarded here as 2×2 matrix). Thus we have found the form of the Dirac operator up to two constants. But Theorem III.2 of [14] says that the following are equivalent:

1. $\partial\!\!\!/^2 = \Box$.
2. $\gamma_i \gamma_j + R_{ji,lk} \gamma_k \gamma_l = 2g_{ji}\mathbf{1}$, $\quad i,j \in \mathcal{S}$.
3. $ab = 1$.

Moreover, the remaining freedom in the choice of a results in a trivial scaling of the undotted spinor and we can set $a = b = 1$. Thus we have obtained

$$\gamma_i = \begin{pmatrix} 0 & A_i \\ \sigma_i & 0 \end{pmatrix}$$

where A_i are given by (5). Now the form of the Dirac equation $(i\partial\!\!\!/ + m)\varphi = 0$ is determined.

In the next step we find (formal) solutions of Klein–Gordon and Dirac equations in two important cases (Section 4 of [12]). In one of them we use (as a tool during calculations) an additional algebra $\mathcal{F}$ and its representations. Specific calculations are made in [14] (including the form of metric tensor, gamma matrices, representations of $\mathcal{F}$ and the solutions of Klein–Gordon and Dirac equations). For spin 0 particles the momenta $P_j = i\partial_j$. For spin 1/2 particles we set [14] the momenta as

$$\tilde{P}_j = i\tilde{\partial}_j, \qquad (6)$$

$$\tilde{\partial}_j = (Y_j \otimes \text{id})\tilde{\Psi} : \tilde{\mathcal{C}} \longrightarrow \tilde{\mathcal{C}} \tag{7}$$

(motivation: (6)–(7) remain true if we omit tildas everywhere). We get four objects $\tilde{P}^k = g_{kj}\tilde{P}_j$ which have good transformation properties, commute with the Dirac operator and are hermitian w.r.t. (indefinite) inner product in the space of bispinors. However, in many cases their spectral properties are not satisfactory. The problem of further improvement in this matter remains open. We also study certain expressions like the deformed Lagrangian. They transform themselves in a similar way as in the standard theory.

It turns out (cf. Theorem 1.13 of [11]) that there exist invertible matrices W such that $W(\mathcal{P} \otimes \mathcal{P}) = (\mathcal{P} \otimes \mathcal{P})W$ and the Yang–Baxter equation

$$(W \otimes \mathbf{1})(\mathbf{1} \otimes W)(W \otimes \mathbf{1}) = (\mathbf{1} \otimes W)(W \otimes \mathbf{1})(\mathbf{1} \otimes W)$$

is satisfied. Namely, up to a constant they are given by the unit matrix and

$$R_Q = \begin{pmatrix} R & Z & -R \cdot Z & (R - \mathbf{1}^{\otimes 2})T + b \cdot g \\ 0 & 0 & \mathbf{1} & 0 \\ 0 & \mathbf{1} & 0 & 0 \\ 0 & 0 & 0 & \mathbf{1} \end{pmatrix},$$

where $b \in \mathbf{C}$.

In [13] we show the existence of $\mathcal{R} \in (\mathcal{B} \otimes \mathcal{B})'$ such that $(\mathcal{B}, \Delta, \mathcal{R})$ is a coquasitriangular (CQT) Hopf algebra. In other words, for any $v, z \in \text{Rep } G$ we define

$$R^{vz} \in \text{Lin}(\mathbf{C}^{\dim v} \otimes \mathbf{C}^{\dim z}, \mathbf{C}^{\dim z} \otimes \mathbf{C}^{\dim v})$$

by

$$(R^{vz})_{ij,kl} = \mathcal{R}(v_{jk} \otimes z_{il}), \ j,k = 1,\ldots,\dim v, \ i,l = 1,\ldots,\dim z,$$

and prove

$$R^{1v} = R^{v1} = \mathbf{1},$$
$$R^{v_1 \otimes v_2, z} = (R^{v_1 z} \otimes \mathbf{1})(\mathbf{1} \otimes R^{v_2 z}),$$
$$R^{v, z_1 \otimes z_2} = (\mathbf{1} \otimes R^{v z_2})(R^{v z_1} \otimes \mathbf{1}),$$
$$(z \otimes v)R^{vz} = R^{vz}(v \otimes z),$$

for all $v, v_1, v_2, z, z_1, z_2 \in \text{Rep } G$ ($\mathbf{1} = (I)$ is the trivial representation).

The classification of all CQT Hopf algebra structures (for all quantum Poincaré groups) is done in Theorem 3 of [13]. In particular, $R^{\mathcal{PP}} = R_Q$,

$$R^{v\mathcal{P}} = \begin{pmatrix} G_v, & H_v \\ 0, & \mathbf{1} \end{pmatrix},$$

$R^{\mathcal{P}v} = (R^{v\mathcal{P}})^{-1}$, $R^{ww} = kL$, $R^{w\bar{w}} = kX$, $R^{\bar{w}w} = qkX^{-1}$, $R^{\bar{w}\bar{w}} = k\tau L\tau$, for all representations v of the corresponding quantum Lorentz group H (these data determine $\mathcal{R}$ uniquely), where $L = sq^{1/2}(\mathbf{1} + qEE')$, $k = \pm 1$ (two possible $\mathcal{R}$ for each $b \in \mathbf{C}$).

We have to do with CQT Hopf *-algebra iff $\overline{\mathcal{R}(y^* \otimes x^*)} = \mathcal{R}(x \otimes y)$, $x, y \in \mathcal{B}$, iff $q = 1$ and $b \in \mathbf{R}$. We have cotriangular (CT) Hopf algebra iff $(R^{vz})^{-1} = R^{zv}$ for all $v, z \in \text{Rep } G$ iff $q = 1$ and $b = 0$ (so then it is also a CT Hopf *-algebra).

Using the above results, universal enveloping algebras for quantum Poincaré groups are introduced. Their commutation relations are investigated. Moreover, we classify C(Q)T Hopf (*-)algebra structures for quantum Lorentz groups. We also show some general statements concerning coquasitriangularity. The results of [13] are used in Section 5 of [12] to define the Fock space for non-interacting particles of spin 0 on a quantum Minkowski space. Then we take a CT Hopf *-algebra structure $\mathcal{R}$ as above and introduce the particles interchange operator $K : \mathcal{C} \otimes \mathcal{C} \longrightarrow \mathcal{C} \otimes \mathcal{C}$ by

$$K(x \otimes y) = \mathcal{R}(y^{(1)} \otimes x^{(1)})(y^{(2)} \otimes x^{(2)}),$$

where $\Psi(x) = x^{(1)} \otimes x^{(2)}$, $\Psi(y) = y^{(1)} \otimes y^{(2)}$. It defines the action of the permutation group Π_n in $\mathcal{C}^{\otimes n}$ ($\Pi_n \ni \sigma \longrightarrow \pi_\sigma$) which agrees with the action of the quantum Poincaré group G. Thus G acts in the boson subspace $\mathcal{C}^{\otimes_s n}$. If $W : \mathcal{C} \to \mathcal{C}$ is an operator related to a single particle then the corresponding n-particle operator is given by

$$W^{(n)} = \sum_{m=1}^{n} \pi_{(1,m)}(W \otimes \mathbf{1}^{\otimes(n-1)})\pi_{(1,m)} : \mathcal{C}^{\otimes_s n} \to \mathcal{C}^{\otimes_s n}$$

(the m-th term is the operator in $\mathcal{C}^{\otimes n}$ corresponding to the m-th particle). We can also define the Fock space $F = \oplus_{n=0}^{\infty} \mathcal{C}^{\otimes_s n}$ and the operator $\oplus_{n=0}^{\infty} W^{(n)}$ acting in F.

For particles of mass m we should consider $\ker(\Box + m^2)$ instead of $\mathcal{C}$ and a scalar product there (heuristically e.g. $W = P^k$, $k \in \mathcal{S}$, would be hermitian operators in such a space).

Results of [12] and [13] are proven also for general inhomogeneous quantum groups (satisfying certain conditions). References to the existing literature are given in [11], [12], [13], [14].

Concluding, quantum Minkowski spaces and quantum Poincaré groups have a lot of properties similar to that of the classical ones. It suggests a possibility of building more advanced physical models which use those objects. However, it would need further studies concerning deformed quantum field theory and interaction of particles. There is also another advantage of quantum Minkowski spaces: the fact that there are many possibilities in the choice of parameters somehow forces us to find the proofs which have good geometric meaning. In particular, the invariance of the Dirac operator turns out to be equivalent to the fact that some object built from the gamma matrices intertwines two specific representations of the quantum Poincaré group. Then the form of gamma matrices is easy to find (without using the Lie algebra at all).

REFERENCES

1. Carow–Watamura, U., Schlieker, M., Scholl, M. and Watamura, S., Tensor representation of the quantum group $SL_q(2,\mathbf{C})$ and quantum Minkowski space, *Z. Phys. C – Particles and Fields* **48** (1990), 159–165.
2. Carow–Watamura, U., Schlieker, M. and Watamura, S., $SO_q(N)$ covariant differential calculus on quantum space and quantum deformation of Schrödinger equation, *Z. Phys. C – Particles and Fields* **49** (1991), 439–446.
3. Chaichian, M. and Demichev, A.P., Quantum Poincaré group, *Phys. Lett.* **B304** (1993), 220–224.
4. Dobrev, V.K., Canonical q-deformations of noncompact Lie (super-) algebras, *J.Phys.A: Math. Gen.* **26**(1993), 1317–1334.
5. Lukierski, J., Nowicki, A. and Ruegg, H., New quantum Poincaré algebra and κ-deformed field theory, *Phys. Lett.* **B293** (1992), 344–352; Zakrzewski, S., Quantum Poincaré group related to the κ-Poincaré algebra, *J. Phys. A: Math. Gen.* **27** (1994), 2075–2082.
6. Majid, S., Braided momentum in the q-Poincaré group, *J. Math. Phys.* **34** (1993), 2045–2058.
7. Ogievetsky, O., Schmidke, W.B., Wess, J. and Zumino, B., q-Deformed Poincaré algebra, *Commun. Math. Phys.* **150** (1992), 495–518.
8. Podleś, P., Quantum spheres, *Lett. Math. Phys.* **14** (1987), 193–202; The classification of differential structures on quantum 2-spheres, *Commun. Math. Phys.* **150** (1992), 167–179; Quantization enforces interaction. Quantum mechanics of two particles on a quantum sphere, *Int. J. Mod. Phys. A*, **7**, Suppl. 1B (1992), 805–812.
9. Podleś, P. and Woronowicz, S.L., Quantum deformation of Lorentz group, *Commun. Math. Phys.* **130** (1990), 381–431.
10. Podleś, P. and Woronowicz, S.L., On the structure of inhomogeneous quantum groups, hep-th/9412058, *Commun. Math. Phys.* **185** (1997), 325–358.
11. Podleś, P. and Woronowicz, S.L., On the classification of quantum Poincaré groups, *Commun. Math. Phys.* **178** (1996), 61–82.
12. Podleś, P., Solutions of Klein–Gordon and Dirac equations on quantum Minkowski spaces, *Commun. Math. Phys.* **181** (1996), 569–585.
13. Podleś, P., Quasitriangularity and enveloping algebras for inhomogeneous quantum groups, *J. Math. Phys.* **37** (1996), 4724–4737.
14. Podleś, P., The Dirac operator and gamma matrices for quantum Minkowski spaces, *J. Math. Phys.* **38** (1997), 4474–4491.
15. Reshetikhin, N. Yu., Takhtadzyan, L. A. and Faddeev, L. D., Quantization of Lie groups and Lie algebras, *Leningrad Math. J.* **1:1** (1990), 193–225. Russian original: *Algebra i analiz* **1:1** (1989), 178–206.
16. Schlieker, M., Weich, W. and Weixler, R., Inhomogeneous quantum groups, *Z. Phys. C. – Particles and Fields* **53** (1992), 79–82.
17. Woronowicz, S.L., Twisted $SU(2)$ group. An example of a non-commutative differential calculus, *Publ. RIMS, Kyoto University* **23** (1987), 117–181; Compact matrix pseudogroups, *Commun. Math. Phys.* **111** (1987), 613–665.
18. Woronowicz, S.L. and Zakrzewski, S., Quantum deformations of the Lorentz group.

The Hopf *-algebra level, *Comp. Math.* **90** (1994), 211–243.
19. Zakrzewski, S., Geometric quantization of Poisson groups – diagonal and soft deformations, *Contemp. Math.* **179** (1994), 271–285.

Deformation Quantization: Twenty Years After[1]

Daniel Sternheimer

Laboratoire Gevrey de Mathématique Physique, CNRS ESA 5029
Département de Mathématiques, Université de Bourgogne
BP 400, F-21011 Dijon Cedex France
e-mail: dastern@u-bourgogne.fr

Abstract. We first review the historical developments, both in physics and in mathematics, that preceded (and in some sense provided the background of) deformation quantization. Then we describe the birth of the latter theory and its evolution in the past twenty years, insisting on the main conceptual developments and keeping here as much as possible on the physical side. For the physical part the accent is put on its relations to, and relevance for, "conventional" physics. For the mathematical part we concentrate on the questions of existence and equivalence, including most recent developments for general Poisson manifolds; we touch also noncommutative geometry and index theorems, and relations with group theory, including quantum groups. An extensive (though very incomplete) bibliography is appended and includes background mathematical literature.

I. BACKGROUND

In this Section we briefly present the fertile ground which was needed in order for deformation quantization to develop, even if from an abstract point of view one could have imagined it on the basis of Hamiltonian classical mechanics. Indeed there are two sides to "deformation quantization". The philosophy underlying the rôle of *deformations in physics* has been consistently put forward by Flato since more than 30 years and was eventually expressed by him in [66] (see also [58,67]). In short, the passage from one level of physical theory to another, more refined, can be understood (and might even have been predicted) using what mathematicians call deformation theory. For instance one passes from Newtonian physics to special relativity by deforming the invariance group (the Galilei group $SO(3) \cdot \mathbb{R}^3 \cdot \mathbb{R}^4$) to the Poincaré group $SO(3,1) \cdot \mathbb{R}^4$ with deformation parameter c^{-1}, where c is the velocity of light. There are many other examples among which *quantization* is perhaps the most seminal.

---

[1] *This review is dedicated to the memory of our good friend Ryszard Rączka, in whose honor the meeting in Łódz "Particles, Fields and Gravitation", was held in April 1998.*

As a matter of fact it seems that the idea that quantum mechanics is some kind of deformed classical mechanics has been, almost from the beginning of quantum theory, "in the back of the mind of many physicists" (after we came out with the preprint of [17], a scientist even demanded that we quote him for that!). This is attested by the notion of classical limit and even more by that of semi-classical approximation, a good presentation of which can be found in [157]. But the idea remained hidden "in the back of the minds" for a long time, in particular due to the apparently insurmountable "quantum jump" in the nature of observables – and probably also because the mathematical notion of deformation and the relevant cohomologies were not available. A long maturation was needed which eventually gave birth to full-fledged deformation quantization about 20 years ago [17].

A word of caution may be needed here. It is possible to intellectually imagine new physical theories by deforming existing ones. Even if the mathematical concept associated with an existing theory is mathematically rigid, it may be possible to find a wider context in which nontrivial deformations exist. For instance the Poincaré group may be deformed to the simple (and therefore rigid in the category of Lie groups) anti De Sitter group SO(3,2) very popular recently – though it had been studied extensively by us 15-20 years ago [3], resulting in particular in a formulation of *QED with photons dynamically composed of two singletons* [70] in AdS universe. As is now well-known, there exist deformations of the Hopf algebras associated with a simple Lie group (these "quantum groups" [55], which are in fact an example of deformation quantization [25], have been extensively studied and applied to physics). Nevertheless such intellectual constructs, even if they are beautiful mathematical theories, need to be somehow confronted with physical reality in order to be taken seriously in physics. So some physical intuition is still needed when using deformation theory in physics.

I.1 Weyl Quantization and Related Developments

We assume the reader somewhat familiar with (classical and) quantum mechanics. In an "impressionist" fashion we only mention the names of Planck, Einstein, de Broglie, Heisenberg, Schrödinger and finally Hermann Weyl. What we call here "deformation quantization" is related to Weyl's quantization procedure. In the latter [161], starting with a classical observable $u(p,q)$, some function on phase space $\mathbb{R}^{2\ell}$ (with $p, q \in \mathbb{R}^\ell$), one associates an operator (the corresponding quantum observable) $\Omega(u)$ in the Hilbert space $L^2(\mathbb{R}^\ell)$ by the following general recipe:

$$u \mapsto \Omega_w(u) = \int_{\mathbb{R}}^{2\ell} \tilde{u}(\xi,\eta) \exp(i(P.\xi + Q.\eta)/\hbar) w(\xi,\eta) \, d^\ell\xi d^\ell\eta \tag{1}$$

where $\tilde{u}$ is the inverse Fourier transform of u, P_α and Q_α are operators satisfying the canonical commutation relations $[P_\alpha, Q_\beta] = i\hbar\delta_{\alpha\beta}$ ($\alpha, \beta = 1,...,\ell$), w is a weight function and the integral is taken in the weak operator topology. What is now called normal ordering corresponds to choosing the weight $w(\xi,\eta) = \exp(-\frac{1}{4}(\xi^2 \pm \eta^2))$, standard ordering (the case of the usual pseudodifferential operators in mathematics) to $w(\xi,\eta) = \exp(-\frac{i}{2}\xi\eta)$ and the original Weyl (symmetric) ordering to $w = 1$. An inverse formula was found shortly afterwards by Eugene Wigner [162] and maps an operator into what mathematicians call its symbol by a kind of trace formula. For example Ω_1 defines an isomorphism of Hilbert spaces between $L^2(\mathbb{R}^{2\ell})$ and Hilbert-Schmidt operators on $L^2(\mathbb{R}^{\ell})$ with inverse given by

$$u = (2\pi\hbar)^{-\ell}\text{Tr}[\Omega_1(u)\exp((\xi.P+\eta.Q)/i\hbar)] \qquad (2)$$

and if $\Omega_1(u)$ is of trace class one has $\text{Tr}(\Omega_1(u)) = (2\pi\hbar)^{-\ell}\int u\,\omega^{\ell}$ where ω^{ℓ} is the (symplectic) volume dx on $\mathbb{R}^{2\ell}$. Numerous developments followed in the direction of phase-space methods, many of which can be found described in [2]. Of particular interest to us here is the question of finding an interpretation to the classical function u, symbol of the quantum operator $\Omega_1(u)$; this was the problem posed (around 15 years after [162]) by Blackett to his student Moyal. The (somewhat naïve) idea to interpret it as a probability density had of course to be rejected (because u has no reason to be positive) but, looking for a direct expression for the symbol of a quantum commutator, Moyal found [126] what is now called the Moyal bracket:

$$M(u,v) = \nu^{-1}\sinh(\nu P)(u,v) = P(u,v) + \sum_{r=1}^{\infty}\nu^{2r}P^{2r+1}(u,v) \qquad (3)$$

where $2\nu = i\hbar$, $P^r(u,v) = \Lambda^{i_1 j_1}\ldots\Lambda^{i_r j_r}(\partial_{i_1\ldots i_r}u)(\partial_{j_1\ldots j_r}v)$ is the rth power ($r \geq 1$) of the Poisson bracket bidifferential operator P, $i_k, j_k = 1,\ldots,2\ell$, $k = 1,\ldots,r$ and $(\Lambda^{i_k j_k}) = \begin{pmatrix} 0 & -I \\ I & 0 \end{pmatrix}$. To fix ideas we may assume here $u,v \in C^{\infty}(\mathbb{R}^{2\ell})$ and the sum taken as a formal series (the definition and convergence for various families of functions u and v was also studied, including in [17]). A similar formula for the symbol of a product $\Omega_1(u)\Omega_1(v)$ had been found a little earlier [95] and can now be written more clearly as a (Moyal) *star product*:

$$u *_M v = \exp(\nu P)(u,v) = uv + \sum_{r=1}^{\infty}\nu^r P^r(u,v). \qquad (4)$$

Several integral formulas for the star product have been introduced and the Wigner image of various families of operators (including bounded operators on $L^2(\mathbb{R}^{\ell})$) were studied, mostly after deformation quantization was developed (see e.g. [44,123,107]). An adaptation to Weyl ordering of the mathematical notion of pseudodifferential operators (ordered, like differential operators, "first q, then p") was done in [96] – and the converse in [102]. Starting from field theory, where normal (Wick) ordering is essential (the rôle of q and p above is played by $q \pm ip$), Berezin [18,19] developed in the mid-seventies an extensive study of what he called "quantization", based on the correspondence principle and Wick symbols. It is essentially based on Kähler manifolds and related to pseudodifferential operators in the complex domain [32]. However in his theory (which we noticed rather late), as in the studies of various orderings [2], the important concepts of *deformation* and *autonomous* formulation of quantum mechanics in general phase space are absent.

I.2 Classical Mechanics on General Phase Space and its Quantization

Initially classical mechanics, in Lagrangean or Hamiltonian form, assumed implicitly a "flat" phase space $\mathbb{R}^{2\ell}$, or at least considered only an open connected set thereof. Eventually more general configurations were needed and so the mathematical notion of manifold, on which mechanics imposed some structure, was needed. This has lead in particular to using the notions of symplectic and later of Poisson manifolds, which have been introduced also for purely mathematical reasons. One of these reasons has to do with families of infinite-dimensional Lie algebras, which date back to works by Élie Cartan at the beginning of this century and regained a lot of popularity (including in physics) in the past 30 years.

A typical example can be found with Dirac constraints [50]: second class Dirac constraints restrict phase space from some $\mathbb{R}^{2\ell}$ to a symplectic manifold W imbedded in it (with induced symplectic form), while first class constraints further restrict to a Poisson manifold with symplectic foliation (see e.g. [72]). Some of the references where one can find detailed information on the symplectic approach to classical (Hamilton) mechanics are [118,1,98] and (which includes the derivation of symplectic manifolds from Lagrangean mechanics) [149]. The question of quantization on such manifolds was certainly treated by many authors (including in [50]) but did not go beyond giving some (often useful) recipes and hoping for the best.

A first systematic attempt started around 1970 with what was called soon afterwards *geometric quantization* [113], a by-product of Lie group representations theory where it gave significant results [10,109]. It turns out that it is geometric all right, but its scope as far as quantization is concerned has been rather limited since few classical observables could be quantized, except in situations which amount essentially to the Weyl case considered above. In a nutshell one considers phase-spaces W which are coadjoint orbits of some Lie groups (the Weyl case corresponds to the Heisenberg group with the canonical commutation relations $\mathfrak{h}_\ell$ as Lie algebra); there one defines a "prequantization" on the Hilbert space $L^2(W)$ and tries to halve the number of degrees of freedom by using polarizations (often complex ones, which is not an innocent operation as far as physics is concerned) to get a Lagrangean submanifold $\mathcal{L}$ of dimension half that of W and quantized observables as operators in $L^2(\mathcal{L})$; "Moyal quantization" on a symplectic groupoid $\mathbb{R}_1^{2\ell} \times \mathbb{R}_2^{2\ell}$ was obtained therefrom in [92]. A recent exposition can be found in [163].

I.3 Pseudodifferential Operators and Index Theorems

One may argue that physicists had invented the theory of distributions (with Dirac's δ) and symbols of pseudodifferential operators (with standard ordering) much before mathematicians developed the corresponding theories. These may also be considered as belonging to the large family of examples of a fruitful interaction between physics and mathematics, even if in the latter case (symbols) it seems that the two developments were largely independent at the beginning, and in fact converged only with the advent of deformation quantization.

In this connection one should not forget that there is a significant difference in attitude to Science (with notable exceptions): in physics a very good idea may be enough to earn you a Nobel prize (with a little bit of luck, enough PR and provided you live long enough to see it well recognized); in mathematics one usually needs to have, young enough, several good ideas and prove that they are really good (often with hard work, because "problems worthy of attack prove their worth by hitting back") in order to be seriously considered for the Fields medal. This rather ancient difference may explain Goethe's sentence (much before Nobel and Fields): "Mathematicians are like Frenchmen: They translate everything into their own language and henceforth it is something completely different".

In the fifties [38] the notion of Fourier integral operators was introduced, generalizing and making precise the sometimes heuristic calculus of "differential operators of noninteger order". Soon it evolved into what are now called pseudodifferential operators, defined on general manifolds [101] and as indispensable to theories of partial differential equations as distributions (with which they are strongly mixed). But what gained to this tool fame and respectability among all mathematicians was the proof, in 1963 (and following years for various generalizations) of the so-called *index theorem* for elliptic (pseudo)differential operators on manifolds by Atiyah, Singer, Bott, Patodi and others [9,136,39,89]. The (analytical) index of a linear map between two vector spaces is defined (when both terms are finite) as the dimension of its kernel minus the codimension of its image. Elliptic partial differential operators d on compact manifolds X have an index $i(d)$, equal (and this is the original theorem) to a "topological index" which depends only on topological invariants associated with the manifold (the Todd class $\tau(X)$ of the complexified cotangent bundle of X and the fundamental class $[X]$) and on a cohomological invariant $\mathrm{ch}d$ (a Chern character) associated with the symbol of the principal part of the operator d. To give the flavor of the result we write a precise formula (see e.g. Atiyah's lecture in [39]), valid for compact manifolds (without or with boundary):

$$i(d) = \langle (\mathrm{ch}d)\tau(X), [X] \rangle. \tag{5}$$

The existence of such a formula had also been conjectured by Gel'fand. The proof is very elaborate and one cannot avoid doing it also for pseudodifferential operators. Topological arguments and factorization (which imposes consideration of continuous symbols – this was my share in [39]) permit eventually to reduce the proof of the equality to the cases of the Dirac operator on even dimensional manifolds and one particular (convolution) operator on the circle. There have been numerous developments in a wide array of mathematical domains provoked by this seminal result. The formula itself has been very much generalized, including to "algebraic" index theorems where the algebra of pseudodifferential operators is replaced by more abstract algebras (this is an major ingredient in noncommutative geometry [41] and is strongly related to star products [42,131]). Throughout the theory a capital rôle is played by the symbol $\sigma(d)$ (the classical function associated with the standard-ordered operator d). Note that the principal part of a differential operator is independent of the ordering, but eventually the whole symbol was used. In the proof of the reduction one needs an expression for the symbol of a product of operators, given by an integral formula analogous to a star product. So mathematicians had been using star products (albeit corresponding to a different ordering and without formal series development in some parameter like $\hbar$) before they were systematically defined. This permitted eventually to give original proofs of existence of star products on quite general manifolds [31,97] by adapting techniques and results developed [32] in the theory of pseudodifferential operators.

I.4 Cohomologies and Deformation Theory

In an often ignored section of a paper, I.E. Segal [146] and (independently) a little later Wigner and Inonü [104,143] have introduced in the early fifties a kind of inverse [117] to the mathematical notion of *deformation* of Lie groups and algebras, notion which was precisely defined only in 1964 by Murray Gerstenhaber [86]. That inverse was called *contraction* and typical examples (mentioned at the beginning of this Section) are the passage from De Sitter to Poincaré groups (by taking the limit of zero curvature in space-time) or from Poincaré to Galilei (by taking $c^{-1} \to 0$). Intuitively speaking a contraction is performed by neglecting in symmetries, at some level of physical reality, a constant (like c^{-1}) which has negligible impact at this level but significant effects at a more "refined" level. Note that this may be realized mathematically in varying generality (e.g. [143] is more general than [104] but both have for inverse a Gerstenhaber deformation). The notion of deformation of algebras, which may be seen as an outcome of the notion (introduced a few years before) of deformations of complex analytic structures [110], gives rise to a better defined mathematical theory which, for completeness, we shall briefly present in the following two subsections. All this is by now well-known, even to physicists, and we shall keep details to a minimum, referring the reader interested in more details to papers and textbooks cited here and references quoted therein.

It also turns out that recently we had to introduce deformations which are even more general than those introduced by Gerstenhaber (see [53,139,128] and [67] in these Proceedings) in order e.g. to quantize Nambu mechanics. They are still inverse to some contraction procedure, applied (like deformation quantization, where the algebra is that of classical observables and the parameter Planck's constant) to algebras which are not geometrical symmetries. So one should keep an open mathematical mind, let physics be a guide and develop if needed completely new mathematical tools. This is true *physical mathematics*, in contradistinction with standard mathematical physics where one mainly applies existing tools or with theoretical physics where mathematical rigor is too often left aside and a good physical intuition (which Dirac certainly had e.g. when he worked with his δ "function") is then required as a guide.

I.4.1 Hochschild and Chevalley-Eilenberg cohomologies

Let first A be an *associative* algebra (over some commutative ring $\mathbb{K}$) and for simplicity we consider it as a module over itself with the adjoint action (algebra multiplication); the generalization to cohomology valued in a general module is straightforward. A *p-cochain* is a p-linear map C from A^p into (the module) A and its *coboundary* bC is given by

$$bC(u_0, \ldots, u_p) = u_0 C(u_1, \ldots, u_p) - C(u_0 u_1, u_2, \ldots, u_p) + \cdots$$
$$+ (-1)^p C(u_0, u_1, \ldots, u_{p-1} u_p) + (-1)^{p+1} C(u_0, \ldots, u_{p-1}) u_p. \quad (6)$$

One checks that we have here what is called a complex, i.e. $b^2 = 0$. We say that a p-cochain C is a *p-cocycle* if $bC = 0$. We denote by $\mathcal{Z}^p(A, A)$ the space of p-cocycles and by $\mathcal{B}^p(A, A)$ the space of those p-cocycles which are coboundaries (of a $(p-1)$-cochain). The pth *Hochschild cohomology* space (of A valued in A) is defined as $\mathcal{H}^p(A, A) = \mathcal{Z}^p(A, A)/\mathcal{B}^p(A, A)$. *Cyclic cohomology* is defined using a bicomplex which includes the Hochschild complex and we shall briefly present it at the end of this review in the example of interest for us here.

For *Lie algebras* (with bracket $\{\cdot, \cdot\}$) one has a similar definition, due to Chevalley and Eilenberg [40]. The p-cochains are here skew-symmetric, i.e. linear maps $B: \wedge^p A \longrightarrow A$, and the Chevalley coboundary operator ∂ is defined on a p-cochain B by (where $\hat{u}_j$ means that u_j has to be omitted):

$$\partial C(u_0, \ldots, u_p) = \sum_{j=0}^{p} (-1)^j \{u_j, C(u_0, \ldots, \hat{u}_j, \ldots, u_p)\}$$
$$+ \sum_{i<j} (-1)^{i+j} C(\{u_i, u_j\}, u_0, \ldots, \hat{u}_i, \ldots, \hat{u}_j, \ldots, u_p). \quad (7)$$

Again one has a complex ($\partial^2 = 0$), cocycles and coboundaries spaces Z^p and B^p (resp.) and by quotient the *Chevalley cohomology* spaces $H^p(A, A)$, or in short $H^p(A)$; the collection of all cohomology spaces is often denoted H^*.

I.4.2 Gerstenhaber theory of deformations of algebras

Let A be an algebra. By this we mean an *associative*, *Lie* or *Hopf* algebra, or a *bialgebra*. Whenever needed we assume it is also a *topological* algebra, i.e. endowed with a locally convex topology for which all needed algebraic laws are continuous. For simplicity we may think that the base (commutative) ring $\mathbb{K}$ is the field of complex numbers $\mathbb{C}$ or that of the real numbers $\mathbb{R}$. Extending it to the ring $\mathbb{K}[[\nu]]$ of formal series in some parameter ν gives the module $\tilde{A} = A[[\nu]]$, on which we can consider the preceding various algebraic (and topological) structures.

I.4.2.1 Deformations and cohomologies. A concise formulation of a Gerstenhaber deformation of an algebra (which we shall call in short a *DrG-deformation* whenever a confusion may arise with more general deformations) is [86,87,25]:

DEFINITION 1 *A deformation of such an algebra A is a* $\mathbb{K}[[\nu]]$-*algebra* $\tilde{A}$ *such that* $\tilde{A}/\nu\tilde{A} \approx A$. *Two deformations* $\tilde{A}$ *and* $\tilde{A}'$ *are said equivalent if they are isomorphic over* $\mathbb{K}[[\nu]]$ *and* $\tilde{A}$ *is said trivial if it is isomorphic to the original algebra A considered by base field extension as a* $\mathbb{K}[[\nu]]$-*algebra.*

Whenever we consider a topology on A, $\tilde{A}$ is supposed to be topologically free. For associative (resp. Lie algebra) Definition 1 tells us that there exists a new product $*$ (resp. bracket $[\cdot,\cdot]$) such that the new (deformed) algebra is again associative (resp. Lie). Denoting the original composition laws by ordinary product (resp. $\{\cdot,\cdot\}$) this means that, for $u, v \in A$ (we can extend this to $A[[\nu]]$ by $\mathbb{K}[[\nu]]$-linearity) we have:

$$u * v = uv + \sum_{r=1}^{\infty} \nu^r C_r(u,v) \qquad (8)$$

$$[u,v] = \{u,v\} + \sum_{r=1}^{\infty} \nu^r B_r(u,v) \qquad (9)$$

where the C_r are Hochschild 2-cochains and the B_r (skew-symmetric) Chevalley 2-cochains, such that for $u, v, w \in A$ we have $(u*v)*w = u*(v*w)$ and $\mathcal{S}[[u,v],w] = 0$, where $\mathcal{S}$ denotes summation over cyclic permutations. At each level r we therefore need to fulfill the equations ($j, k \geq 1$):

$$D_r(u,v,w) \equiv \sum_{j+k=r} (C_j(C_k(u,v),w) - C_j(u,C_k(v,w))) = bC_r(u,v,w) \qquad (10)$$

$$E_r(u,v,w) \equiv \sum_{j+k=r} \mathcal{S}B_j(B_k(u,v),w) = \partial B_r(u,v,w) \qquad (11)$$

where b and ∂ denote (respectively) the Hochschild and Chevalley coboundary operator. In particular we see that for $r = 1$ the driver C_1 (resp. B_1) must be a 2-cocycle. Furthermore, assuming one has shown that (10) or (11) are satisfied up to some order $r = t$, a simple calculation shows that the left-hand sides for $r = t+1$ are then 3-cocycles, depending only on the cochains C_k (resp. B_k) of order $k \leq t$. If we want to extend the deformation up to order $r = t+1$ (i.e. to find the required 2-cochains C_{t+1} or B_{t+1}), this cocycle has to be a coboundary (the coboundary of the required cochain): *The obstructions to extend a deformation from one step to the next lie in the 3-cohomology.* In particular (and this was Vey's trick) if one can manage to pass always through the null class in the 3-cohomology, a cocycle can be the driver of a full-fledged (formal) deformation.

For a (topological) *bialgebra* (an associative algebra A where we have in addition a coproduct $\Delta : A \longrightarrow A \otimes A$ and the obvious compatibility relations), denoting by $\otimes_\nu$ the tensor product of $\mathbb{K}[[\nu]]$-modules, we can identify $\tilde{A} \hat{\otimes}_\nu \tilde{A}$ with $(A \hat{\otimes} A)[[\nu]]$, where $\hat{\otimes}$ denotes the algebraic tensor product completed with respect to some operator topology (e.g. projective for Fréchet nuclear topology), we similarly have a deformed coproduct $\tilde{\Delta} = \Delta + \sum_{r=1}^{\infty} \nu^r D_r$, $D_r \in \mathcal{L}(A, A\hat{\otimes}A)$ and in this context appropriate cohomologies can be introduced. Here we shall not elaborate on these, nor on the additional requirements for Hopf algebras, referring for more details to original papers and books; there is a huge literature on the subject, among which we may mention [55,88,23,25,26,148,151] and references quoted therein.

I.4.2.2 Equivalence means that there is an isomorphism $T_\nu = I + \sum_{r=1}^{\infty} \nu^r T_r$, $T_r \in \mathcal{L}(A, A)$ so that $T_\nu(u *' v) = (T_\nu u * T_\nu v)$ in the associative case, denoting by $*$ (resp. $*'$) the deformed laws in $\tilde{A}$ (resp. $\tilde{A}'$); and similarly in the Lie case. In particular we see (for $r = 1$) that a deformation is trivial at order 1 if it starts with a 2-cocycle which is a 2-coboundary. More generally, exactly as above, we can show [17] that if two deformations are equivalent up to some order t, the condition to extend the equivalence one step further is that a 2-cocycle (defined using the T_k, $k \leq t$) is the coboundary of the required T_{t+1} and therefore *the obstructions to equivalence lie in the 2-cohomology*. In particular, if that space is null, all deformations are trivial.

I.4.2.3 Unit. An important property is that a *deformation of an associative algebra with unit* (what is called a unital algebra) is again unital, and *equivalent to a deformation with the same unit*. This follows from a more general result of Gerstenhaber (for deformations leaving unchanged a subalgebra) and a proof can be found in [87].

I.4.2.4. In the case of (topological) *bialgebras* or *Hopf* algebras, *equivalence* of deformations has to be understood as an isomorphism of (topological) $\mathbb{K}[[\nu]]$-algebras, the isomorphism starting with the identity for the degree 0 in ν. A deformation is again said *trivial* if it is equivalent to that obtained by base field extension. For Hopf algebras the deformed algebras may be taken (by equivalence) to have the same unit and counit, but in general not the same antipode.

I.4.3 Examples of special interest: the differentiable cases

Consider the algebra $N = C^\infty(X)$ of functions on a differentiable manifold X. When we look at it as an associative algebra acting on itself by pointwise multiplication, we can define the corresponding Hochschild cohomologies. Now let Λ be a skew-symmetric contravariant two-tensor (possibly degenerate) defined on X, satisfying $[\Lambda, \Lambda] = 0$ in the sense of the supersymmetric Schouten-Nijenhuis bracket [132,145] (a definition of which, both intrinsic and in terms of local coordinates, can be found in [17,71]). Then the inner product $P(u,v) = i(\Lambda)(du \wedge dv)$ of Λ with the 2-form $du \wedge dv$, $u, v \in N$, defines a *Poisson bracket* P: it is obviously skew-symmetric, satisfies the Jacobi identity because $[\Lambda, \Lambda] = 0$ and the Leibniz rule $P(uv, w) = P(u, w)v + uP(v, w)$. It is a bidifferential 2-cocycle for the (general or differentiable) Hochschild cohomology of N, skewsymmetric of order $(1,1)$, therefore [17] nontrivial and thus defines an infinitesimal deformation of the pointwise product on N. We say that X, equipped with such a P, is a *Poisson manifold* [17,120].

When Λ is everywhere nondegenerate (X is then necessarily of even dimension 2ℓ), its inverse ω is a closed everywhere nondegenerate 2-form ($d\omega = 0$ is then equivalent to $[\Lambda, \Lambda] = 0$) and we say that X is *symplectic*; ω^ℓ is a volume element on X. Then one can in a consistent manner work with differentiable cocycles [17,71] and the differentiable Hochschild p-cohomology space $\mathcal{H}^p(N)$ is [103,155] that of all skew-symmetric contravariant p-tensor fields, and therefore is infinite-dimensional. Thus, except when X is of dimension 2 (because then necessarily $\mathcal{H}^3(N) = 0$), the obstructions belong to an infinite-dimensional space where they may be difficult to trace. On the other hand, when $2\ell = 2$, any 2-cocycle can be the driver of a deformation of the associative algebra N: "anything goes" in this case; some examples for $\mathbb{R}^2$ can be found in [155].

Now endow N with a Poisson bracket: we get a Lie algebra and can look at its Chevalley cohomology spaces. Note that P is bidifferential of order $(1,1)$ so it is important to check whether the Gerstenhaber theory is *consistent* when restricted to *differentiable* cochains (both of arbitrary order and of order at most 1), especially since the general (Gelfand-Fuks) cohomology is very complicated [85] (but in fact the pathology arises only when non-continuous cochains are allowed). This is a nontrivial question and we gave it a positive answer [71,17]; in brief, if a coboundary is differentiable, it is the coboundary of a differentiable cochain.

Again, since P is of order $(1,1)$, we first studied the 1-differentiable cohomologies. When the cochains are restricted to be of order $(1,1)$ with no constant term (then they annihilate constant functions, which we write "n.c." for "null on constants") it was found [119] that the Chevalley-Eilenberg cohomology $H^*_{1-\text{diff,n.c.}}(N)$ of the Lie algebra N (acting on itself with the adjoint representation) is exactly the de Rham cohomology $H^*(X)$. Thus $\dim H^p_{1-\text{diff,n.c.}}(N) = b_p(X)$, the pth Betti number of the manifold X. Without the n.c. condition one gets a slightly more complicated formula [119]; in particular if X is symplectic with an exact 2-form $\omega = d\alpha$, one has here $H^p_{1-\text{diff}}(N) = H^p(X) \oplus H^{p-1}(X)$.

All this allowed us "three musketeers" to study in 1974 what we called in [71] *1-differentiable deformations* of the Poisson bracket Lie algebra N and to give some applications [72]. In particular we noticed that the "pure" order $(1,1)$ deformations correspond to a deformation of the 2-tensor Λ; allowing constant terms and taking the deformed bracket in Hamilton equations instead of the original Poisson bracket gave (at this classical level) a kind of friction term.

Shortly afterwards, triggered by our works, a "fourth musketeer" J. Vey [155] noticed that in fact $\dim H^2_{\text{diff,n.c.}}(N) = 1 + \dim H^2_{1-\text{diff,n.c.}}(N)$ and that $H^3_{\text{diff,n.c.}}(N)$ is also finite-dimensional, which allowed him to study differentiable deformations. Incidentally, in the $\mathbb{R}^{2\ell}$ case, Vey rediscovered (independently, because he ignored it) the Moyal bracket. The latter was then rather "exotic" and few authors (except for a number of physicists, like [2] or J. Plebański who described it in Polish lecture notes [140] he gave us later) paid any attention to it. In *Mathematical Reviews* this bracket, for which [126] is nowadays often quoted, is not even mentioned in the review! We then came back to the problem [73], this time with differentiable deformations, and deformation quantization was conceived.

II. DEFORMATION QUANTIZATION AND ITS RAMIFICATIONS

II.1 The Birth of Deformation Quantization

Though we had mentioned the main features in 1976 and 1977 in short papers [72,16], meetings [78] and a long preprint, it is only with the publication of the latter [17], our first major (and often quoted) contribution in this new domain, that what eventually became known as *deformation quantization* [159] took off. Incidentally, as for the two other parts of our "deformation trilogy" (see e.g. [69,70] and [74,75,78]) which deal (resp.) with singleton physics and with nonlinear evolution equations, true recognition was slow to come (it took about twenty years). Let me stress once more (and this will become evident with the section on physical applications), that the important and most original conceptual aspect is that *quantization* is here an *autonomous theory* based on a *deformation* of the composition law of classical observables, not on a radical change in the nature of the observables. In addition to this important *conceptual advantage*, our approach is more general (simple examples can be given); it can be shown to coincide with the conventional (operatorial) approach in known applications (see below) whenever a (possibly generalized) Weyl mapping can be defined; it also paves the way to better conventional quantizations in field theory (e.g. on the infinite-dimensional symplectic manifold of initial conditions for nonlinear evolution equations) via a kind of cohomological renormalization.

II.1.1 Differentiable deformations and star products

Let X be a differentiable manifold (of finite, or possibly infinite, dimension; to be precise, in the former case, we assume it is locally of finite dimension, paracompact and Hausdorff, and by differentiable we mean infinitely differentiable; the base field may be $\mathbb{R}$ or $\mathbb{C}$). We assume given on X a *Poisson structure* (a Poisson bracket P).

DEFINITION 2 *A star product is a deformation of the associative algebra of functions $N = C^\infty(X)$ of the form $* = \sum_{n=0}^\infty \nu^n C_n$ where for $u, v \in N$, $C_0(u,v) = uv$, $C_1(u,v) - C_1(v,u) = 2P(u,v)$ and the C_n are bidifferential operators (locally of finite order). We say a star product is strongly closed if $\int_X (u*v - v*u)dx = 0$ where dx is a volume element on X.*

Remark 1. a. The parameter ν of the deformation is in physical applications taken to be $\nu = \frac{i}{2}\hbar$.

b. Using equivalence one may take $C_1 = P$. The latter is the case of Moyal, but other orderings like standard or normal do not verify this condition (only the skew-symmetric part of C_1 is P). Again by equivalence, in view of Gerstenhaber's result mentioned in I.4.2.3, we may take cochains C_r which are without constant term (what we called n.c. or null on constants). In fact, in the original paper [17], we considered only this case and we also concentrated on "Vey products" [121] for which the cochains C_r have the same parity as r and have P^r for principal symbol in any Darboux chart, with X symplectic; when X is symplectic of dimension 2ℓ with symplectic form ω, the (Liouville) volume element is $dx = \omega^\ell$.

c. It is also possible to consider star products for which the cochains C_n are allowed to be slightly more general. Allowing them to be *local* ($C_n(u,v) = 0$ on any open set where u or v vanish) gives nothing new [36]. (Note that this is not the same as requiring the whole associative product to be local; in fact [142] the latter condition is very restrictive and, like true pseudodifferential operators, a star product is a nonlocal operation). In some cases (e.g. for star representations of Lie groups) it may be practical to consider pseudodifferential cochains. As far as the cohomologies are concerned, it has been recently shown [94,138] that as long as one requires at least continuity for the cochains, the theory is the same as in the differentiable case. Incidentally this indicates that one has to go beyond continuous cochains to get the pathological features of the general Gelfand-Fuks cohomology of infinite Lie algebras.

Also, due to formulas like (2) and the relation with Lie algebras (see II.4.1), it is sometimes convenient [131,63,30] to take $\mathbb{K}[\nu^{-1}, \nu]]$ (Laurent series in ν, polynomial in ν^{-1} and formal series in ν) for the ring on which the deformation is defined. Again, this will not change the theory.

d. By taking the corresponding commutator $[u,v]_\nu = (2\nu)^{-1}(u*v - v*u)$, since the skew-symmetric part of C_1 is P, we get a deformation of the Poisson bracket Lie algebra (N, P). This is a crucial point because (at least in the symplectic case) we know the needed Chevalley cohomologies and (in contradistinction with the Hochschild cohomologies) they are small [155,48]. The interplay between both structures gives existence and classification; in addition it will explain why (in the symplectic case) the classification of star products is based on the 1-differentiable cohomologies, hence ultimately on the de Rham cohomology of the manifold.

II.1.2 Invariance and covariance

Since the beginning [17] we realized that there is a big difference between Poisson brackets and their deformations, from the point of view of geometric invariance. Indeed, while a Poisson bracket P is (by definition) invariant under all symplectomorphisms, i.e. transformations of the manifold X which preserve the symplectic form ω (generated by the flows $x_u = i(\Lambda)(du)$ defined by Hamiltonians $u \in N$), already on $\mathbb{R}^{2\ell}$ one sees easily that its powers P^r, hence also the Moyal bracket (3), are invariant only under flows generated by Hamiltonians u which are polynomials of maximal order 2, forming the "affine" symplectic Lie algebra $\mathfrak{sp}(\mathbb{R}^{2\ell}) \cdot \mathfrak{h}_\ell$. For other orderings the invariance is even smaller (only $\mathfrak{h}_\ell$ remains). For general Vey products the first terms of a star product are [17,121] $C_2 = P_\Gamma^2 + bH$ and $C_3 = S_\Gamma^3 + T + 3\partial H$. Here H is a differential operator of maximal order 2, T a 2-tensor corresponding to a closed 2-form, ∂ the Chevalley coboundary operator.

P_Γ^2 is given (in canonical coordinates) by an expression similar to P^2 in which usual derivatives are replaced by covariant derivatives with respect to a given symplectic connection Γ (a torsionless connection with totally skew-symmetric components when all indices are lowered using Λ). S_Γ^3 is a very special cochain given by an expression similar to P^3 in which the derivatives are replaced by the relevant components of the Lie derivative of Γ in the direction of the vector field associated to the function (u or v). Fedosov's algorithmic construction [61] shows that the symplectic connection Γ plays a rôle at all orders. Therefore the invariance group of a star product is a subgroup of the *finite-dimensional* group of symplectomorphisms preserving a connection. Its Lie algebra is $\mathfrak{g}_0 = \{a \in N; [a, u]_\nu = P(a, u) \, \forall u \in N\}$, *preferred observables* Hamiltonians for which the classical and quantum evolutions coincide. We are thus lead to look for a weaker notion and shall call a star product *covariant* under a Lie algebra $\mathfrak{g}$ of functions if $[a,b]_\nu = P(a,b) \, \forall a, b \in \mathfrak{g}$. It can be shown [8] that $*$ is $\mathfrak{g}$-covariant iff there exists a representation τ of the Lie group G whose Lie algebra is $\mathfrak{g}$ into $\mathrm{Aut}(N[[\nu]]; *)$ such that $\tau_g u = (Id_N + \sum_{r=1}^\infty \nu^r \tau_g^r)(g.u)$ where $g \in G, u \in N$, G acts on N by the natural action induced by the vector fields associated with $\mathfrak{g}$, $(g \cdot u)(x) = u(g^{-1}x)$, and where the τ_g^r are differential operators on W. Invariance of course means that the geometric action preserves the star product: $g \cdot u * g \cdot v = g \cdot (u * v)$. This is the basis for the theory of star representations which we shall briefly present below (II.4.1).

II.2 Existence and Classification of Deformation Quantizations

II.2.1 Symplectic finite-dimensional manifolds; reduction

As early as 1975, Vey [155] had shown that on a symplectic manifold X with $b_3(X) = 0$, there exists a globally defined deformation of the Poisson bracket P. He did this by a careful study allowing him to show (by induction) that at each order of deformation (each order of $\hbar$), he can manage to pass via the zero class of the finite-dimensional obstructions space $H^3_{\mathrm{diff,n.c.}}(N)$. This was later easily extended to star products [130] and in essence tells us that we can in this case "glue" Moyal products defined on local charts (equivalence will take care of intersection of two charts and the vanishing of $b_3(X)$ permits to do it in a way compatible with multiple intersections).

The restriction $b_3(X) = 0$ seemed purely technical and indeed already in [17], with the important case of the hydrogen atom ($X = T^*(S^3)$), we showed that it is not essential. The latter case was generalized by Gutt to $X = T^*(M)$ where M is a Lie group [93] (see also regular star representations in [35]) or more generally a parallelizable manifold. Shortly afterwards De Wilde and Lecomte were able to find a proof of existence, first for a general cotangent bundle, then for exact symplectic manifolds and finally for a general symplectic manifold. The latter required at first a very abstract proof which the authors eventually made more "palatable" [47] along the lines used by the Japanese group (gluing Moyal on Darboux charts), and the question of invariant star products was also studied [49].

What is behind the scene for cotangent bundles $T^*(M)$, is that there one has globally defined "momentum coordinates" which permits to work with globally defined differentiations on M and polynomials in them (differential operators). From there a natural step forward is to localize everything on a general symplectic manifold X and to work with a bundle $\mathcal{W}(X)$ of *Weyl algebras*; a Weyl algebra W_ℓ is the enveloping algebra of the Heisenberg canonical commutation relations Lie algebra $\mathfrak{h}_\ell$, possibly completed to formal series, and the product there is the $\mathrm{Sp}(2\ell)$-invariant Moyal product. The "miracle" is that this bundle has a flat connection and a global section; therefore locally defined (Moyal) star products on polynomials (in any given Darboux chart) can be glued together to a globally defined star product. (A Darboux chart on a symplectic manifold is a chart with local coordinates the q's and p's of physicists, i.e. where $\omega = \sum_{j=1}^{2\ell} dq_j \wedge dp_j$). This line of conduct was taken by Fedosov since 1985 (in an obscure paper [60] which was made detailed and precise later [61,63]), using a "germ" approach (infinitesimal neighborhoods), symplectic connections and an algorithmic construction of the flat connection on the Weyl bundle, canonically constructed starting from a given symplectic connection on X. Independently a Japanese group [134] obtained also a general proof of existence, gluing together Moyal products defined on Darboux charts by "projecting" from the Weyl bundle, and the method could also be easily adapted to give the existence [134] of *closed star products* [42]. For more details on these (especially Fedosov's) constructions we refer to the original papers and to the reviews in [20,99,159]. The relation between the "Russian" approach and the "Belgian" one was made [46] by a famous Belgian mathematician (with a Russian wife), translating both into his own language of "gerbes". Eventually the Fedosov construction was shown (cf. e.g. [131]) to be "generic" in the sense that any differentiable star product is equivalent to a Fedosov star product.

Now an important tool in symplectic mechanics is that of *reduction* [124] caused by the action of an invariance group G and subsequent reduction of the algebra of observables. In fact, already in [17], a reduction of this general type was used in connection with the hydrogen atom. Fedosov has recently showed [64] that the classical reduction theory can be "quantized" in the same conditions, i.e. that reduction commutes with G-invariant deformation quantization (at least for G compact); note however that, as shown in a simple example [158], reductions may give nonequivalent star products.

Remark 2: connection with 1-differentiable deformations. We have indicated that equivalence classes of star products are in one-to-one correspondence with formal series in the deformation parameter ν with coefficients in $H^2(X)$, i.e. series of the form $[\Lambda] + \nu[H^2(X)][[\nu]]$, where Λ is the (nondegenerate) Poisson 2-tensor given on X. Since $H^2(X)$ classifies 1-differentiable deformations of the Lie algebra (N, P), one expects that *any Fedosov deformation can be obtained by a sequence of successive 1-differentiable deformations of the initial Poisson bracket,* starting with any given one, at least for X symplectic. This is indeed true. Intuitively, one starts – if needed by equivalence – with a star product n.c. at order 1, $(u,v) \mapsto uv + \nu P(u,v) + O(\hbar^2)$. Associativity is satisfied modν^2. Then one takes a 1-differentiable infinitesimal deformation of P of the form $P + \nu P_1 + \cdots$, corresponding in fact to an infinitesimal deformation $\omega + \nu\omega_1 + O(\nu^2)$. Fedosov tells us that we can find P_1' and higher order terms so that the new product $(u,v) \mapsto uv + \nu(P(u,v) + \nu P_1(u,v)) + \nu^2 P_1'(u,v) + \cdots$ is associative to order 2 in ν. The classes of the 2-tensors associated with P_1 give all possible choices to order 2. One does the same at the next step with $(P + \nu P_1 + \nu^2 P_2)$ and P_2', and so on. Indeed, in [24], where at every order in ν the effect of adding a de Rham 2-cocycle was traced in the star product, Bonneau gave a detailed proof of this fact. So by "plugging in" 1-differentiable deformations of Poisson brackets one can cover all possible equivalence classes, starting from any given Fedosov star product.

In a more abstract form a similar conclusion is a consequence of results developed in the nice review [99]. The mathematically oriented reader will find there a detailed presentation (in a Čech cohomology context) of the equivalence question and related problems.

II.2.2 Poisson finite-dimensional manifolds

As we explained earlier, physics (e.g. with first class Dirac constraints) requires sometimes manifolds which have a Poisson bracket P, but a degenerate one and are therefore not symplectic. These are called *Poisson manifolds* [17] and they are foliated with symplectic leaves. A typical example is the dual of a finite-dimensional Lie algebra, foliated by coadjoint orbits (see e.g. [109]). As this example shows (even in the case of the Heisenberg Lie algebra $\mathfrak{h}_\ell$, where one must not forget the trivial orbit), Poisson manifolds are in general not regular; a *regular* Poisson manifold is foliated with symplectic leaves of constant even dimension (before introducing Poisson manifolds we had considered [71] "canonical manifolds", regular Poisson manifolds where the leaves have codimension 1). In the regular case the theory we just explained extends in a straightforward manner. Some non conclusive attempts were made in the general case, most notably in [135] following more "traditional" lines and [156] in the direction indicated by Kontsevich's formality conjecture [111].

The solution to this difficult problem was recently given by Maxim Kontsevich [112] and this is in a way "the cherry on the cake" of deformation quantization (and contributed to getting him the Fields medal in 1998). It involves very elaborate constructions, both conceptually and computationally and makes an essential use of ideas coming from string theory. We shall not attempt here to describe it in detail but give the flavor of the development. The reader interested in more details should refer to [112] and follow subsequent developments. The mathematically oriented reader may be interested in the "Bourbaki-style" presentation (by a French mathematician [133], see quotation above...) of the context and results.

Since, as we noted above, a Poisson bracket P is a nontrivial 2-cocycle for the

Hochschild cohomology of the algebra $N = C^\infty(X)$, a natural question is to decide whether this infinitesimal deformation can be extended to a star product on N. Kontsevich answers positively, and more:

THEOREM 1 *Let X be a differentiable manifold and $N = C^\infty(X)$. There is a natural isomorphism between equivalence classes of deformations of the null Poisson structure on X and equivalence classes of differentiable deformations of the associative algebra N; in particular, any Poisson bracket P on X comes from a canonically defined (modulo equivalence) star product.*

With this concise formulation of the result (which gives a positive answer to Kontsevich's formality conjecture) we see that, in this more general context, a main result from the symplectic case is still valid: *classes of star products correspond to classes of deformations of the Poisson structure*. A deformation of the null Poisson structure is a formal series $\Lambda(\hbar) = \sum_{n=1}^\infty \Lambda_n \hbar^n$ having vanishing Schouten-Nijenhuis bracket with itself: $[\Lambda(\hbar), \Lambda(\hbar)] = 0$. Any given Poisson structure Λ_0 can be identified with the series $\Lambda_0 \hbar$.

Remark 3. From Theorem 1 it is natural to conjecture that the relation with 1-differentiable deformations of the Poisson bracket mentioned at the end of (II.2.1) extends to general Poisson manifolds, but a full proof has not yet been given. Furthermore (and this is one of the developments to come) a comparison of the proof with e.g. that of Fedosov in the symplectic case is not done either.

The bulk of the proof is the "affine case", essentially when X is some $\mathbb{R}^\ell$ (or an open set in it), the result being formulated in such a way that "gluing" charts, though still nontrivial, is not too difficult to perform for an experienced mathematician. Doing so Kontsevich gives an interesting explicit universal formula for the star product on such an X where graphs (and Stokes formula) play a crucial rôle. The formula looks like

$$u * v = \sum_{n=0}^\infty \sum_{\Gamma \in G_n} w_\Gamma B_{\Gamma, \Lambda}(u, v) \qquad (12)$$

where Λ is an arbitrary Poisson structure on an open domain in $\mathbb{R}^\ell$ and G_n a set of labeled oriented graphs Γ. The latter are pairs (V_Γ, E_Γ) such that $E_\Gamma \subset V_\Gamma \times V_\Gamma$ with $n+2$ vertices V_Γ and $2n$ edges E_Γ satisfying some additional conditions (see [112], section 2). G_n has $(n(n+1))^n$ elements (1 for $n = 0$). B_Γ is an explicitly defined bidifferential operator (of total order $2n$, as in (4)) and w_Γ a weight defined by an absolutely convergent integral (of the exterior product of the differentials of $2n$ harmonic angular variables associated with Γ, taken over the space of configurations of n numbered pairwise distinct points on the Lobatchevsky upper half-plane).

II.2.3 Remarks on the infinite-dimensional case

Poisson structures are known on infinite-dimensional manifolds since a long time and there is an extensive literature on this subject, which alone would require a book. A typical structure, for our purpose, is a symplectic structure such as that defined by Segal [147] (see also [114]) on the space of solutions of a classical field equation like $\Box \Phi = F(\Phi)$, where $\Box$ is the d'Alembertian. Now if one considers scalar valued functionals Ψ over such a space of solutions, i.e. over the phase space of initial conditions $\varphi(x) = \Phi(x,0)$ and $\pi(x) = \frac{\partial}{\partial t}\Phi(x,0)$, one can consider a Poisson bracket defined by

$$P(\Psi_1, \Psi_2) = \int \left(\frac{\delta\Psi_1}{\delta\varphi}\frac{\delta\Psi_2}{\delta\pi} - \frac{\delta\Psi_1}{\delta\pi}\frac{\delta\Psi_2}{\delta\varphi}\right)dx \tag{13}$$

where δ denotes the functional derivative. The problem is that while it is possible to give a precise mathematical meaning to (13), the formal extension to powers of P, needed to define the Moyal bracket, is highly divergent, already for P^2.

The same difficulty is met if one takes e.g. a space N of differentiable functions on a Hilbert space with orthonormal basis $\{p_k, q_k; k = 1, \ldots, \infty\}$ and a Poisson bracket $P(u,v) = \sum_{k=1}^{\infty}\left(\frac{\partial u}{\partial p_k}\frac{\partial v}{\partial q_k} - \frac{\partial u}{\partial q_k}\frac{\partial v}{\partial p_k}\right)$.

This is not so surprising for physicists who know from experience that the correct approach to field theory is via normal ordering, and that there are infinitely many inequivalent representations of the canonical commutation relations (as opposed to the von Neumann uniqueness in the finite-dimensional case, for projective representations). Integral formulas, related to Feynman path integrals, can also be used with some success. The participants at the Łodz meeting may be interested to learn that an analogue of the pseudodifferential calculus in the infinite-dimensional case, and especially the ("Wigner") notion of symbols of operators, has been developed already in 1978 by Paul Krée and Ryszard Rączka [115].

We shall come back to this question in (II.3.2) with more specific examples and give indications showing that with proper care the deformation quantization approach can help making better mathematical sense of field theory calculations done by theoretical physicists.

II.2.4 Generalized deformations, n-gebras and related structures

One of the mathematical reasons we started with the study of deformations of Poisson brackets is related to the fact that it is the only one, among classical infinite-dimensional algebras, which is not rigid, even at the level of 1-differentiable deformations. In particular unimodular structures (defined by a determinant) are rigid. It turns out that in connection with Nambu mechanics, where the Poisson bracket is replaced by an n-bracket, say a functional determinant, one meets structures of this type and it is not a big surprise that a specific quantization (not of Heisenberg type) was difficult to find.

Roughly speaking, a *generalized deformation* of a $\mathbb{K}$-algebra A (associative, Lie or other) is a $\mathbb{K}$-algebra A_ν having A for limit as the deformation parameter $\nu \to 0$. Among the "other" algebras are of course the bialgebras to which we shall come back in (II.4.2) when dealing with quantum groups (incidentally, since 'al' means 'the' in Arabic, applied to a set containing only one element, the French denomination 'bigèbre', imposed by Cartier, is far better).

Here we shall be concerned mostly with the so-called n-gebras, algebras A endowed with a composition law $A^n \to A$ satisfying some conditions including skew-symmetry. Structures of this kind were introduced by Nambu [129] in connection with his "generalized mechanics" and (in a paper published in an obscure journal) by Filippov [65]. Serious interest in them developed only from 1992, when Takhtajan [152] and independently Flato and Frønsdal (unpublished) discovered that Nambu n-brackets satisfy a generalization of Jacobi identity, called the Fundamental Identity (FI); surprisingly enough, this identity had not been discovered before.

The resurgence of *operads* which occurred at the same time [90,122] and are related to n-gebras [91], as well as the new notion of *strong homotopy Lie algebras* introduced then by Stasheff and is also related to deformation theory [150] add to the interest in these structures.

Recently, there have been several works dealing with various generalizations of Poisson structures by extending the binary bracket to an n-bracket. The main point for these generalizations is to look for the corresponding identity which would play the rôle of Jacobi identity for the usual Poisson bracket. Indeed, in view of generalizations, the Jacobi identity for a Lie 2-bracket can be presented in a number of ways [84] among which two have been recently extensively studied. The most straightforward way is to require that the sum over the symmetric group $\mathfrak{S}_3$ of the composed brackets $[[\cdot,\cdot],\cdot]$ is zero. When extended to n-brackets, leads to the notion of *generalized Poisson structures* studied e.g. in [11]; the corresponding identity is obtained by complete skew-symmetrization of the $2n-1$ composed brackets when n is even; this is equivalent to require that the Schouten bracket of the n-tensor defining the n-bracket with itself vanishes.

A physically more appealing way is to say that the adjoint map $b \mapsto [a,b]$ is a Lie algebra derivation. Indeed this means that the bracket of conserved quantities is again a conserved quantity. The two formulations coincide only for $n = 2$ and for $n \geq 3$ the latter is a stronger requirement. This *Fundamental Identity* of Nambu Mechanics can be written:

$$[x_1,\ldots,x_{n-1},[y_1,\ldots,y_n]] - \sum_{1\leq i\leq n}[y_1,\ldots,y_{i-1},y_{i+1},\ldots,y_n,[x_1,\ldots,x_{i-1},y_i,x_i,\ldots,x_{n-1}]] = 0. \quad (14)$$

Nambu brackets (like Poisson brackets and commutators such as the Moyal bracket, for $n = 2$) are n-brackets required to satisfy, with respect to the usual algebra multiplication and in addition to skew-symmetry $\{x_1,\ldots,x_n\} = \epsilon(\sigma)\{x_{\sigma_1},\ldots,x_{\sigma_n}\}$ $\forall \sigma \in \mathfrak{S}_n$ and the FI, a *Leibniz rule*:

$$\{x_0 x_1, x_2,\ldots,x_n\} = x_0\{x_1,x_2,\ldots,x_n\} + \{x_0,x_2,\ldots,x_n\}x_1. \quad (15)$$

The related cohomologies are not yet completely known, though a major step in this direction was done in [84] where one can also find a very interesting and detailed study of all intermediate possibilities between the two generalizations described here (generalized Poisson and Nambu). In (II.3.3) we shall nevertheless indicate, specializing to the case $A = N$, how one can quantize Nambu brackets using generalized deformations based on the factorization of polynomials and methods of second quantization. One of the steps there (and this produces a non-DrG-deformation) is an operation, the effect of which is that in products the deformation parameter $\hbar$ behaves as if it was nilpotent (e.g. multiplied by a Dirac γ matrix).

This last fact has very recently induced Pinczon [139] and Nadaud [128] to generalize Gerstenhaber theory to the case of a deformation parameter σ which *does not commute with the algebra*. A similar theory can be done in this case, with appropriate cohomologies. While that theory does not reproduce the above mentioned Nambu quantization, it gives new and interesting results. For instance [139], while the Weyl algebra W_1 (generated by the Heisenberg Lie algebra $\mathfrak{h}_1$) is known [56] to be DrG-rigid, it can be nontrivially deformed in such a *supersymmetric deformation theory* to the supersymmetry enveloping algebra $\mathcal{U}(\mathfrak{osp}(1,2))$; or [128], on the polynomial algebra $\mathbb{C}[x,y]$ in 2 variables, Moyal-like products of a new type were discovered. This is another example of a motivated study which goes beyond a generally accepted framework.

II.3 Physical Applications

In this subsection and the following, I shall present a few of the numerous developments which have made use of deformation quantization and/or are strongly related to it. The presentation made, and therefore the bibliography, is by no means exhaustive – more than a whole volume would be needed for that – and the absence of reference to any specific work does not (in general) reflect a lack of appreciation; I did not even quote all of my publications in the domain. The aim of these two last subsections (in fact, of all this review) is mainly to give the flavor of the many facets of deformation quantization and quite naturally the presentation will be somewhat biased towards the works of our group. Nevertheless the interested reader should be able to complete whatever is missing by a kind of "hyper-referencing", looking at references of references a few times.

II.3.1 Quantum mechanics

Let us start with a phase space X, a symplectic (or Poisson) manifold and N an algebra of classical observables (functions, possibly including distributions if proper care is taken for the product). We shall call *star quantization* a star product on N invariant (or sometimes only covariant) under some Lie algebra $\mathfrak{g}_0$ of "preferred observables". Invariance of the star product ensures that the classical and quantum evolutions of observables under a Hamiltonian $H \in \mathfrak{g}_0$ will coincide [17]. The typical example is the Moyal product on $W = \mathbb{R}^{2\ell}$.

II.3.1.1 Spectrality. Physicists want to get numbers matching experimental results, e.g. for energy levels of a system. That is usually achieved by describing the spectrum of a given Hamiltonian $\hat{H}$ supposed to be a self-adjoint operator so as to get a real spectrum and so that the evolution operator (the exponential of $it\hat{H}$) is unitary (thus preserves probability). A similar spectral theory can be done here, in an *autonomous manner*. The most efficient way to achieve it is to consider [17] the *star exponential* (corresponding to the evolution operator)

$$\mathrm{Exp}(Ht) \equiv \sum_{n=0}^{\infty} \frac{1}{n!} \left(\frac{t}{i\hbar}\right)^n (H*)^n \qquad (16)$$

where $(H*)^n$ means the nth star power of the Hamiltonian $H \in N$ (or $N[[\nu]]$). Then one writes its Fourier-Stieltjes transform $d\mu$ (in the distribution sense) as $\mathrm{Exp}(Ht) = \int e^{\lambda t/i\hbar} d\mu(\lambda)$ and defines *the spectrum of* $(H/\hbar)$ as the *support S of $d\mu$* (incidentally this is the definition given by L. Schwartz for the spectrum of a distribution, out of motivations coming from Fourier analysis). In the particular case when H has discrete spectrum, the integral can be written as a sum (see the top equation in (18) below for a typical example): the distribution $d\mu$ is a sum of "delta functions" supported at the points of S multiplied by the symbols of the corresponding eigenprojectors.

In different orderings with various weight functions w in (1) one gets in general different operators for the same classical observable H, thus different spectra. For $X = \mathbb{R}^{2\ell}$ all orderings are mathematically equivalent (to Moyal under the Fourier transform T_w of the weight function w). This means that every observable H will have the same spectrum under Moyal ordering as $T_w H$ under the equivalent ordering. But this does not imply physical equivalence, i.e. the fact that H will have the same spectrum under both orderings. In fact the opposite is true [34]: if two equivalent star products are isospectral (give the same spectrum for a large family of observables and all $\hbar$), they are identical.

It is worth mentioning that our definition of spectrum permits to define a spectrum even for symbols of non-spectrable operators, such as the derivative on a half-line which has different deficiency indices; this corresponds to an infinite potential barrier (see also [154] for detailed studies of similar questions). That is one of the many advantages of our autonomous approach to quantization.

II.3.1.2 Applications. In quantum mechanics it is preferable to work (for $X = \mathbb{R}^{2\ell}$) with the star product that has maximal symmetry, i.e. $\mathfrak{sp}(\mathbb{R}^{2\ell}) \cdot \mathfrak{h}_\ell$ as algebra of preferred observables: the Moyal product. One indeed finds [17] that the star exponential of these observables (polynomials of order ≤ 2) is proportional to the usual exponential. More precisely, if $H = \alpha p^2 + \beta pq + \gamma q^2 \in \mathfrak{sl}(2)$ with $p, q \in \mathbb{R}^\ell$, $\alpha, \beta, \gamma \in \mathbb{R}$, setting $d = \alpha\gamma - \beta^2$ and $\delta = |d|^{1/2}$ one gets (the sums and integrals appearing in the various expressions of the star exponential being convergent as distributions, both in phase-space variables and in t or λ)

$$\text{Exp}(Ht) = \begin{cases} (\cos\delta t)^{-l} \exp((H/i\hbar\delta)\tan(\delta t)) & \text{for} \quad d > 0 \\ \exp(Ht/i\hbar) & \text{for} \quad d = 0 \\ (\cosh\delta t)^{-l} \exp((H/i\hbar\delta)\tanh(\delta t)) & \text{for} \quad d < 0 \end{cases} \quad (17)$$

hence the Fourier decompositions

$$\text{Exp}(Ht) = \begin{cases} \sum_{n=0}^{\infty} \Pi_n^{(\ell)} e^{(n+\frac{\ell}{2})t} & \text{for} \quad d > 0 \\ \int_{-\infty}^{\infty} e^{\lambda t/i\hbar} \Pi(\lambda, H) d\lambda & \text{for} \quad d < 0 \end{cases} \quad (18)$$

We thus get the discrete spectrum $(n+\frac{\ell}{2})\hbar$ of the *harmonic oscillator* and the continuous spectrum $\mathbb{R}$ for the dilation generator pq. The eigenprojectors $\Pi_n^{(\ell)}$ and $\Pi(\lambda, H)$ are given [17] by known special functions on phase-space (generalized Laguerre and hypergeometric, multiplied by some exponential). Formulas (17) and (18) can, by analytic continuation, be given a sense outside singularities and even (as distributions) for singular values of t.

Other examples can be brought to this case by functional manipulations [17]. For instance the Casimir element C of $\mathfrak{so}(\ell)$ representing *angular momentum*, which can be written $C = p^2 q^2 - (pq)^2 - \ell(\ell-1)\frac{\hbar^2}{4}$, has $n(n+(\ell-2))\hbar^2$ for spectrum. For the *hydrogen atom*, with Hamiltonian $H = \frac{1}{2}p^2 - |q|^{-1}$, the Moyal product on $\mathbb{R}^{2\ell+2}$ ($\ell = 3$ in the physical case) induces a star product on $X = T^*S^\ell$; the energy levels, solutions of $(H-E)*\phi = 0$, are found from (18) and the preceding calculations for angular momentum to be (as they should, with $\ell = 3$) $E = \frac{1}{2}(n+1)^{-2}\hbar^{-2}$ for the discrete spectrum, and $E \in \mathbb{R}^+$ for the continuous spectrum.

We thus have recovered, in a completely autonomous manner entirely within deformation quantization, the results of "conventional" quantum mechanics in those typical examples (and many more can be treated similarly). It is worth noting that the term $\frac{\ell}{2}$ in the harmonic oscillator spectrum, obvious source of divergences in the infinite-dimensional case, disappears if the normal star product is used instead of Moyal – which is one of the reasons it is preferred in field theory.

II.3.1.3 Remark on convergence. We have always considered star products as formal series and looked for convergence only in specific examples, and then generally in the sense of distributions. The same applies to star exponentials, as long as each coefficient in the formal series is well defined. In the case of the harmonic oscillator or more generally the preferred observables H in Weyl ordering, this study was facilitated by the fact that the powers $(H*)^n$ are polynomials in H. Moreover, in the case of star exponentials, a notion of convergence stronger than as distributions would require considerations analogous to the problem of analytic vectors in Lie groups representations [76] and pose problems also when looking at their Fourier decomposition. Nevertheless some authors (see e.g. [141] and references therein) insist in making a stronger parallel with operator algebras and look for domains (in N) where one has convergence, speaking of "strict" deformation quantization. In the Weyl case on $\mathbb{R}^{2\ell}$ this question is related to the "Wigner image" of classes of operators [100,107,123] on $L^2(\mathbb{R}^\ell)$. While this is a perfectly legitimate mathematical problem, we do not feel that it is physically wise. In particular it lacks the flexibility that exists in deformation quantization as compared with physics based on algebras of bounded operators or on the less developed approach of algebras of unbounded operators [106], and puts deformation quantization back into the C^* algebras Procrustean bed. Therefore the question of "strict" convergence shall not be asked, except in examples. This does not mean that one should not look for domains where pointwise convergence can be proved; this was done e.g. for Hermitian symmetric spaces [37]. But it should be clearly understood that one can consider wider classes of observables – in fact, the latter tend to be physically more interesting. It is also worthwhile to adapt to deformation quantization procedures that were successful in operator algebras, like the GNS construction [30,28].

II.3.2 Path integrals, field theory and statistical mechanics

II.3.2.1 Path integrals are intimately connected to star exponentials. In fact, in quantum mechanics the path integral of the action is nothing but the partial Fourier transform of the star exponential (16) with respect to the momentum variables, for $X = \mathbb{R}^{2\ell}$ as phase space with the Moyal star product [137]. For normal ordering the path integral is essentially the star exponential [51] and we shall come back to it in (II.3.2.2).

For compact groups the star exponential E (defined in a similar manner, see below (II.4.1)) can be expressed in terms of unitary characters using a global coherent state formalism [33] based on the Berezin dequantization of compact group representation theory used in [4,125] (it gives star products somewhat similar to normal ordering); the star exponential of any Hamiltonian on G/T (where T is a maximal torus in the compact group G) is then equal to the path integral for this Hamiltonian.

II.3.2.2 Field theory. The deformation quantization of a given classical field theory consists in the giving a proper definition for a star product on the infinite-dimensional manifold of initial data for the classical field equation (see II.2.3) and constructing with it, as rigorously as possible, whatever physical expressions are needed. As in other approaches to field theory, here also one faces serious divergence difficulties as soon as one is considering interacting fields theory, and even at the free field level if one wants a mathematically rigorous theory. But the philosophy in dealing with the divergences is significantly different and one is in position to take advantage of the cohomological features of deformation theory to perform what can be called *cohomological renormalization*.

Starting with some star product $*$ (e.g. an infinite-dimensional version of a Moyal-type product or, better, a star product similar to the normal star product (19)) on the manifold of initial data, one would interpret various divergences appearing in the theory in terms of coboundaries (or cocycles) for the relevant Hochschild cohomology. Suppose that we are suspecting that a term in a cochain of the product $*$ is responsible for the appearance of divergences. Applying the procedure described in (I.4.2.2), we can try to eliminate it, or at least get a lesser divergence, by subtracting at the relevant order a coboundary; we would then get a better theory with a new star product, equivalent to the original one. Furthermore, since in this case we can expect to have at each order an infinity of non equivalent star products, we can try to subtract a cocycle and then pass to a non equivalent star product whose lower order cochains are identical to those of the original one. We would then make an analysis of the divergences up to order $\hbar^r$, identify a divergent cocycle, remove it, and continue the procedure (at the same or hopefully a higher order). Along the way one should preserve the usual properties of a quantum field theory (Poincaré covariance, locality, etc.) and the construction of adapted star products should be done accordingly. The complete implementation of this program should lead to a cohomological approach to renormalization theory.

A very good test for this approach would be to start from classical electrodynamics, where (among others) the existence of global solutions and a study of infrared divergencies were recently rigorously performed [77], and go towards mathematically rigorous QED. Physicists will think that spending so much effort in trying to give complete mathematical sense to recipes that work so well is a waste of time, but I am sure that the mathematical tools needed will prove very efficient. Would De Gaulle have been a mathematician he could have said about this scheme "vaste programme" (supposedly his answer to a minister who wanted to get rid of all stupid bureaucrats); but had he been a scientist he would probably have been a physicist and share the attitude of too many physicists towards mathematics: "l'intendance suivra" ("Supply Corps will follow", needed logistics will be provided).

In the case of free fields, one can write down an explicit expression for a star product corresponding to normal ordering. Consider a (classical) free massive scalar field Φ with initial data (ϕ, π) in the Schwartz space $\mathcal{S}$. The initial data (ϕ, π) can advantageously be replaced by their Fourier modes $(\bar{a}, a)$ which after quantization become the usual creation and annihilation operators, respectively. The normal star product $*_N$ is formally equivalent to the Moyal product and an integral representation for $*_N$ is given by:

$$(F *_N G)(\bar{a}, a) = \int_{\mathcal{S}' \times \mathcal{S}'} d\mu(\bar{\xi}, \xi) F(\bar{a}, a + \xi) G(\bar{a} + \bar{\xi}, a), \qquad (19)$$

where μ is a Gaussian measure on $\mathcal{S}' \times \mathcal{S}'$ defined by the characteristic function $\exp(-\frac{1}{\hbar} \int dk\, \bar{a}(k) a(k))$ and F, G are holomorphic functions with semi-regular kernels. Likewise, Fermionic fields can be cast in that framework by considering functions valued in some Grassmann algebra and super-Poisson brackets (for the deformation quantization of the latter see e.g. [27]).

For the normal product (19) one can formally consider interacting fields. It turns out that the star exponential of the Hamiltonian is, up to a multiplicative well-defined function, equal to Feynman's path integral. For free fields, we have a mathematical meaningful equality between the star exponential and the path integrals as both of them are defined by a Gaussian measure, and hence well-defined. In the interacting fields case, giving a rigorous meaning to either of them would give a meaning to the other.

The interested reader will find in [51] calculations performing some steps in the above direction, for free scalar fields and the Klein-Gordon equation, and an example of cancellation of some infinities in $\lambda\phi_2^4$-theory via a λ-dependent star product equivalent to a normal star product. Finally we mention here for more completeness (though neither is directly related to what precedes) the symbolic calculus of [115] and the Fedosov-like approach to self-dual Yang-Mills and gravity of [83].

II.3.2.3 Statistical mechanics. In view of our philosophy on deformations, a natural question to ask is their *stability*: Can deformations be further deformed, or does "the buck stops there"? As we indicated at the beginning of this review and shall exemplify with quantum groups, the answer to that question may depend on the context. Here is another example.

If one looks for deformations of the Poisson bracket Lie algebra (N, P) one finds (assuming mild technical assumptions on parity of cochains in [13] which, in view of the classification of star products, are not required) that a further deformation of the Moyal bracket, with another deformation parameter ρ, is again a Moyal bracket for a ρ-deformed Poisson structure; in particular, for $X = \mathbb{R}^{2\ell}$, *quantum mechanics viewed as a deformation is unique and stable.*

Now, for the associative algebra N, the only *local* associative composition law is [142] of the form $(u, v) \mapsto ufv$ for some $f \in N$. If we take $f = f_\beta \in N[[\beta]]$ we get a 0-differentiable deformation (with parameter β) of the usual product, which for convenience we shall call here a Rubio product. We were thus lead [14] to look, starting from a $*_\nu$ product, for a new composition law

$$(u, v) \mapsto u \tilde{*}_{\nu,\beta} v = u *_\nu f_{\nu,\beta} *_\nu v \quad \text{with} \quad f_{\nu,\beta} = \sum_{r=0}^{\infty} \nu^{2r} f_{2r,\beta} \in N[[N[[\nu^2]], \beta]] \quad (20)$$

where $f_{0,\beta} \equiv f_\beta \neq 0$ and $f_0 = 1$. The transformation $u \mapsto T_{\nu,\beta} u = f_{\nu,\beta} *_\nu u$ intertwines $*_\nu$ and $\tilde{*}_{\nu,\beta}$ but it is not an equivalence of star products because $\tilde{*}_{\nu,\beta}$ is not a star product: it is a (ν, β)-deformation of the usual product (or a ν-deformation of the Rubio product) with at first order in ν the driver given by $P_\beta(u,v) = f_\beta P(u,v) + uP(f_\beta, v) - P(f_\beta, u)v$, a conformal Poisson bracket associated with a *conformal symplectic structure* given by the 2-tensor $\Lambda_\beta = f_\beta \Lambda$ and the vector $E_\beta = [\Lambda, f_\beta]$.

In view of applications we suppose given a star product, denoted $*$, on some algebra $\mathcal{A}$ of observables (possibly defined on some infinite-dimensional phase-space) and take for $f_{\nu,\beta}$ the exponential $g_\beta \equiv \exp_*(c\beta H) = 1 + \sum_{n=1}^{\infty} \frac{(c\beta)^n}{n!} (H*)^n$ with $c = -\frac{1}{2}$ (we omit ν from now on and write $\tilde{*}$ for $\tilde{*}_{\nu,\beta}$). The star exponential $\mathrm{Exp}(Ht)$ defines an automorphism $u \mapsto \alpha_t(u) = \mathrm{Exp}(-Ht) * u * \mathrm{Exp}(Ht)$. A *KMS state* σ on $\mathcal{A}$ is a state (linear functional) satisfying, $\forall a, b \in \mathcal{A}$, the Kubo–Martin–Schwinger condition $\sigma(\alpha_t(a) * b) = \sigma(b * \alpha_{t+i\hbar\beta}(a))$. Then the (quantum) KMS condition can be written [14], with $[a,b]_\beta = (i/\hbar)(a\tilde{*}b - b\tilde{*}a)$, simply $\sigma(g_{-\beta} * [a,b]_\beta) = 0$: up to a conformal factor, a KMS state is like a trace with respect to this new product. The (static) classical KMS condition is the limit for $\hbar = 0$ of the quantum one. So we can recover known features of statistical mechanics by introducing a new deformation parameter $\beta = (kT)^{-1}$ and the related conformal symplectic structure. This procedure commutes with usual deformation quantization. Finally let us mention that recently several people [29,160] have considered the question of KMS states and related modular automorphisms from a more conventional point of view in deformation quantization.

II.3.3 Nambu mechanics and its quantization

We mention this aspect here mainly for the sake of completeness, as an example of generalized deformation. A somewhat detailed recent review can be found in [68] (see also [67]), so we shall just briefly indicate a few highlights.

Nambu [129] started with a kind of "Hamilton equations" on $\mathbb{R}^3$, of the form $\frac{dr}{dt} = \nabla g(r) \wedge \nabla h(r)$, $r = (x,y,z) \in \mathbb{R}^3$, where x, y, z are the dynamical variables and g, h are two functions of r. Liouville theorem follows directly from the identity $\nabla \cdot (\nabla g(r) \wedge \nabla h(r)) = 0$, which tells us that the velocity field in the above equation is divergenceless. From this we derive the evolution of a function f on $\mathbb{R}^3$:

$$\frac{df}{dt} = \frac{\partial(f,g,h)}{\partial(x,y,z)}, \tag{21}$$

where the right-hand side is the Jacobian of the mapping $\mathbb{R}^3 \to \mathbb{R}^3$ given by $(x,y,z) \mapsto (f,g,h)$. In this "baby model for integrable systems", Euler equations for the angular momentum of a rigid body are obtained when the dynamical variables are taken to be the components of the angular momentum vector $L = (L_x, L_y, L_z)$, g is the total kinetic energy $\frac{L_x^2}{2I_x} + \frac{L_y^2}{2I_y} + \frac{L_z^2}{2I_z}$ and h the square of the angular momentum $L_x^2 + L_y^2 + L_z^2$. Other examples can be given, in particular Nahm's equations for static $\mathfrak{su}(2)$ monopoles, $\dot{x}_i = x_j x_k$ ($i,j,k = 1,2,3$) in $\mathfrak{su}(2)^* \sim \mathbb{R}^3$, with $h = x_1^2 - x_2^2$, $g = x_1^2 - x_3^2$, etc. Here the principle of least action, which states that the classical trajectory C_1 is an extremal of the action functional $A(C_1) = \int_{C_1}(pdq - Hdt)$, is replaced by a similar one [152] with a 2-dimensional cycle C_2 and "action functional" $A(C_2) = \int_{C_2}(xdy \wedge dz - hdg \wedge dt)$ (which bears some flavor of strings and some similitude with the cyclic cocycles of Connes [41]).

Expression (21) was easily generalized to n functions f_i, $i = 1,\ldots,n$. One introduces an n-tuple of functions on $\mathbb{R}^n$ with composition law given by their Jacobian, linear canonical transformations $SL(n,\mathbb{R})$ and a corresponding $(n-1)$-form which is the analogue of the Poincaré-Cartan integral invariant. The Jacobian has to be interpreted as a generalized Poisson bracket: It is skew-symmetric with respect to the f_i's, satisfies the FI which is an analogue of the Jacobi identity (but was discovered much after [129]) and a derivation of the algebra of smooth functions on $\mathbb{R}^n$ (i.e., the Leibniz rule is verified in each argument, e.g. $\{f_1 f_2, f_3, \ldots, f_{n+1}\} = f_1\{f_2, \ldots, f_{n+1}\} + \{f_1, f_3, \ldots, f_{n+1}\}f_2$, etc.). Hence there is a complete analogy with the Poisson bracket formulation of Hamilton equations, including the important fact that the components of the $(n-1)$-tuple of "Hamiltonians" $(f_2, \ldots, f_n)$ are constants of motion.

Shortly afterwards it was shown [15,127] that Nambu mechanics could be seen as a coming from constrained Hamiltonian mechanics; e.g. for $\mathbb{R}^3$ one starts with $\mathbb{R}^6$ and an identically vanishing Hamiltonian, takes a pair of second class constraints to reduce it to some $\mathbb{R}^4$ and one more first-class Dirac constraint, together with time rescaling, will give the reduction. This "chilled" the domain for almost 20 years – and gives a physical explanation to the fact that Nambu could not go beyond Heisenberg quantization.

In order to quantize the Nambu bracket, a natural idea is to replace, in the definition of the Jacobian, the pointwise product of functions by a deformed product. For this to make sense, the deformed product should be Abelian, so we are lead to consider commutative DrG-deformations of an associative and commutative product. Looking first at polynomials (this restriction can be removed [138]) we are lead to the commutative part of Hochschild cohomology called Harrison cohomology, which is trivial [12,87]. Dealing with polynomials, a natural idea is to factorize them and take symmetrized star products of the factors. More precisely we introduce an operation α which maps a product of factors into a symmetrized tensor product (in a kind of Fock space) and an evaluation map T which replaces tensor product by star product. Associativity will be satisfied if α annihilates the deformation parameter $\hbar$ (there are still $\hbar$-dependent terms in a product due to the last action of T); intuitively one can think of a deformation parameter which is $\hbar$ times a Dirac γ matrix. This fact brought us to generalized deformations, but even this was not enough. Dealing with distributivity of the product with respect to addition and with derivatives posed difficult problems. In the end we took for observables Taylor developments of elements of the algebra of the semi-group generated by irreducible polynomials ("polynomials over polynomials", inspired by second quantization techniques) and were then able to perform a meaningful quantization of these Nambu-Poisson brackets (cf. [53] for more details and [52] for subsequent development).

II.4 Related Mathematical Developments

II.4.1 Star representation theory of Lie groups

Let G be a Lie group (connected and simply connected), acting by symplectomorphisms on a symplectic manifold X (e.g. coadjoint orbits in the dual of the Lie algebra $\mathfrak{g}$ of G). The elements $x, y \in \mathfrak{g}$ will be supposed realized by functions u_x, u_y in N so that their Lie bracket $[x,y]_\mathfrak{g}$ is realized by $P(u_x, u_y)$. Now take a G-covariant star-product $*$, that is $P(u_x, u_y) = [u_x, u_y] \equiv (u*v - v*u)/2\nu$, which shows that the map $\mathfrak{g} \ni x \mapsto (2\nu)^{-1} u_x \in N$ is a Lie algebra morphism. The appearance of ν^{-1} here and in the trace (see (I.1)) cannot be avoided and explains why we have often to take into account both ν and ν^{-1}. We can now define the *star exponential*

$$E(e^x) = \text{Exp}(x) \equiv \sum_{n=0}^{\infty} (n!)^{-1}(u_x/2\nu)^{*n} \tag{22}$$

where $x \in \mathfrak{g}$, $e^x \in G$ and the power $*n$ denotes the nth star-power of the corresponding function. By the Campbell-Hausdorff formula one can extend E to a *group homomorphism* $E : G \to (N[[\nu, \nu^{-1}]], *)$ where, in the formal series, ν and ν^{-1} are treated as independent parameters for the time being. Alternatively, the values of E can be taken in the algebra $(\mathcal{P}[[\nu^{-1}]], *)$, where $\mathcal{P}$ is the algebra generated by $\mathfrak{g}$ with the $*$-product (a representation of the enveloping algebra).

We call *star representation* [17,82] of G a distribution $\mathcal{E}$ (valued in ImE) on X defined by $D \ni f \mapsto \mathcal{E}(f) = \int_G f(g)E(g^{-1})dg$ where D is some space of testfunctions on G. The corresponding *character* χ is the (scalar-valued) distribution defined by $D \ni f \mapsto \chi(f) = \int_X \mathcal{E}(f)d\mu$, $d\mu$ being a quasi-invariant measure on X.

The character is one of the tools which permit a comparison with usual representation theory. For semi-simple groups it is singular at the origin in irreducible representations, which may require caution in computing the star exponential (22). In the case of the harmonic oscillator that difficulty was masked by the fact that the corresponding representation of $\mathfrak{sl}(2)$ generated by (p^2, q^2, pq) is integrable to a double covering of $SL(2,\mathbb{R})$ and decomposes into a sum $D(\frac{1}{4}) \oplus D(\frac{3}{4})$: the singularities at the origin cancel each other for the two components.

This theory is now very developed, and parallels in many ways the usual (operatorial) representation theory. It is not possible here to give a detailed account of all of them, but among notable results one may quote:

i) An exhaustive treatment of *nilpotent* or solvable exponential [5] and even *general solvable* Lie groups [6]. The coadjoint orbits are there symplectomorphic to $\mathbb{R}^{2\ell}$ and one can lift the Moyal product to the orbits in a way that is adapted to the Plancherel formula. Polarizations are not required, and "star-polarizations" can always be introduced to compare with usual theory. Wavelets [45], important in signal analysis, are manifestations of star products on the (2-dimensional solvable) affine group of $\mathbb{R}$ or on a similar 3-dimensional solvable group [21].

ii) For *semi-simple* Lie groups an array of results is already available, including [4,125] a complete treatment of the *holomorphic discrete series* (this includes the case of compact Lie groups) using a kind of Berezin dequantization, and scattered results for specific examples. Similar techniques have also been used [37,108] to find invariant star products on Kähler and Hermitian symmetric spaces (convergent for an appropriate dense subalgebra). Note however, as shown by recent developments of unitary representations theory (see e.g. [144]), that for semi-simple groups the coadjoint orbits alone are no more sufficient for the unitary dual and one needs far more elaborate constructions.

iii) For semi-direct products, and in particular the Poincaré and Euclidean groups, an autonomous theory has also been developed (see e.g. [7]).

Comparison with the usual results of "operatorial" theory of Lie group representations can be performed in several ways, in particular by constructing an invariant Weyl transform generalizing (1), finding "star-polarizations" that always exist, in contradistinction with the geometric quantization approach (where at best one can find complex polarizations), study of spectra (of elements in the center of the enveloping algebra and of compact generators) in the sense of (II.3.1), comparison of characters, etc. Note also in this context that the pseudodifferential analysis and (non autonomous) connection with quantization developed extensively by Unterberger, first in the case of $\mathbb{R}^{2\ell}$, has been recently extended to the above invariant context [154]. But our main insistence is that the theory of star representations is an *autonomous* one that can be formulated completely within this framework, based on coadjoint orbits (and some additional ingredients when required).

II.4.2 Quantum groups

Around 1980 Kulish and Reshetikhin [116], for purposes related to inverse scattering and 2-dimensional models, discovered a strange modification of the $\mathfrak{sl}(2)$ Lie algebra, where the commutation relation of the two nilpotent generators is a sine in the semi-simple generator instead of being a multiple of it – this in fact requires some completion of the enveloping algebra $\mathcal{U}(\mathfrak{g})$. The theory was developed in the first half of the 80's by the Leningrad school of L. Faddeev [59], systematized by V. Drinfeld who developed the Hopf algebraic context and coined the extremely effective (though somewhat misleading) term of *quantum group* [55] and from the enveloping algebra point of view by Jimbo [105]. Shortly afterwards, Woronowicz [164] realized these models in the context of the noncommutative geometry of Alain Connes [41] by matrix pseudogroups, with coefficients (satisfying some relations) in C^* algebras. A typical example of such Hopf algebras is a Poisson Lie group, a Lie group G with compatible Poisson structure i.e. a Poisson bracket P on $N = C^\infty(G)$, considered as a bialgebra with coproduct defined by $\Delta u(g, g') = u(gg')$, $g, g' \in G$, satisfying $\Delta P(u, v) = P(\Delta u, \Delta v)$, $u, v \in N$.

Now the topological dual of N is the space N' of distributions with compact support on G; it includes G (Dirac's δs at the points of G) and a completion of $\mathcal{U}(\mathfrak{g})$ (differential operators). Taking an adequate subspace N_0 of N (generated by the coefficients of suitably chosen representations, e.g. the "well-behaved" vectors of Harish Chandra) will give a dual $N_0' \supset N'$. All these are reflexive (the bidual coincides with the original space; the algebraic dual of a Hopf algebra is in general not a Hopf algebra). This is the basis of the theory of topological Hopf algebras developed recently, first for G compact [26,25] and then for G semi-simple and in general [22]. In the compact or semi-simple case the quantum group is obtained by giving a star product on N or N_0 and keeping unchanged the coproduct (what is called a preferred deformation) or equivalently by deforming the coproduct in the dual (and keeping the product unchanged). Associativity of the star product corresponds to the Yang-Baxter equation, and the Faddeev-Reshetikhin-Takhtajan and Jimbo models of quantum groups can be seen in this way. Also, all Poisson-Lie groups can be quantized [57,22], though not necessarily with preferred deformations. We have therefore shown that quantum groups are in fact a special case of star products. For more details see e.g. the original papers and [79,80].

II.4.3 Noncommutative geometry and index theorems

Noncommutative geometry arose by a kind of "distillation" from the works of Connes on C^*-algebras and the use in that connection of methods and results of algebraic geometry. It involves in particular *cyclic cohomology* which was introduced by A. Connes in connection with trace formulas for operators (cyclic homology was introduced independently by Tsygan [153]). In particular cyclic cocycles are higher analogues of traces (see [54] for a generalization of the notion of trace). Thus they facilitate (by setting it algebraically) the computation of the index, which can obviously be viewed as the trace of some operator, and permit to generalize the index theorem, producing *algebraic index formulas* [41] of which the Atiyah-Singer formula (5) is a special case. As a matter of fact, Fedosov worked first in problems related to the index theorem and this brought him naturally to star product algebras of functions and to the index question in that context [62] as a fruitful alternative to algebras of pseudodifferential operators. Recently Nest and Tsygan [131] gave a nice proof of general algebraic index theorems in the framework of deformation quantization; doing so they show the existence of a "formal trace" (for X symplectic of dimension 2ℓ) given by $\mathrm{Tr}_{\mathrm{NT}}(u) = \frac{1}{\ell!\nu^\ell} \int_X (u\omega^\ell + \nu\tau_1(u) + \nu^2\tau_2(u) + \cdots)$ where the τ_k are local expressions in u. That trace satisfies $\mathrm{Tr}_{\mathrm{NT}}(u*v - v*u) = 0$; thus the integrand will give an equivalence, over $\mathbb{K}[\nu^{-1}, \nu]]$, between any given n.c. star product and a strongly closed one.

Cyclic cohomology is based on a bicomplex containing a Hochschild complex with coboundary operator b of degree 1 and another one with operation B of degree -1 anticommuting with b. For a precise definition and properties, see [41]. The concept does not require to make reference to operator algebras; formulated abstractly, it applies even better to star products algebras provided the star products considered are closed (see [42], where a explanation of cyclic cohomology in this context can be found). Indeed, if $*$ is closed (see Def. 2) and a trace τ is defined on $u = \sum_{r=0}^\infty \nu^r u_r \in N[[\nu]]$ by $\tau(u) = \int u_\ell \omega^\ell$, we can consider the quasi-homomorphism (that measures the noncommutativity of the $*$-algebra and is also a Hochschild 2-cocycle) $\theta(u_1, u_2) = u_1 * u_2 - u_1 u_2$; then $\varphi_{2k}(u_0, \ldots, u_{2k}) = \tau(u_0 * \theta(u_1, u_2) * \ldots * \theta(u_{2k-1}, u_{2k}))$ defines the *components of a cyclic cocycle* φ in the (b, B) bicomplex on N that is called the *character* of the closed star product. In particular $\varphi_{2\ell}(u_0, \ldots, u_{2\ell}) = \int u_0 du_1 \wedge \ldots \wedge du_{2\ell}$. The composition of symbols of pseudodifferential operators is [42] a closed star product, the character of which coincides with that defined by the trace on these operators.

A natural extension of the associative algebra context of noncommutative geometry is to Hopf algebras (in the line of [25]) and this indeed permitted now Connes and Moscovici [43] to compute the index of transversally elliptic operators on foliations, a longstanding problem (which among many other tools required hypoelliptic pseudodifferential operators). Another extension, motivated by physics, is to supersymmetric data, and this has been the subject of recent studies by Fröhlich and coworkers [81], first in the context of usual differential geometry and now in that of noncommutative geometry. There are many more developments in this framework, including quantized space, but we shall not develop these further.

Acknowledgements. I want to thank Giuseppe (= Joseph) Dito and Moshé Flato for very useful comments, and Piotr Rączka and the organizers of PFG98 in Lodz (especially Jakub Rembielinski) for excellent hospitality in Poland.

REFERENCES

1. Abraham R. and Marsden J.E., *Foundations of mechanics*, Benjamin/Cummings Publ. Co., Advanced Book Program, Reading, Mass. (Second edition, 1978).
2. Agarwal G.S. and Wolf E. "Calculus for functions of noncommuting operators and general phase-space methods in quantum mechanics I, II, III", Phys. Rev.**D2** (1970), 2161-2186, 2187-2205, 2206-2225.
3. Angelopoulos E., Flato M., Fronsdal, C. and Sternheimer D. "Massless particles, conformal group and De Sitter universe", Phys. Rev. **D23** (1981), 1278-1289.
4. Arnal D., Cahen M. and Gutt S. "Representations of compact Lie groups and quantization by deformation", Bull. Acad. Royale Belg. **74** (1988), 123-141; "Star exponential and holomorphic discrete series", Bull. Soc. Math. Belg. **41** (1989), 207-227.
5. Arnal D. and Cortet J.C. "Nilpotent Fourier transform and applications", Lett. Math. Phys. **9** (1985), 25-34; "Star-products in the method of orbits for nilpotent Lie groups", J. Geom. Phys. **2** (1985), 83-116; "Représentations star des groupes exponentiels", J. Funct. Anal. **92** (1990), 103-135.
6. Arnal D., Cortet J.C. and Ludwig J. "Moyal product and representations of solvable Lie groups", J. Funct. Anal. **133** (1995), 402-424.
7. Arnal D., Cortet J.C. and Molin P. "Star-produit et représentation de masse nulle du groupe de Poincaré", C.R. Acad. Sci. Paris Sér. A **293**, 309-312 (1981).
8. Arnal D., Cortet J.C., Molin P. and Pinczon G. "Covariance and geometrical invariance in star-quantization", J. Math. Phys. **24** (1983), 276-283.
9. Atiyah M.F. and Singer I.M. "The index of elliptic operators on compact manifolds", Bull. Amer. Math. Soc. **69** (1963) 422-433; "The index of elliptic operators I,III and IV,V", Ann. of Math. **87** (1968) 484-530, 546-604 and **93** (1971), 119-138, 139-149; Atiyah M., Bott R. and Patodi V.K. "On the heat equation and the index theorem", Invent. Math. **19** (1973), 279-330 and **28** (1975), 277-280.
10. Auslander L. and Kostant B. "Polarization and unitary representations of solvable groups", Inv. Math. **14** (1971), 255-354.
11. de Azcárraga J.A., Perelomov A.M. and Pérez Bueno J.C. "New generalized Poisson structures", J. Phys. A **29** (1996), L151–L157.

12. Barr M. "Harrison Homology, Hochschild Homology, and Triples", J. Alg. **8** (1968), 314-323.
13. Basart H. and Lichnerowicz A. "Déformations d'un star-produit sur une variété symplectique". C. R. Acad. Sci. Paris **293** I (1981), 347-350; "Conformal Symplectic Geometry, Deformations, Rigidity and Geometrical (KMS) Conditions", Lett. Math. Phys. **10** (1985), 167-177.
14. Basart H., Flato M., Lichnerowicz A. and Sternheimer D. "Deformation theory applied to quantization and statistical mechanics", Lett. Math. Phys. **8** (1984), 483-494; "Mécanique statistique et déformations", C.R. Acad. Sci. Paris Sér. I **299** (1984), 405-410.
15. Bayen F. and Flato M. "Remarks concerning Nambu's generalized mechanics", Phys. Rev. **D11** (1975), 3049-3053.
16. Bayen F., Flato M., Fronsdal C., Lichnerowicz A. and Sternheimer D. "Quantum mechanics as a deformation of classical mechanics". Lett. Math. Phys. **1** (1977), 521-530.
17. Bayen F., Flato M., Fronsdal C., Lichnerowicz A. and Sternheimer D. "Deformation theory and quantization I, II", Ann. Phys. (NY) (1978) **111**, 61-110, 111-151.
18. Berezin F. A. "General concept of quantization", Comm. Math. Phys. **40** (1975), 153-174; "Quantization", Izv. Akad. Nauk SSSR Ser. Mat. **38** (1974), 1116-1175; "Quantization in complex symmetric spaces", Izv. Akad. Nauk SSSR Ser. Mat. **39** (1975), 363-402, 472. [English translations: Math. USSR-Izv. **38** 1109-1165 (1975) and **39**, 341–379 (1976)].
19. Berezin F. A. and Šubin M. A. "Symbols of operators and quantization" in: *Hilbert space operators and operator algebras* (Proc. Internat. Conf., Tihany, 1970), Colloq. Math. Soc. Janos Bolyai **5**, 21-52. North-Holland, Amsterdam (1972).
20. Bertelson M., Cahen M. and Gutt S. "Equivalence of star products", Classical and Quantum Gravity **14** (1997), A93-A107.
21. Bertrand J. and Bertrand P. "Symbolic calculus on the time-frequency half-plane", J. Math. Phys. **39** (1998), 4071-4090.
22. Bidegain F. and Pinczon G. "A star product approach to noncompact quantum groups", Lett. Math. Phys. **33** (1995), 231-240; "Quantization of Poisson-Lie Groups and Applications", Comm. Math. Phys. **179** (1996), 295-332.
23. Bonneau P. "Cohomology and associated deformations for not necessarily coassociative bialgebras", Lett. Math. Phys. **26** (1992), 277-280.
24. Bonneau P. "Fedosov star-products and 1-differentiable deformations", Lett. Math. Phys. (in press; expanded version in `math/980....`).
25. Bonneau P., Flato M., Gerstenhaber M. and Pinczon G. "The hidden group structure of quantum groups: strong duality, rigidity and preferred deformations". Comm. Math. Phys. **161** (1994), 125-156.
26. Bonneau P., Flato M. and Pinczon G. "A natural and rigid model of quantum groups". Lett. Math. Phys. **25** (1992), 75-84.
27. Bordemann M. "On the Deformation Quantization of super-Poisson Brackets", `q-alg/9605038`.

28. Bordemann M., Neumaier N. and Waldmann S., "Homogeneous Fedosov Star Products on Cotangent Bundles" I, II, q-alg/9711016, q-alg/9707030.
29. Bordemann M., Roemer H. and Waldmann S., "A Remark on Formal KMS States in Deformation Quantization", math/9801139.
30. Bordemann M. and Waldmann S. "Formal GNS Construction and States in Deformation Quantization", q-alg/9607019, Comm. Math. Phys. **195** (1998), 549-583.
31. Boutet de Monvel L. "Star products on conic Poisson manifolds of constant rank", Mat. Fiz. Anal. Geom. **2** (1995), 143-151; "Star produit associé à un crochet de Poisson de rang constant", in: *Partial differential equations and functional analysis*, 111-119, Progr. Nonlinear Diff. Eqs. Appl., **22**, Birkhäuser Boston, (1996).
32. Boutet de Monvel L. and Guillemin V. *The spectral theory of Toeplitz operators.* Annals of Mathematics Studies **99**, Princeton University Press (1981).
33. Cadavid C. and Nakashima M. "The star-exponential and path integrals on compact groups", Lett. Math. Phys. **23** (1991), 111-115.
34. Cahen M., Flato M., Gutt S. and D. Sternheimer "Do different deformations lead to the same spectrum?", J. Geom. Phys. **2** (1985), 35-48.
35. Cahen M. and Gutt S. "Regular *-representations of Lie algebras", Lett. Math. Phys. **6** (1982), 395-404.
36. Cahen M., Gutt S. and De Wilde M. "Local cohomology of the algebra of C^∞ functions on a connected manifold", Lett. Math. Phys. **4** (1980), 157-167.
37. Cahen M., Gutt S. and Rawnsley J. "Quantization of Kähler manifolds IV, Lett. Math. Phys. **34** (1995), 159-168.
38. Calderon A.P. and Zygmund A. "Singular integral operators and differential equations", Amer. J. Math. **77** (1957), 901-921.
39. Cartan H. and Schwartz L. *Séminaire Henri Cartan 1963/64*, "*Théorème d'Atiyah-Singer sur l'indice d'un opérateur différentiel elliptique*", fasc. 1 & 2, Secrétariat mathématique, Paris (1965).
40. Chevalley C. and Eilenberg S. "Cohomology theory of Lie groups and algebras", Trans. Amer. Math. Soc. **63** (1948), 85-124.
41. Connes A. *Noncommutative Geometry*, Academic Press, San Diego (1994).
42. Connes A., Flato M. and Sternheimer D. "Closed star-products and cyclic cohomology", Lett. Math. Phys. **24** (1992), 1-12.
43. Connes A. and Moscovici H. "Hopf algebras, cyclic cohomology and the transverse index theorem", preprint math.DG/9806109 (1998).
44. Daubechies I. "On the distributions corresponding to bounded operators in the Weyl quantization", Comm. Math. Phys. **75** (1980), 229-238; Daubechies I. and Grossmann A. "An integral transform related to quantization", J. Math. Phys. **21** (1980), 2080-2090.
45. Daubechies I. "Wavelets and other phase space localization methods", in: *Proceedings of the International Congress of Mathematicians*, (Zürich, 1994), 56-74, Birkhäuser, Basel (1995).
46. Deligne P. "Déformations de l'Algèbre des Fonctions d'une Variété Symplectique: Comparaison entre Fedosov et De Wilde, Lecomte", Selecta Math. N.S. **1** (1995), 667-697.

47. De Wilde M. and Lecomte P.B.A. "Existence of star-products and of formal deformations of the Poisson Lie algebra of arbitrary symplectic manifolds", Lett. Math. Phys. **7** (1983), 487-496; "Star-products on cotangent bundles", Lett. Math. Phys. **7** (1983), 235-241 and **8** (1984), 79; "Existence of star-products on exact symplectic manifolds", Ann. Inst. Fourier (Grenoble) **35** (1985), 117-143; "Existence of star-products revisited", dedicated to the memory of Professor Gottfried Köthe, Note Mat. **10** (1990), suppl. 1, 205-216 (1992); "Formal deformations of the Poisson Lie algebra of a symplectic manifold and star-products. Existence, equivalence, derivations" in *Deformation Theory of Algebras and Structures and Applications* (M. Hazewinkel and M. Gerstenhaber Eds.), NATO ASI Ser. C **247**, 897-960, Kluwer Acad. Publ., Dordrecht (1988).
48. De Wilde M., Lecomte P.B.A. and Gutt S. "A propos des deuxième et troisième espaces de cohomologie de l'algèbre de Lie de Poisson d'une variété symplectique", Ann. Inst. H. Poincaré Sect. A (N.S.) **40** (1984), 77-83.
49. De Wilde M., Lecomte P.B.A. and Mélotte D. "Invariant star-products on symplectic manifolds", J. Geom. Phys. **2** (1985), 121-129; "Invariant cohomology of the Poisson Lie algebra of a symplectic manifold", Comment. Math. Univ. Carolin. **26** (1985), 337-352.
50. Dirac P.A.M. *Lectures on Quantum Mechanics*, Belfer Graduate School of Sciences Monograph Series No. 2 (Yeshiva University, New York, 1964). "Generalized Hamiltonian dynamics", Canad. J. Math. **2** (1950), 129-148.
51. Dito J. "Star-product approach to quantum field theory: the free scalar field", Lett. Math. Phys. **20** (1990), 125-134; "Star-products and nonstandard quantization for Klein-Gordon equation", J. Math. Phys. **33** (1992), 791-801; "An example of cancellation of infinities in star-quantization of fields", Lett. Math. Phys. **27** (1993), 73-80.
52. Dito G. and Flato M. "Generalized Abelian Deformations: Application to Nambu Mechanics", Lett. Math. Phys. **39** (1997), 107-125.
53. Dito G., Flato M., Sternheimer D. and Takhtajan L. "Deformation Quantization and Nambu Mechanics", Comm. Math. Phys. **183** (1997), 1-22.
54. Dixmier J. "Existence de traces non normales", C.R. Acad. Sci. Paris Sér. A **262** (1966), 1107-1108.
55. Drinfeld V.G. "Quantum Groups", in: *Proc. ICM86, Berkeley*, **1**, 101-110, Amer. Math. Soc., Providence (1987); "Quasi-Hopf algebras", Leningrad Math. J. **1** (1990), 1419-1457; "On almost co-commutative Hopf algebras", *ibid.*, 321-431.
56. Du Cloux F. "Extensions entre représentations unitaires irréductibles des groupes de Lie nilpotents", Astérisque **125** (Soc. Math. Fr. 1985), 129-211.
57. Etingof P. and Kazhdan D. "Quantization of Lie Bialgebras" I. Selecta Math. (N.S.) **2** (1996), 1-41 (q-alg/9506005); II. q-alg/9701038; III. q-alg/9610030; IV math/9801043; "Quantization of Poisson algebraic groups and Poisson homogeneous spaces" q-alg/9510020.
58. Faddeev L. D. "On the relation between mathematics and physics". in: *Integrable systems* (Tianjin, 1987), 3–9, Nankai Lectures Math. Phys., World Sci. Pub. (1990).
59. Faddeev L.D., Reshetikhin N.Yu. and Takhtajan L.A. "Quantization of Lie groups and Lie algebras", Leningrad Math. J. **1** (1990), 193-225.

60. Fedosov B.V. "Formal quantization" in *Some topics of modern mathematics and their applications to problems of mathematical physics*, 129-139 (Moscow, 1985); "Quantization and index", Dokl. Akad. Nauk SSSR **291**, 82-86 (1986).
61. Fedosov B.V. "A simple geometrical construction of deformation quantization", J. Diff. Geom. **40** (1994), 213-238.
62. Fedosov B.V. "Analytical formulas for the index of elliptic operators", Trans. Moscow Math. Soc. **30** (1974), 159-240; "An index theorem in the algebra of quantum observables, Dokl. Akad. Nauk SSSR **305** (1989), 835-838; "The Index Theorem for Deformation Quantization", in: *Boundary Value Problems, Schrödinger Operators, Deformation Quantization*, 206-318, Math. Top. **8**, Akademie Verlag, Berlin (1995); "A trace density in deformation quantization", *ibid.*, 319-333;
63. Fedosov B.V. *Deformation Quantization and Index Theory*, Mathematical Topics **9**, Akademie Verlag, Berlin (1996).
64. Fedosov B.V. "Non-abelian reduction in deformation quantization", Lett. Math. Phys. **43** (1998), 137-154; "Reduction and eigenstates in deformation quantization", in: *Pseudo-differential calculus and mathematical physics*, 277-297, Math. Top. **5**, Akademie Verlag, Berlin (1994).
65. Filippov V.T. "n-Lie algebras", Siberian Math. J. **26** (1985) 875–879.
66. Flato, M. "Deformation view of physical theories", Czechoslovak J. Phys. **B32** (1982), 472-475.
67. Flato, M. "Two Disjoint Aspects of the Deformation Programme: Quantizing Nambu Mechanics; Singleton Physics", these Proceedings.
68. Flato M., Dito G. and Sternheimer D. "Nambu mechanics, n-ary operations and their quantization" in *Deformation theory and symplectic geometry, Proceedings of Ascona meeting, June 1996* (D. Sternheimer, J. Rawnsley and S. Gutt, Eds.), Math. Physics Studies **20**, 43-66, Kluwer Acad. Publ., Dordrecht (1997).
69. Flato M. and Fronsdal C. "One massless particle equals two Dirac singletons", Lett. Math. Phys. **2** (1978), 421-426.
70. Flato M. and Fronsdal C. "Composite Electrodynamics". J. Geom. Phys. **5** (1988), 37-61; "Interacting Singletons", Lett. Math. Phys. **44** (1998), 249-259.
71. Flato M., Lichnerowicz A. and Sternheimer D. "Déformations 1-différentiables d'algèbres de Lie attachées à une variété symplectique ou de contact", C.R. Acad. Sci. Paris Sér. A **279** (1974), 877-881 and Compositio Mathematica, **31** (1975), 47-82; "Algèbres de Lie attachées à une variété canonique", J. Math. Pures Appl. **54** (1975), 445-480.
72. Flato M., Lichnerowicz A. and Sternheimer D. "Deformations of Poisson brackets, Dirac brackets and applications". J. Math. Phys. **17** (1976), 1754-1762.
73. Flato M., Lichnerowicz A. and Sternheimer D. "Crochets de Moyal-Vey et quantification", C.R. Acad. Sci. Paris Sér. A **283** (1976), 19-24.
74. Flato M., Pinczon G. and Simon J. "Non-linear representations of Lie groups", Ann. Sci. Éc. Norm. Sup. (4) **10** (1977), 405-418.
75. Flato M. and Simon J., "Non-linear wave equations and covariance", Lett. Math. Phys. **2** (1977), 115-160.
76. Flato M., Simon J., Snellman H. and Sternheimer D. "Simple facts about analytic vectors and integrability", Ann. Sci. Éc. Norm. Sup. (4) **5** (1972), 432-434.

77. Flato M., Simon J.C.H. and Taflin E. *The Maxwell-Dirac equations: the Cauchy problem, asymptotic completeness and the infrared problem*, Memoirs of the American Mathematical Society **127** (number 606, May 1997).
78. Flato M. and Sternheimer D. "Deformations of Poisson brackets, separate and joint analyticity in group representations, nonlinear group representations and physical applications." in: *Lectures at the Advanced Summer Institute on Harmonic Analysis* (Liège 1977), Mathematical Physics and Applied Mathematics **5**, 385-448. D. Reidel, Dordrecht (1980).
79. Flato M. and Sternheimer D. "Closedness of star products and cohomologies", in: *Lie Theory and Geometry: In Honor of B. Kostant* (J.L. Brylinski et al., eds.), 241-259. Progress in Mathematics, Birkhäuser, Boston (1994).
80. Flato M. and Sternheimer D. "Topological Quantum Groups, star products and their relations", St. Petersburg Mathematical Journal (Algebra and Analysis) **6** (1994), 242-251.
81. Fröhlich J., Grandjean O. and Recknagel A. "Supersymmetric quantum theory and differential geometry", Comm. Math. Phys. **193** (1998), 527-594; "Supersymmetric quantum theory and non-commutative geometry", `math-ph/9807006` (1998).
82. Fronsdal C. "Some ideas about quantization", Rep. Math. Phys. **15** (1978), 111-145.
83. Garcia-Compean H., Plebański J. and Przanowski M. "Geometry Associated with Self-dual Yang-Mills and the Chiral Model Approaches to Self-dual Gravity", Acta Phys.Polon. **B29** (1998), 549-571.
84. Gautheron Ph. "Some remarks concerning Nambu mechanics", Lett. Math. Phys. **37** (1996), 103-116; "Simple facts concerning Nambu algebras", Comm. Math. Phys. **195** (1998), 417-434.
85. Gel'fand I.M., Kalinin D.I. and Fuks D.B. "The cohomology of the Lie algebra of Hamiltonian formal vector fields", Funkcional. Anal. i Priložen. **6** (1972), 25-29.
86. Gerstenhaber M. "On the deformation of rings and algebras", Ann. Math. **79** (1964), 59-103; and (IV), *ibid.* **99** (1974), 257-276.
87. Gerstenhaber M. and Schack S.D. "Algebraic cohomology and deformation theory", in *Deformation Theory of Algebras and Structures and Applications* (M. Hazewinkel and M. Gerstenhaber Eds.), NATO ASI Ser. C **247**, 11-264, Kluwer Acad. Publ., Dordrecht (1988).
88. Gerstenhaber M. and Schack S.D. "Bialgebra cohomology, deformations and quantum groups", Proc. Nat. Acad. Sci. USA **87** (1990), 478-481; "Algebras, bialgebras, quantum groups and algebraic deformations", in *Deformation Theory and Quantum Groups with Applications to Mathematical Physics* (M. Gerstenhaber and J. Stasheff, eds.), Contemporary Mathematics **134**, 51-92, American Mathematical Society, Providence (1992).
89. Gilkey P. *Invariance theory, the heat equation and the Atiyah-Singer index theorem*, Second edition, Studies in Advanced Mathematics, CRC Press, Boca Raton FL (1995).
90. Ginzburg G. and Kapranov M. "Koszul duality for operads", Duke Math. J. **76** (1994), 203-272.
91. Gnedbaye A.V. "Les algèbres k-aires et leurs opérades", C. R. Acad. Sci. Paris Sér.I **321** (1995), 147-152.

92. Gracia-Bondía J.M. and Várilly J.C. "From geometric quantization to Moyal quantization", J. Math. Phys. **36** (1995), 2691-2701.
93. Gutt S. "An explicit $*$-product on the cotangent bundle to a Lie group", Lett. Math. Phys. **7** (1983), 249-258.
94. Gutt S. "On some second Hochschild cohomology spaces for algebras of functions on a manifold", Lett. Math. Phys. **37** (1997), 157-162.
95. Groenewold A. "On the principles of elementary quantum mechanics", Physica **12** (1946), 405-460.
96. Grossman A., Loupias G. and Stein E.M. "An algebra of pseudodifferential operators and quantum mechanics in phase space", Ann. Inst. Fourier Grenoble **18** (1968), 343-368.
97. Guillemin V. "Star products on compact pre-quantizable symplectic manifolds". Lett. Math. Phys. **35** (1995), 85-89.
98. Guillemin V. and Sternberg S. *Symplectic techniques in physics*. Cambridge University Press (1984).
99. Gutt S. and Rawnsley J. "Equivalence of star products on a symplectic manifold: an introduction to Deligne's Čech cohomology classes", preprint (June 1998).
100. Hansen F. "The Moyal product and spectral theory for a class of infinite-dimensional matrices", Publ. RIMS, Kyoto Univ., **26** (1990), 885-933; "Quantum Mechanics in Phase Space", Reports On Math. Phys. **19** (1984), 361-381.
101. Hörmander L. "Pseudo-differential operators and non-elliptic boundary problems", Ann. Math. **83** (1966), 129-209. "Fourier Integral Operators I", Acta Math. **127** (1971), 79-183.
102. Hörmander L. "The Weyl calculus of pseudodifferential operators", Comm. Pure Appl. Math. **32** (1979), 360-444; "Symbolic calculus and differential equations" in *18th Scandinavian Congress of Mathematicians (Aarhus, 1980)*, 56-81, Progr. Math. **11**, Birkhäuser Boston (1981).
103. Hochschild G., Kostant B. and Rosenberg A. "Differential forms on regular affine algebras", Trans. Am. Math. Soc. **102** (1962), 383-406.
104. Inonü E. and Wigner E.P. "On the contraction of groups and their representations", Proc. Nat. Acad. Sci. U. S. A. **39** (1953), 510-524.
105. Jimbo M. "A q-difference algebra of $\mathcal{U}(\mathfrak{g})$ and the Yang-Baxter equation", Lett. Math. Phys. **10** (1985), 63-69.
106. Jurzak J.P. *Unbounded noncommutative integration*, Mathematical Physics Studies **7**, D. Reidel Publ. Co., Dordrecht (1985).
107. Kammerer J.B. "Analysis of the Moyal product in a flat space", J. Math. Phys. **27** (1986), 529-535.
108. Karabegov, A. V. "Cohomological classification of deformation quantizations with separation of variables", Lett. Math. Phys. **43** (1998), 347-357; "Berezin's quantization on flag manifolds and spherical modules", Trans. Amer. Math. Soc. **350** (1998), 1467-1479.
109. Kirillov A.A. *Elements of the theory of representations*, Springer, Berlin (1976).
110. Kodaira K. and Spencer D.C. "On deformations of complex analytic structures" I, II, Ann. of Math. **67** (1958), 328-466; III "Stability theorems for complex structures" *ibid.* **71** (1960), 43-76.

111. Kontsevich M. "Formality conjecture", in: *Deformation Theory and Symplectic Geometry*, 139-156, Math. Phys. Stud. **20**, Kluwer Acad. Publ., Dordrecht (1997).
112. Kontsevich M. "Deformation quantization of Poisson manifolds, I" `q-alg/9709040` and private communications; "Deformation quantization", Arbeitstagung (1997).
113. Kostant B. *Quantization and unitary representations*, in: Lecture Notes in Math. **170**, 87-208, Springer Verlag, Berlin (1970); "On the definition of quantization", in: *Géométrie symplectique et physique mathématique* (Colloq. Int. CNRS, N° 237), 187-210, Éds. CNRS, Paris (1975); "Graded manifolds, graded Lie theory, and prequantization" in *Differential geometrical methods in mathematical physics*, 177-306, Lecture Notes in Math. **570**, Springer, Berlin (1977).
114. Kostant B. "Symplectic spinors", Symposia Mathematica **14** (1974), 139-152.
115. Krée P. and Rączka R. "Kernels and symbols of operators in quantum field theory", Ann. Inst. H. Poincaré Sect A (N.S.) **28** (1978), 41-73.
116. Kulish P.P. and Reshetikhin N.Yu. "Quantum linear problem for the sine-Gordon equation and higher representations". Zap. Nauch. Sem. LOMI **101** (1981), 101-110 (English translation in Jour. Sov. Math. **23** (1983), 24-35).
117. Lévy-Nahas M. "Deformations and contractions of Lie algebras", J. Math. Phys. **8** (1967), 1211-1222. "Déformations du groupe de Poincaré", in: *L'extension du groupe de Poincaré aux symétries internes des particules élémentaires* (Colloque Int. CNRS N° 159), 25-45, Éds CNRS, Paris (1968).
118. Libermann P. and Marle C.-M. *Symplectic geometry and analytical mechanics*, Mathematics and its Applications **35**. D. Reidel Publ. Co., Dordrecht (1987).
119. Lichnerowicz A. "Cohomologie 1-différentiable des algèbres de Lie attachées à une variété symplectique ou de contact", J. Math. Pures Appl. **53** (1974), 459-483.
120. Lichnerowicz A. "Les variétés de Poisson et leurs algèbres de Lie associées", J. Diff. Geom. **12** (1977), 253-300.
121. Lichnerowicz A. "Déformations d'algèbres associées à une variété symplectique (les $*_\nu$-produits)", Ann. Inst. Fourier, Grenoble, **32** (1982), 157-209.
122. Loday J.-L. "La renaissance des opérades", *Séminaire Bourbaki*, Exposé 792 (Novembre 1994).
123. Maillard J.M. "On the twisted convolution product and the Weyl transform of tempered distributions", J. Geom. Phys. **3** (1986), 231-261.
124. Marsden, J. E. and Weinstein, A. "Reduction of symplectic manifolds with symmetry', Rep. Math. Phys. **5** (1974), 121-130.
125. Moreno C. "Invariant star products and representations of compact semi-simple Lie groups", Lett. Math. Phys. **12** (1986), 217-229.
126. Moyal J.E. "Quantum mechanics as a statistical theory", Proc. Cambridge Phil. Soc. **45** (1949), 99-124.
127. Mukunda, N. and Sudarshan, E.C.G. "Relation between Nambu and Hamiltonian mechanics", Phys. Rev. **D 13** (1976), 2846-2850.
128. Nadaud F. "Generalized deformations, Koszul resolutions, Moyal Products", Reviews Math. Phys. **10** (5) (1998).
129. Nambu Y. "Generalized Hamilton dynamics", Phys. Rev **D7** (1973), 2405-2412.
130. Neroslavsky O.M. and Vlasov A.T. "Sur les déformations de l'algèbre des fonctions d'une variété symplectique", C.R. Acad. Sc. Paris Sér. I **292** (1981), 71-76.

131. Nest R. and Tsygan B. "Algebraic Index Theorem", Comm. Math. Phys. **172** (1995), 223-262; "Algebraic Index Theorem For Families", Advances in Mathematics **113** (1995), 151-205; "Formal deformations of symplectic manifolds with boundary", J. Reine Angew. Math. **481** (1996), 27-54.
132. Nijenhuis A. "Jacobi-type identities for bilinear differential concomitants of certain tensor fields. I, II ", Indag. Math. **17** (1955), 390-397, 398-403.
133. Oesterlé J. "Quantification formelle des variétés de Poisson [d'après Maxim Kontsevich]", Séminaire Bourbaki, exposé 843 (mars 1998).
134. Omori H., Maeda Y. and Yoshioka A. "Weyl manifolds and deformation quantization", Adv. in Math. **85** (1991), 225-255; "Existence of a closed star product", Lett. Math. Phys. **26** (1992), 285-294;
135. Omori H., Maeda Y. and Yoshioka A. "Deformation Quantizations of Poisson Algebras", Contemporary Mathematics Amer. Math. Soc. **179** (1994), 213-240 and Proc. Japan Acad. **68** (1992), 97-101; "Deformation quantizations of the Poisson algebra of Laurent polynomials", Lett. Math. Phys. (in press).
136. Palais R.S. *Seminar on the Atiyah-Singer index theorem*, Annals of Mathematics Studies No. 57, Princeton University Press (1965).
137. Pankaj Sharan "Star-product representation of path integrals", Phys. Rev. D **20** (1979), 414-418.
138. Pinczon G. "On the equivalence between continuous and differential deformation theories", Lett. Math. Phys. **39** (1997), 143-156.
139. Pinczon G. "Non commutative deformation theory", Lett. Math. Phys. **41** (1997), 101-117.
140. Plebański J. "Naviasy Poissona i komutatory", Torun preprint Nr 69 (1969).
141. Rieffel M. "Questions on quantization", `quant-ph/9712009`, to be published in *Proceedings of the International Conference on Operator Algebras and Operator Theory* (Shanghai, July 1997).
142. Rubio R. "Algèbres associatives locales sur l'espace des sections d'un fibré en droites", C.R. Acad. Sci. Paris Sér. I **299** (1984), 699-701.
143. Saletan E.J. "Contraction of Lie groups", J. Math. Phys. **2** (1961), 1-21 and 742.
144. Schmid W., "Character formulas and localization of integrals", in *Deformation theory and symplectic geometry, Proceedings of Ascona meeting, June 1996* (D. Sternheimer, J. Rawnsley and S. Gutt, Eds.), Math. Physics Studies **20**, 259-270, Kluwer Acad. Publ., Dordrecht (1997).
145. Schouten J.A. "On the differential operators of first order in tensor calculus", Convegno Internazionale di Geometria Differenziale, Italia, 1953, 1-7, Edizioni Cremonese, Roma (1954).
146. Segal I.E. "A class of operator algebras which are determined by groups", Duke Math. J. **18** (1951), 221-265.
147. Segal I.E. "Symplectic structures and the quantization problem for wave equations", Symposia Mathematica **14** (1974), 79-117.
148. Shnider S. and Sternberg S. *Quantum Groups*, Graduate Texts in Mathematical Physics vol. II, International Press (Boston and Hong-Kong, 1993). [With an extended reference list of 1264 items!].

149. Souriau, J.-M. *Structure of dynamical systems. A symplectic view of physics.* Progress in Mathematics **149**. Birkhäuser Boston (1997): Edited translation from the French *Structure des systèmes dynamiques*, Dunod, Paris (1970); "Des particules aux ondes: quantification géométrique" in: *Huygens' principle 1690–1990: theory and applications* (The Hague and Scheveningen, 1990), 299–341, Stud. Math. Phys. **3**, North-Holland, Amsterdam (1992); "La structure symplectique de la mécanique décrite par Lagrange en 1811", Math. Sci. Humaines **94** (1986), 45-54.
150. Stasheff J. "Deformation theory and the Batalin-Vilkovisky master equation", in *Deformation theory and symplectic geometry (Ascona, 1996)*, 271-284, Math. Phys. Stud. **20**, Kluwer Acad. Publ., Dordrecht (1997).
151. Takhtajan L.A. *Introduction to quantum groups* in: Springer Lecture Notes in Physics, **370**, 3-28 (1990); *Lectures on quantum groups* in: Nankai Lectures on Math. Phys. (M. Ge and B. Zhao eds.) 69-197, World Scientific, Singapore (1990).
152. Takhtajan L.A. "On foundation of the generalized Nambu mechanics", Comm. Math. Phys. **160** (1994), 295-315.
153. Tsygan B. "Homology of matrix Lie algebras over rings and Hochschild homology", Uspekhi Math. Nauk **38** (1983), 217-218.
154. Unterberger A. and Upmeier H. *Pseudodifferential analysis on symmetric cones*, Studies in Advanced Mathematics, CRC Press, Boca Raton FL (1996); "The Berezin transform and invariant differential operators", Comm. Math. Phys. **164** (1994), 563-597; Unterberger A. and J. "Quantification et analyse pseudodifférentielle", Ann. Sci. Éc. Norm. Sup. (4) **21** (1988), 133-158.
155. Vey J. "Déformation du crochet de Poisson sur une variété symplectique", Comment. Math. Helv. **50** (1975), 421-454.
156. Voronov A. "Quantizing Poisson manifolds", `q-alg/9701017`.
157. Voros A. *Développements semi-classiques*, Thèse (Orsay et Saclay), mai 1977. "Semiclassical approximations", Ann. Inst. Henri Poincaré **24** (1976), 31-90.
158. Waldman S. "A Remark on Non-equivalent Star Products via Reduction for $\mathbb{CP}$" `math/9802078`.
159. Weinstein A. "Deformation quantization", Séminaire Bourbaki, exposé 789 (juin 1994), Astérisque **227**, 389-409.
160. Weinstein A. "The modular automorphism group of a Poisson manifold", J. Geom. Phys. **23** (1997), 379-394.
161. Weyl, H. *The theory of groups and quantum mechanics*, Dover, New-York (1931), translated from *Gruppentheorie und Quantenmechanik*, Hirzel Verlag, Leipzig (1928); "Quantenmechanik und Gruppentheorie", Z. Physik **46** (1927), 1-46.
162. Wigner, E.P. "Quantum corrections for thermodynamic equilibrium", Phys. Rev. **40** (1932), 749-759.
163. Woodhouse N. M. J. *Geometric quantization.* Oxford Science Publications, Oxford Mathematical Monographs. Oxford University Press, New York (2nd edition, 1992).
164. Woronowicz S.L. "Compact matrix pseudogroups", Commun. Math. Phys. **111** (1987), 613-665; "Differential calculus on compact matrix pseudogroups (quantum groups)", Commun. Math. Phys. **160** (1989), 125-170; "Quantum E(2) group and its Pontryagin dual", Lett. Math. Phys. **23** (1991), 251-263.

q-Derivatives, Quantization Methods and q-Algebras

Reidun Twarock[1]

Abstract. Using the example of Borel quantization on S^1, we discuss the relation between quantization methods and q-algebras. In particular, it is shown that a q-deformation of the Witt algebra with generators labeled by $\mathbb{Z}$ is realized by q-difference operators. This leads to a discrete quantum mechanics. Because of $\mathbb{Z}$, the discretization is equidistant. As an approach to a non-equidistant discretization of quantum mechanics one can change the Witt algebra using not the number field $\mathbb{Z}$ as labels but a quadratic extension of $\mathbb{Z}$ characterized by an irrational number τ. This extension is denoted as quasi-crystal Lie algebra, because this is a relation to one-dimensional quasicrystals. The q-deformation of this quasicrystal Lie algebra is discussed. It is pointed out that quasicrystal Lie algebras can be considered also as a "deformed" Witt algebra with a "deformation" of the labeling number field. Their application to the theory is discussed.

INTRODUCTION

The use of q-algebras and q-derivatives to construct a discrete quantum mechanics on S^1 has been discussed in connection with Borel quantization in [1–3] or independent of quantization procedures e.g. in [4,5]. These approaches make use of the fact that a discretization of S^1 can be achieved in terms of roots of unity and they are hence leading to an equidistant discretization of S^1. The essential ingredient in the construction is a q-deformed version of the Witt algebra $W^{\mathbb{Z}}$ with generators labeled by the number field $\mathbb{Z}$, where the deformation parameter q is itself a root of unity.

After a review of the above results, we address the question, if also a non-equidistant, e.g. aperiodic discretization, is possible. In particular, we present another change of the Witt algebra $W^{\mathbb{Z}}$ by using another number field $\mathbb{Z}[\tau]$ with τ an irrational number. This "deformation" of the Witt algebra is referred to as QUASICRYSTAL LIE ALGEBRAS $W^{\mathbb{Z}[\tau]}$. They resemble the Witt algebra, but are linked to an aperiodic set, also called quasicrystal or cut and project quasicrystal. The possibility of a link of Borel quantization to an aperiodically discretized quantum mechanics based on quasicrystal Lie algebras is discussed.

---

[1] supported by the Land of Lower Saxony in the framework of the Dorothea-Erxleben program

REVIEW OF BOREL QUANTIZATION ON S^1

In this section the main ideas of Borel quantization on the configuration space S^1 are briefly summarized according to [6].

For a system localized and moving on the classical configuration space S^1 (with parametrization $\phi \in [0, 2\pi)$), the position and momentum observables are modeled by the **kinematical algebra**

$$\mathcal{S}(S^1) = C^\infty(S^1, \mathbb{R}) \mathbin{\textcircled{s}} \mathrm{Vect}(S^1), \qquad (1)$$

which is given by smooth functions $f \in C^\infty(S^1, \mathbb{R})$ and smooth vector fields $\tilde{X}(\phi) = X(\phi)\frac{d}{d\phi}$ with $X \in C^\infty(S^1, \mathbb{R})$. A realization of this object in the set $SA(\mathcal{H})$ of self-adjoint operators in the Hilbert space $L^2(S^1, d\phi)$ of $\mathbb{C}$-valued wave functions ψ together with some more technical physically motivated assumptions leads to the **quantum Borel kinematics** on S^1:

$$\begin{aligned}(Q(f)\psi)(\phi) &= f(\phi)\psi(\phi) \\ (P(\tilde{X})\psi)(\phi) &= \left(-i\tilde{X}\right) + \left(-\frac{1}{2}i + D\right)\left(\mathrm{div}\tilde{X}\right) + \omega\left(\tilde{X}\right)\psi(\phi).\end{aligned} \qquad (2)$$

where $\omega = \theta d\phi$ with $\theta \in [0,1)$ denotes a closed one-form. Unitarily inequivalent quantum kinematics are labeled by

$$(\theta, D) \in \pi_1^*(S^1) \times \mathbb{R}. \qquad (3)$$

The connection with the inhomogeneous Witt algebra becomes apparent if one introduces the coordinates $z := \exp(i\phi)$ and expresses $f(\phi)$ and $X(\phi)$ via a Fourier expansion as

$$\begin{aligned} f(\phi) &= \sum_{n=-\infty}^{\infty} f_n z^n \\ X(\phi) &= \sum_{n=-\infty}^{\infty} X_n z^n \end{aligned} \qquad (4)$$

with $f_n = \bar{f}_{-n}$ and $X_n = \bar{X}_{-n}$. Then one obtains:

$$\begin{aligned} Q(f) &= \sum_{n=-\infty}^{\infty} f_n z^n \\ P(X) &= \sum_{n=-\infty}^{\infty} X_n z^n \left(z\frac{d}{dz} + \frac{n}{2} + \theta + iDn\right). \end{aligned} \qquad (5)$$

With the notation

$$\begin{aligned} T_n &= z^n \\ L_n^\theta &= z^n \left(z\frac{d}{dz} + \frac{n}{2} + \theta\right) \end{aligned} \qquad (6)$$

(5) can be expressed as

$$Q(f) = \sum_{n=-\infty}^{\infty} f_n T_n$$
$$P(X) = \sum_{n=-\infty}^{\infty} X_n \left(L_n^\theta + iDnT_n \right).$$
(7)

For fixed $\theta \in \mathbb{R}$, i. e. for a particular quantization, the generators T_n and $L_n \equiv L_n^\theta$ fulfill the commutation relations:

$$\begin{aligned} [T_m, T_n] &= 0 \\ [L_n, T_m] &= mT_{m+n} \\ [L_m, L_n] &= (n-m)L_{m+n} \end{aligned}$$
(8)

and are thus generators of the inhomogeneous Witt algebra.

A corresponding dynamics can be obtained via [7]. An application to S^1 leads to the Schrödinger equation

$$i\partial_t \psi = -\frac{1}{2}\frac{d}{d\phi}^2 \psi - i\theta \frac{d}{d\phi}\psi + i\frac{D}{2\rho}\left(\frac{d}{d\phi}^2 \rho\right)\psi + \underbrace{\Re G[\bar\psi,\psi]\psi}_{R[\psi]},$$
(9)

in which the real part $R[\psi]$ of the nonlinear term $G[\bar\psi,\psi]$ remains undetermined by the formalism.

The results in this section lead to a deformed version of the quantum kinematics, if a q-deformed version of the Witt algebra as introduced in the next section, is implemented.

q-DEFORMATION OF THE WITT ALGEBRA $W^{\mathbb{Z}}$

In order to replace the derivative $z\partial_z$ by q-difference operators which satisfy an algebraic relation $W_q^{\mathbb{Z}}$ that in the limit $q \to 1$ reduces to the Witt algebra $W^{\mathbb{Z}}$, we need some additional assumptions, called q-assumptions, since the process of q-deformation is not unique:

- $W_q^{\mathbb{Z}}$ is a Lie algebra
- $W_q^{\mathbb{Z}}$ introduces a minimum of extra parameters
- $W_q^{\mathbb{Z}}$ is realized via q-difference operators

The physical motivation for the assumptions are:

- Stay as close as possible at the undeformed structures
- Mathematically necessary extra parameters are assumed to be more likely to have a physical interpretation.

- Discretization

It is shown in [1,2] that the following generators form a Lie algebra $W_q^{\mathbb{Z}}$ with the required properties:

$$L_n = z^n \left(z\frac{d}{dz} + \frac{n}{2} + \theta \right) \Rightarrow \mathcal{L}_n^j := z^n \frac{\left[j\left(z\partial_z + \frac{n}{2} + \theta \right) \right]}{[j]} \tag{10}$$

where $[a] := \dfrac{q^a - q^{-a}}{q - q^{-1}}$ with $\lim_{q \to 1}[a] = a$ denotes a q-number.

q-QUANTUM MECHANICS

On $L^2(S^1, d\phi)$ [1,2] the above deformation has led to the **q-quantum kinematics**:

$$\begin{aligned} Q_q(f) &= \sum_{n=-\infty}^{\infty} f_n T_n \ (= Q(f)) \\ P_q^j(X) &= \sum_{n=-\infty}^{\infty} X_n \left(\mathcal{L}_n^{(j)} \theta + i \frac{[jn]}{[j]} DT_n \right). \end{aligned} \tag{11}$$

which has been shown to be self-adjoint in [2].

A dynamics compatible with the q-quantum kinematics can be derived via a deformed version of the first Ehrenfest relation (see [1,2]. One obtains a family of nonlinear q-Schrödinger equations:

$$\begin{aligned} (i\partial_t \psi)\bar{\psi} &= \frac{1}{[j]} \left(\frac{[jz\partial_z]}{[j]} \right) \left[\frac{z\partial_z}{2} \right] \psi \right) S\bar{\psi} \\ &\quad - \frac{i}{[j]} \left[j\frac{\theta}{2} \right] \left\{ q^{-\epsilon\frac{1}{2}\theta} \left(i\frac{[jz\partial_z]}{[j]} q^{-\epsilon\frac{1}{2}z\partial_z} \psi \right) \right. \\ &\quad \left. + q^{\epsilon\frac{1}{2}\theta} \left(i\frac{[jz\partial_z]}{[j]} q^{\epsilon\frac{1}{2}z\partial_z} \psi \right) q^{-\epsilon j z\partial_z} \right\} S\bar{\psi} + F_{NL} \end{aligned} \tag{12}$$

with $S = \frac{1}{2}\left(q^{jz\partial_z} + q^{-jz\partial_z} \right)$. F_{NL} denotes a nonlinear term which is fixed in dependence on the shifts used in the ansatz for a nonlinear Schrödinger equation.

THE LIMIT $q \to 1$

The limit $q \to 1$ is of particular interest, because it allows to deduce information about the undeformed situation **from** the deformed results. A detailed discussion of this issue is presented in [1,2]. The linear part and the imaginary part of the nonlinearity of the Schrödinger equation derived from Borel quantization without deformation are recovered. **In addition**, information on the real part of the nonlinearity is gained.

For S^1 one obtains (' denotes $\partial_\phi = iz\partial_z$, $\psi = \psi(\phi, t)$):

$$i\partial_t\psi = -\frac{1}{2}\psi'' - i\theta\psi' + i\frac{D}{2\rho}\rho''\psi + \Re G[\bar\psi, \psi]\psi \tag{13}$$

with three types of real parts $\Re G[\bar\psi, \psi]$, depending on the shifts $R = R_1 + iR_2$:

1. If $R_1 = 0$ or $R_2 = 0$ $A\dfrac{\psi''\bar\psi' - \bar\psi''\psi'}{\psi\bar\psi' - \bar\psi\psi'}$
($A \in \mathbb{R}$).

2. If $R_1 \neq 0$ and $R_2 \neq 0$ $B\dfrac{D\rho''}{2\rho}$
($B \in \mathbb{R}$).

3. For R proportional to the identity (trivial shift operator) $\Re G[\bar\psi, \psi]$ remains undetermined.

Correspondingly, the result derived via (undeformed) Borel quantization is reproduced correctly and **in addition** information of the real part of the nonlinearity is obtained.

qc-DEFORMATION OF THE WITT ALGEBRA

As an alternative to the q-deformed Witt algebra presented above, quasicrystal Lie algebras are introduced. The deformation process in not a q-deformation in the sense of Gerstenhaber or quantum groups, but rather consists in a replacement of the indexing set $\mathbb{Z}$ of the usual Witt algebra by an aperiodic lattice Σ, where Σ can be viewed as a cut and project quasicrystal, or quasicrystal in short.

The structure of one-dimensional quasicrystal by a cut and project method is given by the following construction: we start with the quadratic extension $\mathbb{Q}[\sqrt{5}]$ of the rational number field $\mathbb{Q}$ and its ring of integers

$$\mathbb{Z}[\tau] := \{a + \tau b | a, b \in \mathbb{Z}\}, \tag{14}$$

where τ, given by $\tau := \frac{1}{2}(1 + \sqrt{5})$, is an irrational number. Since $\mathbb{Z}[\tau]$ is dense, we need some operation to select a discrete set in $\mathbb{Z}[\tau]$. For this we use the automorphism

$$* : c + \sqrt{5}d \mapsto c - \sqrt{5}d \tag{15}$$

with $c, d \in \mathbb{Q}$ which – acting on τ – gives $\tau^* := \frac{1}{2}(1 - \sqrt{5})$. It is important to note for later purposes that τ and τ^* fulfill the relation

$$\tau + \tau^* = 1 \text{ and } \tau\tau^* = -1. \tag{16}$$

Now we select those elements of $\mathbb{Z}[\tau]$ which have under the automorphism $*$ an image in a bounded interval with nonempty interior in $\mathbb{R}$ called **acceptance window** Ω, i. e. a one-dimensional quasicrystal is given by

$$\Sigma(\Omega) := \{x \in \mathbb{Z}[\tau] | x^* \in \Omega\}. \tag{17}$$

As a special example, let us consider the quasicrystal $\Sigma([0,1])$:
Here the condition

$$0 \leq a + \tau^* b \leq 1 \tag{18}$$

translates into

$$\frac{b}{\tau} \leq a \leq 1 + \frac{b}{\tau}. \tag{19}$$

With this one obtains

$$\Sigma([0,1]) := \{\left[1 + \frac{b}{\tau}\right] + b\tau | b \in \mathbb{N}\} \cup \{0\} \tag{20}$$

and the points nearest to the origin are given by:

$$\cdots, -\tau, 0, 1, 1+\tau, 2+2\tau, 2+3\tau, 3+4\tau, 4+5\tau, 4+6\tau, 5+7\tau, 5+8\tau, 6+9\tau, \cdots \tag{21}$$

Some special features of the quasicrystal are:

- $\Sigma([0,1])$ has a GLOBAL REFLECTION SYMMETRY IN THE POINT $1/2$, because $x \in \Sigma([0,1])$ implies $y = 1 - x \in \Sigma([0,1])$.

- The DISTANCES BETWEEN ADJACENT POINTS are of the length 1, τ or τ^2. 1 is an exceptional tile which occurs only once.

Quasicrystal Lie algebras

We now come back to our quantization problem. Because we used $W^{\mathbb{Z}}$ the q-deformation has led to a quantum mechanics on the discretized S^1 with a (for q N-th root of unity) N equidistant points. To have instead of equidistant points aperiodic ones it is tempting to choose the aperiodicity of one-dimensional quasicrystals and to try to define a "deformation" of the Witt algebra $W^{\mathbb{Z}}$ by choosing $\Sigma(\Omega)$ instead of $\mathbb{Z}$.

Let $\Sigma(\Omega)$ be a one-dimensional quasicrystal with an acceptance window of the form $[a,b]$, $(a,b]$, $[a,b)$, or (a,b), where $a \neq b \in \mathbb{R}$ satisfy $0 \leq ab < \infty$. Then we define:

For a number field F with $F \supset \mathbb{Q}[\tau]$ the quasicrystal Lie algebra $Q(\Omega)$ over F is the F-span of its basis

$$B(Q(\Omega)) = \{L_n \mid n \in \Sigma(\Omega)\} \tag{22}$$

with the commutation relations of the basis elements given by

$$[L_n, L_m] = \begin{cases} (m-n)L_{n+m} & \text{if } n+m \in \Sigma(\Omega) \\ 0 & \text{if } n+m \notin \Sigma(\Omega). \end{cases} \tag{23}$$

Or, equivalently, using the characteristic function χ_Ω of the interval Ω:

$$[L_n, L_m] = (m-n)\chi_\Omega(n^* + m^*)L_{n+m}. \tag{24}$$

The **antisymmetry** of the commutators is obvious, the **Jacobi identity** holds provided $ab \geq 0$.

Although these algebras look similar to the Witt algebra $W^{\mathbb{Z}}$, their properties differ substantially, since here, in contrast to the Witt algebra $W^{\mathbb{Z}}$, many commutators are equal to zero, because

$$m, n \in \Sigma(\Omega) \Rightarrow m + n \in \Sigma(\Omega). \tag{25}$$

We list some properties of quasicrystal Lie algebras relevant for Borel quantization, further properties can be found in [8].

- The algebras $Q(\Omega)$ admit only a trivial central extension, i.e. the direct sum $Q(\Omega) \oplus Fc$.

- Quasicrystal Lie algebras admit an ABUNDANCE OF FINITE DIMENSIONAL SUBALGEBRAS, because the closure of any finite set of generators from $B(Q(\Omega))$ under the commutation relations is a finite dimensional subalgebra of $Q(\Omega)$.

- Their realization is LOCAL, i.e. it depends on the objects it is acting on.
  Example: $L_n z^k = z^n z \partial_z \chi_\Omega(n^* + k^*) z^k$

THE CORRESPONDING qc-QUANTUM MECHANICS

Here we address the question if the quasicrystal Lie algebras, or qc-deformation of the Witt algebra $W^{\mathbb{Z}}$, are suitable to construct a qc-quantum mechanics along the lines used before. We have no full answer at this stage, but only discuss the first steps in this direction and comment on the status quo.

A substitution of the generators of the Witt algebra $W^{\mathbb{Z}}$ by the qc-Witt algebra $W^{\mathbb{Z}[\tau]}$ leads to the qc-quantum kinematics:

$$\begin{aligned} Q(f) &= \sum_{n \in \Sigma} f_n T_n \\ P(X) &= \sum_{n = \in \Sigma} X_n (L_n + iDnT_n). \end{aligned} \tag{26}$$

with $(n \in \Sigma(\Omega))$

$$T_n z^k = z^{n+k} \chi_\Omega(n^* + k^*)$$
$$L_n z^k = z^n \left(z\frac{d}{dz} + \frac{n}{2} + \theta \right) \chi_\Omega(n^* + k^*) z^k \qquad (27)$$

and $\theta \in [0, 1)$, $D \in \mathbb{R}$ labeling unitarily inequivalent quantizations as before.

In comparison with the construction related to the q-Witt algebra $W_q^\mathbb{Z}$, a difficulty arises: the operators are local and not global, i.e. they depend on the functions they are acting on. To obtain a realization with discrete operators, and hence a discretization on the operator level, one would have to introduce discrete operators as before. However, a q-derivative with one deformation parameter cannot be sufficient, since at least two different tiles (spacings between adjacent points) occur in quasicrystals, in contrast to equidistant discretizations which have only 1 tile. A solution to this problem is not yet at hand and requires further investigation.

CONCLUSION

It has been demonstrated in the first part how a deformation of the Witt algebra $W^\mathbb{Z}$ can be successfully implemented in the framework of Borel quantization on S^1 to construct a discrete quantum mechanics related to an equidistant discretization of S^1. The first steps for the extension of this concept to a non-equidistant discretization, i.e. the introduction of quasicrystal Lie algebras, have been presented and the implementation of the latter has been discussed. The derivation of a realization in terms of discrete operators, e.g. via a 2 parameter deformation of quasicrystal Lie algebras, and the derivation of a corresponding dynamics, is still under investigation.

ACKNOWLEDGEMENTS

The part of the work related to the q-deformation of the Witt algebra $W^\mathbb{Z}$ and a derivation of a quantum mechanics related to an equidistant discretization has been derived in collaboration with Prof. Dobrev from Sofia and Prof. Doebner from Clausthal. The results on quasicrystal Lie algebras where obtained together with Prof. Patera from Montreal and Prof. Pelantova from Prague.

REFERENCES

1. Dobrev, V. K., Doebner, H.-D., and Twarock, R. *J. Phys. A: Math. Gen.*, **38**, 1161 (1997).
2. Twarock, R. *Quantum Mechanics on S^1 with q-Difference Operators*. Ph.D. thesis, TU Clausthal (1997).

3. Twarock, R. In: Dobrev, V. K. and Doebner, H.-D., eds., *q-Group 21*. Sofia: Heron Press (1997), pp. 158–164.
4. Kobayashi, T. and Suzuki, T. *Phys. Lett.*, **B317**, 359 (1993).
5. Kobayashi, T. and Suzuki, T. *J. Phys. A: Math. Gen.*, **26**, 6055 (1993).
6. Doebner, H.-D. and Tolar, J. *Ann. Phys. Leipzig*, **47**, 116 (1990).
7. Doebner, H.-D. and Hennig, J. D. In: Gruber, B., ed., *Symmetries in Science VIII*. New York: Plenum Publ. (1995), pp. 85–90.
8. Patera, J., Pelantova, E., and Twarock, R. *to appear in Phys. Lett. A* (1998).

Non-commutative Space-Time and Quantum Groups

Julius Wess

*Sektion Physik der Ludwig-Maximilians-Universität,
Theresienstraße 17, D-80333 München and
Max-Planck-Institut für Physik (Werner-Heisenberg-Institut),
Föhringer Ring 6, D-80805 München*

We have in mind a noncommutative space time structure that is based on an algebra where some of the elements can be identified with position operators - we shall call them X^a - and some of the elements with momentum operators P^a.

This is best illustrated by our usual space time structure:

$$\text{Position} \quad x^a : \quad x^a x^b = x^b x^a \tag{1}$$

$$\text{Momentum} \quad p^a : \quad p^a p^b = p^b p^a \tag{2}$$

The noncommutative part of this algebra is defined by the Heisenberg relation:

$$x^a p^b - p^b x^a = i\eta^{ab}; \quad \eta^{ab} \in \mathbb{C} \tag{3}$$

If η^{ab} is an invariant tensor of some symmetric group ($SO(3)$ or Lorentz group):

$$\Lambda^a{}_c \Lambda^b{}_d \eta^{cd} = \eta^{ab}, \tag{4}$$

then eqns (1), (2) and (3) are covariant under this group. The position operators and the momentum operators belong to representations of this group.

Let L^i be the generators of this group:

$$[L^i, L^j] = i f^{ij}{}_k L^k. \tag{5}$$

Belonging to a representation means:

$$[L^i, x^a] = -l^{ia}{}_b x^b$$
$$[L^i, p^a] = -l^{ia}{}_b p^b \tag{6}$$

where $l^{ia}{}_b$ is the a, b matrix element of the generator L^i in the respective representation. We have assumed that position and momenta are in the same representation.

$$[l^i, l^j] = i f^{ij}{}_k l^k \tag{7}$$

It follows that L^i can be represented as an ordered element of the algebra:

$$L^i = l^{ia}{}_b \eta_{ac} x^c p^b; \quad \eta_{ac}\eta^{cd} = \delta_a{}^d \qquad (8)$$

such that

$$[L^i, L^j] = if^{ij}{}_k L^k \qquad (9)$$

By ordered we mean that the x elements in (8) are all to the left of the p elements.

This is the algebraic structure of our usual space time. In addition we have conjugation properties:

$$\overline{x^a} = x^a, \qquad \overline{p^a} = p^a \qquad (10)$$

They are consistent with the algebra defined by eqns (1), (2) and (3) if $\eta^{ab} \in \mathbb{R}$.

The formalism of quantum mechanics tells us how to proceed from the algebra to physics. We have to find Hilbertspace representations of the algebra. Observables have to be represented by selfadjoint operators. Their real eigenvalues are the possible results of a precise measurement and a physical state is prepared by such a measurement.

The algebraic conjugation (10) has to be represented by conjugation of linear operators in a Hilbert space. Algebraic selfadjoint has to become selfadjoint of linear operators.

In this context the algebraic structure defined above could be seen as the defining algebra for our space-time structure and it is the basis for the cinematics of quantum theory. For a dynamic we have to add the Schroedinger equation.

We would like to generalize the above picture by bringing a noncommutative structure already to the relations (1) and (2) of position operators and momentum operators. This can be achieved by generalizing the symmetric structure (5) to a quantum group structure. We shall discuss the quantum groups $SO_q(3)$ and the q-Lorentz group. We demand that the space variables and the momentum variables be (co-) modules of the quantum group.... This generalizes eqns (6). Then we ask for generalizations of (1) and (2) that are compatible with the (co-) module structure. At a next step we have to find a conjugation that is consistent with the algebraic structure. This then would be the starting point of a physical interpretation along the lines of quantum mechanics.

There is, however, one more property of the algebra that we would like to maintain. From the relations (1) or (2) follows that the ordered monomials in x (or p) of fixed degree form a basis for the elements of the x (or p) algebra of this degree. This property we would like to keep also in the generalized case. We shall call it the Poincare-Birkhoff-Witt (PBW) property. Though it is formulated very mathematically it has important physical implications. A measurement of the observables x or p should not be more restricted then in the commutative case. Relation (3) allows us to extend this PBW property to the full x, p algebra.

The above requirements restrict the possible algebras very much. Indeed we were not able to find any algebra in terms of x and p that would satisfy all our demands. In order to generalize (3) in accordance with PBW and selfadjointness of the observables we have to enlarge the algebra by a new element Λ. Then it turns out that the generators L^i of the quantum group cannot be expressed as ordered polynomials in x and p as it was the case in eqn (8). This has as a consequence that the ordered elements of the x, p algebra do not form a basis of the full algebra.

Now we have to learn a few facts about quantum groups. This will also explain the notation in the following formulas. The representations of a quantum group $SU_q(2)$ or $SL_q(2, \mathbb{C})$ have for real q exactly the same pattern as the corresponding non-deformed ($q = 1$) groups.

The product of two vector representations of $SU_q(2)$ (two triplets) decomposes into a singlet, a triplet and a quintuplet.

$$3 \otimes 3 = 1 \oplus 3 \oplus 5 \tag{11}$$

The Clebsch-Gordan coefficients that combine two triplets to a singlet define a metric:

$$X \circ Y = g_{AB} X^A Y^B \tag{12}$$

$$g_{33} = 1, \quad g_{+-} = -q, \quad g_{-+} = -\frac{1}{q}$$

The notation X^3, X^+, X^- is adapted to the quantum group notation where it is natural to introduce $X^\pm = X^1 \pm iX^2$.

The Clebsch-Gordan coefficients that combine two triplets to a triplet define a generalized ε-tensor:

$$Z^A = X^C Y^B \varepsilon_{BC}{}^A = (X \times Y)^A$$
$$\varepsilon_{+-}{}^3 = q, \quad \varepsilon_{-+}{}^3 = -q, \quad \varepsilon_{33}{}^3 = 1 - q^2,$$
$$\varepsilon_{+3}{}^+ = 1, \quad \varepsilon_{3+}{}^+ = -q^2, \tag{13}$$
$$\varepsilon_{-3}{}^- = -q^2, \quad \varepsilon_{3-}{}^- = 1.$$

More generally we can compute the projectors on the invariant subspaces:

$$P_1 + P_3 + P_5 = 1 \tag{14}$$

These are 9 by 9 matrices.

There is a combination of these matrices that satisfies the Yang-Baxter equation. This combination is called $\hat{R}$ matrix. We need it to guarantee the PBW property for the algebra we want to define.

$$\hat{R} = P_5 - \frac{1}{q^4} P_3 + \frac{1}{q^6} P_1 \tag{15}$$

With $\hat{R}$ the matrix $\hat{R}^{-1}$ will always satisfy the Yang-Baxter equation as well. For 9 by 9 matrices that can be decomposed into the three projectors these are the only solutions of the Yang-Baxter equation.

We are now ready to present the result of our construction of a generalized space algebra based on $SO_q(3)$ as a symmetry.

The relations (1) and (2) are generalized to:

$$X \times X = 0, \qquad P \times P = 0 \tag{16}$$

More explicitely:

$$\begin{aligned} X^3 X^+ &= q^2 X^+ X^3 \\ X^3 X^- &= q^{-2} X^- X^3 \\ X^- X^+ &= X^+ X^- + (q - q^{-1}) X^3 X^3 \end{aligned} \tag{17}$$

This demonstrates the noncommutative space structure. The position variables as well as the momentum variables remain subalgebras, both having the PBW property.

The quantum group algebra $SO_q(3)$ that replaces (5) and that was the starting point of our construction can be cast in the form:

$$\begin{aligned} L \times L &= -\frac{1}{q^2} W L \\ WL &= LW \\ W^2 &= 1 + q^4 (q^2 - 1)^2 L \circ L \end{aligned} \tag{18}$$

The module structure of X and P that generalizes (6) becomes:

$$\begin{aligned} L^A X^B &= -\frac{1}{q^4} \varepsilon^{ABC} X_C W - \frac{1}{q^2} \varepsilon_{KC}{}^A \varepsilon^{KBD} X^C L_D \\ W X^A &= (q^2 - 1 + q^{-2}) X^A W + (q^2 - 1)^2 \varepsilon^{ABC} X_C L_B \end{aligned} \tag{19}$$

For the P module X has to be replaced by P in (19). Thus the elements X^A, L^A, W generate a subalgebra as well as the elements P^A, L^A, W.

If we now try to complete the algebra by defining q-deformed Heisenberg relations as a generalization of (3) we find that we have to add an additional element Λ to our algebra in order to satisfy all our requirements, essentially selfadjointness and PBW. We find:

$$\begin{aligned} P^A X^B - \hat{R}^{-1\,AB}{}_{CD} X^C P^D &\\ = -\frac{i}{2} \Lambda^{-\frac{1}{2}} \Big\{ (1 + q^{-6}) g^{AB} W - (1 - q^{-4}) \varepsilon^{ABC} L_C \Big\} \end{aligned} \tag{20}$$

The 9 by 9 $\hat{R}$ matrix in (20) reflects the PBW property. On the right hand side of (20) L^A appears. By projecting out the singlet part in (20) (the $g^{AB} W$ part) we see that we can represent L as a polynomial in X and P but not as an ordered one. The ordering of the XP algebra uses L explicitely.

The new element $\Lambda^{-\frac{1}{2}}$ had to be introduced:

$$\Lambda^{-\frac{1}{2}} X^A = q^{-2} X^A \Lambda^{-\frac{1}{2}}$$
$$\Lambda^{-\frac{1}{2}} P^A = q^2 P^A \Lambda^{-\frac{1}{2}} \qquad (21)$$
$$\Lambda^{-\frac{1}{2}} L^A = L^A \Lambda^{-\frac{1}{2}}$$
$$\Lambda^{-\frac{1}{2}} W = W \Lambda^{-\frac{1}{2}}$$

This completes the definition of our algebra. If we consider X^A, P^A, L^A, W and $\Lambda^{-\frac{1}{2}}$ to be generators of the algebra we divide the freely generated algebra by the ideal generated by the above relations.

We find that the algebra allows the following conjugation:

$$\overline{X^A} = g_{AB} X^B, \quad \overline{P^A} = g_{AB} P^B \qquad (22)$$
$$\overline{L^A} = g_{AB} L^B, \quad \overline{W} = W, \quad \overline{\Lambda^{\frac{1}{2}}} = q^{-6} \Lambda^{-\frac{1}{2}}.$$

The representations of this algebra have been studied in ref [3]. He choose $L \circ L$, L^3 and $X^2 = X \circ X$ as a complete set of commuting variables. Note that for $q = 1$ the eigenvalues of these operators would label a state completely. The same is true for $q \neq 1$, $q \in \mathbb{R}$. For the $SU_q(2)$ part W and L^A the representations are well known, the eigenvalues of $L \circ L$ and L^3 are

$$L \circ L |j, m, n\rangle = \frac{q^{-6}}{(q^2 - q^{-2})^2} \left(q^{4j+2} + q^{-4j-2} - (q^2 + q^{-2}) \right) |j, m, n\rangle \qquad (23)$$

$$L^3 |j, m, n\rangle = \frac{q^{-3}}{(q - q^{-1})} \left\{ q^{2m} - \frac{q^{2j+1} + q^{-2j-1}}{q + q^{-1}} \right\} |j, m, n\rangle,$$

where $j, m \in \mathbb{Z}$, $j \geq 0$, $-j \leq m \leq +j$.

For X^2 we have the following eigenvalues

$$X^2 |j, m, n\rangle = l_0^2 q^{4n} |j, m, n\rangle, \qquad n \in \mathbb{Z}. \qquad (24)$$

A parameter l_0 appears in the eigenvalues of X^2. It is a parameter with the dimension of a length and it labels inequivalent representations.

As q is a dimensionless parameter no new parameter with a dimension of a length that would relate to a lattice is in the algebra. It however occurs as a parameter that labels inequivalent representations.. A physical system "chooses" a irreducible representation and thus a constant l_0. All the elements of the algebra do not change this value of l_0. If the Hamiltonian is part of the algebra l_0 will not change in time either.

To incorporate time into the noncommutative structure we start with a q-deformed Lorentz algebra. Again, there is the same Clebsch-Gordan structure

$$4 \times 4 = 1 + 9 + 6, \qquad (25)$$

where 6 decomposes into a selfdual and an antiselfdual antisymmetric tensor. The singlet defines a metric η. The projectors can again be combined to a solution of the Yang-Baxter equation. This time, however, there are two independent combinations $\hat{R}_I$ and $\hat{R}_{II}$ with their inverse.

$$\hat{R}_I = P_S + P_T - q^2 P_+ - q^{-2} P_- \qquad (26)$$
$$\hat{R}_{II} = q^{-2} P_S + q^2 P_T - P_+ - P_-$$

Here the notation is P_T (trace) for the singlet, P_S (symmetric, traceless) for the nonet, P_+ (antisymmetric, selfdual) for the one triplet and P_- (antisymmetric, antiselfdual) for the other triplet.

We only list a few of the algebraic relations. A complete description of the algebra is given in ref [1].

$$X^0 X^A = X^A X^0 \qquad (27)$$
$$X^C X^D \varepsilon_{DC}{}^A = (1-q^2) X^0 X^A.$$

The "time" X^0 commutes with the space variables X^A ($A = 3, +, -$). The space variables, however, produce the time variable.

The momenta are subject to the identical relations (27).

The q-Lorentz group can be split into the Pauli decomposition:

$$R^C R^D \varepsilon_{DC}{}^A = (1+q^2)^{-1} U R^A$$
$$S^C S^D \varepsilon_{DC}{}^A = -(1+q^2)^{-1} U S^A \qquad (28)$$
$$R^A S^B = q^2 \hat{R}^{AB}{}_{CD} S^C R^D$$

The $\hat{R}$ matrix here is the 9×9 matrix defined by eqn (15).

U is related to the Casimir:

$$U^2 = 1 + \frac{1}{2}(q^4 - 1)^2 (R \circ R + S \circ S) \qquad (29)$$

The important relation is the Heisenberg relation:

$$P^a X^b - q^{-2} \hat{R}_{II}^{-1\,ab}{}_{cd} X^c P^d = -\frac{i}{2} \Lambda^{-\frac{1}{2}} \left\{ (1+q^4) \eta^{ab} U + q^2 (1-q^4) V^{ab} \right\} \qquad (30)$$

Again a scaling operator $\Lambda^{-\frac{1}{2}}$ had to be introduced. The indices a,b take the values $(0, 3, +, -)$. V^{ab} is the tensor notation for the generators of the q-Lorentz algebra.

Finally the conjugation properties:

$$\overline{X^0} = X^0, \quad \overline{X^A} = g_{AB} X^B$$
$$\overline{P^0} = P^0, \quad \overline{P^A} = g_{AB} P^B$$
$$\overline{R^A} = -g_{AB} S^B \qquad (31)$$
$$\overline{U} = U \quad \overline{\Lambda^{\frac{1}{2}}} = q^4 \Lambda^{-\frac{1}{2}}.$$

We are now ready to discuss Hilbertspace representations of this algebra. A complete set of commuting observables is:

$$X^0, \quad r^2 = g_{AB}X^A X^B, \quad L \circ L, \quad L^3 \tag{32}$$

L is the three-dimensional angular momentum

$$L^A = \frac{q^2+1}{q^2}(US^A - UR^A + (q^4-1)\varepsilon_{CB}{}^A R^B S^C) \tag{33}$$

that commutes with the time variable X^0. The eigenvalues of these observables (32) completely characterize a state in the Hilbertspace as they would do in the commuting case.

The representations of the algebra have been constructed in ref [2].

Here I first would like to plot the eigenvalues of the observable X^0 (time) and r (three-dimensional radius) for a particular value of q ($q = 1.1$). A scale has to be chosen, it characterizes the representations. We chose $t_0 = 1$.

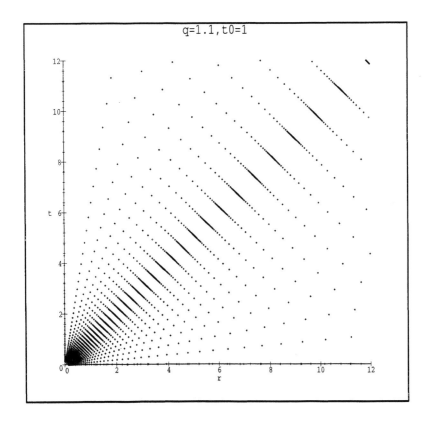

FIGURE 1.

The representations have an interesting pattern. In fig. 1 the forward time-like, backward time-like and space-like regions are clearly visible. Each of these furnishes a representation of the algebra. If, however, X^a and P^a have to be represented by selfadjoint linear operators then we have to glue together all three representations [4]. The light cone consists of limit points that again have a limit point at the origin. The hyperbola corresponding to the invariant length $\eta_{ab}X^aX^b$ are clearly visible. For $q \to 1$ the points come closer and closer and have the continuum as a limit.

We learn that a q-deformation of the Minkowski space latticizes the space-time manifold.

We finally give the formulas for the eigenvalulues of X^0 and r^2 and the range of the eigenvalues of $L \circ L$. As mentioned above there are inequivalent representations for the forward and for the backward time-like regions as well as for the space-like region.

Space-like: $s^2 = (X^0)^2 - r^2 < 0$: 8

$$X^0|j,m,n,M\rangle = \frac{l_0}{q+q^{-1}}q^M(q^n - q^{-n})|j,m,n,M\rangle \tag{34}$$

$$r^2|j,m,n,M\rangle = \frac{l_0^2}{(q+q^{-1})^2}q^{2M}(q^{n+1} + q^{-n-1})(q^{n-1} + q^{-n+1})|j,m,n,M\rangle$$

The eigenvalues j and m refer to $L \circ L$ and L^3. In the space-like region we have

$$j = 0, \ldots \infty, \quad n = -\infty \ldots \infty, \quad M = -\infty \ldots \infty \tag{35}$$

Time-like: $s^2 = (X^0)^2 - r^2 > 0$, $t_0 > 0$ or $t_0 < 0$:

$$X^0|j,m,n,M\rangle = \frac{t_0}{q+q^{-1}}q^M(q^{n+1} + q^{-n-1})|j,m,n,M\rangle \tag{36}$$

$$r^2|j,m,n,M\rangle = \frac{t_0^2}{(q+q^{-1})^2}q^{2M}(q^{n+2} - q^{-n-2})(q^n - q^{-n})|j,m,n,M\rangle$$

In this case we have

$$M = -\infty \ldots \infty, \quad n = 0, \ldots \infty, \quad j = 0, \ldots, n. \tag{37}$$

Angular momentum is limited by the eigenvalue n.

The q-space time lattice is labelled by "points" that are linear combinations of q^{M+n} and q^{M-n} for X^0 and of $q^{2(M+n)}$, q^{2M} and $q^{2(M-n)}$ for r^2 where M and n are integers. This seems to be the general pattern of q-lattices.

Note the M labels the hyperboles:

$$s^2|j,m,n,M\rangle = \begin{cases} -l_0^2 q^{2M}|j,m,n,M\rangle & \text{space-like} \\ t_0^2 q^{2M}|j,m,n,M\rangle & \text{time-like} \end{cases} \tag{38}$$

and n labels the points on the hyperboles.

REFERENCES

1. Lorek, A., Weich, W., Wess, J., *Z. Phys. C* **76**, 375–386 (1997).
2. Cerchiai, B. L., Wess, J., Preprint math.QA/9801104, accepted for publication *Eur. Phys. J. C* (1998).
3. Schraml, S. *Diplomarbeit, Ludwig-Maximilians-Universität München* (1998)
4. Hebecker, A., Schreckenberg, S., Schwenk, J., Weich, W., Wess, J., *Z.. Phys. C*bf 64, 355-359 (1994)
    Cerchiai, B., Wess, J., to be published

QUANTUM MECHANICS, SOLVABLE

AND QUASI-SOLVABLE PROBLEMS

The Complex Pendulum

Carl M. Bender

Department of Physics, Washington University, St. Louis, MO 63130

Abstract. In this talk we propose to broaden the conventional notion of quantum mechanics. In conventional quantum mechanics one imposes the condition $H^\dagger = H$, where † represents complex conjugation and matrix transpose, to ensure that the Hamiltonian has a real spectrum. Replacing this mathematical condition by the weaker and more physical requirement $H^\ddagger = H$, where $\ddagger = \mathcal{PT}$ represents combined parity reflection and time reversal, one obtains new infinite classes of complex Hamiltonians whose spectra are also real and positive. These $\mathcal{PT}$-symmetric theories may be viewed as analytic continuations of conventional theories from real to complex phase space. This talk describes the unusual classical and quantum properties of $\mathcal{PT}$-symmetric quantum mechanical and quantum field theoretic models.

INTRODUCTION

Several years ago, D. Bessis conjectured on the basis of numerical studies that the spectrum of the Hamiltonian $H = p^2 + x^2 + ix^3$ is *real and positive* [1]. To date there is no rigorous proof of this conjecture. The reality of the spectrum of H is due to $\mathcal{PT}$ symmetry. Note that H is invariant *neither* under parity $\mathcal{P}$, whose effect is to make spatial reflections, $p \to -p$ and $x \to -x$, *nor* under time reversal $\mathcal{T}$, which replaces $p \to -p$, $x \to x$, and $i \to -i$. However, $\mathcal{PT}$ symmetry is crucial. For example, the Hamiltonian $p^2 + ix^3 + ix$ has $\mathcal{PT}$ symmetry, and our numerical studies indicate that its entire spectrum is positive definite; the Hamiltonian $p^2 + ix^3 + x$ is not $\mathcal{PT}$-symmetric, and the entire spectrum is complex.

The connection between $\mathcal{PT}$ symmetry and the reality of spectra can be understood as follows: We know that if two linear diagonalizable operators commute, then they can be simultaneously diagonalized. This theorem does not hold in general if the operators are not linear, and indeed $\mathcal{PT}$ is not a linear operator because it involves complex conjugation. However, if $\mathcal{PT}$ does commute with the Hamiltonian H and *if we assume that we can simultaneously diagonalize these two operators*, then we have $E = E^*$, where E is an eigenvalue of H. Thus E, is real.

The connection between $\mathcal{PT}$ symmetry and reality of spectra is nicely illustrated by some exactly solvable models. Consider the harmonic oscillator $H = p^2 + x^2$, whose energy levels are $E_n = 2n+1$. Adding ix to H does not break $\mathcal{PT}$ symmetry, and the spectrum remains positive definite: $E_n = 2n + \frac{5}{4}$. Adding $-x$ also does not break $\mathcal{PT}$ symmetry if we define $\mathcal{P}$ as reflection about $x = \frac{1}{2}$, $x \to 1-x$, and again the spectrum remains positive definite: $E_n = 2n + \frac{3}{4}$. However, adding $ix - x$ *does* break $\mathcal{PT}$ symmetry, and the spectrum is now complex: $E_n = 2n + 1 + \frac{1}{2}i$.

A QUANTUM MECHANICAL MODEL

The Hamiltonian studied by Bessis is just one example of a huge and remarkable class of complex Hamiltonians whose energy levels are real and positive. The purpose of this talk is to describe the properties of such Hamiltonians. We begin by examining the one-parameter class of quantum-mechanical Hamiltonians

$$H = p^2 - (ix)^N \quad (N \text{ real}). \tag{1}$$

We find that as a function of N, there are two phases with a transition point at $N = 2$ at which the entirely real spectrum begins to develop complex eigenvalues. A full description of the behavior of the eigenvalues of H is given in [2,3].

We conjecture that the underlying reason that the spectrum of a $\mathcal{PT}$-symmetric theory is real is that the Hamiltonian is actually self-adjoint with respect to a new definition of adjoint: $H^\ddagger = H$, where $\ddagger = \mathcal{PT}$. The Hilbert space consists of the set of vectors that can be represented as *real* linear combinations of the eigenfunctions $\phi_n(x)$ of H, which are also simultaneous eigenfunctions of $\mathcal{PT}$. [Note that $\phi_n(x)$ are themselves complex functions because they solve a complex differential equation.] The eigenfunctions of H are orthonormal: $\int dx \, \phi_m(x)\phi_n(x) = \delta_{mn}$. We may define the norm of a vector $\Phi(x)$ in the Hilbert space as $\int dx \, [\Phi(x)]^2$; this norm is positive. The path of integration in the definition of the norm is a complex contour, as we will explain later. As the parameter N is varied, this path may cease to be continuous because of the presence of cuts in the complex-x plane. When this happens, we find that the eigenvalues of H become complex.

We have studied the Hamiltonian (1) extensively using both numerical and analytical methods. As shown in Fig. 1, the spectrum of H exhibits three distinct behaviors as a function of N: When $N \geq 2$, the spectrum is infinite, discrete, and entirely real and positive. (This region includes the case $N = 4$ for which $H = p^2 - x^4$; the spectrum of this Hamiltonian is positive and discrete and $\langle x \rangle \neq 0$ in the ground state because H breaks parity symmetry!) At the lower bound $N = 2$ of this region lies the harmonic oscillator. A phase transition occurs at $N = 2$; when $1 < N < 2$, there are only a *finite* number of real positive eigenvalues and an infinite number of complex conjugate pairs of eigenvalues. In this region $\mathcal{PT}$ symmetry is *spontaneously broken* [3]. As N decreases from 2 to 1, adjacent energy levels merge into complex conjugate pairs beginning at the high end of the spectrum; ultimately, the only remaining real eigenvalue is the ground-state energy, which diverges as $N \to 1^+$ [4]. When $N \leq 1$, there are no real eigenvalues.

The Schrödinger eigenvalue differential equation for the Hamiltonian (1) is

$$-\psi''(x) - (ix)^N \psi(x) = E\psi(x). \tag{2}$$

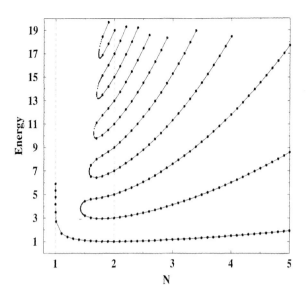

FIGURE 1. Energy levels of the Hamiltonian $H = p^2 - (ix)^N$ as a function of the parameter N. There are three regions: When $N \geq 2$ the spectrum is real and positive. The lower bound of this region, $N = 2$, corresponds to the harmonic oscillator, whose energy levels are $E_n = 2n + 1$. When $1 < N < 2$, there are a finite number of real positive eigenvalues and an infinite number of complex conjugate pairs of eigenvalues. As N decreases from 2 to 1, the number of real eigenvalues decreases; when $N \leq 1.42207$, the only real eigenvalue is the ground-state energy. As N approaches 1^+, the ground-state energy diverges. For $N \leq 1$ there are no real eigenvalues.

Ordinarily, the boundary conditions that give quantized energy levels E are that $\psi(x) \to 0$ as $|x| \to \infty$ on the real axis; this condition suffices when $1 < N < 4$. However, for arbitrary real N we must continue the eigenvalue problem for (2) into the complex-x plane. Thus, we replace the real-x axis by a contour in the complex plane along which the differential equation holds and we impose the boundary conditions that lead to quantization at the endpoints of this contour. (Eigenvalue problems on complex contours are discussed in Ref. [5].)

The regions in the cut complex-x plane in which $\psi(x) \to 0$ exponentially as $|x| \to \infty$ are *wedges* (see Fig. 2); these wedges are bounded by the *Stokes lines* of the differential equation [6]. The center of the wedge, where $\psi(x)$ vanishes most rapidly, is called an *anti-Stokes line*.

There are many wedges in which $\psi(x) \to 0$ as $|x| \to \infty$. Thus, there are many eigenvalue problems associated with a given differential equation [5]. However, we choose to continue the eigenvalue equation (2) away from the harmonic oscillator problem at $N = 2$. The wave function for $N = 2$ vanishes in wedges of angular opening $\frac{1}{2}\pi$ centered about the negative- and positive-real x axes. For arbitrary N the anti-Stokes lines at the centers of the left and right wedges lie at the angles

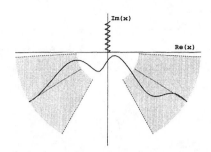

FIGURE 2. Wedges in the complex-x plane containing the contour on which the eigenvalue problem for the differential equation (2) for $N = 4.2$ is posed. In these wedges $\psi(x)$ vanishes exponentially as $|x| \to \infty$. The wedges are bounded by *Stokes lines* of the differential equation. The center of the wedge, where $\psi(x)$ vanishes most rapidly, is an anti-Stokes line.

$$\theta_{\text{left}} = -\pi + \frac{N-2}{N+2}\frac{\pi}{2} \quad \text{and} \quad \theta_{\text{right}} = -\frac{N-2}{N+2}\frac{\pi}{2}. \tag{3}$$

The opening angle of these wedges is $\Delta = 2\pi/(N+2)$. The differential equation (2) may be integrated on any path in the complex-x plane so long as the ends of the path approach complex infinity inside the left wedge and the right wedge [7]. Note that these wedges contain the real-x axis when $1 < N < 4$.

As N increases from 2, the left and right wedges rotate downward into the complex-x plane and become thinner. At $N = \infty$, the differential equation contour runs up and down the negative imaginary axis and thus there is no eigenvalue problem at all. Indeed, Fig. 1 shows that the eigenvalues all diverge as $N \to \infty$. As N decreases below 2 the wedges become wider and rotate into the upper-half x plane. At $N = 1$ the angular opening of the wedges is $\frac{2}{3}\pi$ and the wedges are centered at $\frac{5}{6}\pi$ and $\frac{1}{6}\pi$. Thus, the wedges become contiguous at the positive-imaginary x axis, and the differential equation contour can be pushed off to infinity. Consequently, there is no eigenvalue problem when $N = 1$ and, as we would expect, the ground-state energy diverges as $N \to 1^+$ (see Fig. 1).

To ensure the accuracy of our numerical computations of the eigenvalues in Fig. 1, we have solved the differential equation (2) using two independent procedures. The most accurate and direct method is to convert the complex differential equation to a system of coupled, real, second-order equations that we solve using the Runge-Kutta method; the convergence is most rapid when we integrate along anti-Stokes lines. We then patch the two solutions together at the origin. We have verified those results by diagonalizing a truncated matrix representation of the Hamiltonian in Eq. (1) in harmonic oscillator basis functions.

Semiclassical analysis. WKB gives a good approximation to the eigenvalues in Fig. 1 when $N \geq 2$. The novelty of this WKB calculation is that it must be performed in the complex plane. The turning points $x_\pm$ are those roots of $E + (ix)^N = 0$ that *analytically continue* off the real axis as N moves away from $N = 2$ (the harmonic oscillator):

$$x_- = E^{1/N} e^{i\pi(3/2 - 1/N)}, \quad x_+ = E^{1/N} e^{-i\pi(1/2 - 1/N)}. \tag{4}$$

These turning points lie in the lower (upper) x plane in Fig. 2 when $N > 2$ ($N < 2$).

The leading-order WKB phase-integral quantization condition is $(n + 1/2)\pi = \int_{x_-}^{x_+} dx \sqrt{E + (ix)^N}$. It is crucial that this integral follow a path along which the *integral is real*. When $N > 2$, this path lies entirely in the lower-half x plane and when $N = 2$ the path lies on the real axis. When $N < 2$ the path is in the upper-half x plane; it crosses the cut on the positive-imaginary axis and thus is *not a continuous path joining the turning points*. Hence, WKB fails when $N < 2$.

When $N \geq 2$, we deform the phase-integral contour so that it follows the rays from x_- to 0 and from 0 to x_+: $(n + 1/2)\pi = 2\sin(\pi/N) E^{1/N+1/2} \int_0^1 ds \sqrt{1 - s^N}$. We then solve for E_n:

$$E_n \sim \left[\frac{\Gamma(3/2 + 1/N) \sqrt{\pi}(n + 1/2)}{\sin(\pi/N) \Gamma(1 + 1/N)} \right]^{\frac{2N}{N+2}} \quad (n \to \infty). \tag{5}$$

This result is quite accurate. The fourth exact eigenvalue (obtained using Runge-Kutta) for the case $N = 3$ is 11.3143 while WKB gives 11.3042, and the fourth eigenvalue for the case $N = 4$ is 18.4590 while WKB gives 18.4321.

The spectrum of the $|x|^N$ potential is like that of the $-(ix)^N$ potential. The WKB quantization condition is like Eq. (5) except that $\sin(\pi/N)$ is absent. However, as $N \to \infty$, the spectrum of $|x|^N$ approaches that of the square-well potential $[E_n = (n+1)^2 \pi^2/4]$, while the energies of the $-(ix)^N$ potential diverge (see Fig. 1).

Asymptotic study of the ground-state energy near $N = 1$: When $N = 1$, the differential equation (2) can be solved exactly in terms of Airy functions. The anti-Stokes lines lie at 30° and at 150°. We find the solution that vanishes exponentially along each of these rays and then rotate back to the real-x axis:

$$\psi_{\text{left,right}}(x) = C_{1,2} \, \text{Ai}(\mp x e^{\pm i\pi/6} + E e^{\pm 2i\pi/3}). \tag{6}$$

We must patch these solutions together at $x = 0$ according to the patching condition $\frac{d}{dx} |\psi(x)|^2 \big|_{x=0} = 0$. But for real E, the Wronskian identity for the Airy function is

$$\frac{d}{dx} |\text{Ai}(x e^{-i\pi/6} + E e^{-2i\pi/3})|^2 \big|_{x=0} = -\frac{1}{2\pi} \tag{7}$$

instead of 0. Hence, there is no real eigenvalue.

Next, we perform an asymptotic analysis for $N = 1 + \epsilon$, $-\psi''(x) - (ix)^{1+\epsilon} \psi(x) = E\psi(x)$, and take $\psi(x) = y_0(x) + \epsilon y_1(x) + O(\epsilon^2)$ as $\epsilon \to 0+$. We find that $E \to \infty$ as $\epsilon \to 0$ roughly like $E \propto (-\ln \epsilon)^{2/3}$. The explicit asymptotic formula is [2]

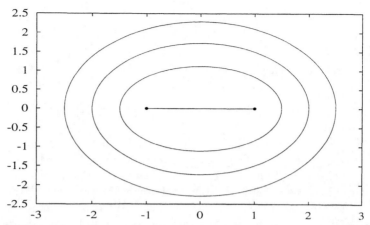

FIGURE 3. Classical paths in the complex-x plane for the $N = 2$ oscillator. The paths form a set of nested ellipses.

$$1 \sim \epsilon e^{\frac{4}{3}E^{3/2}} E^{-3/2}[\sqrt{3}\ln(2\sqrt{E}) + \pi - (1-\gamma)\sqrt{3}]/8. \qquad (8)$$

To test the accuracy of this formula we have compared the exact ground-state energy E near $N = 1$ with the asymptotic results in Eq. (8). For $N = 1.1$, $E_{\text{exact}} = 1.6837$ while $E_{\text{asymptotic}} = 2.0955$; for $N = 1.001$, $E_{\text{exact}} = 3.4947$ while $E_{\text{asymptotic}} = 3.6723$; for $N = 1.00001$, $E_{\text{exact}} = 4.7798$ while $E_{\text{asymptotic}} = 4.8776$; for $N = 1.0000001$, $E_{\text{exact}} = 5.8943$ while $E_{\text{asymptotic}} = 5.9244$.

Behavior near $N = 2$: The most intriguing aspect of Fig. 1 is the transition that occurs at $N = 2$. To describe quantitatively the merging of eigenvalues that begins when $N < 2$, we let $N = 2 - \epsilon$ and study the asymptotic behavior of the determinant of the matrix as $\epsilon \to 0+$. (A conventional Hermitian perturbation of a Hamiltonian causes adjacent energy levels to repel, but in this case the complex perturbation of the harmonic oscillator $(ix)^{2-\epsilon} \sim x^2 - \epsilon x^2[\ln(|x| + \frac{1}{2}i\pi \, \text{sgn}(x)]$ causes the levels to merge.) A complete description of this asymptotic study is given elsewhere [3]. The onset of eigenvalue merging is a phase transition. This transition occurs even at the *classical* level. We discuss the classical problem below.

CLASSICAL VERSION OF THE THEORY

The classical equation of motion (Newton's law) for the Hamiltonian (1) describes a particle of energy E subject to the *complex* forces. The trajectory $x(t)$ of the particle obeys $\pm dx[E + (ix)^N]^{-1/2} = 2dt$. While E and dt are real, $x(t)$ is a path in the complex plane in Fig. 2. Consider the simple case $N = 2$. Here, there are two turning points. There is one classical path that terminates at the classical turning points $x_\pm$ in (4). Other paths are nested ellipses with foci at the turning points (see Fig. 3). Cauchy's theorem implies that all these paths have the same period.

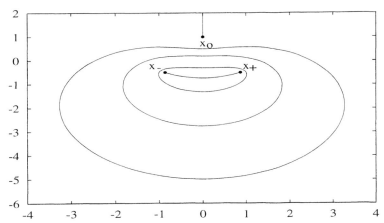

FIGURE 4. Classical paths in the complex-x plane for the $N = 3$ oscillator. In addition to the periodic orbits there is one path that runs off to $i\infty$ from the turning point on the imaginary axis.

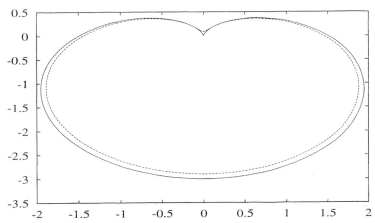

FIGURE 5. Classical paths complex-x plane for the $N = 3$ oscillator. As the paths get larger, they approach a limiting shape that resembles a cardioid. We have plotted the *rescaled* paths.

When $N = 3$ there is again a classical path that joins the left and right turning points and an infinite class of paths enclosing the turning points (see Fig. 4). As these paths increase in size they approach a cardioid shape. The indentation in the limiting cardioid occurs because paths may not cross, and thus all periodic paths must avoid the path that runs up the imaginary axis (see Fig. 5). When N is noninteger, we obtain classical paths that move off onto *different sheets* of the Riemann surface (see Fig. 6).

In general, whenever $N \geq 2$, the trajectory joining $x_\pm$ is a smile-shaped arc in the lower complex plane. The motion is *periodic*; thus, we have a *complex pendulum* whose (real) period T is

173

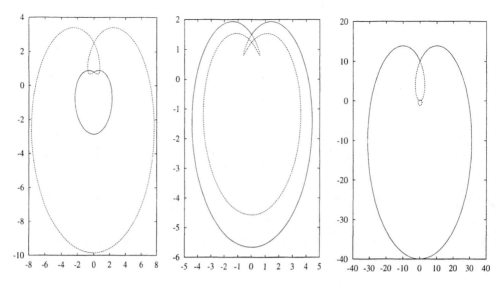

FIGURE 6. Classical paths for the case $N = 2.5$. These paths do not intersect; the graph shows the projection of the parts of the path that lie on different sheets of the Riemann surface. As the size of the paths increases a limiting cardioid appears on the principal sheet of the Riemann surface. On the remaining sheets of the surface the path exhibits a knot-like topological structure.

$$T = 2E^{\frac{2-N}{2N}} \cos\left[\frac{(N-2)\pi}{2N}\right] \frac{\Gamma(1+1/N)\sqrt{\pi}}{\Gamma(1/2+1/N)}. \quad (9)$$

At $N = 2$ there is a global change. Below $N = 2$ a path starting at one turning point, say x_+, moves toward but *misses* the turning point x_-. This path spirals outward crossing from sheet to sheet on the Riemann surface, and eventually veers off to infinity asymptotic to the angle $\frac{N}{2-N}\pi$. Hence, the period abruptly becomes infinite. The total angular rotation of the spiral is finite for all $N \neq 2$ and as $N \to 2^+$ but becomes infinite as $N \to 2^-$ (see Fig. 7). The path passes many turning points as it spirals clockwise from x_-. [The nth turning point lies at the angle $\frac{3N-2-4n}{2N}\pi$ (x_- corresponds to $n = 0$).] As N approaches 2 from below, when the classical trajectory passes a new turning point, there corresponds an additional merging of the quantum energy levels as shown in Fig. 1). This correspondence becomes exact in the limit $N \to 2^-$ and is a manifestation of Ehrenfest's theorem.

APPLICATIONS OF COMPLEX HAMILTONIANS

There appear to be many applications of complex $\mathcal{PT}$-invariant Hamiltonians in physics. Hamiltonians having an imaginary external field have been introduced recently to study delocalization transitions in condensed matter systems such as vortex flux-line depinning in type-II superconductors [8] or even to study population

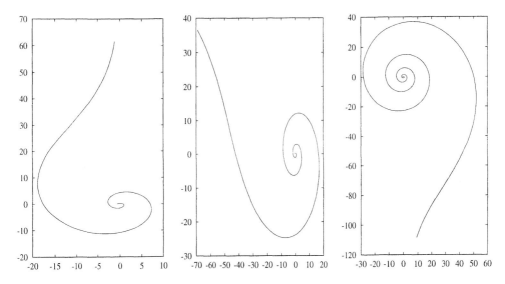

FIGURE 7. Classical paths in the complex-x plane for $N = 1.8$, $N = 1.85$ and $N = 1.9$. These paths are not periodic. The paths spiral outward to infinity. As $N \to 2$ from below, the number of turns in the spiral increases. The lack of periodic orbits corresponds to a broken $\mathcal{PT}$ symmetry.

biology [9]. Here, initially real eigenvalues bifurcate into the complex plane due to the increasing external field, indicating the unbinding of vortices or the growth of populations. We believe that one can also induce dynamic delocalization by tuning a physical parameter (here N) in a self-interacting theory.

The $\mathcal{PT}$-symmetric Hamiltonian in Eq. 1 may be generalized to include a mass term $m^2 x^2$. The massive case is more elaborate than the massless case; phase transitions appear at $N = 0$ and at $N = 1$ as well as at $N = 2$ [2]. Replacing the condition of Hermiticity by the weaker constraint of $\mathcal{PT}$-symmetry allows one to construct new kinds of quasi-exactly solvable quantum theories [10].

Quantum field theories analogous to the quantum-mechanical theory in Eq. (1) have astonishing properties. The Lagrangian $\mathcal{L} = (\nabla \phi)^2 + m^2 \phi^2 - g(i\phi)^N$ (N real) possesses $\mathcal{PT}$ invariance, the fundamental symmetry of local self-interacting scalar quantum field theory [11]. While the Hamiltonian for this theory is complex, the spectrum appears to be positive definite. Also, as $\mathcal{L}$ is explicitly not parity invariant, the expectation value of the field $\langle \phi \rangle$ is nonzero even when $N = 4$ [12]. Thus, one can in principle calculate directly (using the Schwinger-Dyson equations, for example [13]) the (real positive) Higgs mass in a renormalizable theory such as $-g\phi^4$ or $ig\phi^3$ in which symmetry breaking occurs naturally (without introducing a symmetry-breaking parameter).

Replacing conventional $g\phi^4$ or $g\phi^3$ theories by $-g\phi^4$ or $ig\phi^3$ theories reverses signs in the beta functions. Thus, theories that are not asymptotically free become asymptotically free and theories lacking stable critical points develop such points.

We believe that $-g\phi^4$ in four dimensions is nontrivial. Furthermore, $\mathcal{PT}$-symmetric massless electrodynamics has a nontrivial stable critical value of the fine-structure constant α [14].

We have examined supersymmetric $\mathcal{PT}$-invariant Lagrangians [15] and find that the breaking of parity symmetry does not induce a breaking of the apparently robust global supersymmetry. We have investigated the strong-coupling limit of $\mathcal{PT}$-symmetric quantum field theories [16]; the correlated limit in which the bare coupling constants g and $-m^2$ both tend to infinity with the renormalized mass M held fixed and finite is dominated by solitons. (In parity-symmetric theories the corresponding limit, called the Ising limit, is dominated by instantons.)

I thank my coauthors S. Boettcher, H. Jones, P. Meisinger, and K. Milton for their contributions to this research and D. Bessis, A. Wightman, and Y. Zarmi for useful conversations. This work was supported by the U.S. Department of Energy.

REFERENCES

1. D. Bessis, private discussion. This problem originated from discussions between Bessis and J. Zinn-Justin, who was studying Lee-Yang singularities using renormalization group methods. An $i\phi^3$ field theory arises if one translates the field in a ϕ^4 theory by an imaginary term.
2. C. M. Bender and S. Boettcher, *Phys. Rev. Lett.* (to be published).
3. C. M. Bender, S. Boettcher, and P. N. Meisinger *Phys. Rev. D* (submitted).
4. It is known that the spectrum of $H = p^2 - ix$ is null. See I. Herbst, *Commun. Math. Phys.* **64**, 279 (1979).
5. C. M. Bender and A. Turbiner, *Phys. Lett. A* **173**, 442 (1993).
6. C. M. Bender and S. A. Orszag, *Advanced Mathematical Methods for Scientists and Engineers*, New York: McGraw-Hill, 1978.
7. In the case of a Euclidean path integral representation for a quantum field theory, the (multiple) integration contour for the path integral follows the same anti-Stokes lines. See Ref. [12].
8. N. Hatano and D. R. Nelson, *Phys. Rev. Lett.* **77**, 570 (1996), and *Phys. Rev. B* **56**, 8651 (1997).
9. D. R. Nelson and N. M. Shnerb, condmat/9708071.
10. C. M. Bender and S. Boettcher, *J. Phys. A: Math. Gen.* **31**, L273 (1998).
11. R. F. Streater and A. S. Wightman, *PCT, Spin & Statistics, and all that*, New York: Benjamin, 1964. Ultimately we may define ‡ as $\mathcal{PCT}$, the fundamental symmetry of the world. There is no analog of the $\mathcal{C}$ operator in quantum mechanical systems having one degree of freedom and in scalar field theories.
12. C. M. Bender and K. A. Milton, *Phys. Rev. D* **55**, R3255 (1997).
13. C. M. Bender and K. A. Milton (submitted).
14. C. M. Bender and K. A. Milton (submitted).
15. C. M. Bender and K. A. Milton, *Phys. Rev. D* **57**, 3595 (1998).
16. C. M. Bender, S. Boettcher, H. F. Jones, and P. N. Meisinger, *Phys. Rev. D* (submitted).

Quasi Exactly Solvable Operators and Abstract Associative Algebras

Y. Brihaye

Dep. of Mathematical Physics, University of Mons-Hainaut, Mons, Belgium

P. Kosinski

Dep. of Theoretical Physics, University of Lodz, Lodz, Poland

Abstract. We consider the vector spaces consisting of direct sums of polynomials of given degrees and we show how to classify the linear differential operators preserving these spaces. The families of operators so obtained are identified as the envelopping algebras of particular abstract associative algebras. Some of these operators can be transformed into quasi exactly solvable Schroedinger operators which, having a hidden algebra, can be partially solved algebraically; we exhibit however a series of Schoedinger equations which, while completely solvable algebraically, do not possess a hidden algebra.

INTRODUCTION

In this report, we present a catalogue of the abstracts algebras which can be represented in terms of the differential operators acting onto a direct sum of vector spaces of polynomials of given degrees. We introduce the topic by means of two fundamental cases i) and ii) presented below. More sophisticated cases will them be treated in sections II-V.

Preliminary example i) Let x be a real variable, n be a positive integer and $P(n)$ denote the set of polynomials $p(x)$ degree at most n in x. We then pose the question: "what are the differential operators $A(x, \frac{d}{dx})$ preserving $P(n)$?". The answer is given in [1,2]: the operator A is (up to an element of the kernel of $P(n)$) an element of the envelopping algebra generated by the three operators

$$j_+ = x^2 \frac{d}{dx} - nx \;,\; j_0 = x\frac{d}{dx} - \frac{n}{2} \;,\; j_- = \frac{d}{dx} \tag{1}$$

These operators obey the (n-independant) commutation relations of the Lie algebra $sl(2)$

$$[j_\pm, j_0] = \mp j_\pm \quad , \quad [j_+, j_-] = -2j_0 \tag{2}$$

and constitute an $n+1$-dimensional irreducible representation of this algebra. This idea is at the basis of the theory of quasi-exactly-solvable (QES) operators. They refer to a class of spectral problems for which a finite number of eigenvalues and eigenvectors can be computed by solving an algebraic equation. Several examples of QES operators are given in [1,3], using the approach by representation of Lie algebra that we will follow here too. It is worth to mention that a different approach of QES problem was developped by A. Ushveridze and is, namely, the object of a very complete monograph [4].

The relations (2) reveal the deep connexion between the operators preserving the vector space $P(n)$ and group theory; we now present another example.

Preliminary example ii)   We consider the vector space $P(n-1,n) \equiv P(n-1) \oplus P(n)$, i.e.

$$P(n-1,n) = \{(p(x), q(x)) \mid \deg_x p \leq n-1, \deg_x q \leq n\}. \tag{3}$$

and ask the same question as above : "What are the linear differential operators A preserving $P(n-1,n)$?".

The answer is that A is an element of the envelopping algebra generated by the four operators

$$Q_0 = \sigma_- \quad , \quad Q_1 = x\sigma_- \tag{4}$$

$$\bar{Q}^0 = (x\frac{d}{dx} - n)\sigma_+ \quad , \quad \bar{Q}^1 = \frac{d}{dx}\sigma_+ \tag{5}$$

with $\sigma_\pm = (\sigma_1 \pm i\sigma_2)/2$. The four operators (4),(5), supplemented by the four defined by $J_a^b \equiv Q_a \bar{Q}^b + \bar{Q}^b Q_a$ constitute an irreductible representation of the graded Lie algebra $spl(1,2)$ (also isomorphic to $osp(2,2)$) with dimension $2n+1$. The structure relations of this algebra can be obtained from (25) below with $q=1$.

Many generalizations of the above examples can be considered, here are a few possibilities.

1) Consider the examples i) (resp. ii)) above for polynomials of several variables $x_1, x_2, ..., x_v$. The relevant operators are then related to the Lie algebra $sl(V+1)$ (resp. the graded Lie algebra $spl(V+1,1)$) [5].

2) Classify the operators preserving the direct sum $P(n-\delta, n) \equiv P(n-\Delta) \oplus P(n)$. The underlying algebra, say $\mathcal{A}$., is a non linear algebra. It will be presented in Sect.2 and its representations will be discussed in Sect.3.

3) Extend the case ii) above to more general sums $P(n_i) \oplus P(n_2) \oplus ... \oplus P(n_k)$. This was considered in [6].

4) Consider all the problems evoked above but for finite difference (rather than differential) operators. As illustrated in Sect.4, the underlying operators are related to deformations of the algebras mentionned above.

The differential operators produced by these different extentions of the case i) can be transformed, after a suitable change of the variable and a change of function, into quasi exactly solvable Schroedinger operators which, having a hidden algebra, can be partially solved algebraically. This does not exclude, however, the existence of Schoedinger operators which, while completely solvable algebraically, do not possess a hidden algebra; we construct a family of such operators in Sect.5.

COUPLES OF POLYNOMIALS

In order to construct the operators preserving the vector space $P(n - \Delta, n)$, it is convenient to introduce a completely symmetric multi-index $[A] \equiv a_1, a_2..., a_\Delta$, $a_k = 0$ or 1. We first define two sets of $\Delta + 1$ operators, $Q_{[A]}$ and $\bar{Q}^{[A]}$, as follows

$$Q_{[A]} = x^\delta \sigma_- \tag{6}$$

where δ is the number of "1" in [A]; and (defining $D = x\frac{d}{dx}$)

$$\bar{Q}^{11...11} = (\frac{d}{dx})^\Delta \sigma_+$$

$$\bar{Q}^{01...11} = (D - n + \Delta - 1)(\frac{d}{dx})^{\Delta-1} \sigma_+$$

$$\bar{Q}^{001...11} = (D - n + s - 1)(D - n + \Delta - 2)(\frac{d}{dx})^{\Delta-2} \sigma_+$$

...

$$\bar{Q}^{00...00} = (D - n + \Delta - 1)(D - n + \Delta - 2)...(D - n)\sigma_+ \tag{7}$$

Then, we further introduce four operators J_a^b $(a, b = 0, 1)$ according to

$$J_a^b = \begin{pmatrix} j_a^b(n - \Delta) & 0 \\ 0 & j_a^b(n) \end{pmatrix} + \delta_a^b \begin{pmatrix} \kappa_1 & 0 \\ 0 & \kappa_2 \end{pmatrix} \tag{8}$$

$$j_0^0(n) = D - n \quad , \quad j_1^1 = -D \quad , \quad j_1^0(n) = -x(D - n) \quad , \quad j_0^1 = \frac{d}{dx} \tag{9}$$

where κ_1, κ_2 are arbitrary constants. We then have [7] the
Proposition 1. The operator preserving $P(n - \Delta, n)$ are the element of the envelopping algebra generated by the $2\Delta + 6$ opertors $Q_{[A]}, \bar{Q}^{[A]}, j_a^b$.

We now discuss the commutation relations obeyed by the operators above. First, it can be checked easily that the J_a^b obey the commutation relations of $gl(2)$.

$$[J_a^b, J_c^d] = \delta_a^d J_c^b - \delta_c^b J_a^d \tag{10}$$

and that the $Q_{[A]}$ (and $\bar{Q}^{[A]}$) transform as the tensorial operators under the adjoint action of the $gl(2)$ subalgebra

$$[J_a^b, Q_{[C]}] = \kappa \delta_a^b Q_{[C]} - \sum_{k=1}^{\Delta} \delta_{C_k}^b Q_{[\hat{C}_k, a]} \tag{11}$$

$$[J_a^b, \bar{Q}^{[C]}] = -\kappa \delta_a^b \bar{Q}^{[C]} + \sum_{k=1}^{\Delta} \delta_a^{C_k} \bar{Q}^{[\hat{C}_k, a]} \tag{12}$$

The notation $[\hat{C}_k, a]$ means that the index c_k has been removed from $[C] = c_1, c_2, ..., c_\Delta$ and the index a has been added. The following relations lead to a complete set of the normal ordering rules between the operators

$$\{Q_{[A]}, Q_{[B]}\} = 0 \quad , \quad \{\bar{Q}^{[A]}, \bar{Q}^{[B]}\} = 0 \tag{13}$$

and

$$\{Q_{[A]} \bar{Q}^{[B]}\} = \sum_{k=0}^{\Delta} \alpha_a J_{[A]}^{[B]}(k) \tag{14}$$

with the definition

$$J_{[A]}^{[B]}(k) \equiv (\frac{1}{\Delta!})^2 S[A].S[B](J_{a_1}^{b_1} J_{a_2}^{b_2} ... J_{a_k}^{b_k} \delta_{a_{k+1}}^{b_{k+1}} \cdots \delta_{a_\Delta}^{b_\Delta}) \tag{15}$$

and $S[A]$ means the sum over all permutations of the indices a_k.

The constant κ in (11), (12) and the coefficient α_k in (14) depend on the choice κ_1, κ_2 in (8). The choice

$$\kappa_1 = -\frac{1+\Delta}{2} \quad , \quad \kappa_2 = -\frac{1-\Delta}{2} \tag{16}$$

leads to $\kappa = \Delta$ and leads to the following equations for the α_k

$$\sum_{k=0}^{\Delta} \alpha_k (y + \frac{\Delta-1}{2})^k = \Pi_{j=0}^{\Delta-1}(y+j) \tag{17}$$

The set relations given above define ordering rules among the operators (6),(7),(8) but they do not obey the generalized Jacobi identities which would promote these rules to the rank of structure relations defining an abstract algebra. We have tried to modify the relations above minimally in order to render then compatible with the Jacobi identities. It appears that (13) is very stringent because it implies $\frac{(\Delta+1)(\Delta+2)}{2}$ relations among the anti-commutators (trivially satisfied by the particular representation (6),(7)).

After trials to replace (13) by some weaker conditions we came to the conditions

$$S[A, B]\{Q_{[A]}, Q_{[B]}\} = 0 \tag{18}$$

$$S[A,B]\{\bar{Q}^{[A]}, \bar{Q}^{[B]}\} = 0 \tag{19}$$

which result into $2\Delta + 1$ independent conditions for the anticommutators among the Q (and similarly for the $\bar{Q}$). Our result [5] is the summarized by

Proposition 2

The set of rules (10), (11), (12), (13), (18), (19) fulfill all Jacobi identities provided $\kappa = \Delta$. The parameters α_k in (14) are free.

So we end up with a family of associative algebras, say $\mathcal{A}_\Delta(\alpha_k)$. In fact, only $\Delta - 1$ of the parameters are non trivial, we can set $\alpha_\Delta = 1$ by normalizing the Q and $\alpha_{\Delta-1} = 0$ by translating the J. To finish this section, we mention that, when Δ is even, an alternative choice to (18) is also compatible with associativity, this is discussed in [5].

REPRESENTATIONS

In order to construct the representation of the algebra $\mathcal{A}(\Delta)$ obtained in the previous section it is usefull to observe that the operator

$$T = \frac{1}{\Delta}(J_0^0 + J_1^1) \tag{20}$$

is a grading operator:

$$[T, J_a^b] = 0 \;,\; [T, Q_{[A]}] = Q_{[A]} \;,\; [T, \bar{Q}^{[A]}] = -\bar{Q}^{[A]} \tag{21}$$

The space $\mathcal{V}$ of the representation can be decomposed into eigenspace of T. Starting from the subspace $\mathcal{V}_0$ which is annihilated by the $\bar{Q}$'s, the other levels of the representation space are acceeded by applying successive powers of Q of $\mathcal{V}_0$. We further assume $Q^N = 0$ for some $N \in \mathbb{N}$ to guarantee the representation to be finite dimensional. After these generalities we discuss the cases $\Delta = 1$ and $\Delta = 2$.

Case $\Delta = 1$.

Applying the reasonning on the case $\Delta = 1$, we find the space of the generic representation to be

$$\mathcal{V} = P(n) \oplus P(n+1) \oplus P(n-1) \oplus P(n) \tag{22}$$

whose T eigenvalue are respectively $t, t-1, t+1, t+2$. The representation is the labelled by $n \in \mathbb{N}, t \in \mathbb{C}$. It is irreducible for generic values of t but if $t = n+2$, the subspace $P(n-1) \oplus P(n)$ decouple and provide an atypical representation; it is precisely the one we started with in the preliminary problem ii). In fact this is just a reformulation of the result of [8] in terms of polynomials.

Case $\Delta = 2$.

We now illustrate the generic irreducible representations of the algebra $\mathcal{A}_2(\alpha_0)$. The representations are characterized by two integers L, n, respectively the number of levels of the operator T and the dimension of the space corresponding to the highest

T-eigenvalue	polynomial content
t	$P(n)$
$t+1$	$P(n-1) \oplus P(n) \oplus P(n+2)$
$t+2$	$P(n-2) \oplus P(n) \oplus P(n) \oplus P(n+2)$
...	...
$t+L-3$	$P(n-2) \oplus P(n) \oplus P(n) \oplus P(n+2)$
$t+L-2$	$P(n-2) \oplus P(n) \oplus P(n+2)$
$t+L-1$	$P(n)$

eigenvalue of T, say $t \in \mathbb{C}$. The T-eigenspaces organize as follows

The only condition implied by the (anti) commutation relations is $t + \alpha_0 = \frac{L-1}{2}$ In fact, several extra conditions on t, L, m, α_0 lead to five types of atypical representations which are discussed at length in [9].

FINITE DIFFERENCE OPERATORS

We finally come to the topic of finite difference operators preserving spaces of polynomials. In this purpose, it is convenient to introduce the Jackson derivative

$$D_q f(x) = \frac{f(x) - f(qx)}{x(1-q)} \quad , \quad q \in \mathbb{R}^+ \quad , \quad q \neq 0, 1 \tag{23}$$

$$D_q x^n = [n]_q X^n \quad , \quad [n]_q = \frac{1-q^n}{1-q} \tag{24}$$

A few years ago, it was observed that the finite difference operators preserving $P(n)$ are related to the particular deformation of $sl(2)$ known as the Witten-Woronowicz deformation [10,11].

Owing that the majority of physically relevant examples of quasi-exactly solvable matrix operators are related to $osp(2,2)$, it was natural to try to construct a deformation of this graded algebra which admit representation expressible in terms of the finite difference operator (23). This is on of the result presented in [12], the relevant deformation of $osp(2,2)$ reads

$$\{Q_a, Q_b\}_{q^{b-a}} = 0 \quad , \quad \{\bar{Q}^a, \bar{Q}^b\}_{q^{b-a}} = 0 \quad , \quad \{Q_a, \bar{Q}^b\} = J_a^b$$
$$[J_a^b, Q_c] = q^{\frac{c-b-1}{2}}(\delta_a^b Q_c - \delta_a^b Q_a)$$
$$[J_a^b, \bar{Q}^c] = q^{\frac{a-b-1}{2}}(\delta_a^c \bar{Q}^b - \delta_a^b \bar{Q}^c)$$
$$[J_a^b, \bar{J}_c^d] = q^{\frac{S-r-1}{2}}(\delta_a^d J_c^b - q^r \delta_c^b J_a^d) + \frac{q-1}{q^2}(Q_0 \bar{Q}^0 + Q_1 \bar{Q}^1)\delta_a^d \delta_c^b (1 - \delta_a^b \delta_c^d) \tag{25}$$

with $s \equiv b + c - a - d$, $r \equiv (a-c)(b-d)$ and

$$[A, B]_q \equiv AB - qBA \quad , \quad \{A, B\} - q = AB + qBA \tag{26}$$

Irreducible representations preserving $P(n-1) \oplus P(n)$ can be constructed in the form

$$Q_0 = q^{\frac{-n}{2}}\sigma_- \quad , \quad Q_1 = -x\sigma_-$$
$$\bar{Q}^0 = q^{\frac{-n}{2}}(xD_1 - [n]_1)\sigma_+ \quad , \quad \bar{Q}^1 = D_q\sigma_+. \tag{27}$$

SOME COUNTEREXAMPLES

Let us now turn back to the QES problems related to the algebra (1). They are defined by one dimensional Schroedinger equation

$$-\frac{d^2\psi}{dy^2} + V(y)\psi(y) = E\psi(y) \tag{28}$$

having the property that, under an appropriate change of the wave function and of the independent variable

$$\psi(y) = \psi_0(y)F(y) \quad , \quad y = y(x) \tag{29}$$

the Schroedinger operator becomes an element of the enveloping algebra of sl(2) represented by the operators (1) for some n. The n+1 analytically accessible wave functions have then the form of the product of some fixed function ψ_0 times a polynomial of degree less or equal to n in the new variable.

There is a long-standing problem whether for any QES onedimensional problem there is an underlying sl(2) hidden symmetry. This problem, as it stands, is somewhat vague because it strongly depends on what we mean by "analytic methods". For example, one can view as QES all problems that can be, by any method available, transformed to the one which has an underlying hidden sl(2) symmetry. In the examples given in [13], [14], respectively, the supersymmetry and inverse scattering method were used to enlarge the number of problems with analytically accessible part of energy spectrum. These new QES equations do not, in general, share the property that, in appropriate variable, the available wave functions are, up to a common factor, polynomials. Therefore, within this wider definition of QES problems there are ones which do not fit in general scheme of sl(2) hidden symmetry. However, as we are going to show below, there are even exactly solvable problems, with all eigenfunctions of the (up to a common factor) polynomial form, with no underlying sl(2) hidden symmetry.

Let us consider the reflectionless potential V which has N bound states of energies

$$E_n = -(N-n)^2\kappa^2 \quad , \quad n = 0, 1, ..., N-1 \tag{30}$$

Using the inverse scattering method (in the form given in [15]) one can construct the most general potential V. It reads

$$V(y) = -2\frac{d^2}{dy^2}\log \det A(y) \tag{31}$$

where $A(y)$ is an $N \times N$ matrix defined by

$$A_{mn}(y) = \delta_{mn} + \frac{\beta_m}{m+n}\exp(\kappa(m+n)y) \tag{32}$$

where β_m are arbitrary real numbers. Moreover, it is easy to check that the bound state wave functions can be written as

$$\psi_n(y) = \frac{\det A^{(n)}(y)}{\det A(y)} \tag{33}$$

where $A^{(n)}$ denotes the matrix A of Eq.(32) with the elements of the nth column replaced by $\beta_m \exp(\kappa m y)$. It is obvious from the above equations that, up to the common factor $1/\det A$, all bound state wave functions are polynomials in $w \equiv \exp(\kappa y)$. Let us now pose the question whether there is an sl(2) symmetry present here. There is just one case where the answer is yes. One can check easily that $V(y)$ reduces to the reflectionless Poschl-Teller potential if the following constraint on the parameters β_m is imposed

$$\det A = (1 + w^2)^{N(N+1)/2} \tag{34}$$

This equation can be always solved with respect to the β's, yielding real positive solutions, because eq.(31) gives the most general form of $V(y)$ which covers the Poschl-Teller case. Our conjecture is that, for any N, the only sl(2)-symmetric potentials of the kind considered here are those obtained from Poschl-Teller potential by space translation. It is not hard to see that, in order to prove our conjecture we have to show only that all ratios ψ_n/ψ_0, $n = 2, 3, ..., N-1$ are polynomially expressible in terms of ψ_1/ψ_0 if and only if the parameters beta belong to one-parameter family yielding shifted Poschl-Teller potential. We have checked explicitly that this holds true for N=3,4,5.

Acknowledgments. This work was supported by the Belgian F.N.R.S. and carried out under the grant n⁰ 2 P03B 076 10.

REFERENCES

1. Turbiner, A. V., *Comm. Math. Phys.* **119**, 467 (1988).
2. Turbiner, A. V., *J. Phys. A* **25**, L 1087 (1992).
3. Shifman, M. A., and Turbiner, A. V., *Comm. Math. Phys.* **120**, 347 (1989).
4. Ushveridze, A. G., *Quasi exact solvability in quantum mechanics* (Institute of Physics Publishing, Bristol and Philadelphia, 1993).
5. Brihaye, Y., and Nuyts, J., 'The hidden symmetry algebras of a class of quasi exactly solvable multi dimensional equation', *Commun. Math. Phys.* in press, q-alg:9701016.

6. Brihaye, Y., Giller, S., Gonera, C., and Kosinski, P., *J. Math. Phys.* **36**, 4340 (1995).
7. Brihaye, Y., and Kosinski, P., *J. Math. Phys.* **35**, 3089 (1994).
8. Scheunert, M., Nahm, W., and Rittenberg, V., *J. Math. Phys.* **18**, 155 (1977).
9. Brihaye, Y., Giller, S., Kosinski, P., and Nuyts, J., *Commun. Math. Phys.* **187**, 201 (1997).
10. Witten, E., *Nucl. Phys. B* **330**, 285 (1990).
11. Woronowicz, S., *Commun. Math. Phys.* **111** 613 (1987).
12. Brihaye, Y., Giller, S., and Kosinski, P., *spl(p,q) superalgebra and differential operators*, Lodz-Mons preprint in Memory of Stanislaw Malinowski (1997), q-alg:9710018.
13. Shifman, M. A., *Int. Journ. Mod. Phys.* **A4** 3305 (1989).
14. Ushveridze, A. G., *Mod. Phys. Lett.* **A6** 739 (1991).
15. Abraham, P. B., and Moses, H. E., *Phys. Rev.* **A22** 1333 (1980).

On Conformal Reflections in Compactified Phase Space

P. Budinich

International School for Advances Studies and International Center for Theoretical Physics, Strada Costiera 13, 34014 Trieste,Italy

Abstract. Some results from arguments of research dealt with R. Raczka are exposed and extended. In particular new arguments are brought in favor of the conjecture, formulated with him, that both space-time and momentum may be conformally compactified, building up a compact phase space of automorphism for the conformal group, where conformal reflections determine a convolution between space-time and momentum space which may have consequence of interest for both classical and quantum physics.

INTRODUCTION

In 1931 P.M.A. Dirac wrote [1]: *"There are at present fundamental problems in theoretical physics ... whose solution will presumably require a more drastic revision of our fundamental concepts than any that have gone before. Quite likely these changes will be so great that it will be beyond the power of human intelligence to get the necessary new ideas by direct attempts to formulate the experimental data in mathematical terms. The theoretical work in the future will therefore have to proceed in a more indirect way. The most powerful method of advance that can be suggested at present is to employ all the resources of pure mathematics in attempts to perfect and generalize the mathematical formalism that forms the existing basis of theoretical physics, and after each success in this direction, to try to interpret the new mathematical features in terms of physical entities ..."*. Later he added: *"It seems that, if one is working from the point of view of getting beauty in one's equations, and if one has really sound insight, one is on a sure line of progress"*.

Dirac started himself the process of research in abstract mathematics, aimed at the understanding of physical quantum phenomena, which brought him to discover his beautiful spinor field equation which, not only explained the origin of electron spin, but even anticipated the discovery of new, unexpected physical phenomena; as those deriving from the existence of antimatter. He was certainly a forerunner and now, after 67 years, when looking at the recent developments in theoretical physics, where some branches like quantum groups, noncommutative geometry, deal mainly with subjects of pure mathematics, his words sound like a prophecy.

Having in mind his recommendations, an undeniable source of beauty, of relevance for quantum physical phenomena, may certainly be found in the E. Cartan work on simple spinor geometry [2]. E. Cartan stressed specially two concepts.

1. Vectors of euclidean spaces may be conceived as bilinearly composed by spinors.

2. Rotations in vector spaces may be decomposed in products of reflections.

Notoriously, the main geometrical tool for the description of the phenomena of classical mechanics is euclidean geometry in pseudo euclidean vector spaces and rotation therein. Following E. Cartan, spinors and reflection may be conceived as the elementary constituents of these.

The study of covariance of the equations of motion with respect to rotation groups has been one of the main mathematical instrument of research in classical mechanics in the last and also in the present century, when covariance of Maxwell's equations with respect to the Lorentz-Poincaré groups has brought Einstein to the discovery of special relativity and of the geometrical structure of space-time: $\mathbb{M} = \mathbb{R}^{3,1}$.

The importance of reflection groups, instead, has been only recognized after the advent of quantum mechanics. In fact space-time reflection group play an important role in the explanation of quantum phenomena. Precisely space reflections allow us to define the concept of parity (not conserved in weak decays) while time inversion allows us to understand the existence of antimatter. The most appropriate space for dealing with such reflections seems to be spinor space.

It is known since 1909 that Maxwell's equations are also covariant with respect to the conformal group. A discovery which has also brought to the conjecture that Minkowski space-time may be conformally compactified, and represented by a particular realization of an homogeneous space of the conformal group. With R. Raczka, somehow in line with Dirac's recommendations, we have observed that another realization of that homogeneous space could well represent conformally compactified momentum space which, together with compactified space-time, could then build up a compact phase space where both the concept of infinity and that of infinitesimal, of difficult, if not impossible, self consistent mathematical definition, would not be needed, allowing then not only a rigorous mathematical formulation of theoretical physics, but also the elimination of the main difficulties encountered by quantum physics: the ones of so called infrared and ultraviolet divergences.

The conformal group, in which the Lorentz-Poincaré group is contained as a subgroup, has a reflection group which contains, beside space-time reflections, mentioned above, also two additional reflections, called conformal reflections. However, somehow surprisingly, despite the relevance of conformal covariance for the understanding of several physical phenomena concerning massless systems, these reflections have failed, up to now, to manifest their role in physics.

One of purposes of this paper is to review some of the results obtained in the work with R. Raczka and to outline some follow up of our thinking, in particular on the possible role in physics of conformal reflections, obtained after he left us.

COMPACT PHASE SPACE.

In a similar way as, from Maxwell's equations Lorentz covariance, Minkowski derived the pseudo euclidean structure of space-time: $\mathbb{M} = \mathbb{R}^{3,1}$, from their conformal covariance Veblen derived and adopted [3] the conformally compactified structure $\mathbb{M}_c$ of $\mathbb{M}$:

$$\mathbb{M}_c = \frac{S^3 \times S^1}{\mathbb{Z}_2} , \qquad (1)$$

in which Minkowski space-time $\mathbb{M}$ is densely contained (which means: to every point of $\mathbb{M}_c$ there correspond one point of $\mathbb{M}$; to a submanifold of $\mathbb{M}_c$, of dimension 3, there corresponds the points of $\mathbb{M}$ at infinity) afterwards also adopted by several authors [4].

We will show now that the same argument which induces to postulate that $\mathbb{M}$ is compactified induces also to postulate that its Fourier dual momentum space $\mathbb{P} = \mathbb{R}^{3,1}$ is conformally compactified as well.

In fact the conformal group may be linearly represented by $O(4,2)$ acting in $\mathbb{R}^{4,2}$. Let us consider the equations for the corresponding Weyl spinors or twistors:

$$\sum_{a=1}^{6} p_a \gamma^a (1 \pm \gamma_7) \pi = 0 , \qquad (2)$$

where $p_a \in \mathbb{R}^{4,2}$, γ^a are the generators of the Clifford algebra $\mathbb{C}\ell(4,2)$, γ_7 its volume elements and π is a vector of the spinor space S defined by $\text{End} S = \mathbb{C}\ell(4,2)$.

Eq. (2) for non zero twistors $\pi_\pm = \frac{1}{2}(1 \pm \gamma_7) \pi$, implies $p_a p^a = 0$, and therefore the directions of p_a form the projective quadric $\mathbb{P}_c$:

$$\mathbb{P}_c = \frac{S^3 \times S^1}{\mathbb{Z}_2}. \qquad (3)$$

It may be easily seen that eq. (2) contains the equations:

$$\sum_{\mu=0}^{3} p_\mu \gamma^\mu \varphi_\pm = 0 , \qquad (4)$$

where $\varphi_\pm$ represent Weyl spinors associated with $\mathbb{C}\ell(3,1)$, of which γ^μ are the generators. Eq. (4) is the Weyl equation in momentum space $\mathbb{P} = \mathbb{R}^{3,1}$, Fourier dual of Minkowski space $\mathbb{M} = \mathbb{R}^{3,1}$, from it is easy to obtain [5] the equations:

$$p_\mu F_+^{\mu\nu} = 0 ; \qquad p_\mu F_-^{\mu\nu} = 0 , \qquad (5)$$

where
$$F_\pm^{\mu\nu} = \langle \varphi_\pm^\dagger [\gamma^\mu, \gamma^\nu] \varphi_\pm \rangle .$$

Eq.s (5) represent the homogeneous Maxwell's equations, in momentum space $\mathbb{P} = \mathbb{R}^{3,1}$, which then results densely contained in $\mathbb{P}_c$ of eq. (3).

This short-cut derivation of Maxwell's equations in momentum space from twistors equations indicates that, in so far, Maxwell's equations conformal covariance implies the conformal compactification $\mathbb{M}_c$, as given in eq. (1), of Minkowski space $\mathbb{M}$, their derivability from twistors equations in momentum space implies the conformal compactification $\mathbb{P}_c$ of momentum space $\mathbb{P}$, as given in eq. (3), as well.

The resulting phase space will then be compact and consequently any field theory formulated in such a compact phase space should, a priori, expected to be free from both infrared and ultraviolet divergences.

The main problem will be to define, for every function $f(x)$ taking values in $\mathbb{M}_c$, a transform to a function $F(k)$, taking values in $\mathbb{P}_c$, such that in the flat limit (radiuses of S^3 and S^1 going to infinity) it identifies with the standard Fourier transform correlating $\mathbb{M}$ and its dual $\mathbb{P}$.

The problem may be solved [6] for the two-dimensional case $\mathbb{M} = \mathbb{R}^{1,1} = \mathbb{P}$, for which:

$$\mathbb{M}_c = \frac{S^1 \times S^1}{\mathbb{Z}_2} = \mathbb{P}_c ; \qquad (6)$$

one needs only to inscribe in each S^1 a regular polygon with $2N = 2\pi RK$ vertices, where R (of dimension $[\ell]$: length) and K (of dimension $[\ell^{-1}]$) are the radiuses of the $\mathbb{M}_c$ and $\mathbb{P}_c$ circles respectively. They define in $\mathbb{M}_c$ and $\mathbb{P}_c$ two lattices: $M_L \subset \mathbb{M}_c$; $P_L \subset \mathbb{P}_c$ which are Fourier dual. Indicating in fact with $f(x_{nm})$ a function taking values in M_L and with $F(k_{\rho\tau})$ a function taking values in P_L we have:

$$f(x_{nm}) = \frac{1}{2\pi R^2} \sum_{\rho,\tau=-N}^{N-1} \varepsilon^{(n\rho-m\tau)} F(k_{\rho\tau}) ,$$

$$F(k_{\rho\tau}) = \frac{1}{2\pi K^2} \sum_{n,m=-N}^{N-1} \varepsilon^{-(n\rho-m\tau)} f(x_{nm}) ,$$

(7)

where $\varepsilon = e^{i\frac{\pi}{N}}$ is the $2N$-root of unity.

Eqs. (7), in the limit $R, K \to \infty$ may be easily identified with the standard Fourier transforms in $\mathbb{M} = \mathbb{R}^{1,1} = \mathbb{P}$.

It is obvious that in M_L and P_L any field theory will be free from both infrared and ultraviolet divergences.

In the realistic, four dimensional case, since, in principle, the concept of infinity and infinitesimal should not be realizable, one could expect again that phase space should restrict to discrete and fine lattices which however do not seem to be obtainable with standard mathematical algorithms [7], through which, instead one may anticipate some aspects of the convergences of field theories in conformally compactified space-time and momentum space.

In fact it is known that the space $\mathbb{P}_c$, given by eq. (3), is conformally flat i.e

$$g_{\mu\nu}(p) = \Omega^2(p)\, \eta_{\mu\nu},$$

where $\Omega(p)$ is the conformal factor and $\eta_{\mu\nu}$ is the metric tensor of flat $\mathbb{R}^{3,1}$. $\Omega(p)$ may be obtained [1] by adopting the Dirac six-dimensional formalism: $p_\mu = P_\mu/(P_5 + P_6)$ and, as shown in reference [8], it provides a convergence factor since:

$$\lim_{|p|^2 \to \infty} \Omega(p) = \frac{M^4}{p^4}, \tag{8}$$

where M is a mass scale ($c = 1$).
In particular for the de Sitter subgroup $SO(4,1)$ (obtained for $P_6 = 1$):

$$\Omega(p) = \frac{4M^2}{4M^2 + p^2}, \tag{9}$$

identical to the Pauli-Villars regularizing factor, often adopted in relativistic field theories for the elimination of ultraviolet divergences in perturbation expansions. For the anti-de Sitter group $SO(3,2)$ (for $P_5 = 1$) we obtained:

$$\Omega(p) = \frac{4M^2}{4M^2 - p^2}. \tag{10}$$

The analogous procedure starting from space-time $\mathbb{M}$ will provide convergence factors $\Omega(x)$ which will eliminate infrared divergences. That conformally covariant theories may be free from divergences was also shown by Mack and Todorov [9].

THE HOMOGENEOUS SPACE, ACTION OF CONFORMAL REFLECTIONS.

In reference [8] it was shown how $\mathbb{M}_c$ and $\mathbb{P}_c$ may be represented as homogeneous spaces of the conformal group $C = L \otimes D \rtimes P^{(4)} \rtimes S^{(4)}$ (where L, D, $P^{(4)}$, $S^{(4)}$ stand for Lorentz-Dilatation-Poincaré-Special conformal-transformations, respectively).
Precisely:

$$\mathbb{M}_c = \frac{C}{c_1} \; ; \quad \mathbb{P}_c = \frac{C}{c_2}, \tag{11}$$

---

[1] It is interesting to observe that $\Omega(p)$ may be rigorously set in the form $\Omega(p) = M^2/p_W^2$, where p_W^2 is the Wick rotated p^2, that is $p_W^2 = p_1^2 + p_2^2 + p_3^2 + p_0^2$ which ensures the non singularity of $\Omega(p)$, which instead is not guaranteed by (9) and (10) for p^2 time-like and space-like respectively.

where $c_1 = L \otimes D \rtimes S^{(4)}$; $c_2 = L \otimes D \rtimes P^{(4)}$. $\mathbb{M}_c$ and $\mathbb{P}_c$ in eq. (11) are both isomorphic to $(S^3 \times S^1)/\mathbb{Z}_2$. Furthermore if we represent with I a conformal reflection (a reflection with respect to a plane orthogonal to the 5th or 6th axis) then:

$$I \mathbb{M}_c I^{-1} = \mathbb{P}_c \ . \tag{12}$$

If we now consider the conformal group C inclusive of reflections (represented in $\mathbb{R}^{4,2}$ by $O(4,2)$), then $\mathbb{M}_c$ and $\mathbb{P}_c$ are two copies of the same homogeneous space of C, transformed in each other by conformal reflections. And then neither $\mathbb{M}_c$ nor $\mathbb{P}_c$ are automorphism spaces for C, but only the two taken together; that is conformally compactified phase space. A conformal reflection I determines a convolution between $\mathbb{M}_c$ and $\mathbb{P}_c$ and then also between $\mathbb{M}$ and $\mathbb{P}$. This duality, which could be named conformal duality, to distinguished it from the quite different Fourier one, could be of relevance for physics.

The action of I in space-time $\mathbb{M}$, densely imbedded in $\mathbb{M}_c$, is well known; for $x_\mu \in \mathbb{M}$:

$$I: \quad x_\mu \to I(x_\mu) = \pm \frac{x_\mu}{x^2} \ , \tag{13}$$

which, for x_μ space like, is often interpreted as the map of every point, inside a unit sphere S^2 in ordinary space[2], at a distance x form its centre to a point (on the same ray) at a distance x^{-1}:

$$I: \quad x \to I(x) = \frac{1}{x} \ . \tag{14}$$

For the physical interpretation x is thought to be dimensionless that is represented by x/L where L is an arbitrary unit at length (and this breaks conformal covariance which is already broken together with Lorentz covariance in (14)). With this interpretation a micro world ($x/L \ll 1$) is transformed by I to the macro world ($x/L \gg 1$) in ordinary 3D space.

The corresponding interpretation may be also adopted [8] for the action of I in momentum space $\mathbb{P} = \mathbb{R}^{3,1}$ densely imbedded in $\mathbb{P}_c$, where, for $k_\mu \in \mathbb{P}$:

$$I: \quad k_\mu \to I(k_\mu) = \pm \frac{k_\mu}{k^2} \tag{15}$$

and correspondingly

$$I: \quad k \to I(k) = \frac{1}{k} \ . \tag{16}$$

---

[2] It is interesting to observe that, for $\mathbb{M} = \mathbb{R}^{2,1}$, for which the conformal group is represented by the anti-de Sitter group $O(3,2)$, the sphere S^2 reduces to a circle S^1 and then eq. (14) reminds the Target Space duality in string theory [10], which then might be correlated with conformal duality advocated in this paper.

If instead we take into account of (12) eq. (14) must be interpreted as bringing a point of $\mathbb{M}$ to a point of $\mathbb{P}$ and therefore we do not need to interpret x as dimensionless, we may give it the meaning of a length and then eq. (14) becomes:

$$I : x \to I(x) = \frac{1}{x} = k \in \mathbb{P} \tag{17}$$

and similarly eq. (16) becomes

$$I : k \to I(k) = \frac{1}{k} = x \in \mathbb{M} . \tag{18}$$

Reminding that C is an automorphism group for phase space and that the physical momentum p is obtained multiplying k by an unit of action $H : p = H \cdot k$, taking together the above equations we arrive to the following interpretation for the action of I in physical phase space:

$$I : \frac{xp}{H} \to I\left(\frac{xp}{H}\right) = \frac{H}{xp} ; \tag{19}$$

which means: in phase space conformal inversion I brings from regions of where the action is $\ll H$ to those where it is $\gg H$; that is from those appropriate for the description of quantum phenomena (in the micro world) to those appropriate for the description of classical phenomena (in the macro world). It could then represent a sort of geometrical prerequisite for the realization of the correspondence principle.

Now, since the conformal inversion I brings also from $\mathbb{M}_c$ to $\mathbb{P}_c$ and vice versa, it would appear that, since obviously $\mathbb{M}_c$ is appropriate for the description of classical mechanics with the geometrical instrument of euclidean geometry, momentum space could be the most appropriate for the description of quantum mechanics, and in this space the most appropriate geometrical instrument for its description seems to be spinor geometry.

SPINOR REPRESENTATION OF QUANTUM MECHANICS IN MOMENTUM SPACE.

Fermions are the most elementary constituents of matter. Their properties may be ideally described in the frame of spinor geometry, discovered by E. Cartan [2], which, for what concerns us, may be summarized as follows [11].

Given a real, $2n$ dimensional, vector space V with scalar product g with signature $(k, l); k + l = 2n$, the corresponding Clifford algebra $\mathbb{C}\ell(k, l)$, is central simple and has one, up to equivalence, representation:

$$\gamma : \mathbb{C}\ell(k, l) = \text{End } S_D \tag{20}$$

in a complex, 2^n dimensional space S_D of Dirac spinors. If γ_a, obeying $[\gamma_a, \gamma_b]_+ = 2g_{ab}$, are the generators of $\mathbb{C}\ell(k, l)$ and p_a the components of a vector $p \in V$, a Dirac spinor ψ may be defined through the Cartan's equation

$$\sum_{a=1}^{2n} p_a \gamma^a \psi = 0. \tag{21}$$

For $\psi \neq 0$, we have that the vector p is null: $p_a p^a = 0$ which implies that the directions of p_a form the compact, projective quadric:

$$P_c = \frac{S^{k-1} \times S^{l-1}}{\mathbb{Z}_2}, \tag{22}$$

of which eq. (3) is a particular case for the signature $(4, 2)$.

Eq. (21) associates to each spinor ψ a totally null plane in V defined by all null, mutually orthogonal vectors $p \in V$ satisfying it. When such a plane has dimension n, that is maximal, the spinor ψ was named simple by E. Cartan (and pure by C. Chevalley).

E. Cartan has further shown how vectors of euclidean geometry in V may be conceived as bilinearly composed of spinors. In fact if we represent the generators γ_a of $\mathbb{C}\ell(k,l)$ with $2^n \times 2^n$ matrices acting on spinor space S, also the transposed matrices γ_a^t, defined by

$$\gamma_a^t = B \gamma_a B^{-1}, \tag{23}$$

will generate $\mathbb{C}\ell(k,l)$ and B is uniquely defined since $\mathbb{C}\ell(k,l)$ is simple, and they will act on the dual of S_D. We will have then, for ψ and $\phi \in S_D$ [12]:

$$\psi \otimes B\phi = \frac{1}{2^n} \sum_{k=0}^{2n} \gamma_{\mu_1} \gamma_{\mu_2} \ldots \gamma_{\mu_k} B_k^{\mu_1 \mu_2 \ldots \mu_k} (\psi, \phi) \tag{24}$$

where

$$B_k^{\mu_1 \mu_2 \ldots \mu_k}(\psi, \phi) = \langle B\psi, \gamma^{\mu_1} \gamma^{\mu_2} \ldots \gamma^{\mu_k} \phi \rangle,$$

in which:

$$1 \leq \mu_1 < \mu_2 \ldots \leq 2n.$$

From (24) we may then obtain:

$$(\gamma_a \psi \otimes B\phi) \gamma^a \psi = p_a \gamma^a \psi, \tag{25}$$

where

$$p_a = \langle B\phi, \gamma_a \psi \rangle. \tag{26}$$

Now $p_a p^a = 0$ for either ϕ or ψ simple [12] and then form (25) we obtain identically Cartan's eq. (21) where the p_a are bilinearly expressed in terms of spinors.

Define now with γ_{2n+1} the volume element of $\mathbb{C}\ell(2n) = \mathbb{C}\ell(k,l)$:

$$\gamma_{2n+1} = i^l \gamma_1 \gamma_2 \ldots \gamma_{2n}, \tag{27}$$

it anti commutes with all the γ_a, and it may be considered as the $(2n+1)^{th}$ generator of $\mathbb{C}\ell(k+1,l)$ which is a non simple algebra while its even sub algebra $\mathbb{C}\ell_0(2n+1)$ is simple: $\mathbb{C}\ell_0(2n+1) = EndS_P$ and the associated spinors are named Pauli spinors.

The volume element γ_{2n+1} defines the Weyl spinors $\psi_\pm$ of opposite helicity of $\mathbb{C}\ell(2n)$ corresponding to each Dirac spinor ψ:

$$\psi_\pm = \frac{1}{2}(1 \pm \gamma_{2n+1})\psi; \quad \psi_+ + \psi_- = \psi, \tag{28}$$

building up the endomorphism spaces of the even subalgebra $\mathbb{C}\ell_0(2n)$ of $\mathbb{C}\ell(2n)$, which is non-simple.

For physical applications we need the vectors p_a given in eq. (26) to be real. To this end we introduce the charge-conjugate spinor $\psi_c = C\bar{\psi}$ where $\bar{\psi}$ means ψ complex conjugate and C is defined by $C\gamma_a = \bar{\gamma}_a C$. Then we have [13] that for the signature $(k,l) = (m+1, m-1)$ the vectors:

$$p_a^\pm = \langle B\psi_c, \gamma_a(1 \pm \gamma_{2n+1})\psi\rangle, \tag{29}$$

are real (or imaginary) for m even while complex for m odd. That is p_a will be real for the signatures $(3,1)$, $(5,3)$, $(7,4)$ ... (for $(4,2)$, that is for twistors, p_a will be complex)[3].

It is remarkable that in this way one obtains [13] from Cartan's eq. (21), (where also the vectors p_a are conceived as bilinearly composed by spinors) not only the elementary equations of quantum mechanics in first quantization, however in momentum space; including Maxwell's equations (which somehow constitute a bridge between quantum and classical physics), but also those manifesting the so called internal symmetry.

In fact consider the following isomorphisms of algebras:

$\mathbb{C}\ell(2n)$ is isomorphic to $\mathbb{C}\ell_0(2n+1)$ – both simple
$\mathbb{C}\ell_0(2n)$ is isomorphic to $\mathbb{C}\ell(2n+1)$ – both non-simple

which allows to consider a Dirac spinor associated with $\mathbb{C}\ell(2n)$ as a direct sum of Weyl spinor or of Pauli spinor which in turn may be conceived as Dirac spinors of $\mathbb{C}\ell(2n-2)$ and so on. An elementary and historical example is the space-time Dirac spinor, direct sum of right- and left-handed Weyl spinors, which may also be considered as a doublet of Pauli spinors associated with $\mathbb{C}\ell(3)$ (for non relativistic motions).

In this way from the Cartan's eq. (21) for Weyl spinors associated with $\mathbb{C}\ell(5,3)$, taking into account of (29) the following equation is obtained [13]:

$$(p_\mu \gamma^\mu \cdot \mathbb{I} + \vec{\pi} \cdot \vec{\sigma} \otimes \gamma_5 + m \cdot \mathbb{I})N = 0, \tag{30}$$

where: $\vec{\pi} = \langle \tilde{N}, \vec{\sigma} \otimes \gamma_5 N\rangle$; $N = \begin{bmatrix}\psi_1 \\ \psi_2\end{bmatrix}$; $\tilde{N} = [\tilde{\psi}_1, \tilde{\psi}_2]$; $\tilde{\psi}_j = \psi_j^\dagger \gamma_0$,

with ψ_1, ψ_2—space-time Dirac spinors, and $\vec{\sigma} = (\sigma_1, \sigma_2, \sigma_3)$ Pauli matrices.

---

[3] It may be shown that they are also real for the lorenzian signature $(2n-1, 1)$.

Eq. (30) is formally identical to the proton-neutron equation interacting with the pseudoscalar isotriplet $\vec{\pi}$ representing the pion, however in momentum space, and the internal isospin symmetry appears as generated by the conformal reflections with respect to the planes orthogonal to the 5th, 6th and 7th axis, ans proton-neutron equivalence might represent a natural realization of quaternion algebra.

In fact it is known that a reflection with respect to a plane orthogonal to γ_a is represented in spinor space by $\psi \to \gamma_a \psi$; and, if γ_a is time-like, it has to be substituted by $i\gamma_a$, if we impose that the square of a reflection equals the identity, which is the case for γ_6 in $\mathbb{C}\ell(5,3)$, which brings to eq. (30). Furthermore also the pseudoscalar nature of the pion triplet is uniquely determined from spinor geometry since the representation in which the eighth component spinor N is a doublet of equivalent Dirac spinors, imposes for the gamma matrices to have the form:

$$\Gamma_\mu = \mathbb{I} \otimes \gamma_\mu \; ; \; \Gamma_5 = \sigma_1 \otimes \gamma_5 \; ; \; \Gamma_6 = i\sigma_2 \otimes \gamma_5 \; ; \; \Gamma_7 = \sigma_3 \otimes \gamma_5. \tag{31}$$

As we have seen the 8-component spinor N of eq. (30) may be also considered as a doublet of Weyl spinors associated with $\mathbb{C}\ell_0(4,2)$, or twistors: $\Psi = \begin{bmatrix} \pi_+ \\ \pi_- \end{bmatrix}$ obeying the Cartan's equation:

$$\left(p^a \tilde{\Gamma}_a + p^8 \cdot \mathbb{I} \right) \Psi = 0 \;, \tag{32}$$

where $\tilde{\Gamma}_a$ have the form:

$$\tilde{\Gamma}_\mu = \sigma_1 \otimes \gamma_\mu \; ; \; \tilde{\Gamma}_5 = \sigma_1 \otimes \gamma_5 \; ; \; \tilde{\Gamma}_6 = i\sigma_2 \otimes \mathbb{I} \; ; \; \tilde{\Gamma}_7 = \sigma_3 \otimes \mathbb{I}. \tag{33}$$

Let us now define the 8×8 matrix

$$U = \begin{bmatrix} L & R \\ R & L \end{bmatrix} = U^{-1} \;,$$

where $L = \frac{1}{2}(1+\gamma_5)$; $R = \frac{1}{2}(1-\gamma_5)$. It is easily seen that

$$U\Gamma_a U^{-1} = \tilde{\Gamma}_a \; ; \quad U = \Psi \; ; \quad U\Psi = N \;,$$

from which we have that $(1 \otimes L)N = N_L \equiv (1 \otimes L)\Psi = \Psi_L$.

But then eq.s (32) and (30) may be summed to give[4]

$$\left(p_\mu + \vec{A}_\mu \cdot \vec{\sigma} \right) \gamma^\mu N_L + B N_R = 0 \;, \tag{34}$$

where $\vec{A}_\mu = \langle \tilde{N}, \vec{\sigma} \otimes \gamma_\mu N \rangle$ and $N_R = (1 \otimes R) N$. If we now suppose that, of the two Dirac spinors ψ_1, ψ_2, the first represents the electron: e and the second the left handed neutrino: ν_L then eq. (34) becomes:

---

[4] The charged vector bosons derived from the 6-vector Z_a, bilinearly generated by twistors π_+, π_-, which are complex. They identically satisfy the equation $Z_a \gamma^a \pi_+ = 0$ and $\bar{Z}_a \gamma^a \pi_- = 0$, out of which the charged part of eq. (34) is obtained.

$$\left(p_\mu + \vec{A}_\mu \cdot \vec{\sigma}\right) \gamma^\mu \begin{bmatrix} e_L \\ \nu_L \end{bmatrix} + \alpha e_R = 0 \;, \tag{35}$$

where α is a free parameter.

Eq. (35) is the equation of the electroweak model, here derived from eq. s (32) and (30) both obtained from eq. (21) for $\mathbb{C}\ell(5,3)$. It has been shown [15], [16] that if one considers the triplet e_L, e_L^c, ν_L (or the corresponding 2-component Pauli spinors) to transform with $SU(3)$, the mixing angle Θ results determined such that $sin^2\Theta = 0.25$. Further details and consequences of these computations will be given elsewhere.

These examples, naturally derived from Cartan's eq. (21), representing some of the basic equations of quantum physics in momentum space, in the frame of spinor geometry, may induce to think that the method could be extended also to higher dimensional spinor spaces, e.g. associated with $\mathbb{C}\ell(9,1)$, in order to explain the multiplicities of elementary fermions (and bosons). In such spinor spaces the problem of dimensional reduction (from 10 to 4,say) of pseudo-euclidean vector spaces through *ad hoc* compactifications, could be avoided [15] [17].

FURTHER ASPECTS OF CONFORMAL DUALITY.

Conformally compactified phase space, conceived as an automorphism space of the extended conformal group, implies the conformal duality between space-time and momentum space which appears as a convolution, determined by conformal reflections. This duality might have two aspects of interest for physics.

The first follows from its comparison with Fourier duality which is defined trough functions, or physical fields, which may be defined in space time and its Fourier-dual momentum space. As we have seen in general, and in particular in the soluble two dimensional case, for a compact phase space such Fourier dual spaces may be only discrete and finite. As such they might be named the "physical" spaces, to be distinguished from the homogeneous or "mathematical" spaces $\mathbb{M}_c$ and $\mathbb{P}_c$ which are also finite but continuous. The "physical" discrete spaces will be both Fourier and conformally-dual while the "mathematical" spaces will be only conformally dual, and only the former should be the appropriate ones for description of physical phenomena.

The second derives from the possible correlation of conformal duality with the correspondence principle in so far it could be, as shown above, a sort of geometrical prerequisite for the realization of the correspondence principle, once one has found the motivation for identifying the unit of action H introduced in eq. (6) with the Planck's constant $\hbar$ (as in the de Broglie equality $p = \hbar \cdot k$). But it could perhaps, in any case, throw some light on some of the still somehow mysterious aspects of the correspondence principle, as we will try to show elsewhere. Furthermore one could expect that some of the geometrical aspects, that is of the topological and symmetry properties which are common to both $\mathbb{M}_c$ and $\mathbb{P}_c$ could be manifested by both classical and quantum physical systems independently (and above) of the correlation, specifically predicated by the correspondence principle (identification of wave functions with classical orbits for high quantum numbers). One of them is the $SO(4)$ symmetry which could be identified as the maximal compact subgroup of $SO(4,2)$ and manifested by the presence of its isometry sphere S^3 in both $\mathbb{M}_c$ and in its conformally dual $\mathbb{P}_c$; and then to be expected in both classical and quantum mechanical stationary (non relativistic) systems in ordinary- and momentum-space respectively. These systems exist they are the planetary motions in space-time and the H-atom in momentum space. In fact it is remarkable that the $SO(4)$ symmetry of the H-atom was discovered by V. Fock in the S^3 compactification of momentum space [18], which suggests that this $SO(4)$ symmetry might be a consequence of conformal duality, rather than being "accidental", as it was named by W. Pauli when discovered.

With R. Raczka [19] we have also conjectured an eigenvibration of the S^3 sphere of the Robertson Walker universe represented by

$$M_{RW} = S^3 \times R^1, \tag{36}$$

(which is often considered as a natural realization of $\mathbb{M}_c$ given by eq. (1) where R^1 is the infinite covering of S^1), in order to explain a remarkable regularity in the distribution of distant galaxies [20] (more than eleven peaks equally spaced by about $4 \cdot 10^8$ light years, in the direction of the North and South galactic poles). We have shown [21] how the astronomical data are well represented by the most symmetric spherical harmonic of S^3:

$$Y_{n,0,0} = k_n \frac{\sin{(n+1)\chi}}{\sin \chi} \tag{37}$$

where χ is the geodesic distance from center of the eigenvibration.

If further astronomical observations will confirm this model it would represent a remarkable test of conformal duality since eq. (37) is exactly the eigenfunction of the stationary S-states of the H-atoms found by V. Fock precisely in S^3 compactification of momentum space (eq. (26) of ref. [18]). Then the Universe and the H-atom would constitute an example of realization of conformal duality, representing two conformally dual systems having the same eigenfunction : the first in ordinary space and the second in the conformally dual momentum space.

REFERENCES

1. Dirac, P. A. M., *Proc. Roy. Soc.* **133**, 60 (1931).
2. Cartan, E., *Lecons sur la theorie des spineurs*, Paris: Hermann, 1937.
3. Veblen, O., *Proc. Math. Acad. Sci. USA* **90**, 503 (1933).
4. Penrose, R., and MacCallum, M. A. H., *Phys. Rep.* **6c**, 242 (1973); Flato, M., et al., *Ann. Phys.* **61**, 78 (1970); Michelson, J., and Niederle, J., *Ann. Inst. H. Poincaré*, **XXIII A**, 277 (1975); Segal, I. E., *Nuovo Cimento* **79B**, 187 (1984).

5. Budinich, P., in *Symmetry in Nature*, Pisa: Scuola Normale Superiore, 1989, p. 141.
6. Budinich, P., *Acta Phys. Pol.* **B29**, 905 (1998).
7. Budinich, P., Dabrowski, L., and Heidenreich, F., *Nuovo Cimento* **110B**, 1035 (1995).
8. Budinich, P., and Raczka, R., *Found. Phys.* **23**, 599 (1993).
9. Mack, G., and Todorov, I. T., *Phys. Rev.* **D8**, 1764 (1973).
10. Giveon, A., Poratti, M., and Rabinovici, E., *Phys. Rep.* **244**, 77 (1994).
11. Budinich, P., and Trautman, A., *The Spinorial Chessboard*, New York: Springer, 1989.
12. Budinich, P., and Trautman, A., *J. Math. Phys.* **30**, 2125 (1989).
13. Budinich, P., *Nuovo Cimento* **53A**, 31 (1979); Budinich, P., *Found. Phys.* **23**, 949 (1993).
14. Bandyopadhyay, P., *Geometry, Topology and Quantization*, Dordrecht: Kluver Acad. Pub., 1996.
15. Budinich, P., and Furlan, P., in *Proceedings of the 2nd Adriatic Meeting*, Zagreb, 1976, p. 259.
16. Adler, S., *Phys. Lett.* **B225**, 143 (1989).
17. Manogue, C. A., and Bray, T., Oregon St. Univ. preprint, (hep-th 9807044).
18. Fock, V., *Z. Phys.* **98**, 145 (1935).
19. Budinich, P., and Raczka, R., *Found. Phys.* **23**, 225 (1993).
20. Broadhurst, T. J., et. al., *Proc. VI Marcel Grossman Meeting*, Singapore: World Scientific, 1992, p. 17.
21. Barut, A. O., Budinich, P., Niederle, J., and Raczka, R., *Found. Phys.* **24**, 1461 (1994); Budinich, P., Nurowski, P., Raczka, R., and Ramella, M., *Astrophys. Journ.* **451**, 10 (1995).

Localization Problem in Quantum Mechanics and Preferred Frame[1]

Paweł Caban and Jakub Rembieliński

*Department of Theoretical Physics, University of Łódź,
Pomorska 149/153, 90-236 Łódź, Poland*

Abstract. In this paper the localization problem in the relativistic quantum mechanics is considered in the framework of a nonstandard synchronization scheme for clocks. Such a synchronization preserves Poincaré covariance but (at least formally) distinguishes an inertial frame. Our analysis has been focused mainly on the problem of existence of a proper position operator for massive particles. We have found the explicit form of the position operator and have demonstrated that in the preferred frame our operator coincides with the Newton–Wigner one. Moreover, full algebra of observables consisting of position operators and fourmomentum operators is manifestly Poincaré covariant in this framework. Our results support expectations of other authors (Bell [1], Eberhard [3]) that a consistent formulation of quantum mechanics demands existence of a preferred frame.

INTRODUCTION

According to Bell [1] (see also Eberhard [3]) consistent formulation of relativistic quantum mechanics may be necessarily with a preferred frame at the fundamental level. Following these suggestions we try to construct here a quantum mechanics which has built-in the preferred frame and which is at the same time Poincaré covariant. The key point is to reformulate theory in a way which preserves the Poincaré covariance but abandons the relativity principle and consequently allows us to introduce a preferred frame. Such a formulation has been given by one of the authors (J.R.) in [8,9]. It was shown there, that, using a nonstandard synchronization procedure for clocks (named in [9] as the Chang–Tangherlini synchronization), it is possible to obtain such a form of transformations of coordinates beetwen inertial observers, that they realize Lorentz transformations, the time coordinate is only rescaled by a positive factor and the space coordinates do not mix to it. A price for this is existence of a preferred frame in the theory and dependence of Lorentz group transformations on the additional parameter — the fourvelocity of the preferred frame.

---

[1] Supported by University of Łódź grant No. 621.

The main problem which is solved in this context is the localisation problem. Various aspects of localizability of particles have been studied from the early days of quantum mechanics, but, in the relativistic case (in contradiction to the non-relativistic one) the fully satisfactory position operator has not been found up to now. Let us explain at this point what we mean by the satisfactory position operator. Such an operator should be Hermitean, have commuting components (for massive particles), fulfill the canonical commutation relations with the momentum operators, be covariant and have covariant eigenstates (localized states). Operator constructed in the framework of our approach fulfills all of the stated above conditions.

THE CHANG–TANGHERLINI (CT) SYNCHRONIZATION

In this chapter we briefly describe main results connected with the CT synchronization scheme which we shall use in the following. Derivation of these results one can find in [9,2]. The idea applied there is based on well known facts that the definition of the time coordinate depends on the choice of the synchronization scheme for clocks and that this choice is a convention [4,5,7,11,12]. Using this freedom of choice one can try to find a synchronization procedure resulting in the desired form of the Lorentz transformations. Performing such a program we have to distinguish, at least formally, one inertial frame — so called preferred frame. Thus, at least formally, the relativity principle may be broken. Now, each inertial frame is determined by the fourvelocity of this frame with respect to the distinguished one. We denote fourvelocity of the preferred frame as seen by an inertial observer by u^μ. Hereafter, quantities in the Einstein-Poincaré (EP) synchronization are denoted by the subscript (or superscript) E. Quantities in the CT synchronization will have no index. We use the natural units ($\hbar = c = 1$).

Now we provide a geometric description of the special relativity in the CT synchronization scheme in the language of frame bundles. To do this, let us denote:
M – the Minkowski space–time,
$L_+^\uparrow$ – ortochronous Lorentz group (the group of space–time transformations);
$F(L_+^\uparrow)$ – the set of all frames in the space M obtained by action of $L_+^\uparrow$ on one particular (but arbitrary) frame in the space M; thus $F(L_+^\uparrow)$ is isomorphic to the group $L_+^\uparrow$. An element of $F(L_+^\uparrow)$ corresponding, by means of this isomorphism, to the element $g \in L_+^\uparrow$ is designated by $e(g)$.
Now let us consider the following structure

$$M_w = \left[L_+^\uparrow, (F(L_+^\uparrow) \times M, M, pr_2), \pi_w, \psi_w\right], \qquad (1)$$

where pr_2 is the canonical projection on the second factor of the cartesian product; therefore $(F(L_+^\uparrow) \times M, M, pr_2)$ is a frame bundle with the typical fibre $F(L_+^\uparrow)$. π_w is a projection on a fixed time-like fourvector w, while ψ_w is the action of the group

$L_+^\uparrow$ on the bundle $(F(L_+^\uparrow) \times M, M, pr_2)$ fulfilling the following conditions:

$$(e'(g), x') = (e(kg), x), \tag{2}$$

$$e^\mu(kg, x) = D(k,g)^\mu{}_\nu e^\nu(g, x), \tag{3}$$

$$D^{T-1}(k,g)\pi_w D^{-1}(k,g) = \pi_w. \tag{4}$$

where $k \in L_+^\uparrow$, $x \in M$. It is clear that we consider here Lorentz transformations as passive transformations — the action of the Lorentz group changes the observer, not the physical state. In our language it means that the action of the Lorentz group changes the frame $e(g)$. The condition (2) means that the action ψ_w is trivial on the manifold M; the group $L_+^\uparrow$ acts only in the fiber. The second condition (3) says that the action ψ_w is linear on frames. Now, we associate the time direction with w which means that the projector $\pi_w = \frac{w \otimes w}{w^2}$ is equal to π_{e^0}

$$\pi_{e^0} = \pi_w. \tag{5}$$

This construction defines a time-orientation of M along w. The condition (4) means that the direction of the fourvector w is invariant under the action of the group $L_+^\uparrow$.

Thus we have a collection of time-oriented space-times M_w, where w is the arbitrary time-like fourvector. The objects M_w and $M_{w'}$ corresponding to different w and w' are evidently connected by the action of another Lorentz group $L_+^{\uparrow(S)}$ (the so called synchronization group — see [9]). The whole family of time-oriented space-times M_w together with the transformations φ of the synchronization group, treated as morphisms, form a category

$$\mathcal{A} = (M_w, \varphi). \tag{6}$$

The action φ of the synchronization group $L_+^{\uparrow(S)}$ is defined in the most natural way

$$\varphi(M_w) = M_{\Lambda^S \circ w}, \qquad \Lambda^S \in L_+^{\uparrow(S)}. \tag{7}$$

From the physical point of view all choices of the element of the category $\mathcal{A}$ are equivalent provided that the relativity principle holds. However, if we want to introduce covariant canonical formalism for a relativistic free particle on the classical level or to define a proper position operator for such a particle on the quantum level we have to give up the relativity principle; a consistent description is possible only if we use a fixed element of the category $\mathcal{A}$. This construction shows, that some notions (like localizability) are simultaneously compatible with quantum mechanics and Poincaré covariance only if we resign with democracy between inertial frames, i.e. if a privileged frame is distinguished. Usually it is claimed that existence of a preferred frame violates the Poincaré covariance. This is really the case when we restrict ourselves to the Minkowski space-time. But in our approach the additional set of parameters (the mentioned above fourvelocity) allows us to solve this difficulty. In our framework, average light velocity over closed paths is still constant and equals to c, so Michelson-Morley like experiments do not distinguish such a possibility from the standard one [4,9].

The transformation of coordinates between inertial frames in the CT synchronization has the following form (for contravariant coordinates)

$$x'(u') = D(\Lambda, u)x(u), \tag{8}$$

where $D(\Lambda, u)$ is a Λ and u dependent 4×4 matrix, Λ – element of the Lorentz group and u^μ is the fourvelocity of the preferred frame with respect to considered frame, so (8) is accompanied by

$$u' = D(\Lambda, u)u. \tag{9}$$

Now the matrix $D(k, g)$ can be expressed by $D(\Lambda, u)$ given in eqs. (8–9) as follows: let $\Lambda_1 = k$, $\Lambda_2 = g$, then

$$D(k, g) = D(\Lambda_1, \Lambda_2 \tilde{u}), \tag{10}$$

and $\tilde{u} = (1, \vec{0})$ fix an observer in the preferred frame. Matrices $D(\Lambda, u)$ fulfill the following group composition rule

$$D(\Lambda_2, D(\Lambda_1, u)u)D(\Lambda_1, u) = D(\Lambda_2 \Lambda_1, u) \tag{11}$$

so

$$D^{-1}(\Lambda, u) = D(\Lambda^{-1}, D(\Lambda, u)u), \qquad D(I, u) = I. \tag{12}$$

Explicitely $D(\Lambda, u)$ is given for rotations ($R \in SO(3)$) by

$$D(R, u) = \left(\begin{array}{c|c} 1 & 0 \\ \hline 0 & R \end{array}\right); \tag{13}$$

while for boosts by

$$D(W, u) = \left(\begin{array}{c|c} \frac{1}{W^0} & 0 \\ \hline -\vec{W} & I + \frac{\vec{W} \otimes \vec{W}^T}{\left(1+\sqrt{1+(\vec{W})^2}\right)} - u^0 \vec{W} \otimes \vec{u}^T \end{array}\right), \tag{14}$$

where W^μ denotes the fourvelocity of the primed frame $O_{u'}$ with respect to the frame O_u. The explicit relationship beetwen coordinates in EP and CT is given by

$$\begin{array}{ll} x_E^0 = x^0 + u^0 \vec{u}\vec{x}, & \vec{x}_E = \vec{x}, \\ u_E^0 = \frac{1}{u^0}, & \vec{u}_E = \vec{u}. \end{array} \tag{15}$$

We see that only the time coordinate changes. Note also, that in the same space point we have $\Delta x_E^0 = \Delta x^0$ so the time lapse is the same in both synchronizations. One can easily see that the line element

$$ds^2 = g_{\mu\nu}(u)dx^\mu dx^\nu, \tag{16}$$

is invariant under the Lorentz transformations if

$$[g_{\mu\nu}] = \left(\begin{array}{c|c} 1 & u^0 \vec{u}^T \\ \hline u^0 \vec{u} & -I + (u^0)^2 \vec{u} \otimes \vec{u}^T \end{array} \right). \tag{17}$$

The contravariant metric tensor has the form

$$g^{-1}(u) = \left(\begin{array}{c|c} (u^0)^2 & u^0 \vec{u}^T \\ \hline u^0 \vec{u} & -I \end{array} \right), \tag{18}$$

so the space line element is the Euclidean one: $dl^2 = d\vec{x}^2$.

Let us notice here that the triangular form of the boost matrix (14) implies, that under the Lorentz transformations the time coordinate is only rescaled by a positive factor ($x'^0 = \frac{1}{W^0} x^0$); the space coordinates do not mix to it.

Heareafter the three vector part of a covariant (contravariant) fourvector a_μ (a^μ) will be denoted by $\underline{a}$ ($\vec{a}$) respectively.

Now, as was showen in [10,2] it is possible to construct covariant canonical formalism of massive particle starting with the action:

$$S_{12} = -m \int_{\lambda_1}^{\lambda_2} \sqrt{ds^2}, \tag{19}$$

The corresponding covariant Poisson bracket is given by [10,2]:

$$\{A, B\} = -\left(\delta^\mu_{\ \nu} - \frac{k^\mu u_\nu}{uk} \right) \left(\frac{\partial A}{\partial x^\mu} \frac{\partial B}{\partial k_\nu} - \frac{\partial B}{\partial x^\mu} \frac{\partial A}{\partial k_\nu} \right), \tag{20}$$

where all variables x^μ, k_ν are treated as independent; in particular k_0 is not *a priori* connected with k_i via the disperssion relation $k^2 = m^2$. However the bracket (20) is consistent with this constraint. Thus we do not need Dirac bracket.

ALGEBRA OF QUANTUM OBSERVABLES IN THE CT SYNCHRONIZATION

In our approach we are able, in analogy to the classical Poisson algebra, to introduce a Poincaré covariant algebra of momentum and position operators satisfying all fundamental physical requirements. In the CT synchronization the following point of view is the most natural one: with each inertial observer O_u we connect his own Hilbert space H_u (space of states). The states vectors from H_u are denoted by u: $|u, \ldots\rangle$. In other words we have a bundle of Hilbert spaces corresponding to the bundle of frames described in the previous section. In such an interpretation we have to distinguish carefully active and passive transformations, because in our approach active transformations are represented by operators acting in one Hilbert space while passive ones by operators acting beetwen different Hilbert spaces. So, in particular, the Lorentz group transformations are considered as passive ones.

Now, let $U(\Lambda)$ be an operator representing a Lorentz group element Λ. We postulate the following, standard, transformation law for a contravariant fourvector operator

$$U(\Lambda)\hat{A}(u)^\mu U^{-1}(\Lambda) = \left(D^{-1}(\Lambda, u)\right)^\mu_{\ \nu} \hat{A}(u')^\nu; \tag{21}$$

where $D(\Lambda, u)$ is given by eqs. (13,14) and $u' = D(\Lambda, u)u$; for a covariant fourvector $\hat{A}(u)_\mu$ we have to replace D^{-1} by D^T on the right hand side of (21).

Now we can introduce in each space H_u the Hermitean fourmomentum operators $\hat{p}_\lambda(u)$ (generators of translations). These operators are interpreted as observables in the corresponding reference frame. In the CT synchronization we can also introduce in each space H_u Hermitean position operators $\hat{x}^\mu(u)$. According to the Poisson bracket on the classical level (20) we postulate the following commutators between $\hat{x}^\mu(u)$ and $\hat{p}_\lambda(u)$

$$[\hat{x}^\mu(u), \hat{p}_\lambda(u)] = i \left(\frac{u_\lambda \hat{p}^\mu(u)}{u\hat{p}(u)} - \delta^\mu_{\ \lambda} \right), \tag{22}$$

$$[\hat{p}_\mu(u), \hat{p}_\nu(u)] = 0, \tag{23}$$

$$[\hat{x}^\mu(u), \hat{x}^\nu(u)] = 0. \tag{24}$$

In particular

$$\left[\hat{x}^0(u), \hat{p}_\lambda(u)\right] = 0, \tag{25}$$

$$\left[\hat{x}^i(u), \hat{p}_j(u)\right] = -i\, \delta^i_{\ j}. \tag{26}$$

We see that $\hat{x}^0$ commutes with all observables, it allows us to interprete $\hat{x}^0$ as a parameter just like in the standard nonrelativistic quantum mechanics.

We have to stress here once again that above commutation relations defining position operators *are covariant* in the CT synchronization. It can be checked directly; one simply has to use the eqs. (13,14,21) and to transform the eqs. (22,23,24) to another reference frame.

One can also check that

$$[\hat{x}^\mu(u), \hat{p}^2] = 0, \tag{27}$$

which means that localized states have definite mass.

UNITARY ORBITS OF THE POINCARÉ GROUP IN MASSIVE CASE

According to our interpretation we deal with a bundle of Hilbert spaces H_u rather than with a single space of states. Therefore transformations of the Lorentz group induce an orbit in this bundle. In this section we construct and classify unitary orbits of the Poincaré group in the above mentioned bundle of Hilbert spaces. As we will see, the unitary orbits are classified with help of mass and spin, similarly as for the standard unitary representations of the Poincaré group.

As in the standard case we assume that the eigenvectors $|k, u, \ldots\rangle$ of the four-momentum operators

$$\hat{p}_\mu(u) |k, u, \ldots\rangle = k_\mu |k, u, \ldots\rangle \qquad (28)$$

with $k^2 = m^2$, form a base in the Hilbert space H_u. We adopt the following Lorentz-covariant normalization

$$\langle k', u, \ldots | k, u, \ldots \rangle = 2 k^0 \delta^3(\underline{k}' - \underline{k}), \qquad (29)$$

where $\underline{k}$ denotes the space part of the covariant fourvector k_μ and $k^0 = g^{0\mu} k_\mu$ is positive. Energy k_0 is the solution of the dispersion relation $k^2 = m^2$ and is given by

$$k_0 = \frac{1}{u^0} \left(-\vec{u}\underline{k} + \sqrt{(\vec{u}\underline{k})^2 + (\underline{k})^2 + m^2} \right), \qquad (30)$$

so

$$k^0 \equiv \omega(\underline{k}) = u^0 \sqrt{(\vec{u}\underline{k})^2 + (\underline{k})^2 + m^2}. \qquad (31)$$

In the construction of the unitary irreducible orbits we use the operator $e^{-iq\hat{\tilde{x}}(u)}$. Action of this operator on the basis states can be determined by using its unitarity, normalization of the basis vectors (29) and the commutation relations (22). Its final form is

$$e^{-i\underline{q}\hat{\tilde{x}}(u)} |k, u, \ldots\rangle = e^{iq_0 \hat{x}^0(u)} \sqrt{\frac{uk}{u(k+q)}} |k+q, u, \ldots\rangle, \qquad (32)$$

where on the right hand side of (32) q_0 is determined by $\underline{k}$, $\underline{q}$ and u; namely

$$q_0 = \frac{1}{u^0} \left(-\vec{u}\underline{q} - \sqrt{(\vec{u}\underline{k})^2 + (\underline{k})^2 + m^2} + \sqrt{(\vec{u}\underline{q} + \vec{u}\underline{k})^2 + (\underline{k} + \underline{q})^2 + m^2} \right). \qquad (33)$$

The basis vectors of the space H_u can be generated from a vector representing a particle at rest with respect to the preferred frame. Firstly we act $U(L_u)$ on such a vector; the resulting state has fourmomentum mu_μ and belongs to H_u. Next, by means of the formula (32), we generate in H_u a vector with fourmomentum k_μ. Therefore

$$|k, u, \ldots\rangle = \sqrt{\frac{uk}{m}} e^{-i(k_\mu - mu_\mu)\hat{x}^\mu(u)} U(L_u) |\underline{k}, \tilde{u}, \ldots\rangle, \qquad (34)$$

where:

$$\tilde{u} = (1, \vec{0}), \qquad \underline{k} = (m, \underline{0}), \qquad u = D(L_u, \tilde{u})\tilde{u}. \qquad (35)$$

The above mentioned orbit induced by the action of the operator $U(\Lambda)$ in the bundle of Hilbert spaces is fixed by the following covariant conditions
— $k^2 = m^2$;
— $\varepsilon(k^0) = inv.$, for physical representations $k^0 > 0$, $\varepsilon(k^0) = 1$.
As a consequence there exists a positive defined, Lorentz invariant measure

$$d\mu(k, m) = d^4k \, \theta(k^0) \, \delta(k^2 - m^2). \qquad (36)$$

Now, applying the Wigner method and using eq. (21) one can easily determine the action of the operator $U(\Lambda)$ on the basis vector. We find

$$U(\Lambda) |k, u, m; s, \sigma\rangle = \mathcal{D}^{s}_{\sigma\lambda}{}^{-1}(R_{\Lambda,u}) |k', u', m; s, \lambda\rangle, \qquad (37)$$

where

$$u' = D(\Lambda, u)u = D(L_{u'}, \tilde{u})\tilde{u}, \qquad (38)$$

$$k' = D^{T-1}(\Lambda, u)k, \qquad (39)$$

$$R_{\Lambda,u} = D(R_{\Lambda,u}, \tilde{u}) = D^{-1}(L_{u'}, \tilde{u})D(\Lambda, u)D(L_u, \tilde{u}) \subset SO(3) \qquad (40)$$

and $\mathcal{D}^{s}_{\sigma\lambda}(R_{\Lambda,u})$ is the standard spin s rotation matrix $s = 0, \frac{1}{2}, 1, \ldots$; $\sigma, \lambda = -s, -s+1, \ldots, s-1, s$. $D(R_{\Lambda,u}, \tilde{u})$ is the Wigner rotation belonging to the little group of a vector $\tilde{u}$. Let us stress that in this approach, contrary to the standard one, representations of the Poincaré group are induced from the little group of a vector $\tilde{u}$, not $\underline{k}$.

Finally, the normalization (29) takes the form

$$\langle k, u, m; s, \lambda \, | k', u, m; s', \lambda'\rangle = 2k^0 \delta^3(\underline{k}' - \underline{k}) \delta_{s's} \delta_{\lambda'\lambda}. \qquad (41)$$

LOCALIZED STATES

In this section we find localized states in the Schrödinger picture. Let $\left|\vec{\xi}, \tau, u, m; s, \lambda\right\rangle$ denotes a state localized at the time τ in the space point $\vec{\xi}$

$$\hat{\vec{x}}(u)\left|\vec{\xi}, \tau, u, m; s, \lambda\right\rangle = \vec{\xi}\left|\xi, \tau, u, m; s, \lambda\right\rangle. \quad (42)$$

The state $\left|\vec{\xi}, \tau, u, m; s, \lambda\right\rangle$ can be expressed with help of the invariant measure (36) in terms of the basis vectors $|k, u, m; s, \lambda\rangle$, namely

$$\left|\vec{\xi}, \tau, u, m; s, \lambda\right\rangle = \frac{1}{(2\pi)^{3/2}} \int \frac{d^3\underline{k}}{2\omega(\underline{k})} \sqrt{uk}\, e^{i\underline{k}\cdot\vec{\xi}} |k, u, m; s, \lambda\rangle. \quad (43)$$

Now, after an arbitrary time t this state evolves to

$$\left|\vec{\xi}, \tau, u, m; s, \lambda; t\right\rangle = \frac{1}{(2\pi)^{3/2}} \int \frac{d^3\underline{k}}{2\omega(\underline{k})} \sqrt{uk}\, e^{ik_\mu \xi^\mu} |k, u, m; s, \lambda\rangle \quad (44)$$

with $\xi^0 = \tau - t$. One can easily check that these states are normalized as follows

$$\left\langle \vec{\xi'}, \tau, u, m; s', \lambda'; t \,\middle|\, \vec{\xi}, \tau, u, m; s, \lambda; t \right\rangle = \frac{1}{2u^0}\delta^3(\vec{\xi} - \vec{\xi'})\delta_{ss'}\delta_{\lambda\lambda'}. \quad (45)$$

It is worthwhile to notice here that the states given by the eq. (43) are covariant in the CT synchronization, i.e. a state localized in the time $t = \tau$ for the observer O_u is localized in the time $t' = \tau' = D^0{}_0(\Lambda, u)\tau$ for the observer $O_{u'}$ too.

Now, in the momentum representation wave functions are defined in the standard way

$$\psi_\lambda^{m,s}(k, u) = \langle k, u, m; s, \lambda | \psi \rangle. \quad (46)$$

So the scalar product is defined by the wave functions as follows

$$\langle \varphi | \psi \rangle = \sum_\lambda \int \frac{d^3\underline{k}}{2\omega(\underline{k})} \varphi_\lambda^{*m,s}(k,u)\psi_\lambda^{m,s}(k,u). \quad (47)$$

Now we can identify the wave functions related to the localized states (43); namely, we have

$$\chi_\lambda^{m,s}(\vec{\xi}, \tau, k, u; \sigma; t) = \frac{1}{(2\pi)^{3/2}}\sqrt{uk}\, e^{ik_\mu \xi^\mu}\delta_{\sigma\lambda}. \quad (48)$$

It follows that in this realisation

$$\hat{x}^i = -i\frac{\partial}{\partial k_i} + \frac{1}{2}i\left(\frac{u^i}{uk} - \frac{k^i}{(uk)^2}\right). \quad (49)$$

Evidently, for $\xi^0 = 0$ (i.e. for $t = \tau$) the functions χ are eigenvectors of $\hat{x}^i$. It can be easily demonstrated that in the preferred frame ($u = (1, \vec{0})$) the function (48) reduces to the Newton–Wigner localized state; then also the operator (49) coincides with the Newton–Wigner position operator for a spinless particle (see [6]).

SUMMARY

According to suggestions of some authors (Bell [1], Eberhard [3]), that a consistent formulation of quantum mechanics demands existence of a preferred frame, we constructed here a quantum mechanics for a free massive particle which has built-in the preferred frame and which is at the same time Poincaré covariant. In this formulation the boost matrix has the lower-triangular form so the time coordinate rescales only under Lorentz transformations. Such a realisation corresponds to a nonstandard synchronization of clocks i.e. to a different than standard coordinate time definition. Clasically such a scheme is operationally indistinguishable from the standard one.

We constructed and classified unitary orbits of the Poincaré group in the appriopriate bundle of Hilbert spaces. The unitary orbits are classified with help of mass and spin, similary as for the standard unitary representations of the Poincaré group, howeover are induced differently from $SO(3)$.

The introduced Poincaré covariant algebra of momentum and position operators satisfy all fundamental physical requirements. In particular the position operator fulfills all the physical conditions: it is Hermitean and covariant, it has commuting components and moreover its eigenvectors (localised states) are also covariant. We found the explicit functional form of the position operator and demonstrated that in the preferred frame our operator coincides with the Newton–Wigner one.

REFERENCES

1. Bell J. S., "Quantum mechanics for cosmologists" in "Quantum Gravity", eds. Isham, Penrose and Sciama, Oxford 1981, p. 611.
2. Caban P., Rembieliński J., "Lorentz-covariant quantum mechanics and preferred frame", quant-ph/9808013.
3. Eberhard P. H., *Nuovo Cim.*, **46B**, 392 (1978).
4. Jammer M. in *Problems in the Foundations of Physics*, North-Holland, Bologne, 1979.
5. Mansouri R., Sexl R. V., *Gen. Relativ. Gravit.* **8**, 497 (1977).
6. Newton T. D., Wigner E. P., *Rev. Mod. Phys.*, **21**, 400 (1949).
7. Reinchenbach H., *Axiomatization of the Theory of Relativity* University of California Press, Berkeley, CA, 1969.
8. Rembieliński J., *Phys. Lett.* **A78**, 33 (1980).
9. Rembieliński J., *Int. J. Mod. Phys.* **A12**, 1677, (1997), hep-th/9607232.
10. Rembieliński J., Caban P., "The Preferred Frame and Poincaré Symmetry" in *Physical Applications and Mathematical Aspects of Geometry, Groups and Algebras* eds. H. D. Doebner, W. Scherer, P. Nattermann, World Scientific, Singapore 1997, p. 349, hep-th/9612072.
11. Will C. M., *Phys. Rev.* **D45**, 403 (1992).
12. Will C. M., *Theory and Experiment in Gravitational Physics*, Cambridge University Press, 1993.

Solvable Potentials, Non-Linear Algebras, and Associated Coherent States

F. Cannata[1]*, G. Junker[2]† and J. Trost[3]†

*Dipartmento di Fisica and INFN, Via Irnerio 46, I-40126 Bologna, Italy
†Institut für Theoretische Physik, Universität Erlangen-Nürnberg,
Staudtstr. 7, D-91058 Erlangen, Germany

Abstract. Using the Darboux method and its relation with supersymmetric quantum mechanics we construct all SUSY partners of the harmonic oscillator. With the help of the SUSY transformation we introduce ladder operators for these partner Hamiltonians and shown that they close a quadratic algebra. The associated coherent states are constructed and discussed in some detail.

INTRODUCTION

Since the early days of quantum mechanics there has been enormous interest in exactly solvable quantum systems. In fact, Schrödinger himself initiated a program [1] which resulted in the famous Schrödinger-Infeld-Hull factorization method [2]. In the last 10-15 years this program has been revived in connection with supersymmetric (SUSY) quantum mechanics [3]. To be a little more precise, it has been found [4] that the so-called property of shape-invariance of a given Schrödinger potentials, which is in fact equivalent to the factorization condition, is sufficient for the exact solvability of the eigenvalue problem of the associated Schrödinger Hamiltonian. However, SUSY quantum mechanics has also been shown to be an effective tool in finding new exactly solvable systems. Here in essence one utilizes the fact that SUSY quantum mechanics consists of a pair of essentially isospectral Hamiltonians whose eigenstates are related by SUSY transformations. This is the basic idea of a recent construction method for so-called conditionally exactly solvable potentials [5]. Here one constructs a SUSY quantum system for which, under certain conditions imposed on its parameters, one of the SUSY partner Hamiltonians reduces to that of an exactly solvable (shape-invariant) one. Other approaches,

---

[1] cannata@bo.infn.it
[2] junker@theorie1.physik.uni-erlangen.de
[3] jtrost@theorie1.physik.uni-erlangen.de

which are also based on the presence of pairs of essentially isospectral Hamiltonians, go back to an idea formulated by Darboux [6], are based on the inverse scattering method [7], or on the factorization method [8]. Clearly, these approaches are closely connected to each other and to the SUSY approach.

In this paper we will construct with the help of the Darboux method all possible SUSY partners of the harmonic oscillator Hamiltonian on the real line and discuss their algebraic properties in some detail. In doing so we review in the next section the Darboux method and explicitly show its equivalence to the supersymmetric approach. Section 3 then briefly presents the basic idea for the construction of conditionally exactly solvable (CES) potentials. Section 4 is devoted to a detailed discussion of the harmonic oscillator case. Here we first present all possible SUSY partners of the harmonic oscillator and give explicit expressions for the corresponding eigenstates. Secondly, with the help of the standard ladder operators of the harmonic oscillator we introduce similar ladder operators for the SUSY partners and show that they close a quadratic algebra, which is also briefly discussed. Finally, we introduce so-called non-linear coherent states which are associated with this non-linear algebra. The properties of these coherent states are discussed in some detail.

THE DARBOUX METHOD

In this section we briefly review the Darboux method [6] and show its connection to supersymmetric quantum mechanics [3]. In doing so we start with considering a pair of standard Schrödinger Hamiltonians acting on $L^2(\mathbb{R})$,

$$H_\pm = -\frac{\hbar^2}{2m}\frac{\partial^2}{\partial x^2} + V_\pm(x),\tag{1}$$

and a linear operator

$$A = \frac{\hbar}{\sqrt{2m}}\frac{\partial}{\partial x} + \Phi(x), \quad \Phi : \mathbb{R} \to \mathbb{R},\tag{2}$$

obeying the intertwining relation

$$H_+ A = A H_-.\tag{3}$$

It is obvious that this intertwining relation cannot be obeyed for arbitrary functions $V_\pm$ and Φ. In fact, the relation (3) explicitly reads

$$\left(-\frac{\hbar^2}{2m}\Phi''(x) + V_+(x)\Phi(x) - \frac{\hbar}{\sqrt{2m}}V_-'(x) - \Phi(x)V_-(x)\right)1 = \\ \left(\frac{\hbar^2}{m}\Phi'(x) + \frac{\hbar}{\sqrt{2m}}V_-(x) - \frac{\hbar}{\sqrt{2m}}V_+(x)\right)\frac{\partial}{\partial x}.\tag{4}$$

As the unit operator **1** and the momentum operator (i.e. $\partial/\partial x$) are linearly independent, their coefficients have to vanish. In other words, we are left with two conditions between the three functions $V_\pm$ and Φ:

$$V_-(x) = V_+(x) - \frac{2\hbar}{\sqrt{2m}}\Phi'(x), \tag{5}$$

$$-\frac{\hbar^2}{2m}\Phi''(x) + V_+(x)\Phi(x) - \frac{\hbar}{\sqrt{2m}}V'_-(x) - \Phi(x)V_-(x) = 0. \tag{6}$$

Inserting the first one into the second one and integrating once we find

$$\frac{\hbar}{\sqrt{2m}}\Phi'(x) - V_+(x) + \Phi^2(x) = -\varepsilon, \tag{7}$$

where ε is an arbitrary real integration constant sometimes called factorization energy [3]. With this relation and with (5) we can express the two potentials under consideration in terms of the function Φ:

$$V_\pm(x) = \Phi^2(x) \pm \frac{\hbar}{\sqrt{2m}}\Phi'(x) + \varepsilon. \tag{8}$$

At this point one realizes that these are so-called SUSY partner potentials [3]. In fact, using relations (8) we note that

$$H_+ = AA^\dagger + \varepsilon, \qquad H_- = A^\dagger A + \varepsilon. \tag{9}$$

These supersymmetric partner Hamiltonians are due to the intertwining relation (3) essentially isospectral, that is,

$$\text{spec } H_+ \setminus \{\varepsilon\} = \text{spec } H_- \setminus \{\varepsilon\}. \tag{10}$$

Their eigenstates are related via SUSY transformations. To make this more explicit, let us denote by $|\phi_n^\pm\rangle$ the eigenstates of $H_\pm$ for eigenvalues $E_n > \varepsilon$,

$$H_\pm|\phi_n^\pm\rangle = E_n|\phi_n^\pm\rangle. \tag{11}$$

Then these states are related by SUSY transformations [3]

$$|\phi_n^+\rangle = \frac{1}{\sqrt{E_n - \varepsilon}}A|\phi_n^-\rangle, \quad |\phi_n^-\rangle = \frac{1}{\sqrt{E_n - \varepsilon}}A^\dagger|\phi_n^+\rangle. \tag{12}$$

In addition to the states in (11) one of the two Hamiltonians $H_\pm$ may have an additional eigenstate $|\phi_\varepsilon^\pm\rangle$ with eigenvalue ε obeying the first-order differential equation $A|\phi_\varepsilon^-\rangle = 0$ and $A^\dagger|\phi_\varepsilon^+\rangle = 0$, respectively. In terms of the function Φ they explicitly read

$$\phi_\varepsilon^\pm(x) = N_\pm \exp\left\{\pm\frac{\sqrt{2m}}{\hbar}\int dx\, \Phi(x)\right\}, \tag{13}$$

where $N_\pm$ stands for a normalization constant. Clearly, only one of the two solutions (13) may be square integrable. This situation corresponds to an unbroken SUSY. If none of them is square integrable then SUSY is said to be broken [3].

The Darboux method reviewed in this section can now be used to find for a given potential, say V_+, all its possible SUSY partners V_-. Firstly, one has to solve equation (7), that is, finding all possible SUSY potentials Φ. This in fact corresponds to find all possible factorizations for the corresponding Hamiltonian H_+. Finally, the corresponding SUSY partner V_- can be obtained via (5). In this way one can construct new exactly solvable potentials. The parameters involved in the SUSY potential turn out to obey certain conditions and therefore these new potentials are more precisely called conditionally exactly solvable (CES) potentials. Let us note that the Darboux method may be generalized to intertwining operators containing higher orders of the momentum operator [9].

MODELLING OF CES POTENTIALS

In this section we give some more details on the construction of CES potentials using the Darboux method. As just mentioned above we start with a given potential V_+ and try to find all its associated SUSY potentials. That is, we have to find the most general solution of the generalized Riccati equation (7). In doing so we will first linearize this non-linear differential equation via the substitution $\Phi(x) = (\hbar/\sqrt{2m})u'(x)/u(x)$,

$$-\frac{\hbar^2}{2m}u''(x) + V_+(x)u(x) = \varepsilon u(x), \qquad (14)$$

which is actually a Schrödinger-like equation for V_+. Note, however, that we are not restricted to normalizable solution of (14). In other words, the energy-like parameter ε is up to now still arbitrary.

In terms of u the linear operator A reads

$$A = \frac{\hbar}{\sqrt{2m}}\left(\frac{\partial}{\partial x} + \frac{u'(x)}{u(x)}\right) \qquad (15)$$

and thus is only a well-defined operator on $L^2(\mathbb{R})$ if u does not have any zeros on the real line. As a consequence we may admit only those solutions of (14) which have no zeros. Form Sturmian theory we know that this is only possible if ε is below the ground-state energy of H_+ which we will denote by E_0. Hence, we obtain a first condition on the parameter ε, which reads $\varepsilon < E_0$. This also implies that ε does not belong to the spectrum of H_+. In fact, the associated eigenfunction (13) would read $\phi_\varepsilon^+(x) = N_+ u(x)$, which is not normalizable due to condition put on ε.

The above condition on ε is still not sufficient to guarantee a nodeless solution. Being a second-order linear differential equation (14) has two linearly independent fundamental solutions denoted by u_1 and u_2. Hence, the most general solution for $\varepsilon < E_0$ is given by a linear combination of the fundamental ones:

$$u(x) = \alpha\, u_1(x) + \beta\, u_2(x) \,. \tag{16}$$

Therefore, the condition that u does not vanish also imposes conditions on the parameters α and β, which have to be studied case by case [5].

Let us now assume that H_+ is an exactly solvable Hamiltonian, which means that its eigenvalues E_n and eigenstates $|\phi_n^+\rangle$ are exactly known in closed form. For simplicity we have assumed that H_+ has a purely discrete spectrum enumerated by $n = 0, 1, 2, \ldots$ such that $\varepsilon < E_0 < E_1 < \ldots$. Then via the method outlined above one can construct all its SUSY partners H_- which are conditionally exactly solvable due to the conditions which have to be imposed on the parameters α, β and ε. By construction the eigenvalues of H_+ are also eigenvalues of H_- and the corresponding eigenfunctions are obtained via the SUSY transformation (12). In the case of unbroken SUSY H_- has one additional eigenvalue ε which belongs to its ground state given by $\phi_\varepsilon^-(x) = N_-/u(x)$. Finally, we note that in terms of u the partner potentials read

$$V_-(x) = \frac{\hbar^2}{m}\left(\frac{u'(x)}{u(x)}\right)^2 - V_+(x) + 2\varepsilon \tag{17}$$

and form a two-parameter family label by ε and β/α. Note that only the quotient β/α or its inverse is relevant for (17). For various examples of CES potentials found by this method see [5]. Here we limit our discussion to those related to the harmonic oscillator.

THE HARMONIC OSCILLATOR

In this section we will now construct all possible SUSY partner potentials for the harmonic oscillator $V_+(x) = (m/2)\omega^2 x^2$, $\omega > 0$, via the Darboux method. The corresponding Schrödinger-like equation (14) reads in this case[4]

$$-\frac{1}{2} u''(x) + \frac{1}{2} x^2 u(x) = \varepsilon u(x) \tag{18}$$

and has as general solution a linear combination of confluent hypergeometric functions

$$u(x) = e^{-x^2/2}\left[\alpha\,_1F_1(\tfrac{1-2\varepsilon}{4}, \tfrac{1}{2}, x^2) + \beta\, x\,_1F_1(\tfrac{3-2\varepsilon}{4}, \tfrac{3}{2}, x^2)\right] \,. \tag{19}$$

The condition that u does not have a real zero implies that α must not vanish and thus can be set equal to unity without loss of generality. Furthermore, β has to obey the inequality [5,10]

---

[4] From now on we will use dimensionless quantities, that is, x is given in units of $\sqrt{\hbar/m\omega}$ and all energy-like quantities are given in units of $\hbar\omega$.

$$|\beta| < \beta_c(\varepsilon) := 2\frac{\Gamma(\frac{3}{4} - \frac{\varepsilon}{2})}{\Gamma(\frac{1}{4} - \frac{\varepsilon}{2})}. \quad (20)$$

The corresponding partner potentials of the harmonic oscillator then read according to (17)

$$V_-(x) = \left(\frac{u'(x)}{u(x)}\right)^2 - \frac{1}{2}x^2 + 2\varepsilon. \quad (21)$$

We note that for the above u SUSY remains unbroken and therefore, the spectral properties of H_- are given by

$$\text{spec } H_- = \{\varepsilon, E_0, E_1, \ldots\} \quad \text{with} \quad E_n = n + \tfrac{1}{2}, \quad n = 0, 1, 2, \ldots,$$

$$\phi_\varepsilon^-(x) = \frac{N_- e^{x^2/2}}{{}_1F_1(\frac{1-2\varepsilon}{4}, \frac{1}{2}, x^2) + \beta x \,{}_1F_1(\frac{3-2\varepsilon}{4}, \frac{3}{2}, x^2)}, \quad (22)$$

$$\phi_n^-(x) = \frac{\exp\{-x^2/2\}}{[\sqrt{\pi}\, 2^{n+1} n!(n+1/2-\varepsilon)]^{1/2}} \left[H_{n+1}(x) + \left(\frac{u'(x)}{u(x)} - x\right) H_n(x)\right],$$

where H_n denotes the Hermite polynomial of degree n. Figures of the potential family (21) for various values of ε and β can be found in [5]. Here let us stress that one can even allow for complex valued $\beta \in \mathbb{C}\setminus[-\beta_c(\varepsilon), \beta_c(\varepsilon)]$ which in turn will give rise to complex potentials generating the same real spectrum [10]. We also note that the present CES potential (21) contains as special cases those previously obtain by Abraham and Moses [7] and by Mielnik [8]. See also [5] for a detailed discussion.

Algebraic Structure

We will now analyse the algebraic structure for the partner Hamiltonians of the harmonic oscillator. In fact, using the standard raising and lowering operators of the harmonic oscillator $H_+ = AA^\dagger + \varepsilon = a^\dagger a + 1/2$,

$$a = \frac{1}{\sqrt{2}}\left(\frac{\partial}{\partial x} + x\right), \quad a^\dagger = \frac{1}{\sqrt{2}}\left(-\frac{\partial}{\partial x} + x\right), \quad (23)$$

which close the linear algebra

$$[H_+, a] = -a, \quad [H_+, a^\dagger] = a^\dagger, \quad [a, a^\dagger] = \mathbf{1}, \quad (24)$$

one may introduce via the SUSY transformation (12) similar ladder operators for the SUSY partners [11]

$$B = A^\dagger a A, \quad B^\dagger = A^\dagger a^\dagger A, \quad (25)$$

which act on the eigenstates of H_- in the following way

$$B|\phi_{n+1}^-\rangle = \sqrt{(n+\tfrac{1}{2}-\varepsilon)(n+1)(n+\tfrac{3}{2}-\varepsilon)}|\phi_n^-\rangle ,$$
$$B^\dagger|\phi_n^-\rangle = \sqrt{(n+\tfrac{3}{2}-\varepsilon)(n+1)(n+\tfrac{1}{2}-\varepsilon)}|\phi_{n+1}^-\rangle , \qquad (26)$$
$$B|\phi_0^-\rangle = 0 , \quad B|\phi_\varepsilon^-\rangle = 0 , \quad B^\dagger|\phi_\varepsilon^-\rangle = 0 .$$

The last two relations explicate that the ground state $|\phi_\varepsilon^-\rangle$ of H_- is isolated in the sense that it cannot be reached via B from any of the excited states and, vice versa, the excited states cannot be constructed with $B^\dagger$ from $|\phi_\varepsilon^-\rangle$. These ladder operators close together with the Hamiltonian H_- the quadratic, hence non-linear, algebra

$$[H_-, B] = -B , \quad [H_-, B^\dagger] = B^\dagger , \quad [B, B^\dagger] = 3H_-^2 - 4\varepsilon H_- + \varepsilon^2 . \qquad (27)$$

This quadratic algebra belongs to the class of so-called W_2 algebras and may be viewed as a polynomial deformation of the $su(1,1)$ Lie algebra. Such deformations have been discussed by Roček [12] and, within a more general context, by Karassiov [13] and Katriel and Quesne [14]. The quadratic Casimir operator associated with the algebra (27) reads

$$C = BB^\dagger - \Psi(H_-) , \quad \Psi(H_-) - \Psi(H_- - 1) = 3H_-^2 - 4\varepsilon H_- + \varepsilon^2 . \qquad (28)$$

In the Fock space representation (26) we have the following explicit expression

$$\Psi(H_-) = (H_- - \varepsilon)(H_- + \tfrac{1}{2})(H_- + 1 - \varepsilon) \qquad (29)$$

and the relations $BB^\dagger = \Psi(H_-)$ and $B^\dagger B = \Psi(H_- - 1)$. Hence the Casimir (28) vanishes within this representation as expected [13,14].

Non-linear coherent states

Let us now construct the non-linear coherent states [15] associated with the quadratic algebra (27). There are several ways to define such states [16]. Here we will define them as eigenstates of the "non-linear" annihilation operator B, leading essentially to so-called Barut-Girardello coherent states [17]. We also note that the construction procedure presented below is very similar to that of coherent states associated with quantum groups [18].

Let us note that the ground state $|\phi_\varepsilon^-\rangle$ of H_- is isolated and therefore we may construct the coherent states over the excited states $\{|\phi_n^-\rangle\}_{n\in\mathbb{N}_0}$ only. For this reason we make the ansatz

$$|\mu\rangle = \sum_{n=0}^\infty c_n \mu^n |\phi_n^-\rangle , \qquad (30)$$

where μ is an arbitrary complex number and the real coefficients c_n are to be determined from the defining relation

$$B|\mu\rangle = \mu|\mu\rangle = \sum_{n=0}^{\infty} c_n \mu^n B|\phi_n^-\rangle . \tag{31}$$

Using relations (26) we obtain the following recurrence relation for the c_n's,

$$c_{n+1} = c_n \left[(n + \tfrac{1}{2} - \varepsilon)(n+1)(n + \tfrac{3}{2} - \varepsilon)\right]^{-1/2} . \tag{32}$$

That is, the coefficients c_n for $n \geq 1$ can be expressed in terms of c_0,

$$c_n = c_0 \left[n!(\tfrac{1}{2} - \varepsilon)_n (\tfrac{3}{2} - \varepsilon)_n\right]^{-1/2} \tag{33}$$

where $(z)_n = \Gamma(z+n)/\Gamma(z)$ denotes Pochhammer's symbol. The remaining coefficient $c_0 = c_0(\mu)$ is determined via the normalization of the coherent states

$$\langle\mu|\mu\rangle = c_0^2(\mu) \sum_{n=0}^{\infty} \frac{|\mu|^{2n}}{n!} \frac{1}{(\tfrac{1}{2} - \varepsilon)_n (\tfrac{3}{2} - \varepsilon)_n} = 1 . \tag{34}$$

Thus, we can express c_0 in terms of a generalized hypergeometric function [19]

$$c_0^{-2}(\mu) = {}_0F_2\left(\tfrac{1}{2} - \varepsilon, \tfrac{3}{2} - \varepsilon; |\mu|^2\right) . \tag{35}$$

Let us now discuss some properties of these non-linear coherent states. First we note that these states are not orthogonal for $\mu \neq \nu$ as expected:

$$\langle\mu|\nu\rangle = c_0(\mu) c_0(\nu) \, {}_0F_2\left(\tfrac{1}{2} - \varepsilon, \tfrac{3}{2} - \varepsilon; \mu^*\nu\right) . \tag{36}$$

Secondly, let us investigate whether these states form an overcomplete set. In other words, we consider the question: Can these states generate a resolution of the unit operator? For this we have to recall that the non-linear coherent states have been constructed over the excited states of H_-. Therefore, we start with postulating a positive measure ρ on the complex μ-plane obeying the following resolution of unity:

$$\int_{\mathbb{C}} d\rho(\mu^*, \mu) |\mu\rangle\langle\mu| = \mathbf{1} - |\phi_\varepsilon^-\rangle\langle\phi_\varepsilon^-| . \tag{37}$$

Within the polar decomposition $\mu = \sqrt{x}\, e^{i\varphi}$ we make the ansatz

$$d\rho(\mu^*, \mu) = \frac{d\varphi\, dx\, \sigma(x)}{2\pi c_0^2(\sqrt{x})} , \tag{38}$$

with a yet unknown positive density σ on the positive half-line. Inserting this ansatz into (37) we obtain the following conditions on σ

$$\int_0^\infty dx\, \sigma(x)\, x^n = \Gamma(n+1) \frac{\Gamma(\tfrac{1}{2} - \varepsilon + n)\Gamma(\tfrac{3}{2} - \varepsilon + n)}{\Gamma(\tfrac{1}{2} - \varepsilon)\Gamma(\tfrac{3}{2} - \varepsilon)} , \quad n = 0, 1, 2, \ldots . \tag{39}$$

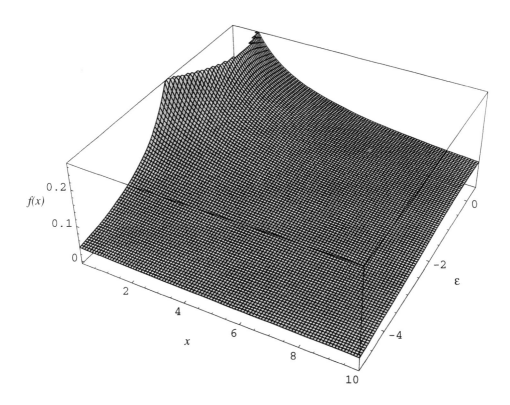

FIGURE 1. The radial density $f(x) = \sigma(x)/c_0^2(\sqrt{x})$ giving rise to the resolution of unity (37) with (38) as a function of $x = |\mu|^2$ and for various parameters $\varepsilon < \frac{1}{2}$.

Hence, σ is a probability density on the positive half-line defined by its moments given on the right-hand side of (39). Let us note that the integral in (39) may be viewed as a Mellin transformation [20] of σ and in turn the latter is given by the inverse Mellin transformation of the moments. This inverse Mellin transformation turns out to lead to the integral representation of Meijer's G-function [19]. In other words, we have the explicit form:

$$\sigma(x) = \frac{1}{\Gamma(\frac{1}{2} - \varepsilon)\Gamma(\frac{3}{2} - \varepsilon)} G_{03}^{30}\left(x|0, -\frac{1}{2} - \varepsilon, \frac{1}{2} - \varepsilon\right) . \tag{40}$$

In Figure 1 a plot of the radial density $f(|\mu|^2) = 2\pi \, d\rho(\mu^*, \mu)/(d\varphi d|\mu|^2)$ is given showing that it leads to a well-behaved positive measure on the complex μ-plane.

Finally, let us point out that similar non-linear coherent states associated with the CES potentials of the radial harmonic oscillator have been constructed in [15]. In that case broken as well as unbroken SUSY can be considered and the corresponding symmetry algebra is a cubic one. In analogy to the discussion in [15] one can show that the coherent states discussed here are also minimum uncertainty states.

ACKNOWLEDGEMENTS

One of us (G.J.) would like to thank the organizers for their kind invitation to this very stimulating meeting. In particular, he has enjoyed valuable discussions with C.M. Bender, P.P. Kulish and A. Odzijewicz during this conference.

REFERENCES

1. Schrödinger E., *Proc. Roy. Irish Acad.* **46A**, 9–14 (1940); **46A**, 183–206 (1941); **47A**, 53–54 (1941).
2. Infeld L. and Hull T.E., *Rev. Mod. Phys.* **23**, 21–68 (1951).
3. Junker G., *Supersymmetric Methods in Quantum and Statistical Physics*, Berlin: Springer-Verlag, 1996.
4. Gendenshteîn L.É., *JETP Lett.* **38**, 356–358 (1983).
5. Junker G. and Roy P., *Conditionally exactly solvable potentials: A supersymmetric construction method*, preprint quant-ph/9803024.
6. Darboux G., *Comptes Rendus Acad. Sci. (Paris)* **94**, 1456–1459 (1882).
7. Abraham P.B. and Moses H.E., *Phys. Rev. A* **22**, 1333–1686 (1980).
   Luban M. and Pursey D.L., *Phys. Rev. D* **33**, 431–436 (1986).
   Pursey D.L., *Phys. Rev. D* **33**, 1048–1055 (1986).
8. Mielnik B., *J. Math. Phys.* **25**, 3387–3389 (1984).
9. Andrianov A.A., Ioffe M.V. and Spiridonov V.P., *Phys. Lett. A* **174**, 273–179 (1993).
   Andrianov A.A., Ioffe M.V., Cannata F. and Dedonder J.-P., *Int. J. Mod. Phys. A* **10**, 2683–2702 (1995).
   Samsonov B.F., *J. Math. Phys. A* **28**, 6989–6998 (1995).
10. Cannata F., Junker G. and Trost J., *Schrödinger operators with complex potentials but real spectrum*, preprint quant-ph/9805085.
11. Junker G. and Roy P., *Phys. Lett. A* **232**, 155–161 (1997).
    Junker G. and Roy P., *Supersymmetric construction of exactly solvable potentials and non-linear algebras*, preprint quant-ph/9709021.
12. Roček M., *Phys. Lett. B* **255**, 554–557 (1991).
13. Karassiov V.P., *J. Phys. A* **27**, 153–165 (1994).
14. Katriel J. and Quesne C., *J. Math. Phys.* **37**, 1650–1661 (1996).
15. Junker G. and Roy P., *Non-linear coherent states associated with conditionally exactly solvable problems*, preprint (1998).
16. Zhang W.-M., Feng D.H. and Gilmore R., *Rev. Mod. Phys.* **62**, 867–927 (1990).
17. Barut A.O. and Girardello L., *Commun. Math. Phys.* **21**, 41–55 (1971).
18. Spiridonov V., *Phys. Rev. A* **52**, 1909–1935 (1995).
    Odzijewicz A., *Commun. Math. Phys.* **192**, 183–215 (1998).
19. Erdélyi A., Magnus W., Oberhettinger F. and Tricomi F.G., *Higher Transcedental Functions*, Volume I, New York: McGraw-Hill, 1953.
20. Erdélyi A., Magnus W., Oberhettinger F. and Tricomi F.G., *Tables of Integral Transforms*, Volume I, New York: McGraw-Hill, 1954.

On the Inverse Variational Problem in Classical Mechanics

J. Cisło*, J.T. Łopuszański* and P.C. Stichel[†]

*Institute of Theoretical Physics
University of Wrocław,
PL 50-204 Wrocław, pl. M. Borna 9
[†]Fakultät für Physik
Università Bielefeld
Bielefeld, Germany

Im memoriam – Ryszard Rączka

Our approach to the problems formulated below is rather elementary, this means the main mathematical tool we are going to use is the <u>Lagrange formalism</u> based on Lagrange functions $L = L(\underline{x}, \underline{\dot{x}}, t)$ and the Lagrange–Euler Equations, equations of motion. The setting is a Euclidean space, in most cases 1– or 3–dimensional one, in which the variables $\underline{x}^{(i)}(t), \underline{\dot{x}}^{(i)} = \frac{dx(t)}{dt}$ and $\underline{\ddot{x}}^{(i)} = \frac{d^2x(t)}{dt^2}, i = 1, 2, 3, ..., n$, describe the position, velocity and acceleration of the i–th particle. These variables depend on time t, while t is independent variable. In most cases, investigated by us, we choose the Lagrange function to be autonomous for simplicity reasons. To solve our problems we make an extensive use of the <u>inverse lemma of Poincaré</u> and the theory of the linear partial differential equations of first order <u>(method of characteristics)</u>, with which the theoretical physicists are well acquainted.

The basic problem to be solved is: assume that we are given the equations of motion of one or two point particles in a 1 – or 3 – dimensional Euclidean space; the question which has to be answered is: does there exist such a Lagrange function for which these equations of motion coincide with the Euler–Lagrange Equations of this Lagrange function, and, if so, is there only one <u>nontrivial</u> Lagrange function or there are more. This is our <u>inverse variational problem</u>.

A comment is in order as far as the dimensions of the Euclidean space are concerned. Many of our results can be easily extended to spaces of dimension $n = 2$ or $n > 3$. We shall also explain below, what we mean by nontrivial Lagrange functions. It should be also mentioned that the Lagrange functions as well as the Euler–Lagrange Equations can display certain symmetry properties; if this is the case the symmetry of the Lagrange function can <u>never exceed</u> the symmetry of the equations.

The inverse variational problem in classical mechanics is an old problem. The necessary[1] and sufficient conditions for existence of a Lagrange function for given equations of motion were give first by H. Helmholtz [1] in 19-th century. His conditions are of great value but rather for abstract mathematical considerations and for solving simple problems. For more involved mechanical problems one is compelled to try other approaches.

In our considerations we shall restrict ourselves to regular Lagrange functions i.e. the canonical momenta

$$p_i \equiv \frac{\partial L}{\partial \dot{x}_i} \quad (i = 1, ..., 3n) \tag{1}$$

are well defined, so that the system of variables $(\underline{x}, \underline{\dot{x}})$ is equivalent to the system of variables $(\underline{x}, \underline{p})$. This implies

$$\det\left(\frac{\partial L}{\partial \dot{x} \partial \dot{x}}\right) \neq 0 \quad \text{a.e..} \tag{2}$$

Keeping this in mind we are going to give some definitions:

i) two systems of equations are called <u>equivalent</u> if they have the same set of solutions,

ii) two Lagrange functions are called <u>s–equivalent</u> if their Euler–Lagrange Equations are equivalent,

iii) two Lagrange functions, L_1 and L_2 are called equivalent if they are s–equivalent and

$$\left.\begin{array}{l} [x_i, \dot{x}_j]_{L_1} = \alpha [x_i, \dot{x}_j]_{L_2} \quad \alpha \in \mathbb{R}^1, \neq 0 \\ [\dot{x}_i, \dot{x}_j]_{L_1} = \alpha [\dot{x}_i, \dot{x}_j]_{L_2} \\ [x_i, x_j]_{L_1} = \alpha [x_i, x_j]_{L_2}, \end{array}\right\} \tag{3}$$

where

$$[a, b] = \sum_{s=1}^{n} \left(\frac{\partial a}{\partial x_s}\frac{\partial b}{\partial p_s} - \frac{\partial a}{\partial p_s}\frac{\partial b}{\partial x_s}\right)$$

denotes the Poisson Bracket[2].

Since we deal here with Euler–Lagrange Equations the systems of equivalent equations have to be linear with respect to the variable $\ddot{x}$.

A <u>theorem of M. Henneaux</u> [2] gives the necessary and sufficient conditions for two Lagrange functions, L_1 and L_2, to be equivalent. This condition reads

---

[1] Helmholtz proved only that his system of relations is necessary for the existence of the Lagrange function. The sufficiency of them were shown later by A. Mayer in 1896.
[2] One can use also Lagrange instead of Poisson Brackets.

$$L_2 = \alpha L_1 + \frac{d}{dt} g(\underline{x}, t) = \alpha L_1 + \sum_{i=1}^{n} \frac{\partial g}{\partial x_i} \dot{x}_i + \frac{\partial g}{\partial t}. \tag{4}$$

Here $\alpha \in \mathbb{R}^1 \neq 0$ is the same as in (3) and g is an arbitrary function.

We are not interested in equivalent Lagrange functions. What we are interested in, is the set of s–equivalent Lagrange functions, whose subsets consist of equivalent elements, which are, however, inequivalent to elements of other subsets. The set of equivalent Lagrange functions we shall call <u>trivial</u>.

The only case where we are interested in equivalent Lagrange functions is the case of symmetry. If the Euler–Lagrange Equations display a certain symmetry it can happen that the corresponding Lagrange function L', although not strictly invariant under the symmetry transformation, goes over into an equivalent L', satisfying (4) with $\underline{\alpha} = 1$. We say then that L is <u>weakly invariant</u> under the symmetry transformations.

The term $\frac{dg}{dt}$ in (4) we call the <u>gauge term</u>.

The main reason of our interest in inequivalent classical Lagrange functions is that, when used in the Feynman method of "integral over all paths", they can lead to different quantization schemes.

We shall illustrate our problem, our method and goal on a very simple example, given by J. Cisło.

Let us consider the equation

$$\ddot{x} - \alpha \dot{x} = 0, \qquad \alpha \in \mathbb{R}^1, \neq 0, \tag{5}$$

in (1+1) dimensional spacetime. There does not exist any Lagrange function yielding (5) as its Euler–Lagrange Equation (friction!). But to the equation

$$\frac{1}{\dot{x}} (\ddot{x} - \alpha \dot{x}) = 0 \tag{6}$$

equivalent to (5), corresponds already a Lagrange function, namely

$$L = \dot{x} \ln |\dot{x}| + \alpha x. \tag{7}$$

The most general set of Lagrange functions, s–equivalent to (7), is

$$L = \int_c^{\dot{x}} \frac{1}{u} \sigma(x, u, t)(\dot{x} - u) du + g(x, t) \tag{8}$$

where

$$\sigma = \sigma(a, b), \qquad a \equiv \alpha t - \ln |\dot{x}|, \qquad b \equiv \alpha x - \dot{x}, \tag{9}$$

is an arbitrary function, and $\underline{\sigma} \neq 0, \infty$ a.e.. The quantities $\underline{a}$ and $\underline{b}$ are constants of motion which implies also

$$\frac{d\sigma}{dt} = 0 \tag{10}$$

along the trajectory. The constant c takes care of the convergence of the integral. The function $g(x,t)$ can be evaluated from the equation

$$\frac{\partial g(x,t)}{\partial x} = \int_c^{\dot{x}} \frac{\alpha}{u}\left(\frac{\partial \sigma}{\partial a} + \frac{\partial \sigma}{\partial b}u\right) du + \alpha\sigma \tag{11}$$

as soon as $\underline{\sigma}$ is chosen. The Euler–Lagrange Equations of (8) are

$$\frac{\sigma}{\dot{x}}(\ddot{x} - \alpha\dot{x}) = 0. \tag{12}$$

The Lagrange Brackets for $\underline{x}$ and $\underline{\dot{x}}$ read for two s-equivalent Lagrange functions, L_1 and L_2,

$$\{x,\dot{x}\}_{L_1} = \frac{1}{\dot{x}}\sigma_1, \qquad \{x,\dot{x}\}_{L_2} = \frac{1}{\dot{x}}\sigma_2 = \frac{\sigma_2}{\sigma_1}\{x_1,\dot{x}\}_{L_1}. \tag{13}$$

Since $\tilde{\sigma} \equiv \dfrac{\sigma_2}{\sigma_1}$ is, in general, not constant, the Lagrange functions, L_1 and L_2 are inequivalent to each other, by virtue of Henneaux's theorem (see (3)), barring the case $\tilde{\sigma} = const.$. As the equivalence sets are marked by σ each equivalence set for given non constant σ consists only of one element.

Another example worth to quote is due to J. Douglas [3]. It concerns equations

$$\ddot{x}_1 = \dot{x}_2, \qquad \ddot{x}_2 = x_2 \tag{14}$$

in a (2+1)–dimensional spacetime. For equations (14) as well as equivalent to them equations no Lagrange function can be found.

The generalization of Cisło's example is the problem of one point particle in (1+1)–dimensional spacetime [4]. We assume the existence of an (for simplicity) autonomous Lagrange function L_0, the structure of which we need not specify. Our goal is to find the set of Lagrange functions, s–equivalent to L_0, and investigate, what is their mutual equivalence.

The starting point is the identity

$$\frac{\partial^2 L'}{\partial \dot{x}^2}\ddot{x} + \frac{\partial^2 L'}{\partial \dot{x}\partial x}\dot{x} - \frac{\partial L'}{\partial x} \equiv \sigma(x,\dot{x})\left(\frac{\partial^2 L_0}{\partial \dot{x}^2}\ddot{x} + \frac{\partial^2 L_0}{\partial \dot{x}\partial x}\dot{x} - \frac{\partial L_0}{\partial x}\right), \tag{15}$$

where $\sigma(x,\dot{x}) \neq 0, \infty$ a.e. and $L'(x,\dot{x})$ is an autonomous Lagrange function. Our task is to find the structure of σ as well as of L'. The most general form of the Lagrange function is

$$L' = \dot{x} \int_c^{\dot{x}} G(x,u)du + \sum(H_0)$$

where

$$G(x,\dot{x}) \equiv \sigma(H_0)\frac{\partial^2 L_0}{\partial \dot{x}^2}$$

$$\frac{d\sum(z)}{dz} \equiv \sigma(z)$$

$$H_0 = \frac{\partial L_0}{\partial \dot{x}}\dot{x} - L^0.$$

(16)

H_0 and consequently σ are constants of motion.

The Lagrange Brackets for x and $\dot{x}$ are

$$\{x,\dot{x}\}_{L'} = \sigma(H_0)\frac{\partial^2 L_0}{\partial \dot{x}^2} \qquad \{x,\dot{x}\}_{L_0} = \frac{\partial^2 L_0}{\partial \dot{x}^2}$$

$$\{x,\dot{x}\}_{L'_2} = \sigma_1 \sigma_2^{-1} \{x,\dot{x}\}_{L'_1}.$$

(17)

So we conclude that L' are not equivalent among themselves and with respect to L^0, barring the case σ is a constant.

These introductory comments and examples were meant to make the reader familiar with the problem. We list now our results.

We investigate thoroughly the case of two point particles in (1+1)-dimensional spacetime, where the equations of motion are assumed to be forminvariant under Galilei transformations and should be Euler–Lagrange Equations for a certain Lagrange function, the structure of which has to be found [5]. We establish the most general form of the Lagrange function and its Galilei covariant Euler–Lagrange Equations. We also present the centrally extended Lie algebra of the Galilei group. The extension of the algebra is due to the fact that we do not assume the 3-rd Newton Law to be valid.

After that is done we choose a certain Lagrange function L of the form we found before and establish the set of Lagrange functions L', s–equivalent among themselves and with respect to L, the Euler–Lagrange Equations of which are Galilei covariant. The set of s–equivalent Lagrange functions L' decomposes into subsets built out of equivalent elements, which are not equivalent with respect to other subsets.

We did not succeed in extension of the set of Lagrange functions L' by all Lagrange functions, which, although s–equivalent to L', yield Euler–Lagrange Equations <u>not</u> forminvariant under the Galilei transformation. A quite large set of such Lagrange functions was nevertheless found.

We may score a partial success as far as one or two particles in (3+1) dimensional spacetime are concerned [6].

For one point particle we assume rotational forminvariance. For some special cases, where the Euler–Lagrange Equations coincide with the Newton Equations, we get the theory of a charged particle in an electromagnetic field (Dirac's magnetic monopole), as well as a complete solution with nonequivalent subsets of Lagrange functions of the set of s–equivalent Lagrange functions. As far as the general case is considered we succeeded in evaluating the most general form of the Lagrange function and of the Euler–Lagrange Equations. We did not succeed in evaluating the set of Lagrange functions, s–equivalent to a chosen Lagrange function established before for one particle with rotationally covariant Euler–Lagrange Equations. This problem seems to be a difficult task.

For two point particles we assumed that the Euler–Lagrange Equations are forminvariant under the Galilei group. We succeeded in establishing the most general form of the Lagrange function and of the Euler–Lagrange Equations. We presented the centrally extended Lie algebra of the Galilei group. We did not try to solve the problem of evaluating the set of s–equivalent Lagrange functions to a given one.

To summarize: the general case of one and two particles in (1+1) dimensional spacetime is completely solved and can be readily used in quantization procedure of Feynman's "integral over all paths". The same is true for the particular case of one particle in (3+1) dimensional spacetime for the Lagrange function, the Euler Lagrange Equations of which coincide with the Newton Equations.

To end the story we present the results obtained for one particle in a (3+1) dimensional spacetime, the equation of motion of which are subjected to rotational covariance. These results were mentioned before. The most general form of our antonomous Lagrange function L reads

$$L = K\left(\underline{x}^2, \underline{x}\,\underline{\dot{x}}, \dot{x}^2\right) + \sum_i \dot{x}_i \frac{\partial}{\partial x_i}\left[W(\underline{x}^2)P(\underline{x} \mid \underline{\dot{x}}, \underline{b})\right] \tag{18}$$

where K and W are arbitrary functions of their variables, $\underline{b} \neq 0$ is an arbitrary constant vector and

$$P(\underline{x} \mid \underline{\dot{x}}, \underline{b}) \equiv \frac{1}{|\underline{x}|} \arctan\left[\frac{|\underline{x}|\sum_i (\underline{b} \times \underline{\dot{x}})_i x_i}{\sum_i (\underline{x} \times \underline{b})_i (\underline{x} \times \underline{\dot{x}})_i}\right] . \tag{19}$$

The Euler–Lagrange Equations, displaying the rotational covariance, are

$$\left.\begin{aligned}&\sum_j \frac{\partial^2 K}{\partial \dot{x}_i \partial \dot{x}_j}\ddot{x}_j + \sum_j \frac{\partial^2 K}{\partial \dot{x}_i \partial x_j}\dot{x}_j - \frac{\partial K}{\partial x_i} + \\ &- \left(2\underline{x}^2 W' - W\right) \sum_j \frac{\partial}{\partial x_j}\left[\frac{(\underline{x}\times\underline{\dot{x}})_i}{(\underline{x}\times\underline{\dot{x}})^2}\right]\ddot{x}_j + \\ &+ \left(2\underline{\dot{x}}^2 W' + 4\left(\underline{x}\,\underline{\dot{x}}\right)^2 W''\right)\frac{(\underline{x}\times\underline{\dot{x}})_i}{(\underline{x}\times\underline{\dot{x}})^2}\end{aligned}\right\} \qquad (20)$$

where
$$W'(\alpha) \equiv \frac{dW(\alpha)}{d\alpha}.$$

These formulae, (19) and (20), seem to us to have a rather unexpected, not conventional form.

REFERENCES

1. H. Helmholtz, *Journ. Reine und Angewandte Mathematik*, **100**, 137 (1887).
2. M. Henneaux, *Annales of Physics (N.Y.)* **140**, 45 (1982).
3. J. Douglas, *Trans. Am. Math. Soc.* **50**, 71 (1941).
4. see e.g. E. Engels, *Il N.C.*, **26B**, 481 (1975),
   J. Cisło, J. Łopuszański, P.C. Stichel, *Fortschr. Phys.* **43**, 745 (1995).
5. J. Łopuszański, P.C. Stichel, *Fortschr. Phys.* **45**, 79 (1997),
   J. Cisło, J. Łopuszański, P.C. Stichel, *Fortschr. Phys.* **46**, 45 (1998).
6. J. Cisło, J. Łopuszański, P.C. Stichel, *Fortschr. Phys.* **43**, 733 (1995),
   J. Cisło, J. Łopuszański, P.C. Stichel, *Fortschr. Phys.* **43**, 745(1995),
   J. Cisło, J. Łopuszański, submitted for publication in *Commun. Math. Phys.*,
   J. Łopuszański, P.C. Stichel, in : *From Field Theory to Quantum Groups*. W.Sc. 1996, p. 243,
   J. Łopuszański, P.C. Stichel, *Fortschr. Phys.* **45**, 79 (1997).

Change of Variables, Fundamental Solutions and Borel Resummation

Stefan Giller†[1] and Piotr Milczarski‡[2]

Theoretical Physics Department II, University of Łódź,
Pomorska 149/153, 90-236 Łódź, Poland
e-mail: † *sgiller@krysia.uni.lodz.pl*
‡ *jezykmil@krysia.uni.lodz.pl*

Abstract. It is shown that a change of variable in 1-dim Schrödinger equation applied to the Borel summable fundamental solutions [8,9] is equivalent to Borel resummation of the fundamental solutions multiplied by suitably chosen $\hbar$-dependent constant. This explains why change of variable can improve JWKB formulae [11]. It is shown also that a change of variable alone cannot provide us with the exact JWKB formulae.

A change of variable in the 1-dim Schrödinger equation (1-DSE) is one of the basic technic used to solve 1-dim problems (see [12] for example). In the context of semiclassical (JWKB) approximation the procedure is in fact a main ingredient of Fröman and Fröman (F-F) approach to 1-DSE [3,4] with an aim of getting improved JWKB quantization formulae [4–6]. Sometimes, a suitable change of variable provides us with JWKB-like formulae solving the problem of energy spectra even exactly [4]. No doubts, however, that the latter possibility depends totally on a potential considered and changing variable plays in such cases only an auxiliary role [11].

An improvement of the standard JWKB formulae achieved by the changing variable procedure appears as corrections having typically forms of additional $\hbar$-dependent term in emerging effective potentials [1,4–6]. Since in all these cases of variable changing the standard JWKB formulae can be easily restored simply by $\hbar$-expansions of the improved ones the latter seems to be a kind of some hidden resummation of a part (in the case of improvements only) or a full (when exact formulae emerge) standard semiclassical expansion corresponding to considered cases.

It is the aim of this paper to show that indeed this hidden resummation mentioned above really takes place and that a class of applied changes of variable in 1-DSE results as the Borel resummation of suitably chosen standard semiclassical solutions

---

[1] Supported by KBN 2PO3B 07610
[2] Supported by the Łódź University Grant No 580

to SE multiplied by appropriately chosen $\hbar$-dependent constants which can always be attached to any of such semiclassical solution.

As it has been shown by Milczarski and Giller [7] (see also [8]) such specific Borel summable solutions to SE are provided for meromorphic potentials by the F-F construction [3] in the form of so called fundamental solutions (FS) [8,9]. These are the only solutions with the Borel summability property among all the F-F-like solutions [7]. Despite their rareness the FS's when collected into a full set of them allow us to solve any 1-dim problem [8,9] (see also a discussion below).

A standard way to introduce FS's is a construction of a Stokes graph (SG) [7–9] for a given (meromorphic) potential $V(x)$.

SG consists of Stokes lines (SL) emerging from roots (turning points) of the equation:

$$V(x) + \hbar^2 \delta(x) = E \qquad (1)$$

with E as energy as well as from simple poles of the considered potential $V(x)$.

The presence and role of the δ-term in (1) is explained below. It contributes to (1) only when $V(x)$ contains simple and second order poles. The δ-term is constructed totally from these poles.

The points of SL's satisfy one of the following equations:

$$\Re \int_{x_i}^{x} \sqrt{V(y) + \hbar^2 \delta(y) - E}\, dy = 0 \qquad (2)$$

with x_i being a root of (1) or a simple pole of $V(x)$.

SL's which are not closed end at these points of the x-plane (i.e. have the latter as the boundaries) for which an action integral in (2) becomes infinite. Of course such points are singular for the potential $V(x)$ and can be finite poles, higher than the simple ones, or poles of $V(x)$ lying at the infinity.

Each such a singularity x_0 of $V(x)$ defines a domain called a sector. This is the connected domain of the x-plane bounded by the SL's and x_0 itself with the latter point being also a boundary for the SL's or being an isolated boundary point of the sector (as it is in the case of the second order pole).

In each sector the LHS in (2) is only positive or negative.

Consider now the Schrödinger equation:

$$\Psi''(x) - \hbar^{-2} q(x) \Psi(x) = 0 \qquad (3)$$

where $q(x) = V(x) - E$ (we have put the mass m in (3) to be equal to 1/2).

Following Fröman and Fröman one can define in each sector k having x_0 at its boundary a solution of the form:

$$\Psi_k(x) = \tilde{q}^{-\frac{1}{4}}(x) \cdot e^{\frac{z}{\hbar} W(x)} \cdot \chi_k(x) \quad k = 1, 2, \ldots \qquad (4)$$

where:

$$\chi_k(x) = 1 + \sum_{n\geq 1} \left(-\frac{\sigma\hbar}{2}\right)^n \int_{x_0}^x d\xi_1 \int_{x_0}^{\xi_1} d\xi_2 \ldots \int_{x_0}^{\xi_{n-1}} d\xi_n \omega(\xi_1)\omega(\xi_2)\ldots\omega(\xi_n) \quad (5)$$

$$\times \left(1 - e^{-\frac{2\sigma}{\hbar}(W(x) - W(\xi_1))}\right)\left(1 - e^{-\frac{2\sigma}{\hbar}(W(\xi_1) - W(\xi_2))}\right)\ldots\left(1 - e^{-\frac{2\sigma}{\hbar}(W(\xi_{n-1}) - W(\xi_n))}\right)$$

with

$$\omega(x) = \frac{\delta(x)}{\tilde{q}^{\frac{1}{2}}(x)} - \frac{1}{4}\frac{\tilde{q}''}{\tilde{q}^{\frac{3}{2}}(x)} + \frac{5}{16}\frac{\tilde{q}'^2}{\tilde{q}^{\frac{5}{2}}(x)} \quad (6)$$

and

$$W(x, E) = \int_{x_i}^x \sqrt{\tilde{q}(\xi, E)}\, d\xi \quad (7)$$

$$\tilde{q}(x, E) = V(x) + \hbar^2 \delta(x) - E$$

In (4) and (5) a sign $\sigma(=\pm 1)$ and an integration path are chosen in such a way to have:

$$\sigma \Re\left(W(\xi_j) - W(\xi_{j+1})\right) \leq 0 \quad (8)$$

for any ordered pair of integration variables (with $\xi_0 = x$). Such a path of integration is then called canonical.

The term $\delta(x)$ appearing in (6) and in (7) is necessary to ensure all the integrals in (5) to converge when x_0 is a first or a second order pole of $V(x)$ or when solutions (4) are to be continued to such poles. Each such a pole x_0 demands a contribution to $\delta(x)$ of the form $(2(x-x_0))^{-2}$ so that $\delta(x)$ collects all of them and its final form depends of course on the corresponding singular structure of $V(x)$.

Note that the effect of introducing the δ-term is completely equivalent to making some change of variable in the SE, a possibility which in this context shall however not be discussed in the paper.

In a domain D_k of the x-plane where the condition (8) is satisfied (so called canonical domain) the series in (5) defining χ_k is uniformly convergent. χ_k itself satisfies the following initial conditions:

$$\chi_k(x_0) = 1 \text{ and } \chi_k'(x_0) = 0 \quad (9)$$

corresponding to the equation:

$$\chi_k(x) = 1 - \frac{\sigma\hbar}{2}\int_{x_0}^x dy\, \omega(y)\chi_k - \frac{\sigma\hbar}{2}\tilde{q}^{-\frac{1}{2}}(x)\chi_k'(x) \quad (10)$$

this function has to obey as a consequence of SE (3) and the initial conditions (9).

In the canonical domain D_k and the sector $S_k(\subset D_k)$ where the solution (4) is defined the latter has two following basic properties:

1^0 It can be expanded in D_k into a standard semiclassical series obtained by iterating Eq.(10) and taking into account the initial conditions (9);

2^0 The emerging semiclassical series is Borel summable in S_k to the solution itself.

The solutions (4) defined in the above way are known as the fundamental ones [8,9]. They are pairwise independent and collected into a full set of them they allow to solve any one-dimensional problem. They are distinguished by the property 2^0 above i.e. they are the unique solutions to SE with this property [7].

By a standard semiclassical expansion for χ we mean the following series:

$$\chi(x) \sim C(\hbar) \sum_{n \geq 0} \left(-\frac{\sigma\hbar}{2}\right)^n \chi_n(x)$$

$$\chi_0(x) = 1$$

$$\chi_n(x) = \int_{x_0}^{x} d\xi_n \tilde{D}(\xi_n) \times \quad (11)$$

$$\times \int_{x_0}^{\xi_n} d\xi_{n-1} \tilde{D}(\xi_{n-1}) \ldots \int_{x_0}^{\xi_3} d\xi_2 \tilde{D}(\xi_2) \int_{x_0}^{x} d\xi_1 \left(\tilde{q}^{-\frac{1}{4}}(\xi_1)\left(\tilde{q}^{-\frac{1}{4}}(\xi_1)\right)'' + \tilde{q}^{-\frac{1}{2}}(\xi_1)\delta(\xi_1)\right)$$

$$n = 1, 2, \ldots$$

$$\tilde{D}(x) = \tilde{q}^{-\frac{1}{4}}(x) \frac{d^2}{dx^2} \tilde{q}^{-\frac{1}{4}}(x) + \tilde{q}^{-\frac{1}{2}}(x)\delta(x)$$

$$C(\hbar) = \sum_{n \geq 0} C_n \left(-\frac{\sigma\hbar}{2}\right)^n$$

where a choice of a point x_0 and constants C_k, $k = 1, 2, \ldots$, is arbitrary. However, for the particular χ_k (as defined by (5), for example) this choice is of course definite (if x_0 is given by the lower limit of the integrations in the expansion (5) then $C(\hbar) \equiv 1$). Nevertheless, even in such cases the choice of x_0 can be arbitrary. Only the constants C_k accompanied to the choice are definite depending on the choice [7].

The representation (11) is standard in a sense that any other one can be brought to (11) by redefinitions of the constants C_k. Therefore, any semiclassical expansion can be uniquely given by fixing x_0 and the constants C_k.

And conversely, multiplying a given semiclassical expansion by an asymptotic series as defined by the last series in (11) with other constants $C_k, k = 1, 2, \ldots$, one can obtain any other semiclassical expansion.

We have mentioned above that the semiclassical series for χ is Borel summable for x staying in the sector S_k where χ is defined. In fact it is as such at least inside a circle $Re(\hbar^{-1})^* = (2R)^{-1}$ of the $\hbar$-plane satisfying sufficient conditions of the Watson-Sokal-Nevanlinna (WSN) theorem [10].

Construct now a new semiclassical series by multiplying the one for χ by a $\hbar$-dependent constant $C(\hbar)$ with an *analytic* behaviour at $\hbar = 0$. Expand $C(\hbar)$ into a power series in $\hbar$, the latter being simultanuously an *asymptotic* expansions for the constant. Multiply with this power series the corresponding semiclassical expansion for χ.

A resulting semiclassical series can be now Borel resummed leading us again to another solution to SE. However, this new solution can have now two representations: the one being the solutions (4) multiplied by $C(\hbar)$, and the second being a solution provided by the performed Borel resummation i.e. there is no a priori a necessity for these two representations to coincide.

This is exactly what is observed when a change of variable in SE is performed.

Consider therefore a change of variable in (3) putting $y = y(x)$ and assuming $y'(x)$ to be meromorphic. Such a change of variable preserves the SE (3) if simultanuously we make a substitution: $\Phi(y(x)) \equiv y'^{\frac{1}{2}}(x)\Psi(x)$ and $Q(y)$ corresponding to $\Phi(y)$ in its Schrödinger-like equation is given by:

$$y'^2(x)Q(y(x)) = q(x) - \hbar^2 \left(\frac{3}{4} \frac{y''^2(x)}{y'^2(x)} - \frac{1}{2} \frac{y'''(x)}{y'(x)} \right) \qquad (12)$$

Therefore, the above change of variable provide us with a new potential differing from the old one by the term which depends totaly on $y(x)$. It follows from the form of this term that since $y'(x)$ is assumed to be meromorphic this dependence can introduce to the new potential at most second order poles not cancelling the ones of the original potential $V(x)$ if the latter poles do not depend on $\hbar$. It then follows further that the new second order poles can introduce to the corresponding SG additional sectors and SL's not cancelling the old ones built around the old infinite points of the actions. The old sectors of course change their boundaries and enviroments (having possibly as their neighbours some new sectors).

Consider now therefore the old sector S_k and its new modified form $\tilde{S}_k$. Both the sectors have a common part containing x_0 at its boundary. Using $\Phi(y)$ and $Q(y)$ we can construct in the $\tilde{S}_k$ a solution $\tilde{\Psi}_k(x)$ to SE (3). Namely, we have:

$$\tilde{\Psi}_k = \left(y'^2 \tilde{Q}(y(x)) \right)^{-\frac{1}{4}} e^{\frac{\sigma}{\hbar} \int_{x_i}^x \sqrt{y'^2(\xi)\tilde{Q}(y(\xi))}d\xi} \tilde{\chi}_k(y(x)) \qquad (13)$$

$$k = 1, 2, \ldots$$

where $\tilde{\chi}_k(y)$ is constructed according to (5) - (7) by making there substitutions: $x \to y(= y(x))$, $\delta(x) \to \tilde{\delta}(y)$, $\tilde{q}(x) \to \tilde{Q}(y)$, $\omega(x) \to \tilde{\omega}(y)$, $W(x) \to \tilde{W}(y)$ and $x_0 \to y_0(= y(x_0))$.

Note that the new second order poles introduced to (13) by $y'(x)$ being not present in the original potential $V(x)$ are not real singularities of $\tilde{\Psi}_k(x)$. They are only singularities of the representation (13).

To the solution (13) there correspond a domain $\tilde{D}_k$ (an obvious analogue of D given by the inequality (8)) in which the solution has the same properties $1^0, 2^0$ above as the previous ones defined by (4)-(7). In particular the solutions (13) is Borel summable to itself in $\tilde{S}_k$

Let us note further that because the sectors S_k and $\tilde{S}_k$ have a common part with x_0 at its boundary then the solutions (4) and (13) defined in the corresponding sectors have to coincide with each other up to a muliplicative constant C_k i.e.

$$\tilde{\Psi}_k(x) = C_k(\hbar)\Psi_k(x) \quad k = 1, 2, \ldots \qquad (14)$$

with $C_k(\hbar)$ given by

$$C_k(\hbar) = exp\left[\sigma\hbar \int_{x_i}^{x_0} \frac{\tilde{\delta}(x) - f(x)}{\sqrt{\tilde{q}(x)} + \sqrt{q(x) + \hbar^2\tilde{\delta}(x) - \hbar^2 f(x)}} dx\right] \qquad (15)$$

where

$$f(x) = \frac{3}{4}\frac{y''^2(x)}{y'^2(x)} - \frac{1}{2}\frac{y'''(x)}{y'(x)} \qquad (16)$$

The coefficient C_k was calculated by taking a limit $x \to x_0$ on both sides of (14).

From (14) and (15) we get the following relation between $\tilde{\chi}_k$ and χ_k:

$$\tilde{\chi}_k(x) = \left(1 + \hbar^2\frac{\tilde{\delta}(x) - f(x)}{\tilde{q}(x)}\right)^{\frac{1}{4}} \times$$

$$\times exp\left[-\sigma\hbar \int_{x_0}^{x} \frac{\tilde{\delta}(\xi) - f(\xi)}{\sqrt{\tilde{q}(\xi)} + \sqrt{\tilde{q}(\xi) + \hbar^2\tilde{\delta}(\xi) - \hbar^2 f(\xi)}} d\xi\right] \chi_k(x) \qquad (17)$$

Note that the two factors in (17) staying in front of $\tilde{\chi}_k$ are holomorphic with respect to $\hbar$ at $\hbar = 0$.

We shall now show that the solution (13) as well as its $\tilde{\chi}_k$-function are just the Borel sums of the corresponding semiclassically expanded right hand sights in (14) and (17), respectively.

This is an immediate consequence of the holomorphicity of the coefficient $C_k(\hbar)$ and of the two factors in (17) at $\hbar = 0$ due to which their semiclassical expansions coincide with their convergent power series expansion in $\hbar$. Therefore, due to our earlier discussion the WSN conditions for Borel summability of the semiclassical series emerging from RHS in (14) and (17) are satisfied and $\tilde{\Psi}_k(x)$ and $\tilde{\chi}_k(x)$ are obtained by taking these Borel sums.

Summarizing the above discussion we can conclude that the effect of variable changing leading us to the solutions (13) can be obtained also as a result of Borel resummations of the standard semiclassical expansions for the solutions (5) multiplied by a suitably chosen $\hbar$-dependent constants.

A choice of the constants in (14) can be even done in such a way to produce simultaneously fundamental solutions for which the series in (6) start with an arbitrary high power of $\hbar$ [6]. Such a choice corresponds to a total effect of repeating changes of variable when for each subsequent Schrödinger-like equation a new independent variable is the action i.e. $y'^2(x) = \tilde{q}(x,\hbar)$. The 'lacking' powers of $\hbar$ are collected then in $(y'^2(x,\hbar)\tilde{Q}(x,\hbar))^{\frac{1}{4}}$ and in the corresponding exponential factors of the solutions (5). These two factors are then the sources of new JWKB approximations generalizing the conventional ones [6].

To conclude we have shown that the Borel summable fundamental solutions to SE can be modified by appropriate Borel resummations of the latter multiplied by properly chosen $\hbar$-dependent constans. Sometimes effects of such resummations can be recognized as a proper change of variable in SE. But the latter can always be considered as an effect of such resummations. This justifies certainly all the improvements and sometimes exact results provided by the change-of-variable procedure applied in JWKB calculations. The latter possibility (i.e. the exact results), however, can realize only due to particular properties of considered potentials reflected in global structures of their respective Stokes graphs [11]. Some further consequences of this Borel resummability feature of the change-of-variable procedure shall be discussed elsewhere [13].

REFERENCES

1. Langer, R. E., *Phys. Rev.* **51**, 669-76 (1937).
2. Berry, M. V., and Mount, K. E., *Rep. Prog. Phys.* **35**, 315-397 (1972).
3. Fröman, N., and Fröman, P. O., *JWKB Approximation. Contribution to the Theory*, Amsterdam: North-Holland, 1965.
4. Rozenzweig, C., and Krieger, J. B., *J. Math. Phys.* **9**, 849-860 (1968).
5. Krieger, J. B., *J. Math. Phys.* **10**, 1455-1458 (1969).
6. Giller, S., *J. Phys. A: Math. Gen.* **21**, 909 (1988).
7. Giller, S., and Milczarski, P., "Borel summable solutions to 1-dim Schrödinger equation", quant-ph/9801031, to be published.
8. Giller, S., *Acta Phys. Pol.* **B23**, 457-511 (1992).
9. Giller, S., *Acta Phys. Pol.* **B21**, 675-709 (1990).
10. Watson, G. N., *Philos. Trans. Soc. London Ser.A* **211**, 279 (1912).
    Sokal, A. D., *J. Math. Phys.* **21**, 261-263 (1980).
    Nevanlinna, F., *Ann. Acad. Sci. Fenn. Ser. A* **12**, No.3, (1918-19).
11. Milczarski, P., "Exactness of Conventional and Supersymmetric JWKB Formulae

and Global Symmetries of Stokes Graphs", quant-ph/9807039, *Ph.D. thesis, to be published.*
12. Bose, A. K., *Nuo. Cim.* **32**, 679 (1964).
13. Giler, S., and Milczarski, P., "Change of variable as Borel ressummation of semiclassical series", quant-ph/9712039, *to be published.*

Einstein-Podolski-Rosen Experiment from Noncommutative Quantum Gravity

Michael Heller

Vatican Observatory, V-12000 Vatican City State[1]

Wiesław Sasin

Institute of Mathematics, Warsaw University of Technology
Plac Politechniki 1, 00-661 Warsaw, Poland

Abstract. It is shown that the Einstein–Podolski–Rosen type experiments are the natural consequence of the groupoid approach to noncommutative unification of general relativity and quantum mechanics. The geometry of this model is determined by the noncommutative algebra $\mathcal{A} = C_c^\infty(G, \mathbf{C})$ of complex valued, compactly supported, functions (with convolution as multiplication) on the groupoid $G = E \times \Gamma$. In the model considered in the present paper E is the total space of the frame bundle over space-time and Γ is the Lorentz group. The correlations of the EPR type should be regarded as remnants of the totally non-local physics below the Planck threshold which is modelled by a noncommutative geometry.

INTRODUCTION

One of the greatest challenges of contemporary physics is to explain the non-local effects of quantum mechanics theoretically predicted (in the form of a gedanken experiment) by Einstein, Podolski and Rosen (EPR, for short) [1] and experimentally verified by Aspect et al. [2–4] (for a comprehensive review see [5]). Although non-local effects of this type logically follow from the postulates of quantum mechanics, it seems strange and against our "realistic common sense" to accept that two particles separated in space could be so strongly correlated (provided they once interacted with each other) that they seem to "know" about each other irrespectively of the distance separating them. In spite of long lasting discussions, so far no satisfactory explanation of this effect has been offered. In the present paper we shall argue that effects of the EPR type are remnants of the totally non-local physics of the fundamental level (below the Planck threshold). We substantiate

---

[1] Correspondence address: ul. Powstańców Warszawy 13/94, 33-110 Tarnów, Poland. E-mail: mheller@wsd.tarnow.pl

our argument by explaining the EPR experiment in terms of a quantum gravity model, based on a noncommutative geometry, proposed by us in [6] (see also [7]), although the explanation itself does not depend on particulars of the model.

The main physical idea underlying our model is that below the Planck threshold (we shall speak also on the "fundamental level") there is no space-time but only a kind of pregeometry which is modeled by a suitable noncommutative space, and that space-time emerges only in the transition process to the classical gravity regime. Accordingly, we start our construction not from a space-time manifold M, but rather from a groupoid $G = E \times \Gamma$ where E is a certain abstract space and Γ a suitable group of "fundamental symmetries". In the present paper, for the sake of simplicity, we shall assume that E is the total space of the frame bundle over space-time M and $\Gamma = SO(3,1)$. We define, in terms of this geometry, the noncommutative algebra $\mathcal{A} = C_c^\infty(G, \mathbf{C})$ of smooth, compactly supported, complex-valued functions on the groupoid G with the usual addition and convolution as multiplication. We develop a noncommutative differential geometry basing on this algebra, and define a noncommutative version of Einstein's equation (in the operator form). The algebra $\mathcal{A}$ can be completed to become a C^*-algebra, and this subalgebra of $\mathcal{A}$ which satisfies the generalized Einstein's equation is called *Einstein C^*- algebra*, denoted by $\mathcal{E}$ (for details see [6]). And now quantization is performed in the standard algebraic way. Since the explanation of the EPR type experiments depends on the noncommutative structure of the groupoid G rather than on details of our field equations and the quantization procedure, we shall not review them here; the reader interested in the particulars of our model should consult the original paper [6].

It can be shown that the subalgebra $\mathcal{A}_{proj}$ (elements of $\mathcal{A}_{proj}$ are called *projectible*) of functions which are constant on suitable equivalence classes of fibres $\pi_E^{-1}(p)$, π_E being the projection $G = E \times \Gamma \to \Gamma$ and $p \in E$, is isomorphic to the algebra $C^\infty(M)$ of smooth functions on M. Consequently, by making the restriction of $\mathcal{A}$ to $\mathcal{A}_{proj}$ we recover the ordinary space-time geometry and the standard general relativity. In our model, to simplify calculations, we have assumed that the noncommutative differential geometry is determined by the submodule V of the module Der$\mathcal{A}$ of all derivations of $\mathcal{A}$, and that V has the structure adapted to the product structure of the groupoid $G = E \times \Gamma$, i. e., $V = V_E \oplus V_\Gamma$, where V_E and V_Γ are "parts" parallel to E and Γ, correspondingly. It can be seen that in our model the geometry "parallel" to E is responsible for generally relativistic effects, and that "parallel" to Γ for quantum effects. In general, "mixed terms" should appear, and then one would obtain stronger interaction between general relativity and quantum physics. This remains to be elaborated in the future.

The crucial point is that the geometry as determined by the noncommutative algebra $\mathcal{A}$ is non-local, i. e., there are no maximal ideals in $\mathcal{A}$ which could determine points and their neighborhoods in the corresponding space, and consequently neither space points nor time instants can be defined in terms of $\mathcal{A}$. Physical states of a quantum gravitational system are identified with states on the algebra $\mathcal{A}$, i. e., with the set of positive linear functionals (normed to unity) on $\mathcal{A}$, and pure states in the mathematical sense are identified with pure states in the physical sense. Let $a \in \mathcal{A}$ be a quantum gravitational observable, i. e., a projectible and Hermitian element of $\mathcal{A}$ (a must be projectible to leave traces in the macroscopic world), and φ a state on $\mathcal{A}$. Then $\varphi(a)$ is the expectation value of the observable a when the system is in the state φ. The fact that a is an element of a "non-local" (noncommutative) algebra $\mathcal{A}$ implies that when a is projected to the space-time M it becomes a real-valued (since a is Hermitian) function on M, and the results of a measurement corresponding to a are values of this function. Consequently, one should expect correlations between various measurement results even if they are performed at distant points of space-time M. We shall see that this is indeed the case.

The organization of our material is the following. First, we consider the eigenvalue equation for quantum gravitational observables. In the following section, we show that correlations of the EPR type between distant events in space-time are consequences of non-local (noncommutative) physics of the quantum gravitational regime, and in the next section we present details of the EPR experiment in terms of the noncommutative approach. The last section contains concluding remarks.

MEASUREMENT ON QUANTUM GRAVITATIONAL SYSTEM

Let $\varphi : \mathcal{A} \to \mathbf{C}$ be a state on the algebra $\mathcal{A}$, i. e., $\varphi(1) = 1$ and $\varphi(aa^*) \geq 0$ for every $a \in \mathcal{A}$. It can be easily seen that $\varphi|\mathcal{A}_{proj} : \mathcal{A}_{proj} \to \mathbf{C}$ is a state on the subalgebra $\mathcal{A}_{proj}$.

Let now $a \in \mathcal{A}_{proj}$ be Hermitian, then there exists a function $\bar{a} \in C^\infty(M)$ with $\bar{a} \circ pr = a$, where $pr : G \to M$ is the projection, and the state $\bar{\varphi} : C^\infty(M) \to \mathbf{R}$ on the algebra $C^\infty(M)$, such that $\varphi(a) = \bar{\varphi}(\bar{a})$. Since the algebras $\mathcal{A}_{proj}$ and $C^\infty(M)$ are isomorphic, the spaces of states of these algebras are isomorphic as well.

To make a contact with the standard formulation of quantum mechanics we represent the noncommutative algebra $\mathcal{A}$ in a Hilbert space by defining, for each $p \in E$, the representation

$$\pi_p : \mathcal{A} \to \mathcal{B}(\mathcal{H}),$$

where $\mathcal{B}(\mathcal{H})$ is the algebra of operators on the Hilbert space $\mathcal{H} = L^2(G_p)$ of square integrable functions on the fibre $G_p = \pi_E^{-1}(p)$, $\pi_E : G \to E$ being the natural projection, with the help of the formula

$$\pi_p(a)\psi = \pi_p(a) * \psi$$

or more explicitly

$$(\pi_p(a)\psi)(\gamma) = \int_{G_p} a(\gamma_1)\psi(\gamma_1^{-1}\gamma),$$

$a \in \mathcal{A}$, $\gamma = \gamma_1 \circ \gamma_2$, $\gamma, \gamma_1, \gamma_2 \in G_p$, $\psi \in L^2(G_p)$, and the integration is with respect to the Haar measure. This representation is called the *Connes representation* (see [8, p.102], [6]).

Now, let us suppose that a is an observable, i. e., $a \in \mathcal{A}_{proj}$, and we perform a measurement of the quantity corresponding to this observable. The eigenvalue equation for a is

$$\int_{G_P} a(\gamma_1)\psi(\gamma_1^{-1}\gamma) = r_p\psi(\gamma) \tag{1}$$

where the eigenvalue r_q is the expected result of the measurement when the system is in the state ψ. Here we must additionally assume that the "wave function" ψ is constant on fibres of G to guarantee for equation (1) to have its usual meaning in the non-quantum gravity limit. If this is the case, equation (1) can be written in the form

$$\psi(\gamma_1^{-1}\gamma) \int_{g_p} a(\gamma_1) = r_p\psi(\gamma)$$

and consequently

$$r_p = \int_{G_p} a(\gamma_1).$$

Let us notice that the measurement result is a measure in the mathematical sense.

It is obvious that if we define the "total phase space" of our quantum gravitational system

$$L^2(G) := \bigoplus_{p \in E} L^2(G_p)$$

and the operator

$$\pi(a) := (\pi_p(a))_{p \in E}$$

acting on $L^2(G)$ then the eigenvalue equation becomes

$$\pi(a)\psi = r\psi$$

where $r : M \to \mathbf{R}$ is a function on space-time M given by

$$r(x) = \int_{G_p} a(\gamma_1) \tag{2}$$

where x is a point in M to which the frame p is attached. Let us notice that the function r is equal to the function $\bar{a} : M \to \mathbf{R}$ (see the beginning of the present Section). Let us now consider a composed quantum system the state of which is described by the single vector in the Hilbert space, and let us perform a measurement on its parts when they are at a great distance from each other. Formula (2) asserts that in such a case the results of the measurement are not independent but are the values of the same function defined on space-time. This can be regarded as a "shadow" of a non-local character of the observable a projected down to space-time M.

EPR NON-LOCALITY

So far we were mainly concerned with what happens when we project the algebra $\mathcal{A}$ onto the "horizontal component" E of the groupoid G. This, of course, gives us the transition to the classical space-time geometry (general relativity). In the present Section, we shall be interested in projecting $\mathcal{A}$ onto the "vertical component" Γ of G. This gives us quantum effects of our model.

Let us consider functions *projectible* to Γ. We define

$$\mathcal{A}_\Gamma := \{f \circ pr_\Gamma : f \in C_c^\infty(\Gamma, \mathbf{C})\} \subset \mathcal{A}.$$

The reasoning similar to that in the beginning of the present section shows that if $s \in \mathcal{A}_\Gamma$ and $\psi : \mathcal{A}_\Gamma \to \mathbf{C}$ is a state on $\mathcal{A}_\Gamma$ then $\psi(s) = \underline{\psi}(\underline{s})$, where $s = \underline{s} \circ pr_\Gamma$, $pr_\Gamma : G \to \Gamma$ is the projection, and $\underline{\psi} : C_c^\infty(\Gamma, \mathbf{C}) \to \mathbf{C}$ is a state on $C_c^\infty(\Gamma, \mathbf{C})$.

Let now Φ be a state on $\overline{C}_c^\infty(\Gamma, \mathbf{C})$. We say that the state $\varphi : \mathcal{A} \to \mathbf{C}$ is Γ-*invariant associated to* Φ on $\mathcal{A}$ if

$$\varphi(s) = \begin{cases} \Phi(\underline{s}), & \text{if } s \in \mathcal{A}_\Gamma, \\ 0, & \text{if } s \notin \mathcal{A}_\Gamma. \end{cases}$$

Since all fibres of G_p, $p \in E$, of G are isomorphic, the number $\varphi(s) = \Phi(\underline{s})$, for $s \in \mathcal{A}_\Gamma$, is the same in each fibre G_p. If additionally s is a Hermitian element of $\mathcal{A}$, and if a measurement performed at a certain point of space-time M gives the number $\varphi(s)$ as its result, then this result is immediately "known" at all other fibres G_p, $p \in E$, of G, and consequently at all other points of space-time $x = \pi_M(p) \in M$, where $\pi_M : E \to M$ is the canonical projection.

This can be transparently seen if we consider the problem in the Hilbert space by using the Connes representation of the algebra $\mathcal{A}$. Let $a \in \mathcal{A}_\Gamma$, and let us consider the following Connes representations

$$\pi_p(a)(\xi_p) = a_p * \xi_p, \tag{3}$$

and

$$\pi_q(a)(\xi_q) = a_q * \xi_q, \tag{4}$$

where $\xi_p \in L^2(G_p)$, $\xi_q \in L^2(G_q)$, $p, q \in E$, $p \neq q$. Since G_p and G_q are isomorphic, we can choose ξ_p and ξ_q to be isomorphic with each other, which implies that $\pi_p(a)$ and $\pi_q(a)$ are isomorphic as well. We have the following important

Lemma. If $a \in \mathcal{A}_\Gamma$ then its image under the Connes representation π_p does not depend of the choice of $p \in E$ (up to isomorphism).

Since $p \in E$ projects down to the space-time point $\pi_M(p) \in M$, $\pi_M : E \to M$, the above result should be interpreted as stating that all points of M "know" what happens in the fiber G_g, $g \in \Gamma$. This, together with the fact that vectors ξ_p, upon which the observable $\pi_p(a)$, $a \in \mathcal{A}_\Gamma$ acts, also do not depend of p, in principle, explains the EPR type experiments. However, let us go deeper into details.

EPR EXPERIMENT IN TERMS OF NONCOMMUTATIVE GEOMETRY

In this section we consider a group Γ such that $\Gamma_0 = SU(2)$ is its compact subgroup. We look for an element $s \in \mathcal{A}_\Gamma$ such that

$$\pi_p(s) : L^2(\Gamma_0) \to L^2(\Gamma_0).$$

Of course, $\mathbf{C}^2 \subset \mathbf{L}^2(\mathbf{\Gamma_0})$. We define two linearly independent functions on the group Γ_0, for instance the constant function

$$\mathbf{1} : \Gamma_0 \to \mathbf{C},$$

and

$$\det : \Gamma_0 \to \mathbf{C},$$

which span the linear space $\mathbf{C}^2$, i. e., $\mathbf{C}^2 = \langle \mathbf{1}, \det \rangle_{\mathbf{C}}$. Let $\hat{S}_z = \pi_p(s)|_{\mathbf{C}^2}$ be the usual z-component spin operator. We have

$$\pi_p(s)\psi = \hat{S}_z \psi,$$

for $\psi \in \mathbf{C}^2$ or, by using the Connes representation and the fact that $\hat{S}_z \psi = \pm\frac{\hbar}{2}\psi$,

$$\int_{\Gamma_0} s_p(\gamma_1)\psi(\gamma_1^{-1}\gamma) = \pm\frac{\hbar}{2}\psi.$$

Since $s_p =$const, one obtains

$$\int_{\Gamma_0} \psi(\gamma_1^{-1}\gamma) \sim \psi(\gamma).$$

One of the solutions of this equation is $\psi = \mathbf{1}_{\Gamma_0}$. Therefore

$$\frac{\hbar}{2} = \pm \int_{\Gamma_0} s_p(\gamma_1).$$

Hence

$$(s_p)_1 = +\frac{\hbar}{2}\frac{1}{\mathrm{vol}\Gamma_0},$$

$$(s_p)_2 = -\frac{\hbar}{2}\frac{1}{\mathrm{vol}\Gamma_0},$$

and consequently

$$\pi_p((s_p)_1)\psi = +\frac{\hbar}{2}\psi \text{ for } \psi \in \mathbf{C}^+,$$

$$\pi_p((s_p)_2)\psi = -\frac{\hbar}{2}\psi \text{ for } \psi \in \mathbf{C}^-,$$

where $\mathbf{C}^+ := \mathbf{C} \times \{0\}$, and $\mathbf{C}^- := \{0\} \times \mathbf{C}$. To summarize these results we can define

$$\hat{S}_z\psi = \pi_p(s_1, s_2)\psi := \begin{cases} (s_1)_p * \psi & \text{if } \psi \in \mathbf{C}^+, \\ (s_2)_p * \psi & \text{if } \psi \in \mathbf{C}^-. \end{cases}$$

Now, the analysis of the "EPR paradox" proceeds in the same way as in the standard textbooks on quantum mechanics (see for instance [9, pp. 179-181]). An observer A, situated at $\pi_M(p) = x_A \in M$, measures the z-spin component of the one of the electrons[2], i. e., he applies the operator $\hat{S}_z \otimes \mathbf{1}|_{\mathbf{C}^2}$ to the vector $\xi = \frac{1}{\sqrt{2}}(\psi \otimes \varphi - \varphi \otimes \psi)$ where $\psi \in \mathbf{C}^+$ and $\varphi \in \mathbf{C}^-$. Let us suppose that the result of the measurement is $\frac{\hbar}{2}$. This means that the state vector $\xi = \frac{1}{\sqrt{2}}(\psi \otimes \varphi - \varphi \otimes \psi) \in \mathbf{C}^2 \otimes \mathbf{C}^2 \subset L^2(G_r) \otimes L^2(G_r)$, $r \in E$, has collapsed to $\xi_0 = \frac{1}{\sqrt{2}}(\psi \otimes \varphi)$, and that immediately after the measurement the system is in the state ξ_0 which is the same (up to isomorphism) for all fibres G_r whatever $r \in E$, and consequently it does not depend of the point in space-time to which r is attached (see formulae (3) and (4) which are obviously valid also for tensor products). In particular, the vector ξ_0 is the same for the fibres G_p and G_q where p is such that $\pi_M(p) = x_A$ and q is such that $\pi_M(q) = x_B$ ($x_A \neq x_B$). It is now obvious that if an observer B, situated a x_B measures the z-spin component of the second electron, i. e., if he applies the operator $\mathbf{1}|_{\mathbf{C}^2} \otimes \hat{S}_z$ to the vector ξ_0, he will obtain the value $-\frac{\hbar}{2}$ as the result of his measurement.

CONCLUDING REMARKS

To conclude our analysis it seems suitable to make the following remarks.

It should be emphasized that our scheme for noncommutative quantum gravity does not "explain" quantum mechanical postulates. In the very construction of our scheme it has been assumed that the known postulates which, in the standard formulation of quantum mechanics are valid for the algebra of observables, can be extended to a more general noncommutative algebra. However, the very fact that these postulates are valid in the conceptual framework of noncommutative geometry gives them a new flavour. For instance, since in the noncommutative regime there is no time in the usual sense, the sharp distinction between the continuous unitary evolution and the non-continuous process of measurement ("collapse of the wave function") disappears. This distinction becomes manifest only when time emerges (see [10]) in changing from the noncommutative regime to the usual space-time geometry.

What our approach does explain is the fact that some quantum effects are strongly correlated even if they occur at great distances from each other. These

---

[2] Let us notice that when A measures the spin of the electron, he simultaneously determines the position of the electron (at least roughly), i. e., the position x_A at which he himself is situated (spin and position operators commute).

effects are "projections" from the fundamental level at which all concepts have purely global meanings.

This explanation does not depend on "details" of our model, such as some particulars of the construction of noncommutative differential geometry, the concrete form of generalized Einstein's equation, or the dynamical equation for quantum gravity. However, it does depend on (or even more, it is deeply rooted in) the noncommutative character of the algebra $\mathcal{A} = C_c^\infty(G, \mathbf{C})$ and the product structure of the groupoid $G = E \times \Gamma$.

REFERENCES

1. A. Einstein, B. Podolski and N. Rosen, "Can quantum description of physical reality be considered complete?" *Phys. Rev.* **47**, 777–780 (1935).
2. A. Aspect, P. Grangier and G. Roger, "Experimental tests of realistic local theories via Bell's theorem", *Phys. Rev. Lett.*, **47**, 460–463 (1981).
3. A. Aspect, P. Grangier and G. Roger, "Experimental realization of Einstein-Podolsky-Rosen-Bohm Gedankenexperiment: A new violation of Bells inequalities", *Phys. Rev. Lett.*, **49**, 91–94 (1982).
4. A. Aspect, J. Dalibard and G. Roger, "Experimental tests of Bell inequalities using time-varying analyzers", *Phys. Rev. Lett.*, **49**, 1804–1807 (1982).
5. M.L.G. Redhead, *Incompleteness, Nonlocality, and Realism: A Prolegomenon to the Philosophy of Quantum Mechanics*, Clarendon Press, Oxford, 1987.
6. M. Heller, W. Sasin and D. Lambert, "Groupoid approach to noncommutative quantization of gravity, *J. Math. Phys.*, **38**, 5840–5853 (1997).
7. M. Heller, and W. Sasin, "Towards noncommutative quantization of gravity, gr-qc/9712009.
8. A. Connes, *Noncommutative Geometry*, Academic Press, San Diego-New York, 1994.
9. C. J. Isham, *Lectures on Quantum Theory*, Imperial College Press, London, 1995.
10. M. Heller and W. Sasin, "Emergence of Time", gr-qc/9711051.

Wigner Quantization Problem for External Forces

Edward Kapuścik and Andrzej Horzela

Department of Theoretical Physics
Henryk Niewodniczański Institute of Nuclear Physics
ul. Radzikowskiego 152, 31-342 Kraków, Poland

Abstract. The Wigner problem of finding most general quantum mechanical commutation relations compatible with equations of motion is discussed for particles influenced by quantized external forces. The problem is solved here under an assumption that Wigner commutation relations determine minimal Lie algebras with initial observables as generators.

INTRODUCTION

In 1950 E.P. Wigner [1] raised the question of the interrelation between the quantum mechanical commutation relations and equations of motion. For the harmonic oscillator he found a one parameter class of commutation relations alternative to the standard Heisenberg one.

The Wigner approach to quantization is now extensively discussed in literature [2], [3]. Here we present a new way of treating Wigner commutation relations. The novelty consists in constructing Lie algebras determined by the Wigner commutation relations.

LIE ALGEBRAS DETERMINED BY WIGNER COMMUTATION RELATIONS

As a starting point we take classical solutions of equations of motion for one dimensional motion of a single mass point influenced by driving forces. We assume that a sufficiently large class of driving forces may be obtained as superpositions of forces with simple time dependence. We shall therefore consider driving forces given in the following form

$$F(t) = \sum_{k=1}^{N} F_k \, f_k(t) \tag{1}$$

where the "amplitudes" F_k are some parameters which will undergo quantization while $f_k(t)$ are known functions of time which describe simple "elementary" forces $f_k(t)$. For the forces (1) the solutions of classical equations of motion may be written in the form

$$x(t) = x_0 + \frac{p_0}{M}(t - t_0) + \sum_{k=1}^{N} F_k\, x_k(t) \tag{2}$$

where x_0 and p_0 are initial values of the coordinate and momentum at the instant of time t_0 while $x_k(t)$ are solutions of the equations

$$M\ddot{x}_k(t) = f_k(t) \tag{3}$$

which vanish at $t = t_0$ together with their first derivatives. The momentum of the particle is equal to

$$p(t) = p_0 + M\sum_{k=1}^{N} F_k \dot{x}_k(t). \tag{4}$$

Let us now assume that as a result of quantization all parameters in (1), (2) and (4), except time, are treated as operators in some Hilbert space. The coordinate and momentum of the particle, as well as the force acting on it, are then operators with the following commutation relations

$$[x(t), p(t)] = i\hbar\Lambda + iM\sum_{k=1}^{N} \Delta_k \dot{x}_k(t) +$$

$$+i\sum_{k=1}^{N} \Gamma_k (t-t_0)^2 \frac{d}{dt}\frac{x_k(t)}{t-t_0} + +iM\sum_{j,k=1}^{N} \Theta_{jk} x_j(t)\, \dot{x}_k(t), \tag{5}$$

$$[x(t), F(t)] = i\sum_{k=1}^{N} \Delta_k f_k(t) + \sum_{k=1}^{N} \Gamma_k f_k(t)(t-t_0) +$$

$$+i\sum_{k,j=1}^{N} \Theta_{kj}\, x_k(t)\, f_j(t), \tag{6}$$

$$[p(t), F(t)] = i\sum_{k=1}^{N} \Gamma_k f_k(t) + iM\sum_{k,j=1}^{N} \Theta_{kj} \dot{x}_k(t)\, f_k(t) \tag{7}$$

where we have introduced the notations

$$[x_0, p_0] = i\hbar\Lambda, \tag{8}$$

$$[x_0, F_k] = i\Delta_k, \qquad (9)$$

$$[p_0, F_k] = i\Gamma_k \qquad (10)$$

and

$$[F_k, F_j] = i\,\Theta_{kj} = -i\,\Theta_{jk}. \qquad (11)$$

In the standard Heisenberg canonical quantization we simply have

$$\Lambda = I, \qquad \Delta_k = \Gamma_k = \Theta_{jk} = 0. \qquad (12)$$

It is clear that the nature of the above commutation relations and consequently the quantum nature of all physical observables depends on the choice of the operators Λ, Δ_k, Γ_k and Θ_{jk} ($k, j = 1, 2, 3$). Physically, these operators are defined by the initial (at the instant of time t_0) meaning given to all quantized observables. They determine the compatibility of all measurements performed on the considered physical system. It is therefore natural to require that all uncertainties of these measurements are mutually compatible because otherwise they could not be simultaneously determined. This means that we should have

$$[\Lambda, \Delta_k] = [\Lambda, \Gamma_k] = [\Lambda, \Theta_{kj}] = [\Delta_k, \Delta_j] =$$

$$[\Delta_k, \Gamma_j] = [\Delta_k, \Theta_{jl}] = [\Gamma_k, \Theta_{jl}] = [\Theta_{kj}, \Theta_{lm}] = 0 \qquad (13)$$

for all values of the indices.

It is clear from (5) - (7) that in the case when

$$\Delta_k \neq 0, \qquad \Gamma_k \neq 0, \qquad \Theta_{kj} \neq 0 \qquad (14)$$

all basic commutation relations acquire specific time dependence. This is the most important difference between the Wigner and Heisenberg quantizations.

The choice of the uncertainties Λ, Δ_k, Γ_k and Θ_{jk} determines the quantum nature not only of the position and momentum but also the quantum nature of the driving forces. Among all possible ways of choosing them we prefer the way based on Lie algebras determined by the Wigner commutators. Such an approach selects those Wigner quantum mechanical systems which most closely remind the standard Heisenberg ones. In our method we construct only algebras which are invariant under Galilean transformations, space reflection and time reversal. As a result we get that in such algebras all the operators Λ, Δ_k, Γ_k and Θ_{jk} commute with x_0, p_0 and F_k. All the operators Λ, Δ_k, Γ_k and Θ_{jk} belong therefore to the centers of the corresponding Lie algebras. The story changes however radically when we extend the algebra of initial operators by the Hamiltonian of the system defined by the commutation relations

$$[x_0, H] = i\hbar \frac{p_0}{M}, \tag{15}$$

$$[p_0, H] = i\hbar \sum_{k=1}^{N} F_k \, f_k(0) \tag{16}$$

and

$$[F_k, H] = i\hbar \sum_{j=1}^{N} \alpha_{kj} F_j \tag{17}$$

where the coefficients α_{kj} depend on the particular dynamics of the system. In the extended algebras the uncertainties cease to commute with the Hamiltonian. This means that the energy content of the Wigner systems is different from the energy content of systems described by the standard quantum mechanical Heisenberg algebra.

To proceed further we shall illustrate our method on two simple examples [4], [5].

THE CASE OF A CONSTANT DRIVING FORCE

In the case of a constant in time force we have only one force parameter F_1 and

$$f_1(t) = 1, \qquad x_1(t) = \frac{1}{2M}(t - t_0)^2. \tag{18}$$

The Wigner commutation relations are of the form

$$[x(t), p(t)] = i\hbar\Lambda + i(t - t_0)\,\Delta + \frac{i(t - t_0)^2}{2M}\Gamma, \tag{19}$$

$$[x(t), F] = i\Delta + \frac{i(t - t_0)}{M}\Gamma, \tag{20}$$

$$[p(t), F] = i\Gamma \tag{21}$$

where for obvious reason we skipped the unessential single subscript. Clearly in the present case we have

$$\Theta_{jk} = 0. \tag{22}$$

The commutation relations (16) and (17) for the Hamiltonian are

$$[p_0, H] = i\hbar F \tag{23}$$

and
$$[F, H] = 0 \tag{24}$$

while for the uncertainties we get
$$[\Lambda, H] = i\hbar\Delta, \tag{25}$$

$$[\Delta, H] = i\hbar\Gamma \tag{26}$$

and
$$[\Gamma, H] = 0. \tag{27}$$

From these results we see that the operator Γ remains to be in the center of the whole algebra. In each irreducible representation Γ is therefore realized as a dimensional number. The relation (26) provides us the only pair of canonically conjugated operators with the constant value of Γ as a new fundamental quantization constant.

THE CASE OF A HARMONIC DRIVING FORCE

As a less trivial example we consider here the case of a harmonic driving force of the form
$$F(t) = F_1 \cos \omega (t - t_0) + F_2 \frac{\sin \omega (t - t_0)}{\omega} \tag{28}$$

which is the superposition of two "elementary" harmonic forces
$$f_1(t) = \cos \omega (t - t_0) \tag{29}$$

and
$$f_2(t) = \frac{\sin \omega (t - t_0)}{\omega}. \tag{30}$$

From the equation of motion we get
$$x_1(t) = \frac{1 - \cos \omega (t - t_0)}{M\omega^2} \tag{31}$$

and
$$x_2(t) = \frac{\omega(t - t_0) - \sin \omega (t - t_0)}{M\omega^3}. \tag{32}$$

The Wigner commutation relations are now much more complicated than in the previous case. Among the terms we find such which increase in time and such which are bounded. It turns out that the unrestricted in time terms can be cancelled by passing to a particular subalgebra of the whole algebra of Wigner commutation relations. This subalgebra is selected by the conditions

$$\Gamma_2 = 0, \qquad \Theta_{12} = \omega^2 \Gamma_1. \tag{33}$$

Under these conditions the commutation relations are

$$[x(t), p(t)] = i\hbar\Lambda + i\Delta_1 \frac{\sin\omega(t-t_0)}{\omega} +$$

$$+ i\Delta_2 \frac{1-\cos\omega(t-t_0)}{\omega^2} + i\Gamma_1 \frac{1-\cos\omega(t-t_0)}{M\omega^2}, \tag{34}$$

$$[x(t), F(t)] = i\Delta_1 \cos\omega(t-t_0) +$$

$$+ i\Delta_2 \frac{\sin\omega(t-t_0)}{\omega} + i\Gamma_1 \frac{\sin\omega(t-t_0)}{M\omega^2}, \tag{35}$$

$$[p(t), F(t)] = i\Gamma_1. \tag{36}$$

The Hamiltonian, in addition to the commutation relation (15), satisfies now the relations

$$[p_0, H] = i\hbar F_1, \tag{37}$$

$$[F_1, H] = i\hbar F_2, \tag{38}$$

$$[F_2, H] = -i\hbar\omega^2 F_1. \tag{39}$$

The Jacobi identities for the complete Lie algebra require now that the uncertainties must satisfy the relations

$$[\Lambda, H] = i\hbar\Delta_1, \tag{40}$$

$$[\Delta_1, H] = i\hbar\left(\Delta_2 + \frac{1}{M}\Gamma_1\right), \tag{41}$$

$$[\Delta_2, H] = -i\hbar\Delta_1, \tag{42}$$

$$[\Gamma_1, H] = [\Theta_{12}, H] = 0. \tag{43}$$

In the present case the center of the algebra consists of two operators Γ_1 and Θ_{12}. We do not get however any pair of canonically conjugated operators because on the right hand side of (41) apart from Γ_1 we have also the operator Δ_2.

CONCLUSIONS

We have shown that Wigner commutation relations may form new Lie algebras of quantum mechanical observables. In these algebras operators responsible for uncertainties are treated dynamically. As a result the observed uncertainties of physical measurements may change in time.

REFERENCES

1. Wigner,E.P, *Phys. Rev.* **77**, 711 (1950).
2. Man´ko,V.I, Marmo,G., Sudarshan,E.C.G., Zaccaria,F., *Phys. Scr.***55**, 528 (1997).
3. López-Pena,R., Man´ko,V.I., Marmo,G., *Phys. Rev. A* **56**, 1126 (1997).
4. Kapuścik,E., "Wigner Time Dependent Commutation Relations for Particles in External Force Fields", in *Proceedings of the Conference Fifth International Conference on Squeezed States and Uncertainty Relations, Balatonfüred, Hungary, 27-31 May, 1997*, D. Han, J. Janszky, Y.S. Kim, and V.I. Man´ko, Eds., NASA/CP - 1998 - 206855, p.275 (1998).
5. Horzela,A., "Wigner Time Dependent Quantization for External Harmonic Forces", in *Proceedings of the Conference Fifth International Conference on Squeezed States and Uncertainty Relations, Balatonfüred, Hungary, 27-31 May, 1997*, D. Han, J. Janszky, Y.S. Kim and V.I. Man´ko, Eds., NASA/CP - 1998 -206855, p.251 (1998).

Coherent States for a Quantum Spin 1/2 Particle on a Sphere[1]

K. Kowalski and J. Rembieliński

Department of Theoretical Physics, University of Łódź, ul. Pomorska 149/153, 90-236 Łódź, Poland

L.C. Papaloucas

Department of Mathematics, University of Athens, Panepistemiopolis, 157 84 Athens, Greece

Abstract. The coherent states for a quantum spin 1/2 particle on a sphere are introduced. These states are marked with points of the classical phase space, that is the position and the angular momentum of a particle.

INTRODUCTION

As is well known the celebrated spin coherent states introduced by Radcliffe [1] and Perelomov [2] are labelled by points of a sphere, that is the elements of the configuration space. On the other hand, it seems that as with the standard coherent states, the coherent states for a quantum particle on a sphere should be marked with points of the phase space.

In this work we introduce the coherent states for a quantum spin 1/2 particle on a sphere S_2, labelled by points of the phase space $S_2 \times TS_2$.

DEFINITION OF THE COHERENT STATES FOR A QUANTUM PARTICLE ON A SPHERE

In order to study the coherent states for a quantum particle on a sphere we should first identify the corresponding algebra. It seems that the most natural choice is the $e(3)$ algebra such that

$$[J_i, J_j] = i\varepsilon_{ijk}J_k, \quad [J_i, Y_j] = i\varepsilon_{ijk}Y_k, \quad [Y_i, Y_j] = 0, \quad i, j, k = 1, 2, 3. \quad (1)$$

---

[1] Supported by University of Łódź grant No. 621.

Indeed, let us recall that the Casimir operators for (1) are

$$\mathbf{Y}^2 = r^2, \qquad \mathbf{J}\cdot\mathbf{Y} = \kappa. \tag{2}$$

From (2) it follows that the generators Y_i, $i = 1, 2, 3$, can be interpreted as quantum counterparts of the Cartesian coordinates of the points of the sphere S_2 with radius r. Furthermore, the second relation of (2) for $\kappa = 0$ and $\mathbf{J} = \mathbf{L}$ can be regarded as the orthogonality condition for the orbital momentum $\mathbf{L}$ and the position $\mathbf{Y}$ of a quantum particle on the sphere. We now restrict to the case of the spin 1/2 particle on a unit sphere, so we have

$$\mathbf{J} = \mathbf{L} + \tfrac{1}{2}\boldsymbol{\sigma}, \tag{3}$$

where σ_i, $i = 1, 2, 3$, are the Pauli matrices, and

$$\mathbf{Y}^2 = 1, \qquad \mathbf{L}\cdot\mathbf{Y} = 0. \tag{4}$$

We seek the coherent states as the solution of the eigenvalue equation

$$X|\xi\rangle = |\xi\rangle. \tag{5}$$

What is X ? Based on the experience with the coherent states for a quantum particle on a circle [3] we are looking for the polar decomposition of X such that

$$X = f(\mathbf{J})U(\mathbf{Y}), \tag{6}$$

where $f(\mathbf{J})$ is Hermitian and the unitary operator $U(\mathbf{Y})$ should describe the position of a quantum particle on a sphere. It seems that the most natural choice of $U(\mathbf{Y})$ preserving the covariance under rotations is to set

$$U(\mathbf{Y}) = \boldsymbol{\sigma}\cdot\mathbf{Y}, \tag{7}$$

Such a form of $U(\mathbf{Y})$ indicates that f should be the function of $\boldsymbol{\sigma}\cdot\mathbf{J}$. Further, the elegant commutation relation

$$[K, \boldsymbol{\sigma}\cdot\mathbf{Y}]_+ = 0. \tag{8}$$

satisfied by the Dirac operator [4] defined by

$$K := -(\boldsymbol{\sigma}\cdot\mathbf{L} + 1). \tag{9}$$

suggests that it is more convenient to use operator K instead of $\boldsymbol{\sigma}\cdot\mathbf{J}$. Finally, the form of the operator X in the case with the coherent states for a quantum particle on a circle indicates that the eigenvalue equation (5) should be written as

$$X|\mathbf{z}\rangle = \mathbf{z}\cdot\boldsymbol{\sigma}|\mathbf{z}\rangle, \tag{10}$$

where

$$X = e^{-K}\boldsymbol{\sigma}\cdot\mathbf{Y}, \tag{11}$$

and $\mathbf{z} \in \mathbf{C}^3$ obeys $\mathbf{z}^2 = 1$. Of course, the vectors $\mathbf{z}$ should parametrize the points of the classical phase space. Taking into account (11) we put

$$\mathbf{z}\cdot\boldsymbol{\sigma} = e^{\boldsymbol{\sigma}\cdot\mathbf{l}}\boldsymbol{\sigma}\cdot\mathbf{x}, \tag{12}$$

where the vectors $\mathbf{l}, \mathbf{x} \in \mathbf{R}^3$, satisfy $\mathbf{x}^2 = 1$ and $\mathbf{l}\cdot\mathbf{x} = 0$, that is we assume that $\mathbf{l}$ is the classical momentum and $\mathbf{x}$ is the radius vector of a particle on a sphere. From (12) it follows that

$$\mathbf{z} = \cosh|\mathbf{l}|\mathbf{x} + i\frac{\sinh|\mathbf{l}|}{|\mathbf{l}|}\mathbf{l}\times\mathbf{x}, \tag{13}$$

where $\mathbf{z}^2 = 1$. It thus appears that the vector $\mathbf{z}$ really parametrizes the points $(\mathbf{x}, \mathbf{l})$ of the classical phase space $S_2 \times TS_2$.

CONSTRUCTION OF THE COHERENT STATES

The experience with the circular motion suggests that the most convenient representation for the study of the eigenvalue equation (10) is that with diagonal (Hermitian) operator K. We find

$$K|l,m\rangle_+ = l|l,m\rangle_+, \qquad K|l,m\rangle_- = -(l+1)|l,m\rangle_-, \tag{14}$$

where

$$|l,m\rangle_+ = \frac{1}{\sqrt{2l+1}}\begin{pmatrix}\sqrt{l-m}\,|l,m\rangle \\ -\sqrt{l+m+1}\,|l,m+1\rangle\end{pmatrix}, \qquad -l \leq m < l, \tag{15a}$$

$$|l,m\rangle_- = \frac{1}{\sqrt{2l+1}}\begin{pmatrix}\sqrt{l+m+1}\,|l,m\rangle \\ \sqrt{l-m}\,|l,m+1\rangle\end{pmatrix}, \qquad -l-1 \leq m \leq l. \tag{15b}$$

and $|l,m\rangle$ are the usual common eigenvectors of the operators $\mathbf{L}^2$ and L_3. The vectors $|l,m\rangle_\pm$ are the spherical spinors [5]. They refer to the parallel ($|l,m\rangle_-$) and anti-parallel ($|l,m\rangle_+$) orientation of the orbital momentum and the spin of a quantum particle. The orthogonality and the completeness condition satisfied by the vectors (15) can be writen as

$$_+\langle l,m|l',m'\rangle_+ = \delta_{ll'}\delta_{mm'}, \quad _-\langle l,m|l',m'\rangle_- = \delta_{ll'}\delta_{mm'}, \quad _+\langle l,m|l',m'\rangle_- = 0, \tag{16}$$

$$\sum_{l=0}^{\infty}\sum_{m=-l-1}^{-1}|l,m\rangle_{--}\langle l,m| + \sum_{l=0}^{\infty}\sum_{m=0}^{l}|l+1,m\rangle_{++}\langle l+1,m| = I. \tag{17}$$

The action of the operator $\boldsymbol{\sigma}\cdot\mathbf{Y}$ on the vectors (15) is of the following simple form:

$$\boldsymbol{\sigma}\cdot\mathbf{Y}|l+1,m\rangle_+ = |l,m\rangle_-, \qquad \boldsymbol{\sigma}\cdot\mathbf{Y}|l,m\rangle_- = |l+1,m\rangle_+, \qquad -l-1 \le m \le l. \tag{18}$$

We now return to the eigenvalue equation (10). On projecting (10) on the basis vectors (15) we arrive at the system of linear difference equations satisfied by the Fourier coefficients of the expansion of the coherent state $|\mathbf{z}\rangle$ in the basis $|l,m\rangle_\pm$. In spite of its linearity the direct solution of such system in the general case with an arbitrary $\mathbf{z}$ satisfying (13) seems to be difficult task. Therefore, we first solve the eigenvalue equation for $\mathbf{z} = \mathbf{n}_3 = (0,0,1)$, and then generate the coherent states from the "vacuum vector" $|\mathbf{n}_3\rangle$. The solution of the eigenvalue equation

$$X|\mathbf{n}_3\rangle = \sigma_3|\mathbf{n}_3\rangle. \tag{19}$$

is

$$|\mathbf{n}_3\rangle = \sum_{l=0}^{\infty}(e^{-\frac{1}{2}(l+1)(l+2)}\sqrt{l+1}|l+1,0\rangle_+ + e^{-\frac{1}{2}l(l+1)}\sqrt{l+1}|l,0\rangle_-). \tag{20}$$

Now, using the relation

$$[\mathbf{w}\cdot\mathbf{J}, X] = 0, \tag{21}$$

where $\mathbf{w} \in \mathbf{C}^3$, we find that the coherent states $|\mathbf{z}\rangle$ given by (10) can be defined as

$$|\mathbf{z}\rangle := e^{\mathbf{w}\cdot\mathbf{J}}|\mathbf{n}_3\rangle, \tag{22}$$

where

$$\mathbf{z}\cdot\boldsymbol{\sigma} = e^{\mathbf{w}\cdot\mathbf{J}}\sigma_3 e^{-\mathbf{w}\cdot\mathbf{J}} = e^{\frac{1}{2}\mathbf{w}\cdot\boldsymbol{\sigma}}\sigma_3 e^{-\frac{1}{2}\mathbf{w}\cdot\boldsymbol{\sigma}}, \tag{23}$$

and $\mathbf{z}^2 = 1$. From (22) we obtain after some algebra the following relation:

$$|\mathbf{z}\rangle = \exp\left[\frac{\operatorname{arccosh} z_3}{\sqrt{1-z_3^2}}(\mathbf{z}\times\mathbf{n}_3)\cdot\mathbf{J}\right]|\mathbf{n}_3\rangle, \tag{24}$$

where $\mathbf{z}^2 = 1$, so the coherent states $|\mathbf{z}\rangle$ can be really generated from the "vacuum vector" $|\mathbf{n}_3\rangle$. On the other hand, taking into account (24) we get the expansion of the coherent states in the basis vectors $|l,m\rangle_\pm$ such that

$$|\mathbf{z}\rangle = \sum_{l=0}^{\infty}\left(e^{-\frac{1}{2}(l+1)(l+2)}\sum_{m=0}^{l}\frac{\nu^m(l+m+1)!}{m!\,(l-m)!}e^{\frac{\nu}{2}(2m+1)}\sum_{k=0}^{l+m+1}\frac{\mu^k}{k!}\sqrt{\frac{(l-m+k)!}{(l+m-k+1)!}}|l+1,m-k\rangle_+\right.$$
$$\left.+ e^{-\frac{1}{2}l(l+1)}\sum_{m=0}^{l}\frac{\nu^m(l+m+1)!}{m!\,(l-m)!}e^{\frac{\nu}{2}(2m+1)}\sum_{k=0}^{l+m+1}\frac{\mu^k}{k!}\sqrt{\frac{(l-m+k)!}{(l+m-k+1)!}}|l,m-k\rangle_-\right), \tag{25}$$

, where

$$\mu = \frac{z_1 + iz_2}{1 + z_3}, \quad \nu = \frac{-z_1 + iz_2}{1 + z_3}, \quad \gamma = \ln\frac{1 + z_3}{2}. \tag{26}$$

Using elementary properties of the hypergeometric function $_2F_1$ and the Jacobi polynomials $P_n^{(\alpha,\beta)}(x)$, and rescaling $|\mathbf{z}\rangle \to \left(\frac{1+z_3}{2}\right)^{-\frac{1}{2}}|\mathbf{z}\rangle$, we obtain from (25) the following relations:

$$_+\langle l+1, m|\mathbf{z}\rangle = \frac{1+z_3}{2} e^{-\frac{1}{2}(l+1)(l+2)} \sqrt{\frac{(l+m+1)!(l-m)!}{l!^2}} \left(\frac{-z_1 + iz_2}{2}\right)^m P_{l-m}^{(m,m+1)}(z_3), \tag{27a}$$

$$_-\langle l, m|\mathbf{z}\rangle = \frac{1+z_3}{2} e^{-\frac{1}{2}l(l+1)} \sqrt{\frac{(l+m+1)!(l-m)!}{l!^2}} \left(\frac{-z_1 + iz_2}{2}\right)^m P_{l-m}^{(m,m+1)}(z_3), \tag{27b}$$

where $0 \leq m \leq l$, and

$$_+\langle l+1, m|\mathbf{z}\rangle = e^{-\frac{1}{2}(l+1)(l+2)} \sqrt{\frac{(l+m+1)!(l-m)!}{l!^2}} \left(\frac{z_1 + iz_2}{2}\right)^{-m} P_{l+m+1}^{(-m,-m-1)}(z_3), \tag{28a}$$

$$_-\langle l, m|\mathbf{z}\rangle = e^{-\frac{1}{2}l(l+1)} \sqrt{\frac{(l+m+1)!(l-m)!}{l!^2}} \left(\frac{z_1 + iz_2}{2}\right)^{-m} P_{l+m+1}^{(-m,-m-1)}(z_3), \tag{28b}$$

where $-l-1 \leq m < 0$.

Thus, it turns out that as with the standard coherent states the projection of the coherent states for a quantum spin 1/2 particle on a sphere on the discrete basis vectors is polynomial in a complex variable parametrizing these states.

COHERENT STATES AND THE CLASSICAL PHASE SPACE

We now discuss the parametrization (13) in a more detail. It follows from computer simulations that

$$\frac{\langle \mathbf{z}|\mathbf{L}|\mathbf{z}\rangle}{\langle \mathbf{z}|\mathbf{z}\rangle} \approx \mathbf{l}, \tag{29}$$

where $|\mathbf{l}| \geq 10$. The relative error for $|\mathbf{l}| \sim 10$ is of order 1 per cent. We stress that the condition $|\mathbf{l}| \geq 10\,\hbar$ is not the same as the classical limit $|\mathbf{l}| \to \infty$. We only point out that we measure $|\mathbf{l}|$ in units of $\hbar$ and $10\,\hbar \approx 10^{-33}$ J·s. Further, the computer simulations indicate that

$$\frac{\langle \mathbf{z}|\mathbf{Y}|\mathbf{z}\rangle}{\langle \mathbf{z}|\mathbf{z}\rangle} \approx e^{-\frac{1}{4}}\mathbf{x}. \tag{30}$$

The range of applicability of (30) is the same as with (29). Because of the term $e^{-\frac{1}{4}}$, it turns out that the expectation value of $\mathbf{Y}$ does not belong to the unit sphere.

Proceeding analogously as in the case of the circle we introduce the relative average value of **Y** of the form

$$\frac{\langle Y_i \rangle_{\mathbf{z}}}{\langle Y_i \rangle_{\mathbf{w}_i}} := \frac{\langle \mathbf{z}|Y_i|\mathbf{z} \rangle}{\langle \mathbf{w}_i|Y_i|\mathbf{w}_i \rangle}, \qquad i = 1, 2, 3, \tag{31}$$

where $|\mathbf{z}\rangle$ and $|\mathbf{w}_i\rangle$ are normalized coherent states, and

$$\mathbf{w}_k = \cosh|\mathbf{l}|\mathbf{n}_k + i\frac{\sinh|\mathbf{l}|}{|\mathbf{l}|}\mathbf{l} \times \mathbf{n}_k, \qquad k = 1, 2, 3, \tag{32}$$

where $\mathbf{n}_k$ is the unit vector along the k coordinate axis. We have

$$\frac{\langle Y_i \rangle_{\mathbf{z}}}{\langle Y_i \rangle_{\mathbf{w}_i}} \approx x_i, \qquad i = 1, 2, 3, \tag{33}$$

so, the relative expectation value $\langle Y_i \rangle_{\mathbf{z}}/\langle Y_i \rangle_{\mathbf{w}_i}$ seems to be adequate to describe the average position of a quantum spin $1/2$ particle on a sphere. We interpret the approximate relations (29) and (33) as the best possible approximation of the classical phase space. In our opinion the approximative nature of these relations is connected with the quantum fluctuations which are immanent feature of the quantum mechanics on a sphere.

MINIMALIZATION OF THE HEISENBERG UNCERTAINTY RELATIONS

The introduced coherent states minimalize the Heisenberg uncertainty relations. Indeed, let us introduce the following Hermitian operators playing the similar role as the usual position and momentum operators for the standard coherent states:

$$Q := \frac{1}{2}(X + X^\dagger), \tag{34a}$$

$$P := \frac{1}{2i}(X - X^\dagger), \tag{34b}$$

where X is given by (11). Consider the Heisenberg uncertainty relation satisfied by operators (34)

$$\Delta_\phi Q \Delta_\phi P \geq \frac{1}{2}\frac{|\langle \phi|[Q,P]|\phi \rangle|}{\langle \phi|\phi \rangle}. \tag{35}$$

It follows that

$$\Delta_\mathbf{z} Q \Delta_\mathbf{z} P = \frac{1}{2}\frac{|\langle \mathbf{z}|[Q,P]|\mathbf{z} \rangle|}{\langle \mathbf{z}|\mathbf{z} \rangle}. \tag{36}$$

That is the (normalized) coherent states minimalize the Heisenberg uncertainty relation (35).

EXAMPLE: FREE MOTION ON A SPHERE

For an easy illustration of the introduced formalism we now discuss the free quantum spin 1/2 particle on a sphere, i.e. the rotator. Let us remind that in the Pauli approximation the expression on the energy of the rotator is the same as that referring to the case of the spinless particle, so

$$E = \tfrac{1}{2}l(l+1). \tag{37}$$

The normalized wave functions are given by spinors (15). It follows from computer simulations that the distribution of energies in the coherent states described by

$$p_{l,m}^{(+)}(\mathbf{x},\mathbf{l}) = \frac{|\langle_+ \langle l+1,m|\mathbf{z}\rangle|^2}{\langle \mathbf{z}|\mathbf{z}\rangle}, \qquad p_{l,m}^{(-)}(\mathbf{x},\mathbf{l}) = \frac{|\langle_- \langle l,m|\mathbf{z}\rangle|^2}{\langle \mathbf{z}|\mathbf{z}\rangle}, \qquad -l-1 \le m \le l, \tag{38}$$

has the following behaviour. For fixed integer $m = l_3$ the function $p_{l,m}^{(\pm)}$ has a maximum at $l_{\max}^{(\pm)}$ coinciding with the integer nearest to the positive root of the equation

$$(l+1)(l+2) = \mathbf{l}^2, \tag{39a}$$

and

$$l(l+1) = \mathbf{l}^2, \tag{39b}$$

respectively. If the root $l_*^{(\pm)}$ of (39) is half-integer then the function $p_{l,m}^{(\pm)}$ reaches its maximum value at $l_{\max 1,2}^{(\pm)} = l_*^{(\pm)} \pm \tfrac{1}{2}$. Thus it turns out that the parameter $\tfrac{1}{2}\mathbf{l}^2$ can be identified with the energy of the particle. Further, for fixed integer l in $p_{l,m}^{(\pm)}$ such that (39a) and (39b) holds, respectively, the function $p_{l,m}^{(+)}$ has a maximum at $m_{\max}^{(+)} = [l_3]$, if $l_3 < 0$, and it reaches a maximum value at $m_{\max 1}^{(+)} = [l_3]$ and $m_{\max 2}^{(+)} = [l_3] - 1$, if $l_3 > 0$. Accordingly, the function $p_{l,m}^{(-)}$ has a maximum at $m_{\max}^{(-)} = [l_3]$, if $l_3 > 0$ and its maximum value is reached at points $m_{\max 1}^{(-)} = [l_3]$ and $m_{\max 2}^{(-)} = [l_3] - 1$ for $l_3 < 0$. For $l_3 = 0$ the maximum value of $p_{l,m}^{(\pm)}$ refers to $m_{\max 1}^{(\pm)} = 0$ and $m_{\max 2}^{(\pm)} = -1$. It thus appears that the parameter l_3 can be regarded as the projection of the momentum on the x_3 axis.

REFERENCES

1. J.M. Radcliffe, J. Phys. A **4**, 313 (1971).
2. A.M. Perelomov, Commun. Math. Phys. **26**, 222 (1972); A.M. Perelomov, *Generalized Coherent States and Their Applications* (Springer, Berlin, 1986).

3. K. Kowalski, J. Rembieliński and L.C. Papaloucas, J. Phys. A **29**, 4149 (1996).
4. L.C. Biedenharn and J.D. Louck, *Angular Momentum in Quantum Physics. Theory and Application.* (Addison-Wesley, Massachusetts, 1981).
5. A.A. Sokolov, I.M. Ternov and V.C. Zhukovskii, *Quantum Mechanics* (Nauka,Moscow,1979) (in Russian).

On Time-Dependent Quasi-Exactly Solvable Problems[1]

Dieter Mayer

*Institute of Theoretical Physics, TU Clausthal,
Arnold Sommerfeld Str. 6, 38-678 Clausthal, Germany*[2]

Alexander Ushveridze and Zbigniew Walczak

*Department of Theoretical Physics, University of Łódź,
ul. Pomorska 149/153, 90-236 Łódź, Poland*[3]

Abstract. In this paper we demonstrate that there exists a close relationship between quasi-exactly solvable quantum models and two special classes of classical dynamical systems. One of these systems can be considered a natural generalization of the multi-particle Calogero-Moser model and the second one is a classical matrix model.

INTRODUCTION

A quantum mechanical model is called *quasi-exactly solvable* (QES) if a finite number of energy levels and the corresponding wavefunctions of the model can be constructed explicitly. One possible way of studying the QES models and their solutions in the one-dimensional case is based on a reformulation of the spectral equations in terms of the wavefunction zeros. It can be shown that, in all the cases when the number of wavefunction zeros is finite (and this is just the case of QES models), the problem of reconstructing solutions of the stationary Schroedinger equation becomes purely algebraic and is reduced to the determination of their positions in the complex plane. These positions (which hereafter we shall denote by ξ_i) obey the system of algebraic equations

$$\sum_{k=1,k\neq i}^{M} \frac{\hbar}{\xi_i - \xi_k} + F(\xi_i) = 0, \quad i = 1,\ldots, M \tag{1}$$

---

[1] This work was partially supported by DFG grant No. 436 POL 113/77/0 (S)
[2] E-mail address: dieter.mayer@tu-clausthal.de
[3] E-mail addresses: alexush@mvii.uni.lodz.pl and walczak@mvii.uni.lodz.pl

where $F(\xi)$ is a rational function in which all the information of a QES model is contained. In turns out that equations of the type (1) are typical not only for QES models but appear in many branches of mathematical physics and this fact enables one to establish a deep relationship between QES models and many other seemingly unrelated models. For example, QES models turn out to be equivalent to completely integrable Gaudin spin chains [1,10] (for which system (1) plays the role of the Bethe ansatz equations for the coordinates of elementary spin excitations), to random matrix models [4,5] (for which system (1) determines the distribution of eigenvalues of large random matrices), and also to purely classical models of 2-dimensional elctrostatics [1,2,6] respectively hydrostatics of point vortices [3]. For both the latter models system (1) determines the equilibrium positions of pointwise classical objects (the charged Coulomb particles or resp. vortices) in an external (electrostatic or resp. hydrostatic) field.

Up to now only the stationary solutions of QES Schroedinger equations have been considered from this point of view. The aim of the present paper is to consider time dependent solutions for QES models and to show how they can equivalently be described as dynamical equations for the motion of wavefunction zeros [4]. This enables one to reveal three classical (complex) dynamical systems closely related to quantum QES models. One of these classical systems does not have an immediate physical interpretation, but the two remaining ones are very interesting from both the physical and mathematical point of view. The point is that one of these two systems is a natural generalization of the famous classical Calogero-Moser multi-particle system and the second one is a classical matrix model from which the first system can be obtained by means of Olshanetsky and Perelomov's projection method [13].

Strictly speaking, the idea of studying classical dynamical systems describing the motion of zeros of solutions of linear differential equations belongs to Calogero and is far from being new. Papers devoted to the investigation of such systems appear in the literature rather frequently. The goal of our paper is to apply Calogero and Olshanetsky-Perelomov methods to a concrete class of quantum QES models and to derive the associated classical models. The most interesting and quite unexpected result which we intend to present here is that the potentials of the resulting classical matrix models turn out to almost coincide with the potentials of the initial quantum QES ones (up to terms of order $\hbar$). This fact may hint to the existence of a certain non-standard "quantization procedure" relating the classical multi-particle systems of Calogero-Moser type to QES models of one-dimensional quantum mechanics.

The paper is organized as follows: in the following section we remind the reader of the basic facts concerning the simplest QES model in the stationary case. In the next section we derive the explicit form of solutions for this model in the time-dependent case and the corresponding system of evolution equations for the

---

[4] We mean here just the solutions of evolution equations for standard QES models with time-independent potentials. Note that QES models with time-dependent potentials were discussed (from different point of view) in paper [7]

wavefunction zeros. In the fourth section we show thatsolutions of this system can be considered as "soliton like solutions" of a multi-particle classical system of the Calogero-Moser type. The matrix version of this system is discussed in the next section. In sixth section we discuss the limiting case when the number of algebraically calculable states in the QES model tends to infinity. The last section is devoted to a discussion of our results.

THE SIMPLEST QES MODEL

The simplest QES model is the sextic anharmonic oscillator. Its potential has the following form

$$V(x) = \frac{x^2}{2}\left(ax^2 + b\right)^2 - \hbar a \left(M + \frac{3}{2}\right)x^2, \qquad (2)$$

where a and b are real parameters with $a > 0$, and M is an arbitrary non-negative integer. It can be shown [8] that for any given $M \geq 0$, the stationary Schroedinger equation

$$\left(-\frac{\hbar^2}{2}\frac{d^2}{dx^2} + V(x)\right)\Psi(x) = E\Psi(x) \qquad (3)$$

for model (2) admits solutions of the form

$$\Psi(x) = \prod_{i=1}^{M}(x - \xi_i)\exp\left[-\frac{1}{\hbar}\left(\frac{bx^2}{2} + \frac{ax^4}{4}\right)\right], \qquad (4)$$

$$E = \hbar b\left(M + \frac{1}{2}\right) + \hbar a \sum_{i=1}^{M}\xi_i^2, \qquad (5)$$

where the numbers ξ_i, $i = 1, \ldots, M$ (playing the role of the wavefuntion zeros) satisfy the system of numerical equations

$$\sum_{k=1, k \neq i}^{M}\frac{\hbar}{\xi_i - \xi_k} = b\xi_i + a\xi_i^3, \quad i = 1, \ldots, M \qquad (6)$$

with the following additional condition

$$\sum_{i=1}^{M}\xi_i = 0. \qquad (7)$$

Note that the form of system (6) coincides exactly with that of the famous Bethe Ansatz equations appearing in the theory of completely integrable Gaudin models [9,10]. Therefore we hereafter will call (6) the Bethe Ansatz equations.

It is not difficult to show that for any fixed M the Bethe Ansatz equations (6)-(7) have

$$N_{sol} = \left[\frac{M}{2}\right] + 1 \tag{8}$$

solutions. This means that for the sextic anharmonic oscillator we can find N_{sol} wavefunctions $\Psi_i(x)$ and N_{sol} corresponding energy levels E_i by means of purely algebraic methods.

Note also that model (2) can be considered a deformation of the simple harmonic oscillator with the potential

$$V_0(x) = \frac{b^2 x^2}{2}. \tag{9}$$

The role of the deformation parameter is played by a. It is not difficult to see that after taking $a = 0$ in formulas (4)-(6) they reduce to the well known formulas describing the solution of the simple harmonic oscillator in terms of wavefunction zeros [1,12]. In this case the number M can be considered a free parameter because it is not any longer correllated with the form of the potential.

EVOLUTION EQUATIONS

Let us next consider the time-dependent Schroedinger equation

$$i\hbar \frac{\partial \Psi(x,t)}{\partial t} = \left\{-\frac{\hbar^2}{2}\frac{\partial^2}{\partial x^2} + V(x)\right\} \Psi(x,t). \tag{10}$$

for potential (2). It is obvious that any linear combination of the stationary solutions

$$\Psi(x,t) = \sum_{n=1}^{N_{sol}} c_n \Psi_n(x) e^{-iE_n t/\hbar} =$$

$$= \left(\sum_{n=1}^{N_{sol}} c_n \prod_{i=1}^{M} (x - \xi_i) e^{-iE_n t/\hbar}\right) \exp\left[-\frac{1}{\hbar}\left(\frac{bx^2}{2} + \frac{ax^4}{4}\right)\right] \tag{11}$$

gives a certain dynamical solution of equation (10). Remember that the pre-exponential factor in (4) is always a polynomial of degree M. Therefore also the pre-exponential factor in (11) is a certain polynomial and hence its zeros are functions of time. This enables one to write

$$\Psi(x,t) = C(t) \prod_{i=1}^{M} (x - \xi_i(t)) \exp\left[-\frac{1}{\hbar}\left(\frac{bx^2}{2} + \frac{ax^4}{4}\right)\right]. \tag{12}$$

Formula (12) enables one to derive the evolution equations for the functions $\xi_i(t)$ and $C(t)$. For this it is convenient to rewrite equation (10) in the form

$$V(x) = i\hbar \frac{\partial}{\partial t} \ln \Psi(x,t) + \frac{\hbar^2}{2} \left\{ \left(\frac{\partial}{\partial x} \ln \Psi(x,t) \right)^2 + \frac{\partial^2}{\partial x^2} \ln \Psi(x,t) \right\}. \tag{13}$$

Substituting expressions (2) and (12) into equation (13) we obtain after some algebra the relation

$$i\hbar \frac{\dot{C}(t)}{C(t)} - \hbar a \sum_{i=1}^{M} \xi_i^2(t) - \hbar a x \sum_{i=1}^{M} \xi_i(t) - \hbar b \left(M + \frac{1}{2} \right) +$$
$$- \sum_{i=1}^{M} \frac{\hbar}{x - \xi_i(t)} \left(i\dot{\xi}_i(t) - \sum_{k=1, k \neq i}^{M} \frac{\hbar}{\xi_i(t) - \xi_k(t)} + b\xi_i(t) + a\xi_i^3(t) \right) = 0 \tag{14}$$

from which it immediately follows that the functions $\xi_i(t)$ must satisfy the following system of first order differential equations

$$i\dot{\xi}_i(t) = \sum_{k=1, k \neq i}^{M} \frac{\hbar}{\xi_i(t) - \xi_k(t)} - b\xi_i(t) - a\xi_i^3(t), \quad i = 1, \ldots, M \tag{15}$$

supplemented by the condition

$$\sum_{i=1}^{M} \xi_i(t) = 0. \tag{16}$$

For the function $C(t)$ we obtain on the other hand

$$C(t) = \exp \left[-i \left(a \sum_{i=1}^{M} \int \xi_i^2(t) dt + b \left(M + \frac{1}{2} \right) t \right) \right]. \tag{17}$$

The system (15)-(16) is the dynamical extension of the stationary system (6)-(7). Since the functions $\xi_i(t)$ are assumed to be complex, we face a typical example of a complex dynamical system of dimension M. It is remarkable that this system can be rewritten in the following "potential" form

$$i\dot{\xi}_i(t) = -\frac{\partial}{\partial \xi_i(t)} U(\xi(t)), \tag{18}$$

where

$$U(\xi) = -\hbar \sum_{k=1, k \neq i}^{M} \ln(\xi_i(t) - \xi_k(t)) + \frac{b}{2} \sum_{i=1}^{M} \xi_i^2(t) + \frac{a}{4} \sum_{i=1}^{M} \xi_i^4(t) \tag{19}$$

is playing the role of a complex potential.

COMPLEX MULTI-PARTICLE SYSTEMS

Despite the fact that system (18) cannot be directly derived from the Lagrange principle, it is possible to relate it to a certain Lagrangian system in the following way. For this, consider the complex "Lagrangian"

$$L(\xi,\dot\xi) = \frac{1}{2}\sum_{i=1}^{M}\dot\xi_i^{\,2}(t) - \frac{1}{2}W(\xi) \tag{20}$$

with

$$W(\xi) = \sum_{i=1}^{M}\left(\frac{\partial U(\xi)}{\partial \xi_i}\right)^2. \tag{21}$$

It is not difficult to see that this Lagrangian (20) can be represented in the form

$$L(\xi,\dot\xi) = -\frac{1}{2}\sum_{i=1}^{M}\Lambda_i^2(\xi,\dot\xi) + i\frac{d}{dt}U(\xi), \tag{22}$$

where

$$\Lambda_i(\xi,\dot\xi) = i\dot\xi_i(t) - \sum_{k=1, k\neq i}^{M}\frac{\hbar}{\xi_i(t)-\xi_k(t)} + b\xi_i(t) + a\xi_i^3(t). \tag{23}$$

The total time derivative in (22) can be omitted because it does not affect the form of the equations of motion. The form of these equations derived from the action principle reads then

$$\sum_{i=1}^{M}\left[\frac{d}{dt}\left(\Lambda_i(\xi,\dot\xi)\frac{\partial \Lambda_i(\xi,\dot\xi)}{\partial \dot\xi_n}\right) - \Lambda_i(\xi,\dot\xi)\frac{\Lambda_i(\xi,\dot\xi)}{\partial \xi_n}\right] = 0, \quad n=1,\ldots,M. \tag{24}$$

From (24) it immediately follows that the solutions

$$\Lambda_i(\xi,\dot\xi) = 0, \quad i=1,\ldots,M \tag{25}$$

are automatically solutions of the dynamical equations for the Lagrangian (20). Note that system (25) exactly coincides with equations (15). To derive the form of the Lagrangian (20) it is sufficient to substitute expression (19) into (20). This gives

$$L(\xi,\dot\xi) = \frac{1}{2}\sum_{i=1}^{M}\dot\xi_i^2(t) - \sum_{i=1}^{M}\sum_{k=1,k>i}^{M}\frac{\hbar^2}{(\xi_i(t)-\xi_k(t))^2} + $$
$$- \sum_{i=1}^{M}\frac{\xi_i^2(t)}{2}\left(a\xi_i^2(t)+b\right)^2 + \hbar a\left(M-\frac{3}{2}\right)\sum_{i=1}^{M}\xi_i^2(t). \tag{26}$$

We have obtained a complex Lagrangian system supplemented with the additional constraint (16). This system can obviously be regarded a deformation of a complexified version of the famous Calogero-Moser system with Lagrangian

$$L_0(\xi, \dot\xi) = \frac{1}{2}\sum_{i=1}^{M} \dot\xi_i^2(t) - \sum_{i=1}^{M}\sum_{k=1, k>i}^{M} \frac{\hbar^2}{(\xi_i(t) - \xi_i(t))^2} - \frac{b^2}{2}\sum_{i=1}^{M} \xi_i^2(t). \qquad (27)$$

Note that the above construction is very similar to the one usually used in constructing the soliton solutions of some classical dynamical equations. Therefore it is natural to interpret the solutions of equation (25) describing the motion of wavefunction zeros in quantum QES model (2) as solitons of the classical multi-particle system with Lagrangian (20). The relation between the undeformed theories of the simple harmonic oscillator and the Calogero-Moser system is well known in the literature [11,12].

CLASSICAL MATRIX MODEL OF THE SEXTIC ANHARMONIC OSCILLATOR

Let X be a $M \times M$ traceless complex matrix whose entries are considered as dynamical variables of a certain complex dynamical system. The Lagrangian of this system can be chosen in the form

$$\mathcal{L}(X, \dot X) = \frac{1}{2}\text{Tr}\dot X^2 - \text{Tr}V(X) \qquad (28)$$

with $V(X)$ given by formula (2). The corresponding dynamical equations then read

$$\ddot X = -\frac{\partial}{\partial X} V(X). \qquad (29)$$

It is known that any complex matrix with non-coinciding eigenvalues can be reduced to diagonal form by means of an appropriate similarity transformation

$$X = S \Xi S^{-1} \qquad (30)$$

where we used the notation $\Xi = \text{diag}\{\xi_1(t), \ldots, \xi_M(t)\}$ with $\xi_1(t) + \cdots + \xi_M(t) = 0$. Taking the time derivative of (30) we obtain

$$\dot X = SLS^{-1}, \qquad (31)$$

where

$$L = \dot\Xi + [M, \Xi] \qquad (32)$$

and

$$M = S^{-1}\dot{S}. \tag{33}$$

Differentiating (31) once more and using (29) and (30) we get the equation

$$\dot{L} + [M, L] = -\frac{\partial}{\partial \Xi} V(\Xi) \tag{34}$$

which together with (32) is equivalent to system (29). Note that (34) has the form of a deformed Lax equation. A relation between equations (29) and (34) can also be established in the opposite direction: Assume that we have two matrices M and L satisfying (34) and (32). Let S be a solution of the linear evolution equation

$$\dot{S} = SM. \tag{35}$$

Then the function X in (30) is a solution of equation (29). It is easy to check that the matrices L and M with components

$$L_{ii} = \dot{\xi}_i(t), \quad L_{ik} = \frac{i\hbar}{\xi_i(t) - \xi_k(t)},$$

$$M_{ii} = \sum_{k=1, k \neq i}^{M} \frac{-i\hbar}{(\xi_i(t) - \xi_k(t))^2}, \quad M_{ik} = \frac{-i\hbar}{(\xi_i(t) - \xi_k(t))^2} \tag{36}$$

satisfy the system (32), (34) provided the functions $\xi_i(t)$ are solutions of the deformed Calogero-Moser system (26). As mentioned above for this it is sufficient that $\xi_i(t)$ be solutions of system (15)-(16) describing the motion of the wavefunction zeros in the quantum QES model with potential (2).

From the above reasonings it follows that we essentially established a relationship between the quantum QES model with Hamiltonian

$$H = \frac{p^2}{2} + \frac{x^2}{2}\left(ax^2 + b\right)^2 - \hbar a \left(M + \frac{3}{2}\right) x^2 \tag{37}$$

and the classical matrix model with Hamiltonian

$$\mathcal{H} = \text{Tr}\left(\frac{P^2}{2} + \frac{X^2}{2}\left(aX^2 + b\right)^2 - \hbar a \left(M - \frac{3}{2}\right) X^2\right). \tag{38}$$

THE LARGE M LIMIT

It is remarkable that the hamiltonians of both the quantum and classical models (37) and (38) essentially coincide. The difference is of order $\hbar a$. Let us demonstrate now that this difference also disappears if we take a proper large M limit of these models.

The neccessity for taking the limit $M \to \infty$ in formulas (37) and (38) arises from the following reasonings. For finite M it is not correct to speak of an equivalence of models (37) and (38). The point is that for finite M model (37) describes the evolution of only a certain finite superposition of quantum states (because the model (37) is quasi-exactly solvable). However, if M tends to infinity, then the number of stationary states in this superposition also tends to infinity and fills all the spectrum of the model. Only in this case we can say that models (37) and (38) are equivalent. But how to proceed to the large M limit? It is clear that if we simply take $M = \infty$ in formulas (37) and (38) then we obtain a meaningless (minus) infinity. In order to get a finite expression we must take into account that the parameters a and b may also depend on M. Choosing this dependence according to the conditions

$$ab = g, \quad \frac{b^2}{2} - \hbar a M = \frac{\omega^2}{2} \qquad (39)$$

in which g and ω are fixed (positive) numbers, we find that in the large M limit the parameter a must behave as $a \sim M^{-1/3}$. This means that in the limit $M \to \infty$ the terms containing a^2 and a (i.e. the sextic term and also the harmonic term responsible for the difference between the models) disappear and we obtain the two models

$$H = \frac{p^2}{2} + \frac{\omega^2 x^2}{2} + gx^4 \qquad (40)$$

and

$$\mathcal{H} = \mathrm{Tr}\left(\frac{P^2}{2} + \frac{\omega^2 X^2}{2} + gX^4\right) \qquad (41)$$

with exactly coinciding classical and quantum hamiltonians. Note however that model (40) is a one-particle quantum model while model (41) is a classical model of traceless infinite matrices. The models are equivalent in the sense that the soliton-like solutions in the classical model (41) describe the evolution of the wavefunctions in the quantum model (40) and vice verso.

DISCUSSION

When speaking of a relationship between quantum and classical mechanics one usually thinks of a pair of quantum and classical models related to each other by a certain quantization – dequantization procedure. The hamiltonians of these models, considered as functions of coordinates and momenta, formally coincide up to terms of order $\hbar$. However, the mathematical meaning of the coordinates and momenta is essentially different in the quantum and classical case. In the quantum case they are operators in Hilbert space (infinite matrices) while in the classical case – they are simply numbers. Correspondingly, the number of quantities (degrees of freedom) neccessary to fix uniquely the state of a dynamical system is infinitely larger in the quantum case compared to the classical one. In this sense the transition to the classical limit is equivalent to freezing infinitely many degrees of freedom of the quantum system.

There are however several examples where the correspondence between the quantum and classical model is not approximate but exact and is not related to any limiting procedure and any lose of degrees of freedom. The idea underlying such examples is based on a proper parametrization of the wavefunction by an infinite number of parameters depending on time and considered as canonically conjugated dynamical variables. If such a parametrization is found then we can associate with a given quantum model a certain classical one which in this case should neccessarily be infinite-dimensional. It is quite clear that a priori there are no reasons for any relation between the forms of the corresponding quantum and classical hamiltonians. In general, they may be of absolutely different nature. Consider a simple but most instructive example. Let us take the evolution equation for a certain quantum model with hamiltonian H

$$i\hbar \partial_t \Psi = H\Psi \qquad (42)$$

Since the wavefunction is complex, $\Psi = Q + iP$, we can rewrite equation (42) in real form

$$\hbar \partial_t Q = HP, \quad \hbar \partial_t P = -HQ, \qquad (43)$$

which, after introducing the functional

$$\mathcal{H}(Q, P) = \frac{1}{\hbar}(Q, HQ) + \frac{1}{\hbar}(P, HP), \qquad (44)$$

can in turn be rewritten in the form of the classical Hamilton-equations:

$$\partial_t Q = \frac{\delta \mathcal{H}(Q, P)}{\delta P} \quad \partial_t P = -\frac{\delta \mathcal{H}(Q, P)}{\delta Q}. \qquad (45)$$

We see that irrespective of the form of the initial quantum model, the resulting classical one describes an infinite-dimensional coupled harmonic oscillator.

In this paper we have found examples for an exact relationship between quantum and classical models. The main distinguished feature of these examples is that the potentials of the initial quantum and the resulting classical models exactly coincide. In this case the construction of the classical counterpart of a given quantum model (in our case the quartic or harmonic oscillator) is extremely simple. One should simply replace the operators of coordinate and momentum entering into the quantum model by infinite traceless complex matrices and, after this, take the trace of this matrix hamiltonian. What we obtain will be just the hamiltonian of a classical model whose solutions contain the complete information of the dynamics of the wavefunctions in the initial quantum model. The origin of this coincidence is not clear to us at the moment but it is quite obvious that it cannot be accidental. It would be tempting to conjecture that any one-dimensional quantum model with hamiltonian

$$H = \frac{p^2}{2} + V(x) \tag{46}$$

is somehow equivalent to the classical infinite and traceless complex matrix model with the same hamiltonian

$$\mathcal{H} = \text{Tr}\left(\frac{P^2}{2} + V(X)\right) \tag{47}$$

We have an idea how to check this conjecture at least for models with polynomial anharmonicity and hope to publish the results in the near future.

REFERENCES

1. Ushveridze, A., *Sov. J. of Particles and Nuclei* **20**, 504 (1989).
2. Ushveridze, A., *Sov. Phys. - Lebedev Inst. Rep.* **2**, 50; 54 (1988).
3. Chorin, A., J. Marsden, J., *A mathematical introduction to fluid mechanics*, New York: Springer, 1993.
4. Cicuta, G. M., Stramaglia, S., and Ushveridze, A., *Mod. Phys. Lett.* **A11**, 119 (1996).
5. Cicuta, G. M., and Ushveridze, A., *Phys. Lett.* **A215**, 167 (1996).
6. Shifman, M. A., *Int. J. Mod. Phys.* **A4**, 2897 (1989).
7. Finkel, F., and Kamran, N., "Quasi-exactly solvable time-dependent potentials", (hep-physics/9705022).
8. Krajewska, A., Ushveridze, A., and Walczak, Z., *Mod. Phys. Lett.* **A12**, 1225 (1997).
9. Gaudin, M., *La Fonction d'Onde de Bethe*, Paris: Masson, 1983.
10. Ushveridze, A., *Quasi-Exactly Solvable Problems in Quantum Mechanics*, Bristol: IOP Publishing, 1994.
11. Calogero, F., "Solvable many-body problems and related mathematical findings", *Bifurcation Phenomena in Mathematical Physics and Related Topics*, Dordrecht: Reidel, 1980, pp. 371-384.
12. Calogero, F., *Nuovo Cimento*, **43B**, 177 (1978).
13. Olshanetsky, M. A., and Perelomov, A. M., *Lett. Nuovo Cim.* **16**, 333 (1976).

Discrete Symmetries and Supersymmetries of Quantum–Mechanical Systems[1,2]

J. Niederle

Institute of Physics, Academy of Sciences of Czech Republic, Prague,
E-mail: niederle@fzu.cz

Dear Ola, distinguished colleagues, ladies and gentlemen,

I am very pleased to give a talk in this Memorial Conference devoted to Ryszard Raczka. There are several reasons for that. First, Professor Raczka was an outstanding theoretical physicist which I have admired for his knowledge, his steady interest in problems of contemporary science, in general, and of physics, in particular, and for his originality in handling and solving these problems. Second, it was he who essentially influenced my research activity and career. Third Ryszard Raczka was a Christian in the full meaning of the word with lofty ideals and good intentions and last but not least he was my very good friend indeed.

I met Professor Ryszard Raczka for the first time as a PhD student at the International Center for Theoretical Physics in Trieste in 1964, i.e., thirty–four years ago. I attended his stimulating lectures and seminars devoted to a very fascinating subject - to symmetries - or, more precisely, to a mathematical tool for their systematic studies, to group theory.

I soon started to collaborate with him, first together with Nedzad Limić and then with Jan Fischer, Robert Anderson and others. We wrote together several highly cited papers on the most degenerate irreducible representations of the pseudo–orthogonal groups, on the complete sets of functions on arbitrary hyperboloids and cones, on separation of variables in the Laplace–Beltrami operators and a few papers devoted to new approaches to the Bethe–Salpeter equation, to the role of conformal symmetry in physics and, especially, in astrophysics, and so on.

Ladies and gentlemen, there are and always will be many open questions in science and many generous activities for well–being of the people and for solving the problems threatening the world today. It is difficult for me to become reconciled with the fact that Ryszard Raczka is no longer with us to answer the questions of science in his witty and original way and to help us with activities beneficial to mankind, as was typical for him: effectively and *"without din and glory"*. Surely, we shall greatly miss him.

---

[1] An invited talk at the International Conference "Particles, Fields and Gravitation". Devoted to the memory of Professor Ryszard Rączka, Łódz, April 15 - 19, 1998.
[2] This publication is based on works sponsored by the Grant Agency of the Academy of Sciences of the Czech Republic undre the project number A1010711

Ladies and gentlemen, in my today's lecture I want to talk about symmetries - about the subject which many Raczka's works dealt with. I shall confine to discrete symmetries and supersymmetries of various quantum–mechanical systems and, moreover, rather to a few illustrative examples of them, since their full exposition can be found in the recent papers by Nikitin and myself [1-4].

INTRODUCTION

We shall start with some comments and general remarks:

(i) It is well known that quantum-mechanical systems are usually described in terms of (ordinary or partial, linear or nonlinear) differential equations and that symmetries of these equations form powerful tools for their studies. They are used to separate variables in these equations (see e.g. [5]), to find their solutions as well as to solve their labelling problem (see e.g. [5-9]), to derive spectra and related complete sets of functions of linear differential operators (see e.g. [5,10,11]), to derive the corresponding conservation laws (see e.g. [12]), to construct of new theories, i.e., to find out differential equations invariant with respect to a given symmetry (see e.g. [13]), and so on.

(ii) Let us recall that in quantum mechanics the statement: *"The physical system S has a symmetry group G"* means that there is a group of transformations leaving both the equation of motion of system S as well as the rules of quantum mechanics invariant. In particular no transformation from symmetry group G is allowed to produce an observable effect. Thus if system S is described by an observable A in states $|\psi>, |\varphi>, \ldots$ and system S' obtained by a symmetry transformation $g \in G, g : S \to S'$, is described by the corresponding observable A' in the states $|\psi'>, |\varphi'>, \ldots$, then the equality

$$|\langle\psi'|A'|\phi'\rangle|^2 = |\langle\psi|A|\phi\rangle|^2 \tag{1}$$

must hold. As shown by E. Wigner [14], equality (1) implies that to any symmetry g there exists a unitary or anti-unitary operator U_g (representing g in the Hilbert space of the states of system S) such that

$$|\psi'> \, = \, U_g|\psi> \quad \text{and} \quad A' = U_g A U_g^+ \tag{2}$$

describe the effect of g, i.e., the change $S \to S'$. It is obvious that unitary and anti-unitary operators preserve (1). It is, however, more difficult to show that they exhaust all possibilities.

(iii) There are two types of symmetries: *continuous* (e.g. rotations) and *discrete* (e.g. reflections). For continuous symmetries any $g \in G$ is a function of one or more continuous parameters α^i, $i = 1, 2, \ldots, n$, i.e., $g = g(\alpha^1, \alpha^2, \ldots \alpha^n)$, and any U_g can be expressed in terms of Hermitian operators, i.e., observables

$B_1, B_2, \ldots$ via $e^{i\alpha^j B_j}$. Due to continuity of parameters α^j any B_j is a constant of motion, since for a given quantum-mechanical system described by Hamiltonian H

$$\left[e^{i\alpha^j B_j}, H\right] = 0 \Leftrightarrow \sum_{n=0}^{\infty} \frac{(i\alpha^j)^n}{n!} [B_j^n, H] = 0 \Leftrightarrow [B_j, H] = 0 . \qquad (3)$$

Now, if $g \in G$ is a discrete symmetry, it does not depend on continuous parameters. The corresponding operator U_g can still be written as e^{iB} or Ke^{iB}, where B is an observable and K is an anti-unitary operator, but the equality $[B, H] = 0$ is only a sufficient condition for $\sum_{n=0}^{\infty} \frac{(i)^n}{n!} [B^n, H] = 0$ but not necessary. Consequently, B is not, in general, a constant of motion. However, all discrete symmetries in physics fulfil the condition $U_g^2 = 1$, i.e., they are involutive. Then if U_g is unitary ($U_g U_g^+ = U_g^+ U_g = 1$), it is also Hermitian $U_g^+ = U_g$, and therefore an observable. This is not true for $U_g^2 \neq 1$.

Now we are ready to discuss some of the results derived in [1-4].

First we shall describe new symmetry algebras of various quantum–mechanical systems made of discrete involutive symmetries.

NEW SYMMETRY ALGEBRAS OF QUANTUM–MECHANICAL SYSTEMS

For the sake of definiteness and of simplicity we shall consider the free Dirac equation

$$L_0 \psi = (i\gamma^\mu \partial_\mu - m)\psi = 0 \qquad (4)$$

with

$$\gamma_0 = \begin{pmatrix} 0 & 1_2 \\ 1_2 & 0 \end{pmatrix}, \quad \gamma_a = \begin{pmatrix} 0 & -\sigma_a \\ \sigma_a & 0 \end{pmatrix}, \quad a = 1, 2, 3, \quad \gamma_5 = \begin{pmatrix} 1_2 & 0 \\ 0 & 1_2 \end{pmatrix}$$

(for the new symmetry algebras of the other systems see [1]). It is invariant w.r.t the complete Lorentz group. Discrete involutive symmetries form a finite subgroup of the Lorentz group consisting of 4 reflections of x_μ, 6 reflections of pairs of x_μ, 4 reflections of triplets of x_μ, reflection of all x_μ and the identity transformation.

If the coordinates x_μ in (4) are transformed by these involutive symmetries, function $\psi(x)$ contransforms according to a projective representation of the symmetry group, i.e., either via $\psi(x) \to R_{kl}\psi(x)$ or via $\psi(x) \to B_{kl}\psi(x)$ (for details see [1]). Here R_{kl} and $B_{kl} = CR_{kl}$ are linear and anti-linear operators respectively which commute with L_0 and consequently transform solutions of (4) into themselves. The operators $R_{kl} = -R_{lk}$ form a representation of the algebra $so(6)$ and C is the operator of charge conjugation $C\psi(x) = i\gamma_2\psi^*(x)$. Among the operators B_{kl} there are six which satisfy the condition that $(B_{kl})^2 = -1$ and nine for which $(B_{kl})^2 = 1$. We shall consider further only B_{kl} fulfilling the last condition (for the reason mentioned in the Introduction and since otherwise B_{kl} cannot be diagonalised to real γ_5 and consequently used for reduction of the system). As shown in [1] the operators R_{kl}, B_{kl} and C form a 25-dimensional Lie algebra. It can be extended to a 64-dimensional real Lie algebra or via non-Lie symmetries, to a 256-dimensional Lie algebra. In this connection, first let me recall that a linear operator Q is the symmetry operator of the system described by operator L if there exists an operator α_Q such that the relation $[Q, L] = \alpha_Q L$ holds.

Operator Q is a Lie symmetry operator provided it is of the form $ia^\mu \partial_\mu + b$, with a^μ and b being a function and matrix of x_μ respectively. It is a non–Lie symmetry operator if its form is more general than that of a Lie one. It is possible to show that for Lie symmetry operators the maximal Lie symmetry algebra of system (4) is isomorphic to the Poincaré algebra (with $Q = P_\mu$, $J_{\mu\nu}$, where $P_\mu = p_\mu$, $J_{\mu\nu} = x_\mu p_\nu - x_\nu p_\mu + \frac{i}{4}[\gamma_\mu, \gamma_\nu]$ and $\alpha_Q = 0$) and that the simplest non–Lie symmetry algebra of eq.(4) (constructed in terms of non–Lie operators $Q = ia^\mu \partial_\mu + b$ with both a^μ and b being matrices of x) is given by

$$\Sigma_{\mu\nu} = \frac{1}{2}[\gamma_\mu, \gamma_\nu] + \frac{1}{m}(1 - i\gamma_4)(\gamma_\mu p_\nu - \gamma_\nu p_\mu)$$

$$\Sigma_1 = \gamma_4 - \frac{i}{m}(1 - i\gamma_4)\gamma^\mu p_\mu \quad (5)$$

and the corresponding α_Q are of the form

$$\alpha_{\Sigma_{\mu\nu}} = \frac{1}{m}(\gamma_\mu p_\nu - \gamma_\nu p_\mu),$$

$$\alpha_{\Sigma_1} = \frac{1}{m}\gamma_4 \gamma^\mu p_\mu. \quad (6)$$

The new symmetry algebras of various quantum–mechanical systems do not provide the essentially new conservation laws but can be used to reduce the considered systems to simpler ones or to find their supersymmetries and explain structures of their spectra.

REDUCTION OF QUANTUM–MECHANICAL SYSTEMS

Let us show now only one example how to use this new symmetry algebra to reduce the physical system into uncoupled subsystems (for the other examples see [1]). Let our system be a spin $\frac{1}{2}$ particle interacting with a magnetic field A_μ described by the Dirac equation

$$L\psi(x) = (\gamma^\mu(i\partial_\mu - eA_\mu) - m)\psi(x) = 0 . \quad (7)$$

Eq. (7) is invariant w.r.t. discrete symmetries provided $A_\mu(x)$ contransforms appropriately. For instance,

$$A_\mu(-x) = -A_\mu(x) \tag{8}$$

for $x \to -x$ and $\psi(x) \to \hat{R}\psi(x) = \gamma_5\hat{\theta}\psi(x) = \gamma_5\psi(-x)$. Then, diagonalising symmetry operator $\hat{R}$ by means of the operator

$$W = \frac{1}{\sqrt{2}}\left(1 + \gamma_5\gamma_0\right) \frac{1}{\sqrt{2}}\left(1 + \gamma_5\gamma_0\hat{\theta}\right), \tag{9}$$

i.e., $W\hat{R}W^+ = \gamma_5$, and applying the same transformation to L, equation (7) is reduced to the block diagonal form

$$\left(-\mu(i\partial_0 - eA_0) - \vec{\sigma}\cdot\left(i\vec{\partial} - e\vec{A}\right)\hat{\theta} - m\right)\psi_\mu(x) = 0, \tag{10}$$

where $\mu = \pm 1$ and ψ_μ are two-component spinors satisfying $\gamma_5\psi_\mu = \mu\psi_\mu$.

If equations (10) admit a discrete symmetry again, then they can be further reduced to one-component uncoupled subsystems.

The other treated examples in [1] include: the Dirac equation with the anomalous Pauli interaction, the Dirac and Weyl oscillators, and so on.

SUPERSYMMETRIES OF QUANTUM–MECHANICAL SYSTEMS

It was shown in [3] that extended, generalized and reduced supersymmetries appear rather frequently in many quantum-mechanical systems. Here I shall illustrate it only on two examples. Let us begin with a familiar case – the Schrödinger equation with a matrix superpotential W:

$$H\psi = \frac{1}{2}(p^2 + W^2 + \sigma_3 W')\psi = E\,\psi, \tag{11}$$

where $p = -i\frac{d}{dx}, W' = \frac{d}{dx}W(x)$. It is well known that system (11) admits supercharges

$$Q_1 = \frac{1}{\sqrt{2}}(\sigma_1 p + \sigma_2 W) \quad \text{and} \quad Q_2 = \frac{1}{\sqrt{2}}(\sigma_2 p - \sigma_1 W) \tag{12}$$

satisfying the $N = 2$ superalgebra

$$\{Q_a, Q_b\} = 2\delta_{ab}H, \quad [Q_a, H] = 0, \quad a,b = 1,2. \tag{13}$$

There is, however, an additional supersymmetry of (11) provided superpotential W is an even function of x! Indeed, if $W(-x) = W(x)$, then there exists the third supercharge

$$Q_3 = i\sigma_1 R Q_1 \tag{14}$$

with R defined by

$$R\psi(x) = \psi(-x) \tag{15}$$

such that the relations (13) hold for $a, b = 1, 2, 3$.

If W is an odd function of x, then Q_1 and Q_2 transform according to a reducible representation of the $N = 2$ superalgebra. This follows from the fact that for $W(-x) = -W(x)$ there exists an invariant operator namely, $I = \sigma_3 R$, which commutes with H and Q_a, $a = 1, 2$, and which can be used to diagonalise H and Q_a to a block diagonal form. Let us remark that the irreducible components of Q_1 and Q_2 are expressed in terms of p, R and W, i.e., without the usual fermionic variables (for details see [3])!

As a second example let us consider a spin $\frac{1}{2}$ particle interacting with a constant and homogeneous magnetic field $\vec{H}$. It is described by the Schrödinger-Pauli equation:

$$H\psi(x) = \left(\left(-i\vec{\partial} - e\vec{A}\right)^2 - \frac{1}{2}\, eg\vec{\sigma} \cdot \vec{H}\right)\psi(x) = 0\,. \tag{16}$$

This system is exactly solvable and admits the $N = 4$ extended supersymmetry (for details see [3]). One of its supercharges is of the standard form

$$Q_1 = \vec{\sigma}\left(-i\vec{\partial} - e\vec{A}\right),\ Q_1^2 = H\,, \tag{17}$$

three additional ones can be constructed by means of space reflections R_a of x^a, $a = 1, 2, 3$, since equation (14) is invariant w.r.t. them. As found in [3], they are of the form:

$$Q_2 = iR_3\vec{\sigma} \cdot \left(-i\vec{\partial} - e\vec{A}\right),\ Q_3 = iCR_4\vec{\sigma} \cdot \left(-i\vec{\partial} - e\vec{A}\right),\ Q_4 = iCR_2\vec{\sigma} \cdot \left(-i\vec{\partial} - e\vec{A}\right)\,.$$

Supercharges Q_a are integrals of motion for system (16) (notice that without the usual "fermionic" variables) and responsible for the eight-fold degeneracy of the energy spectrum of this system. For details, many other examples and appropriate references see [3].

CONCLUSIONS

- The new symmetry algebra for the Dirac equation (or the Schrödinger-Pauli equation) describing various quantum-mechanical systems was derived. This algebra is based on discrete involutive symmetries of the considered systems and is of the biggest dimension from all known invariant algebras of the considered systems.

- The new symmetry algebra was used for two purposes: i) to reduce the considered quantum-mechanical system to simpler subsystems and ii) to search for its supersymmetries (extended, reducible or generalized) and to explain degeneracy of its energy spectrum.

- Since the required transformation properties of electromagnetic field $A_\mu(x)$ for the considered quantum-mechanical systems to be supersymmetric seem to be physically realisable (for details see [3]), these systems give strong indications that supersymmetry is indeed a symmetry of Nature.

Thank you for your attention.

REFERENCES

1. J. Niederle and A.G. Nikitin, *J. Phys. A: Math. Gen.* **30**, 999 (1997).
2. J. Niederle and A.G. Nikitin, *Nonlin. Math. Phys.* **4**, 436 (1997).
3. J. Niederle and A.G. Nikitin, "Extended supersymmetries for the Schrödinger-Pauli equation", *J. Math. Phys.* (1998) (in press).
4. J. Niederle, in the Proc. of the Second International Conference "Symmetries in Nonlinear Mathematical Physics", Kiev, 1997, Vol. 2, pp.495-498;
   A.G. Nikitin, in the Proc. GROUP21 Physical Applications and Mathematical Aspects of Geometry, Groups and Algebras, Vol. 1, Eds. H.-D. Doebner et al. (World Scientific, Singapore, 1997) pp. 509-514.
5. H.Ya. Vilenkin and A.U. Klimyk, *Representations of Lie Groups and Special Functions*, Vol. 1-3. *Recent Advances* (Kluwer, Dordrecht, 1991-1995);
   U. Müller, *Symmetry and Separation of Variables* (Reading, MA: Addison-Wesley, 1997).
6. P. Olver, *Application of Lie Groups to Differential Equations* (Springer, New York, 1986);
   N.Ch. Ibragimov, *Transformation Groups in Mathematical Physics* (Nauka, Moscow, 1983).
7. E. Taflin, *Pacific J. Math.* **108**, 203 (1983);
   M. Flato and J. Simon, *J. Math. Phys.* **21**, 913 (1980); *Phys. Lett.* **94B**, 518 (1980) and references cited therein.
8. W.I. Fushchich, W.M. Shtelen and N.I. Serov, *Symmetry Analysis and Exact Solutions of Equations of Nonlinear Mathematical Physics* (Reidel, Dordrecht, 1993).
9. B.G. Konopelchenko, *Nonlinear Integrable Equations* (Lecture Notes in Physics 270), (Springer, Berlin, 1987).
10. N. Limic, J. Niederle and R. Raczka, *J. Math. Phys.* **7**, 1861, 2026 (1966); *J. Math. Phys.* **8**, 1079 (1967).
11. W.I. Fushchich and A.G. Nikitin, *Symmetries of Equations of Quantum Mechanics* (Allerton, New York, 1994).
12. N.N. Bogolubov, D.V. Shirkov, *An Introduction to the Theory of Quantised Fields* (Nauka, Moscow, 1976);
    N.P. Konoplyova, V.N. Popov, *Gauge Fields* (Atomizdat, Moscow, 1972);
    see also:
    C. Gardner et al., *Phys. Rev. Lett.* **19**, 109 (1967);
    J. Mickelsson and J. Niederle, *Lett. Math. Phys.* **8**, 195 (1984).

13. I.M. Gel'fand, A.M. Yaglom, *Z. Eksp. Teor. Fiz.* **8**, 703 (1948);
    also in:
    I.M. Gel'fand, R.A. Minlos, Z.Ya. Shapiro, *Representations of the Rotation and Lorentz Groups and their Applications* (Pergamon, New York, 1963);
    See also
    R. Kotecký, J. Niederle, *Czech. J. Phys.* **B25**, 123 (1975);
    R. Kotecký, J. Niederle, *Reports Math. Phys.* **12**, 237 (1977);
    J. Mickelsson, J. Niederle, *Ann. Inst. Henri Poincaré* **XXIII A**, 277 (1975).
14. E.P. Wigner, *Group Theory and its Application to the Quantum Mechanics of Atomic Spectra* (Addison-Wesley, New York, 1959).

Foundations of Quantum Theory and Thermodynamics

Victor Olkhov

United Institute of High Temperatures,
Krasnokazarmenaia 17-a, Moscow, Russia.
e-mail: olkhov@dpc.asc.rssi.ru

Abstract. Physical reasons to support the statement that Quantum theory (Quantum Gravity in particular as well as Classical Gravity) loose applicability due to Thermodynamical effects are presented. The statement is based on several points: 1. N.Bohr requirement that measuring units must have macro size is one of common fundamentals of Quantum theory. 2. The Reference System - the base notion of Classical and Quantum theory and of any observation process as well, must be protected from any external Thermal influence to provide precise measurements of Time and Distance. 3. No physical screen or process, that can reduce or reflect the action of Gravity is known and hence nothing can cool or protect the measuring units of the Reference System from heating by Thermal Gravity fluctuations. 4. Thermal Gravity fluctuations - Thermal fluctuations of Gravity free fall acceleration, are induced by Thermal behavior of matter and Thermal properties of Electromagnetic fields, but usually are neglected as near zero values. Matter heat Gravity and Gravity heat Matter. Thermal fluctuations of Gravity free fall acceleration act as a Universal Heater on any kind of Matter or Field. 5. Nevertheless the usual Thermal properties of Gravity are negligible, they can be dramatically increased by Gravity Blue Shift (near Gravitational Radius) or usual Doppler effects. 6. If Thermal action of Gravity become significant all measurements of Time and Distance that determine the Reference System notion, must depend on the Thermal properties of Gravity, like Temperature or Entropy, and that violate applicability of the Reference System notion and Quantum and Classical theories as well. If so, Thermal notions, like Temperature or Entropy, become more fundamental than common Time and Distance characters. The definition of the Temperature of the Gravity fluctuations and it's possible measurements are suggested.

INTRODUCTION

The foundations of any physical theory and Quantum theory in particular, can be divided into two parts: the basement of theoretical scheme and the measurement procedures. The significance of the measurements theory on the foundation of Quantum theory was studied in distinguished works of N.Bohr, von Neumann, L. de Brogle, D. Bom and others. Till now the measurement theory and corresponding problems of interpretation of different aspects of Quantum and Classical theories are of great interest [1], [2], [3], [4], [5], [6], [7], [8].

One regard the Quantum description as more fundamental than Classical one. At the same time the Quantum theory is based on the measurements provided by the macro size equipment. Only principle ability to measure the observables of certain theoretical scheme might make the theory reasonable. Otherwise, if there are no way to provide any measurements of the observables, the theory is useless.

The necessary part of any measurement procedure is the Reference System (RS) - the macro size physical equipment to measure Time, Distance and Direction. Any other physical measuring units are assumed to be replaced at certain RS. RS and its mathematical images - Frames or System of Coordinates - are the necessary part and the origin of mathematical formalism of physical theory. The RS is assumed to be the most essential and obvious part and notion of modern theories.

We would like to attract the attention to the problems that seems to be simple and obvious. We are going to study the meaning of the RS notion and it's function in the foundation of Quantum (and Classical) theory. The simplicity of this notion is accompanied with the fundamental dependance of the basement of modern Physics on the RS notion. It creates the root of all theoretic schemes, the root of modern understanding of space-time, and on the other hand it is the root of all measuring procedures. Both theory scheme, measurement procedures and interpretations are founded on the RS notion.

Meanwhile, RS is a definite approximation used to describe the Reality and to measure it's properties. RS is not a Nature phenomena. As any other notion used to describe the Nature, RS notion must have certain domain of applicability. One must to determine the physical conditions for which the RS can be applied and when it has no use. The study of the applicability of the RS notion must be done far ago, but we failed to find any investigations of this problem.

But if for certain case due to definite physical reasons the RS notion loose applicability, hence all modern physical theories (Quantum and Classical) also have no use for this case. All modern theories are based on RS and any measuring procedures are based on RS. If one shows that due to certain physical reasons RS loose applicability - hence all modern physical theories loose applicability too.

In our work we present a brief study of the problem of applicability of the RS notion. Due to our research Thermodynamics and it's properties are exactly those that can reduce applicability of the RS notion. Hence, applicability of modern Quantum and Classical theories (Gravity theories in particular) is reduced due to Thermodynamical reasons.

THE REFERENCE SYSTEM NOTION

First, to avoid misunderstanding, we determine our treatment of the RS. We treat RS as a collection of a macro size solid bars with physical instruments to measure Time, Distance and Direction (TDD) and additional equipment to receive and to transmit information. RS can be added with other devices to measure any additional properties like mass, charge, color etc. So, we treat here RS as a macro-size physical measuring equipment. System of Coordinates, Frames, Observers and so on are mathematical images of RS that are used in the formalism of physical models. The use of these notions definitely depends on the applicability of the origin RS notion.

The significance of the RS notion is based on the assumption: Time, Distance and Direction (TDD) quantities measured by the macrosize measuring units of the RS - are the most fundamental characters of the Universe (the origin of Space-Time formalism). The second significant assumption: it is possible to transmit and to receive information from one RS to another. These statements are fulfilled with high accuracy. They seems to be so obvious and simple, that no studies of the applicability of the RS notion are known. Meanwhile, such problem must be studied to determine the applicability of RS in Nature.

Indeed, to assume, that TDD are base parameters that determine the most important properties of the Reality one must to prove, that any other properties, Thermodynamical properties in particular, can be made careless and insignificant and hence can be derived on the base of TDD. To do so one must be able to protect the equipment of the RS from any external influence to reduce the action of Thermal properties on the measurement of TDD.

Indeed, universality of Thermodynamics means that Thermal properties are attached to all macro bodies. Thus any measuring devices of the RS must have certain Thermodynamical properties like Temperature, for example. So, why Thermal properties are not treated as fundamental similar to TDD? Why the influence of Thermal properties of the equipment on the measuring of TDD is not taken into account? Why one assumes that Thermal properties (Temperature , Entropy, etc.) can be derived on base of TDD?

The common answer is: it is possible to protect the equipment of the RS from any external Thermal action and to reduce it so, that it is possible to neglect the Thermal influence on the results of the measurements of TDD. All our experience and precision predictions of modern physical theories based on the RS notion prove that this statement is exact enough.

But if for certain physical conditions it will be impossible to protect RS from Thermal action and impossible to avoid it in any way, and if Thermal properties are significant for the measurements of TDD, than RS have no use there. For such case Thermal properties, like Temperature or Entropy, must be treated as more fundamental or at least as same base notions as TDD. So, if we present the physical reasons, that demonstrate why RS equipment and measurements of TDD can not be protected by any means from significant Thermal influence and dependence - than RS loose applicability in this case.

To do so we regard the Nature of Thermal behavior. Due to common treatment Thermal action is induced by particles movement and by the radiation action. One knows that it is possible to protect the RS from particles action and from the Electromagnetic Thermal action by creating corresponding screens and coolers.

Meanwhile there exist additional source of Thermal influence, that can play significant role and lead to unmovable action on RS measuring units.

ON THERMAL PROPERTIES OF GRAVITY

We attract the attention to the Thermal properties of Gravity.

Indeed, Universality of Thermodynamics means that all macro size bodies have Thermal properties, like Temperature. These Thermal properties must be reflected by the behavior of Gravity, and Gravity free fall acceleration in particular. As Thermal behavior of macro body corresponds, due to modern description, with Thermal fluctuations of it's particles position, energy fluctuations and so on, than corresponding fluctuations of the free fall acceleration must be induced. Thermal properties of Electro magnetic radiation can be reflected by fluctuations of energy and that also induce corresponding fluctuations of the free fall acceleration. It is important to outline that the scales of Thermal fluctuations of matter and radiation induce similar scales of free fall accelerations. It can be easy seen, that the values of such free fall accelerations are very small and definitely can be neglected.

Nevertheless such small fluctuations of the free fall acceleration exist. As we shall show below there exist the mechanism that increase the value of such fluctuations up to infinity. Let us study the influence of such strong fluctuations of the free fall acceleration.

We shall mention them as Thermal Gravity Fluctuations (TGF). Assume, that in certain area such fluctuations are strong enough. What can we say about the physics of such area? Our description is pure phenomenological. We can not suggest now any developed theory. We just present some problems and their consequences.

1. The protection problem. No screens or physical effects that can protect or reflect the action of Gravity are known. Hence, it is impossible to reduce or to reflect the action of such TGF and to reduce their action on the equipment of the RS. Any equipment, matter or field will be under the same action of TGF in that area. Nothing can be created in that area without the action of TGF. Nothing can be moved in that area from outside without the influence of TGF. Thus it is forbidden to regard here any physical system or process that assumed to be isolated from the action of TGF. TGF and their Thermal properties become the characters of this area.

2. The problem of Temperature of Gravity. TGF are induced by Thermal behavior of matter and field. TGF induce similar effect on any matter and field: TGF heat any matter and field and act as a Universal Heater. Indeed, if one regard the body in the area with fluctuations of free fall accelerations with the space and time scales small enough (exactly with Thermal scales) than different particles of the body will be under the action of different fluctuations of free fall acceleration. That will cause the fluctuating movement of particles of the body - exactly the same as Thermal movement. Due to dissipation processes the properties of the movement of the particles under the action of TGF will reach certain stationary

state. The Temperature of such state can be treated as Temperature T_G of TGF. It is obvious, that any body in this area will reach the same Temperature T_G. As nothing can protect matter or field from the action of TGF - than any attempt to cool the object will fail, as it is impossible to isolate the cooling agent from TGF and it will be heated up to T_G.

3. The problem of Min Temperature. The Thermal action of TGF produce a new phenomena: if the T_G is high enough (the value of TGF is great and the scales of TGF are small) than T_G determines the Min Temperature that exist in such area. Indeed, any matter or field with Temperature lower that T_G that are moved inside the area will be heated up to T_G, nothing can protect it from the action of Gravity. It is impossible to decrease the temperature of any body, field or process lower than T_G, as it is impossible to protect the object from the action of TGF and hence it will compensate the decreasing of temperature immediately. So, the Min Temperature T_{min} that equals the Temperature of TGF $T_{min} = T_G$ becomes the necessary the character of the area under consideration. Present physical models that are based on the RS notion can be treated as $T_{min} = 0$ approximation.

4. The problem of observation and identification. If $T_{min} = T_G$ is significant any measurements of TDD will depend on Thermal properties of the surrounding TGF. These Thermal characters, like Temperature T_G must be treated as more fundamental, than TDD. Moreover, it is not clear, how any measurements of Time or Distance can be proceeded. Indeed, as one unable to protect any device or process from the action of TGF, than any oscillator will be Thermalized and strongly connected with the surrounding TGF. So, nothing similar to common "repeated processes" like oscillator, can be created. If TGF are strong enough, than the problem of identification of the physical events must be treated. Indeed, our experience makes the identification problem trivial. But it is not clear how to "mark" the event (to "calculate" the number of points), as all characters are fluctuating under the action of TGF. Common Space-Time relations seems to be unusable in such case. Even the question: What is the dimension of the area ? - have no answer.

5. The problem of creating, transmitting and receiving information. The necessary part of the RS notion are devices for creating, transmitting and receiving information. In the area with strong thermal action of TGF these necessary properties of the RS notion can not be arranged. Indeed, any informative signal will be Thermalized under the action of TGF and thus lose information. To create information, as well as to receive it, one need the identification process, and that also the problem (see above). The lack of exchange information, except the Thermal one, makes usual RS notion unusable. At the same time one may suppose, that Thermal information will be sufficient for the description of physics in such exotic area.

6. The problem of foundation of Statistical Physics. Applicability of RS is based on the assumption: Thermal properties of Gravity are sufficiently small and can be neglected. Thus Dynamics (Classical and Quantum), is founded on the Low-Thermal-Gravity approximation. Hence, Thermodynamics and Statistical Physics can not be derived exactly on base of pure Dynamical description, based on the RS

notion, as RS and Dynamics itself loose applicability in the area with strong TGF, where Thermal and Statistical effects of Gravity become more significant and must be treated as more fundamental, than TDD.

7. The problem of measurements of T_G. The treatments of the Thermal Gravity problems are of physical interest if possible scheme of the measurement of T_G is suggested. As the nature of Thermal Gravity and it's description are not clear we are able to present the phenomenological reasons of the measurements of T_G only.

We assume, that in the area with TGF certain Thermal Equilibrium conditions are fulfilled. If one moves the body in this area, the Temperature of this body $T = T_G$ is constant (nevertheless we unable to explain exactly, how one can measure the Temperature in different points at different time, how to measure Time, how to exchange information etc.). If this body is a big box, than it seems reasonable to assume, that equilibrium Black-Body Radiation (BBR), generated by the walls, with the same Temperature $T = T_G$ will fill the inside area of the box.

We assume, that Thermal properties of TGF must be reflected by the presence of the corresponding BBR with temperature $T = T_G$. This statement can not be proved till the theory of the phenomena of TGF will be created.

The existence of TGF contradict with applicability of the RS notion and with applicability of Relativity anywhere. Hence, our Universe can not be described on the base of Relativity models (Classical or Quantum). Thus, the Universe Evolution models are incorrect. The common explanation of the nature of the Cosmological Microwave radiations discovered by [9] at 1965 must be reviewed. So, instead of common explanation, based on the Universe Evolution model, we suggest to treat the observed Microwave radiation with temperature $T_B \sim 2.7 K^o$ as factor, that reflect Thermal properties of the surrounding Gravity and it temperature is equal $T_G = T_B \sim 2.7 K^o$.

Now we present some reasons to prove the existence of strong TGF.

STRONG TGF AREAS

The assumption, that Thermal properties of Gravity may exist immediately lead to the statement, that strong TGF areas must exist due to Gravity Blue Shift (GBS) and usual Doppler effects, at least.

The GBS corresponds with the Black Holes physics [10], [11], [12], [13], [14], [15] only few references. GBS near the Gravitational Radius R_G of the Black Hole increase up to infinity. Hence if there is BBR with Temperature T_B far from R_G thus, due to common results the Temperature of the measured BBR near R_G trend to infinity. If due to our considerations Microwave radiation reflect the Thermal properties of TGF with Temperature $T_G = T_B \sim 2.7 K^o$, hence $T_G \to \infty$ near R_G due to the rise of GBS. That violate applicability of the RS notion and of Relativity as a whole. Near and inside R_G physics can not be described on the base of Classical or Quantum Relativity.

Doppler effect. It is known, that due to Doppler effect the Temperature measured in the direction of the movement increase up. So, if Thermal properties of Gravity exist and are reflected by Microwave radiation, thus the RS that moves with respect to Microwave radiation will measure the Temperature $T > T_B \sim 2.7 K^o$. When the speed v of the RS with respect to Microwave radiation trend to speed of light c : $v \to c$ the temperature measured in the direction of the movement increase up to infinity. As Doppler increase any observed Temperature, it also increase $T_G \to \infty$. Thus, when measured Temperature of TGF $T >> 1$ the applicability of the RS notion is vanished.

The existence of Thermal properties of Gravity reduce applicability of the RS notion and hence applicability of Classical and Quantum Relativity models by pure Thermodynamical conditions by the region out of R_G and by the speed of the RS with respect to Microwave radiation $v < v_T < c$. If so, modern Universe Evolution models are incorrect.

DISCUSSION

We failed to find any publications on the subject similar to our work.

At the same time there are a lot of studies of different aspects of Thermal effects for Classical and Quantum models in curved space and near R_G in particular [16], [17], [15], [18], [19] and references there in. The main difference from our treatment: in these works Thermal effects in the given space-time are studied and it is assumed that Thermal processes do not change the REGULAR properties of the initial model. No self-Thermal properties of Gravity are studied or discussed. RS or Observer are assumed to have applicability anywhere.

On the other hand the problems named "Chaos in General Relativity" are of great interest last years [20], [21], [22] and ref. there in. These studies of chaotic behavior of matter and Gravity also miss the problem of possible Thermal influence of fluctuating Gravity on the RS and the measurements. The main interest of the similar studies is to obtain certain averaged procedure and to derive closed equations for the corresponding momentum values. We know no attempts to mention, that fluctuations of Gravity may induce Thermal action on any physical processes and on the RS in particular.

The reasons presented in our work may be treated as not too satisfactory. Nevertheless the central problems of the research still exist and must be studied: 1.The problem of applicability of the RS notion. 2.What principles can be treated as primary and what properties can be measured as fundamental for the case, when RS has no use and Thermal properties of Gravity are significant? 3.How to describe physics in the STG fluctuation area? And many other questions.

CONCLUSIONS

We repeat the main points of our study. The sources of Thermal fluctuations of free fall accelerations exist. The factors that can increase the Temperature of such fluctuations up to infinity exist. No physical processes that can protect, reduce or cool the Thermal action of Gravity on the measuring units of the RS are known . Hence, the RS notion can be used only till the Thermal action of Gravity is insignificant and can be neglected. Otherwise RS notion loose applicability and hence modern Quantum and Classical physics loose applicability too. For such case Thermal notions like Temperature or Entropy must be treated as more fundamental notions, then TDD.

1. Applicability of the RS notion - the base notion of the Quantum and Classical description - is reduced by Thermal properties of Gravity. One use RS notion if $T_{min} = T_G = 0$ is a good approximation. Otherwise RS notion has no use and Thermal parameters like Temperature and Entropy must be treated as more fundamental notions then TDD. For that case modern Quantum and Classical theories have no use.

2. Areas with strong TGF with $T_G \gg 1$ must exist due to GBS and Doppler effects. Physics near and inside R_G can not be described by Quantum or Classical Relativity models. Universe Evolution models must be reviewed.

3. Cosmological Microwave Radiation assumed to reflect Thermal properties of the surrounding Gravity and $T_G = T_B \sim 2.7 K^o$.

4. Statistical Physics and Thermodynamics can not be derived on base of pure Dynamics.

ACKNOWLEDGMENTS

My exclusive thanks to Prof. Vladimir G. Kurt for his moral support and help. As well I would like to thank Andrey Postuhov for his pictures used during the presentation of this work at the Conference.

REFERENCES

1. D.Bom Quantum theory, Prentice Hall, N.Y. (1952)
2. J. von Neumann Mathematical foundations of Quantum Mechanics, Princeton Univ.Press, New Jersey, (1955)
3. P.A.M.Dirac The Principles of Quantum Mechanics, Clarendon Press, Oxford, (1958)
4. E.P.Wigner The problem of measurements, Amer. J. Phys. , 31,6-15, (1963)
5. M.Jammer, The conceptual development of Quantum Mechanics, McGraw-Hill , (1967)
6. C.W.Helstrom Quantum detection and estimation theory, Academic Press, N.Y. (1976)

7. L. de Brogle Les incertitudes dHeisenberg et linterpretation probaliste de la mecanique ondulatoire, Gauthier-Villars, (1982)
8. A.Sudbery Quantum Mechanics and the particles of nature , Cambridge Univ.Press, (1986)
9. A.A.Penzias, R.W.Wilson Astrophys.J., 142, 419, (1965)
10. L.D.Landaw Phys.Zs.Sowjet, 2, 46, (1932)
11. J.R.Oppenheimer, G.M.Volkoff Phys.Rev., 55, 374, (1939).
12. J.R.Oppenheimer, H.Snider Phys.Rev., 56, 455, (1939).
13. R.C.Tolman Phys.Rev., 55, 364, (1939).
14. S.Chandrasekhar Astrophys.J., 140, 417, (1964).
15. Black Holes: The Membrane Paradigm, Ed. K.S.Torne, R.H.Price, D.A.Macdonald, Yale Univ.Press, (1986).
16. S.W.Hawking Phys.Rev.D 13, 191, (1976).
17. D.W. Sciama Vistas Astron., 19, 385, (1976)
18. T.Jacobson , Phys.Rev.Lett. 75, 1260, (1995)
19. E.A.Martinez , Phys.Rev.D54, 6302, (1996)
20. R.Penrose Singularities and Time assimetry, In. An Einstein Centerary Survey, Ed. S.W.Hawking and W.Israel, Cambridge Univ.Press, (1979)
21. Barrow J.D. Chaotic behavior in General Relativity, Phys.Rep. 85,1,1-49, (1982)
22. R.M. Zalaletdinov, Gen. Rel. Grav. 28, 953 (1996)

Integrability of the $N=3$ Supersymmetric KdV Equation

Z. Popowicz

*Institute of Theoretical Physics, University of Wroclaw,
pl. M. Borna 9, 50-205 Wroclaw, Poland*

Abstract. We present the Lax operator for the $N=3$ Supersymmetric KdV equation. In that manner we prove the integrability of this equation.

INTRODUCTION

The famous Kortewg de Vries equation [1,2] colud be written in the bihamiltonian form as follows

$$u_t = P_1 \frac{\delta H_1}{\delta u} = P_2 \frac{\delta H_2}{\delta u} = -u_{xxx} + 6uu_x, \tag{1}$$

where the Hamiltonian operators are

$$P_1 := \partial, \tag{2}$$

$$P_2 := -\partial^3 + 2u\partial + 2\partial u, \tag{3}$$

and

$$H_1 = \int dx \left(u_x^2 + 3u^2 \right), \tag{4}$$

$$H_2 = \int dx \, u^2, \tag{5}$$

The Korteweg de Vries equation is usually written down in the so called Lax pair representation

$$\frac{\partial L}{\partial t} = \left[L, (L_+)^{\frac{3}{2}} \right] \tag{6}$$

where Lax operator L is

$$L = \partial^2 + u \qquad (7)$$

and (+) denotes the projection onto purely differential part in the algebra of the pseudo-differential algebra G

$$G \in g = \sum_{n=-\infty}^{\infty} a_n \partial^n, \qquad (8)$$

where a_n are an arbitray functions. The infinite number of conserved currents could be obtained from Lax operator using

$$H_n = TrL^{\frac{n}{2}} \qquad (9)$$

where Tr denotes the a_{-1} element in the pseudo-differential algebra.

The second Hamiltonian operator P_2 is connected with the Virasoro algebra. It can be easily seen if we use the Fourier expansion of the field u

$$u(x) := \frac{6}{c} \sum_{n=-\infty}^{\infty} exp(-inx) L_n - \frac{1}{4} \qquad (10)$$

where c is the central element. Substituting this expansion to the Hamiltonian operator we obtain that the generators L_n constitue the Virasoro algebra (conformal algebra)

$$\{L_n, L_m\} := (n-m)L_{n+m} + cn(n^2-1)\delta_{n+m,0} \qquad (11)$$

The observation that Korteweg de Vries equation is connected with the Virasoro algebra allowed to carry out the supersymmetrization of this equation to the non extended $N = 1$ as well as to the extended $N = 2, 3, 4$ supersymmetric case [3,4,5,6,7].

The supersymmetrization of the soliton equation appeared almost in parallel to the usage of the supersymmetry in the quantum field theory. In order to get a supersymmetric theory we have to add to a system of k bosonic equations kN fermions and $k(N-1)$ boson fields $k = 1, 2.., N = 1, 2, ..$ in such a way that a final theory becomes a suspersymmetric invariant. From a soliton point of view we can distinguish two important classes of the supersymmetric equations: the non-extened ($N = 1$) and extended ($N > 1$) cases. Consideration of the extended case may imply new bosonic equations whose properties need further investigation. This may be viewed as a bonus, but this extended case is in no way more fundamental than the non-extended one.

We know all supersymmetric extensions of the Virasoro algebra [4,8] and hence we are able to construct the Hamiltonian operator for $N = 1, 2, 3, 4$ extensions. For example for the $N = 3$ case it has the following form

$$P_2 := -D_1 D_2 D_3 + \Phi \partial + (D_i \Phi) D_i + 2\Phi_x \qquad (12)$$

where the summation convention over repeated indexes is assumed ($i = 1, 2, 3$) and

$$D_i := \frac{\partial}{\partial \theta_i} + \theta_i \partial, \tag{13}$$

Φ is the superfermionic field which can be considered as

$$\Phi := \eta + \theta_i f_i + \sum_{i<j} \theta_i \theta_j h_{i,j} + \theta_1 \theta_2 \theta_3 u, \tag{14}$$

where θ are Majorana spinors $\theta_k^2 = 0$, $\eta, h_{i,j}$ are fermions fields while f_i, u are bosons fields. $D_i \Phi$ denotes the action of the supersymmetric derivate on the superfield Φ while $(D_i \Phi)$ the autcome.

If we assume the most general supersymmetric Hamiltonian of the gradation 2.5 (the field Φ has gradation 0.5)

$$H := \int dx \, d^3\theta \Big(\Phi(D_1 D_2 D_3 \Phi) + \frac{1}{3} \Phi(D_i \Phi)(D_i \Phi) \Big) \tag{15}$$

and use the opertaor P_2 (8) we obtain the supersymmetric $N=3$ extension of the Korteweg de Vries equation

$$\Phi_t := P_2 \frac{\delta H}{\delta \Phi} = \tag{16}$$

$$-\Phi_{xxx} + (\Phi(D^3\Phi))_x - \frac{5}{2}(D^3(\Phi \Phi_x)) + \frac{5}{2}(D^3(D_i\Phi)(D_i\Phi)) + \tag{17}$$

$$10(D_i\Phi)(D^{3-i}\Phi) + \frac{5}{3}\Phi((D_i\Phi)(D_i\Phi))_x, \tag{18}$$

where

$$D^3 := D_1 D_2 D_3 \tag{19}$$

$$D^{3-i} := \frac{1}{2} \epsilon_{i,j,k} D_j D_k \tag{20}$$

The bosonic sector of the supersymmetric KdV equaition in which all fermionic fields vanishes reads

$$u_t = -u_{xxx} + 3\Big(u^2 - f_i f_{ixx} + u f_i f_i\Big), \tag{21}$$

$$f_{it} = -f_{ixxx} + 3\Big(uf_i\Big)_x + 3f_i f_j f_{jx}. \tag{22}$$

As we see this bosonic sector describes the interaction of the Korteweg de Vries field with three fields.

Let us mention that this supersymmetric extension has been know for physicist since 10 years. However nothing up to now has been know on its integrability.

The supersymmetric Lax operator has been constarcted only for $N = 1, 2$ supersymmetric KdV case only [5,9]. Moreover one can easily prove that it is impossible to construct such for the $N = 3, 4$ [9] case. In the next section we show how it is possible to overcome this problem and prove the integrability of $N = 3$ supersymmetric extension of Korteweg de Vries equation.

LAX REPRESENTATION FOR $N = 3$ SUPER KDV EQUATION

In order to find the proper Lax pair representation of this equation let us notice that it is possible to rewritte this equation in terms of $N = 2$ superfields [10]. This can be written as the following coupled system of evolution equations for the general bosonic spin 1 superfield $J(Z)$ and general fermionic spin 1/2 superfield $g(Z)$:

$$\frac{\partial}{\partial t} J = -J''' - (J^3)' + 3(J\,[D,\overline{D}]J)' + 3(g'\,[D,\overline{D}]g)' - $$
$$12(J\,Dg\,\overline{D}g)' + 6(\overline{D}J\,g\,Dg)' + 6(DJ\,g\,\overline{D}g)' ,$$
$$\frac{\partial}{\partial t} g = -g''' + 6J(\overline{D}J\,Dg + DJ\,\overline{D}g) + 3(J\,[D,\overline{D}]g)' - 3J^2\,g' - $$
$$6g'\,Dg\,\overline{D}g - 6(g\,Dg\,\overline{D}g)', \qquad (23)$$

where $Z = (z,\theta,\overline{\theta})$ is a coordinate of the $N = 2$ superspace, $dZ \equiv dzd\theta d\overline{\theta}$ and the fermionic covariant derivatives D and $\overline{D}$ are defined as

$$D = \frac{\partial}{\partial \theta} - \frac{1}{2}\overline{\theta}\frac{\partial}{\partial z}, \quad \overline{D} = \frac{\partial}{\partial \overline{\theta}} - \frac{1}{2}\theta\frac{\partial}{\partial z}, \quad D^2 = \overline{D}^2 = 0, \quad \{D,\overline{D}\} = -\frac{\partial}{\partial z} \equiv -\partial. \qquad (24)$$

One can check that the equations (23) are covariant with respect to the following transformations of an extra hidden supersymmetry:

$$\delta J = \varepsilon g', \quad \delta g = \varepsilon J , \qquad (25)$$

where ε is a Grassmann parameter. Just this additional supersymmetry together with explicit $N = 2$ ones form $N = 3$ supersymmetry.

The $N = 3$ KdV equation (23) can be obtained from the Hamiltonian

$$H = -3\int dZ \left(J[D,\overline{D}]J + \frac{1}{3}J^3 - g[D,\overline{D}]g' + 2Jgg' + 8JDg\overline{D}g \right) , \qquad (26)$$
$$\frac{\partial}{\partial t} J \equiv \{H, J\}, \quad \frac{\partial}{\partial t} g \equiv \{H, g\},$$

if we use the $N = 3$ superconformal algebra Poisson brackets, which read in terms of $N = 2$ supercurrents $J(Z)$ and $g(Z)$ as follows:

$$\{J(Z_1), J(Z_2)\} = \left(\frac{1}{2}[D,\overline{D}]\partial + \partial J + J\partial + \overline{D}JD + DJ\overline{D}\right)\Delta(Z_1 - Z_2),$$
$$\{J(Z_1), g(Z_2)\} = \left(\partial g + \frac{1}{2}g\partial - \overline{D}gD - Dg\overline{D}\right)\Delta(Z_1 - Z_2),$$
$$\{g(Z_1), g(Z_2)\} = \frac{1}{2}\left(J + [D,\overline{D}]\right)\Delta(Z_1 - Z_2) . \qquad (27)$$

Here $\Delta(Z_1 - Z_2) = (\theta_1 - \theta_2)(\bar\theta_1 - \bar\theta_2)\delta(z_1 - z_2)$ is $N = 2$ superspace delta function and all operators in r.h.s. are evaluated in the second point.

Now we turn to the basic item of this section, the construction of the Lax operator for the $N = 3$ KdV hierarchy [10]. Keeping in mind that in the limit $g \to 0$ the $N = 3$ KdV equation reduces to the $N = 2$ $a = 1$ KdV one, which can be described by the following Lax operator [9]

$$L_{a=1} = \partial - \left[D, \overline{D}\right]\partial^{-1} J \qquad (28)$$

we propose the following Lax operator for $N = 3$ KdV hierarchy

$$L = \partial - \left[D, \overline{D}\right]\partial^{-1} J - \left[D, \overline{D}\right]\partial^{-1} g \left[D, \overline{D}\right]\partial^{-1} g . \qquad (29)$$

Now, one can check that the Lax operator (29) indeed gives rise to the $N = 3$ KdV flows via the Lax equation

$$\frac{\partial}{\partial t_{2n+1}} L = \left[\left(L^{2n+1}\right)_+, L\right] , \qquad (30)$$

where $\{+\}$ denotes the differential part of the Lax operator. The conserved currents for the Lax operator (29) are defined by the standard $N = 2$ residue form as

$$H_n = \int dZ\, tr\left(L^{2n+1}\right) , \qquad (31)$$

where tr denotes the coefficient standing before $[D,\overline{D}]\partial^{-1}$.

Thus, we have proved the integrability of the $N = 3$ KdV equation.

REFERENCES

1. Faddev, L., and Takhtadjan, L., *Hamiltonian Methods in the Theory of Solitons*, Berlin: Springer, 1987.
2. Ablowitz, M., and Segur, H., *Solitons and the Inverse Scattering Transforms*, Philadelphia (PA): SIAM, 1981.
3. Kupershmidt, B., *Elements of Superintegrable Systems*, Dordrecht: Kluwer, 1987.
4. Chaichian, M., and Lukierski, J., *Phys. Lett.* **B183**, 169 (1987).
5. Laberge, C., and Mathieu, P., *Phys. Lett.* **B215**, 718 (1988).
6. Yung, C., *Mod. Phys. Lett.* **A8**, 1161 (1993).
7. Bellucci, S., Ivanov, E., and Krivonos, S., *J. Math. Phys.* **34**, 3087 (1993).
8. Schouten, K., *Nucl. Phys.* **B295**, 634 (1988).
9. Popowicz, Z., *Phys. Lett.* **A174**, 411 (1993).
10. Krivons, S., Pashnev, A., and Popowicz, Z., "Lax pairs for N=2,3 supersymmetric Kdv equation and their extensions", preprint solv-int/9802003 (to appear in *Mod. Phys. Lett. A*).

DEVELOPMENTS IN QUANTUM AND TOPOLOGICAL FIELD THEORIES

Curved Domain Wall/Vortex Solutions in Relativistic Field Theories [1]

Henryk Arodź

Institute of Physics, Jagellonian University, Reymonta 4, 30-059 Cracow, Poland

Abstract. We briefly review two methods for construction of curved domain wall or vortex solutions: classical effective action method and improved expansion in the width.

INTRODUCTION

Domain walls and vortices appear in nonrelativistic effective field theoretic models of condensed matter physics, as well as in relativistic field models of particle physics. In spite of the fact that the models describe totally different physics, e. g., cosmology [1], quantum chromodynamics [2], superfluids [3,4], liquid crystals [5] and magnetic materials [6], theoretical analysis of dynamics of domain walls/vortices is to large extent model independent [2]. The similarities begin with the common prerequisite for the existence of domain walls or vortices: the degeneracy of the vacuum state such that the manifold of vacuum states has nontrivial π_0 or π_1 homotopy groups. Next, line-like or surface-like physical shape common for all vortices or domain walls, respectively, leads to similarities in mathematical description of dynamics of these objects.

Mathematically, domain walls and vortices are represented by particular solutions of pertinent nonlinear field equations. Only rarely the solutions are known explicitly. This is true even in cases distinguished by the largest possible symmetry, like a planar domain wall or a straight-linear vortex, which have been studied for quite a long time. On the other hand, there is little doubt about existence of these highly symmetric solutions. They have been studied with the help of numerical methods. One could also use analytical approximations, like the polynomial approximation [7], [8]. Therefore, for most purposes one may assume that the planar

---

[1] The research supported in part by KBN grant 2 PO3B 095 13.
[2] We use the term *vortex* in a rather broad sense, including, e.g., line defects in liquid crystals or a flux-tube in quantum chromodynamics. The defining feature is that the fields have values different from those corresponding to a vacuum (a ground state) only in certain vicinity of a line (closed or infinite) in space. In the domain wall case the line is replaced by a surface.

domain wall and the straight-linear vortex solutions are known with any reasonable accuracy.

Curved domain wall and vortex solutions pose a much harder problem even at the level of approximate solutions. It seems that the first significant step towards such solutions has been taken while investigating a relationship between relativistic Nambu-Goto string and a vortex in the Abelian Higgs model [9], [10], [11]. The action functional for the string follows from the field theoretical action of the model in the limit of negligibly small ratio of the width of the vortex to its curvature radius. As a byproduct of those investigations it was shown that the simplest solution corresponding to a straight-linear vortex could serve as a basis for construction of a class of approximate solutions representing the curved vortex. Later on, the work in this direction has been resumed [12], [13] in connection with the cosmic string hypothesis [3]. It has also been realised that analogous problems and formalism appear in the case of curved domain wall — here the string is replaced by a membrane.

For the single curved domain wall or vortex solution so called classical effective action method has been worked out. The main idea has been given already in the old papers [9], [10], and elaborated in many papers, see, e. g., [12], [13], [14], [15], [16], [17]. Below we shall briefly describe the method, and we shall point out its deficiences. Next, we shall outline another method, called by us the improved expansion in the width. This perturbative scheme has been proposed for domain walls in the papers [18], [19]. As for the curved vortex, in the paper [20] it has been shown that an analogous perturbative scheme can be constructed — a corrected and expanded version of it will be presented elsewhere [21].

THE CLASSICAL EFFECTIVE ACTION METHOD

The first step in the effective action method, as well as in the improved expansion in the width, consists in introducing a coordinate system comoving with the domain wall in the three dimensional space. Let us introduce a smooth surface S attached to the domain wall (comoving with it) — we shall call it *the membrane*. The world-volume of S is denoted by Σ and it is parametrised as follows

$$\Sigma \ni (Y^\mu)(u^a) = (\tau, Y^i(u^a)). \tag{1}$$

We use here the notation $(u^a)_{a=0,1,2} = (\tau, \sigma^1, \sigma^2)$, where τ coincides with the laboratory frame time x^0, while σ^1, σ^2 parametrise the comoving membrane S at each instant of time. The index $i = 1, 2, 3$ refers to the spatial components of the four-vector. The points of the comoving membrane S at the instant τ are given by

---

[3] This hypothesis, is it true or not, has inspired extensive investigations of dynamics of topological defects in relativistic field theory as well as in condensed matter physics, including such complex phenomena as, e.g., evolution of a network of vortices, or production of topological defects in phase transitions.

$(Y^i)(\tau, \sigma^1, \sigma^2)$. The coordinate system $(\tau, \sigma^1, \sigma^2, \xi)$ comoving with the domain wall is defined by the formula

$$x^\mu = Y^\mu(u^a) + \xi\, n^\mu(u^a), \qquad (2)$$

where x^μ are Cartesian laboratory frame coordinates in Minkowski space-time, and (n^μ) is a normalised space-like four-vector orthogonal to Σ,

$$n_\mu(u^a) Y^\mu_{,a}(u^a) = 0, \qquad n_\mu n^\mu = 1,$$

where $Y^\mu_{,a} \equiv \partial Y^\mu / \partial u^a$. The three four-vectors $Y_{,a}$ are tangent to Σ. The definition (2) implies that ξ and u^a are Lorentz scalars. In the comoving coordinates the comoving membrane is described by the simple Lorentz invariant condition $\xi = 0$.

Let us introduce the extrinsic curvature coefficients K_{ab} and the induced metrics g_{ab} on Σ:

$$K_{ab} \stackrel{df}{=} n_\mu Y^\mu_{,ab}, \qquad g_{ab} \stackrel{df}{=} Y^\mu_{,a} Y_{\mu,b},$$

where $a, b = 0, 1, 2$. The covariant metric tensor transformed to the comoving coordinates has the following components

$$[G_{\alpha\beta}] = \begin{bmatrix} G_{ab} & 0 \\ 0 & 1 \end{bmatrix},$$

where $\alpha, \beta = 0, 1, 2, 3$, with $\alpha, \beta = 3$ corresponding to the ξ coordinate, and

$$G_{ab} = N_{ac} g^{cd} N_{db}, \qquad N_{ac} \stackrel{df}{=} g_{ac} - \xi K_{ac}.$$

Thus, $G_{\xi\xi} = 1$, $G_{\xi a} = 0$. Furthermore,

$$\sqrt{-G} = \sqrt{-g}\, h(\xi, u^a),$$

where $g = det[g_{ab}]$, $G = det[G_{\alpha\beta}]$, and

$$h(\xi, u^a) = 1 - \xi K^a_a + \frac{1}{2}\xi^2(K^a_a K^b_b - K^b_a K^a_b) - \frac{1}{3}\xi^3 K^a_b K^b_c K^c_a.$$

For raising and lowering the Latin indices of the extrinsic curvature coefficients we use the induced metric tensors g^{ab}, g_{ab}.

The inverse metric tensor $G^{\alpha\beta}$ has the following form

$$[G^{\alpha\beta}] = \begin{bmatrix} G^{ab} & 0 \\ 0 & 1 \end{bmatrix},$$

where

$$G^{ab} = (N^{-1})^{ac} g_{cd} (N^{-1})^{db},$$

and

$$(N^{-1})^{ac} = \frac{1}{h}\left\{ g^{ac}[1 - \xi K^b_b + \frac{1}{2}\xi^2(K^b_b K^d_d - K^d_b K^b_d)] \right.$$

$$+ \xi(1 - \xi K_b^b)K^{ac} + \xi^2 K_d^a K^{dc}\}.$$

In general, the comoving coordinates (u^a, ξ) are defined in a vicinity of the worldvolume Σ of the comoving membrane. Roughly, the allowed range of the ξ coordinate is determined by the smaller of the two main curvature radia of the membrane in a local rest frame. We assume that the curvature radia are sufficiently large so that on the outside of the region of validity of the co-moving coordinates fields differ from their vacuum values only by exponentially small terms.

As a relatively simple example we shall consider a domain wall in the model with the following Lagrangian

$$\mathcal{L} = -\frac{1}{2}\eta_{\mu\nu}\partial^\mu \Phi \partial^\nu \Phi - \frac{\lambda}{2}(\Phi^2 - \frac{M^2}{4\lambda})^2, \qquad (3)$$

where Φ is a single real scalar field, $(\eta_{\mu\nu})$=diag(-1,1,1,1) is the space-time metric, and λ, M are positive constants. Two vacuum values of Φ are equal to $\pm M/2\sqrt{\lambda}$. The domain wall is given by the solution Φ of the corresponding Euler-Lagrange equation smoothly interpolating between the two vacuum values. Then, at each instant of time the field Φ vanishes somewhere in the interior of the domain wall. The locus of these zeros is assumed to be a smooth connected surface $\tilde{S}$ in the space, called *the core* of the domain wall. The transverse width of such a domain wall is of the order M^{-1}. Energy density is exponentially localised around the core.

In all known to us implementations of the effective action method the membrane has been identified with the core. In a sense this is quite natural a choice because otherwise one would have to provide an additional definition of the membrane and this would make the scheme more complicated. Nevertheless, in the improved expansion in the width the membrane and the core in general do not coincide.

In the co-moving coordinates Lagrangian (3) has the following form

$$\mathcal{L} = -\frac{M^4}{32\lambda}\left[\partial_s\phi\partial\phi_s + (\phi^2 - 1)^2 + \frac{4}{M^2}G^{ab}\partial_a\phi\partial_b\phi\right], \qquad (4)$$

where $\partial_s = \partial/\partial s, \partial_a = \partial/\partial u^a$. The field Φ and the coordinate ξ have been expressed by dimensionless ϕ and s,

$$\Phi(x^\mu) = \frac{M}{2\sqrt{\lambda}}\phi(s, u^a), \quad \xi = \frac{2}{M}s.$$

The corresponding Euler-Lagrange equation has the following form

$$\frac{2}{M^2}\frac{1}{\sqrt{-g}}\partial_a(\sqrt{-g}hG^{ab}\partial_b\phi) + \frac{1}{2}\partial_s(h\partial_s\phi) + h\phi(1 - \phi^2) = 0. \qquad (5)$$

In the second step of the effective action method the field equation (5) is used to determine dependence of the ϕ field on the ξ coordinate for arbitrarily fixed trajectory of the comoving membrane. The resulting ϕ we shall denote by $\phi(\xi; Y^i(u^a))$

— the dependence on ξ is supposed to be explicit while $Y^i(u^a)$ is not known as yet. This step is a bottleneck in applications of the effective action method to domain walls and vortices. The point is that we are not able to find the ξ dependence exactly, and moreover, approximate formulas presented in the literature in most cases are not fully satisfactory. Such approximate formulas can be divided in two classes: *ad hoc* composed formulas which are meant to be only quolitatively correct, and formulas claimed to be obtained with the help of expansion in the width. However, the version of expansion in the width which has been used in that context is not satisfactory because it misses the consistency conditions described in the next Section.

Presented in the next Section the improved expansion in the width gives results which in general are different from the ones obtained with the previous version of the expansion. In the case of domain walls the differences appear already in the first order correction ($\sim 1/M$) to the zeroth order approximation, and in the case of vortex in the Abelian Higgs model in the third order correction $((1/m_H)^3$ [21], where m_H is the mass of the Higgs field — hence in this case the old and the improved versions of the expansion in the width agree up to the second order.

In order to obtain the actual curved domain wall solution we still have to determine the physical trajectory of the comoving membrane. To this end, the function $\phi(\xi; Y^i(u^a))$ obtained in the second step is inserted in the action integral of the model (3). Because the dependence on the transverse variable ξ is explicit we can integrate over it. The resulting expression is a functional of the membrane trajectory $Y^i(u^a)$ — this is the classical effective action $S_{eff}[Y^i]$ for the membrane. The physical trajectory Y^i_{phys} of the membrane is determined from the Euler-Lagrange equations following from this effective action. Inserting Y^i_{phys} in the function $\phi(\xi; Y^i(u^a))$ we obtain the curved domain wall solution.

The classical effective action method is correct in principle, as noted in the paper [14]. In the paper [22] this method has been checked on simple examples from classical mechanics. We have obtained effective actions which contain higher derivatives and nonlocalities in time. Moreover, initial data for the Euler-Lagrange equations derived from the effective actions have to be appropriately restricted in order to eliminate some spurious trajectories which do not correspond to solutions of the initial equations of motion. All these complications notwithstanding, the method reproduces the original equations of motion. Nevertheless, it has become clear that it requires a great care, and that it can lead to quite unpleasant effective models.

THE IMPROVED EXPANSION IN THE WIDTH

This expansion yields approximate solutions of the curved domain wall type without employing the classical effective action. This is possible because the consistency conditions yield all necessary equations of motion.

Let us first solve Eq.(5) in the leading approximation obtained by putting $1/M = 0$. The equation is then reduced to

$$\frac{1}{2}\partial_s^2 \phi^{(0)} + \phi^{(0)}[1 - (\phi^{(0)})^2] = 0. \tag{6}$$

Equation (6) does not contain derivatives with respect to time in spite of the fact that it is supposed to approximate the evolution equation (5). This annoying fact reflects the singular character of the perturbation given by the first term on the l.h.s. of Eq.(5).

Equation (6) has the following well-known domain wall type solution

$$\phi^{(0)} = \tanh s. \tag{7}$$

The solution (7) is given in the co-moving coordinates and it does not determine the field ϕ in the laboratory frame because we do not know yet the position of the comoving membrane with respect to the laboratory frame, but we shall shortly see that the first order terms following from Eq.(5) in the perturbative expansion imply Nambu-Goto equation for the membrane — no effective action will be needed.

It is convenient to split the field ϕ in two components

$$\phi(s, u^a) = B(u^a)\psi_0(s) + \chi(s, u^a), \tag{8}$$

where

$$B(u^a) \stackrel{df}{=} \phi(0, u^a)$$

is the component of the scalar field living on the co-moving membrane, and

$$\chi \stackrel{df}{=} \phi(s, u^a) - B(u^a)\psi_0(s)$$

is the remaining part. The auxiliary, *fixed* function $\psi_0(s)$ depends on the variable s only. It is smooth, concentrated around $s = 0$, and

$$\psi_0(0) = 1.$$

It follows that

$$\chi(0, u^a) = 0. \tag{9}$$

The best choice for $\psi_0(s)$ is given by formula [19]

$$\psi_0(s) = \frac{1}{\cosh^2 s}.$$

The expansion in the width has the form

$$\chi(s, u^a) = \tanh s + \frac{1}{M}\chi^{(1)}(s, u^a) + \frac{1}{M^2}\chi^{(2)}(s, u^a) + \frac{1}{M^3}\chi^{(3)}(s, u^a) + ..., \tag{10}$$

$$B(u^a) = \frac{1}{M}B^{(1)}(u^a) + \frac{1}{M^2}B^{(2)}(u^a) + \frac{1}{M^3}B^{(3)}(u^a) + ..., \tag{11}$$

where we have taken into account the zeroth order result (7). The expansion parameter is $1/M$ and not $1/M^2$ because $1/M$ in the first power appears in h and G^{ab} after passing to the s variable. In order to obey the condition (9), and to ensure the proper asymptotics of χ at large $|s|$ we assume that for $n \geq 1$

$$\chi^{(n)}(0, u^a) = 0, \quad \lim_{s \to \pm\infty} \chi^{(n)} = 0. \tag{12}$$

Inserting the perturbative Ansatz (10, 11) in Eq.(5), expanding the l.h.s.'s of it in powers of $1/M$, and equating to zero coefficients in front of the powers of $1/M$ we obtain a sequence of linear, inhomogeneous equations for $\chi^{(n)}(s, u^a)$ with $n \geq 1$. They can be written in the form

$$\hat{L}\chi^{(n)} = f^{(n)}, \tag{13}$$

where the source term $f^{(n)}$ is determined by the lower order terms in χ and B, and

$$\hat{L} \stackrel{df}{=} \frac{1}{2}\partial_s^2 + 1 - 3(\phi^{(0)})^2.$$

$B^{(n)}$ is absent on the l.h.s. of Eq.(13) because $\hat{L}\psi_0 = 0$. Explicit formulas for $f^{(n)}$ with $n \leq 4$ can be found in [19]. In particular,

$$f^{(1)} = K_a^a \partial_s \phi^{(0)}$$

— we shall need this formula below. Solutions of Eqs.(13) are obtained in a standard manner [19]. They are given by the formula

$$\chi^{(n)}(s, u^a) = \int_{-\infty}^{+\infty} dx\, G(s, x) f^{(n)}(x, u^a), \tag{14}$$

where $G(s, x)$ is a Green's function adjusted to the boundary conditions (12).

Obviously, we assume that all proportional to positive powers of $1/M$ terms in Eq.(5) are small. For this it is not sufficient that the extrinsic curvatures are small, that is that $K_b^a/M \ll 1$. We have also to assume that the derivatives $\chi_{,a}^{(n)}, B_{,a}^{(n)}$ are not proportional to M. It will not be the case, for example, if χ and B contain modes oscillating with a frequency $\sim M$. Such modes would give positive powers of M upon differentiation with respect to u^a, and then the counting of powers of $1/M$ would no longer be so straightforward as we have assumed. This excludes radiation modes as well as massive excitations of the domain wall. Therefore, the approximate solution we obtain gives what we may call *the basic curved domain wall*. The expansion yields domain walls of concrete transverse profile — the dependence on s is explicitly given by the approximate solution even at the initial instant of time. We may choose the initial position and velocity of points of the membrane, but the dependence of the scalar field on the variable s at the initial time is given by formula (14). This unique profile is characteristic for the basic curved domain wall.

The most important point in our approach is the observation that operator $\hat{L}$ has the normalizable zero-mode:

$$\hat{L}\psi_0 = 0, \tag{15}$$

where $\psi_0(s)$ is given below the condition (9). Notice that $\psi_0 = \partial_s \phi^{(0)}$ — this means that the zero-mode ψ_0 is related to invariance of Eq.(6) with respect to the translations $s \to s + \text{const}$.

The presence of the zero-mode implies the consistency conditions. For example, let us multiply the first order equation

$$\hat{L}\chi^{(1)} = K_a^a \partial_s \phi^{(0)} \tag{16}$$

by ψ_0 and integrate over s. It is easy to see that $\int \psi_0 \hat{L}\chi^{(1)}$ vanishes because of (15), and we obtain the following condition

$$K_a^a \int ds\, \psi_0(s) \partial_s \phi^{(0)}(s) = 0,$$

which is equivalent to

$$K_a^a = 0. \tag{17}$$

Eq.(17) coincides with the well-known Nambu-Goto equation for the relativistic membrane. It determines the motion of the comoving membrane, that is the functions $Y^i(u^a)$, i=1,2,3, once appropriate initial data are fixed. Subsequently, it also determines the extrinsic curvature K_{ab} and the metric g_{ab}.

The r.h.s. of Eq.(16) vanishes due to Nambu-Goto equation (17), and the resulting homogeneous equation

$$\hat{L}\chi^{(1)} = 0$$

with the boundary conditions (12) taken into account has only the trivial solution

$$\chi^{(1)} = 0.$$

Hence, the first order correction to the total field ϕ is equal to $B^{(1)}\psi_0/M$. In general, it does not vanish on the comoving membrane that is at $s = 0$.

Equation (16) does not give any restriction on the function $B^{(1)}$. However, we have also analogously obtained consistency conditions for equations (13) with $n > 1$. For some n (e.g. for $n = 2$) we obtain only the trivial identity $0 = 0$, but for other n the consistency conditions have the form of equations for $B^{(k)}$. For example, the consistency condition for Eq.(13) with $n = 3$ gives the following equation for $B^{(1)}$ [18]:

$$\frac{1}{\sqrt{-g}} \partial_a(\sqrt{-g} g^{ab} \partial_b B^{(1)}) + (\frac{\pi^2}{4} - 1) K_b^a K_a^b B^{(1)} + \frac{9}{35}(B^{(1)})^3 = (\frac{\pi^2}{6} - 1) K_b^a K_c^b K_a^c. \tag{18}$$

At this point we have the complete set of equations determining the evolution of the domain wall up to the $1/M$ order. Each of Eqs.(16), (17), (18) describes different aspect of the dynamics of the curved domain wall. Eq.(16) determines dependence of $\chi^{(1)}$ on s, trivial in this case. Because the term $B\psi_0$ in formula (8) has the explicit dependence on s, we may say that Eq.(16) for $\chi^{(1)}$ fixes the transverse profile of the domain wall.

Equation (18) determines the $B^{(1)}(u^a)$ function which can be regarded as a (2+1)-dimensional scalar field defined on Σ and having nontrivial nonlinear dynamics. The extrinsic curvature K_{ab} of Σ acts as an external source for this field. The field $B^{(1)}$ can propagate along Σ. One may regard this effect as causal propagation of deformations which are introduced by the extrinsic curvature.

Finally, Nambu-Goto equation (17) for the comoving membrane determines the evolution of the shape of the domain wall.

Equation (17) for the comoving membrane and equations for $B^{(n)}$ obtained from the consistency conditions are of the evolution type — we have to specify initial data for them, otherwise their solutions are not unique. Equations (13) for the perturbative contributions $\chi^{(n)}$ are of different type — in order to ensure uniqueness of their solutions it is sufficient to adopt the boundary conditions (12). The initial data for $B(u^a)$ and $Y^i(u^a)$ follow from initial data for the original field ϕ. From such data for ϕ we know in particular the initial position and velocity of the core. We may assume, for example, that at the initial instant τ_0 the comoving membrane and the core have the same position and velocity. Hence,

initial data for the membrane = initial data for the core.

Using formula (8) one can show that then

$$B^{(n)}(\tau_0, \sigma^1, \sigma^2) = 0, \quad \partial_\tau B^{(n)}(\tau_0, \sigma^1, \sigma^2) = 0. \tag{19}$$

In order to find the domain wall solution one should first solve the collective dynamics of the domain wall, that is to compute evolution of the comoving membrane and of the $B^{(n)}$ fields. The profile χ of the domain wall is calculated in the next step with the help of formula (14). In our perturbative scheme the profile of the domain wall can not be chosen arbitrarily even at the initial time — it is fixed uniquely once the initial data for the membrane and for the B field are given.

Evolution of the core can be determined afterwards, from the explicit expression for the scalar field ϕ [19]. It is easy to see that the core and the membrane coincide for $\tau > \tau_0$ only if the r.h.s. of Eq.(18) vanishes. This is not the case for, e.g., a spherical domain wall [18].

In paper [19] the perturbative solution has been constructed up to the order $(1/M)^4$.

In paper [23] second order perturbative solutions for cylindrical and spherical domain walls have been compared with numerical solutions of the field equation (5). The results of the comparison are quite encouraging.

The perturbative approach to dynamics of the domain walls we have just sketched can be generalised to models involving several fields. Also the requirement of relativistic invariance can be dropped out. As an example, we have applied the perturbative scheme to find evolution of curved domain walls and disclination lines in nematic liquid crystals, [24].

Acknowledgment. It is a pleasure to thank the Organizers for the invitation to this very instructive and stimulating Conference.

REFERENCES

1. T.W.B. Kibble, *J.Phys.* **A9**, 1387 (1976); A.L. Vilenkin, *Phys. Rep.* **121**, 263 (1985).
2. See, e.g., M. Baker, J.S. Ball and F. Zachariasen, *Phys.Rep.* **209**, 73 (1991).
3. R.P. Huebener, *Magnetic Flux Structures in Superconductors*, Berlin - Heidelberg - New-York: Springer-Verlag, 1979.
4. R.J. Donnelly, *Quantized Vortices in HeliumII*, Cambridge: Cambridge University Press,
5. See, e.g., S. Chandrasekhar and G.S. Ranganath, *Adv. Phys.* **35**, 507 (1986).
6. See, e.g., A. M. Kosevich, B. A. Ivanov and A. S. Kovalev, *Phys. Rep.* **194**, 117 (1990).
7. H. Arodź, *Phys. Rev.* **D52** , 1082 (1995).
8. H. Arodź and L. Hadasz, *Phys. Rev.* **D54**, 4004 (1996).
9. H.B. Nielsen and P. Olesen, *Nucl.Phys.* **B61**, 45 (1973).
10. D. Förster, *Nucl.Phys.* **B81**, 84 (1974).
11. J.-L. Gervais and B. Sakita, *Nucl.Phys.* **B91**, 301 (1975).
12. K. Maeda and N. Turok, *Phys.Lett.* **B202**, 376 (1988).
13. R. Gregory, *Phys.Lett.* **B206**, 199 (1988).
14. S.M. Barr and D. Hochberg, *Phys.Rev.* **D39**, 2308 (1989).
15. R. Gregory, D. Haws and D. Garfinkle, *Phys.Rev.* **D42**, 343 (1990).
16. B. Carter and R. Gregory, *Phys. Rev.* **D51**, 5839 (1995).
17. M. Anderson, F. Bonjour, R. Gregory and J. Stewart, *Phys. Rev.* **D56** , 8014 (1997).
18. H. Arodź, *Nucl. Phys.* **B450**, 174 (1995).
19. H. Arodź, *Nucl. Phys.* **B509** , 273 (1998).
20. H. Arodź, *Nucl. Phys.* **B450**, 189 (1995).
21. H. Arodź and A. L. Larsen, in preparation.
22. H. Arodź and P. Węgrzyn, *Phys.Lett.* **B291**, 251 (1992).
23. J. Karkowski and Z. Świerczyński, *Acta Phys. Pol.* **B30** , 234 (1996).
24. H. Arodź, in preparation.

Three-Dimensional Topological Quantum Field Theory of Witten Type[1]

Małgorzata Bakalarska* and Bogusław Broda†

Department of Theoretical Physics
University of Łódź, Pomorska 149/153
PL-90-236 Łódź, Poland
*gosiabak@krysia.uni.lodz.pl
†bobroda@krysia.uni.lodz.pl

Abstract. Description of two three-dimensional topological quantum field theories of Witten type as twisted supersymmetric theories is presented. Low-energy effective action and a corresponding topological invariant of three-dimensional manifolds are considered.

INTRODUCTION

The aim of our work is to present a step made in the direction of understanding of three-dimensional (3d) topological quantum field theory (TQFT) in the spirit Witten and Seiberg have done in the four-dimensional case. Actually, we will consider two "microscopic" 3d TQFT's and sketch their common low-energy consequences for topology of 3d manifolds.

First of all, let us recall the milestones in the development of four-dimensional TQFT. In 1988, Witten proposed a description of Donaldson's theory ("topological" invariants of four-dimensional manifolds) in terms of an appropriately twisted $\mathcal{N} = 2$ SUSY SU(2) pure gauge theory [1]. But not so many mathematical consequences had followed from this approach until 1994 when Seiberg-Witten (SW) theory entered the scene. It has appeared that a newly-discovered dual description of the $\mathcal{N} = 2$ theory in low-energy limit [2] provides us a new, alternative (but essentially equivalent and simpler) formulation of "topological" invariants of four-dimensional manifolds [3].

In the first of our previous papers [4], a physical scenario has been proposed to reach our present goal using a geometric-topological construction with scalar curvature distribution "compatible" with surgery. Though the scalar curvature distribution used "agrees" with the surgery procedure, nevertheless it is unclear why

---

[1] Talk delivered by B. B.

such a distribution should be privileged. In the second paper [5], we have proposed a simpler and more natural mechanism without any reference to curvature. In fact, we have directly applied known results concerning low-energy limit of 3d $\mathcal{N} = 4$ SUSY gauge theory [6].

TOPOLOGICAL FIELD THEORY

In 3d, we have the two important topological quantum field theories (of "cohomological" type): topological SU(2) gauge theory of flat connection and 3d version of the (topological) SW theory. The former is a 3d twisted $\mathcal{N} = 4$ SUSY SU(2) pure gauge theory or a 3d version of the Donaldson-Witten (DW) theory, and "by definition" it describes the Casson invariant that appropriately counts the number of flat SU(2) connections [7] (see, Section A). The latter (3d SW) is a 3d twisted version of $\mathcal{N} = 4$ SUSY U(1) gauge theory with a matter hypermultiplet [2], [3] (see, Section B). It is interesting to note that the both theories can be derived from 4d $\mathcal{N} = 2$ SUSY SU(2) pure gauge theory corresponding via twist to DW theory.

Donaldson-Witten and Casson theory

Let us consider first the usual $\mathcal{N} = 2$ supersymmetric SU(2) Yang-Mills theory in flat Euclidean space R^4 [8]. The Yang-Mills gauge field A_μ is embedded in the $\mathcal{N} = 2$ chiral multiplet $\mathcal{A}$ consisting of one $\mathcal{N} = 1$ chiral multiplet $\Phi = (B, \psi_\alpha)$ and one $\mathcal{N} = 1$ vector multiplet $W_\alpha = (\lambda_\alpha, A_\mu)$.[2] The $\mathcal{N} = 2$ chiral multiplet can be arranged in a diamond form,

$$A^a_\mu$$
$$\lambda^a_\alpha \quad \psi^a_\alpha,$$
$$B^a$$

to exhibit the $SU(2)_I$ symmetry which acts on the rows. This theory is described by the action

$$S = \int_{R^4} d^4 x \, \mathrm{Tr} \left(\frac{1}{4} F_{\mu\nu} F_{\mu\nu} + D_\mu \bar{B} D_\mu B + \frac{1}{2} [B, \bar{B}]^2 - \bar{\lambda}_{\dot{\alpha} i} \bar{\sigma}^{\dot{\alpha}\alpha}_\mu D_\mu \lambda_\alpha{}^i + \right.$$
$$\left. - \frac{i}{\sqrt{2}} \bar{B} \varepsilon_{ij} [\lambda^i, \lambda^j] + \frac{i}{\sqrt{2}} B \varepsilon^{ij} [\bar{\lambda}_i, \bar{\lambda}_j] \right), \qquad (1)$$

---

[2] We will use the conventions of Wess and Bagger [8]. For instance, doublets of the $SU(2)_L$ (or $SU(2)_R$) rotation symmetries are represented by spinor indices $\alpha, \beta, \ldots = 1, 2$ (or $\dot{\alpha}, \dot{\beta}, \ldots = 1, 2$). Doublets of the internal $SU(2)_I$ symmetry will be denoted by indices $i, j, \ldots = 1, 2$. These indices are raised and lowered with the antisymmetric tensor $\varepsilon_{\alpha\beta}$ (or $\varepsilon_{\dot{\alpha}\dot{\beta}}, \varepsilon_{ij}$) with sign convention such that $\varepsilon_{12} = 1 = \varepsilon^{21}$. Tangent vector indices are denoted as $\mu, \nu, \ldots = 1, \ldots, 4$. The spinor and tangent vector indices are related with the tensors $\sigma_{\mu\alpha\dot{\beta}} = (-1, -i\tau^a)$, $\bar{\sigma}^{\dot{\alpha}\beta}_\mu = (-1, i\tau^a)$ described in the Appendices A and B of [8] by letting $\eta_{\mu\nu} \to -\delta_{\mu\nu}$ and $\varepsilon^{\mu\nu\rho\sigma} \to -i\varepsilon_{\mu\nu\rho\sigma}$ with $\varepsilon_{1234} = 1$. Covariant derivatives are defined by $D_\mu B = \partial_\mu B - [A_\mu, B]$, and the Yang-Mills field strength is $F_{\mu\nu} = \partial_\mu A_\nu - \partial_\nu A_\mu - [A_\mu, A_\nu]$.

where $\lambda_\alpha{}^i = (\lambda_\alpha, \psi_\alpha)$.

Now we will construct twisted TQFT. The rotation group K in four-dimensional Euclidean space is locally $SU(2)_L \times SU(2)_R$. In addition, the connected component of the global symmetry group of the $\mathcal{N} = 2$ theory is $SU(2)_I$. The theory, when formulated on a flat R^4, therefore has a global symmetry group [9]

$$H = SU(2)_L \times SU(2)_R \times SU(2)_I.$$

Let $SU(2)_{R'}$ be a diagonal subgroup of $SU(2)_R \times SU(2)_I$ obtained by sending $SU(2)_I$ index "i" to dotted index "$\dot{\alpha}$", and let

$$K' = SU(2)_L \times SU(2)_{R'}.$$

Then, the transformations of the fields under $SU(2)_L \times SU(2)_{R'}$ are

$$A_{\alpha\dot\alpha} : \left(\frac{1}{2}, \frac{1}{2}, 0\right) \longrightarrow A_{\alpha\dot\alpha} : \left(\frac{1}{2}, \frac{1}{2}\right),$$

$$\lambda_{\alpha i} : \left(\frac{1}{2}, 0, \frac{1}{2}\right) \longrightarrow \lambda_{\alpha\dot\beta} : \left(\frac{1}{2}, \frac{1}{2}\right),$$

$$\bar\lambda_{\dot\alpha i} : \left(0, \frac{1}{2}, \frac{1}{2}\right) \longrightarrow \bar\lambda_{\dot\alpha\dot\beta} : (0,0) \oplus (0,1),$$

$$B : (0,0,0) \longrightarrow B : (0,0).$$

And, we decompose the gaugino doublet $\lambda_\alpha{}^i$ into K' irreducible representations as [10]:

$$\lambda_{\alpha\dot\beta} = \frac{1}{\sqrt{2}} \sigma_{\mu\alpha\dot\beta} \psi_\mu,$$

$$\bar\lambda_{\dot\alpha\dot\beta} = \frac{1}{\sqrt{2}} \left(\bar\sigma_{\mu\nu\dot\alpha\dot\beta} \chi_{\mu\nu} + \varepsilon_{\dot\alpha\dot\beta} \eta \right). \tag{2}$$

Substituting (2) into the Lagrangian (1), we have

$$S = \int_{R^4} d^4x \left(\mathcal{L}_1 + \mathcal{L}_2 + \mathcal{L}_3 \right), \tag{3}$$

where

$$\mathcal{L}_1 = \text{Tr} \left(\frac{1}{4} F_{\mu\nu} F_{\mu\nu} - \chi_{\mu\nu} (D_\mu \psi_\nu - D_\nu \psi_\mu) + \frac{i}{\sqrt{2}} B \{\chi_{\mu\nu}, \chi_{\mu\nu}\} \right),$$

$$\mathcal{L}_2 = \text{Tr} \left(-\eta D_\mu \psi_\mu + D_\mu \bar{B} D_\mu B - \frac{i}{\sqrt{2}} \bar{B} \{\psi_\mu, \psi_\mu\} \right),$$

$$\mathcal{L}_3 = \text{Tr} \left(\frac{1}{2} [B, \bar{B}]^2 + \frac{i}{\sqrt{2}} B \{\eta, \eta\} \right).$$

Thus, we have obtained a TQFT — Donaldson-Witten theory.

Now we simply assume all the fields to be independent of the fourth coordinate and discard all mention of the fourth coordinate from the Lagrangian. The dimensionally reduced version of (3) is

$$S = \int_{R^3} d^3x \, \text{Tr} \left\{ \frac{1}{4} F_{mn} F_{mn} + \frac{1}{2} D_m \varphi D_m \varphi + 2\chi_m [\varphi, \psi_m] - 2\omega D_m \chi_m + \right.$$
$$- 2\varepsilon_{mnk} \chi_k D_m \psi_n + \frac{4i}{\sqrt{2}} B \{\chi_m, \chi_m\} + \eta[\phi, \omega] - \eta D_m \psi_m + \quad (4)$$
$$+ D_m \bar{B} D_m B + [\varphi, \bar{B}][\varphi, B] - \frac{i}{\sqrt{2}} \bar{B} \{\psi_m, \psi_m\} - \frac{i}{\sqrt{2}} \bar{B} \{\omega, \omega\} +$$
$$\left. + \frac{1}{2} [B, \bar{B}]^2 + \frac{i}{\sqrt{2}} B \{\eta, \eta\} \right\}.$$

We can obtain a supersymmetric action in three dimensions from (1) by the process of dimensional reduction,

$$S = \int_{R^3} d^3x \, \text{Tr} \left\{ \frac{1}{4} F_{mn} F_{mn} + \frac{1}{2} D_m \varphi D_m \varphi + D_m \bar{B} D_m B + \right.$$
$$+ [\varphi, \bar{B}][\varphi, B] + \frac{1}{2} [B, \bar{B}]^2 - i \bar{\lambda}_{\alpha\beta} \sigma_m^{\alpha\gamma} D_m \lambda_\gamma{}^\beta + \quad (5)$$
$$\left. + \bar{\lambda}_{\alpha\beta} [\varphi, \lambda^{\beta\alpha}] - \frac{i}{\sqrt{2}} \bar{B} \varepsilon_{ij} [\lambda^i, \lambda^j] + \frac{i}{\sqrt{2}} B \varepsilon^{ij} [\bar{\lambda}_i, \bar{\lambda}_j] \right\}.$$

And then substituting the above-mentioned twist

$$\lambda_{\alpha\beta} = \frac{1}{\sqrt{2}} (\sigma_{4\alpha\beta} \omega - i \sigma_{m\alpha\beta} \psi_m),$$
$$\bar{\lambda}_{\alpha\beta} = \frac{1}{\sqrt{2}} (2i\sigma_{k\alpha\beta} \chi_k + \varepsilon_{\alpha\beta} \eta), \quad (6)$$

into the action (5) we get the same action (4).

Dropping fermions and topologically trivial bosons in equation (5) we obtain

$$S = \int_{R^3} d^3x \, \text{Tr} \left(\frac{1}{4} F_{mn} F_{mn} \right).$$

It is a part of the bosonic action, with the absolute minima equation,

$$F_{mn}^a = 0,$$

corresponding to the Casson invariant of 3d manifolds [7].

Seiberg-Witten theory

We start again from the $\mathcal{N} = 2$ SUSY SU(2) Yang-Mills theory. We recall that the moduli space of the $\mathcal{N} = 2$ SUSY SU(2) Yang-Mills theory [2] contains two singular points. At these points the low energy effective theory is $\mathcal{N} = 2$ SUSY U(1) theory coupled to an additional massless matter (monopoles or dyons) in the form of the $\mathcal{N} = 2$ hypermultiplet (sometimes called the scalar multiplet):[3]

$$\begin{array}{c} \psi_A \\ A \quad \tilde{A}^\dagger. \\ \psi_{\tilde{A}}^\dagger \end{array}$$

The action of an $\mathcal{N} = 2$ supersymmetric abelian gauge theory coupled to a massless hypermultiplet is given by

$$S = \int_{R^4} d^4x \left\{ \frac{1}{4} F_{\mu\nu} F_{\mu\nu} + \partial_\mu \bar{B} \partial_\mu B - \bar{\lambda}_{\dot{\alpha} i} \bar{\sigma}_\mu^{\dot{\alpha}\alpha} \partial_\mu \lambda_\alpha{}^i + D_\mu \bar{A}^i D_\mu A_i + \right.$$
$$+ \frac{1}{2} (\bar{A}^i A_i)^2 + i\bar{\psi}\gamma^\mu D_\mu \psi + i\sqrt{2} \bar{A}^i \bar{\lambda}_i \psi - i\sqrt{2} \bar{\psi} \lambda^i A_i - \bar{\psi}(B - \gamma_5 \bar{B})\psi + \quad (7)$$
$$\left. - \bar{A}^i (B^2 + \bar{B}^2) A_i \right\}.$$

We know that the twist consists of considering as the rotation group the group, $K' = \mathrm{SU}(2)_L \times \mathrm{SU}(2)_{R'}$ and this implies that the hypermultiplet field content is modified as follows:

$$A^i : \left(0, 0, \frac{1}{2}\right) \longrightarrow M^{\dot{\alpha}} : \left(\frac{1}{2}, 0\right),$$
$$\psi_{A\alpha} : \left(\frac{1}{2}, 0, 0\right) \longrightarrow u_\alpha : \left(\frac{1}{2}, 0\right),$$
$$\bar{\psi}_{\tilde{A}\dot{\alpha}} : \left(0, \frac{1}{2}, 0\right) \longrightarrow \bar{v}_{\dot{\alpha}} : \left(0, \frac{1}{2}\right),$$
$$A_i^\dagger : \left(0, 0, \frac{1}{2}\right) \longrightarrow \bar{M}_{\dot{\alpha}} : \left(\frac{1}{2}, 0\right),$$
$$\bar{\psi}_{A\dot{\alpha}} : \left(0, \frac{1}{2}, 0\right) \longrightarrow \bar{u}_{\dot{\alpha}} : \left(0, \frac{1}{2}\right),$$
$$\psi_{\tilde{A}\alpha} : \left(\frac{1}{2}, 0, 0\right) \longrightarrow v_\alpha : \left(\frac{1}{2}, 0\right).$$

Substituting equation (2) into the action (7) and taking into account that field content we get the following twisted euclidean action (compare to [11])

$$S = \int_{R^4} d^4x \left\{ \frac{1}{4} F_{\mu\nu} F_{\mu\nu} + \partial_\mu \bar{B} \partial_\mu B - 2\chi_{\mu\nu} \partial_\mu \psi_\nu - \eta \partial_\mu \psi_\mu + D_\mu \bar{M}^{\dot{\alpha}} D_\mu M_{\dot{\alpha}} + \right.$$
$$+ \frac{1}{2} (\bar{M}^{\dot{\alpha}} M_{\dot{\alpha}})^2 - i \left(\bar{M}^{\dot{\beta}} \psi_{\dot{\alpha}\dot{\beta}} u^\alpha - v^\alpha \psi_{\alpha\dot{\beta}} M^{\dot{\beta}} \right) + i\eta \left(\bar{u}^{\dot{\alpha}} M_{\dot{\alpha}} - \bar{M}^{\dot{\beta}} \bar{v}_{\dot{\beta}} \right) + \quad (8)$$
$$+ i\chi_{\dot{\alpha}\dot{\beta}} (\bar{M}^{\dot{\alpha}} \bar{v}^{\dot{\beta}} - \bar{u}^{\dot{\alpha}} M^{\dot{\beta}}) - B^2 \bar{M}^{\dot{\alpha}} M_{\dot{\alpha}} - \bar{B}^2 \bar{M}^{\dot{\alpha}} M_{\dot{\alpha}} + v^\alpha \bar{B} u_\alpha +$$
$$\left. - B v^\alpha u_\alpha - B \bar{u}_{\dot{\alpha}} \bar{v}^{\dot{\alpha}} - \bar{B} \bar{u}_{\dot{\alpha}} \bar{v}^{\dot{\alpha}} + i v^\alpha D_{\alpha\dot{\alpha}} \bar{v}^{\dot{\alpha}} + i \bar{u}^{\dot{\alpha}} D_{\alpha\dot{\alpha}} u^\alpha \right\}.$$

---

[3] Here the low energy fields are: the vector multiplet (gauge field A_μ, SU(2)$_I$ doublet of fermions λ_α and ψ_α, a complex scalar B) and the hypermultiplet (consisting of two Weyl fermions ψ_A and $\psi_{\tilde{A}}^\dagger$ and complex bosons A and $\tilde{A}^\dagger$).

The dimensionally reduced version of (8) is

$$S = \int_{R^3} d^3x \left\{ \frac{1}{4} F_{mn} F_{mn} + \frac{1}{2} \partial_m \varphi \partial_m \varphi + \partial_m \bar{B} \partial_m B - 2\varepsilon_{mnk} \chi_k \partial_m \psi_n + \right.$$
$$- 2\omega \partial_m \chi_m - \eta \partial_m \psi_m + D_m \bar{M}^\alpha D_m M_\alpha + [\varphi, \bar{M}^\alpha][\varphi, M_\alpha] +$$
$$+ \frac{1}{2} (\bar{M}^\alpha M_\alpha)^2 + i\omega \left(\bar{M}^\beta u_\beta - v^\alpha M_\alpha \right) + v^\alpha \psi_{\alpha\beta} M^\beta - \bar{M}^\beta \psi_{\alpha\beta} u^\alpha + \quad (9)$$
$$+ i\eta \left(\bar{u}^\alpha M_\alpha - \bar{M}^\beta \bar{v}_\beta \right) + 2\chi_{\alpha\beta} \left(\bar{u}^\alpha M^\beta - \bar{M}^\alpha \bar{v}^\beta \right) - B^2 \bar{M}^\alpha M_\alpha +$$
$$- \bar{B}^2 \bar{M}^\alpha M_\alpha + \bar{B} v^\alpha u_\alpha - B v^\alpha u_\alpha - B \bar{u}_\alpha \bar{v}^\alpha - \bar{B} \bar{u}_\alpha \bar{v}^\alpha + v^\alpha D_{\alpha\beta} \bar{v}^\beta +$$
$$+ \bar{u}^\beta D_{\alpha\beta} u^\alpha + iv^\alpha [\varphi, \bar{v}_\alpha] + i\bar{u}^\alpha [\varphi, u_\alpha] \right\}.$$

As a result of the dimensional reduction of (7) we get

$$S = \int_{R^3} d^3x \left\{ \frac{1}{4} F_{mn} F_{mn} + \frac{1}{2} \partial_m \varphi \partial_m \varphi + \partial_m \bar{B} \partial_m B - i\bar{\lambda}_{\alpha i} \sigma_m^{\alpha\beta} \partial_m \lambda_\beta{}^i + \right.$$
$$+ D_m \bar{A}^i D_m A_i + [\varphi, \bar{A}^i][\varphi, A_i] + \frac{1}{2} (\bar{A}^i A_i)^2 - \bar{A}^i (B^2 + \bar{B}^2) A_i +$$
$$+ i\sqrt{2} \bar{A}^i \bar{\lambda}_{\alpha i} \bar{\psi}_{\bar{A}}^\alpha - i\sqrt{2} \bar{\psi}_{A\alpha} \bar{\lambda}^{\alpha i} A_i - i\sqrt{2} \bar{A}^i \lambda_{\alpha i} \psi_A^\alpha + i\sqrt{2} \psi_{\bar{A}}^\alpha \lambda_{\alpha i} A^i + \quad (10)$$
$$+ \bar{B} \psi_{\bar{A}}^\alpha \psi_{A\alpha} - B \psi_{\bar{A}}^\alpha \psi_{A\alpha} - B \bar{\psi}_{A\alpha} \bar{\psi}_{\bar{A}}^\alpha - \bar{B} \bar{\psi}_{A\alpha} \bar{\psi}_{\bar{a}}^\alpha + \psi_A^\alpha D_{\alpha\beta} \bar{\psi}_{\bar{A}}^\beta +$$
$$+ \bar{\psi}_A^\beta D_{\alpha\beta} \psi_A^\alpha + i\psi_A^\alpha [\varphi, \bar{\psi}_{\bar{A}\alpha}] + i\bar{\psi}_{\bar{A}}^\alpha [\varphi, \psi_{A\alpha}] \right\}.$$

Using the twist

$$\lambda_{\alpha\beta} = \frac{1}{\sqrt{2}} \left(\sigma_{4\alpha\beta} \omega - i\sigma_{m\alpha\beta} \psi_m \right),$$
$$\bar{\lambda}_{\alpha\beta} = \frac{1}{\sqrt{2}} \left(2i\sigma_{k\alpha\beta} \chi_k + \varepsilon_{\alpha\beta} \eta \right),$$
$$A^i \to M^\alpha,$$
$$\psi_{A\alpha} \to u_\alpha,$$
$$\bar{\psi}_{\bar{A}\alpha} \to \bar{v}_\alpha,$$
$$\bar{A}_i \to \bar{M}_\alpha,$$
$$\bar{\psi}_{A\alpha} \to \bar{u}_\alpha,$$
$$\psi_{\bar{A}\alpha} \to v_\alpha,$$

we get the action (9).

Dropping fermions and topologically trivial bosons we obtain

$$S = \int_{R^3} d^3x \left(\frac{1}{4} F_{mn} F_{mn} + \frac{1}{2} (\bar{A}^i A_i)^2 + D_m \bar{A}^i D_m A_i \right). \quad (11)$$

It is a part of the bosonic matter action.

LOW-ENERGY CONSEQUENCES

Interestingly, it follows from [6] that low-energy effective theory for both 3d $\mathcal{N}=4$ SUSY $SU(2)$ case as well as for the 3d $\mathcal{N}=4$ SUSY abelian one with matter hypermultiplet is pure gauge abelian, and the moduli spaces are smooth. More precisely, we have the so-called Atiyah-Hitchin manifold (interpreted as the two-monopole moduli space), a complete hyper-Kähler manifold, in the first case, and the Taub-NUT manifold, in the second one.

3d $\mathcal{N}=4$ supersymmetric abelian gauge theory is described by the following action[4]

$$S = \int_{R^3} d^3x \left(\frac{1}{4} F_{mn} F_{mn} + \frac{1}{2} \partial_m \varphi \partial_m \varphi + \partial_m \bar{B} \partial_m B + \right.$$
$$\left. - i \bar{\lambda}_{\alpha i} \sigma_m^{\alpha\beta} \partial_m \lambda_\beta{}^i - \frac{1}{2} (\vec{G})^2 \right). \tag{12}$$

Introducing infinitesimal parameters $\xi_\alpha{}^i$ and $\bar{\xi}_{\alpha i}$, the supersymmetry transformations are given by

$$\delta A_m = i \bar{\lambda}_{\alpha i} \sigma_m^{\alpha\beta} \xi_\beta{}^i - i \bar{\xi}_{\alpha i} \sigma_m^{\alpha\beta} \lambda_\beta{}^i,$$
$$\delta \varphi = \bar{\xi}_i^\alpha \lambda_\alpha{}^i - \bar{\lambda}_{\alpha i} \xi^{\alpha i},$$
$$\delta B = -\sqrt{2} \varepsilon_{ij} \xi^{\alpha i} \lambda_\alpha{}^j,$$
$$\delta \bar{B} = -\sqrt{2} \varepsilon^{ij} \bar{\xi}_i^\alpha \bar{\lambda}_{\alpha j}, \tag{13}$$
$$\delta \lambda_\alpha{}^i = -\frac{i}{2} \varepsilon_{mnk} \sigma_{k\alpha}{}^\beta \xi_\beta{}^i F_{mn} - i\sqrt{2} \varepsilon^{ij} \sigma_{m\alpha\beta} \bar{\xi}_j^\beta \partial_m B +$$
$$- i \sigma_{m\alpha}{}^\beta \xi_\beta{}^i \partial_m \varphi,$$
$$\delta \bar{\lambda}_{\alpha i} = i\sqrt{2} \varepsilon_{ij} \sigma_{m\alpha\beta} \xi^{\beta j} \partial_m \bar{B} + \frac{i}{2} \varepsilon_{mnk} \sigma_{k\alpha}{}^\beta \bar{\xi}_{\beta i} F_{mn} +$$
$$- i \sigma_{m\alpha}{}^\beta \bar{\xi}_{\beta i} \partial_m \varphi + 2 i \sigma_{m\alpha}{}^\beta \bar{\xi}_{\beta i} G_m,$$
$$\delta \vec{G} = i \bar{\xi}_{\alpha i} \vec{\tau}_j^i \sigma_m^{\alpha\beta} \partial_m \lambda_\beta{}^j.$$

Using the equations of motion for all the fields one can easily show that this action is invariant under the supersymmetric transformations (13). Taking into account the following substitution

$$\lambda_{\alpha\beta} = \frac{1}{\sqrt{2}} \left(\sigma_{4\alpha\beta} \omega - i \sigma_{m\alpha\beta} \psi_m \right),$$
$$\bar{\lambda}_{\alpha\beta} = \frac{1}{\sqrt{2}} \left(2 i \sigma_{k\alpha\beta} \chi_k + \varepsilon_{\alpha\beta} \eta \right),$$

and knowing the auxiliary field $\vec{G}$ transforms in the $(1,0)$ representation (and hence becomes a vector H_k), we get the following 3d twisted version of the action (12)

---

[4] $\vec{G}$ is an auxiliary field.

$$S = \int_{R^3} d^3x \left(\frac{1}{4} F_{mn} F_{mn} + \frac{1}{2} \partial_m \varphi \partial_m \varphi + \partial_m \bar{B} \partial_m B + 2\varepsilon_{mnk} \psi_n \partial_m \chi_k + \right.$$
$$\left. - 2\omega \partial_m \chi_m - \eta \partial_m \psi_m - 2 H_k H_k \right). \tag{14}$$

The topological BRST transformation δ_B induced by the $\mathcal{N} = 4$ supertransformations is found by putting $\xi = 0$, $\bar{\xi} = \frac{\varepsilon_{\alpha\beta}\cdot\rho}{\sqrt{2}}$, which reads $\delta = -\rho \cdot \delta_B$.[5] The BRST transformations are

$$\begin{aligned}
\delta_B A_m &= \psi_m, \\
\delta_B \varphi &= -\omega, \\
\delta_B B &= \delta_B \omega = \delta_B \eta = 0, \\
\delta_B \bar{B} &= -\sqrt{2}\eta, \\
\delta_B \psi_m &= -\sqrt{2} \partial_m B, \\
\delta_B \chi_k &= -\frac{1}{2} \partial_k \varphi - \frac{1}{2} \varepsilon_{kmn} \partial_m A_n - H_k, \\
\delta_B H_k &= \frac{1}{2} \partial_k \omega - \frac{1}{2} \varepsilon_{kmn} \partial_m \psi_n.
\end{aligned} \tag{15}$$

These BRST transformations are off-shell nilpotent up to a gauge transformation $\delta_B^2 = -\sqrt{2}\delta_{\text{gauge}}$; e.g., $\delta_B^2 A_m = -\sqrt{2}\partial_m B$. The action (14) is also nilpotent but on-shell.

Confining ourselves to the contribution of the Coulomb branch coming from abelian flat connections on the 3d manifold $\mathcal{M}^3$ we should consider a mathematical object akin to the Casson invariant. We have argued in [4] that such an object should count (algebraically) the number of abelian flat connection on a cover of $\mathcal{M}^3$. If $\mathcal{M}^3$ arises from $\mathcal{S}^3$ via 0-framed surgery on a knot [12] the 3d invariant is directly related to the Alexander invariant of $\mathcal{M}^3$ [13].

CONCLUSIONS

In this paper, we have indicated that both 3d TQFT of SU(2) flat connection as well as 3d version of topological SW theory can be described in low-energy regime by the Alexander invariant. In a future work, it will be necessary to improve the analysis to include all contribution coming from the Coulomb branch.

ACKNOWLEDGMENTS

The work has been supported by KBN grant 2 P03B 084 15 and by University of Łódź grant 621.

---

[5] ρ is an anticommuting parameter and the lower index B denotes BRST transformation.

REFERENCES

1. Witten E., *Commun. Math. Phys.* **117**, 353 (1988).
2. Seiberg N., and Witten E., *Nucl. Phys.* B **426**, 19 (1994).
3. Witten E., *Math. Res. Lett.* **1**, 769 (1994).
4. Broda B., in *New Developments in Quantum Field Theory*, NATO ASI Series, 1997, ed. P. H. Damgaard, Plenum Press, pp. 261-268.
5. Broda B., and Bakalarska M., Acta Phys. Polon. B **4**, 995 (1998).
6. Seiberg N., and Witten E., "Gauge Dynamics And Compactification To Three Dimensions", *E-print* hep-th/9607163.
7. Blau M., and Thompson G., *Commun. Math. Phys.* **152**, 41 (1993).
8. Wess J., and Bagger J., *Supersymmetry and Supergravity*, Princeton University Press, 1982, pp. 171-179.
9. Witten E., *J. Math. Phys.* **35**, 10 (1994).
10. Yoshida Y., "On Instanton Calculations of $\mathcal{N} = 2$ Supersymmetric Yang-Mills Theory", *E-print* hep-th/9610211.
11. Labastida J. M. F., and Lozano C., "Lectures on Topological Quantum Field Theory", E-print hep-th/9709192.
12. Rolfsen D., *Knots and Links*, Publish or Perish, Inc., Wilmington, 1976.
13. Meng G., and Taubes C., *Math. Res. Lett.* **3**, 661 (1996).
    Hutchings M., and Lee Y., "Circle-valued Morse theory, Reidemeister torsion, and Seiberg-Witten invariants of 3-manifolds", *E-print* math.DG/9612004;

On Notivarg Propagator

M. Bakalarska, W. Tybor[1]

Department of Theoretical Physics
University of Łódź
ul. Pomorska 149/153, 90-236 Łódź, Poland

Abstract. The covariant propagator of the notivarg is found. It has the Feynmann-like form.

1. The essential difficulty, we are faced with on calculating the gauge field propagator, is that the Lagrangian operator of the gauge theory is singular. The origin of the singularity is the gauge freedom of the theory. To remove it we impose the suitable gauge conditions. That allow us to obtain an effective action leading to the regular Lagrangian operator. The propagator is defined as the inverse operator to the Lagrangian one. The method works very well, for example, in the electrodynamics. It has been verified in the theory of the notoph [1,2]. However, the method is rather troublesome if the gauge field is a many component tensor. Such a situation is in the case of the notivarg that is described by the field with symmetry properties of a Riemann tensor.

We can follow another way to obtain the form of propagators. We consider the interaction of the gauge field with an external current that obeys some conservation law ensuring the gauge invariance for the interacting term. In the fixed Lorentz frame we perform the canonical analysis of the theory. Solving constraints we obtain the physical Hamiltonian. Its free part describes only the physical degrees of freedom. The interaction part is a sum of two terms. The first one describes the interaction of the physical components of the gauge field with the current components that are not restricted by the current conservation law. The second one describes the instant interactions, bilinear in the currents (for example, the Coulomb term $\rho \frac{1}{\Delta} \rho$ in the electrodynamics). Using the standard methods of the S - matrix formalism, with the help of this Hamiltonian, we can calculate the amplitude of the current - current interaction, square in the coupling constant. On the other hand, the amplitude can be calculated using the covariant propagator. Taking into account the Lorentz properties of the field and the current and the law of the current conservation, we can predict the general form of the covariant

---

[1] Supported by Łódź University Grant No. 505/581

propagator. The result of calculations must be independent of the calculation method. So, we can verify correctness of our prediction of the form of the covariant propagator.

2. The notivarg field $K^{\mu\nu\alpha\beta}$ interacting with the external Weyl current $j^{\mu\nu\alpha\beta}$ is described by the Lagrangian density [3]

$$\mathcal{L} = -(\partial_\sigma K^{\sigma\nu\alpha\beta})^2 + (\partial_\sigma K^{\sigma\nu\alpha}{}_\nu)^2 + \frac{1}{4} j^{\mu\nu\alpha\beta} K_{\mu\nu\alpha\beta}, \qquad (1)$$

where $K^{\mu\nu\alpha\beta}$ has the symmetry properties of a Riemann tensor $K^{\mu\nu\alpha\beta} = K^{\alpha\beta\mu\nu} = -K^{\nu\mu\alpha\beta}$, $\varepsilon_{\mu\nu\alpha\beta} K^{\mu\nu\alpha\beta} = 0$. The Weyl current $j^{\mu\nu\alpha\beta}$ is the Riemann tensor obeying

$$j^{\mu\nu\alpha}{}_\nu = 0. \qquad (2)$$

The conservation law for the Weyl current is

$$\partial_\mu \partial_\alpha j^{\mu\nu\alpha\beta} = 0. \qquad (3)$$

The action determined by the Lagrangian (1) is invariant under the gauge transformations

$$\begin{aligned}\delta K^{\mu\nu\alpha\beta} &= \varepsilon^{\mu\nu\sigma\lambda} \varepsilon^{\alpha\beta\varphi\kappa} \partial_\sigma \partial_\varphi \omega_{\lambda\kappa} + g^{\mu\alpha}(\partial^\nu \eta^\beta + \partial^\beta \eta^\nu) + \\ &+ g^{\nu\beta}(\partial^\mu \eta^\alpha + \partial^\alpha \eta^\mu) - g^{\mu\beta}(\partial^\nu \eta^\alpha + \partial^\alpha \eta^\nu) + \\ &- g^{\nu\alpha}(\partial^\mu \eta^\beta + \partial^\beta \eta^\mu) - 2(g^{\mu\alpha}g^{\nu\beta} - g^{\mu\beta}g^{\nu\alpha})\partial_\sigma \eta^\sigma,\end{aligned} \qquad (4)$$

where $\omega_{\alpha\beta} = \omega_{\beta\alpha}$ and η_α are gauge tensors.

In the covariant gauge

$$K^{\mu\nu\alpha}{}_\nu = 0, \qquad \partial_\mu \partial_\alpha K^{\mu\nu\alpha\beta} = 0 \qquad (5)$$

the field equation has the form [2]

$$\Box K^{\mu\nu\alpha\beta} = -\frac{1}{2} j^{\mu\nu\alpha\beta}. \qquad (6)$$

3. Let us consider the exchange of the notivarg between two external currents. The general structure of the amplitude describing the process in the second order of the perturbation theory is

$$A = (-i)^2 \left(\frac{a}{k^2} j^{\mu\nu\alpha\beta}(-k) j_{\mu\nu\alpha\beta}(k) + \frac{b}{k^4} k_\mu j^{\mu\nu\alpha\beta}(-k) k^\sigma j_{\sigma\nu\alpha\beta}(k) + \right.$$
$$\left. + \frac{c}{k^6} k_\mu k_\alpha j^{\mu\nu\alpha\beta}(-k) k^\sigma k^\kappa j_{\sigma\nu\kappa\beta}(k) \right),$$

---

[2] In Ref. [3] the right hand side of Eq. (7) reads $-\frac{1}{2} j^{\mu\nu\alpha\beta}$.

where a, b, c are number factors. The last term vanishes due to the conservation law (3). The second one has, in fact, the structure of the first one due to the following identity for the Weyl tensor

$$j^{\mu\nu\alpha\beta}j_{\sigma\nu\alpha\beta} = \frac{1}{4}\delta^\mu_\sigma j^{\kappa\nu\alpha\beta}j_{\kappa\nu\alpha\beta}.$$

Assuming the following form of the notivarg propagator

$$D_{\mu\nu\alpha\beta,\sigma\lambda\gamma\delta}(k) = -\frac{1}{k^2}\frac{1}{8}(g_{\mu\sigma}g_{\nu\lambda}g_{\alpha\gamma}g_{\beta\delta} + g_{\mu\lambda}g_{\nu\sigma}g_{\alpha\delta}g_{\beta\gamma} +$$
$$+ g_{\mu\gamma}g_{\nu\delta}g_{\alpha\sigma}g_{\beta\lambda} + g_{\mu\delta}g_{\nu\gamma}g_{\alpha\lambda}g_{\beta\sigma} - g_{\mu\lambda}g_{\nu\sigma}g_{\alpha\gamma}g_{\beta\delta} +$$
$$- g_{\mu\sigma}g_{\nu\lambda}g_{\alpha\delta}g_{\beta\gamma} - g_{\mu\gamma}g_{\nu\delta}g_{\alpha\lambda}g_{\beta\sigma} - g_{\mu\delta}g_{\nu\gamma}g_{\alpha\sigma}g_{\beta\lambda}) \qquad (7)$$

we obtain the amplitude

$$A = (-i)^2(-\frac{1}{8})j^{\mu\nu\alpha\beta}(-k)D_{\mu\nu\alpha\beta,\sigma\lambda\gamma\delta}(k)j^{\sigma\lambda\gamma\delta}(k). \qquad (8)$$

The number factor $-\frac{1}{8}$ follows from Eqs. (1) and (6)

$$\frac{1}{4}j^{\mu\nu\alpha\beta}K_{\mu\nu\alpha\beta} \to \frac{1}{4}j^{\mu\nu\alpha\beta}D_{\mu\nu\alpha\beta,\sigma\lambda\gamma\delta}(-\frac{1}{2}j^{\sigma\lambda\gamma\delta}).$$

Using
(i) the current conservation law (3),
(ii) the decomposition of the Weyl current

$$j^{\mu\nu\alpha\beta} = (\tau^{ij}, \sigma^{ij}), \qquad i,j = 1,2,3;$$

where τ^{ij} and σ^{ij} are symmetric and traceless tensors defined by

$$j^{0i0j} = \tau^{ij},$$
$$j^{0ijk} = \varepsilon^{jkp}\sigma^i_p,$$
$$j^{ijkl} = -(g^{ik}\tau^{jl} + g^{jl}\tau^{ik} - g^{il}\tau^{jk} - g^{jk}\tau^{il}).$$

(iii) the helicity decomposition of the symmetric traceless tensor a^{ij}

$$a^{ij} = a^{ij}(0) + a^{ij}(\pm 1) + a^{ij}(\pm 2)$$

where

$$a^{ij}(\pm 1) = -\frac{1}{\Delta}(\partial^i a^j_T + \partial^j a^i_T),$$
$$a^{ij}(0) = \frac{3}{2}(\frac{1}{\Delta}\partial^i\partial^j + \frac{1}{3}g^{ij})a_L,$$
$$a^i \equiv \partial_j a^{ji},$$
$$a_L \equiv \frac{1}{\Delta}\partial_i a^i,$$

we obtain

$$A = \frac{3}{2}\frac{\sigma_L(-k)\sigma_L(k)}{k^2} + 2|\vec{k}|^{-4}\tau_{Ti}(-k)\tau_T^i(k) + \\ + \frac{1}{4}k^2 k_0^{-2}|\vec{k}|^{-2}\tau_{ij}(\pm 2, -k)\tau^{ij}(\pm 2, k). \quad (9)$$

From the canonical analysis [4] we obtain the physical Hamiltonian density describing interaction

$$\mathcal{H}_{int} = \sqrt{\frac{3}{2}}\varphi\sigma_L + \frac{1}{2}\sigma_m^n(\pm 2)\frac{1}{\Delta}\sigma_n^m(\pm 2) + \\ + \frac{1}{8}(\frac{1}{\Delta}\partial^0\tau^{ij}(\pm 2))(\frac{1}{\Delta}\partial^0\tau_{ij}(\pm 2)) - \frac{3}{8}\tau^{ij}(\pm 2)\frac{1}{\Delta}\tau_{ij}(\pm 2) + \quad (10) \\ - (\frac{1}{\Delta}\tau_T^i)(\frac{1}{\Delta}\tau_{Ti}).$$

Two remarks are necessary:
1. The term (Eq. (34) in Ref. [4])

$$-\frac{1}{4}\sigma_m^n(\pm 2)\frac{1}{\Delta}\sigma_n^m(\pm 2)$$

is wrong. The right form is

$$-\frac{1}{2}\sigma_m^n(\pm 2)\frac{1}{\Delta}\sigma_n^m(\pm 2).$$

2. The new field $\varphi = \sqrt{6}S_L$ is introduced because the free physical Lagrangian has then the standard form

$$\mathcal{L}_{free} = \frac{1}{2}(\partial_\mu \varphi)^2.$$

We recall that S_L is a scalar component of the symmetric traceless tensor

$$S^{im} = -\frac{1}{4}(\varepsilon^m{}_{jk}K^{0ijk} + \varepsilon^i{}_{jk}K^{0mjk}).$$

The Hamiltonian density (11) in the momentum space is

$$\mathcal{H}_{int} = \sqrt{\frac{3}{2}}\sigma_L(-k)\varphi(k) + \frac{1}{2}\left[2|\vec{k}|^{-4}\tau_{Ti}(-k)\tau_T^i(k) + \\ + \frac{1}{4}k^2 k_0^{-2}|\vec{k}|^{-2}\tau_{ij}(\pm 2, -k)\tau^{ij}(\pm 2, k)\right]. \quad (11)$$

The current conservation law (3) is included in Eq. (11).
With the help of this Hamiltonian, using standard methods of the S - matrix formalism [5], we can obtain the amplitude of the current - current interaction via one notivarg exchange. We get exactly the amplitude (9). So, the Feynmann - like form (7) of the notivarg propagator is confirmed.

We are grateful to Prof. J. Rembieliński for interesting discussion.

REFERENCES

1. Papaloucas, L. C., Rembieliński, J., Tybor, W., *Acta Phys. Pol.* **B22**, 429 (1991).
2. Tybor, W., *Acta Phys. Pol.* **B18**, 369 (1987).
3. Rembieliński, J., Tybor, W., *Acta Phys. Pol.* **B22**, 439 (1991).
4. Rembieliński, J., Tybor, W., *Acta Phys. Pol.* **B22**, 447 (1991).
5. Bjorken, J. D., Drell, S. D., *Relativistic Quantum Mechanics*, New York: McGraw-Hill, 1964.
6. Bjorken, J. D., Drell, S. D., *Relativistic Quantum Fields*, New York: McGraw-Hill, 1965.

Lorentz Covariance, Higher-Spin Superspaces and Self-Duality [1]

Chandrashekar Devchand[a] and Jean Nuyts[b]

[a] Max-Planck-Institut für Mathematik in den Naturwissenschaften
Inselstraße 22-26, 04103 Leipzig, Germany
[b] Physique Théorique et Mathématique, Université de Mons-Hainaut
20 Place du Parc, 7000 Mons, Belgium
devchand@mis.mpg.de, nuyts@umh.ac.be

Abstract. Lorentz covariant generalisations of the notions of supersymmetry, superspace and self-duality are discussed. The essential idea is to extend standard constructions by allowing tangent vectors and coordinates which transform according to more general Lorentz representations than solely the spinorial and vectorial ones of standard lore. Such superspaces provide model configuration spaces for theories of arbitrary spin fields. Our framework is an elegant one for handling higher-dimensional theories in a manifestly SO(3,1) covariant fashion. A further application is the construction of a hierarchy of solvable Lorentz covariant systems generalising four-dimensional self-duality.

INTRODUCTION

There exist various forms of higher dimensional fundamental building blocks for the observed four dimensional 'real world'. An important question is the recovery of four-dimensional Lorentz invariance, the most important observed symmetry in physics. It often seems desirable to have a framework in which the higher dimensional objects transform covariantly under $SO(3,1)$ and therefore have some intrinsic four-dimensional nature. Recent developments in string physics, moreover, seem to require an extension of the standard notions of supersymmetry and superspace. Over the past couple of years, we have been investigating $SO(4, \mathbb{C})$-covariant superalgebras which provide generalisations of supersymmetry and a manifestly Lorentz-covariant framework for the investigation of spaces of dimensions greater than four.

Standard superspace is a homogeneous space constructed as the quotient of the super Poincaré group by the Lorentz group, with four bosonic coordinates $\{Y^{\alpha\dot\alpha}\}$ and four fermionic coordinates $\{Y^\alpha, Y^{\dot\alpha}\}$, transforming according to the vectorial

---

[1] Presented by Jean Nuyts.

($\frac{1}{2}, \frac{1}{2}$) and spinorial $(0, \frac{1}{2})$ and $(\frac{1}{2}, 0)$ representations of the Lorentz group. Superspace can therefore be considered as a direct sum of 4d Minkowski space with two 2d spinorial spaces. Now, once the restriction to coordinates transforming solely according to the vectorial $(\frac{1}{2}, \frac{1}{2})$ representation has been abandoned, one can ask whether one can build more general superspaces from direct sums of more general sets of representation spaces than the usual spinorial and vectorial ones; namely, spaces with coordinates from the set of general Lorentz tensors $\{Y^{\alpha_1...\alpha_{2s}\dot\alpha_1...\dot\alpha_{2\dot s}}\}$, symmetrical in the α indices and in the $\dot\alpha$ indices, for some collection of representations $\Lambda_p = \{(s, \dot s)\}$, where p denotes the maximum value of $s+\dot s$ occurring in the chosen set of representations.

This is clearly a way of parametrising a higher-dimensional space in a manifestly 4d Lorentz covariant fashion. Usually, in higher-dimensional theories, the recovery of 4d Lorentz covariance is achieved by having the extra coordinates transform as a bunch of Lorentz scalars. The idea here is to start with a set of Lorentz representations and hence fix the 4d Lorentz structure to the coordinate system. For instance a simple bosonic extension of 4d space with coordinates $\{Y^{\alpha\dot\alpha}, Y^{\alpha\beta\gamma\dot\alpha}\}$ has dimension 4+8=12. For the sake of generality, we consider a graded vector space with the coordinates $\{Y(s, \dot s)\}$ spanning a supercommutative $\mathbb{Z}_2$-graded algebra, $\mathcal{V}=\mathcal{V}_0+\mathcal{V}_1$. $\mathcal{V}_0$ (resp. $\mathcal{V}_1$) contains bosonic (resp. fermionic) coordinates if $2(s+\dot s)$ is even (resp. odd). Each representation $Y(s, \dot s)$ included, increases the bosonic (resp. fermionic) dimension of the hyperspace $\mathcal{M}$ by $(2s+1)(2\dot s+1)$.

SPIN-P HEISENBERG SUPERALGEBRAS

To describe such hyperspaces as the homogeneous spaces of some algebra, the super-Poincaré algebra needs to be extended to some algebra including higher-spin generators. Such enhancement of customary supersymmetry, going beyond the Haag-Lopusanski-Sohnius barrier, has been extensively studied by Fradkin and Vasiliev [1]. These authors were motivated by physical considerations to realise such higher-spin algebras on 4d de Sitter space fields. Consistency of the dynamics required the inclusion of *all* spins, yielding infinite dimensional algebras realised on infinite chains of fields having spins all the way up to infinity.

Our approach [2–4] has been more abstract and does *not* require infinite dimensionality of the algebra. We consider some basically finite set of higher-spin generators (in addition to the spin 1 and spin $\frac{1}{2}$ super-Poincaré generators); and we interpret the higher-spin generators $X_{\alpha_1...\alpha_{2s}\dot\alpha_1...\dot\alpha_{2\dot s}}$ as 'momenta' in extra dimensions parametrised by higher-spin coordinates. We do not make any a priori field theoretical or dynamical requirements; and we realise our algebras in flat space. The grading, however, remains a $\mathbb{Z}_2$ one, with all integer-spin representations in an even-statistics (bosonic) subspace $\mathcal{A}_0$ and all half-integer-spin representations in an odd-statistics (fermionic) subspace $\mathcal{A}_1$. Thus the Heisenberg superalgebra associated to standard (complexified) superspace with $\Lambda_1 = \{(\frac{1}{2}, \frac{1}{2}), (\frac{1}{2}, 0), (0, \frac{1}{2})\}$

and non-zero canonical supercommutation relations,

$$\{X_\alpha, X_{\dot\beta}\} = X_{\alpha\dot\beta}, \quad [X_\alpha, Y^{\beta\dot\beta}] = \delta_\alpha^\beta Y^{\dot\beta}, \quad [X_{\dot\alpha}, Y^{\beta\dot\beta}] = \delta_{\dot\alpha}^{\dot\beta} Y^\beta$$

$$[X_{\alpha\dot\alpha}, Y^{\beta\dot\beta}] = \delta_\alpha^\beta \delta_{\dot\alpha}^{\dot\beta}, \quad \{X_\alpha, Y^\beta\} = \delta_\alpha^\beta, \quad \{X_{\dot\alpha}, Y^{\dot\beta}\} = \delta_{\dot\alpha}^{\dot\beta},$$

is generalised to a *spin p Heisenberg superalgebra* $\mathcal{G}=\mathcal{A}+\mathcal{V}$, based on a set of representations $\Lambda_p = \{(s, \dot s)\}$ appearing in $\mathcal{A}+\mathcal{V}$.

For any specific $(s, \dot s) \in \Lambda_p$, we label the $(2s+1)(2\dot s+1)$ components of the co-ordinate tensor $Y(s, \dot s)$ as $Y(s, s_3; \dot s, \dot s_3)$, where s_3 (resp. $\dot s_3$) run from $-s$ to s (resp. from $-\dot s$ to $\dot s$) in integer steps. These are in one-to-one correspondence with the components in standard 2-spinor index notation, $Y^{\alpha_1...\alpha_{2s}\dot\alpha_1...\dot\alpha_{2\dot s}}$. The span of coordinates $\{Y(s, s_3; \dot s, \dot s_3)\}$, for all $(s, \dot s)$ in the chosen set Λ_p, is defined to be a supercommutative basis of the vector space $\mathcal{V}$. The corresponding tangent space algebra $\mathcal{A}$ is spanned by components $\{X(s, s_3; \dot s, \dot s_3)\}$ of vector fields $\{X(s, \dot s); (s, \dot s) \in \Lambda_p\}$ in one-to-one correspondence with the coordinate tensors. Denoting

$$S = (s, \dot s), \quad \overline{S} = \{s, s_3; \dot s, \dot s_3\}, \quad R = (r, \dot r), \quad \overline{R} = \{r, r_3; \dot r, \dot r_3\}, \quad \text{etc.,}$$

and defining the sign of the graded bracket as

$$S \bullet R = R \bullet S = (-1)^{4(s+\dot s)(r+\dot r)+1},$$

the most general Lorentz covariant superalgebra $\mathcal{A}$ of vector fields takes the form

$$\left[X(\overline{S}), X(\overline{R}) \right]_{S \bullet R}$$

$$= \sum_{(v,\dot v)\in \Gamma(S,R)\cap \Lambda_p} C(s, s_3, r, r_3; v, s_3+r_3) \, C(\dot s, \dot s_3, \dot r, \dot r_3; \dot v, \dot s_3+\dot r_3)$$

$$\times t(s, \dot s, r, \dot r, v, \dot v) \, X(v, s_3+r_3; \dot v, \dot s_3+\dot r_3) . \qquad (1)$$

Here $C(s, s_3, r, r_3; v, s_3+r_3)$ are $SU(2)$ Clebsch-Gordan coefficients and $\Gamma(S, R)$ denotes the indices in the double Clebsch-Gordon decomposition:

$$\Gamma(S, R) = \{ (v, \dot v); \ v \in \gamma(s, r), \ \dot v \in \gamma(\dot s, \dot r)\};$$

$$\gamma(s, r) = \{ s+r, s+r-1, \ldots, |s-r| \}.$$

The $t(s, \dot s, r, \dot r, v, \dot v)$'s are structure constants. Subject to superskewsymmetry requirements inherited from the graded bracket and quadratic constraints implied by the super Jacobi identities, they parametrise the moduli space of superalgebras with relations (1).

The vector fields $X \in \mathcal{A}=\mathcal{A}_0+\mathcal{A}_1$ act as superderivations on functions of the Y's. We require that the action of $\mathcal{A}$ on $\mathcal{V}$ corresponds to a linear transformation. The combined vector space thus has supercommutation relations of a generalised

Heisenberg superalgebra, with the most general Lorentz covariant relations between the X's and the Y's taking the form

$$\left[X(\overline{S}), Y(\overline{R}) \right]_{S \bullet R}$$

$$= \sum_{(v,\dot{v}) \in \Gamma(S,R) \cap \Lambda_p} C(s, s_3, r, r_3 ; v, s_3 + r_3) \, C(\dot{s}, \dot{s}_3, \dot{r}, \dot{r}_3 ; \dot{v}, \dot{s}_3 + \dot{r}_3)$$

$$\times u(s, \dot{s}, r, \dot{r}, v, \dot{v}) \, Y(v, s_3 + r_3 ; \dot{v}, \dot{s}_3 + \dot{r}_3)$$

$$+ \; C(s, s_3, s, -s_3 ; 0, 0) \, C(\dot{s}, \dot{s}_3, \dot{s}, -\dot{s}_3 ; 0, 0) \, c(s, \dot{s}) \, \delta_{sr} \, \delta_{\dot{s}\dot{r}} \, \delta_{s_3 + r_3, 0} \, \delta_{\dot{s}_3 + \dot{r}_3, 0} \quad . \quad (2)$$

The u's are further structure constants and the 'central' parameters $c(s, \dot{s})$ determine a bilinear pairing between the X's and the Y's,

$$< ., . > : \mathcal{A} \times \mathcal{V} \to \mathcal{C} = \{ c(s, \dot{s}) \; ; \; (s, \dot{s}) \in \Lambda_p \}$$

given by

$$< X(\overline{S}), Y(\overline{R}) >$$
$$= c(s, \dot{s}) \, C(s, s_3, s, -s_3 ; 0, 0) \, C(\dot{s}, \dot{s}_3, \dot{s}, -\dot{s}_3 ; 0, 0) \, \delta_{sr} \, \delta_{\dot{s}\dot{r}} \, \delta_{s_3 + r_3, 0} \, \delta_{\dot{s}_3 + \dot{r}_3, 0} \quad ,$$

where the CG coefficients $C(s, s_3, s, -s_3 ; 0, 0)$ denote Wigner's 'metric' invariant in the representation space of fixed spin s.

Therefore, requiring Lorentz covariance determines the space of a priori allowed structure constants $\{t, u, c\}$. The combined vector space $\mathcal{G} = \mathcal{A} + \mathcal{V}$ forms a superalgebra if these structure constants are subject to quadratic equations, the satisfaction of which make the super Jacobi identities amongst the X's and the Y's automatic. These quadratic equations define the moduli space of spin p Lorentz covariant Heisenberg superalgebras. For any fixed set Λ_p, these equations allow determination using explicit values for Clebsch-Gordan coefficients and $6j$ symbols (from e.g. [4]). Particular solutions of these equations provide concrete examples of Lorentz covariant spin p superspaces; and they exist for any finite or infinite p.

SPIN 2 SUPERSPACES

The hyperspaces $\mathcal{M}$ serve as models for configuration spaces, or for moduli spaces of solutions, of Lorentz invariant field theories; and the supercommutations relations for $\mathcal{G}$ provide canonical supercommutation relations for the corresponding phase spaces. They therefore provide an algebraic description of the local symplectic structure. With this application in mind, it is clear that algebras including spins up to two are of possible significance for the canonical quantisation of gravity and supergravity theories. By including generators transforming according to every Lorentz representation (with unit multiplicity) having spin up to two, the complete set of quadratic equations for the Lorentz covariant structure constants were determined in [4]. These defining equations for *spin two Heisenberg superalgebras* are highly overdetermined. Nevertheless, non-trivial solutions can indeed be found. Explicit examples of algebras $\mathcal{G}$ were presented for spins $s+\dot{s}$ up to $\frac{3}{2}$ in [2,3]; and for spins up to 2 in [4]. The latter have a representation content reminiscent of simple (super)gravity theories: a metric represented by canonically conjugate variables X, Y transforming as $(0,0) \oplus (1,1)$ coupled to single copies of other (spin < 2) representations, including Rarita-Schwinger representations $(1, \frac{1}{2}) \oplus (\frac{1}{2}, 1)$ corresponding to gravitino phase-space variables. Concrete physical application to the canonical quantisation of supergravity theories, however, remains for future investigation.

GENERALISED SELF-DUALITY

A further application of these hyperspaces considers gauge fields on $\mathcal{M}$ [2,3] : Associating a gauge potential A (depending on the Y's) to each of these generalised derivatives, we define gauge-covariant derivatives $D(s,\dot{s}) = X(s,\dot{s}) + A(s,\dot{s})$. The curvature components $F(s, s_3; \dot{s}, \dot{s}_3)$ which are a priori non-zero are in turn defined in a consistent fashion by:

$$\left[D(\overline{S}), D(\overline{R}) \right]_{S \bullet R}$$
$$- \sum_{(v,\dot{v}) \in \widehat{\Gamma}(S,R) \cap \Lambda_p} C(s, s_3, r, r_3; v, s_3 + r_3)\, C(\dot{s}, \dot{s}_3, \dot{r}, \dot{r}_3; \dot{v}, \dot{s}_3 + \dot{r}_3)$$
$$\times t(s, \dot{s}, r, \dot{r}, v, \dot{v})\, D(v, s_3 + r_3; \dot{v}, \dot{s}_3 + \dot{r}_3)$$
$$= \sum_{(v,\dot{v}) \in \widehat{\Gamma}(S,R)} C(s, s_3, r, r_3; v, s_3 + r_3)\, C(\dot{s}, \dot{s}_3, \dot{r}, \dot{r}_3; \dot{v}, \dot{s}_3 + \dot{r}_3)$$
$$\times F(v, s_3 + r_3; \dot{v}, \dot{s}_3 + \dot{r}_3) \quad , \tag{3}$$

where the sums denote Clebsch-Gordon series restricted to terms having the super-skewsymmetry of the graded bracket by choosing

$$\widehat{\Gamma}(S, R) = \{(v, \dot{v}) \in \Gamma(S, R) \mid v + \dot{v} = (s + \dot{s}) + (r + \dot{r}) - 4(s+\dot{s})(r+\dot{r}) - 1 \bmod 2\}.$$

Curvature components thus defined form curvature tensors $F(S, R)$ for all pairs $S, R \in \Lambda_p$.

Generalised self-duality [2,3] corresponds to setting the following curvature representations to zero,

$$F(v, \dot{s}+\dot{r}) = 0 \quad \text{for all} \quad v \in \gamma(s, r) \quad \text{and} \quad (s, \dot{s}), (r, \dot{r}) \in \Lambda_p.$$

Similarly, generalised anti-self-duality corresponds to the imposition of the constraints

$$F(s+v, \dot{v}) = 0 \quad \text{for all} \quad \dot{v} \in \gamma(\dot{s}, \dot{r}) \quad \text{and} \quad (s, \dot{s}), (r, \dot{r}) \in \Lambda_p.$$

A third class of interesting constraints are the *light-like integrable systems*

$$F(s+r, \dot{s}+\dot{r}) = 0 \quad \text{for all} \quad (s,\dot{s}), (r,\dot{r}) \in \Lambda_p \ .$$

All three classes are $SO(4,\mathbb{C})$-covariant equations. They are, moreover, amenable to generalised twistor-like transforms (for example, on the lines of the discussion in [6]). These systems therefore provide an infinitely large hierarchy of gauge- and Lorentz-covariant solvable systems. The corresponding linear (Lax) systems are given in [3].

For example, the simple super-Poincaré set, $\Lambda_1 = \{(\frac{1}{2},\frac{1}{2}),(\frac{1}{2},0),(0,\frac{1}{2})\} \equiv \{V, S, \bar{S}\}$, has the set of a priori non-zero curvature tensors,

$$\widehat{\Gamma}(V,V) \Leftrightarrow \{F_{VV}(0,1), F_{VV}(1,0)\}$$
$$\widehat{\Gamma}(V,S) \Leftrightarrow \{F_{VS}(1,\tfrac{1}{2}), F_{VS}(0,\tfrac{1}{2})\}$$
$$\widehat{\Gamma}(V,\bar{S}) \Leftrightarrow \{F_{V\bar{S}}(\tfrac{1}{2},1), F_{V\bar{S}}(\tfrac{1}{2},0)\}$$
$$\widehat{\Gamma}(S,S) \Leftrightarrow \{F_{SS}(1,0)\}$$
$$\widehat{\Gamma}(S,\bar{S}) \Leftrightarrow \{F_{S\bar{S}}(\tfrac{1}{2},\tfrac{1}{2})\}$$
$$\widehat{\Gamma}(\bar{S},\bar{S}) \Leftrightarrow \{F_{\bar{S}\bar{S}}(0,1)\}$$

For this choice, the above three sets of zero curvature tensors correspond to the following:

$F(v, \dot{s}+\dot{r})$	$F(s+v, \dot{v})$	$F(s+r, \dot{s}+\dot{r})$
$F_{VV}(0,1)$	$F_{VV}(1,0)$	
$F_{VS}(1,\tfrac{1}{2}), F_{VS}(0,\tfrac{1}{2})$	$F_{VS}(1,\tfrac{1}{2})$	$F_{VS}(1,\tfrac{1}{2})$
$F_{V\bar{S}}(\tfrac{1}{2},1)$	$F_{V\bar{S}}(\tfrac{1}{2},1), F_{V\bar{S}}(\tfrac{1}{2},0)$	$F_{V\bar{S}}(\tfrac{1}{2},1)$
$F_{SS}(1,0)$	$F_{SS}(1,0)$	$F_{SS}(1,0)$
$F_{S\bar{S}}(\tfrac{1}{2},\tfrac{1}{2})$	$F_{S\bar{S}}(\tfrac{1}{2},\tfrac{1}{2})$	$F_{S\bar{S}}(\tfrac{1}{2},\tfrac{1}{2})$
$F_{\bar{S}\bar{S}}(0,1)$	$F_{\bar{S}\bar{S}}(0,1)$	$F_{\bar{S}\bar{S}}(0,1)$

These correspond respectively to the supersymmetrisation of the standard four-dimensional self-duality condition, $F_{VV}(0,1) = 0$, the supersymmetrisation of the anti-self-duality condition, $F_{VV}(1,0) = 0$, and the 'conventional constraints' of $N=1$ supersymmetric gauge theory. Further examples are discussed in [3], including the celebrated $N=3$ constraints which are equivalent to the full equations of motion.

CONCLUDING REMARKS

Although we remain in the realm of supercommutative geometry, with $[\mathcal{V},\mathcal{V}]_\bullet = 0$, a generalisation to non-supercommutative geometry is clearly a further possibility, with the simplest superalgebra variant having $[.,.] : \mathcal{V} \times \mathcal{V} \to \mathcal{V}$ such that

$[\mathcal{V}_\alpha, \mathcal{V}_\beta]_\bullet \subset \mathcal{V}_{\alpha+\beta}$. Further generalisations, replacing this superalgebra structure, for instance, by q-deformed supercommutation relations, may also be considered along the lines of the present investigation.

We have considered an element of $\mathcal{A}, \mathcal{V}$ to be of bosonic type if its spin $(s+\dot{s})$ is an integer and of fermionic type if its spin is a genuine half-integer; and we have assumed the corresponding statistics. We note, however, that this assignment is a purely conventional one, motivated by the spin-statistics theorem. This can indeed be lifted, if required, to yield Lie algebra (rather than superalgebra) extensions of the Poincaré algebra containing integer and half-integer spin elements, all of even statistics. Such algebras maintain, nevertheless, their $\mathbb{Z}_2$-graded nature [7]. Such a variant of the supersymmetry algebra was recently shown to be the target space symmetry of the $N=2$ string [8] and the space of string physical states was shown to be elegantly and compactly describable in terms of a self-dual field on a hyperspace with a vectorial and an even-spinorial coordinate. On such hyperspaces with even spinorial coordinates, the generalised self-duality equations are the same as those given above; the only difference being that the a priori set of curvature components is determined from commutators, rather than supercommutators, between covariant derivatives.

REFERENCES

1. E.S. Fradkin and M.A. Vasiliev, *Candidate for the role of higher spin symmetry*, Ann.Phys. **177** (1987) 63; M.A. Vasiliev, *Consistent equations for interacting massless fields of all spins in the first order in curvatures*, Ann.Phys. **190** (1989) 59.
2. C. Devchand and J. Nuyts, *Self-duality in generalised Lorentz superspaces*, hep-th/9612176, Phys. Lett. **B404** (1997) 259.
3. C. Devchand and J. Nuyts, *Supersymmetric Lorentz-covariant hyperspaces and self-duality equations in dimensions greater than (4|4)*, hep-th/9704036, Nucl. Phys. **B503** (1997) 627.
4. C. Devchand and J. Nuyts, *Lorentz covariant spin two superspaces*, hep-th/9804052, Nucl. Phys. **B** (1998).
5. D.A. Varshalovitch, A.N. Moskalev, V.K. Khersonskii, *Quantum Theory of Angular Momentum*, World Scientific Publishing Company, Singapore (1988).
6. C. Devchand and V. Ogievetsky, *Four-dimensional integrable theories*, hep-th/9410147, in Lect. Notes in Physics **447**, (Springer-Verlag, 1995).
7. D.V. Alekseevsky and V. Cortés, *Classification of N-(super)-extended Poincaré algebras and bilinear invariants of the spinor representation of Spin(p,q)*, Commun. Math. Phys. **183** (1997) 477.
8. C. Devchand and O. Lechtenfeld, *Extended self-dual Yang-Mills from the $N=2$ string*, hep-th/9712043, Nucl. Phys. **B516** (1998) 255.

Deformation Stability of BRST-Quantization

M. Dütsch[1] and K. Fredenhagen

II. Institut für Theoretische Physik
Universität Hamburg
Luruper Chaussee 149
D-22761 Hamburg, Germany

Abstract. To avoid the problems which are connected with the long distance behavior of perturbative gauge theories we present a *local* construction of the observables which does not involve the adiabatic limit. First we construct the interacting fields as formal power series by means of causal perturbation theory. The observables are defined by BRST invariance where the BRST-transformation $\tilde{s}$ acts as a graded derivation on the algebra of interacting fields. Positivity, i.e the existence of Hilbert space representations of the local algebras of observables is shown with the help of a local Kugo-Ojima operator Q_{int} which implements $\tilde{s}$ on a local algebra and differs from the corresponding operator Q of the free theory. We prove that the Hilbert space structure present in the free case is stable under perturbations. All assumptions are shown to be satisfied in QED in a finite spatial volume with suitable boundary conditions. As a by-product we find that the BRST-quantization is not compatible with periodic boundary conditions for massless free gauge fields.

PACS. 11.15.-q Gauge field theories, 11.15.Bt General properties of perturbation theory

INTRODUCTION

The long distance behavior of nonabelian perturbative gauge theories is plagued by serious problems. In *massless* theories there appear infrared divergences in the adiabatic limit $g \to$ const. of the S-matrix, where g is a space-time dependent coupling 'constant'. In QED these divergences are logarithmic and cancel in the cross section. (This is proven at least at low orders of the perturbation series [15].) Moreover, Blanchard and Seneor [2] proved that the adiabatic limit of Green's and Wightman functions exists for QED. But in nonabelian gauge theories the divergences are worse. Perturbation theory seems to be not able to describe the long distance properties of these models ("confinement"). In *massive* theories the

---

[1] Work supported by the Alexander von Humboldt Foundation

infrared divergences are absent, but e.g. in the electroweak theory an S-matrix formalism suffers from the instability of some particles, e.g. the W-, Z-bosons and the muons and taus. States containing such particles belong to the physical state space, but they cannot appear as asymptotic states of the S-matrix for $t \to \pm\infty$.

Our way out is to *construct the observables locally*. We consider a fixed, open double cone $\mathcal{O} \subset \mathbf{R}^4$. The coupling 'constant' g has compact support and takes a constant value on $\mathcal{O}$

$$g \in \mathcal{D}(\mathbf{R}^4), \qquad g(x) = e = \text{const.}, \qquad \forall x \in \mathcal{O}. \tag{1.1}$$

The *interacting fields* are defined by Bogoliubov's formula [4]

$$A_{\text{int }\mathcal{L}}(x) \stackrel{\text{def}}{=} \frac{\delta}{i\delta h(x)} S(\mathcal{L})^{-1} S(\mathcal{L} + hA)|_{h=0}, \tag{1.2}$$

and the time ordered products $T(\mathcal{L}(x_1)...\mathcal{L}(x_n))$, which appear in the S-matrix

$$S(\mathcal{L}) = \sum_{n=0}^{\infty} \frac{i^n}{n!} \int d^4 x_1 ... d^4 x_n \, T(\mathcal{L}(x_1)...\mathcal{L}(x_n)), \tag{1.3}$$

are constructed by means of causal perturbation theory [4,6,12,15,17]. The interacting fields $A_{\text{int }\mathcal{L}}(x)$ (A is a Wick polynomial of incoming fields) are formal power series of operator valued distributions on a dense invariant domain $\mathcal{D}$ in the Fock space of incoming fields. They depend on an interaction Lagrangian $\mathcal{L}$ which is a Wick polynomial of incoming fields with testfunctions $g \in \mathcal{D}(\mathbf{R}^4)$ as coefficients.

The crucial observation is that the dependence of the interacting fields on the interaction Lagrangian is *local*, in the sense that Lagrangians $\mathcal{L}_1$ and $\mathcal{L}_2$ which differ only within a closed region which does not intersect the closure of $\mathcal{O}$, lead to unitarily equivalent fields within $\mathcal{O}$, i.e. there exists a unitary formal power series V of operators on $\mathcal{D}$ such that

$$V A_{\text{int }\mathcal{L}_1}(x) V^{-1} = A_{\text{int }\mathcal{L}_2}(x), \qquad \forall x \in \mathcal{O}, \tag{1.4}$$

and V does not depend on A [6]. The proof of (1.4) relies on the causal factorization of the time ordered products.

The field algebra $\tilde{\mathcal{F}}(\mathcal{O})$ which is generated by

$$\{A_{\text{int }\mathcal{L}}(f) = \int d^4 x \, A_{\text{int }\mathcal{L}}(x) f(x) | f \in \mathcal{D}(\mathcal{O})\} \tag{1.5}$$

is up to unitary equivalence uniquely determined by $g|_{\mathcal{O}}$. Since $\mathcal{O}$ is arbitrary, the full net of local algebras can be constructed without ever performing the adiabatic limit $g \to$ constant.

In gauge theories the (local) algebras of interacting fields contain unphysical fields like vector potentials and ghosts. They can be eliminated by the BRST formalism. But it remains to show that the algebra of observables can be (nontrivially) represented on a Hilbert space.

In the free theory positivity can be verified by an explicit calculation. Formally, in the adiabatic limit, the positivity is hence valid also for the interacting theory. We show, that for a localized interaction (i.e. before the adiabatic limit), the physical Hilbert space can be obtained as a deformation of the free one (sect. 2). The construction relies on some assumptions, which are verified for the example of QED (sect. 3). To avoid a volume divergence in Q_{int} we embed our double cone $\mathcal{O}$ into the cylinder $\mathbf{R} \times C_L$, where $\mathbf{R}$ denotes the time axis and C_L is a cube of length L. In sect. 4 we point out the importance of a suitable choice of boundary conditions for the BRST-quantization of massless free gauge fields on $\mathbf{R} \times C_L$.

We hope that due to its local character, our construction can be generalized to curved space-times, continuing the program of [6,7].

CONNECTION OF OBSERVABLE ALGEBRAS AND FIELD ALGEBRAS IN PERTURBATIVE GAUGE THEORIES

Local construction of observables in gauge theories and representation in the physical pre Hilbert space

Let $\mathcal{F}$ be a $\mathbf{Z}_2$-graded *-algebra, e.g. the algebra of fields of a gauge theory where the $\mathbf{Z}_2$-gradiation is $(-1)^{\delta(F)}$ with the ghost number $\delta(F)$. To get rid of the unphysical fields, we use the BRST-transformation s [1], which is a graded derivation on $\mathcal{F}$ with $s^2 = 0$ and $s(F^*) = -(-1)^{\delta(F)} s(F)^*$.

The kernel of s, $\mathcal{A}_0 := s^{-1}(0)$, is a *-subalgebra of $\mathcal{F}$ and $\mathcal{A}_{00} := s(\mathcal{F})$ is a 2-sided ideal in $\mathcal{A}_0$. Hence we may define the *algebra of observables* as the quotient

$$\mathcal{A} \stackrel{\text{def}}{=} \frac{\mathcal{A}_0}{\mathcal{A}_{00}}. \tag{2.1}$$

We ask now under which conditions $\mathcal{A}$ has a nontrivial *-representation by operators on a pre Hilbert space. For this purpose we work with the Kugo-Ojima formalism [14]. We assume that $\mathcal{F}$ has a faithful representation on an inner product space $(\mathcal{K}, <.,.>)$ such that $<F^*\phi, \psi> = <\phi, F\psi>$, $\forall F \in \mathcal{F}$, and that s is implemented by an operator Q on $\mathcal{K}$, i.e.

$$s(F) = QF - (-1)^{\delta(F)} FQ, \tag{2.2}$$

such that

$$<Q\phi, \psi> = <\phi, Q\psi> \quad \text{and} \quad Q^2 = 0. \tag{2.3}$$

Note that if the inner product on $\mathcal{K}$ is positive definite, we find $<Q\phi, Q\phi> = <\phi, Q^2\phi> = 0$, hence $Q = 0$ and thus also $s = 0$. Hence for nontrivial s the inner product must necessarily be indefinite.

Let $\mathcal{K}_0 \stackrel{\text{def}}{=} \text{Ke} \, Q$ be the kernel and $\mathcal{K}_{00}$ the range of Q. Because of $Q^2 = 0$ we have $\mathcal{K}_{00} \subset \mathcal{K}_0$. We assume:

(**Positivity**)   (i) $<\phi, \phi> \geq 0 \quad \forall \phi \in \mathcal{K}_0,$

and   (ii) $\phi \in \mathcal{K}_0 \wedge <\phi, \phi> = 0 \implies \phi \in \mathcal{K}_{00}.$ (2.4)

Then

$$\mathcal{H} \stackrel{\text{def}}{=} \frac{\mathcal{K}_0}{\mathcal{K}_{00}}, \qquad <[\phi_1], [\phi_2]>_{\mathcal{H}} \stackrel{\text{def}}{=} <\psi_1, \psi_2>_{\mathcal{K}}, \quad \psi_j \in [\phi_j] := \phi_j + \mathcal{K}_{00} \quad (2.5)$$

is a pre Hilbert space and

$$\pi([A])[\phi] \stackrel{\text{def}}{=} [A\phi] \qquad (2.6)$$

is a representation on $\mathcal{H}$ (where $A \in \mathcal{A}_0$, $\phi \in \mathcal{K}_0$, $[A] := A + \mathcal{A}_{00}$) [8].

Stability under deformations

It is gratifying that the described structure is *stable under deformations*, e.g. by turning on the interaction. Let $\mathcal{K}$ be fixed and replace $F \in \mathcal{F}$ by a formal power series $\tilde{F} = \sum_n g^n F_n$ with $F_0 = F$ and $F_n \in \mathcal{F}$, $\delta(F_n) = \text{const.}$ In the same way replace s and Q by formal power series $\tilde{s} = \sum_n g^n s_n$, $\tilde{Q} = \sum_n g^n Q_n$ with $s_0 = s$, $Q_0 = Q$ and

$$\tilde{s}^2 = 0, \quad \tilde{Q}^2 = 0, \quad <\tilde{Q}\phi, \psi> = <\phi, \tilde{Q}\psi> \quad \text{and} \quad \tilde{s}(\tilde{F}) = \tilde{Q}\tilde{F} - (-1)^{\delta(\tilde{F})}\tilde{F}\tilde{Q}. \quad (2.7)$$

We can then define $\tilde{\mathcal{A}} \stackrel{\text{def}}{=} \frac{\text{Ke}\,\tilde{s}}{\text{Ra}\,\tilde{s}}$. $\mathcal{K}_0$ and $\mathcal{K}_{00}$ have to be replaced by formal power series $\tilde{\mathcal{K}}_0 := \text{Ke}\,\tilde{Q}$ and $\tilde{\mathcal{K}}_{00} := \text{Ra}\,\tilde{Q}$ with coefficients in $\mathcal{K}$. Due to the above result (2.6), the algebra $\tilde{\mathcal{A}}$ has a natural representation on $\tilde{\mathcal{H}} \stackrel{\text{def}}{=} \frac{\tilde{\mathcal{K}}_0}{\tilde{\mathcal{K}}_{00}}$. The inner product on $\mathcal{K}$ induces an inner product on $\tilde{\mathcal{H}}$ which assumes values in the formal power series over $\mathbf{C}$. We adopt the point of view that a formal power series $\tilde{b} = \sum_n g^n b_n$, $b_n \in \mathbf{C}$ is *positive* if there is another formal power series $\tilde{c} = \sum_n g^n c_n$, $c_n \in \mathbf{C}$ with $\tilde{c}^*\tilde{c} = \tilde{b}$, i.e. $b_n = \sum_{k=0}^n \bar{c}_k c_{n-k}$. (cf. [5])

The assumptions concerning the positivity of the inner product are automatically fulfilled for the deformed theory, if they hold true in the undeformed model [8].

Theorem 1: Let the positivity assumption (2.4) be fulfilled in zeroth order. Then
(i) $<\tilde{\phi}, \tilde{\phi}> \geq 0 \quad \forall \tilde{\phi} \in \tilde{\mathcal{K}}_0,$
(ii) $\tilde{\phi} \in \tilde{\mathcal{K}}_0 \wedge <\tilde{\phi}, \tilde{\phi}> = 0 \implies \tilde{\phi} \in \tilde{\mathcal{K}}_{00}.$
(iii) For every $\phi \in \mathcal{K}_0$ there exists a power series $\tilde{\phi} \in \tilde{\mathcal{K}}_0$ with $(\tilde{\phi})_0 = \phi$.
(iv) Let π and $\tilde{\pi}$ be the representations (2.6) of $\mathcal{A}, \tilde{\mathcal{A}}$ on $\mathcal{H}, \tilde{\mathcal{H}}$ respectively. Then $\tilde{\pi}(\tilde{A}) \neq 0$ if $\pi(A_0) \neq 0$.

From parts (i) and (ii) we conclude that $\tilde{\mathcal{H}} = \frac{\tilde{\mathcal{K}}_0}{\tilde{\mathcal{K}}_{00}}$ is a perturbative analog of a (pre) Hilbert where the scalar product assumes values in the formal power series over C. Note that $\phi \to \tilde{\phi}$ is non-unique and this holds also true for the induced relation between $\mathcal{H}$ and $\tilde{\mathcal{H}}$. A consequence of part (i) of the theorem is the *positivity of the Wightman distributions of $\tilde{s}$-invariant fields* [8].

VERIFICATION OF THE ASSUMPTIONS IN MODELS

Kugo-Ojima [14] argue that at asymptotically early times the interacting fields tend to the free incoming fields. Since $Q_{\text{int}} \stackrel{\text{def}}{=} \tilde{Q}$ is conserved, it coincides with the free one $Q = Q_0$. Hence it is sufficient to check the assumptions for the free theory. But the BRST-current (i.e. the current belonging to Q_{int}) is only conserved in regions where g is constant (see below and [3]). Hence, the Kugo-Ojima procedure involves the (partial) adiabatic limit for $t \to -\infty$, which is difficult to control. The argument does certainly not work in nonabelian gauge theories (as can be seen by an explicit calculation of the first order of Q_{int}) [3]. We therefore prefer not to work in the adiabatic limit. The price to pay is that Q_{int} does not agree with Q, so for the construction of the physical Hilbert space we have to check the assumptions of previous section. We do this for QED. We see no principle obstacle for the generalization to nonabelian gauge theories. But the details still need to be worked out [3].

Free QED

The field algebra $\mathcal{F}$ is generated by the free photon fields A^μ in Feynman gauge, the free spinor fields ψ and $\overline{\psi}$, a pair of free ghost fields u and $\tilde{u}$, the Wick monomials $j^\mu =: \overline{\psi}\gamma^\mu\psi :$, $\gamma_\mu A^\mu \psi$, $\overline{\psi}\gamma_\mu A^\mu$, $j_\mu A^\mu$ and the derivated free fields $\partial_\mu A^\mu$, $F^{\mu\nu} = \partial^\mu A^\nu - \partial^\nu A^\mu$. This algebra is faithfully represented on a Krein space which is given by the usual Fock space of free fields and a Krein operator which defines the indefinite inner product. The graded derivation s is determined by the BRST-transformation of free fields

$$s(A^\mu) = i\partial^\mu u, \quad s(\psi) = 0, \quad s(\overline{\psi}) = 0, \quad s(u) = 0, \quad s(\tilde{u}) = -i\partial_\mu A^\mu \quad (3.1)$$

and by translation invariance of s. This transformation is implemented by the free Kugo-Ojima charge [10]

$$Q \stackrel{\text{def}}{=} \int_{x_0=\text{const.}} d^3x \, (\partial_\nu A^\nu(x)) \overset{\leftrightarrow}{\partial}_0 u(x), \quad (3.2)$$

which fulfills[2] $Q^* = Q$, and $Q^2 = 0$. In addition the inner product $<.,.>$ is positive semidefinite on $\text{Ke}\, Q$ and the space of nullvectors in $\text{Ke}\, Q$ is precisely $\text{Ra}\, Q$ ([9,13])

---

[2] We restrict all operators (resp. formal power series of operators) to the dense invariant domain $\mathcal{D}$ and, therefore, there is no difference between symmetric and self-adjoint operators.

Construction of the interacting Kugo-Ojima charge in QED

In QED the interaction is given by

$$\mathcal{L}(x) = g(x) : \bar\psi(x)\gamma_\mu A^\mu(x)\psi(x) :, \qquad g \in \mathcal{D}(\mathbf{R}^4). \tag{3.3}$$

We fix the double cone $\mathcal{O}$ to be the causal completion of the surface $\{(0,\mathbf{x}), |\mathbf{x}| < r\}$ and assume the switching function $g \in \mathcal{D}(\mathbf{R}^4)$ to be constant on a neighbourhood $\mathcal{U}$ of $\bar{\mathcal{O}}$ (1.1). We study the algebra $\tilde{\mathcal{F}}(\mathcal{O})$ (1.5) of interacting fields localized in $\mathcal{O}$. The ghost fields do not couple in QED, hence $u_{\text{int } \mathcal{L}}(x) = u(x)$ and $\tilde{u}_{\text{int } \mathcal{L}}(x) = \tilde{u}(x)$. The interacting fields can be normalized such that they fulfil the *field equations* [8,11]

$$\Box A^\mu_{\text{int } \mathcal{L}}(x) = -g(x) j^\mu_{\text{int } \mathcal{L}}(x), \tag{3.4}$$

$$(i\gamma_\mu \partial^\mu - m)\psi_{\text{int } \mathcal{L}}(x) = -g(x)(\gamma_\mu A^\mu \psi)_{\text{int } \mathcal{L}}(x), \tag{3.5}$$

electric current conservation

$$\partial_\mu j^\mu_{\text{int } \mathcal{L}} = 0 \tag{3.6}$$

and the following commutation relations at points $x, y \in \mathcal{O}$ [8]

$$[\partial_\mu A^\mu_{\text{int } \mathcal{L}}(x), A^\nu_{\text{int } \mathcal{L}}(y)] = i\partial^\nu D(x-y), \qquad [\partial_\mu A^\mu_{\text{int } \mathcal{L}}(x), \partial_\nu A^\nu_{\text{int } \mathcal{L}}(y)] = 0, \tag{3.7}$$

$$[\partial_\mu A^\mu_{\text{int } \mathcal{L}}(x), \psi_{\text{int } \mathcal{L}}(y)] = D(x-y)e\psi_{\text{int } \mathcal{L}}(y), \tag{3.8}$$

where D is the massless Pauli-Jordan distribution.

The abelian BRST-transformation $\tilde{s} = s_0 + gs_1$ [1] is a graded $*$-derivation with zero square which induces the following transformations on the basic fields,

$$\tilde{s}(A^\mu_{\text{int } \mathcal{L}}(x)) = i\partial^\mu u(x), \qquad \tilde{s}(u(x)) = 0, \qquad \tilde{s}(\tilde{u}(x)) = -i\partial_\mu A^\mu_{\text{int } \mathcal{L}}(x),$$

$$\tilde{s}(\psi_{\text{int } \mathcal{L}}(x)) = -e\psi_{\text{int } \mathcal{L}}(x)u(x), \qquad \tilde{s}(\bar\psi_{\text{int } \mathcal{L}}(x)) = e\bar\psi_{\text{int } \mathcal{L}}(x)u(x) \tag{3.9}$$

for $x \in \mathcal{O}$. (The pointwise products above are well defined.)

On $\tilde{\mathcal{F}}(\mathcal{O})$ $\tilde{s}$ is implemented by the operator

$$Q_{\text{int}}(g, k) = \int d^4x\, k(x)(\partial_\nu A^\nu_{\text{int } \mathcal{L}}(x)) \overset{\leftrightarrow x}{\partial}_0 u(x) \tag{3.10}$$

(where $k \in \mathcal{D}(\mathcal{U})$ is a suitably chosen smeared characteristic function of the surface $\{(0, \mathbf{x}), |\mathbf{x}| \leq r\}$). Note that $[Q_{\text{int}}(g), F]_\mp$, $F \in \tilde{\mathcal{F}}(\mathcal{O})$ is independent of k, since the BRST-current $\partial_\mu A^\mu_{\text{int } \mathcal{L}}(x) \overset{\leftrightarrow x}{\partial^\nu} u(x)$ is conserved within $\mathcal{U}$. $Q_{\text{int}}(g, k)$ is hermitian for real valued k and nilpotent,

$$Q_{\text{int}}(g, k)^2 = \frac{1}{2}\{Q_{\text{int}}(g, k), Q_{\text{int}}(g, k)\} =$$

$$= \frac{1}{2}\int d^4x\, h(x) \int d^4y\, h(y) [\partial_\mu A^\mu_{\text{int } \mathcal{L}}(x), \partial_\nu A^\nu_{\text{int } \mathcal{L}}(y)] \overset{\leftrightarrow x \leftrightarrow y}{\partial_0 \partial_0} u(x)u(y) = 0, \tag{3.11}$$

by means of (3.7).

But we need in addition that the zeroth order term $Q_0(k)$ of $Q_{\text{int}}(g,k)$ (3.10) satisfies the positivity assumption (2.4). There seems to be no reason why this should hold for a generic choice of k. One might try to control the limit when k tends to a smeared characteristic function of the $t=0$ hyperplane (in order that $Q_0(k)$ becomes equal to the free charge Q (3.2)), but without an a priori information on the existence of an $\tilde{s}$-invariant state this appears to be a hard problem.

There is a more elegant way to get rid of these problems which relies on the local character of our construction. We may embed our double cone $\mathcal{O}$ isometrically into the cylinder $\mathbf{R} \times C_L$, where C_L is a cube of length L, $L \gg r$, with suitable boundary conditions (see sect. 4), and where the first factor denotes the time axis. If we choose the compactification length L big enough, the properties of the local algebra $\tilde{\mathcal{F}}(\mathcal{O})$ are not changed.

We assume the switching function g to fulfil

$$g(x) = e = \text{constant} \qquad \forall x \in \mathcal{O} \cup \{(x_0, \vec{x}) | \, |x_0| < \epsilon\} \qquad (r \gg \epsilon > 0) \qquad (3.12)$$

on $\mathbf{R} \times C_L$ and to have compact support in timelike directions. Now we may insert

$$k(x) := h(x_0), \qquad \text{where} \qquad h \in \mathcal{D}([-\epsilon, \epsilon]), \qquad \int dx_0 \, h(x_0) = 1 \qquad (3.13)$$

into the expression (3.10) for Q_{int}. The zeroth order Q_0 then agrees with the free charge Q on C_L (3.2), hence we may apply Theorem 1.

We emphasize that our construction shall describe QED also in the *non-compactified* Minkowski space (this is the main concern of the paper) and, therefore, should not depend on the compactification length L. On the level of the algebras this is evident. We conjecture that also the state space (i.e. the set of expectation functionals induced by vectors in the physical Hilbert space) is independent of L, but this remains to be proven.

An open question is the *physical meaning of the remaining normalization conditions* in a local perturbative construction, after the restrictions from gauge invariance and other symmetries were taken into account. The parameters involved may be considered as structure constants of the algebra of observables, but their usual interpretation as charge and mass involve the adiabatic limit.

BOUNDARY CONDITIONS FOR MASSLESS FREE GAUGE FIELDS IN A FINITE VOLUME

The purpose of this section is to demonstrate the *importance of a suitable choice of boundary conditions*. First we show that the BRST-quantization is not compatible with periodic boundary conditions for massless free gauge fields.

Let T_3 be the 3-torus of length L. The algebra of a free massless scalar field φ on $\mathbf{R} \times T_3$ with periodic boundary conditions is the unital *-algebra generated by elements $\varphi(f)$, $f \in \mathcal{D}(\mathbf{R} \times T_3)$ with the relations

$$f \mapsto \varphi(f) \text{ is linear}, \tag{4.1}$$

$$\varphi(\Box f) = 0, \tag{4.2}$$

$$\varphi(f)^* = \varphi(\bar{f}), \tag{4.3}$$

$$[\varphi(f), \varphi(g)] = \int d^4x d^4y f(x) g(y) D_L(x,y), \tag{4.4}$$

where D_L is the fundamental solution of the wave equation on $\mathbf{R} \times T_3$ with periodic boundary conditions, which has the explicit form

$$D_L(x^0, \vec{x}, y^0, \vec{y}) = \sum_{\vec{n} \in \mathbf{Z}^3} D(x^0 - y^0, \vec{x} - \vec{y} - \vec{n}). \tag{4.5}$$

In particular one sees that D_L coincides with D (the massless commutator function on Minkowski space) on $\mathcal{O}$ if the closure of the double cone $\mathcal{O}$ is contained in $\mathbf{R} \times T_3$, considered as a region in Minkowski space. Hence the algebra $\mathcal{F}(\mathcal{O})$ associated to $\mathcal{O}$ is independent of the boundary conditions.

In a mode decomposition of D_L,

$$D_L(x) = \frac{i}{2L^3} \sum_{\vec{n} \in \mathbf{Z}^3, \vec{n} \neq \vec{0}} \frac{1}{\omega_{\vec{n}}} (e^{-i\omega_{\vec{n}} x^0} - e^{i\omega_{\vec{n}} x^0}) e^{i\vec{k}_{\vec{n}} \vec{x}} + \frac{x_0}{L^3}, \tag{4.6}$$

the zero mode plays a special role.[3] The zero mode part of φ (4.1-4) is defined by

$$\varphi_0(t) \stackrel{\text{def}}{=} \frac{1}{L^3} \int_{T_3} d^3x \, \varphi(t, \vec{x}). \tag{4.7}$$

The algebra of the zero mode is isomorphic to the algebra of p and q in quantum mechanics with the free time evolution $\varphi_0(t) = q + pt$. There exists no ground state on this algebra.

In Feynman gauge, the components of the photon field A_μ are quantized as scalar fields, with a minus sign for the commutator of the zero component. The zero mode of the field $\partial_\mu A^\mu$ is then $-p^0 L^{-3}$, which has a trivial kernel. This makes it impossible to impose the Gupta-Bleuler condition on the physical state space.

The BRST formalism is even worse. The ghost fields are quantized by

$$(u(f) + i\tilde{u}(g))^2 = \int d^4x \, d^4y \, f(x) g(y) D_L(x-y). \tag{4.8}$$

Inserting $f(x) = L^{-3}\delta(x_0 - t_1)$ and $g(y) = -L^{-3}\delta'(y_0 - t_2)$ we obtain

$$(u_0(t_1) + i\partial_0 \tilde{u}_0(t_2))^2 = -L^{-3} \tag{4.9}$$

---

[3] To verify (4.6) note that it is a solution of the wave equation and has the same Cauchy data as (4.5) for $x_0 = 0$.

for the zero mode parts. u_0 and $\partial_0 \tilde{u}_0$ are BRST invariant, hence they are observables. In addition $(u_0(t_1) + i\partial_0 \tilde{u}_0(t_2))$ is hermitian. We conclude that there is no nonzero (pre) Hilbert space representation of the algebra of observables.

$(u_0, \partial_0 \tilde{u}_0)$ corresponds to a 'singlet pair' in the terminology of [14], sect. 3.1. Already there it was pointed out that the appearance of such a pair makes a consistent formulation impossible.

The way out is to choose *boundary conditions which exclude the zero mode*. For the electromagnetic field we may use metallic boundary conditions, i.e. the pullback of the 2-form F vanishes at the boundary (which means that the tangential components of the electric field and the normal component of the magnetic field vanish). In addition we assume that the auxiliary Nakanishi-Lautrup field $B = \partial^\mu A_{L\mu}$ (in Feynman gauge) satisfies Dirchlet boundary conditions. Also the ghost and antighost fields are quantized with Dirichlet boundary conditions. The details are worked out in appendix A of [8].

The BRST-quantization requires no restrictions on the boundary conditions for the electron field. For simplicity, we choose periodic boundary conditions. They have the big advantage that they are invariant under charge conjugation, hence the expectation value of the electric current (normal ordered w.r.t. the Minkowski vacuum) vanishes in the groundstate (of the torus) and, therefore, the interaction Lagrangian $\mathcal{L}$ (3.3) keeps the same form as on Minkowski space.

Acknowledgements: We profitted from discussions with Franz-Marc Boas, Izumi Ojima and Marek J. Radzikowski which are gratefully acknowledged.

REFERENCES

1. Becchi, C., Rouet, A., and Stora, R., *Commun. Math. Phys.* **42**, 127 (1975)
   Becchi, C., Rouet, A., and Stora, R., *Annals of Physics (N.Y.)* **98**, 287 (1976)
2. Blanchard, P., and Seneor, R., *Ann. Inst. H. Poincaré A* **23**, 147 (1975)
3. Boas, F.M., Dütsch, M., and K.Fredenhagen, K., "A local (perturbative) construction of observables in gauge theories: nonabelian gauge theories", work in progress
4. Bogoliubov, N.N., and Shirkov, D.V., *"Introduction to the Theory of Quantized Fields"*, New York (1959)
5. Bordemann, M., and Waldmann, S., q-alg/9611004, to appear in *Commun. Math. Phys.*
6. Brunetti, R., and Fredenhagen, K., "Interacting quantum fields in curved space: Renormalization of ϕ^4", gr-qc/9701048, *Proceedings of the Conference 'Operator Algebras and Quantum Field Theory'*, held at Accademia Nazionale dei Lincei, Rome, July 1996.
   Brunetti, R., and Fredenhagen, K., "Microlocal analysis and interacting quantum field theories: Renormalization on physical backgrounds", in preparation.
7. Brunetti, R., Fredenhagen, K., and Köhler, M., *Commun. Math. Phys.* **180**, 312 (1996)

8. Dütsch, M., and Fredenhagen, K., "A local (perturbative) construction of observables in gauge theories: the example of QED", preprint: hep-th/9807078, DESY 98-090
9. Dütsch, M., Hurth, T., and Scharf, G., *N. Cimento A* **108**, 737 (1995)
10. Dütsch, M., Hurth, T., Krahe, F., and Scharf, G., *N. Cimento A* **106**, 1029 (1993)
11. Dütsch, M., Krahe, F., and Scharf, G., *N. Cimento A* **103**, 871 (1990)
12. Epstein, H., and Glaser, V., *Ann. Inst. H. Poincaré A* **19**, 211 (1973)
13. Krahe, F., *Acta Phys. Polonica B* **27**, 2453 (1996)
14. Kugo, T., and Ojima, I., *Suppl. Progr. Theor. Phys.* **66**, 1 (1979)
15. Scharf, G., *"Finite Quantum Electrodynamics. The causal approach"*, 2nd. ed., Springer-Verlag (1995)
16. Stora, R., "Lagrangian field theory", summer school of theoretical physics about *"particle physics"*, Les Houches, 1-79 (1971)
17. Stora, R., "Differential algebras in Lagrangean field theory", ETH-Zürich Lectures, January-February 1993;
    Popineau, G., and Stora, R., "A pedagogical remark on the main theorem of perturbative renormalization theory", unpublished preprint (1982)

Dynamical Symmetry Breaking in Nambu-Jona-Lasinio Model under the Influence of External Electromagnetic and Gravitational Fields

E. Elizalde[a,b,1], Yu. I. Shil'nov[a,c,2]

[a] *Consejo Superior de Investigaciones Científicas,
IEEC, Edifici Nexus-204, Gran Capità 2-4, 08034, Barcelona, Spain*
[b] *Department ECM, Faculty of Physics, University of Barcelona,
Diagonal 647, 08028, Barcelona, Spain*
[c] *Department of Theoretical Physics, Faculty of Physics,
Kharkov State University, Svobody Sq. 4, 310077, Kharkov, Ukraine*

Abstract. Dynamical symmetry breaking is investigated for a four-fermion Nambu-Jona-Lasinio model in external electromagnetic and gravitational fields. An effective potential is calculated in the leading order of the large-N expansion using the proper-time Schwinger formalism.

Phase transitions accompanying a chiral symmetry breaking in the Nambu-Jona-Lasinio model are studied in detail. A magnetic calalysis phenomenon is shown to exist in curved spacetime but it turns out to lose its universal character because the chiral symmetry is restored above some critical positive value of the spacetime curvature.

INTRODUCTION

Different four-fermion models [1], [2] have been considered to be one of the most convenient ways for an investigation of the low-energy physics of strong interactions. A dynamical symmetry breaking phenomenon (DSB) has been proved to take place within those models, partecularly Nambu-Jona-Lasinio (NJL) one, which seems to show up a nontrivial phase structure. Usually the symmetry to be broken under the DSB mechanism is the chiral one. Dynamical version of fermions mass generation and dynamical chiral symmetry breaking have been investigated very carefully and some fruitful applications for the real high-energy physics have been found [3], [4].

However it has turned out to be very difficult to realize the idea of DSB because all of the calculations should be performed out of pertubation theory. This leads to study already simplified models and that is why we have to investigate any possible

---
[1] E-mail: eli@zeta.ecm.ub.es, elizalde@io.ieec.fcr.es
[2] E-mail: shil@kink.univer.kharkov.ua, visit2@ieec.fcr.es

generalizations within these models those of nonzero temperature and chemical potential, arbitrary dimensions, external fields including gravitational one and so on as some kind of laboratory in order to collect as much new information as we can.

Despite of essential difficulties caused by the nonpertubative character of DSB phenomenon, it has been applied successfully to describe the overcritical behavior of quantum electrodynamics, the top quark condensate mechanism of mass generation in the Weinberg- Salam model of electroweak interactions, technicolor models and, especially, to investigate the composite fields generation in the NJL model. In the frameworks of Schwinger proper-time method this model has been studied in external electromagnetic field by many authors for 20 years [5] - [7].

Recently a new sample of papers devoted to DSB in external electromagnetic field have been published [8], [9]. It shed a new light onto the universal character of magnetic catalysis, which means that magnetic field breaks chiral symmetry for any value of its strength. Furthermore it has been shown that this phenomenon occurs in quantum electrodynamics, 2+1, 3+1 dimensional nonsupersymmetrical and 3+1 supersymmetrical NJL models. So the statement about **the universal character of magnetic catalysis** has been made.

Investigations of the influence of a classical gravitational field on the DSB phenomenon in the NJL model have been carried out for some years [10]. It has been shown that curvature-induced phase transitions exist and might play some essential role in more or less realistic early Universe model. It turns out that, in spite of the relatively small value of the curvature-dependent corrections at the low energy scale to be investigated within the NJL model, these corrections appear to be inescapable, in the sense that they must be taken into account when one performs the necessary "fine tuning" of the different cosmological parameters. Furthemore, positive spacetime curvature changes the universal character of magnetic catalysis dramatically.

It has been shown that the early Universe could contain a large primodial magnetic field and have a huge electrical conductivity. The vicinities of magnetized black holes and neutron stars are the other possible points of application of our model. Therefore both classical external gravitational and electromagnetic fields should be taken into account for the description of a wide sample of events in the Universe.

In the present paper we describe our recent results concerning the DSB under the simultaneous influence of both gravitational and electromagnetic fields in NJL model [11]. The phase transitions accompanying the DSB process on the spacetime curvature, as well as the values of electric or magnetic field strength are investigated.

DYNAMICAL SYMMETRY BREAKING BY A MAGNETIC FIELD IN FLAT SPACETIME

In an arbitrary dimensional flat spacetime the NJL model has the following action:

$$S = \int d^d x \left\{ i\bar{\psi}\gamma^\mu D_\mu \psi + \frac{\lambda}{2N} \left[(\bar{\psi}\psi)^2 + (\bar{\psi}i\gamma_5\psi)^2 \right] \right\}, \tag{1}$$

where the covariant derivative D_μ includes the electromagnetic potential A_μ and N is the number of bispinor fields ψ_a.

Introducing the auxiliary fields

$$\sigma = -\frac{\lambda}{N}(\bar{\psi}\psi), \quad \pi = -\frac{\lambda}{N}(\bar{\psi}i\gamma_5\psi) \tag{2}$$

we can rewrite the action as:

$$S = \int d^d x \left\{ i\bar{\psi}\gamma^\mu D_\mu \psi - \frac{N}{2\lambda}(\sigma^2 + \pi^2) - \bar{\psi}(\sigma + i\pi\gamma_5)\psi \right\}. \tag{3}$$

The effective action in the leading $1/N$ order is

$$\frac{1}{N}S_{eff} = -\int d^d x \frac{\sigma^2 + \pi^2}{2\lambda} - i\ln\det[i\gamma^\mu D_\mu - (\sigma + i\gamma_5\pi)] \tag{4}$$

Then the effective potential (EP), defined for the constant configurations of π and σ as $V_{eff} = -S_{eff}/N\!\int d^d x$, is given by the formula

$$V_{eff} = \frac{\sigma^2}{2\lambda} + i\text{Sp}\ln\langle x|[i\gamma^\mu D_\mu - \sigma]|x\rangle \tag{5}$$

Here we put $\pi = 0$ because the final expression will depend on the combination $\sigma^2 + \pi^2$ only within our approximation. This means that we are actually considering the Gross-Neveu model.

But if we take into account kinetic terms of the fields π and σ generated by quantum corrections we will obtain different dynamics of these two fields. It should be noted that σ will be a massive scalar field in the supercritical area while π will be massless Goldstone particle.

By means of the usual Green function (GF), which obeys the equation

$$(i\gamma^\mu D_\mu - \sigma)_x G(x, x', \sigma) = \delta(x - x'), \tag{6}$$

we obtain the following formula

$$V'_{eff}(\sigma) = \frac{\sigma}{\lambda} - i\text{Sp}G(x, x, \sigma). \tag{7}$$

Now we can substitute in this equation the fermion GF in constant magnetic field

$$G(x, x', \sigma) = \Phi(x, x')\tilde{G}(x - x', \sigma), \tag{8}$$

where

$$\Phi(x, x') = \exp\left[ie \int_{x'}^{x} A^\mu(x'')dx''\right] \tag{9}$$

$$\tilde{G}_0(z, \sigma) = e^{-i\frac{\pi}{4}d} \int_0^\infty \frac{ds}{(4\pi s)^{\frac{d}{2}}} e^{-is\sigma^2} exp(-\frac{i}{4s} z_\mu C^{\mu\nu} z_\nu) \times \tag{10}$$
$$\left(\sigma + \frac{1}{2s}\gamma^\mu C_{\mu\nu} z^\nu - \frac{e}{2}\gamma^\mu F_{\mu\nu} z^\nu\right)\left[\tau \coth \tau - \frac{es}{2}\gamma^\mu \gamma^\nu F_{\mu\nu}\right].$$

Let us describe the 3D case to avoid some more complicated expressions. The EP is given by

$$V_{eff}(\sigma) = \frac{\sigma^2}{2\lambda} + \frac{1}{4\pi^{3/2}} \int_{1/\Lambda^2}^\infty \frac{ds}{s^{5/2}} e^{-s\sigma^2}(eBs)\coth(eBs) \tag{11}$$

The most reliable method to keep all of the divergences is the cut-off parameter introduction. So we can make the following trick: write the integral in the EP like

$$\int_{1/\Lambda^2}^\infty \frac{ds}{s^{5/2}} e^{-s\sigma^2}[(eBs)\coth(eBs) - 1] + \int_{1/\Lambda^2}^\infty \frac{ds}{s^{5/2}} e^{-s\sigma^2} \tag{12}$$

and calculate the last one keeping Λ finite while the first integral is finite already and we can put $1/\Lambda^2 = 0$. Then it appears to be possible to calculate it as a limit $\mu \to -1/2$ using the formula

$$\int_0^\infty dx x^{\mu-1} e^{-ax} \coth(cx) = \Gamma(\mu)\left[2^{1-\mu}(c)^{-\mu}\zeta(\mu, \frac{a}{2c}) - a^{-\mu}\right]. \tag{13}$$

Finally the EP has the form

$$V_{eff}(\sigma) = \frac{\sigma^2}{2\lambda} - \left[\frac{\Lambda \sigma^2}{2\pi^{3/2}} + \frac{\sqrt{2}}{\pi}(eB)^{3/2}\zeta\left(-\frac{1}{2}, 1 + \frac{\sigma^2}{2eB}\right) + \frac{1}{2\pi}eB\sigma\right] \tag{14}$$

There are two ways of justifying the introduction of the Λ parameter in the formula for the EP. The first one is the standard renormalization procedure, by means of the UV cut-off method. Then, in the limit $\Lambda \to \infty$, after renormalization of the coupling constant

$$\frac{1}{\lambda_R} = \frac{1}{\lambda} - \frac{\Lambda}{\pi^{3/2}}, \tag{15}$$

we have the expression for the renormalized EP in 3D spacetime

$$V_{eff}^{ren}(\sigma) = \frac{\sigma^2}{2\lambda_R} - \frac{\sqrt{2}}{\pi}(eB)^{3/2}\zeta\left(-\frac{1}{2}, 1 + \frac{\sigma^2}{2eB}\right) - \frac{1}{2\pi}eB\sigma \tag{16}$$

For $B = 0$ dynamical symmery breaking takes place when

$$\lambda > \lambda_c = \frac{\pi^{3/2}}{\Lambda} \tag{17}$$

if only we keep the finite cut-off Λ meanwhile the renormalized NJL model does not admit this phenomenon in general. However any finite value of the external magnetic field changes the situation dramatically and dynamical symmetry breaking occurs for any coupling constant. For $\sigma^2 \ll eB$ nontrivial solution of the gap equation defining a nontrivial minimum of the EP is given by

$$\sigma = \frac{eB\lambda_R}{2\pi} \tag{18}$$

The same calculations have been done for a constant electric field. The nonzero imaginary part appears in this case caused by vacuum instability of the quantum field theory in electric field. But treating the real part of the EP we can find that electrical field restores the chiral symmetry, initially broken for the finite cut-off parameter case.

Fig.1 illustrates the universal character of magnetic catalysis. It is a plot of 3D $V_{eff}^{ren}(\sigma)$ with $\mu = 100$; $\lambda\mu = 100$. Starting from above the curves correspond to the following electromagnetic field configurations: $eE/\mu^2 = 0.0002, B = 0$; $B = E = 0$; $eB\mu^2 = 0.0002, E = 0$.

After renormalization the chiral symmetry exists without external field but the magnetic field creates the non-zero minimum that indicates that DSB takes place. Meanwhile the external electric field works evidently against symmetry breaking. In all figures, an arbitrary dimensional parameter, μ, defining a typical scale in the model, is introduced in order to perform the plots in terms of dimensionless variables.

GENERAL EXPRESSION FOR EFFECTIVE POTENTIAL IN EXTERNAL ELECTROMAGNETIC AND GRAVITATIONAL FIELDS

We have just the same expression for the EP in curved spacetime

$$V'_{eff}(\sigma) = \frac{\sigma}{\lambda} - i\mathrm{Sp}G(x, x, \sigma) \tag{19}$$

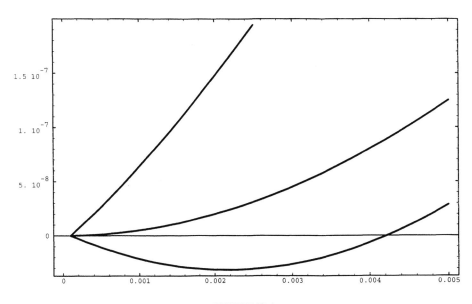

FIGURE 1.

To calculate the linear curvature corrections the local momentum expansion formalism is the most convinient one. Then in the special Riemannian normal coordinate framework

$$g_{\mu\nu}(x) = \eta_{\mu\nu} - \frac{1}{3} R_{\mu\rho\sigma\nu} y^\rho y^\sigma \tag{20}$$

with corresponding formulae for the others values and $y = x - x'$.

Then choosing the vector potential of the external electromagnetic field in the form

$$A_\mu(x) = -\frac{1}{2} F_{\mu\nu} x^\nu, \tag{21}$$

where $F_{\mu\nu}$ is the constant matrix of electromagnetic field strength tensor we find that:

$$G(x, x', \sigma) = \Phi(x, x') \left[\tilde{G}_0(x - x', \sigma) + \tilde{G}_1(x - x', \sigma) \ldots \right], \tag{22}$$

where $G_n \sim R^n$.

Therefore we obtain the iterative sequence of equations for the GF and the linear-curvature corrections are given by

$$\tilde{G}_1(x - x', \sigma) = \int dx'' G_{00}(x - x'', \sigma) \times \tag{23}$$

$$\left[-\frac{i}{6} \gamma^a R^\mu{}_{\rho\sigma a}(x'' - x')^\rho (x'' - x')^\sigma \partial_\mu^{x''} \tilde{G}_0(x'' - x', \sigma) - \tag{24} \right.$$

$$\left. \frac{i}{4} \gamma^a \sigma^{bc} R_{bca\lambda}(x'' - x')^\lambda \right] \tilde{G}_0(x'' - x', \sigma)$$

Here $G_{00}(x - x', \sigma)$ is a free fermion GF.

Substituting an exact flat spacetime GF of fermions in external electromagnetic field into this formula after some algebra we have evident expression for the EP with the linear- curvature accuracy in the constant curvature spacetime.

External constant magnetic field case

For 3D spacetime the EP is given by

$$V_{eff}(\sigma) = \frac{\sigma^2}{2\lambda} + \frac{1}{4\pi^{3/2}} \int_{1/\Lambda^2}^{\infty} \frac{ds}{s^{5/2}} \exp(-s\sigma^2) \tau \coth \tau - \qquad (25)$$

$$\frac{R}{144\pi^{3/2}} \int_{1/\Lambda^2}^{\infty} \int_{1/\Lambda^2}^{\infty} \frac{dsdt}{(t+s)^{5/2}(1+\kappa \coth \tau)^2} \exp[-(t+s)\sigma^2] \times \qquad (26)$$

$$\left[2\kappa(\kappa+\tau) + (9\tau+5\kappa)\coth\tau + \kappa(\tau-3\kappa)\coth^2\tau\right]$$

where $\tau = eBs, \kappa = eBt$.

We can perform the same renormalization procedure as in flat spacetime because no new divergences appear in the linear-curvature corrections. But we keep the cut of scheme here to study the most general situation.

The results are presented on Fig.2. It shows a plot of 3D $V_{eff}^{ren}(\sigma)$ with $\mu = 100$; $\lambda\mu = 100$ and fixed $eB\mu^2 = 0.0002$. Starting from above the curves correspond to the different values of spacetime curvature $R\mu^2 = 0.0025, 0.002, 0.001, 0$. Second-order phase transition ruled by the spacetime curvature takes place.

External constant electrical field case

We have for renormalized EP the following expression

$$V_{eff}^{ren}(\sigma) = \frac{\sigma^2}{2\lambda_R} - \frac{(2ieE)^{3/2}}{4\pi}\left[2\zeta(-\frac{1}{2},\frac{\sigma^2}{2ieE}) - \left(\frac{\sigma^2}{2ieE}\right)^{1/2}\right] + \qquad (27)$$

$$\frac{R\sigma}{24\pi} + \frac{iR(eE)^{1/6}}{2\pi^2 3^{7/3}} \exp(-\pi\frac{\sigma^2}{eE})\Gamma(\frac{2}{3})\sigma^{2/3}.$$

Here we have performed a small electric field expansion in the R-dependent term. A numerical analysis of $\text{Re}V_{eff}(\sigma)$ for negative coupling constant gives the typical behaviour of a first–order phase transition, as shown in Fig. 3. The critical values are defined as usual: R_{c1} corresponds to the spacetime curvature for which a local nonzero minimum appears, R_c, when the real part of EP is equal at zero and at the local minimum, and R_{c2}, when the zero extremum becomes a maximum. There is a plot of 3D $\text{Re}V_{eff}^{ren}/\mu^3$ as a function of σ/μ for fixed $eE/\mu^2 = 0.00005$ and $\lambda\mu = -100$. From above to below, the curves in the plot correspond to the following values of $R/\mu^2 = 0.006; 0.005; 0.004; 0.0032; 0$, respectively. The critical

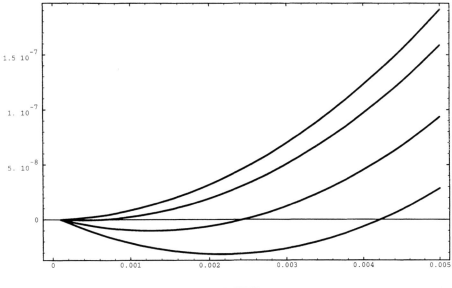

FIGURE 2.

values, defined as usual, are given by: $R_{c1}/\mu^2 = 0.005$; $R_c/\mu^2 = 0.0032$; $R_{c2}/\mu^2 = 0$. Λ obviously does not appear anywhere because after renormalization it must be sent to infinity, $\Lambda \to \infty$.

CONCLUSIONS

We clearly observe that a positive spacetime curvature tries to restore chiral symmetry even in the presence of external magnetic field. Therefore the universal character of magnetic catalysis doesn't survive in curved spacetime. From the other hand electric field increases the critical value of coupling constant as it does in flat spacetime.

It should be noted that for $D < 4$ is renormalizable and these conclusions don't depend already on the cut-off scale Λ.

This work has been partly financed by DGICYT (Spain), project PB96-0095, and by CIRIT (Generalitat de Catalunya), grant 1995SGR-00602. The work of Yu.I.Sh. was supported in part by Ministerio de Educación y Cultura (Spain), grant SB96-AN4620572.

REFERENCES

1. Nambu, Y., Jona-Lasinio G., *Phys. Rev.* **122**, 345 (1961).
2. Gross, D., Neveu, A., *Phys. Rev.* **D10**, 3235 (1974).

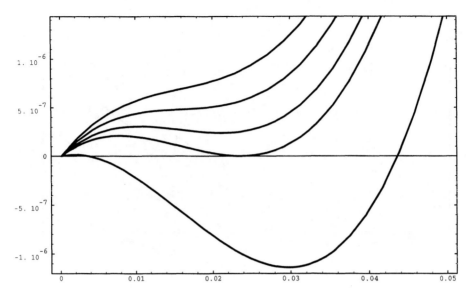

FIGURE 3.

3. Fahri, E., Jackiw, R., Eds., *Dynamical Symmetry Breaking* (World Scientific, Singapore, 1981); Muta, T., Yamawaki K., Eds., *Proceedings of the Workshop on Dynamical Symmetry Breaking* (Nagoya, 1990); Bardeen, W. A., Kodaira, J., Muta, T., Eds., *Proceedings of the International Workshop on Electroweak Symmetry Breaking* (World Scientific, Singapore, 1991)
4. Bando, M., Kugo, T., Yamawaki, K., *Phys. Rep.* **164**, 217 (1988); Rosenstein, B., Warr, B. J., Park, S. H., *ibid.* **205**, 59 (1991); Hatsuda, T., Kinuhiro, T., *ibid.* **247**, 221 (1994); Bijnens, J., *ibid.* **265**, 369 (1996).
5. Schwinger, J., *Phys. Rev.* **82**, 664 (1951).
6. Harrington, B. J., Park, S. Y., Yildiz, A., *Phys. Rev.* **D11**, 1472 (1975); Stone, M. *ibid.* **D14**, 3568 (1976); Kawati, S., Konisi, G., Miyata, H., *ibid.* **D28** 1537 (1983).
7. Klevansky, S. P., Lemmer, R. H., *Phys. Rev.* **D39**, 3478 (1989); Klevansky, S., *Rev. Mod. Phys.* **64**, 649 (1992); Klimenko, K. G., *Theor. Math. Phys.* **89**, 211, 388 (1991); *Z. Phys.* **C54**, 323 (1992); Krive I. V., Naftulin, S., *Phys. Rev.* **D46**, 2337 (1992); Suganuma, H., T. Tatsumi, T., *Ann. Phys. (NY)* **208**, 470 (1991); *Progr. Theor. Phys.* **90**, 379 (1993).
8. Cangemi, D., Dunne, G., D'Hoker, E., *Phys. Rev.* **D51**, R2513 (1995); **D52**, R3163 (1995); Leung, C. N., Ng, Y. J., Ackly, A. W., *ibid.* **D54**, 4181 (1996); Lee, D.-S., Leung, C. N., Ng, Y. J., *ibid.* **D55**, 6504 (1997); Ishi-i, M., Kashiwa, T., Tanemura, N., *Nambu-Jona-Lasinio model coupled to constant electromagnetic fields in D-dimension*, KYUSHU-HET-40, hep-th/9707248; Shushpanov, I. A., Smilga, A. V., *Phys. Lett* **B402**, 351 (1997); Ebert, D., Zhukovsky, V. Ch., *Mod. Phys. Lett.* **A12**, 2567 (1997); Hong, D. K., *Phys. Rev.* **D57**, 3759 (1998); Kanemura, S., Sato H.-T., Tochimura, H., *Nucl. Phys.* **B517**, 567 (1998).

9. Gusynin, V., Miransky, V., Shovkovy, I., *Phys. Rev.* **D52**, 4718 (1995); Gusynin, V., Miransky, V., Shovkovy, I., *Phys. Lett.* **B349**, 477 (1995); Gusynin, V., Miransky, V., Shovkovy, I., *Nucl. Phys.* **B462**, 249 (1996); Elias, V., McKeon, D. G. C., Miransky, V., Shovkovy, I., *Phys. Rev.* **D54**, 7884 (1996); Babansky, A. Yu., Gorbar, E. V., Shchepanyuk, G. V., *Phys. Lett.* **B419**, 272 (1998); Miransky, V. A. *Magnetic catalysis of dynamical symmetry breaking and Aharonov-Bohm effect*, hep-th/9805159.

10. Muta, T., Odintsov, S. D., *Mod. Phys. Lett.* **A6**, 3641 (1991); Hill, C. T., Salopek, D. S., *Ann. Phys. (NY)*, **213**, 21 (1992); Inagaki, T., Muta, T., Odintsov, S. D., *Mod. Phys. Lett.* **A8**, 2117 (1993); Elizalde, E., Odintsov, S. D., Shil'nov, Yu. I., *ibid.* **A9**, 913 (1994); Inagaki, T., Mukaigawa, S., Muta, T., *Phys. Rev.* **D52**, R4267 (1996); Elizalde, E., Leseduarte, S., Odintsov, S. D., Shil'nov, Yu. I., *ibid.* **D53**, 1917 (1996); Kanemura, S., Sato, H.-T., *Mod. Phys. Lett.* **A24**,1777 (1995); Miele, G., Vitale, P., *Nucl. Phys.* **B494**, 365 (1997).

11. Gitman, D. M., Odintsov, S. D., Shil'nov, Yu. I., *Phys. Rev.* **D54**, 2968 (1996); Geyer, B., Granda, L. N., Odintsov, S. D., *Mod. Phys. Lett.* **A11**, 2053 (1996); Elizalde, E., Odintsov, S. D., Romeo, A., *Phys. Rev.* **D54**, 4152 (1996); Inagaki, T., Odintsov, S. D., Shil'nov, Yu. I., *Dynamical symmetry breaking in the external gravitational and constant magnetic fields* KOBE-TH-97-02, hep-th/9709077; Elizalde, E., Shil'nov, Yu. I., Chitov, V. V., *Class. Quant. Grav.* **15**, 735 (1998).

On the Intrerplay between Perturbative and Nonperturbative QCD[1] [2]

Jan Fischer[1] and Ivo Vrkoč[2]

[1] *Institute of Physics, Academy of Sciences of the Czech Republic,*
Na Slovance 2, CZ-18040 Praha 8, Czech Republic
[2] *Mathematical Institute, Academy of Sciences of the Czech Republic,*
Žitná 25, CZ-11467 Praha 1, Czech Republic

Abstract. We discuss the current use of the operator-product expansion in QCD calculations. Treating the OPE as an expansion in inverse powers of an energy-squared variable and assuming a bound on the remainder along the euclidean ray, we observe how the bound develops with increasing deflection from the euclidean ray down to the cut (Minkowski region). We argue that the assumption that the remainder is constant for all angles in the cut complex plane is not justified.

Explicit bounds on the remainder can be obtained under additional conditions, which are still far from the realistic physical situation. Possible generalizations of the scheme are considered.

INTRODUCTION

The operator-product expansion (OPE) [1]

$$\mathrm{i}\int \mathrm{d}x \mathrm{e}^{\mathrm{i}qx} A(x)B(0) \approx \sum_k C_k(q)\mathcal{O}_k, \qquad (1)$$

is an efficient means to investigate the relation between perturbative and nonperturbative effects in quantum chromodynamics. Here, q is the total four-momentum of the system considered, $q^2 = s = -Q^2$. The coefficients $C_k(q)$ are singular at $Q^2 = 0$, and are ordered according to the increasing exponent k in Q^{-2k}.

Our aim is to discuss conditions under which a power expansion of the type

$$f(1/Q^2) \approx \sum_k a_k(q)/Q^{2k} \qquad (2)$$

---

[1] Invited lecture presented by J. Fischer at the International Conference "Particles, Fields and Gravitation" devoted to the memory of Professor Ryszard Rączka, Łódz, Poland, April 15-19 1998.
[2] Based on works supported in part by GAAV and GACR (Czech Republic) under grant numbers A1010711 and 202/96/1616 respectively.

can be extended to angles away from the euclidean semiaxis $Q^2 > 0$ in the complex Q^2 plane. This problem, difficult as it is, has a very practical motivation. In the calculation of many observable quantities in quantum chromodynamics, for instance in the determination of the QCD running coupling constant $\alpha_s(m_\tau^2)$ from the τ lepton hadronic width [2], contour integrals of the type

$$6\pi i \oint_{|s|=m_\tau^2} (1 - s/m_\tau^2)^2 (1 + 2s/m_\tau^2) P(s) \frac{\mathrm{d}s}{m_\tau^2} \qquad (3)$$

occur. (The kinematic factors in the integrand may be different for different observable quantities.) Here, m_τ is the τ-lepton mass and $P(s)$ is some combination of the electromagnetic two-point correlation functions $\Pi(s)$ for the vector and the axial vector colour singlet quark currents j^μ and j^ν, the $\Pi(s)$ being defined by the relations

$$\Pi^{\mu\nu} = (g^{\mu\nu}q^2 - q^\mu q^\nu)\Pi(-q^2) = i \int \mathrm{d}^4 x \, e^{-iqx} \langle 0| \, \mathrm{T}(j^\mu(x) j^\nu(0)) \, |0\rangle \qquad (4)$$

for different currents. As the $\Pi(s)$ are approximated by the corresponding operator-product expansion, it is important to look for methods of estimating the error caused by truncating the expansion. Note that to calculate the integral (3) the knowledge of $P(s)$ is required in all directions of the complex s plane cut along the minkowskian semiaxis $s > 0$, i.e., $Q^2 < 0$.

The precise behaviour of the expansion (1) is not known, not even is known whether exponential terms in Q^{-1} are present on the right-hand side. The symbol $\approx$ is understood differently in different contexts; one can however adopt some additional model assumptions, which make the scheme more or less close to the real physical situation.

In this talk I will report on a result in this field; details will be published separately [3]. We consider the following generic situation. Let $F(s)$ be holomorphic in $\mathcal{C}$, the complex s plane cut along $s \in [0, \infty]$, from which a bounded domain around the origin may be excluded. Let the numbers a_k, $k = 0, 1, 2, ... n-1$, and a positive number A_n exist such that the following inequality

$$|F(s) - \sum_{k=0}^{n-1} a_k/(-s)^k| < A_n/|s|^n \qquad (5)$$

is satisfied for all real negative $s < -b$, with b being a positive number, and n being a positive integer. This problem is of relevance for a number of QCD phenomena; we however make here two simplifying assumptions. As is seen from (5), we assume that, for Q^2 tending to $+\infty$, the n-th order remainder

$$R_n(1/Q^2) = f(1/Q^2) - \sum_{k=0}^{n-1} a_k/Q^{2k}, \qquad (6)$$

where $F(s) \equiv f(1/Q^2)$, tends to zero as the n-th power of $1/Q^2$ for at least one value of n. In addition, we assume that the logarithmic energy dependence of the expansion coefficients a_k can be neglected.

The problem is what inequality (if any) will hold along rays in the complex s plane, away from the negative real semiaxis (euclidean region).

RESULTS

The answer depends on additional assumptions imposed on the function $F(s)$. In the following, we consider two sets of such additional assumptions.

We introduce the variable $z = -1/s$, $z = re^{i\varphi} = x + iy$, and denote $f(z) \equiv F(s)$. Assuming the bound (5), we observe how it varies with the deflection of the ray from euclidean region. The resulting dependence is determined by the form of the additional conditions chosen. We choose the following two sets of them:

(i) Let $f(z)$ be represented in the form

$$f(z) = \int_0^\infty \frac{\rho(t)}{1+zt} dt \qquad (7)$$

with $\rho(t)$ nonnegative for $t \geq 0$, z complex, and let the moment

$$a_n = \int_0^\infty t^n \rho(t) dt \qquad (8)$$

exist for a positive integer n. Then the remainders

$$R_k(z) = (-z)^k \int_0^\infty \frac{\rho(t)}{1+zt} t^k dt \qquad (9)$$

with $k = 1, 2, ..., n$ are bounded by

$$|R_k(z)| \leq a_k |z|^k \qquad (10)$$

and

$$|R_k(z)| \leq a_k |z|^k / |\sin \varphi| \qquad (11)$$

for $\operatorname{Re} z > 0$ and $\operatorname{Re} z < 0$ respectively. Comparing (11) with (10), we see how the factor $1/|\sin \varphi|$ makes the estimate looser when the ray gets closer to the cut, i.e., when $\varphi \to \pm \pi$.

(ii) Note that the above scheme is still far from real physical situations, in which the numerator $\rho(t)$ is usually assumed to tend to a constant for $s \to \infty$, so that the moments (8) do not exist. We therefore consider also a more general scheme, replacing (7) and (8) with the condition that $f(z)$ is bounded by a constant M inside a circle of radius d in the cut z plane. We then obtain the following bound on the remainder $R_n(z)$

$$|R_n(re^{i\varphi})| \leq M_n(r/d_n)^{n(1-|\varphi|/\pi)}, \tag{12}$$

where the constants M_n and d_n are related to the constants M and d. Details of the theorem and its proof will be published elsewhere [3].

The resulting bounds (10) – (12), although derived under simplifying assumptions, nevertheless indicate that the integral (3) can receive essential contributions from the region near $s = m_\tau^2$, where the OPE has little chance appropriately to represent the function expanded.

SUMMARY AND OUTLOOK

1. When approximating $\Pi(s)$ with the operator product expansion, there is no justification to assume that the truncation error is the same in all directions of the complex s plane ranging from the euclidean ($s < 0$) to the minkowskian ($s > 0$)region. There are reasons to expect that the error worsens when proceeding from the former to the latter.

2. It is difficult to make a reliable estimate of the truncation error in the expansions (1) or (2) even in the euclidean region, unless a precise limit of QCD is known in which the OPE becomes exact. Estimates of the type (5) are very optimistic.

3. The combination of (5) with analyticity allows an extension of (5) into the complex s plane, if additional assumptions are made.

4. We have considered two possible additional assumptions, (i) (7) and (8), and (ii) the assumption that $f(z)$ is bounded by a constant inside a circle of the cut z plane. While the case (i) leads to (10) and (11), the case (ii) yields the bound (12), in which the exponent itself depends on the phase φ. In the latter case, the high-energy behaviour of the bound worsens considerably when the Minkowski region is approached.

5. These two sets of assumptions are still very far from physics. On the other hand, it is to be expected that their appropriate modification (e.g., inclusion of the energy dependence of the expansion coefficients a_k) will not imply an essential change of our argument that the truncation error increases with increasing deflection from the euclidean region.

6. In case (i), a significant improvement of the integrated error might be reached by reversing the order of the s-integration and the z-integration. No such chance seems to exist in the case (ii).

7. It is desirable to generalize the scheme by including the logarithmic dependence of the coefficients $a_k(q)$.

Acknowledgements

I am honoured to attend this Conference on "Particles, Fields and Gravitation '98" devoted to the memory of Professor Ryszard Rączka, who was my teacher and co-author in the International Centre of Theoretical Physics, Trieste, in the mid and late sixties, and colleague and close friend since then. Professor Rączka was not only an excellent scientist and teacher; he had also a deep appreciation of friendship, humanity and spiritual values in general, and a very warm attitude to family and children. Science was for him a way to reveal the truth, and also a means of reaching mutual understanding among the peoples of the present world. These attitudes sprang from his genuine Christian faith and were visible from his scientific creativity, his other activities and his whole life.

My personal recollections of this remarkable man are published elsewhere [5].

REFERENCES

1. K.G. Wilson, Phys.Rev. **179** (1969) 1499
2. E. Braaten, S. Narison and A. Pich, Nucl.Phys. **B 373** (1992) 581; A. Pich: QCD tests from tau decays. Invited talk at the 20th Johns Hopkins Workshop (Heidelberg, 27-29 June 1996), hep-ph/9701305; F. LeDiberder and A. Pich, Phys.Lett. **B 289** (1992) 165
3. J. Fischer and I. Vrkoč, to be published
4. M. Shifman, Int.J.Mod.Phys. **A 11** (1996) 3195
5. J. Fischer: Summation of Power Series in Particle Physics. Invited lecture at Professor Rączka's Memorial Day of the XVI Workshop on Geometric Methods in Physics, Białowieża, Poland, June 30 – July 6, 1997; Reports in Mathematical Physics (in print)

Chiral Symmetry Breaking in the Nambu–Jona-Lasinio Model in External Constant Electromagnetic Field

E.V. Gorbar

Bogolyubov Institute for Theoretical Physics, Ukraine

Abstract. Dynamical chiral symmetry breaking (DχSB) is studied in the Nambu–Jona-Lasinio model for an arbitrary external constant electric and magnetic fields. It is shown that the critical coupling constant increases with increasing of the value of the second invariant of electromagentic field $\vec{E} \cdot \vec{B}$, i.e. the second invariant inhibits $D\chi SB$.

As was shown in works [1–3] in the so called ladder approximation (bare vertex and bare photon propagator), quantum electrodynamics (QED) in the regime of strong coupling has a new phase with dynamically broken chiral symmetry. The new phase of QED possesses very interesting properties from the theoretical viewpoint [4–6]. Although at present the problem of occurance of this phase is only of academic interest because one need to have coupling constant that exceed one, there were attempts to use them for modelling electroweak symmetry breaking in technicolor-like models [7].

Recently, it was shown in [8, 9] that the situation may change in the presence of external electromagnetic fields, where $(D\chi SB)$ may occur at the regime of weak coupling. Note also that the problem of studing $D\chi SB$ in external fields is of interest on its own. In [10, 11], by using the Nambu-Jona-Lasinio (NJL) model [12], it was shown that $D\chi SB$ takes place in an external constant magnetic field at however small attraction between fermions (note that the fact that external magnetic field enhances $D\chi SB$ was first noted in the NJL model in [13] (see also [14])).

As shown in [10, 11], in the infrared, the dynamics of fermions in magnetic field 3 + 1 dimensions resembles the dynamics of fermions in 1 + 1 dimensions. Therefore, we have an effective reduction of dimension of space-time by 2 units and as a result the critical value of the coupling coustant in external magnetic field is equal to zero. It was latter shown in [15] that the same effect takes place in QED in external magnetic field. Note that although the critical value of coupling constant is zero extremely strong magnetic fields ($|\vec{B}| \geq 10^{13}G$) are necessary for experimentally significant consequences because the correction to the physical mass of electron is very tiny for weak magnetic fields.

The case of constant electric field was considered in [13] where it was shown that the value of the critical coupling constant is more in this case than in the case without electric field. In the present work we study DχSB in the NJL model in the case of an arbitrary combination of constant electric and magnetic fields. As well known, electromagnetic field has two Lorentz invariants $f_1 = \frac{1}{2}F_{\mu\nu}F^{\mu\nu} = \vec{B}^2 - \vec{E}^2$ and $f_2 = \frac{1}{2}\varepsilon^{\mu\nu\alpha\beta}F_{\mu\nu}F_{\alpha\beta} = \vec{E}\cdot\vec{B}$. Since D$\chi$ SB was already studied in cses of purely electric and magnetic constant external fields when only the first invariant f_1 of electromagnetic field is not equal to zero, in the present work we consider DχSB in the case where $f_2 \neq 0$.

The Lagrangian of the NJL model [12] in an external electromagnetic field reads

$$\mathcal{L} = \sum_{j=1}^{N} i\bar{\Psi}_j \gamma^\mu D_\mu \Psi_j + \frac{G}{2}\sum_{j=1}^{N}\left[(\bar{\Psi}_j\Psi_j)^2 + (\bar{\Psi}_j i\gamma_5 \Psi_j)^2\right], \quad (1)$$

where D_μ is the covariant derivative $D_\mu = \partial_\mu + ieA_\mu$ and $j = 1,2,\ldots,N$ flavor index. Lagrangian (1) is invariant with respect to the $U_L(N)\times U_R(N)$ chiral group. By using auxiliary fields π and σ, we can rewrite (1) in the following form:

$$\mathcal{L} = \sum_{j=1}^{N}\left[i\bar{\Psi}_j\gamma^\mu D_\mu\Psi_j - \bar{\Psi}_j(\sigma_j + i\gamma_5\pi_j)\Psi_j - \frac{1}{2G}\left(\sigma_j^2 + \pi_j^2\right)\right] \quad (2)$$

By taking integrals over fermion fields, we obtain the effective action for π and σ fields

$$\Gamma(\sigma,\pi) = -i\sum_{j=1}^{N}\text{Tr Ln}\left[i\gamma^\mu D_\mu - (\sigma_j + i\gamma_5\pi_j)\right] - \frac{1}{2G}\int d^4x(\sigma_j^2 + \pi_j^2). \quad (3)$$

To obtain the effective potential for σ and π fields, it suffices to consider the case of constant fields $\sigma = const$, $\pi = const$. Since the effective action is invariant with respect to the $U_L(N)\times U_R(N)$ chiral symmetry, the effective potential depends on π and σ fields only through the chirally invariant combination $\rho^2 = \sum_{j=1}^{N}(\sigma_j^2 + \pi_j^2)$. Therefore, in what follows it is sufficient to set $\pi_k = 0$, $\sigma_k = 0$ for $k = 2,\ldots,N$ and consider the effective potential only for the field σ_1 which we simply denote σ. Thus,

$$\Gamma(\sigma) = -i\text{Tr Ln}\left[i\gamma^\mu D_\mu - \sigma\right] - \frac{1}{2G}\int d^4x\sigma^2. \quad (4)$$

By using the method of proper time [16, 17], we represent the first term in (4) as follows:

$$-iTrLn(iD_\mu\gamma^\mu - \sigma) = -\frac{i}{2}TrLn(D^2 + \sigma^2) = \int \frac{i}{2s}tr\langle x|e^{-is(D^2+\sigma^2)}|x\rangle ds d^4x \quad (5)$$

As well known [17], vacuum of QED is not stable in an external electric field and the effective potential has an imaginary part which defines the rate of birth of fermion-antifermion pairs from vacuum per unit volume. Since we study the problem of $D\chi SB$, we can ignore this effect and consider only the real part of effective potential which is equal to

$$V(\sigma) = \frac{\sigma^2}{2G} + \frac{N}{8\pi^2} v.p. \int_{1/\Lambda^2}^{\infty} ds \frac{1}{s} e^{-s\sigma^2} M \coth(Ms) L \cot(Ls), \qquad (6)$$

where $L^2 = e^2 \frac{\sqrt{f_1^2+4f_2^2}-f_1}{2}$, $M^2 = e^2 \frac{\sqrt{f_1^2+4f_2^2}+f_1}{2}$, and $f_1 = \vec{B}^2 - \vec{E}^2$ and $f_2 = \vec{E} \cdot \vec{B}$ are two invariants of electromagnetic field. In (6) we introduced a cut-off $\frac{1}{\Lambda^2}$ and v.p. of the integral in s is present because we consider only the real part of the effective potential (recall that the imaginary part of the effective potential is given by residues in poles of $\cot(Ls)$). The gap equation $\delta V/\delta\sigma|_{\sigma=m} = 0$ has the form

$$\frac{1}{G} - \frac{N}{4\pi^2} v.p. \int_{1/\Lambda^2}^{\infty} ds e^{-sm^2} M \coth(Ms) L \cot(Ls) = 0. \qquad (7)$$

This gap equation was investigated in cases where only the first invariant of electromagnetic field is not equal to zero, i.e. for cases of purely electric and magnetic external fields. In this paper we study how the presence of nonzero electric field parallel to magnetic field ($\vec{E} \cdot \vec{B} \neq 0$) affects $D\chi SB$. By using some inequalities, we first analytically obtain an estimate from below for the critical coupling constant. We add and subtract $1/s$ to $M\coth(Ms)$ in the gap equation (7). Then

$$v.p. \int_{1/\Lambda^2}^{\infty} ds e^{-sm^2} M \coth(Ms) L \cot(Ls) = \Lambda^2 - \frac{\pi}{2}L +$$
$$v.p. \int_{1/\Lambda^2}^{\infty} ds e^{-sm^2} (M \coth(Ms) - 1/s) L \cot(Ls), \qquad (8)$$

where we used the result [13] $v.p. \int_{1/\Lambda^2}^{\infty} ds e^{-sm^2} L \frac{\cot(Ls)}{s} = \Lambda^2 - \frac{\pi}{2}L$. Further, we represent the integral in (8) as a sum of two integrals $\int_{1/\Lambda^2}^{\infty} = \int_{1/\Lambda^2}^{\frac{\pi}{2L}} + \int_{\frac{\pi}{2L}}^{\infty}$ (note that $\frac{\pi}{2L}$ is the first zero of $\cot(Ls)$). We now consider the integral from $\frac{\pi}{2L}$ to infinity. Since $\coth x \leq 1/x + 1$ for $x > 0$, we have

$$v.p. \int_{\frac{\pi}{2L}}^{\infty} ds e^{-sm^2} (M\coth(Ms) - 1/s) L \cot(Ls) \leq v.p. \int_{\frac{\pi}{2L}}^{\infty} ds e^{-sm^2} ML \cot(Ls).$$
$$(9)$$

Integrating by part, we obtain

$$v.p. \int_{\frac{\pi}{2L}}^{\infty} ds e^{-sm^2} ML \cot(Ls) = M \int_{\frac{\pi}{2L}}^{\infty} e^{-m^2 s} d(\ln(2|\sin Ls|))$$
$$= -M\ln(2|\sin\frac{\pi}{2}|) + m^2 M \int_{\frac{\pi}{2L}}^{\infty} e^{-m^2 s} \ln(2|\sin Ls|) ds. \qquad (10)$$

351

By using the formula [18]

$$\int_0^\infty e^{-qx}\ln(2|\sin ax|)dx = -q\sum_{k=1}^\infty \frac{1}{k(q^2+4k^2a^2)}, \operatorname{Re} q > 0, \tag{11}$$

and the fact that the integral $M\int_{1/\Lambda^2}^{\frac{\pi}{2L}}\ln(2|\sin Ls|)ds$ is finite, we conclude that the integral $m^2 M\int_{\frac{\pi}{2L}}^\infty e^{-m^2 s}\ln(2|\sin Ls|)ds$ tends to zero on the critical line (where $m^2 \to 0$). Thus, v.p. $\int_{\frac{\pi}{2L}}^\infty ds e^{-sm^2} ML\cot(Ls) = -M\ln 2$. Therefore, in view of (9), we have

$$\text{v.p.} \int_{\frac{\pi}{2L}}^\infty ds e^{-sm^2}(M\coth(Ms) - 1/s)L\cot(Ls) \leq -M\ln 2. \tag{12}$$

It remains to estimate from below the integral $\int_{1/\Lambda^2}^{\frac{\pi}{2L}} ds(M\coth(Ms)-1/s)L\cot(Ls)$. We have $\int_{1/\Lambda^2}^{\frac{\pi}{2L}} ds(M\coth(Ms)-1/s)L\cot(Ls) \leq \int_{1/\Lambda^2}^{\frac{\pi}{2L}}(M\cot(Ms)-1/s)\frac{ds}{s}$ (because $\cot x \leq 1/x$ for x in the interval from 0 to $\frac{\pi}{2}$). If $M \gg L$, then we use the estimate $\coth x \leq 1/x + 1$ because $\coth(Ms)$ is approximately 1 near the upper limit of integration. Therefore, in this case

$$\int_{1/\Lambda^2}^{\frac{\pi}{2L}} ds(M\coth(Ms) - 1/s)L\cot(Ls) \leq M\ln\frac{\pi\Lambda^2}{2L}. \tag{13}$$

If $M \ll L$, then $\coth(Ms) \ll 1$ in the interval of integration. By using the estimate $\coth x \leq \frac{1}{x} + x/3$ ($\frac{1}{x}$ and $x/3$ are simply two first terms of the Taylor expansion of cothx), we obtain

$$\int_{1/\Lambda^2}^{\frac{\pi}{2L}} ds(M\coth(Ms) - 1/s)L\cot(Ls) \leq \frac{M^2}{3}(\frac{\pi}{2L} - 1/\Lambda^2). \tag{14}$$

We now analyse the obtained results. We assume in what follows that $|f_1| \gg |f_2|$. In the magnetic-type case ($f_1 > 0$), we have $L \approx |e|(\frac{f_2^2}{f_1})^{1/2}$ and $M \approx |e|f_1^{1/2}$. By using (8), (12), and (13), we obtain the following estimate from below for the critical coupling constant in the magnetic-type case:

$$g_{cr} \geq \frac{1}{1 - \frac{L\pi}{2\Lambda^2} + \frac{M}{\Lambda^2}\ln\frac{\Lambda^2}{2L} - \frac{M}{\Lambda^2}\ln 2} \approx \frac{1}{1 + |e|\frac{f_1^{1/2}}{\Lambda^2}\ln\frac{\Lambda^2 f_1^{1/2}}{4|ef_2|}}, \tag{15}$$

where g_{cr} is the dimensionless critical coupling constant $g_{cr} = \frac{4\pi^2 G\Lambda^2}{N}$. It directly follows from (15) that the presence of electric field parallel to magnetic field is very important. Indeed, if $f_2 \neq 0$, then g_{cr} is no longer equal to zero (even if the magnetic field is very strong $|\vec{B}| \sim \Lambda^2$) in contrast to the case of purely magnetic field where $g_{cr} = 0$. If f_2 increases, g_{cr} is also increases. If $f_2 \to 0$, then $g_{cr} \to 0$, i.e. we recover the result obtained by Gusynin, Miransky, and Shovkovy [10]. In the electric-type case ($f_1 < 0$) we have $L \approx |e||f_1|^{1/2}$ and $M \approx |e|(\frac{f_2^2}{|f_1|})^{1/2}$. By using (8), (12), and (14), we obtain

$$g_{cr} \geq \frac{1}{1 - \frac{L\pi}{2\Lambda^2} + \frac{M^2}{3\Lambda^2}(\frac{\pi}{2L} - 1/\Lambda^2) - \frac{M}{\Lambda^2}\ln 2} \approx \frac{1}{1 - \frac{\pi|e||f_1|^{1/2}}{2\Lambda^2} - \frac{|e|}{\Lambda^2}(\frac{f_2^2}{|f_1|})^{1/2}\ln 2}. \quad (16)$$

It follows from (16) that in this case g_{cr} for $f_2 \neq 0$ is more than $g_{cr} = \frac{1}{1 - \frac{\pi|e||f_1|^{1/2}}{2\Lambda^2}}$ [13] in the case $f_2 = 0$. As f_2 goes to zero, our estimate coincides with the result obtained by Klevansky and Lemmer [13] in the electric-type case. Thus, we conclude from the obtained estimates that the second invariant of electromagnetic field inhibits $D\chi SB$. In magnetic-type case it looks rather natural (indeed, if $f_2 \neq 0$, then it means that $\vec{E} \neq 0$ and we know that electric field inhibits $D\chi SB$). However, in the electric-type case it appears unlikely. Indeed, let first $\vec{E} \neq 0, \vec{B} = 0$ (therefore, $f_2 = 0$). It is natural to assume that g_{cr} should decrease in the case $f_2 \neq 0$ because if $f_2 \neq 0$ it means that $\vec{B} \neq 0$ and we know that magnetic field assists $D\chi SB$. We can understand the cause of growth of g_{cr} with increasing of $\vec{E} \cdot \vec{B}$ as follows. Since we study the dependence on the second invariant, we keep the first invariant $\vec{B}^2 - \vec{E}^2$ unchanged. Without loss of generality we can assume that $\vec{E} \parallel \vec{B}$ (if not, one can perform an appropriate Lorentz transformation). If we increase f_2, then in order to keep the first invariant unchanged we have to increase both $\vec{B}$ and $\vec{E}$. Therefore, there is a competition between increasing of $\vec{B}$ and increasing of $\vec{E}$. It turned out that qualitatively increasing of $\vec{E}$ is more significant for g_{cr} than increasing of $\vec{B}$ for any f_1, therefore, g_{cr} always grows with increasing of f_2.

We found a rather rough analytic estimate from below for the critical coupling constant. To obtain a more accurate dependence of the critical coupling constant on f_2, we numerically calculate the integral in (7). A typical dependence of g_{cr} on f_2 in the electric-type case is shown in Fig. 1 (this figure corresponds to $\frac{f_1}{\Lambda^4} = -10^{-4}$) and in the magnetic-type case in Fig. 2 (where $\frac{f_1}{\Lambda^4} = 10^{-4}$).

We see from these figures that the critical coupling constant increases with increasing of f_2 for any value of f_1. In the electric-type case g_{cr} is always more that 1. In the magnetic-type case g_{cr} abruptly drops to zero as f_2 tends to zero. We also numerically calculated g_{cr} in the case where the first invariant is zero $f_1 = 0$ ($|\vec{E}| = |\vec{B}|$) and obtained a dependence which is similar to the electric-type case, i.e. g_{cr} increases with increasing of f_2. Thus, the numerical analysis of the gap equation confirms that the second invariant of electromagnetic field inhibits $D\chi SB$.

As known, the Dirac has an exact solution in the case of constant electric and magnetic fields, in the field of plane wave, and combination of plane wave and constant electric and magnetic fields. Since the case of arbitrary constant electric and magnetic fields has been considered and it is known that plane electromagnetic wave does not influence $D\chi SB$ in the NJL model, it would be of interest to study the last remaining case of combination of plane wave and constant electric field and perhaps more interesting constant magnetic field.

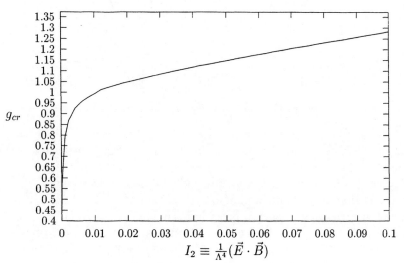

FIGURE 1. The dependence of the critical coupling constant on the second invariant in the magnetic-type case.

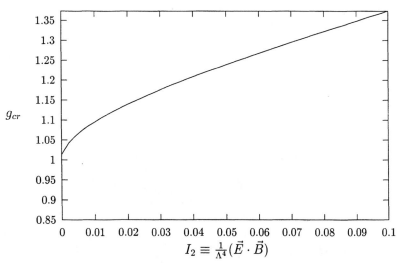

FIGURE 2. The dependence of the critical coupling constant on the second invariant in the electric-type case.

The author is grateful to his collaborators A. Yu. Babansky and G. V. Shchepanyuk on a joint work with whom the present report is based. I would like also to thank Prof. V.P. Gusynin for many fruitful discussions and remarks, Prof. V.A. Miransky for valuable comments, and Dr. I.A. Shovkovy for the help in drawing figures. This work was supported in part through grant INTAS-93-2058-EXT "East-West network in constrained dynamical systems" and by the Foundation of Fundamental Research of the Ministry of Science of the Ukraine through grant No. 2.5.1/003.

REFERENCES

1. T. Maskawa and H. Nakajima, Progr. Theor. Phys. 52, No. 4 (1974) 1326.
2. R. Fukuda and T. Kugo, Nucl Phys. B117, No. 1 (1976) 250.
3. P.I. Fomin and V.A. Miransky, Phys. Lett. B64, No. 2 (1976) 166;
   P.I. Fomin, V.P. Gusynin, V.A. Miransky, and Yu.A. Sitenko, Riv. Nuovo Cim. 6, No. 5 (1983) 1.
4. V.A. Miransky, Nuovo Cim. A90 (1985) 149.
5. C.N. Leung, S.T. Love, and W.A. Bardeen, Nucl. Phys. B273, No. 3 (1986) 649.
6. A. Kocic, E. Daggotto, and J.B. Kogut, Phys. Lett. B213, No. 1 (1988) 56.
7. T. Akiba and T. Yanagida, Phys. Lett. B169, No. 4 (1986) 432;
   K. Yamawaki, M. Bando, and K. Matumoto, Phys. Rev. Lett. 56 (1986) 1335;
   B. Holdom, Phys. Lett. B150 (1985) 301;
   T. Appelquist, D. Karabali, and L.C.R. Wijewardhana, Phys. Rev. Lett. 57 (1986) 957.
8. D. G. Caldi and A. Chodos, Phys. Rev. D36 (1987) 2876.
9. Y.J. Ng and Y. Kikuchi, Phys. Rev. D36 (1987) 2880.
10. V.P. Gusynin, V.A. Miransky, and I.A. Shovkovy, Phys. Rev. Lett. 73 (1994) 3499;
    V.P. Gusynin, V.A. Miransky, and I.A. Shovkovy, Phys. Rev. D52 (1995) 4718.
11. V.P. Gusynin, V.A. Miransky, and I.A. Shovkovy, Phys. Lett. B349 (1995) 477.
12. Y. Nambu and G. Jona-Lasinio, Phys. Rev. 122 (1961) 345.
13. S.P. Klevansky and R.H. Lemmer, Phys. Rev. D39 (1989) 3478.
14. S.P. Klevansky, Rev. Mod. Phys. 64 (1992) 649;
    I.V. Krive and S.A. Naftulin, Phys. Rev. D46 (1992) 2737;
    K.G. Klimenko, Z. Phys. C54 (1992) 323.
15. V.P. Gusynin, V.A. Miransky, and I.A. Shovkovy, Phys. Rev. D52 (1995) 4747;
    C..N. Leung, Y.J. Ng, and A.W. Ackley, Phys. Rev. D54 (1996) 4181;
    D.K. Hong, W. Kim, and S.-J. Sin, Phys. Rev. D54 (1996) 7879.
16. V.A. Fock, Sov. Phys. 12 (1937) 404.
17. J. Schwinger, Phys. Rev. 82 (1951) 664.
18. A.P. Prudnikov, Yu.A. Brychkov, O.I. Marichev, Integrals and series. Elementary functions (Nauka, Moscow, 1981, in Russian).
19. T. Appelquist, M. Bowick, D. Karabali, and L.C.R. Wijewardhana, Phys. Rev. D33 (1986) 3774.

On Quantization of Field Theories in Polymomentum Variables

Igor V. Kanatchikov[*][1]

[*] *Laboratory of Analytical Mechanics and Field Theory*
Institute of Fundamental Technological Research
Polish Academy of Sciences
Świętokrzyska 21, Warsaw PL-00-049, Poland

Abstract. Polymomentum canonical theories, which are manifestly covariant multi-parameter generalizations of the Hamiltonian formulation to field theory, are considered as a possible basis of quantization. We arrive at a multi-parameter hypercomplex generalization of quantum mechanics to field theory in which complex numbers and a time parameter are replaced by the space-time Clifford algebra and space-time variables appearing on equal footing. The corresponding covariant generalization of the Schrödinger equation fulfills several aspects of the correspondence principle: a relation to the Hamilton-Jacobi equation in the classical limit and the Ehrenfest theorem. A relation of the corresponding wave function (over a finite dimensional configuration space of field and space-time variables) to the Schrödinger wave functional in quantum field theory is examined in the ultra-local approximation.

INTRODUCTION

The canonical quantization is based on the Hamiltonian formalism. The conventional Hamiltonian formalism in field theory is an infinite dimensional version of that in mechanics. As a result, the quantum field theory based on it is essentially the quantum mechanics of systems with an infinite number of degrees of freedom. Most of difficulties and ambiguities of quantum field theory are due to this infinite dimensionality. However, should quantum fields always be understood in this way? Does this picture exhaust all aspects of quantum fields? Is there a "genuine quantum field theory" more general that just quantum mechanics applied to fields? It is clear that in pertubative regime, i.e. in the vicinity of a free field theory which can be represented as a continuum of harmonic oscillators, the above picture can work well, and it really does as the experimental triumph of pertubative quantum field theory demonstrates. However, applicability of this picture in non-pertubative domain and in curved space-time, where no natural particle concept exists in general, can be more limited.

---

[1] On leave from Tallin Technical University, Tallin, Estonia

A conceivable approach to the above posed questions can be based on the not yet widely acknowledged fact that the conventional version of the Hamiltonian formalism in field theory is not the only one possible. In fact, there exist different alternative extensions of the Hamiltonian formulation to field theory which all reduce to the Hamilton formalism in mechanics if the number of space-time dimensions equals to one. These extensions originate from the calculus of variations of multiple integrals [1-4]. Unlike the conventional Hamiltonian formalism, all these formulations are constructed in a manifestly covariant way not requiring any singling out of a time dimension. They can be applied even if the signature of the space-time is not Minkowskian. This is achieved by assigning the canonical momentum like variables, which we called *polymomenta* [5], to the whole set of space-time derivatives of a field: $\partial_\mu y^a \rightarrow p_a^\mu$ [2]. An analogue of the phase space is then a finite dimensional phase space of variables (y^a, p_a^μ, x^ν) which we call the *polymomentum phase space*. Corresponding generalizations of the canonical formalism will be referred to as *polymomentum canonical theories*. In the geometric (Cartan's) approach to the calculus of variations these theories (a version of which is also known as the multisymplectic formalism [7]) appear as a result of a certain choice of the so-called Lepagean equivalents of a field-theoretic (multidimensional) analogue of the Poincaré-Cartan form [4,6-9]. Unfortunately, applications of these theories in physics have been so far rather rare (see for references [4,5,7]).

The simplest example of a polymomentum canonical theory is the so-called De Donder-Weyl (DW) theory [1,2,4,7]. Given a Lagrangian density $L = L(y^a, \partial_\mu y^a, x^\nu)$, the polymomenta are introduced by the formula $p_a^\mu := \partial L / \partial(\partial_\mu y^a)$. An analogue of the Hamilton canonical function defined as $H := \partial_\mu y^a p_a^\mu - L$ is referred to as the *DW Hamiltonian function* in what follows. Note, that H is a function of variables $(y^a, p_a^\mu, x^\mu) =: z^M$. In these variables the Euler-Lagrange field equations can be rewritten in the form of *DW Hamiltonian field equations*

$$\partial_\mu y^a = \partial H / \partial p_a^\mu, \quad \partial_\mu p_a^\mu = -\partial H / \partial y^a. \quad (1)$$

Clearly, this formulation reproduces the standard Hamiltonian formulation in mechanics at $n = 1$. At $n > 1$ it provides us with a kind of multi-parameter, or "multi-time", manifestly covariant generalization of the Hamiltonian formalism. In doing so fields are treated not as infinite dimensional mechanical systems evolving with time, but rather as systems varying in space-time, with the DW Hamiltonian function controlling such a variation (similarly to the usual Hamiltonian controlling the time evolution).

The objective of the present contribution is to discuss an approach to quantization of fields based on polymomentum canonical theories. Although we confine ourselves exclusively to the approach based on the DW theory, we believe that basic ideas presented in what follows can be extended to more general polymomentum theories.

---

[2] Throughout the paper y^a denote field variables, x^μ are space-time variables ($\mu = 1, ..., n$), $\partial_\mu y^a$ are space-time derivatives (or first jets) of field variables, p_a^μ denote polymomenta.

GRADED POISSON BRACKET AND QUANTIZATION

The canonical quantization in mechanics is essentially based on the algebraic structure given by the Poisson bracket. One of the reasons why polymomentum canonical theories have not been used as a basis of quantization was the lack of an appropriate generalization of the Poisson bracket. In [5] we proposed such a generalization within the DW theory. The bracket is defined on horizontal differential forms $F = \frac{1}{p!} F_{\mu_1 ... \mu_p}(z^M) dx^{\mu_1} \wedge ... \wedge dx^{\mu_p}$ of various degrees p ($0 \leq p \leq n$), which play the role of dynamical variables (instead of functions in mechanics or functionals in the conventional Hamiltonian formalism in field theory). It leads to graded analogues of the Poisson algebra structure [5,10]. More specifically, the bracket on differential forms in DW theory leads to generalizations of the so-called Gerstenhaber algebra [11] (a graded analogue of the Poisson algebra with the grade of an element of the algebra with respect to the bracket differing by one from its grade with respect to the multiplication). For the purposes of the present paper it suffices to know a small subalgebra of the canonical brackets and a representation of the field equations in terms of the bracket operation[3].

Using the notation $\omega_\mu := (-1)^{(\mu-1)} dx^1 \wedge ... \wedge \widehat{dx^\mu} \wedge ... \wedge dx^n$ the canonical brackets in the (Lie) subalgebra of forms of degree 0 and $(n-1)$ read [5]

$$\{p_a^\mu \omega_\mu, y^b\} = \delta_a^b, \quad \{p_a^\mu \omega_\mu, y^b \omega_\nu\} = \delta_a^b \omega_\nu, \quad \{p_a^\mu, y^b \omega_\nu\} = \delta_a^b \delta_\nu^\mu, \qquad (2a,b,c)$$

with other brackets vanishing. All brackets in (2) reduce to the canonical bracket in mechanics when $n = 1$; in this sense they are canonical and can be viewed as a starting point of quantization.

Let us adopt the Dirac correspondence rule that Poisson brackets go over to commutators divided by $i\hbar$ and apply it to the canonical brackets (2). Note that this is just an assumption: while this principle proved to work well for the usual Poisson bracket its precise form and applicability to graded Poisson bracket in DW theory has to be confirmed. By quantizing (2a) we immediately conclude that

$$\widehat{p_a^\mu \omega_\mu} = i\hbar \partial_a,$$

where ∂_a is a partial derivative with respect to the field variables. The commutator corresponding to (2c) leads to a realization of $\widehat{\omega}_\mu$ and $\widehat{p}_a^\mu$ in terms of Clifford imaginary units, or Dirac matrices, under the assumption that the law of composition of operators is the symmetrized Clifford (=matrix) product [12,13]

$$\widehat{p}_a^\nu = -i\kappa \gamma^\nu \partial_a, \quad \widehat{\omega}_\nu = -\kappa^{-1} \gamma_\nu. \qquad (3)$$

The quantity κ of the dimension $[length]^{n-1}$ appears here on dimensional grounds. Due to the infinitesimal nature of the volume element ω_μ we expect the absolute

---

[3] For the reason of a limited space we avoid discussing properties of graded Poisson bracket in DW theory in details. In what follows we simply chose facts which we need and refer the interested reader for more details to [5,10,13].

value of κ to be "very large". Hence its relation to the ultra-violet cutoff scale [12] can be anticipated (see also the last section before Conclusion).

Note that the realization of operators in terms of Clifford imaginary units implies a certain generalization of the formalism of quantum mechanics. Namely, whereas the conventional quantum mechanics is built up on complex numbers which are essentially the Clifford numbers corresponding to the *one*-dimensional space-time (= the time dimension in mechanics), the present approach to quantization of fields viewed as multi-parameter Hamiltonian systems (of the De Donder-Weyl type) makes use of the hypercomplex (Clifford) algebra of the underlying space-time manifold [14,15].

In order to guess a form of quantum equations of motions within the present approach it is important to know how the field equations are represented in terms of the bracket operation and what is the meaning of the bracket with the DW Hamiltonian function. In fact, the bracket with H exists only for forms of degree higher than $(n-1)$ [5]. Using $(n-1)$-form canonical variables appearing in (2) DW Hamiltonian equations (1) can be written in Poisson bracket formulation as follows [5] (cf. [13])

$$\mathbf{d}(y^a \omega_\mu) = *\{\!\![H, y^a \omega_\mu]\!\!\} = *\partial H/\partial p_a^\mu, \quad \mathbf{d}(p_a^\mu \omega_\mu) = *\{\!\![H, p_a^\mu \omega_\mu]\!\!\} = - *\partial H/\partial y^a, \quad (4)$$

where $*$ is the Hodge duality operator acting on horizontal forms, and $\mathbf{d}$ is the total exterior differential $\mathbf{d}F := \frac{1}{p!}\partial_M F_{\mu_1...\mu_p} \partial_\mu z^M dx^\mu \wedge dx^{\mu_1} \wedge ... \wedge dx^{\mu_p}$, with z^M denoting the set of variables (y^a, p_a^μ, x^μ). For more general dynamical variables represented by p-forms F we need a notion of the bracket with an n-form $H\omega$, where $\omega := dx^1 \wedge ... \wedge dx^n$, which allows us to write the equations of motion in the symbolic form [5]

$$\mathbf{d}F = \{\!\![H\omega, F]\!\!\} + d^h F,$$

where d^h is the exterior differential with respect to the space-time (=horizontal) variables. Hence, we conclude that the DW Hamiltonian "generates" infinitesimal space-time variations of dynamical variables corresponding to the total exterior differentiation, much like the Hamilton function in mechanics generates the infinitesimal evolution along the time dimension.

Now, an analogue of the Schrödinger equation can be expected to have a form $\hat{i}\hat{d}\Psi \sim \widehat{H}\Psi$, where $\hat{i}$ and $\hat{d}$ denote appropriate analogues of the imaginary unit and the exterior differentiation respectively. Keeping in mind the above remark on a hypercomplex generalization of quantum mechanics appearing here, an analogy between the exterior differential and the Dirac operator (in fact, the latter is $d - *^{-1}d*$ [14]), and natural requirements imposed by the correspondence principle, the following generalization of the Schrödinger equation can be formulated [12,13,18]

$$i\hbar\kappa\gamma^\mu \partial_\mu \Psi = \widehat{H}\Psi, \quad (5)$$

where $\widehat{H}$ is the operator corresponding to the DW Hamiltonian function, the constant κ of dimension $[length]^{-(n-1)}$ appears again on dimensional grounds, and

$\Psi = \Psi(y^a, x^\mu)$ is a wave function over the configuration space of field and space-time variables. In the following section we demonstrate that this equations fulfills several aspects of the correspondence principle. Note also that it reproduces the quantum mechanical Schrödinger equation at $n = 1$.

Let us construct the DW Hamiltonian operator for the system of interacting scalar fields y^a in flat space-time given by the Lagrangian density

$$L = \frac{1}{2}\partial_\mu y^a \partial^\mu y_a - V(y). \qquad (6)$$

Then the polymomenta and the DW Hamiltonian function are given by

$$p^a_\mu = \partial_\mu y^a, \quad H = \frac{1}{2}p^a_\mu p^\mu_a + V(y). \qquad (7)$$

DW Hamiltonian field equations take the form

$$\partial_\mu y^a = p^a_\mu, \quad \partial_\mu p^\mu_a = -\partial V/\partial y^a, \qquad (8)$$

which is essentially a first order form of a system of coupled Klein-Gordon equations.

By quantizing the bracket

$$\{\!\!\{p^\mu_a p^a_\mu, y^b \omega_\nu\}\!\!\} = 2p^b_\nu \qquad (9)$$

we obtain [13]

$$\widehat{p^\mu_a p^a_\mu} = -\hbar^2 \kappa^2 \Delta,$$

where $\Delta := \partial_a \partial^a$ is the Laplacian operator in the space of field variables. Thus the DW Hamiltonian operator of the system of interacting scalar fields takes the form

$$\widehat{H} = -\frac{1}{2}\hbar^2 \kappa^2 \Delta + V(y). \qquad (10)$$

Note that for a free scalar field $V(y) = (1/2h^2)m^2 y^2$, so that the DW Hamiltonian operator becomes similar to the Hamiltonian operator of the harmonic oscillator in the space of field variables. Its eigenvalues divided by κ read $m_N = m(N + \frac{1}{2})$. Separating variables $\Psi(y, x^\mu) = \Phi(x) f(y)$ from (5) we obtain

$$\widehat{H} f_N = \kappa m_N f_N, \quad i\hbar \gamma^\mu \partial_\mu \Phi = m_N \Phi.$$

Then for a free scalar field any solution of (5) is a linear combination of

$$\Psi_{N,\mathbf{k},r}(y, \mathbf{x}, t) = u_{N,r}(\mathbf{k}) f_N(y) e^{\epsilon_r(i\omega_{N,\mathbf{k}} t - i\mathbf{k} \cdot \mathbf{x})}, \qquad (11)$$

where $\omega_{N,\mathbf{k}} := \sqrt{\mathbf{k}^2 + m_N^2/\hbar^2}$, $u_{N,r}(\mathbf{k})$ is a properly normalized constant spinor, $\epsilon_r = +1(-1)$ for positive (negative) energy solutions, and f_N are eigenfunctions of the harmonic oscillator in y-space. As a consequence, any Green function of (5) is given by [18]

$$K(y', \mathbf{x}', t'; y, \mathbf{x}, t) = \sum_{N=0}^{\infty} \bar{f}_N(y') f_N(y) D_N(\mathbf{x}' - \mathbf{x}, t' - t), \tag{12}$$

where D_N denotes a Green function of the spinor field of mass m_N. In doing so the type of the Green function D should coincide with the type of the Green function K. Note that at large space-time separations $|x' - x| \gg \hbar/m$ the contribution of the term with $N = 0$ dominates, so that the asymptotic space-time behavior of corresponding Green functions is that of a spinor particle with mass $\frac{1}{2}m$. We hope to present a more detailed analysis elsewhere.

THE CORRESPONDENCE PRINCIPLE

In this section we discuss three properties of Eq. (5) which make it a proper candidate to the Schrödinger equation within the polymomentum quantization. All three are in fact different aspects of the correspondence principle.

Remind first that the DW canonical theory leads to its own field theoretic generalization of the Hamilton-Jacobi theory [2,4]. The corresponding Hamilton-Jacobi equation is a partial differential equation on n functions $S^\mu = S^\mu(y^a, x^\nu)$

$$\partial_\mu S^\mu + H(x^\mu, y^a, p_a^\mu = \partial S^\mu/\partial y^a) = 0.$$

In a simple example of scalar fields (6) the DW Hamilton-Jacobi equation reads

$$\partial_\mu S^\mu = -\frac{1}{2} \partial_a S^\mu \partial_a S_\mu - \frac{1}{2} \frac{m^2}{\hbar^2} y^2. \tag{13}$$

Now, if we substitute (a hypercomplex analogue of) the quasiclassical ansatz

$$\Psi = R \, \exp(i S^\mu \gamma_\mu / \hbar \kappa) \eta, \tag{14}$$

where η is a constant reference spinor, to (5) and (10) we obtain a set of equations which can be transformed to the form [13]

$$\partial_\mu S^\mu = -\frac{1}{2} \partial_a S^\mu \partial_a S_\mu - \frac{1}{2} \frac{m^2}{\hbar^2} y^2 + \frac{1}{2} \hbar^2 \kappa^2 \frac{\Delta R}{R}, \tag{15}$$

$$\partial_a S^\mu \partial^a S_\mu = \partial_a |S| \partial^a |S|, \quad \partial_\mu S^\mu = \frac{S^\mu}{|S|} \partial_\mu |S|. \tag{16a, b}$$

In the first of these we recognize the DW Hamilton-Jacobi equation (13) with an additional term $\frac{1}{2}\hbar^2\kappa^2 \, \Delta R/R$ which is similar to the so-called *quantum potential* known in quantum mechanics [16] and vanishes in the classical limit $\hbar \to 0$. Last two equations are supplementary conditions which appear most likely due to the fact that the quasiclassical ansatz (14) does not represent a most general spinor, thus imposing certain restrictions on dynamics of the wave function. Note that in the case of quantum mechanics, $n = 1$, conditions (16a,b) reduce to trivial identities.

Thus, it is argued that in the classical limit equation (5) leads to the DW Hamilton-Jacobi equation (with two supplementary conditions which are specific to field theory and probably are due to restrictions imposed by the chosen in (14) analogue of the quasiclassical ansatz).

Another aspect of the correspondence principle we are to consider is the Ehrenfest theorem. Let us assume that expectation values of operators are given by

$$\langle \hat{O} \rangle := \int dy\, \overline{\Psi} \hat{O} \Psi, \qquad (17)$$

where $\overline{\Psi}$ is the Dirac conjugate of Ψ. These expectation values depend on space-time points as the averaging is performed only over the field space. Using generalized Schrödinger equation (5) with the DW Hamiltonian (10) we can show that [13]

$$\partial_\mu \langle \hat{p}_a^\mu \rangle = -\langle \widehat{\partial_a H} \rangle, \quad \partial_\mu \langle \widehat{y_a \omega^\mu} \rangle = \langle \widehat{p_a^\mu \omega_\mu} \rangle. \qquad (18)$$

By comparing (18) with DW Hamiltonian field equations (8) we conclude that the latter are fulfilled "in average" as a consequence of the representation of operators (3), generalized Schrödinger equation (5), and the definition of expectation values (17). However, it should be noted that this property is fulfilled only for specially chosen operators (try e.g. to evaluate $\partial_\mu \langle y^a \rangle$ to see that this will not yield the desired result $\langle \hat{p}_\mu^a \rangle$ for scalar fields). Moreover, the scalar product $\int dy\, \overline{\Psi} \Psi$ implied by definition (17) in general is not positive definite and depends on points of the space-time. Therefore, it can not be used for a probabilistic interpretation. These drawbacks urge us to look for a more appropriate version of the Ehrenfest theorem.

An alternative is suggested by the fact that generalized Schrödinger equation (5) possesses a positive definite and time independent scalar product

$$\int d\mathbf{x} \int dy\, \overline{\Psi} \beta \Psi, \qquad (19)$$

where we introduced the notation $\gamma^\mu =: (\gamma^i, \beta)$ ($i, j = 1, ..., n-1$), thus explicitly singling out the time variable $t := x^n$ and the time component of γ-matrices: $\beta := \gamma^t$ ($\beta^2 = 1$). The existence of the satisfactory scalar product of this kind implies that the probabilistic interpretation of the wave function which fulfills generalized Schrödinger equation (5) is possible only upon a time dimension is singled out. The wave function $\Psi(y^a, \mathbf{x}, t)$ is interpreted then as a probability amplitude of obtaining the field value y in the space point $\mathbf{x}$ in the moment of time t. As a result, the theory becomes very much similar to usual quantum mechanics of a fictious (spinor) particle in the space of variables $(y^a, \mathbf{x})$.

Now, new (global) expectation values of operators can be defined by

$$\langle \hat{O} \rangle := \int dy \int d\mathbf{x}\, \overline{\Psi} \beta \hat{O} \Psi. \qquad (20)$$

These expectation values depend only on time. Using definition (20) and generalized Schrödinger equation (5) written in the form

$$i\hbar \partial_t \Psi = -i\hbar \alpha^i \partial_i \Psi + \frac{1}{\kappa}\beta \widehat{H}\Psi, \qquad (21)$$

where $\alpha^i := \beta \gamma^i$, we obtain

$$\partial_t \langle y^a \rangle = \langle \hat{p}_t^a \rangle, \quad \partial_t \langle \hat{p}_a^t \rangle = -\langle \widehat{\partial_i p_a^i} \rangle - \langle \partial_a \widehat{H} \rangle. \qquad (22)$$

Note that in (22) we identified $\widehat{\partial_i p_a^i}$ with $-2i\hbar\kappa\gamma^i \partial_a \partial_i$. This identification is consistent with yet another aspect of the correspondence principle a discussion of which follows.

This aspect is a relation between the classical equations of motion and the Heisenberg equations of motion of operators. From (21) it follows that the time evolution is given by the operator

$$\widehat{\mathcal{E}} := -i\hbar\alpha^i \partial_i + \frac{1}{\kappa}\beta\widehat{H}. \qquad (23)$$

Then, proceeding according to the standard quantum mechanics we obtain

$$\partial_t y^a = \frac{i}{\hbar}[\widehat{\mathcal{E}}, y^a] = \hat{p}_t^a, \quad \partial_t \hat{p}_a^t = \frac{i}{\hbar}[\widehat{\mathcal{E}}, \hat{p}_a^t] = -\widehat{\partial_i p_a^i} - \partial_a \widehat{H}, \qquad (24)$$

where we assumed as before

$$\widehat{\partial_i p_a^i} = -2i\hbar\kappa\gamma^i \partial_a \partial_i. \qquad (25)$$

Hence, as a consequence of generalized Schrödinger equation (5) and the representation of operators (3), the Heisenberg equations of motion have the same form as classical DW Hamiltonian equations (1) written in the form with a singled out time dimension.

RELATION TO THE SCHRÖDINGER WAVE FUNCTIONAL

In this section a possible relationship between the Schrödinger wave functional in quantum field theory [17] and our wave function is examined[4]. We confine ourselves to the simplest example of a free real scalar field. For the seek of simplicity we henceforth put $n = 3 + 1$ and $\hbar = 1$.

The idea is as follows. On the one hand, the Schrödinger wave functional $\Psi([y(\mathbf{x})], t)$ is known to be a probability amplitude of the field configuration $y = y(\mathbf{x})$ to be observed in the moment of time t. On the other hand, our wave

---

[4] The presentation here essentially follows an unpublished preprint by the author [18].

function $\Psi(y, \mathbf{x}, t)$ can be interpreted as a probability amplitude of finding the value y of the field in the point $\mathbf{x}$ in the moment of time t. Hence, the wave functional could in principle be related to a certain composition of single amplitudes given by our wave function.

Let us consider the Schrödinger functional corresponding to the vacuum state of a free scalar field [17]

$$\Psi_0([y(\mathbf{x})], t) = \eta \exp\left(iE_0 t - \frac{1}{2}\int \frac{d\mathbf{k}}{(2\pi)^3} \omega_\mathbf{k}\, \tilde{y}(\mathbf{k})\tilde{y}(-\mathbf{k})\right), \qquad (26)$$

where the Fourier expansion $y(\mathbf{x}) = \int \frac{d\mathbf{k}}{(2\pi)^3} y(\mathbf{k}) e^{i\mathbf{k}\mathbf{x}}$ is used, η is a normalization factor, $\omega_\mathbf{k} := \sqrt{m^2 + \mathbf{k}^2}$, and E_0 is the vacuum state energy

$$E_0 = \lim_{\substack{V\to\infty \\ Q\to\infty}} \frac{1}{2} \int_V d\mathbf{x} \int_Q \frac{d\mathbf{k}}{(2\pi)^3} \omega_\mathbf{k} \qquad (27)$$

which is divergent if either the ultraviolet cutoff Q of the volume of integration in $\mathbf{k}$-space or the infrared cutoff V of the volume of integration over $\mathbf{x}$-space go to infinity. The symbol lim has a formal meaning throughout.

By replacing the Fourier integral by the Fourier series according to the rule $\int \frac{d\mathbf{k}}{(2\pi)^3} \to \lim_{V\to\infty} \frac{1}{V}\sum_{[\mathbf{k}]}$, $[\mathbf{k}] \in \mathbb{Z}^3$, the Schrödinger vacuum state functional can be written in the form of an infinite product of the harmonic oscillator ground state wave functions over all cells in $\mathbf{k}$-space

$$\Psi_0([y(\mathbf{x})], t) = \eta \lim_{V\to\infty} \prod_{[\mathbf{k}]} \exp \frac{1}{2}\left(i\omega_\mathbf{k} t - \frac{1}{V}\omega_\mathbf{k} y^2(|\mathbf{k}|)\right). \qquad (28)$$

Now, let us consider the ground state ($N = 0$) wave functions (cf. Eq. (11)) of generalized Schrödinger equation (5) for a free scalar field

$$\Psi_{N=0,\mathbf{k}}(y_\mathbf{x}, \mathbf{x}, t) = u_{N=0}(\mathbf{k}) e^{i\omega_{0,\mathbf{k}} t - i\mathbf{k}\cdot\mathbf{x}} e^{-\frac{m}{2\kappa}y^2}, \qquad (29)$$

where $\omega_{0,\mathbf{k}} = \sqrt{(\frac{m}{2})^2 + \mathbf{k}^2}$. To simplify a subsequent analysis, which is in any case of preliminary character, we ignore in what follows the spinor nature of the wave function encoded in $u_{N=0}(\mathbf{k})$. Taking into consideration the probabilistic interpretation of solutions (29) and assuming that there are no correlations between the field values in space-like separated points, the amplitude of funding in the vacuum state the whole configuration $y = y(\mathbf{x})$ can be represented as an infinite product of single amplitudes given by the ground state solutions (32) with $y = y(\mathbf{x})$ over all points $\mathbf{x}$ of the space. In order to ensure the spatial isotropy and homogeneity which are expected for the vacuum state we also have to take a product over all possible values of wave numbers because each separate mode with a wave number $\mathbf{k}$ violates these properties. This also agrees with an idea of the vacuum state in which all possible $\mathbf{k}$-states are filled. Hence, the following symbolic formula for the approximate composed vacuum amplitude can be written (up to a normalization)

$$\prod_{\mathbf{k}}\prod_{\mathbf{x}} e^{i\omega_{0,\mathbf{k}}t - i\mathbf{k}\cdot\mathbf{x}} e^{-\frac{m}{2\kappa}y(\mathbf{x})^2}. \qquad (30)$$

This expression can be assigned a meaning if a certain discretization in both **x**- and **k**-spaces is assumed. This discretization can be related to finite values of cutoff parameters V and Q which imply minimal volume elements in **k**-space and in **x**-space to be, respectively, $(2\pi)^3/V =: \xi^3$ and $(2\pi)^3/Q =: \lambda^3$. Then coordinates in **x**- and **k**-space are integers $[\mathbf{x}] \in \mathbb{Z}^3$ and $[\mathbf{k}] \in \mathbb{Z}^3$ such that $\mathbf{x} = [\mathbf{x}]\lambda$, $\mathbf{k} = [\mathbf{k}]\xi$. The continuum limit formally corresponds to $V \to \infty$ and $Q \to \infty$, however, an analysis of its existence in mathematical sense is beyond the scope of the present consideration. Using this discretization, the obvious identity $\prod_{\mathbf{k}} e^{i\mathbf{k}\cdot\mathbf{x}} = 1$, and the Fourier series expansion $y(\mathbf{x}) = \frac{1}{V}\sum_{[\mathbf{k}]} y_{\mathbf{k}} e^{i\mathbf{k}\cdot\mathbf{x}}$, we obtain

$$\prod_{\mathbf{k}} e^{i\omega_{0,\mathbf{k}}t} \prod_{\mathbf{x}} e^{-\frac{m}{2\kappa}y(\mathbf{x})^2}$$

$$= \lim_{\substack{V\to\infty\\Q\to\infty}} \prod_{[\mathbf{k}]} e^{i\omega_{0,\mathbf{k}}t} \prod_{[\mathbf{x}]} \exp\left(-\frac{m}{2\kappa}\frac{1}{V^2}\sum_{[\mathbf{q}']}\sum_{[\mathbf{q}'']} y_{\mathbf{q}'} y_{\mathbf{q}''} e^{i(\mathbf{q}'+\mathbf{q}'')\cdot\mathbf{x}}\right)$$

$$= \lim_{\substack{V\to\infty\\Q\to\infty}} \prod_{[\mathbf{k}]} e^{i\omega_{0,\mathbf{k}}t} \exp\left(-\frac{m}{2\kappa}\sum_{[\mathbf{x}]}\frac{1}{V^2}\sum_{[\mathbf{q}']}\sum_{[\mathbf{q}'']} y_{\mathbf{q}'} y_{\mathbf{q}''} e^{i(\mathbf{q}'+\mathbf{q}'')\cdot\mathbf{x}}\right)$$

$$= \lim_{\substack{V\to\infty\\Q\to\infty}} \prod_{[\mathbf{k}]} e^{i\omega_{0,\mathbf{k}}t} \exp\left(-\frac{m}{2\kappa}\frac{QV}{(2\pi)^3}\frac{1}{V^2}\sum_{[\mathbf{q}]} y_{\mathbf{q}} y_{-\mathbf{q}}\right)$$

$$= \lim_{\substack{V\to\infty\\Q\to\infty}} \prod_{[\mathbf{k}]} \exp\left(i\omega_{0,\mathbf{k}}t - \frac{m}{2\kappa V}\frac{Q}{(2\pi)^3} y_{\mathbf{k}} y_{-\mathbf{k}}\right), \qquad (31)$$

where in passing to the fourth line we have taken into account that the number of cells both in **x**- and **k**-space is equal to $QV/(2\pi)^3$.

Let us compare the composed amplitude (31) with the standard vacuum functional in the form (28). Two additional parameters κ and Q appear in (31): Q is an (infinitely large) ultra-violet cutoff of the volume in **k**-space, while κ is essentially the inverse of an infinitesimal (or very small) volume element in **x**-space (cf. Eq. (3)), i.e a kind of fundamental length to the power 3. Physically it is quite natural to relate the inverse of the fundamental length to the ultraviolet cutoff. We thus identify $\kappa = Q/(2\pi)^3$ obtaining the composed amplitude

$$\lim_{V\to\infty} \prod_{[\mathbf{k}]} \exp\left(i\omega_{0,\mathbf{k}}t - \frac{m}{2V} y_{\mathbf{k}} y_{-\mathbf{k}}\right) \qquad (32)$$

which is similar to (28) except that in (32) the proper mass m appears instead of the frequency $\omega_{\mathbf{k}} = \sqrt{m^2 + \mathbf{k}^2}$ and $\omega_{0,\mathbf{k}}$ replaces $\frac{1}{2}\omega_{\mathbf{k}}$.

It is easy to see that the discrepancy between (28) and (32) disappears in the ultra-local limit $|\mathbf{k}| \ll m$. In this limit the two-point Wightman function $\langle y(\mathbf{x}_1) y(\mathbf{x}_2)\rangle$ between space-like separated points $\mathbf{x}_1$ and $\mathbf{x}_2$ vanishes, so that there are no correlations between the field values in these points. This is, however, exactly the assumption which we made when writing the approximate composed amplitude in the form (30). Hence, in the ultra-local limit the composed amplitude constructed from the the ground state wave functions obeying generalized Schrödinger equation (5) is consistent with the Schrödinger wave functional of the vacuum state (28). Unfortunately, an attempt to extend this correspondence beyond the ultra-local limit leads to a difficulty of writing an expression for the composed amplitude similar to (30) which would account for all relevant correlations between the field values in space-like separated points.

Note, that another important byproduct of our analysis in this section is a conclusion that the constant κ which appeared in (3) and (5) on purely dimensional grounds has to be identified with an ultraviolet cutoff scale quantity.

CONCLUSION

Field theories can be viewed as multi-parameter Hamiltonian-like systems in which space-time variables appear on equal footing as analogues of the time parameter in mechanics. A quantization of such a version of the Hamiltonian formalism leads to an extension of the formalism of quantum mechanics in which the Clifford algebra of underlying space-time manifold plays a key role similar to that of complex numbers in quantum mechanics. The latter thus appears as a special case of a theory with a single (time) parameter. In this formulation a description of quantized fields is achieved in terms of a (spinor) wave function on a finite dimensional analogue of the configuration space (the space of field and space-time variables). The wave function satisfies a multi-parameter covariant generalization of the Schrödinger equation, Eq. (5), which is a partial derivative equation similar to the Dirac equation with the mass term replaced by an operator corresponding to a multi-parameter (polymomentum) analogue of Hamilton's canonical function. Note that despite the dynamics is formulated in a manifestly covariant manner the consideration of scalar products suggests that a proper probabilistic interpretation of the wave function still may require a time parameter to be singled out.

The description outlined above appears to be very different from that known in contemporary quantum field theory. A relation to the latter is a challenge to the theory presented here. In this paper we pointed out a relation to the Schrödinger wave functional which can thus far be followed only in ultra-local approximation. However, the latter is too rough for the real physics. Hence, further efforts are required to clarify possible connections with the standard quantum field theory.

Note in conclusion, that the present approach may have interesting applications to the problem of quantization of gravity and field theories on non-Lorentzian space-times. Further discussion can be found in [19] were a sketch of an approach to quantization of general relativity based on the present framework is presented.

REFERENCES

1. Th. De Donder, *Theorie Invariantive du Calcul des Variations,* Nuov. éd., Paris: Gauthier-Villars, 1935;
   H. Weyl, *Ann. Math. (2)* **36**, 607 (1935).
2. H. Rund, *The Hamilton-Jacobi Theory in the Calculus of Variations,* Toronto: D. van Nostrand, 1966.
3. M. Giaquinta, and S. Hilderbrandt, *Calculus of Variations,* vols. 1/2, Berlin: Springer, 1995/1996.
4. H.A. Kastrup, *Phys. Rep.* **101**, 1 (1983).
5. I.V. Kanatchikov, *Rep. Math. Phys.* **41**, 49 (1998), hep-th/9709229.
6. Th.H.J. Lepage, *Acad. Roy. Belgique Bull. Cl. Sci. (5e Sér)* **27** 27 (1941); *ibid.* **28** 73, 247 (1942)
7. M.J. Gotay, J. Isenberg, and J. Marsden, *Momentum Maps and Classical Relativistic Fields,* Part I: Covariant Field Theory, physics/9801019.
8. M.J. Gotay, in *Symplectic Geometry and Mathematical Physics,* eds. P. Donato, C. Duval, e.a., Boston: Birkhäuser, 1991, p. 160.
9. P. Dedecker, in *Differential Geometrical Methods in Mathematical Physics,* eds. K. Bleuler, and A. Reetz, *Lect. Notes Math.* **570**, Berlin: Springer-Verlag, 1977, p. 395.
10. I.V. Kanatchikov, *Rep. Math. Phys.* **40**, 225 (1997), hep-th/9710069.
11. M. Gerstenhaber, *Ann. Math.* **78**, 267 (1963).
12. I.V. Kanatchikov, *Int. J. Theor. Phys.* **37**, 333 (1998), quant-ph/9712058.
13. I.V. Kanatchikov, *De Donder-Weyl theory and a hypercomplex extension of quantum mechanics to field theory,* preprint, May (1998), submitted to *Rep. Math. Phys.*
14. I.M. Benn and R.W. Tucker, *An Introduction to Spinors and Geometry with Applications in Physics,* Bristol: Adam Hilger, 1987.
15. D. Hestenes, *Space-time algebra,* New York: Gordon and Breach, 1966.
16. D. Bohm, B.J. Hiley, and P.N. Kaloyerou, *Phys. Rep.,* **144**, 349 (1987);
    P.R. Holland, *Phys. Rep.,* **224**, 95 (1993).
17. B. Hatfield, *Quantum Field Theory of Point Particles and Strings,* Redwood City: Addison-Wesley, 1992;
    S.S. Schweber, *An Introduction to Relativistic Quantum Field Theory,* New York: Harper and Row, 1961.
18. I.V. Kanatchikov, *Hypercomplex wave functions and Born-Weyl quantization in field theory,* preprint, March 1996, unpublished.
19. I.V. Kanatchikov, *From the De Donder-Weyl Hamiltonian formalism to quantization of gravity,* to appear in *Proc. Int. Sem. Math. Cosmol.,* Potsdam 1998, eds. M. Rainer, and H.-J. Schmidt, Singapore: World Sci., 1998.

Witten's Integral and the Kontsevich Integral

Louis H. Kauffman

Department of Mathematics, Statistics and Computer Science
University of Illinois at Chicago
851 South Morgan Street
Chicago, IL, 60607-7045

Abstract. This paper is a version of the lecture given by the author in Warsaw in April 1998.

INTRODUCTION

The purpose of this paper is to show how the Kontsevich Integrals, giving rise to Vassiliev invariants in knot theory, arise naturally in the evaluation of Wilson loop contributions to the perturbative expansion of Witten's functional integral. This paper establishes a result that was hitherto folklore among physicists and topologists.

The paper is divided into three sections. Section 2 discusses Vassiliev invariants and invariants of rigid vertex graphs. The section three on the functional integral introduces the basic formalism and shows how the functional integral is related directly to Vassiliev invariants. Finally section 4 shows how the Kontsevich integral arises in the perturbative expansion of Witten's integral in the axial gauge. One feature of section 4 is a new and simplified calcuation of the neccessary correlation functions. We show how the Kontsevich integrals are the Feynman integrals for this theory.

Acknowledgement

It gives the author pleasure to thank Louis Licht, Chris King and Jurg Fröhlich for helpful conversations and to thank the National Science Foundation for support of this research under NSF Grant DMS-9205277 and the NSA for partial support under grant number MSPF-96G-179.

CP453, *Particles, Fields, and Gravitation*
edited by J. Rembieliński
© 1998 The American Institute of Physics   1-56396-837-1/98/$15.00

VASSILIEV INVARIANTS AND INVARIANTS OF RIGID VERTEX GRAPHS

If $V(K)$ is a (Laurent polynomial valued, or more generally - commutative ring valued) invariant of knots, then it can be naturally extended to an invariant of rigid vertex graphs [20] by defining the invariant of graphs in terms of the knot invariant via an "unfolding" of the vertex. That is, we can regard the vertex as a "black box" and replace it by any tangle of our choice. Rigid vertex motions of the graph preserve the contents of the black box, and hence implicate ambient isotopies of the link obtained by replacing the black box by its contents. Invariants of knots and links that are evaluated on these replacements are then automatically rigid vertex invariants of the corresponding graphs. If we set up a collection of multiple replacements at the vertices with standard conventions for the insertions of the tangles, then a summation over all possible replacements can lead to a graph invariant with new coefficients corresponding to the different replacements. In this way each invariant of knots and links implicates a large collection of graph invariants. See [20], [21].

The simplest tangle replacements for a 4-valent vertex are the two crossings, positive and negative, and the oriented smoothing. Let V(K) be any invariant of knots and links. Extend V to the category of rigid vertex embeddings of 4-valent graphs by the formula

$$V(K_*) = aV(K_+) + bV(K_-) + cV(K_0)$$

where K_+ denotes a knot diagram K with a specific choice of positive crossing, K_- denotes a diagram identical to the first with the positive crossing replaced by a negative crossing and K_* denotes a diagram identical to the first with the positive crossing replaced by a graphical node.

This formula means that we define $V(G)$ for an embedded 4-valent graph G by taking the sum

$$V(G) = \sum_S a^{i_+(S)} b^{i_-(S)} c^{i_0(S)} V(S)$$

with the summation over all knots and links S obtained from G by replacing a node of G with either a crossing of positive or negative type, or with a smoothing of the crossing that replaces it by a planar embedding of non-touching segments (denoted 0). It is not hard to see that if $V(K)$ is an ambient isotopy invariant of knots, then, this extension is an rigid vertex isotopy invariant of graphs. In rigid vertex isotopy the cyclic order at the vertex is preserved, so that the vertex behaves like a rigid disk with flexible strings attached to it at specific points.

There is a rich class of graph invariants that can be studied in this manner. The Vassiliev Invariants [41], [6], [4] constitute the important special case of these graph invariants where $a = +1$, $b = -1$ and $c = 0$. Thus $V(G)$ is a Vassiliev invariant if

$$V(K_*) = V(K_+) - V(K_-).$$

Call this formula the *exchange identity* for the Vassiliev invariant V. V is said to be of *finite type* k if $V(G) = 0$ whenever $|G| > k$ where $|G|$ denotes the number of (4-valent) nodes in the graph G. The notion of finite type is of extraordinary significance in studying these invariants. One reason for this is the following basic Lemma.

Lemma. If a graph G has exactly k nodes, then the value of a Vassiliev invariant v_k of type k on G, $v_k(G)$, is independent of the embedding of G.

Proof. The different embeddings of G can be represented by link diagrams with some of the 4-valent vertices in the diagram corresponding to the nodes of G. It suffices to show that the value of $v_k(G)$ is unchanged under switching of a crossing. However, the exchange identity for v_k shows that this difference is equal to the evaluation of v_k on a graph with $k + 1$ nodes and hence is equal to zero. This completes the proof.//

The upshot of this Lemma is that Vassiliev invariants of type k are intimately involved with certain abstract evaluations of graphs with k nodes. In fact, there are restrictions (the four-term relations) on these evaluations demanded by the topology and it follows from results of Kontsevich [4] that such abstract evaluations actually determine the invariants. The knot invariants derived from classical Lie algebras are all built from Vassiliev invariants of finite type. All this is directly related to Witten's functional integral [44].

VASSILIEV INVARIANTS AND WITTEN'S FUNCTIONAL INTEGRAL

In [44] Edward Witten proposed a formulation of a class of 3-manifold invariants as generalized Feynman integrals taking the form $Z(M)$ where

$$Z(M) = \int DA exp[(ik/4\pi)S(M,A)].$$

Here M denotes a 3-manifold without boundary and A is a gauge field (also called a gauge potential or gauge connection) defined on M. The gauge field is a one-form on a trivial G-bundle over M with values in a representation of the Lie algebra of G. The group G corresponding to this Lie algebra is said to be the gauge group. In this integral the "action" $S(M,A)$ is taken to be the integral over M of the trace of the Chern-Simons three-form $A \wedge dA + (2/3)A \wedge A \wedge A$. (The product is the wedge product of differential forms.)

$Z(M)$ integrates over all gauge fields modulo gauge equivalence (See [2] for a discussion of the definition and meaning of gauge equivalence.)

The formalism and internal logic of Witten's integral supports the existence of a large class of topological invariants of 3-manifolds and associated invariants of knots and links in these manifolds.

The invariants associated with this integral have been given rigorous combinatorial descriptions [35], [40], [26], [31], [42], [23], but questions and conjectures arising from the integral formulation are still outstanding. (See for example [3], [10], [12], [13], [36].) Specific conjectures about this integral take the form of just how it implicates invariants of links and 3-manifolds, and how these invariants behave in certain limits of the coupling constant k in the integral. Many conjectures of this sort can be verified through the combinatorial models. On the other hand, the really outstanding conjecture about the integral is that it exists! At the present time there is no measure theory or generalization of measure theory that supports it. Here is a formal structure of great beauty. It is also a structure whose consequences can be verified by a remarkable variety of alternative means.

We now look at the formalism of the Witten integral in more detail and see how it implicates invariants of knots and links corresponding to each classical Lie algebra. In order to accomplish this task, we need to introduce the Wilson loop. The Wilson loop is an exponentiated version of integrating the gauge field along a loop K in three space that we take to be an embedding (knot) or a curve with transversal self-intersections. For this discussion, the Wilson loop will be denoted by the notation $W_K(A) =< K|A >$ to denote the dependence on the loop K and the field A. It is usually indicated by the symbolism $tr(Pexp(\oint_K A))$. Thus

$$W_K(A) =< K|A >= tr(Pexp(\oint_K A)).$$

Here the P denotes path ordered integration - we are integrating and exponentiating matrix valued functions, and so must keep track of the order of the operations. The symbol tr denotes the trace of the resulting matrix.

With the help of the Wilson loop functional on knots and links, Witten writes down a functional integral for link invariants in a 3-manifold M:

$$Z(M,K) = \int DA exp[(ik/4\pi)S(M,A)] tr(Pexp(\oint_K A))$$

$$= \int DA exp[(ik/4\pi)S] < K|A > .$$

Here $S(M, A)$ is the Chern-Simons Lagrangian, as in the previous discussion. We abbreviate $S(M, A)$ as S and write $< K|A >$ for the Wilson loop. Unless otherwise mentioned, the manifold M will be the three-dimensional sphere S^3

An analysis of the formalism of this functional integral reveals quite a bit about its role in knot theory. This analysis depends upon key facts relating the curvature of the gauge field to both the Wilson loop and the Chern-Simons Lagrangian. The idea for using the curvature in this way is due to Lee Smolin [37] (See also [33]). To this end, let us recall the local coordinate structure of the gauge field $A(x)$, where x is a point in three-space. We can write $A(x) = A_k^a(x) T_a dx^k$ where the index a ranges from 1 to m with the Lie algebra basis $\{T_1, T_2, T_3, ..., T_m\}$. The index k goes from 1 to 3. For each choice of a and k, $A_k^a(x)$ is a smooth function defined on three-space. In $A(x)$ we sum over the values of repeated indices. The Lie algebra generators T_a are matrices corresponding to a given representation of the Lie algebra of the gauge group G. We assume some properties of these matrices as follows:

1. $[T_a, T_b] = if^{abc}T_c$ where $[x, y] = xy - yx$, and f^{abc} (the matrix of structure constants) is totally antisymmetric. There is summation over repeated indices.

2. $tr(T_a T_b) = \delta_{ab}/2$ where δ_{ab} is the Kronecker delta ($\delta_{ab} = 1$ if $a = b$ and zero otherwise).

We also assume some facts about curvature. (The reader may enjoy comparing with the exposition in [22]. But note the difference of conventions on the use of i in the Wilson loops and curvature definitions.) The first fact is the relation of Wilson loops and curvature for small loops:

Fact 1. The result of evaluating a Wilson loop about a very small planar circle around a point x is proportional to the area enclosed by this circle times the corresponding value of the curvature tensor of the gauge field evaluated at x. The curvature tensor is written
$$F_{rs}^a(x) T_a dx^r dy^s.$$
It is the local coordinate expression of $A \wedge dA + A \wedge A$.

Application of Fact 1. Consider a given Wilson line $<K|S>$. Ask how its value will change if it is deformed infinitesimally in the neighborhood of a point x on the line. Approximate the change according to Fact 1, and regard the point x as the place of curvature evaluation. Let $\delta <K|A>$ denote the change in the value of the line. $\delta <K|A>$ is given by the formula
$$\delta <K|A> = dx^r dx^s F_a^{rs}(x) T_a <K|A>.$$
This is the first order approximation to the change in the Wilson line.

In this formula it is understood that the Lie algebra matrices T_a are to be inserted into the Wilson line at the point x, and that we are summing over repeated indices. This means that each $T_a <K|A>$ is a new Wilson line obtained from the original line $<K|A>$ by leaving the form of the loop unchanged, but inserting the matrix T_a into that loop at the point x.

Remark. In thinking about the Wilson line $<K|A> = tr(Pexp(\oint_K A))$, it is helpful to recall Euler's formula for the exponential:
$$e^x = lim_{n \to \infty}(1 + x/n)^n.$$
The Wilson line is the limit, over partitions of the loop K, of products of the matrices $(1+A(x))$ where x runs over the partition. Thus we can write symbolically,
$$<K|A> = \prod_{x \in K}(1 + A(x))$$

$$= \prod_{x \in K}(1 + A_k^a(x)T_a dx^k).$$

It is understood that a product of matrices around a closed loop connotes the trace of the product. The ordering is forced by the one dimensional nature of the loop. Insertion of a given matrix into this product at a point on the loop is then a well-defined concept. If T is a given matrix then it is understood that $T < K|A>$ denotes the insertion of T into some point of the loop. In the case above, it is understood from context in the formula that the insertion is to be performed at the point x indicated in the argument of the curvature.

Remark. The previous remark implies the following formula for the variation of the Wilson loop with respect to the gauge field:

$$\delta < K|A > /\delta(A_k^a(x)) = dx^k T_a < K|A >.$$

Varying the Wilson loop with respect to the gauge field results in the insertion of an infinitesimal Lie algebra element into the loop.

Proof.
$$\delta < K|A > /\delta(A_k^a(x))$$

$$= \delta \prod_{y \in K}(1 + A_k^a(y)T_a dy^k)/\delta(A_k^a(x))$$

$$= \prod_{y < x \in K}(1 + A_k^a(y)T_a dy^k)[T_a dx^k] \prod_{y > x \in K}(1 + A_k^a(y)T_a dy^k)$$

$$= dx^k T_a < K|A >.$$

Fact 2. The variation of the Chern-Simons Lagrangian S with respect to the gauge potential at a given point in three-space is related to the values of the curvature tensor at that point by the following formula:

$$F_{rs}^a(x) = \epsilon_{rst}\delta S/\delta(A_t^a(x)).$$

Here ϵ_{abc} is the epsilon symbol for three indices, i.e. it is $+1$ for positive permutations of 123 and -1 for negative permutations of 123 and zero if any two indices are repeated.

With these facts at hand we are prepared to determine how the Witten integral behaves under a small deformation of the loop K.

Theorem. 1. Let $Z(K) = Z(S^3, K)$ and let $\delta Z(K)$ denote the change of $Z(K)$ under an infinitesimal change in the loop K. Then

$$\delta Z(K) = (4\pi i/k)\int dA \exp[(ik/4\pi)S][Vol]T_a T_a < K|A >$$

where $Vol = \epsilon_{rst} dx^r dx^s dx^t$.

The sum is taken over repeated indices, and the insertion is taken of the matrices $T_a T_a$ at the chosen point x on the loop K that is regarded as the "center" of the deformation. The volume element $Vol = \epsilon_{rst} dx_r dx_s dx_t$ is taken with regard to the infinitesimal directions of the loop deformation from this point on the original loop.

2. The same formula applies, with a different interpretation, to the case where x is a double point of transversal self intersection of a loop K, and the deformation consists in shifting one of the crossing segments perpendicularly to the plane of intersection so that the self-intersection point disappears. In this case, one T_a is inserted into each of the transversal crossing segments so that $T_a T_a < K|A>$ denotes a Wilson loop with a self intersection at x and insertions of T_a at $x + \epsilon_1$ and $x + \epsilon_2$ where ϵ_1 and ϵ_2 denote small displacements along the two arcs of K that intersect at x. In this case, the volume form is nonzero, with two directions coming from the plane of movement of one arc, and the perpendicular direction is the direction of the other arc.

Proof.

$$\delta Z(K) = \int DA exp[(ik/4\pi)S][\delta <K|A>]$$

$$= \int DA exp[(ik/4\pi)S][dx^r dy^s F^a_{rs}(x) T_a <K|A>]$$

$$= \int DA exp[(ik/4\pi)S] dx^r dy^s [\epsilon_{rst} \delta S/\delta(A^a_t(x))] T_a <K|A>$$

$$= \int DA [exp[(ik/4\pi)S] \delta S/\delta(A^a_t(x))] \epsilon_{rst} dx^r dy^s T_a <K|A>$$

$$= (-4\pi i/k) \int DA [\delta exp[(ik/4\pi)S]/\delta(A^a_t(x))] \epsilon_{rst} dx^r dy^s T_a <K|A>$$

$$= (4\pi i/k) \int DA exp[(ik/4\pi)S] \epsilon_{rst} dx^r dy^s [\delta[T_a <K|A>]/\delta(A^a_t(x))]$$

(integration by parts and the boundary terms vanish)

$$= (4\pi i/k) \int DA exp[(ik/4\pi)S][Vol] T_a T_a <K|A>.$$

This completes the formalism of the proof. In the case of part 2., a change of interpretation occurs at the point in the argument when the Wilson line is differentiated. Differentiating a self intersecting Wilson line at a point of self intersection is equivalent to differentiating the corresponding product of matrices with respect to a variable that occurs at two points in the product (corresponding to the two places where the loop passes through the point). One of these derivatives gives rise to a term with volume form equal to zero, the other term is the one that is described in part 2. This completes the proof of the Theorem.

In the case of switching a crossing the key point is to write the crossing switch as a composition of first moving a segment to obtain a transversal intersection of the diagram with itself, and then to continue the motion to complete the switch. One then analyses separately the case where x is a double point of transversal self intersection of a loop K, and the deformation consists in shifting one of the crossing segments perpendicularly to the plane of intersection so that the self-intersection point disappears. In this case, one T_a is inserted into each of the transversal crossing segments so that $T^a T^a < K|A>$ denotes a Wilson loop with a self intersection at x and insertions of T^a at $x + \epsilon_1$ and $x + \epsilon_2$ as in part 2. of the Theorem above. The first insertion is in the moving line, due to curvature. The second insertion is the consequence of differentiating the self-touching Wilson line. Since this line can be regarded as a product, the differentiation occurs twice at the point of intersection, and it is the second direction that produces the non-vanishing volume form.

Up to the choice of our conventions for constants, the switching formula is, as shown below

$$Z(K_+) - Z(K_-) = (4\pi i/k) \int DA exp[(ik/4\pi)S] T_a T_a < K_{**}|A>$$

$$= (4\pi i/k) Z(T^a T^a K_{**}),$$

where K_{**} denotes the result of replacing the crossing by a self-touching crossing. We distinguish this from adding a graphical node at this crossing by using the double star notation.

A key point is to notice that the Lie algebra insertion for this difference is exactly what is done (in chord diagrams) to make the weight systems for Vassiliev invariants (without the framing compensation). Here we take formally the perturbative expansion of the Witten integral to obtain Vassiliev invariants as coefficients of the powers of $(1/k^n)$. Thus the formalism of the Witten functional integral takes one directly to these weight systems in the case of the classical Lie algebras. In this way the functional integral is central to the structure of the Vassiliev invariants.

WILSON LINES, AXIAL GAUGE AND THE KONTSEVICH INTEGRALS

In this section we follow the gauge fixing method used by Fröhlich and King [9]. Their paper was written before the advent of Vassiliev invariants, but contains, as we shall see, nearly the whole story about the Kontsevich integral.

Let (x^0, x^1, x^2) denote a point in three dimensional space. Think of $x^2 = t$ as "time". Change to light-cone coordinates

$$x^+ = x^1 + x^2 = x^1 + t$$

and

$$x^- = x^1 - x^2 = x^1 - t.$$

Then the gauge connection can be written in the form

$$A(x) = A_+(x)dx^+ + A_-(x)dx^- + A_0(x)dt.$$

Let $CS(a)$ denote the Chern-Simons integral (over the three dimensional sphere)

$$CS(A) = (1/4\pi) \int tr(A \wedge Aa + (2/3)A \wedge A \wedge A).$$

We define *axial gauge* to be the condition that $A_- = 0$. We shall now work with the functional integral of the previous section under the axial gauge restriction. In axial gauge we have that $A \wedge A \wedge A = 0$ and so

$$CS(A) = (1/4\pi) \int tr(A \wedge dA).$$

Letting $\partial_\pm$ denote partial differentiation with respect to $x^\pm$, we get the following formula in axial gauge

$$A \wedge dA = (A_+\partial_- A_0 - A_0\partial_- A_+)dx^+ \wedge dx^- \wedge dt.$$

Thus, after integration by parts, we obtain the following formula for the Chern-Simons integral:

$$CS(A) = (1/2\pi) \int tr(A_+\partial_- A_0)dx^+ \wedge dx^- \wedge dt.$$

Letting ∂_i denote the partial derivative with respect to x_i, we have that

$$\partial_+\partial_- = \partial_1^2 - \partial_2^2.$$

If we replace x^2 with ix^2 where $i^2 = -1$, then $\partial_+\partial_-$ is replaced by

$$\partial_1^2 + \partial_2^2 = \nabla^2.$$

We now make this replacement so that the analysis can be expressed over the complex numbers.

Letting
$$z = x^1 + ix^2,$$
it is well known that
$$\nabla^2 ln(z) = 2\pi\delta(z)$$
where $\delta(z)$ denotes the Dirac delta function and $ln(z)$ is the natural logarithm of z. Thus we can write
$$(\partial_+\partial_-)^{-1} = (1/2\pi)ln(z).$$

Note that $\partial_+ = \partial_z = \partial/\partial z$ after the replacement of x^2 by ix^2. As a result we have that
$$(\partial_-)^{-1} = \partial_+(\partial_+\partial_-)^{-1} = \partial_+(1/2\pi)ln(z) = 1/2\pi z.$$

Now that we know the inverse of the operator ∂_- we are in a position to treat the Chern-Simons integral as a quadratic form in the pattern

$$(-1/2) < A, LA >= -CS(A)$$

where the operator

$$L = \partial_-.$$

Since we know L^{-1}, we can express the functional integral as a Gaussian integral:
We replace

$$Z(K) = \int DA exp[ikCS(A)] tr(P exp(\oint_K A))$$

by

$$Z(K) = \int DA exp[iCS(A)] tr(P exp(\oint_K A/\sqrt{k}))$$

by sending A to $(1/\sqrt{k})A$. We then replace this version by

$$Z(K) = \int DA exp[(-1/2) < A, LA >] tr(P exp(\oint_K A/\sqrt{k})).$$

In this last equality, we have assumed a replacement of k by $(\sqrt{i})k$ and x^2 by ix^2 so that there are no explicit appearances of i in the functional integral itself.

In this last formulation we can use our knowledge of L^{-1} to determine the the correlation functions and express $Z(K)$ perturbatively in powers of $(1/\sqrt{k})$.

Proposition. Letting

$$< \phi(A) >= \int DA exp[(-1/2) < A, LA >] \phi(A) / \int DA exp[(-1/2) < A, LA >]$$

for any functional $\phi(A)$, we find that

$$< A_+^a(z,t) A_+^b(s,w) >= 0,$$

$$< A_0^a(z,t) A_0^b(s,w) >= 0,$$

$$< A_+^a(z,t) A_0^b(s,w) >= 4\lambda \delta^{ab} \delta(t-s)/(z-w)$$

where λ is a constant.

Proof Sketch. Let's recall how these correlation functions are obtained. The basic formalism for the Gaussian integration is in the pattern

$$< A(z)A(w) >= \int DA exp[(-1/2) < A, LA >] A(z)A(w) / \int DA exp[(-1/2) < A, LA >]$$

$$= (\partial/\partial J(z))((\partial/\partial J(w))|_{J=0} exp[(1/2) < J, L^{-1}J >]$$

Letting $G * J(z) = \int dw G(z-w) J(w)$, we have that when

$$LG(z) = \delta(z)$$

377

($\delta(z)$ is a Dirac delta function of z.) then

$$LG * J(z) = \int dw \, LG(z-w)J(w) = \int dw \, \delta(z-w)J(w) = J(z)$$

Thus $G * J(z)$ can be identified with $L^{-1}J(z)$.
In our case
$$G(z) = 1/2\pi z$$
and
$$L^{-1}J(z) = G * J(z) = \int dw \, J(w)/(z-w).$$

Thus

$$< J(z), L^{-1}J(z) > = < J(z), G * J(z) > = (1/2\pi) \int tr(J(z) \int dw \, J(w)/(z-w))dz.$$

The results on the correlation functions then follow directly from differentiating this expression.

We are now prepared to give an explicit form to the perturbative expansion for

$$< K > = Z(K) / \int DA \exp[(-1/2) < A, LA >]$$

$$= \int DA \exp[(-1/2) < A, LA >] tr(P \exp(\oint_K A/\sqrt{k})) / \int DA \exp[(-1/2) < A, LA >]$$

$$= \int DA \exp[(-1/2) < A, LA >] tr(\prod_{x \in K}(1+(A/\sqrt{k}))) / \int DA \exp[(-1/2) < A, LA >]$$

$$= \sum_n (1/k^{n/2}) \oint_{K_1 < ... < K_n} < A(x_1)...A(x_n) > .$$

The latter summation can be rewrittten (Wick expansion) into a sum over products of pair correlations, and we have already worked out the values of these. In the formula above we have written $K_1 < ... < K_n$ to denote the integration over variables $x_1, ... x_n$ on K so that $x_1 < ... < x_n$ in the ordering induced on the loop K by choosing a basepoint on the loop. After the Wick expansion, we get

$$< K > = \sum_m (1/k^m) \oint_{K_1 < ... < K_n} \sum_{P=\{x_i < x'_i | i=1,...m\}} \prod_i < A(x_i)A(x'_i) > .$$

Now we know that

$$< A(x_i)A(x'_i) > = < A^a_k(x_i) A^b_l(x'_i) > T_a T_b dx^k dx^l.$$

Rewriting this in the complexified axial gauge coordinates, the only contribution is

$$< A^a_+(z,t) A^b_0(s,w) > = 4\lambda \delta^{ab} \delta(t-s)/(z-w).$$

Thus
$$<A(x_i)A(x'_i)>$$
$$=<A^a_+(x_i)A^a_0(x'_i)>T_aT_adx^+\wedge dt+<A^a_0(x_i)A^a_+(x'_i)>T_aT_adx^+\wedge dt$$
$$=(dz-dz')/(z-z')[i/i']$$

where $[i/i']$ denotes the insertion of the Lie algebra elements T_aT_a into the Wilson loop.

As a result, for each partition of the loop and choice of pairings $P=\{x_i<x'_i|i=1,...m\}$ we get an evaluation D_P of the trace of these insertions into the loop. This is the value of the corresponding chord diagram in the weight systems for Vassiliev invariants. These chord diagram evaluations then figure in our formula as shown below:

$$<K>=\sum_m(1/k^m)\sum_P D_P\oint_{K_1<...<K_n}\bigwedge_{i=1}^m(dz_i-dz'_i)/((z_i-z'_i)$$

This is a Wilson loop ordering version of the Kontsevich integral. To see the usual form of the integral appear, we change from the time variable (parametrization) associated with the loop itself to time variables associated with a specific global direction of time in three dimensional space that is perpendicular to the complex plane defined by the axial gauge coordinates. It is easy to see that this results in one change of sign for each segment of the knot diagram supporting a pair correlation where the segment is oriented (Wilson loop parameter) downward with respect to the global time direction. This results in the rewrite of our formula to

$$<K>=\sum_m(1/k^m)\sum_P(-1)^{|P\downarrow|}D_P\int_{t_1<...<t_n}\bigwedge_{i=1}^m(dz_i-dz'_i)/((z_i-z'_i)$$

where $|P\downarrow|$ denotes the number of points (z_i,t_i) or (z'_i,t_i) in the pairings where the knot diagram is oriented downward with respect to global time. The integration around the Wilson loop has been replaced by integration in the vertical time direction and is so inidicated by the replacement of $\{K_1<...<K_n\}$ with $\{t_1<...<t_n\}$

The coefficients of $1/k^m$ in this expansion are exactly the Kontsevich integrals for the weight systems D_P. It was Kontsevich's insight to see (by different means) that these integrals could be used to construct Vassiliev invariants from arbitrary weight systems satsfying the four-term relations. Here we have seen how these integrals arise naturally in the axial gauge fixing of the Witten functional integral.

REFERENCES

1. D. Altschuler and L. Freidel, Vassiliev knot invariants and Chern-Simons perturbation theory to all orders, *Commun. Math. Phys.* 187 (1997), 261-287.
2. M. F. Atiyah, *Geometry of Yang-Mills Fields*, Accademia Nazionale dei Lincei Scuola Superiore Lezioni Fermiare, Pisa ,1979.

3. M. F. Atiyah, *The Geometry and Physics of Knots*, Cambridge University Press, 1990.
4. D. Bar-Natan, On the Vassiliev knot invariants, *Topology* **34** (1995), 423-472.
5. D. Bar-Natan, *Perturbative Aspects of the Chern-Simons Topological Quantum field Theory*, Ph. D. Thesis, Princeton University, June 1991.
6. J. Birman and X. S. Lin, Knot polynomials and Vassiliev's invariants, *Invent. Math.* **111** No. 2 (1993), 225-270.
7. R. Bott and C. Taubes, On the self-linking of knots, *Jour. Math. Phys.* **35** (1994), pp. 5247-5287.
8. P. Cartier, Construction combinatoire des invariants de Vassiliev - Kontsevich des noeuds, *C. R. Acad. Sci. Paris* **316**, Série I, (1993), pp. 1205-1210.
9. J. Fröhlich and C. King, The Chern Simons Theory and Knot Polynomials, *Commun. Math. Phys.* **126** (1989), 167-199.
10. S. Garoufalidis, Applications of TQFT to invariants in low dimensional topology, (preprint 1993).
11. E. Guadagnini, M. Martellini and M. Mintchev, Chern-Simons model and new relations between the Homfly coefficients, *Physics Letters B*, Vol. 238, No. 4, Sept. 28 (1989), pp. 489-494.
12. D. S. Freed and R. E. Gompf, Computer calculation of Witten's three-manifold invariant, *Commun. Math. Phys.*, No. 141, (1991), pp. 79-117.
13. L. C. Jeffrey, Chern-Simons-Witten invariants of lens spaces and torus bundles, and the semi-classical approximation, *Commun. Math. Phys.*, No. 147, (1992), pp. 563-604.
14. V. F. R. Jones, A polynomial invariant of links via von Neumann algebras, *Bull. Amer. Math. Soc.*, 1985, No. 129, pp. 103-112.
15. V. F. R.Jones, Hecke algebra representations of braid groups and link polynomials, *Ann. of Math.*,Vol.126, 1987, pp. 335-338.
16. V. F. R.Jones, On knot invariants related to some statistical mechanics models, *Pacific J. Math.*, Vol. 137, no. 2,1989, pp. 311-334.
17. L. H. Kauffman, *On Knots*, Annals Study No. 115, *Princeton University Press* (1987)
18. L. H. Kauffman, State Models and the Jones Polynomial, *Topology*,Vol. 26, 1987,pp. 395-407.
19. L. H. Kauffman, Statistical mechanics and the Jones polynomial, *AMS Contemp. Math. Series*, Vol. 78,1989, pp. 263-297.
20. L. H. Kauffman, New invariants in the theory of knots, *Amer. Math. Monthly*, Vol.95,No.3,March 1988. pp 195-242.
21. L. H. Kauffman and P. Vogel, Link polynomials and a graphical calculus, *Journal of Knot Theory and Its Ramifications*, Vol. 1, No. 1,March 1992, pp. 59-104.
22. L. H. Kauffman, *Knots and Physics*, World Scientific Pub.,1991 and 1993
23. L. H. Kauffman and S. L. Lins, *Temperley-Lieb Recoupling Theory and Invariants of 3- Manifolds*, Annals of Mathematics Study 114, Princeton Univ. Press,1994.
24. L. H. Kauffman, Functional Integration and the theory of knots, J. Math. Physics, Vol. 36 (5), May 1995, pp. 2402 - 2429.
25. R. Kirby, A calculus for framed links in S^3, *Invent. Math.*, **45**, (1978), pp. 35-56.
26. R. Kirby and P. Melvin, On the 3-manifold invariants of Reshetikhin- Turaev for

sl(2,C), *Invent. Math.* 105, pp. 473-545,1991.
27. M. Kontsevich, Graphs, homotopical algebra and low dimensional topology, (preprint 1992).
28. T. Kohno, Linear representations of braid groups and classical Yang-Baxter equations, *Contemporary Mathematics*, Vol. 78, Amer. Math. Soc., 1988, pp. 339-364.
29. T. Q. T. Le and J. Murakami, Universal finite type invariants of 3-manifolds, (preprint 1995.)
30. R. Lawrence, Asymptotic expansions of Witten-Reshetikhin-Turaev invariants for some simple 3-manifolds. *J. Math. Phys.*, No. 36(11), November 1995, pp. 6106-6129.
31. W. B. R. Lickorish, The Temperley Lieb Algebra and 3-manifold invariants, *Journal of Knot Theory and Its Ramifications*, Vol. 2,1993, pp. 171-194.
32. M. Polyak and O. Viro, Gauss diagram formulas for Vassiliev invariants, *Intl. Math. Res. Notices*, No. 11, (1994) pp. 445-453.
33. P. Cotta-Ramusino,E. Guadagnini, M. Martellini, M. Mintchev, Quantum field theory and link invariants, *Nucl. Phys. B* **330**, Nos. 2-3 (1990), pp. 557-574
34. N. Y. Reshetikhin, Quantized universal enveloping algebras, the Yang-Baxter equation and invariants of links, I and II, *LOMI reprints E-4-87 and E-17-87*, Steklov Institute, Leningrad, USSR.
35. N. Y. Reshetikhin and V. Turaev, Invariants of Three Manifolds via link polynomials and quantum groups, *Invent. Math.*,Vol.103,1991, pp. 547-597.
36. L. Rozansky, Witten's invariant of 3-dimensional manifolds: loop expansion and surgery calculus, In *Knots and Applications*, edited by L. Kauffman, (1995), World Scientific Pub. Co.
37. L. Smolin, Link polynomials and critical points of the Chern-Simons path integrals, *Mod. Phys. Lett. A*, Vol. 4,No. 12, 1989, pp. 1091-1112.
38. V. G. Turaev, The Yang-Baxter equations and invariants of links, LOMI preprint E-3-87, Steklov Institute, Leningrad, USSR., *Inventiones Math.*,Vol. 92, Fasc.3, pp. 527-553.
39. V. G. Turaev and O. Y. Viro, State sum invariants of 3-manifolds and quantum 6j symbols, *Topology 31*, (1992), pp. 865-902.
40. V. G. Turaev and H. Wenzl, Quantum invariants of 3-manifolds associated with classical simple Lie algebras, *International J. of Math.*, Vol. 4, No. 2,1993, pp. 323-358.
41. V. Vassiliev, Cohomology of knot spaces, In *Theory of Singularities and Its Applications*, V.I.Arnold, ed., Amer. Math. Soc.,1990, pp. 23-69.
42. K. Walker, On Witten's 3-Manifold Invariants, (preprint 1991).
43. J. H. White, Self-linking and the Gauss integral in higher dimensions. *Amer. J. Math.*, **91** (1969), pp. 693-728.
44. E. Witten, Quantum field Theory and the Jones Polynomial, *Commun. Math. Phys.*,vol. 121, 1989, pp. 351-399.

Gauge Invariants and Bosonization

J. Kijowski[*], G. Rudolph and M. Rudolph[†]

[*]*Center for Theor. Phys., Polish Academy of Sciences*
al. Lotników 32/46, 02 - 668 Warsaw, Poland
[†]*Institut für Theoretische Physik, Univ. Leipzig*
Augustusplatz 10/11, 04109 Leipzig, Germany

Abstract. We present some results, which are part of our programme of analyzing gauge theories with fermions in terms of local gauge invariant fields. In a first part the classical Dirac–Maxwell system is discussed. Next we develop a procedure which leads to a reduction of the functional integral to an integral over (bosonic) gauge invariant fields. We apply this procedure to the case of QED and the Schwinger model. In a third part we go some steps towards an analysis of the considered models. We construct effective (quantum) field theories which can be used to calculate vacuum expectation values of physical quantities.

I INTRODUCTION

In this contribution we review some results concerning our programme of analyzing gauge theories in terms of physical observables (i.e. gauge invariants). For earlier applications of this programme to non-Abelian Higgs models we refer to [1]–[3]. Recently we have also applied it to theories of gauge fields interacting with fermionic matter fields, see [4] – [6] and [7]. In [4] we have proved that the classical Dirac-Maxwell system can be formulated in a spin-rotation covariant way in terms of gauge invariant quantities. In [5] we have shown that similar constructions work on the level of the (formal) functional integral of QED, where fermion fields are treated as anticommuting (Berezian) quantities, and in [7] we have applied our procedure to the 2–dimensional Schwinger model. As a result we obtained a functional integral completely reformulated in terms of local gauge invariant quantities, which differs essentially from the effective functional integral obtained via the Faddeev-Popov procedure.

We stress, that our approach circumvents any gauge fixing procedure and, therefore, also the Gribov problem. It leads to a natural bosonisation procedure and can be viewed as a general construction scheme for effective field theories. Moreover, we remark that our formulation seems to be well adapted to investigations of nonperturbative aspects of quantum field theories. For a first contribution in the case of QED see [6].

The procedure developed in [4] and [5] can also be applied to Yang–Mills theories. This is done in [8] and [9], where the functional integral of one-flavour chromodynamics is reduced to an integral over purely bosonic invariants.

In this contribution we outline the main ideas of our reduction procedure. We start with a short description of the classical Dirac–Maxwell system. Next we reformulate the full QED as well as the Schwinger model in $(1+1)$ dimension in terms of gauge invariant quantities. We are lead to a functional integral completely written in terms of (commuting) gauge invariant fields. In Section 3 we will go some steps towards the construction of effective theories and their applications within the obtained formulation. Some short remarks about further applications and perspectives will close this contribution.

II THE CLASSICAL DIRAC–MAXWELL–SYSTEM

A field configuration of this model consists of a $U(1)$-gauge potential (A_μ) and a four component spinor field (ψ^a), where $a, b, \ldots = 1, 2, \dot{1}, \dot{2}$ denote bispinor indices and $\mu, \nu, \ldots = 0, 1, 2, 3$ spacetime indices. A bispinor will be represented by a pair of Weyl spinors:

$$\psi^a = \begin{pmatrix} \phi^K \\ \varphi_{\dot K} \end{pmatrix} \equiv \begin{pmatrix} \phi^1 \\ \phi^2 \\ \varphi_{\dot 1} \\ \varphi_{\dot 2} \end{pmatrix}, \qquad (1)$$

where $K, L, \ldots = 1, 2$ denote ordinary spinor indices. The components of (ψ^a_A) are considered as commuting quantities.

The Lagrangian of the classical Dirac–Maxwell–system is given by

$$\mathcal{L} = -\frac{1}{4} F_{\mu\nu} F^{\mu\nu} - m \psi^{a*} \beta_{ab} \psi^b - \mathrm{Im}\left\{ \psi^{a*} \beta_{ab} \left(\gamma^\mu \right)^b{}_c D_\mu \psi^c \right\}, \qquad (2)$$

where

$$F_{\mu\nu} = \partial_\mu A_\nu - \partial_\nu A_\mu, \qquad (3)$$
$$D_\mu \psi^a = \partial_\mu \psi^a + i e A_\mu \psi^a \qquad (4)$$

are the electromagnetic field strength and the covariant derivative. The bar denotes complex conjugation, β_{ab} denotes the Hermitean metric in bispinor space and $(\gamma^\mu)^b{}_c$ are the Dirac matrices. For the representation used for these quantities see [4].

Let us define the following gauge invariant quantities:

$$l^\mu := \tfrac{1}{2} \bar\phi^{\dot K} \phi^L \sigma^\mu{}_{\dot K L}, \qquad r^\mu := \tfrac{1}{2} \varphi_{\dot K} \bar\varphi_L \sigma^{\mu \dot K L},$$
$$h := \bar\varphi_{\dot K} \phi^K, \qquad b_\mu := \mathrm{Im}\left\{ \bar h \left(\bar\varphi_{\dot K}(D_\mu \phi^K) + \phi^K(D_\mu \varphi^*_{\dot K}) \right) \right\}. \qquad (5)$$

Lemma 1 *We have the following identities:*

$$l^\mu l_\mu = 0, \qquad r^\mu r_\mu = 0, \qquad 2\, l^\mu r_\mu = |h|^2. \tag{6}$$

Theorem 1 *Every class $\{A_\mu, \psi^a\}$ of generic gauge equivalent configurations is in one-to-one correspondence with a set $\{l^\mu, r^\mu, \chi, b_\mu\}$, where χ is the phase of the complex field h.*

Proposition 1 *Field dynamics of the classical Dirac–Maxwell-system in terms of the above invariants is given by the Lagrangian*

$$\mathcal{L} = -\frac{1}{4} F_{\mu\nu} F^{\mu\nu} - 2m \sqrt{2(l \cdot r)} \operatorname{Re}\chi - \frac{1}{i}(l^\mu - r^\mu)\chi^* \partial_\mu \chi - 2e\, (l^\mu + r^\mu) B_\mu$$
$$+ \frac{1}{l \cdot r} \epsilon^{\alpha\beta\mu\gamma} l_\alpha r_\beta \nabla_\mu (l_\gamma + r_\gamma), \tag{7}$$

with $B_\mu = \frac{1}{2e|h|^2} b_\mu$ and $F_{\mu\nu} = \partial_{[\mu} B_{\nu]} + \frac{1}{2e(l \cdot r)^2} \epsilon^{\alpha\beta\gamma\delta} l_\alpha r_\beta \left((\nabla_{[\mu} r_\gamma)(\nabla_{\nu]} l_\delta) \right)$.

For the notion "generic" we refer to [4]. The proofs of Theorem 1 and Proposition 1 can be also found in [4].

III THE FUNCTIONAL INTEGRAL OF ABELIAN QUANTUM FIELD THEORIES IN TERMS OF GAUGE INVARIANTS

A Gauge Invariant Formulation of Quantum Electrodynamics

A field configuration of QED consists of a $U(1)$-gauge potential A_μ and a four-component spinor field ψ^a, represented by a pair of Weyl spinors, see (1), where $a, b, \ldots = 1, 2, \dot{1}, \dot{2}$ denote bispinor indices and $\mu, \nu, \ldots = 0, 1, 2, 3$ spacetime indices. The components of (ψ^a) anticommute and (pointwise) build up a Grassmann-algebra generated by 8 elements.

The Lagrangian of the considered model in Minkowski space, with metric $\eta^{\mu\nu} = \operatorname{diag}(1, -1, -1, -1)$, is given by

$$\mathcal{L} = \mathcal{L}_{gauge} + \mathcal{L}_{mat}$$
$$= -\frac{1}{4} F_{\mu\nu} F^{\mu\nu} - m \psi^{a*} \beta_{ab} \psi^b - \operatorname{Im} \left\{ \psi^{a*} \beta_{ab} \left(\gamma^\mu\right)^b{}_c D_\mu \psi^c \right\}, \tag{8}$$

where $F_{\mu\nu}$ and $D_\mu \psi^a$ are given by (3) and (4), respectively.

In the ordinary approach, the reduction of the naive functional integral

$$\mathcal{F} = \int \prod dA\, d\psi\, d\psi^*\, e^{i \int d^4 x\, \mathcal{L}[A, \psi, \psi^*]} \tag{9}$$

to an integral with the correct number of degrees of freedom is done via the Faddeev-Popov gauge fixing procedure. Here we propose a completely different approach.

For this purpose, we define the following Grassmann-algebra valued invariants:

$$L^\mu := \tfrac{1}{2}\phi^{\dot{K}*}\phi^L \sigma^\mu{}_{\dot{K}L}, \qquad R^\mu := \tfrac{1}{2}\varphi^*_L \varphi_{\dot{K}} \sigma^{\mu \dot{K}L},$$
$$H := \varphi^*_K \phi^K, \qquad B_\mu := \operatorname{Im}\left\{H^*\left(\varphi^*_K D_\mu \phi^K + \phi^K D_\mu \varphi^*_K\right)\right\}. \tag{10}$$

Here L^μ and R^μ are real-valued vector fields, H is a complex scalar field, whereas B_μ is a real-valued covector field.

The invariants (10) are subject to the following identities, summarized in

Lemma 2 *The following identities hold:*

$$-2L^\mu R_\mu = HH^* \equiv |H|^2, \tag{11}$$

$$(|H|^2)^2 F_{\mu\nu} = \frac{1}{2e}\left(|H|^2(\partial_{[\mu} B_{\nu]}) - (\partial_{[\mu}|H|^2)B_{\nu]}\right)$$
$$+ \frac{2}{e}\epsilon^{\alpha\beta\gamma\delta} R_\alpha L_\beta (\partial_{[\mu} R_{\gamma})(\partial_{\nu]} L_\delta), \tag{12}$$

$$|H|^2 \mathcal{L}_{mat} = -2m|H|^2 \operatorname{Re}(H) - (L^\mu + R^\mu)B_\mu$$
$$+ (L^\mu - R^\mu)\operatorname{Im}(H^* \partial_\mu H) - 2\epsilon^{\alpha\beta\mu\gamma} L_\alpha R_\beta \partial_\mu (R_\gamma + L_\gamma). \tag{13}$$

The proof of this lemma can be found in [5]. Observe that formula (12) is an identity on the level of elements of maximal rank of the full Grassmann algebra. Moreover, we note that

$$B_\mu = 2e|H|^2 A_\mu + \operatorname{Im}\left\{H^*(\varphi^*_K \partial_\mu \phi^K + \phi^K \partial_\mu \varphi^*_K)\right\}. \tag{14}$$

In a next step we introduce under the functional integral for the invariants L^μ, R^μ, B_μ, H, H^* corresponding c-number mates l^μ, r^μ, b_μ, h, h^*, which by definition are also gauge invariant. For that purpose we make use of the following notion of the δ-distribution on superspace

$$\delta(u - U) = \int e^{2\pi i \xi (u-U)} d\xi = \sum_{n=0}^{\infty} \frac{(-1)^n}{n!} \delta^{(n)}(u) U^n,$$

where u is a c-number variable and U an even-rank combination of Grassmann variables ψ and ψ^*. Due to the nilpotent character of U the above sum is finite. This δ-distribution is a special example of a vector-space-valued distribution in the sense of [10]. For an arbitrary smooth function f one shows easily that

$$f(u)\delta(u-U) = f(U)\delta(u-U).$$

¿From this we get a technique, which frequently will be used:

$$\int f(\alpha)\delta(u-U)\mathrm{d}u = \int f(\alpha\frac{u^n}{\tilde{u}^n})\delta(u-U)\delta(u-\tilde{u})\mathrm{d}u\mathrm{d}\tilde{u}$$
$$= \int f(\frac{\alpha}{\tilde{u}^n}U^n)\delta(u-U)\delta(U-\tilde{u})\mathrm{d}u\mathrm{d}\tilde{u}. \tag{15}$$

Here, α denotes any c-number or Grassmann-algebra-valued quantity and n is a positive integer, such that the rank of αU^n is smaller or equal to the maximal rank.

Thus, we get for the functional integral (9)

$$\mathcal{F} = \int \prod \mathrm{d}A \, \mathrm{d}\psi \, \mathrm{d}\psi^* \int \prod_x \{\mathrm{d}b_\mu \, \mathrm{d}l^\mu \, \mathrm{d}r^\mu \, \mathrm{d}h \, \mathrm{d}h^* \, \delta(h-H)\,\delta(h^*-H^*)$$
$$\times \delta(b_\mu - B_\mu)\,\delta(l^\mu - L^\mu)\,\delta(r^\mu - R^\mu)\} \, e^{i\int \mathrm{d}^4 x\, \mathcal{L}[A,\psi,\psi^*]}. \tag{16}$$

Now, using (15), we rewrite under the functional integral

$$\mathcal{L} = \mathcal{L}_{gauge} + \frac{|h|^2}{|\tilde{h}|^2}\mathcal{L}_{mat},$$

which leads with (12) and (13) to

$$\mathcal{F} = \int \prod_x \{\mathrm{d}\psi\,\mathrm{d}\psi^*\,\mathrm{d}A_\mu\} \int \prod_x \{\mathrm{d}b_\mu\,\mathrm{d}l^\mu\,\mathrm{d}r^\mu\,\mathrm{d}h\,\mathrm{d}h^*\,\delta(h-H)\,\delta(h^*-H^*)$$
$$\times \delta(b_\mu - B_\mu)\,\delta(l^\mu - L^\mu)\,\delta(r^\mu - R^\mu)\} \, e^{i\int \mathrm{d}^4 x\,\mathcal{L}[b_\mu,l^\mu,r^\mu,h,h^*]}, \tag{17}$$

with

$$\mathcal{L}[b_\mu,l^\mu,r^\mu,h,h^*] = -\frac{1}{4}F_{\mu\nu}F^{\mu\nu} - 2m\,\mathrm{Re}\{h\} - \frac{1}{|h|^2}(l^\mu + r^\mu)b_\mu$$
$$-(l^\mu - r^\mu)\frac{\mathrm{Im}\{h^*(\partial_\mu h)\}}{|h|^2} + 2\epsilon^{\alpha\beta\mu\gamma}l_\alpha r_\beta\,\partial_\mu(r_\gamma + l_\gamma),$$
$$F_{\mu\nu} = \frac{1}{2e(|h|^2)^2}\left(|h|^2(\partial_{[\mu}b_{\nu]}) - (\partial_{[\mu}|h|^2)b_{\nu]}\right)$$
$$+\frac{2}{e(|h|^2)^2}\epsilon^{\alpha\beta\gamma\delta}r_\alpha l_\beta\,(\partial_{[\mu}r_\gamma)(\partial_{\nu]}l_\delta).$$

Observe, that the the original field configuration $\{A_\mu,\psi,\psi^*\}$ enters only the δ–distributions. A_μ can be trivially integrated out using (16), giving a factor $2e\,|h|^2$. Inserting the integral representation of the δ–distributions into (17), the interation over ψ and ψ^* is given by a Gaussian integral, which yields a complicated integral kernel.

The final result of this reduction procedure can be summarized in the following

Proposition 2 *The functional integral $\mathcal{F}$ in terms of invariants is given by*

$$\mathcal{F} = \int \prod \mathrm{d}v_\mu\,\mathrm{d}l\,\mathrm{d}r\,\mathrm{d}h\,\mathrm{d}h^*\,K[l^\mu,r^\mu,h,h^*]\,e^{i\int \mathrm{d}^4 x\,\mathcal{L}[v_\mu,l^\mu,r^\mu,h,h^*]}, \tag{18}$$

with the integral kernel

$$K[l^\mu, r^\mu, h, h^*]$$
$$= \frac{1}{2\mathrm{i}} \left\{ \frac{1}{16} \frac{\partial^2}{\partial r^\nu \partial r_\nu} \frac{\partial^2}{\partial l^\mu \partial l_\mu} + \frac{\partial^4}{\partial h^2 \partial h^{*2}} - \frac{1}{2} \frac{\partial^2}{\partial r^\mu \partial l_\mu} \frac{\partial^2}{\partial h \partial h^*} \right\} \delta(h)\delta(h^*)\delta(l)\delta(r), \quad (19)$$

and the (bosonized) Lagrangian

$$\mathcal{L}[v_\mu, l^\mu, r^\mu, h, h^*]$$
$$= -\frac{1}{4} \left(\partial_{[\mu} v_{\nu]} \right)^2 - 2m\mathrm{Re}\{h\} - 2e(l^\mu + r^\mu)v_\mu - 2(l^\mu - r^\mu)\frac{\mathrm{Im}\{h^* \partial_\mu h\}}{2|h|^2}, \quad (20)$$

where $v_\mu := \frac{b_\mu}{2e\,|h|^2}$.

B Gauge Invariant Formulation of the Massless Schwinger Model

As a toy model for some considerations towards effective theories we will treat the gauged Schwinger model in $(1+1)$ dimensions. A field configuration is given by a $U(1)$-gauge potential A_μ, $\mu = 1, 2$, and a spinor field ψ^K, $K = 1, 2$, which is represented by

$$\psi^K = \begin{pmatrix} \phi \\ \varphi \end{pmatrix}, \qquad \psi_K^* = (\phi^*, \varphi^*), \quad (21)$$

with anticommuting components. The functional integral is defined by

$$\mathcal{F} = \int \prod dA\, d\psi\, d\overline{\psi}\, e^{\mathrm{i}\int d^2 x\, \mathcal{L}[A, \psi, \overline{\psi}]}, \quad (22)$$

with the Lagrangian

$$\mathcal{L}[A, \psi, \overline{\psi}] = -\frac{1}{4} F_{\mu\nu} F^{\mu\nu} - \mathrm{Im}\left[\overline{\psi_K} (\gamma^\mu)^K{}_L \left(D_\mu \psi^L\right)\right]. \quad (23)$$

Here we denote $F_{\mu\nu} = \partial_{[\mu} A_{\nu]}$ and $D_\mu \psi^K = \partial_\mu \psi^K + \mathrm{i}e A_\mu \psi^K$ as well as $\overline{\psi_L} := \psi_K^* (\gamma^0)^K{}_L = (\varphi^*, \phi^*)$.

We define the following set of gauge invariant Grassmann algebra valued quantities:

$$H := \overline{\phi}\,\varphi, \qquad B_\mu := \mathrm{Im}\left\{\overline{H}\,\overline{\phi}\,(D_\mu \varphi)\right\},$$
$$J_\mu := \overline{\psi}\,\gamma_\mu\,\psi, \qquad J_\mu^5 := \overline{\psi}\,\gamma_\mu\,\gamma^5\,\psi. \quad (24)$$

Now we introduce for H, B_μ, J_μ, J_μ^5 corresponding c-number mates h, b_μ, j_μ, j_μ^5. Following the lines of our reduction procedure introduced in the last subsection, we are lead to the following

Proposition 3 *The functional integral of the* $(1+1)$-*dimensional Schwinger model in terms of gauge invariant bosonic fields* $v_\mu, j_5^\mu, |h|$ *and* θ, *where* $h := |h|e^{i\theta}$ *and* $v_\mu := \frac{b_\mu}{e|h|^2}$, *is given by:*

$$\mathcal{F} = \int \prod_x \left\{ dv_\mu dj_5^\mu \, d|h|^2 \, d\theta \, K[j_5^\mu, |h|] \right\} e^{i \int d^2x \, \mathcal{L}[v_\mu, j_5^\mu, \theta]}, \qquad (25)$$

with the integral kernel

$$K[j_5^\mu, |h|] = \frac{1}{2\pi} \left(\frac{\delta^2}{\delta j_\mu^5 \delta j_5^\mu} - \frac{1}{4} \frac{\delta^2}{\delta |h|^2} + \frac{1}{4|h|} \frac{\delta}{\delta |h|} \right) \delta(j_\mu^5) \delta(|h|^2), \qquad (26)$$

and the (bosonized) Lagrangian

$$\mathcal{L}[v_\mu, j_\mu^5, \theta] = -\frac{1}{4} \left(\partial_{[\mu} v_{\nu]} \right)^2 - e\, \epsilon^{\mu\nu} v_\mu j_\nu^5 + \frac{1}{2} j_\mu^5 \left(\partial^\mu \theta \right). \qquad (27)$$

We stress that $d\theta$ is the Haar–measure on S^1.

Finally, let us make some remarks. If we consider the massive Schwinger model, where an additional term $-m\overline{\psi}\psi$ occurs in the Lagrangian (23), we get an additional term $-2m\mathrm{Re}(h) = -2m|h|\cos\theta$ in (26). This leads to a field theory of sine–Gordon type, but instead of getting a constant in front of the $\cos(\theta)$–term, we obtain $|h|$. However, since $|h|$ is a non-propagating field, it can be 'averaged" to a constant.

IV TOWARDS EFFECTIVE THEORIES

A The Effective Schwinger Model

In this section we analyse the result of our reduction procedure (Proposition 3) and construct an effective field theoretic model, which can be used to calculate vacuum expectation values of physical quantities. First observe that a nontrivial singular integral kernel $K[j_5^\mu, |h|]$ occurs in the functional integral (25). Obviously, the quantity $|h|$ can be integrated out, because it does not occur in (27). The remaining kernel $K[j_5^\mu] = \frac{1}{2\pi} \left(\frac{\delta^2}{\delta j_\mu^5 \delta j_5^\mu} \right) \delta(j_5^\mu)$ can be treated similarly as in the Faddeev–Popov procedure: it can be averaged with a Gaussian measure, or more generally, with a sum of moments of a Gaussian measure. After this "regularization" a free parameter, say α_j, characterizing the Gaussian measure, enters the theory. In the Faddeev–Popov approach different values of this parameter correspond to different gauge fixings. Here, as will be seen in what follows, this parameter is fixed, e.g. by the physical requirement to give the correct chiral anomaly. After this 'averaging" procedure we get

$$\mathcal{F} = \mathcal{N} \int \prod_x \{ dv_\mu dj_5^\mu \, d\theta \} \, e^{i \int d^2x \, \mathcal{L}[v_\mu, j_5^\mu, \theta]}, \qquad (28)$$

where

$$\mathcal{L}[v_\mu, j^5_\mu, \theta] = -\frac{1}{4}\left(\partial_{[\mu} v_{\nu]}\right)^2 - e\,\epsilon^{\mu\nu} v_\mu j^5_\nu + \frac{1}{2} j^5_\mu (\partial^\mu \theta) - \frac{1}{2\alpha_j} j^\mu_5 j^5_\mu. \tag{29}$$

Here, $\mathcal{N}$ is an infinite normalization constant, which we omit in what follows.

Since j^5_μ does not occur as a dynamical field a further reduction of (28) is possible, namely we can carry out the Gaussian integration over j^5_μ. This leads to:

$$\mathcal{F} = \int \prod_x \{\mathrm{d}v_\mu\, \mathrm{d}\theta\}\, e^{\mathrm{i}\int \mathrm{d}^2 x\, \mathcal{L}[v_\mu,\theta]}, \tag{30}$$

where

$$\mathcal{L}[v_\mu, \theta] = -\frac{1}{4}\left(\partial_{[\mu} v_{\nu]}\right)^2 - \frac{e^2}{2\pi} v_\mu v^\mu - \frac{e}{\sqrt{\pi}} \epsilon^{\mu\nu} v_\mu (\partial_\nu \theta) + \frac{1}{2}(\partial_\mu \theta)(\partial^\mu \theta). \tag{31}$$

Here we performed the reparameterisation $\theta(x) \to 2\sqrt{\pi}\theta(x)$ as well as fixed the parameter $\alpha_j = \frac{1}{\pi}$.

This result has to be compared with other functional integral approaches explaining the bosonisation phenomenon. Essentially, our procedure leads to the same result, but with the following advantages: Within our approach, instead of the electromagnetic potential A_μ in a certain gauge, the gauge invariant field v_μ occurs. It becomes – automatically – massive, with mass $m_v = \frac{e}{\sqrt{\pi}}$. Thus, we see a dynamical Higgs mechanism characteristic for the Schwinger model. Moreover, the field θ, which in other approaches has to be rather introduced by hand, shows up as the 'phase" of a gauge invariant combination of the original fermionic fields. This gives – in our opinion – a deeper insight into the bosonisation phenomenon. Comparing the couplings of v_μ in (29) and (31), we read off the famous 'bosonisation rule"

$$\frac{1}{\sqrt{\pi}}(\partial^\mu \theta) = j^\mu_5. \tag{32}$$

The functional integral (30) together with the Lagrangian (31) can be viewed as an bosonised effective (quantum) theory of interacting gauge invariant – and thus observable – fields. Together with the bosonisation rule (32) it can be used to calculate vacuum expectation values of physical quantities:

1. *chiral anomaly in two dimensions:*

   From (31), (32) and the bosonisation rule we obtain in a straightforward way without using neither perturbative techniques nor regularization techniques of the heat kernel type the chiral anomaly in two dimensions (v_μ is considered as an external field):

$$\partial_\mu < \frac{1}{\sqrt{\pi}}(\partial^\mu \theta) > \equiv < \frac{1}{\sqrt{\pi}}(\partial_\mu \partial^\mu \theta) > = \frac{e}{\pi}\epsilon^{\nu\mu}(\partial_\mu v_\nu). \tag{33}$$

389

2. *current-current propagators*
Using a certain naive regularisation procedure we obtain the well known expressions for the current-current propagators in two dimensions:

$$T^{\mu\nu}_{5\text{reg}}(x_1 - x_2) := <0|Tj_5^\mu(x_1)j^\nu(x_2)|0>_{\text{reg}}$$
$$= \epsilon^{\alpha\nu}\frac{\partial^\mu\partial_\alpha}{\Box}\Pi(x_1 - x_2)$$
$$\equiv \epsilon^{\mu\nu}\Pi(x_1 - x_2) - \epsilon^{\mu\alpha}\frac{\partial_\alpha\partial^\nu}{\Box}\Pi(x_1 - x_2),$$
$$T^{\mu\nu}_{\text{reg}}(x_1 - x_2) := <0|Tj^\mu(x_1)j^\nu(x_2)|0>_{\text{reg}}$$
$$= \eta^{\mu\nu}\Pi(x_1 - x_2) - \frac{\partial^\mu\partial^\nu}{\Box}\Pi(x_1 - x_2), \qquad (34)$$

where $\Pi(x_1 - x_2) = \frac{1}{\pi}\frac{\Box - m_v^2}{\Box}\delta^2(x_1 - x_2)$.

3. *Ward identities*
From (34) we obtain

$$\partial_\mu T^{\mu\nu}_{5\text{reg}}(x_1 - x_2) = \epsilon^{\mu\nu}\partial_\mu\Pi(x_1 - x_2) \neq 0,$$
$$\partial_\nu T^{\mu\nu}_{5\text{reg}}(x_1 - x_2) \equiv 0.$$

The case of the massive Schwinger model is much more complicated, because the additional mass term $-2m|h|\cos\theta$ (see remarks at the end of section III B) can not be treated in an analytic way. Instead we get an power expansion in m which is not yet fully understood.

B Some Comments on Full QED

The reduced functional integral of QED in terms of gauge invariants (Proposition 2) can be treated in a similar way as the Schwinger model in the last subsection. First we introduce the vector current $j^\mu = 2(r^\mu + l^\mu)$ and the axial current $j_5^\mu = 2(r^\mu - l^\mu)$ as well as $h = |h|e^{i\alpha}$. Averaging the singular integral kernel $K[l^\mu, r^\mu, h, h^*]$ with a sum of moments of a Gaussian measure, we obtain an bosonized effective (quantum) field theory for QED:

$$\mathcal{F} = \mathcal{N}\int\prod_x\{dv_\mu\,dj^\mu\,d|h|^2\,d\alpha\}\,e^{i\int d^4x\,\mathcal{L}[v_\mu, j^\mu, |h|, \alpha]} \qquad (35)$$

with the effective Lagrangian

$$\mathcal{L}[v_\mu, j^\mu, |h|, \alpha] = -\frac{1}{4}(\partial_{[\mu} v_{\nu]})^2 - e\,j^\mu v_\mu + \frac{\alpha_{j5}}{16|h|^2}\epsilon^{\alpha\mu\beta\gamma}(\partial_\mu\alpha)j_\alpha(\partial_\beta j_\gamma)$$
$$+\frac{\alpha_{j5}}{128|h|^4}\epsilon^{\alpha\mu\beta\gamma}\epsilon_{\delta\mu\rho\sigma}j_\alpha(\partial_\beta j_\gamma)j^\delta(\partial^\rho j^\sigma) \qquad (36)$$
$$-\frac{1}{2\alpha_j}j^\mu j_\mu + \frac{\alpha_{j5}}{8}(\partial^\mu\alpha)(\partial_\mu\alpha) - \frac{1}{2\alpha_h}|h|^4 - 2m|h|\cos\alpha.$$

We remark that the axial current j_5^μ is a non–dynamical field and, thus, can be integrated out. Moreover, as in the Schwinger model the real scalar field $|h|$ enters the Lagrangian in a non–dynamical way and can – in principle – averaged to a constant. Comparing the couplings proportional to ϵ we obtain the remarkable relation

$$(\partial_\mu \alpha) = \frac{2}{\alpha_{j^5}} j_\mu^5 \tag{37}$$

which can be viewed as the "bosonisation rule" for the four dimensional QED.

Following the same lines as in the last subsection, we can use the effective formulation (35), (36) to calculate vacuum expectation values of physical quantities. In what follows we will restrict to the massless case, i.e. $m \equiv 0$:

1. *current–current propagator*

    The current–current propagator $<0|Tj^\mu(x)\,j^\nu(y)|0>$ for the massless QED is given by the vacuum expectation value

$$<0|Tj^\mu(y_1)\,j^\nu(y_2)|0> = <\alpha_j\,\eta^{\mu\nu}\,\delta^4(y_1-y_2) + m_v^2\alpha_j\,v^\mu(y_1)v^\nu(y_2)> \tag{38}$$

with respect to the functional measure $\prod_x \{dv_\mu\,d|h|^2\}\,e^{i\int d^4x\,\mathcal{L}[v_\mu,|h|]}$ where the Lagrangian is given by

$$\mathcal{L}[v_\mu,|h|] = -\frac{1}{4}(\partial_{[\mu}v_{\nu]})^2 + \frac{m_v^2}{2}v^\mu v_\mu + \frac{\alpha_{j^5}}{128} K^{\mu\perp}[v_\mu,|h|]\,K_\mu^\perp[v_\mu,|h|] - \frac{1}{2\alpha_h}|h|^4. \tag{39}$$

Here $K^{\mu\perp}[\frac{\delta}{v_\mu},|h|]$ denotes the transversal part of the vector field

$$K^\mu[j_\mu,|h|] := \frac{1}{|h|^2}\,\epsilon^{\alpha\mu\beta\gamma}\,j_\alpha(\partial_\beta j_\gamma).$$

This can be viewed as an effective (quantum) theory of a spin-1 field v_μ of mass $m_v^2 = \alpha_j e^2$ with a complicated self-interaction term $\frac{\alpha_{j^5}}{128}K^{\mu\perp}[v_\mu,|h|]\,K_\mu^\perp[v_\mu,|h|]$ of order e^4. This defines a new "perturbation series", which in lowest order yields the result:

$$<0|Tj^\mu(y_1)\,j^\nu(y_2)|0>_0 = \alpha_j\,\frac{\eta^{\mu\nu}\Box - \partial^\mu\partial^\nu}{m_v^2 + \Box}\,\delta^4(y_1-y_2). \tag{40}$$

Since $m_v^2 = \alpha_j e^2$, this leads to the suggestion that within our effective formulation certain quantum correction of higher orders are summed up, leading to a "resummation" of the ordinary perturbation series.

2. *chiral anomaly in four dimensions*

Treating v_μ as an external field and using the bosonisation rule (37) we obtain

$$< \partial^\mu j_\mu^5(y)) >_v \equiv < \frac{\alpha_{j^5}}{2} (\partial_\mu \partial^\mu \alpha(y)) >_v = \frac{e^2 \alpha_j}{8|h|^2} \epsilon^{\mu\nu\alpha\beta} F_{\mu\nu} F_{\alpha\beta}, \qquad (41)$$

where $F_{\mu\nu} := (\partial_{[\mu} v_{\nu]})$ denotes the electromagnetic field strength tensor. This fixes the parameters to $\frac{\alpha_j^2}{|h|^2} = \frac{1}{2\pi^2}$.

V FURTHER APPLICATIONS AND PERSPECTIVES

In this contribution we have presented a description of QED in two and four dimensions in terms of gauge invariant quantities as well as first steps in the analysis of these models with respect to their physical content. It is obvious that standard perturbation techniques are not applicable, because our effective Lagrangians do not decompose into a Gaussian measure and a small perturbation. Thus, we will have to develop new calculational techniques.

As mentioned in the introduction, our reduction procedure can also be applied to nonabelian gauge field theories. For a description of one–flavour QCD we refer to [8] and [9]. There we obtained a description in terms of purely bosonic invariants. The final description contains a set $(j^{ab}, c_{\mu K}{}^L)$ of purely bosonic invariants, where j^{ab} is the c-number mate of $\mathcal{J}^{ab}$ – interpreted as a meson field – and $c_{\mu K}{}^L$ is the c-number mate of a set of complex-valued vector bosons built from the gauge potential and the quark fields. A naive counting of the degrees of freedom encoded in these quantities yields the correct result: The field j^{ab} is Hermitean and carries, therefore, 16 degrees of freedom, whereas $c_{\mu K}{}^L$ is complex-valued and carries, therefore, 32 degrees of freedom. On the other hand, the original configuration $\left(A_{\mu A}{}^B, \psi_A^a\right)$ carries $32 + 24 = 56$ degrees of freedom. Thus, exactly 8 gauge degrees of freedom have been removed. ¿From the physical point of view, this formulation could serve as an effective model of mesons (j^{ab}) interacting via gauge invariantly "dressed" gluons (c_μ^{ab}). This may be considered as a step towards understanding QCD as an effective theory of interacting hadrons. Unfortunately there arise qualitatively new problems related to the fact that nonpolynomial interactions occur in the effective Lagrangian. For that reason, special techniques will have to be developed. One possibility seems to be loop expansion in a nontrivial background field.

Another interesting line of research within our programme consists in doing similar constructions on the level of field operators (Hamiltonian approach) for models formulated on a lattice. In [6] this has been done for QED. In this paper we have given a complete classification of irreducible representations of the algebra of observables. These representations are labeled by eigenvalues of the total charge. Thus, a decomposition of the physical Hilbert space into charge superselection sectors has been obtained. Work on similar constructions for QCD is in progress. On the lattice level both local and nonlocal hadronic invariants containing path ordered exponentials of the gauge potentials will occur in a natural way. As in the case of QED [6], the observable algebra will have a complicated structure due to relations between these invariants. If we were able (as for QED) to classify all representations of this algebra we would get some insight into the structure of hadronic states.

REFERENCES

1. J.Kijowski and G.Rudolph, Nucl. Phys. B325 (1989), 211
2. G.Rudolph, Lett. Math. Phys. 16 (1988), 27
3. J.Kijowski and G.Rudolph, Phys. Rev. D31 (1985), 856,
   G.Rudolph, Annalen der Physik, 7.Folge, Bd. 47, 2/3 (1990) 211
4. J.Kijowski,G.Rudolph, Lett. Math. Phys. **29** (1993), 103
5. J.Kijowski, G.Rudolph and M.Rudolph, *Functional Integral in Terms of Gauge Invariants for QED*, Lett. Math. Phys. **33** (1995), 139
6. J.Kijowski, G.Rudolph and A.Thielmann, *Algebra of Observables and Charge Superselection Sectors for QED on the Lattice*, Commun. Math. Phys., (in print)
   J.Kijowski, G.Rudolph and C.Sliwa, *On the Stucture of the Observable Algebra for QED on the Lattice* Preprint NTZ 11/1997 Univ. Leipzig
7. J.Kijowski, G.Rudolph and M.Rudolph, *Gauge Invariant Formulation and Bosonisation of the Schwinger Model*, Phys. Lett. **B** (1998), (in print)
8. J.Kijowski, G.Rudolph and M.Rudolph, *Effective Bosonic Degrees of Freedom for One-Flavour Chromodynamics*, Ann. Inst. H. Poincaré, Vol. 40 (1997), 565
9. J.Kijowski, G.Rudolph and M.Rudolph, *On the Algebra of Gauge Invariants for One-Flavour Chromodynamics*, Rep. Math. Phys. 40 (1997), 131
10. F.Treves, *Topological Vector Spaces, Distributions and Kernels*, Academic Press, Pure and Applied Math. 25, New York 1967
11. P.H.Damgaard, H.B.Nielsen and R.Sollacher, CERN-TH-6959/93

Conformal Symmetry and Unification

Marek Pawłowski[1]

Soltan Institute for Nuclear Studies, Warsaw, Poland

Abstract.
The Weyl-Weinberg-Salam model is presented. It is based on the local conformal gauge symmetry. The model identifies the Higgs scalar field in SM with the Penrose-Chernikov-Tagirov scalar field of the conformal theory of gravity. Higgs mechanism for generation of particle masses is replaced by the originated in Weyl's ideas conformal gauge scale fixing. Scalar field is no longer a dynamical field of the model and does not lead to quantum particle-like excitations that could be observed in HE experiments. Cosmological constant is naturally generated by the scalar quadric term. Weyl vector bosons can be present in the theory and can mix with photon—Z-boson system.

INTRODUCTION

In 1918, Herman Weyl presented the idea and notion of gauge invariance [1]. It was a consequence of natural generalization of Riemannian geometry used in Einstein's General Relativity theory (GR). Weyl assumed that Einstein's metricity condition

$$\nabla g = 0 \qquad (1)$$

could be replaced by a less restrictive conformal condition

$$\nabla g_{\mu\nu} \sim g_{\mu\nu}. \qquad (2)$$

Thus he supposed that for a vector transported around a closed loop by parallel displacement not only the direction but also the length can change, but the angle between two parallelly transported vectors has to be conserved. Weyl observed that if the Einstein's torsion free condition

$$\Gamma^\lambda_{\mu\nu} - \Gamma^\lambda_{\nu\mu} = T^\lambda_{\mu\nu} = 0. \qquad (3)$$

is kept, then - similarly as in the case of GR - there is a relation between the metric and the affine structure of tangent boundle TM. In contrary to GR case, the Weyl

---

[1] e-mail: pawlowsk@fuw.edu.pl

connection is not given uniquely by the Christophel symbol: it could depend on an arbitrary vector field in principle. This vector field is a compensating potential for a local conformal group of scalar multiplicative transformations conserving the conformal condition (2). Weyl called this group the gauge group as it sets a reference scale from point to point in the space-time. Initially he interpreted the new vector field as the electromagnetic potential and has proposed a dynamics for the model that was based on the bilinear in the generalized curvature Lagrangian.

The dynamics of original Weyl's theory turned out to be much more complicated than the dynamics of Einstein GR. The idea of gauge conformal invariance of the theory was also a subject of intensive criticism. Weyl's conformal theory leaves the freedom for the space-time dependent choice of length standards. It seamed that this gauge freedom clashes with quantum phenomena that provide an absolute standard of length. The point is however, that the freedom to set arbitrary length standards along an atomic path does not mean that atomic frequencies will depend on atomic histories (what was the most popular argument in early literature). In Weyl's theory, an atomic frequency depends on the length standard at a given point but not on a history of the atom. Simultaneously, all other dimensional quantities measured at this point depend on this standard in the same way. Consequently dimensionless ratios are standard independent and experimental predictions do not depend on a particular conformal gauge fixing.

The more fundamental arguments raised against conformal theory were based on the reasonable claim that an acceptable theory should not introduce needles objects and notions. If atomic clocks measure time in an absolute way and velocity of light is an absolute physical quantity (or is definite at least) then the relativism of length is unnatural and redundant. However, we should point out a very essential assumption concerning atomic clocks that is hidden in the above. This is an extrapolation of our *flat and first order experience* that all atomic clocks are proportional always and everywhere. One assumes – roughly speaking – that *the ratios of electron mass to proton mass and to other quantum standards are always and everywhere the same*. One can believe that this statement is true but one should remember (especially when such effects like red shift or other distant signals are interpreted) that at the large scale this statement is only an assumption. It should be (and it could be! [2]) a subject of experimental verification. Conformally invariant gauge theory apparently makes a room to relax from such *a priori* suppositions [4], but in fact it does not predicts itself a dynamics for evolution of fundamental physical "constants".

Weyl conformal theory is a gauge theory of length. It was proposed as a geometrical theory of electromagnetism. Soon after, Dirac proposed his theory of quantum relativistic electron in the flat space [3]. It was a gauge theory of complex electron's phase and it turned out that it provides more adequate framework for description of electromagnetic phenomena. Weyl's proposal was abandoned by the author himself (but still in 1973 Weyl's gauge theory of scale was considered by ... Dirac as a candidate for description of electromagnetism [4]).

The original Dirac's theory of electron was extended to the curved space case [5–7]. Taking a four-dimensional manifold M, a copy of two dimensional complex vector field F_pM can be attached to every point p of M. Then two, in principle independent pairs of affine and metric structures can be implemented on the manifold. The natural tangent boundle TM can be equipped with an affine connection Γ and the field of metric g. Independently a connection γ can be defined in the boundle FM and an arbitrary field ε of Levi-Civita metric can be chosen (for generic two-dimensional complex vector space there is a natural class of antisymmetric Levi-Civita metrics that differ by a complex factor).

The two structures $\{\Gamma, g\}$ and $\{\gamma, \varepsilon\}$ can be naturally correlated. The important observation is that the Levi-Civita metric ε induce Lotentz metric $\varepsilon \otimes \bar{\varepsilon}$ at every fiber of the tensor product boundle $FM \otimes \overline{FM}$ (see e.g. [8] for further details). Thus the real part of $FM \otimes \overline{FM}$ (which is a four dimensional real vector boundle) can be related with the tangent vector boundle TM.

It was found by Infeld and van der Waerden [7] that such correlation of boundles correlates also their metrics and affine structures. Keeping the restrictions of GR (metricity and torsion-free) they have shown that metric structure ε of FM is given by metric structure g of TM up to the arbitrary phase factor. Simultaneously the affine structure γ of FM is given by the affine structure Γ of TM up to an arbitrary vector field. This new vector field is a compensating potential for the $U(1)$ local symmetry group of phase transformations of all Dirac fields in the theory. The authors have identified this new field with electromagnetic potential. Such identification was a subject of criticism as the new vector potential has been coupled universally to all fermions including chargeless neutrino. The modern Weinberg-Salam theory (WS) predicts that all fermions couple to $U(1)$ gauge field. There is a second nonabelian gauge group $SU(2)$ in the theory acting only on the left components of Dirac bispinors. Due to the structure of couplings and the effective mass matrix for gauge bosons the massless field - naturally identified with photon - is a combination of original $U(1)$ and $SU(2)$ bosons. It does not couple to neutrinos despite the fact that the original abelian vector potential does. Thus the Infeld - van der Waerden vector potential can be naturally identified with $U(1)$ gauge group potential of the WS model without any conflict with theory and experiment.

The rest of the present paper is devoted to the description of the version of Weinberg-Salam theory conformally coupled with Weyl's theory of gravity. The first version of the model was proposed in [9] (see also [10]). Similar ideas were also presented in [11]. More comprehensive list of the bibliography of the subject can be found in [12].

Taking into account the roots of the theory it could be called the Weyl-Weinberg-Salam model (WWS).

WEYL-WEINBERG-SALAM MODEL

Let us fix the notation

Weyl's potential will be denoted by S_μ. Let us assume torsion free condition (3). Then the connection in TM is given by

$$\Gamma^\rho_{\mu\nu} = \{^\rho_{\mu\nu}\} + f(S_{m u}g^\rho_\nu + S_\nu g^\rho_\mu - S^\rho g_{\mu\nu}) \tag{4}$$

where f is an arbitrary coupling constant (in principle it could be absorbed at this level by a redefinition of S_μ but it is convenient to keep it here and set its value later). Consequently Weyl's conformal condition (2) gets the form

$$\nabla_\mu \hat{g} = -2f S_\mu \hat{g} \tag{5}$$

Equations (4) and (5) are invariant with respect to Weyl's transformations

$$g_{\mu\nu} \to \Omega^2 g_{\mu\nu} = e^{2\lambda} g_{\mu\nu} \tag{6}$$

$$S_\mu \to S_\mu - \frac{1}{f}\partial_\mu \lambda. \tag{7}$$

Thus metric tensor is covariant with respect to Weyl's transformations with degree 2. The Riemann and Ricci tensors constructed from (4) are conformally invariant objects but their contraction to scalar curvature R is not. R can enter linearly to a conformally invariant expression of dimension of action if it is combined with a scalar Penrose-Chernikov-Tagirov (PCT) field φ_{PCT} [13] that transforms according to

$$\varphi_{PCT} \to e^{-\lambda}\varphi_{PCT}. \tag{8}$$

Then the combination $\varphi^2_{PCT} R$ is conformally invariant. The conformal covariant derivative of φ_{PCT} is given by

$$\nabla_\mu \varphi_{PCT} = (\partial_\mu - fS_\mu)\varphi_{PCT} \tag{9}$$

and it transforms according to (8).

The most general conformally invariant Lagrangian that leads to second order equations of motion for the metric-Weyl-scalar system reads [14]

$$L_g = -\frac{\alpha_1}{12}\varphi^2_{PCT} R + \frac{\alpha_2}{2}\nabla_\mu \varphi_{PCT} \nabla^\mu \varphi_{PCT} - \frac{\alpha_3}{4}H_{\mu\nu}H^{\mu\nu} - \frac{\lambda}{4!}\varphi^4_{PCT} \tag{10}$$

where

$$H_{\mu\nu} = \partial_\mu S_\nu - \partial_\nu S_\mu. \tag{11}$$

The coupling constants α_1, α_2 and α_3 are arbitrary but the last two constants can be absorbed in φ_{PCT} and S_μ by a suitable redefinition of the fields. Observe however, that we are not able to absorb simultaneously α_3 and f. The last coupling remains arbitrary and has to be fixed by experiment.

We can also include the original Weyl Lagrangian being the square of Weyl tensor $L_W = \rho C^2$ where ρ is a coupling constant.

Now we can face the Weinberg-Salam part, or more generally, the full Standard Model of fundamental interactions [15].

First, we should recall [16] that Weyl's vector potential S_μ do not couple directly to Dirac fermions if they transforms according to the rule

$$\Psi \to e^{-\frac{3}{2}\lambda}\Psi. \qquad (12)$$

The conformally invariant part of SM can be written in the following form:

$$\mathcal{L}^c_{SM}[\varphi_H, \mathbf{n}, V, \psi, g] = \mathcal{L}^{SM}_0 + [-\varphi_H F + \varphi_H^2 B - \lambda \varphi_H^4]. \qquad (13)$$

$\mathcal{L}^{SM}_0$ is the conventional SM Lagrangian without the "free" part for the modulus of the Higgs $SU(2)$ doublet φ_H and without the Higgs mass term; B is the mass term of the vector fields generally denoted by V and F is the mass terms of the spinor fields generally denoted by ψ

$$B = D\mathbf{n}(D\mathbf{n})^*\,;\ F = (\bar\psi_L \mathbf{n})\psi_R + h.c.;\quad \mathbf{n} = \begin{pmatrix} n_1 \\ n_2 \end{pmatrix};\ n_1 \overset{*}{n}_1 + n_2 \overset{*}{n}_2 = 1; \qquad (14)$$

$\mathbf{n}$ is the angular component of the Higgs $SU(2)$ doublet.

As there are two abelian gauge groups in the model also a mixed term

$$L_{SB} = \alpha_4 H_{\mu\nu} F^{\mu\nu} \qquad (15)$$

is admitted by all symmetries of the model in general.

The main idea of conformal unification consists in the identification of PCT scalar field φ_{PCT} with the modulus of Higgs doublet φ_H within the rescaling factor χ

$$\varphi_H = \chi \varphi_{PCT}. \qquad (16)$$

The total lagrangian of the conformally unified WWS model can be written as a sum of three terms described above

$$L_T = L_g + L^c_{SM} + L_{SB} \qquad (17)$$

with the constraint (16) resolved.

The rescaling factor χ of (16) is a new coupling constant, which coordinates weak and gravitational scales [17].

SCALE FIXING

The theory given by (17) does not contain any dimensional parameter. This is the necessary condition for it to be conformally invariant. As it was discussed in the Introduction in the context of the Weyl theory alone, dimensional quantities are observed in nature only indirectly. Measuring one of them, we always refer to some other dimensional quantity. We measure ratios of dimensional quantities and we are not able to measure anything more. Our statements express the ratios in the form that carries in the content of its measure an information on the denominator. Thus the dimensional quantities in the half seams to be nothing but only a product of human invention, a logical and a lingual abbreviation representing both the physical information and the chosen convention. There is no doubt that the abbreviation is convenient and useful in practice - in our "flat" surrounding at least (see however the Introduction again). The conformal theory reproduces this conventional abbreviation. It could be done with the help of the most natural mechanism for this purpose, the mechanism of scale fixing which is an example of the gauge fixing of the conformal gauge symmetry group (it is in fact the first historical example of the notion of gauge).

Gauge fixing freedom allows us to impose an additional condition on the theory variables. All lawful conditions (those that can be fulfilled by the fields obtained from a generic configuration by a gauge symmetry transformation) are classically equivalent but not all of them are equally convenient for a given practical purpose. In the case of our conformal theory, we are free to fix the dimensional scale. A natural choice is the one that fixes particle masses in our flat surrounding to theirs conventional space-time independent values. (In fact, nobody will admit in practice that the choice could be a different.) This could be achieved for the conformal symmetry gauge condition that fixes the scalar field in L_{SM}^c (13) to a constant (space-time independent) value. Thus we can demand that

$$\varphi_H = const = v \qquad (18)$$

and it is clear that a generic nonzero scalar field configuration can be conformally transformed to fulfill condition (18).

Choosing $v = 246\mathrm{GeV}$ and choosing ordinary unitary gauge of weak group, we reproduce the whole structure of classical SM masses in WWS model.

It should be stressed here that no mechanism of spontaneous or dynamical symmetry breaking was used in order to produce particle masses. The conformal gauge fixing condition (18) was a sufficient tool. Let us also comment - but without further discussion - that however the condition (18) serves for easy identification the flat space particle content of the model, it needn't be the best toll for other purposes. The condition leads to a massive sigma model that is not perturbatively renormalizable. (The fact does not prejudge the renormalizability of the theory - if we can speak at all about a renormalizability of the theory including gravity. A convenient choice of gauge fixing condition is essential for perturbative analysis of the renormalizability problem. It is known, e.g. that the unitary gauge is not the best choice for this task in SM.)

TOWARD EXPERIMENT

The properties of theory given by (17) depend on the value of coupling constants α_i, ρ, f, λ and χ.

The striking feature of the conformal theory is the lack of ordinary Einstein term in (17). Observe however, that even in the simplest case $\chi = 1$ (the Higgs field identified with PCT scalar field), the condition (18) allows us to reproduce easily the Einstein term [11]. It is sufficient to demand that

$$-\frac{\alpha_1}{12}v^2 = \frac{1}{8\pi G} \tag{19}$$

If the conformal gauge fixing condition (18) is chosen, a mass term for the Weyl's vector field S_μ appears [11] and S_μ acquires mas

$$m_S^2 = \frac{1}{2}f(\alpha_2 - \alpha_1)v^2 \tag{20}$$

The condition (19) leads to Weyl vector mass

$$m_S = 0.5 \cdot 10^{19} f \cdot GeV. \tag{21}$$

In turn the Weyl's mass equals zero only in the special case when $\alpha_2 = \alpha_1$. Then an additional symmetry is realized in the model. Without changing the action we can transform according to the rules of conformal transformations (6), (8) and (12) the metric, the scalar and the all fermion fields leaving the Weyl field unchanged. Similarly we can independently transform S_μ and (17) will not change. In that case the Weyl potential decouples from scalar field and if $\alpha_4 = 0$, it is coupled only to gravity. We get Penrose-Chernikov-Tagirov theory of scalar field conformally coupled with gravity [13]. In order to reproduce appropriate Newtonian limit already at the classical level, we have to demand that χ is very small [9,17]

$$\chi \sim \frac{v}{m_{PLANCK}} \tag{22}$$

In the flat limit approximation (the condition (18) is applied, dynamics of g is frozen and g is chosen to be the metric of Minkowski space) the conformally unified WWS theory leads to the SM-like σ-model. (It holds independently on the values of couplings α_i, ρ, f and λ in (17)). There is still $U(1) \times SU(2)_L \times SU(3)$ gauge symmetry but the feature of perturbative renormalizability is lost. Despite this fact the theory is still predictive. We can reproduce all SM 1-loop results for the processes without external Higgs lines. The SM Higgs mass is replaced in calculations by an effective cutoff that can be expressed (eliminated) by some measured quantity or a combination of observables. 1-loop predictions for 12 LEP observables were given in [18] in reasonable agreement with SM and experiment.

The flat limit of the presented unified WWS model can be a subject of experimental verification and discrimination. The direct verification will be provided (of course!) by LHC. This installation should produce data able to cover all admissible SM Higgs mass range. If no Higgs signal will be found (and we know from LEP that it should be found there if SM is valid) then conformal unified model predicting no dynamical scalar particle at all should be a serious alternative. In turn founding at LHC Higgs particle with the all its SM predicted properties will tell us that the minimal conformal unification is not good.

There was also proposed an indirect method for verification of the flat limit consequences of WWS model [19]. It is based on the observation that while the SM Higgs mass m_H is an energy independent physical constant, the cutoff Λ introduced in the 1-loop analysis of σ-model can depend in principle on the energy of the process considered and on its other parameters. The idea is to derive m_H from two experiments performed at different energy scales. If it will happen that the derived masses disagree it will mean that SM fails while WWS accepts this phenomenon. This is a kind of negative test of SM. It was estimated that the proposed comparison could be made on the base of data given by LEP and CESR B and PEP II if some demanded but realistic luminosity will be achieved.

CONCLUSIONS

The Weyl-Weinberg-Salam model identifies the Higgs scalar field in SM with the Penrose-Chernikov-Tagirov scalar field of the conformally invariant theory of gravity. This identification is very natural and it leads to important physical consequences:

Higgs mechanism for generation of particle masses is replaced by the originated in Weyl's ideas conformal gauge scale fixing. Scalar field is no longer a dynamical field of the model – it is rather a Goldsone direction in field space, the direction that is tangent to the conformal gauge group. Consequently it does not lead to quantum particle-like excitations that could be observed in HE experiments and it does not acquire quantum expectation value in the vacuum. Experimental flat limit consequences of the model could be tested in near future.

No cosmological consequences characteristic for the SM Higgs field can be derived from the present model, but the scalar sector generates cosmological consequences in a different way. The quadric coupling constant λ of thescalar PCT field which in WWS does not play any role in generating particle masses, has its effect in generation of cosmological constant. This constant is dimensional and consequently it is scale choice dependent. In the standard approach, its value is given by λ and by the mass standards fixing gauge condition (18). Thus we get

$$\Lambda = \frac{\lambda}{4!}(\frac{v}{\chi})^4. \tag{23}$$

The very new feature of the Lagrangian (17) is the mixed term (15) that leads to an interaction of Weyl and $U(1)$ Weinberg-Salam vector potentials. At quantum level it would result in a mixing of Weyl boson with photon and weak bosons - the effect in a sense similar to the known $\gamma - Z$ mixing. As the mass of S_μ and the coupling α_4 is not predicted by the theory, the strength of the mixing effect could be small as well as very large. Also the mass m_S cannot be easily estimated from the known data as there is no interaction of fermions with the Weyl potential. Thus definite answers concerning the presence and interactions of Weyl sector should be looked in experiments.

ACKNOWLEDGMENTS

I am indebted to Professor Jakub Rembieliński, Dr. Kordian Smoliński and members of the Organizing Committee for all their efforts in organizing the marvelous meeting in Łódź and for creating the scientific atmosphere. I am indebted to prof. E. Kapuścik, prof. V.N. Pervushin and dr A. Horzela for valuable discussion. I'm also grateful to prof. I. Białynicki-Birula for rendering a copy of English translation of [7]. The work was supported by Polish Committee for Scientific Researches grant no. 2 P03B 183 10.

REFERENCES

1. Weyl, H., *S.-B. Preuss. Akad. Wiss.* 465 (1918); *Math. Z.* **2**, 384 (1918); *Ann. der Phys.* **59**, 101 (1919).
2. Domnin, Yu.S. et. al., *JETP Lett.* **43** 212 (1986); Potekhin, A.Y. et. al., *Testing Cosmological Variability of the Proton to Electron Mass Ratio Using the Spectrum of PKS 0528-250* preprint, astro-ph/9804116; Varshalovich, D.A. and Levshakov, S.A., *JETP Lett.* **58** 237 (1993).
3. Dirac, P.A.M., *Proc. Roy. Soc.* (London) **A117**, 610 (1928).
4. Dirac, P.A.M., *Proc. Roy. Soc.* (London) **A333**, 403 (1973).
5. Schouten, J.A., *Journal of Math. and Phys.* **10**, 239 (1931).
6. Schroedinger, E., *S.-B. Preuss. Akad. Wiss.* 105 (1932).
7. Infeld, L.and van der Waerden, B.L., *S.-B. Preuss. Akad. Wiss.* 380 (1932).
8. Wald, R.M., *General Relativity*, The University of Chicago Press, 1984.
9. Pawłowski, M., *Can gravity do what the Higgs does?* preprint, ICTP-Trieste, IC-90-454; Pawłowski, M. and Rączka, R., *Found. Phys.* **24**, 1305 (1994).
10. Gyngazov, L. N. et. al., *JINR-Preprint* **E-2-98-101** Dubna, 1998; gr-qc/9805083.
11. Cheng, H., *Phys. Rev. Lett.* **61**, 2182 (1988).
12. Hehl, F. W. et. al., *Phys.Rept.* **258**, 1 (1995).
13. Penrose, R., *Relativity, Groups and Topology*, Gordon and Breach, London, 1964; Chernikov, N. A. and Tagirov, E. A., *Ann. Ins. Henri Poincare* **9** 109 (1968).
14. Padmanabhan, T., *Class. Quantum Grav.* **2**, L105 (1985).
15. Particle Data Group, *Phys.Rev.* **D54**, 1 (1996).
16. Hayashi, K. and Shirafuji, T., *Prog. Theor. Phys.* **57**, 302 (1977); Hayashi, K., Kasuya, M. and Shirafuji, T., *ibid*, p. 431
17. Pawłowski, M. and Rączka, R., *A Higgs- Free Model for Fundamental Interactions. Part 1*, preprint SISSA-Trieste, ILAS/EP-3-1995; hep-ph/9503269
18. Pawłowski, M. and Rączka, R., *A Higgs- A Higgs-Free Model for Fundamental Interactions. Part 2*, preprint SISSA-Trieste, ILAS/EP-4-1995; hep-ph/9503270.

19. Pawłowski, M. and Rączka, R., *Consistency test of the Standard Model*, preprint, hep-ph/9610435.

The Zwanziger Action for Electromagnetodynamics Revisited

Dmitri Sorokin*†

*Humboldt-Universität zu Berlin
Institut für Physik
Invalidenstrasse 110, D-10115 Berlin, Germany[1]
and
† Università Degli Studi Di Padova
Dipartimento Di Fisica "Galileo Galilei"
ed INFN, Sezione Di Padova
Via F. Marzolo, 8, 35131 Padova, Italia

Abstract. We demonstrate that a duality–symmetric action proposed by Zwnaziger for describing Maxwell fields interacting with electric and magnetic sources is dual to a duality–symmetric Maxwell action discussed by Deser and Teitelboim, and by Schwarz and Sen.

This paper is a review of results obtained in collaboration with Alexey Maznytsia and Christian Preitschopf [1].

A classical example of duality symmetry of fields is the duality of free Maxwell equations with respect to the interchange of the electric and magnetic field strength. This symmetry can be promoted to the case of coupling Maxwell fields to charged matter if in addition to electrically charged particles there exist magnetically charged particles (monopoles and dyons).

To describe monopoles and dyons on an equal footing with electrically charged particles it is desirable to have a formulation of the theory, where duality symmetry would be a manifest symmetry of the action.

The first action of this kind was proposed by Zwanziger in 1971 [2]. In a source-free case it looks as follows

$$S_1 = \int d^4x [-\frac{1}{8} F^\alpha_{mn} F^{\alpha mn} + \frac{1}{4} n^m \mathcal{F}^\alpha_{mn} \mathcal{F}^{\alpha np} n_p], \tag{1}$$

It contains two abelian vector fields $A^\alpha_m(x)$ ($\alpha = 1, 2$; $m = 0, 1, 2, 3$). The first term in the action is the sum of the standard kinetic terms for these fields ($F^\alpha_{mn} = \partial_{[m} A^\alpha_{n]}$,

---

[1] Alexander von Humboldt Fellow. On leave from Kharkov Institute of Physics and Technology, Kharkov, 310108, Ukraine.

while the second term is bilinear in the self–dual combination of the field strengths of $A_m^\alpha(x)$

$$\mathcal{F}_{mn}^\alpha \equiv \epsilon^{\alpha\beta} F_{mn}^\beta - \frac{1}{2}\epsilon_{mnpq} F^{\alpha pq}; \quad \mathcal{F}_{mn}^\alpha = \frac{1}{2}\epsilon^{\alpha\beta}\epsilon_{mnpq}\mathcal{F}^{\beta pq}, \quad (2)$$

where $\epsilon^{\alpha\beta}$ is the antisymmetric tensor ($\epsilon^{12} = 1$). The self-dual tensor (2) enters the action (1) being contracted with a constant space–like vector n^m ($n^m n_m = 1$). This vector is associated with a rigid Dirac string stemmed from magnetically charged particles. The presence of this vector in the action breaks manifest Lorentz invariance from $SO(1,3)$ to $SO(1,2)$ which acts in the directions transversal to n^m. Breaking of manifest Lorentz invariance is a general situation which one encounters with when trying to construct actions for self–dual fields without the use of auxiliary fields.

Another duality symmetric action for Maxwell theory was discussed in [3,4] and can be written as

$$S_2 = \int d^4x [-\frac{1}{8} F_{mn}^\alpha F^{\alpha mn} - \frac{1}{4} n^m \mathcal{F}_{mn}^\alpha \mathcal{F}^{\alpha np} n_p]. \quad (3)$$

It looks very similar to the Zwanziger action and differs from it only in the sign of the second term.

The purpose of this talk is to demonstrate how these two actions are related to each other by a duality transformation.

But first, let us convince ourselves that both actions produce standard Maxwell equations for a single gauge potential, the second gauge field being related to the first (independent) one by a duality condition. Thus we shall see that both actions describe the dynamics of a single Maxwell field. But the way one derives this result from one action or another is different, because the actions have different symmetry properties and yield *a priori* different equations of motion of A_m^α.

The equations of motion which follow from the Zwanziger action are

$$\partial_m(n^{[m}\mathcal{F}^{n]p\alpha} n_p) = 0, \quad (4)$$

where $\mathcal{F}^{np\alpha}$ is defined in (2). The general solution to this equation is

$$n^{[m}\mathcal{F}^{n]p\alpha} n_p = \epsilon^{mnpq}\partial_p \phi_q^\alpha, \quad (5)$$

where ϕ_q^α are vector functions which depend only on three independent coordinates which parametrize a space-time hypersurface orthogonal to n^m:

$$\phi_q^\alpha = \phi_q^\alpha(y), \quad y^m = x^m - n^m(n_l x^l), \quad y^m n_m \equiv 0. \quad (6)$$

This implies that by choosing appropriate boundary conditions on F_{mn}^α [2] one can restrict the solution of (5) to $\phi_q^\alpha = 0$. Then the right hand side of (5) vanishes and this equation reduces to the duality relation

$$F^1_{mn} = -\frac{1}{2}\epsilon_{mnlp}F^{lp2}, \qquad (7)$$

which reads that only one of the field strengths is independent. Taking a derivative of this expression we recover the Maxwell equations for, say F^1_{mn}

$$\partial_m F^{1mn} = 0, \qquad \partial_m {}^*F^{1mn} = 0. \qquad (8)$$

Thus, the Zwanziger action indeed describes a single electromagnetic field in a duality–symmetric fashion.

Consider now how the duality relation (7) is derived from the action (3). The equations of motion of $A^\alpha_m(x)$ take the form of a Bianchi identity

$$\epsilon^{mnpq}\partial_n(n_p \mathcal{F}^{\alpha r}_q n_r) = 0. \qquad (9)$$

Its general solution is

$$n_{[p}\mathcal{F}^{\alpha r}_{q]} n_r = \partial_{[p}\varphi^\alpha_{q]}. \qquad (10)$$

Analysing eq. (10) one can show that not all components of φ^α_q are independent. They are expressed through a scalar function $\hat{f}^\alpha(x)$ [4,5].

It turns out that in addition to the standard gauge invariance ($\delta A^\alpha_m = \varphi^\alpha$) the action (3) has another local symmetry of the form [4]:

$$\delta A^\alpha_m = n_m f^\alpha(x). \qquad (11)$$

This symmetry is consistently gauge fixed by putting $\hat{f}^\alpha = 0$ and, hence, $\varphi^\alpha_q = 0$ in eq. (10). Then it again reduces to the duality relation (7) and gives rise to the Maxwell equations for one of the gauge fields chosen to be independent.

Note that the Zwanziger action does not have an additional local symmetry analogous to (11). Instead, it is invariant under "semilocal" transformations of A^α_m whose parameters depend on $y^m = x^m - n^m(n_l x^l)$ like ϕ^α_q in (6). Gauge fixing this symmetry is analogous to choosing appropriate boundary conditions.

We have seen that the properties of the two actions are different. However, there exists a duality transformation which relates them to each other. To perform this duality transformation one should consider a manifestly Lorentz–covariant form of the action (3) which was constructed in [5]. This action looks as follows

$$S = \int d^4 x [-\frac{1}{8}\mathcal{F}^\alpha_{mn}\mathcal{F}^{\alpha mn} - \frac{1}{4(\partial^s a)(\partial_s a)}\partial^m a \mathcal{F}^\alpha_{mn}\mathcal{F}^{\alpha np}\partial_p a], \qquad (12)$$

where now $u_m(x) = \partial_m a(x)$ is a derivative of a scalar field which has been introduced to make Lorentz invariance of the action manifest. This scalar field is auxiliary since there is the following local symmetry of the action

$$\delta a = \phi, \qquad \delta A^\alpha_m = \frac{\phi}{(\partial a)^2}\epsilon^{\alpha\beta}\mathcal{F}^\beta_{mn}\partial^n a, \qquad (13)$$

which allows one to gauge fix $a(x) = n_m x^m$ such that $u_m = n_m$ becomes a constant space–like vector. Then manifest Lorentz–invariance is lost, and the action (13) reduces to (3).

Let us now see what happens with the action (12) when the scalar field $a(x)$ is dualized into a two–form field $B_{mn}(x)$.

A standard prescription for dualizing fields in the action consists in following.

Consider u_m in eq. (12) as an independent vector field but take into account its relation to $a(x)$ on the mass shell by adding to eq. (12) the Lagrange multiplier term

$$\int d^4x\, v^m(x)(u_m - \partial_m a(x)). \tag{14}$$

Varying this term with respect to $a(x)$ we get

$$\partial_m v^m = 0$$

whose general solution is

$$v^m = \epsilon^{mnpq} \partial_n B_{pq}(x), \tag{15}$$

and this is how the two–form field $B_{pq}(x)$ appears. The variation of the action (12)+(14) with respect to u_m gives the expression for $v_m(x)$

$$v^m = \frac{1}{2(u^s u_s)^2} \epsilon^{\alpha\beta} \epsilon^{mnpq} u_n \mathcal{F}_p^\alpha \mathcal{F}_q^\beta, \quad (\mathcal{F}_p^\alpha = \mathcal{F}_{pq}^\alpha u^q). \tag{16}$$

Our goal is to use this expression to replace the second, u_m–dependent, term in (12) with a v_m–dependent term.

After some algebra it can be found that the second term in the action (12) is expressed in terms of v^m as follows

$$\frac{1}{(v^s v_s)} v^n \mathcal{F}_{np}^\alpha \mathcal{F}^{\alpha pq} v_q = -\frac{1}{(u^s u_s)} u^n \mathcal{F}_{np}^\alpha \mathcal{F}^{\alpha pq} u_q. \tag{17}$$

Substituting this expression into (12) we get the following Lorentz–covariant action

$$S = \int d^4x \left(-\frac{1}{8} F_{mn}^\alpha F^{\alpha mn} + \frac{1}{4v^s v_s} v^m \mathcal{F}_{mn}^\alpha \mathcal{F}^{\alpha np} v_p \right), \tag{18}$$

where $v^m = \epsilon^{mnpq} \partial_n B_{pq}$.

We observe that the sign of the second term has been changed and become the same as in the Zwanziger action.

To show that the Zwanziger action is indeed a gauge fixed version of the covariant action (18) we should find a local symmetry which allows to choose $\frac{v^m}{\sqrt{v^r v_r}}$ to be a constant space–like vector n^m.

The action (18) does possess this symmetry. The corresponding transformations of fields have the following form

$$\delta B_{mn} = -\frac{1}{2}\epsilon_{mnpq}v^p\Lambda^q \quad \Rightarrow \quad \delta v^m = 2\partial_n(v^{[m}\Lambda^{n]}) = \Lambda^n\partial_n v^m - v^n\partial_n\Lambda^m - v^m\partial_n\Lambda^n,$$

$$\delta A^{m\alpha} = -\frac{1}{v^s v_s}\epsilon^{mnpq}v_n\Lambda_p \mathcal{F}^\alpha_{qt}(A)v^t, \tag{19}$$

where $\Lambda^m(x)$ is a vector parameter.

In conclusion we have shown that the two duality–symmetric formulations of Maxwell theory are dual to each other. This relation has been established by dualizing the auxiliary scalar field $a(x)$ of the Lorentz–covariant generalization of the action (3).

Having performed this dualization we have found the Lorentz–covariant generalization of the Zwanziger action, which implies that the Zwanziger action, though being not manifestly Lorentz invariant, possesses a hidden space–time symmetry which has not been observed before. This symmetry is a combination of Lorentz transformations and a relic of the local symmetry (19) of the covariant action. Under these transformations the vector potentials transform as follows

$$\delta A^\alpha_m = \Omega_m{}^n A^\alpha_n + \Omega^{pq}(x_p\partial_q)A^\alpha_m - \epsilon_{mnpq}n^n\Omega^{pr}x_r\mathcal{F}^{\alpha qt}n_t, \tag{20}$$

The transformations (20) reduce on the mass shell to the standard Lorentz transformations (with parameters $\Omega_m{}^n$), when their last term (proportional to $\mathcal{F}^\alpha_{mn}$) vanishes. This term appeared due to the contribution of the transformations (19) with the parameter $\Lambda^m = \Omega^{mn}x_n$.

Previously Lorentz symmetry was recovered only after quantization of this model [2].

An important point of further analysis of the duality between the two formulations is understanding the difference of their coupling to dyonic sources.

Local gauge symmetries (11), (13) of the action (3) require coupling be non–local and performed through the introduction of a Dirac string [8] as was first proposed in a classical paper by Dirac [6–8], while as was shown in the original paper by Zwanziger [2] the coupling of charged matter to the non–covariant action (1) can be local and minimal. However, this minimal coupling us incompatible with the local symmetry (19) of the covariant version (18) discussed above. The latter requires the coupling to be performed in a non–local way analogous to that in the Dirac formulation. Understanding this discrepancy should be useful when both these formulations are applied for the description of self–dual fields in other theoretical models, and for their quantization.

Acknowledgements. The author would like to thank Paolo Pasti and Mario Tonin for their kind hospitality at the University of Padua, where this work was completed.

REFERENCES

1. A. Maznytsia, C. R. Preitschopf and D. Sorokin, Duality of self–dual actions, hep-th/9805110.
2. D. Zwanziger, *Phys. Rev.* **D3**, 880 (1971).
3. S. Deser and C. Teitelboim, *Phys. Rev.* **D13**, 1592 (1976).
4. J.H. Schwarz and A. Sen, *Nucl. Phys.* **B411**, 35 (1994).
5. P. Pasti, D. Sorokin and M. Tonin, *Phys. Lett* **B352**, 59 (1995); *Phys. Rev.* **D52**, R4277 (1995).
6. P.A.M. Dirac, *Phys. Rev.* **74**, 817 (1948).
7. S. Deser, A. Gomberoff, M. Henneaux and C. Teitelboim, *Phys. Lett.* **B400**, 80 (1997).
8. R. Medina and N. Berkovits, *Phys. Rev.* **D56**, 6388 (1997).

GRAVITATION AND GEOMETRICAL METHODS IN PHYSICS

Non-Generic Symmetries on Extended Taub-NUT metric

Dumitru Baleanu

Bogoliubov Laboratory of Theoretical Physics 141980 Dubna, Moscow region, Russia
and
Institute of Space Sciences, P.O.Box MG-6, Magurele-Bucharest, Romania

Abstract. The geodesic motion of pseudo-classical spinning particles in extended Euclidean Taub-NUT space was analyzed. The non-generic symmetries of Taub-NUT was investigated. We found new non-generic symmetries in the presence of electromagnetic field like a monopole.

INTRODUCTION

The spinning space is an supersymmetric extension of an ordinary Riemannian manifold, parametrized by local coordinates $\{x^\mu\}$, to a graded manifold parametrized by local coordinates $\{x^\mu, \psi^\mu\}$, with the first set of variables being Grassmann even (commuting) and the second set, Grassmann odd (anticommuting). The equation of motion of a spinning particle on a geodesic is derived from the action:

$$S = \int_a^b d\tau \left(\frac{1}{2} g_{\mu\nu}(x) \dot{x}^\mu \dot{x}^\nu + \frac{i}{2} g_{\mu\nu}(x) \psi^\mu \frac{D\psi^\nu}{D\tau} \right). \tag{1}$$

The corresponding world-line Hamiltonian is given by:

$$H = \frac{1}{2} g^{\mu\nu} \Pi_\mu \Pi_\nu \tag{2}$$

where $\Pi_\mu = g_{\mu\nu} \dot{x}^\nu$ is the covariant momentum.

In general, the symmetries of a spinning-particle model can be divided into two classes [1]. First there are four independent *generic* symmetries which exist for any spinning particle model (1) and *non-generic* ones, which depend on the specific background space considered.

To the first class belong: proper-time translations generated by the hamiltonian H; supersymmetry generated by the supercharge

$$Q_0 = \Pi_\mu \psi^\mu, \tag{3}$$

and furthermore chiral and dual supersymmetry, generated respectively by the chiral charge

$$\Gamma_* = \frac{i^{[\frac{d}{2}]}}{d!}\sqrt{g}\epsilon_{\mu_1...\mu_d}\psi^{\mu_1}...\psi^{\mu_d}, \quad (4)$$

and dual supercharge

$$Q^* = i\{\Gamma_*, Q_0\} = \frac{i^{[\frac{d}{2}]}}{(d-1)!}\sqrt{g}\epsilon_{\mu_1...\mu_d}\Pi^{\mu_1}\psi^{\mu_2}...\psi^{\mu_d} \quad (5)$$

where d is the dimension of space-time.

As is well known in the Taub-NUT geometry four Killing-Yano tensors are known to exist [2]. From this point of view, the spinning Taub-NUT space is an exceedingly interesting space to exemplify the effective construction of all conserved quantities in terms of geometric ones, namely Killing-Yano tensors. On the other hand, the Taub-NUT geometry is involved in many modern studies in physics. For example the Kaluza-Klein monopole of Gross and Perry [3] and of Sorkin [4] was obtained by embedding the Taub-NUT gravitational instanton into five-dimensional Kaluza-Klein theory. Remarkably, the same object has re-emerged in the study of monopole scattering. In the long distance limit, neglecting radiation, the relative motion of the BPS monopoles is described by the geodesics of this space [5,6]. The dynamics of well-separated monopoles is completely soluble and has a Kepler type symmetry [7]. In the Taub-NUT case there is a conserved vector, analogous to the Runge-Lenz vector of the Kepler type problem:

$$\vec{K} = \frac{1}{2}\vec{K}_{\mu\nu}\Pi^\mu\Pi^\nu = \vec{p}\times\vec{j} + \left(\frac{q^2}{4m} - 4mE\right)\frac{\vec{r}}{r} \quad (6)$$

where the conserved energy has the following for

$$E = \frac{1}{2}g^{\mu\nu}\Pi_\mu\Pi_\nu = \frac{1}{2}V^{-1}(r)\left[\dot{\vec{r}}^{\,2} + \left(\frac{q}{4m}\right)^2\right]. \quad (7)$$

From these reasons and geodesic motion on extended Taub-NUT metric is very interesting to investigate. The geometrical properties and the existence of a Runge-Lenz like vector for the extended Taub-NUT was analyzed in [8].

The aim of this paper is to investigate the generic and non-generic symmetries of the four dimensional extended Euclidean Taub-NUT manifold and to finding new non-generic symmetries. Therefore, having in mind the importance of the geodesic motion in the presence of spinning variables on Taub-NUT space, we extend the study to the spinning extended Taub-NUT space. The new non-generic symmetries was investigated in the presence of a electromagnetic field like a monopole.

NEW NON-GENERIC SYMMETRIES ON EXTENDED TAUB-NUT METRIC

A generalization of the Euclidean Taub-NUT metric is expressed as [8]

$$ds^2 = f(r)\left(dr^2 + r^2 d\theta^2 + r^2 \sin^2\theta d\varphi^2\right) + g(r)\left(d\psi + \cos\theta d\varphi\right)^2 \tag{8}$$

where $f(r)$ and $g(r)$ are the function of r. It was demonstrated that when

$$f(r) = \frac{a}{r} + b \tag{9}$$

$$g(r) = \frac{ar + br^2}{1 + cr + dr^2} \tag{10}$$

(where a,b,c are constants) the extended metric admits Kepler-type symmetry [8]. If the constants are subjected to the constraints $c = \frac{2b}{a}$, $d = \left(\frac{b}{a}\right)^2$ the extended metric coincides, up to a constant factor, with the original Taub-NUT metric setting $4m = \frac{a}{b}$.

After calculations the Killing vectors have the following form:

$$D^{(\alpha)} = R^{(\alpha)\mu}\partial_\mu, \quad \alpha = 1,\cdots,4 \tag{11}$$

where

$$D^{(1)} = \frac{\partial}{\partial \psi}$$

$$D^{(2)} = \sin\varphi \frac{\partial}{\partial \theta} + \cos\varphi \cot\theta \frac{\partial}{\partial \varphi} - \frac{\cos\varphi}{\sin\theta}\frac{\partial}{\partial \psi}$$

$$D^{(3)} = -\cos\varphi \frac{\partial}{\partial \theta} + \sin\varphi \cot\theta \frac{\partial}{\partial \varphi} - \frac{\sin\varphi}{\sin\theta}\frac{\partial}{\partial \psi}$$

$$D^{(4)} = -\frac{\partial}{\partial \varphi}$$

$$\tag{12}$$

$D^{(1)}$ which generates the $U^{(1)}$ of λ translations, commutes with other Killing vectors. The remaining three vectors obey an $SU(2)$ algebra with

$$\left[D^{(2)}, D^{(3)}\right] = -D^{(4)}, etc... \tag{13}$$

The isometry group is $SU(2) \times U(1)$ and this can be contrasted with the Schwarzschild space-time where the isometry group at spacelike infinity is $SO(3) \times U(1)$ This illustrates the essential topological character of the magnetic mass [9].

In the purely bosonic case we have two constants of the motion corresponding to the invariance given above "relative electric charge" and the angular momentum [10]

$$q = g(r)(\dot{\psi} + \cos\theta\,\dot{\varphi}) \tag{14}$$

$$\vec{j} = \vec{r} \times \vec{p} + q\frac{\vec{r}}{r} \tag{15}$$

We are now in a position to study the extended Taub-NUT metric which admit Killing-Yano tensors [11,12]. An antisimetric tensor $f_{\mu\nu}$ is a Killing-Yano tensor [11,12] if satisfies:

$$\{\mu\nu\lambda\} = D_\mu f_{\nu\lambda} + D_\nu f_{\mu\nu\lambda} = 0 \tag{16}$$

Hence the existence of a Killing Yano tensor of the bosonic manifold is equivalent to the existence of a supersymmetry for the spinning particle with supercharge [1]

$$Q_f = f_a^\mu \Pi_\mu \psi^a - \frac{1}{3} i H_{abc} \psi^a \psi^b \psi^c, \qquad \{Q, Q_f\} = 0. \tag{17}$$

where $H_{\mu\nu\lambda} = f_{\mu\nu;\lambda}$ and semicolon denotes covariant derivative.

After very complicated calculations we found that only three cases of extended Taub-NUT metrics have Killing-Yano tensors.

Case I
For $f(r) = \frac{2m}{r^3}$ and $g(r) = \frac{2m}{r}$ we found from (16) the following expressions for the Killing-Yano tensors in the two-form notations:

$$f^1 = \frac{4m}{r^2}\sin\varphi dr \wedge d\theta - \frac{4m}{r}\sin\theta\sin\varphi d\varphi \wedge \psi - \frac{4m}{r^2}\sin\theta\sin\varphi dr \wedge d\psi$$
$$+ \frac{4m}{r}\cos\varphi d\theta \wedge \varphi + \frac{4m}{r}\cos\theta\cos\varphi d\theta \wedge d\psi$$
$$f^2 = \frac{4m}{r^2}\cos\varphi dr \wedge d\theta - \frac{4m}{r}\sin\theta\cos\varphi d\varphi \wedge d\psi + \frac{4m}{r^2}\sin\theta\sin\varphi dr \wedge d\psi$$
$$+ \frac{4m}{r}\sin\varphi d\theta \wedge d\varphi - \frac{4m}{r}\cos\theta\sin\varphi d\theta \wedge d\psi$$
$$f^3 = \frac{-4m}{r^2}\cos\theta dr \wedge d\psi + \frac{4m}{r^2}d\varphi \wedge dr - \frac{4m}{r}\sin\theta d\theta \wedge d\psi$$
$$f^4 = \frac{-4m}{r^2}\cos\theta dr \wedge d\varphi + \frac{4m}{r^2}d\psi \wedge dr - \frac{4m}{r}\sin\theta d\psi \wedge dr \tag{18}$$

Case II
In the case when $f(r) = \frac{2m}{r}$ and $g(r) = 2mr$, in the two-forms notations, the expressions for the four Killing-Yano tensors are

$$f^1 = 4m\sin\theta\cos\varphi d\psi \wedge dr + 4mr\cos\theta\cos\varphi d\psi \wedge d\theta$$
$$- 4mr\sin\theta\sin\varphi d\psi \wedge d\varphi + 4m\sin\varphi dr \wedge d\theta + 4mr\cos\varphi d\varphi \wedge d\theta$$

$$f^2 = 4m\sin\theta\sin\varphi d\psi \wedge dr + 4mr\cos\theta\sin\varphi d\psi \wedge d\theta$$
$$+ 4mr\sin\theta\cos\varphi d\psi \wedge d\varphi - 4m\cos\varphi dr \wedge d\theta + 4mr\sin\varphi d\varphi \wedge d\theta$$
$$f^3 = 4m\cos\theta d\psi \wedge dr - 4mr\sin\theta d\psi \wedge d\theta + 4md\varphi \wedge dr$$
$$f^4 = 4m\cos\theta d\varphi \wedge dr - 4mr\sin\theta d\varphi \wedge d\theta + 4md\psi \wedge dr \tag{19}$$

Case III

In the Taub-NUT geometry four Killing-Yano tensors are known to exist [13]. In this case in 2-form notation the explicit expression for the f_i are [13].

$$f_i = 4m(d\psi + \cos\theta d\varphi) \wedge dx_i - \epsilon_{ijk}\left(\frac{1+2m}{m}\right)dx_j \wedge dx_k \tag{20}$$

$$Y = 4m(d\psi + \cos\theta d\varphi) \wedge dr + 4r(r+m)\left(1+\frac{r}{m}\right)\sin\theta d\theta \wedge d\varphi \tag{21}$$

When we have two different Killing-Yano tensors of rang two $f_{\mu\nu}$ and $F_{\mu\nu}$ with covariant derivative zero, we can construct an infinite numbers of Killing-Yano tensors using the following procedure

$$F^{(0)}_{\mu\nu} = f_{\mu\nu}, F^{(1)}_{\mu\nu} = F_{\mu\nu}, F^{(2)}_{\mu\nu} = F_{\mu\alpha}f^{\alpha\beta}F_{\beta\nu} \tag{22}$$

$$F^{(n+1)}_{\mu\nu} = F_{\mu\alpha}f^{\alpha\beta}F^{(n)}_{\beta\nu} \tag{23}$$

for $(n = 0, 1, 2, 3, ...)$
We can extract a finite number of independent Killing-Yano tensors and we can construct new non-generic symmetries using (17).

We have many possibilities to construct new independent Killing-Yano tensors in the case of extended Taub-NUT metric because we have many Killing-Yano tensors. Because the expressions of Killing-Yano tensors are long and complicated we prefer to present only some very interesting cases.

Case I.

I.A.
For f^1 and f^2 from (18) we have for $F^{(2)}_{\mu\nu} = f^1_{\mu\alpha}f^{2\alpha\beta}f^1_{\beta\nu}$ the following form

$$F^{(2)}_{\mu\nu} = \frac{2m}{r^2}\cos\theta(\sin^2\varphi + \sin\theta\cos\theta)dr \wedge d\theta + \frac{2m}{r^2}\sin\varphi\sin\theta dr \wedge d\psi$$
$$+ \frac{2m}{r}\sin\varphi[-\sin^2\varphi(\cos\theta+1)+1]d\theta \wedge d\varphi$$
$$+ \frac{2m}{r}\sin\varphi\cos\theta[1+(1-\cos\theta)\cos^2\varphi]d\theta \wedge d\psi$$
$$+ \frac{2m}{r}\sin\theta\cos\varphi(-cos^2\varphi + \cos\theta\sin\varphi)d\varphi \wedge d\psi \tag{24}$$

I.B
For f^3 and f^4 from (18) we found out expressions for the new Killing-Yano tensors $F^{(2)}_{\mu\nu}$:

$$F^{(2)}_{\mu\nu} = \frac{4m}{r^3}\cos\theta dr \wedge d\varphi + \frac{4m}{r^2}(\frac{\cos^2\theta}{r} + \frac{\sin^2\theta}{m})dr \wedge d\psi$$
$$+ \frac{4m}{r^2}\sin\theta d\theta \wedge d\varphi + \frac{4m}{r^2}(\frac{1}{r} - \frac{1}{m})\sin\theta\cos\theta d\theta \wedge d\psi \qquad (25)$$

Case II.
II.A. For f^3 and f^4 from (19) we have for:

$$F^{(2)}_{\mu\nu} = \frac{4m\cos\theta}{\sin\theta}dr \wedge d\varphi + \frac{4m}{\sin\theta}(\cos^2\theta - \sin^2\theta\cos 2\theta)dr \wedge d\psi$$
$$- 4mrd\theta \wedge d\varphi + 4mr\cos\theta(-1 + \sin\theta)d\theta \wedge d\psi \qquad (26)$$

II.B. For f^1 and f^3 from (19) we have the following form for $F^2_{\mu\nu}$

$$F^{(2)}_{\mu\nu} = -4mdr \wedge d\varphi - 4m\cos\theta dr \wedge d\psi + 4mr\sin\theta d\theta \wedge d\psi$$
$$- 4m\sin\theta\cos^2\theta d\varphi \wedge d\psi \qquad (27)$$

The Killing-Yano tensors corresponds to the cases when the extended Taub-NUT metric is conformally flat. In this case we have two different Killing-Yano tensors $f_{\mu\nu}$ and $F_{\mu\nu}$.

After calculations we found that $F^{(n)}$ from (24),(25),(26), and (27), after some finite steps, become equal to one of the Killing-Yano tensors given by in (18) or (19). This means that we have a finite numbers of independent Killing-Yano tensors and a finite numbers of new non-generic symmetries on extended Taub-NUT metric.

Because Killing-Yano tensors in the case of Taub-NUT metric satisfies a quaternionic algebra [13] we can deduce immediately that we have not any new non-generic symmetries in this case.

NEW NON-GENERIC SYMMETRIES IN THE PRESENCE OF A ELECTROMAGNETIC FIELD

Another very interesting case is when we consider an external electromagnetic field [14] on the extended Taub-NUT metric. We investigate now if in this case we have new non-generic symmetries.

A relativistic pointlike fermion, a particle with spin $s = 1/2$ having mass m and electric charge q, can be described in the classical limit $h \to 0$ by a Lagrangian

$$L = \frac{m}{2}g_{\mu\nu}(x)\dot{x}^\mu\dot{x}^\nu + \frac{i}{2}\psi_a\frac{D\psi^a}{D\tau} + q\left(A_\mu\dot{x}^\mu - \frac{i}{2m}F_{ab}(x)\psi^a\psi^b\right) \qquad (28)$$

where x^μ are the particle's space-time coordinates, a dot denoting a derivative with respect to the worldline proprer-time τ. The ψ^a are Grassmann variables, m and q are, respectively, the mass and the charge of a particle, and $A_\mu(x)$ and $F_{\mu\nu}(x)$, respectively, the vector potential and the field strength of the electromagnetic, both of which are considered as external fields.

The trajectories, which make the action stationary under arbitrary variations δx^μ and δx^μ vanishing at the end points, are given by:

$$mg_{\mu\nu}\frac{D^2 x^\mu}{D\tau^2} = qF_{\mu\nu}\dot{x}^\nu - \frac{i}{2}R^{ab}_{\mu\nu}\dot{x}^\nu \psi^a \psi^b - \frac{iq}{2m}D_\mu F^{ab}\psi^a\psi^b \qquad (29)$$

$$\frac{D\psi^a}{D\tau} = \frac{qF^{ab}\psi^b}{m} \qquad (30)$$

The covariant derivative of ψ^a is defined by

$$\frac{D\psi^a}{D\tau} = \dot{\psi}^a - \dot{x}^\mu \omega^{ab}_\mu \psi^b \qquad (31)$$

The Hamiltonian has the following expression

$$H = \dot{x}^\mu p_\mu + \dot{\psi}^a \pi^a - L = \frac{g^{\mu\nu}}{2m}\Pi_\mu \Pi_\nu + \frac{iq}{2m}F^{ab}\psi^a\psi^b \qquad (32)$$

and where Π_μ is defined by

$$\Pi_\mu = p_\mu + \frac{i}{2}\omega^{ab}_\mu \psi^a \psi^b - qA_\mu = mg_{\mu\nu}\dot{x}^\nu \qquad (33)$$

We know that in a spinning particle theory [2] we have a supersymmetry transformation generated by the supercharge Q. In our case it has the following expression

$$Q = \frac{e^\mu_a}{m}\Pi_\mu \psi^a \qquad (34)$$

and satisfies

$$\{Q,H\} = 0, \{Q,Q\} = \frac{-2iH}{m} \qquad (35)$$

We will use the following theorem to investigate the non-generic supersymmetries when the electromagnetic field is present [14].

If the space-time admits a Killing-Yano tensor of valence 2, $f_{\mu\nu}$, and the electromagnetic field $F_{\mu\nu}$ satisfies the condition

$$f^\lambda_\mu F_{\nu\lambda} = f^\lambda_\nu F_{\mu\lambda} \qquad (36)$$

then we have that Y_r is a superinvariant function

$$\{Q, Y_r\} = 0 \qquad (37)$$

(for the form of Y and more detailes see [14]).

For simplicity we consider here a electromagnetic field $F_{\mu\nu} = 4m\cos\theta d\theta \wedge d\varphi$ like a monopole. In this case after calculations we found that only Killing-Yano tensors f^4 from (18) and (19) satisfies (36). This result tell us that we have new non-generic symmetries and in the presence of electromagnetic field like a monopole.

I would like to thank Prof. S. Manoff vor helpful discussions about non-generic symmetries on extended Taub-NUT metric.

REFERENCES

1. Gibbons, G. W., Rietdijk, R. H., and van Holten, J. W., *Nucl. Phys.* **B404** 42 (1993).
2. Baleanu, D., *Helv. Phys. Acta* **67** 405 (1994); *Il Nuovo Cimento B* **109** 845 (1994); *Il Nuovo Cimento B* **111** 973 (1996).
3. Gross, D. J., and Perry, M. J., *Nucl.Phys.* **B226** 29 (1983).
4. Sorkin, R. D., *Phys. Rev. Lett.* **51** 87 (1983).
5. Manton, N. S., *Phys. Lett.* **B110** 54 (1982) 54; id., **B154** 397 (1985); id., (E) **B157** 475 (1985).
6. Atiyah, M. F, and Hitchin, N., *Phys. Lett.* **A107** 21 (1985).
7. Gibbons, G. W., and Ruback, P. J., *Phys. Lett.* **B188** 226 (1987); *Commun. Math. Phys.* **115** 267 (1988).
8. Iwai, T., and Katayama, N., *Journal of Geometry and Physics* **12** 55-75 (1993); *Journal of Physics* **27** 3179 (1994).
9. Feher, L. Gy., and Horvathy, P. A., *Phys. Lett.* **B182** 183 (1987); id., (E) **B188** 512 (1987).
10. Gibbons, G. W., and Manton, N. S., *Nucl. Phys.* **B274** 183 (1986).
11. Yano, K., *Ann. Math.* **55** 328 (1952).
12. Dietz, W., and Rudiger, R., *Proc. Roy. Soc. Lond.* **A375** 361 (1981).
13. van Holten, J. W., *Phys. Lett.* **B342** 47 (1995).
14. Tanimoto, M., *Nucl. Phys.* **B442** 549 (1995).

Some Properties of Light Propagation in Relativity

Stanisław L. Bażański[1]

Institute of Theoretical Physics, University of Warsaw, ul. Hoża 69, 00-681 Warszawa, Poland.

Abstract. An outline of a formalism is proposed that describes relativistic effects associated with the propagation of light rays which after being initially split start to reconverge and intersect each other again. Beside a general geometric description that uses the approach of geometric optics, it has been shown how to compute both the difference of the proper times of arrivals of the two light beams to a measuring apparatus, as well as the frequency shifts of each of the beams taken separately. The formalism used here is applicable to both the special and the general theory of relativity, and it can be used equally well either when the light split is produced by a man-made optical device or when it is caused by the gravity field itself. The geometric description used in the formalism is independent of the physical origin of the frequency shift; of whether it is a Doppler, gravitational or cosmological frequency shift effect. Two simple examples of the application of the formalism have been worked out, which illustrate its results for families of inertial and noninertial observers respectively. Furthermore, a short description of a continuous version of the formalism has been presented in the last section of the article.

INTRODUCTION

The objective of this contribution is to present a geometric formalism that describes some properties of light rays which propagate in a relativistic space-time so that, after being initially split, they start to reconverge and cross each other again. The formalism is universal in the same meaning as the well known unified approach to all frequency shift effects in the theory of relativity, which is independent of the interpretation of an effect under consideration, i.e. of whether it is regarded as a Doppler, gravitational, or cosmological frequency shift. In all these cases, if in a space-time manifold two time like world lines, S (source) and O (observer), are given, and a light ray of frequency ν_s is being emitted by S at a point P_s, then, provided the ray intersects O at a point P_o, the ratio of the frequency ν_o, observed by O at P_o, and of ν_s is given by Schrödinger's formula [1],

---

[1]) This work has been supported in part by the Polish Research Program KBN, contract no PB 1371/P03/97/12, registration no 2 P03B 017 12.

$$\frac{\nu_o}{\nu_s} = \frac{(k_\alpha u^\alpha)_{P_o}}{(k_\beta u^\beta_\parallel)_{P_o}}, \qquad (1)$$

where k^α is the tangent vector to the null geodesic which joins P_s and P_o, expressed in terms of an affine parameter, u^α is the tangent vector to O at P_o, and $u^\alpha_\parallel$ is the tangent vector to S at P_s transported parallelly along the null ray to P_o.

From its derivation, it follows, cf. e.g. [2], that formula (1) can also be applied to a more general situation when a light ray sent by S and received by O meets along its way a number, say n, of time like world lines $\mathcal{R}_i, i = 1, 2, ..., n$, at each of which a change of the light propagation direction takes place. Physically, such lines may be interpreted as representing optical devices like mirrors, prisms etc. The following conclusion can be drawn from the standard derivation of Eq. (1):

Corollary 1 *The relativistic redshift formula (1) is applicable not only in the case when the rays which connect the source and the observer are null geodesics, but also when they are piecewise null geodesics.*

Moreover, formula (1) has been generalized [2] to a situation in which the time like world lines S and O are connected by a one parameter family of null world lines, which need not be null geodesics. These null lines generate a time like two dimensional world sheet that can be interpreted as representing the history of an optical fiber. It turns out [2] that in such a situation Eq. (1) requires only a minute adaptation.

GENERAL APPROACH

Although the geometric approach behind Eq. (1) explains the mechanism of the frequency shift and is very convenient for general consideration, Eq. (1) itself is not very handy for computing the shifts in particular situations. For this purpose, an alternative procedure has been proposed [2].

To describe the procedure, let us assume that in a coordinate system x^α the world line S of the source is described by four equations $x^\alpha = \xi^\alpha(s_s)$, and the observer's world line O is given by $x^\alpha = \eta^\alpha(s_o)$, in terms of the proper times s_s along S and s_o along O, respectively. The light cone emanating from a point P_s with coordinates $\xi^\alpha(s_s)$ is described by four equations of the form $x^\alpha = \kappa^\alpha(p, z_1, z_2, \xi^\mu(s_s))$, where p, z_1, and z_2 are three parameters labeling the points on the surface of the cone. Suppose that the world lines S and O, as well as the light cone with its vertex at a fixed point P_s of S are given, i.e. both the two functions ξ, η, as well as the function κ are known. Eliminating then from the four algebraic equations

$$\eta^\alpha(s_o) = \kappa^\alpha(p, z_1, z_2, \xi^\mu(s_s)). \qquad (2)$$

the parameters $p, z_A, A = 1, 2$, one obtains a single equations that, in general, admits a solution of the form

$$s_o = f(s_s). \tag{3}$$

The function f obtained above determines the value of the proper time interval Δs_o which is the projection, by the light rays, of the interval Δs_s. It can be shown [2] that the redshift formula (1) has to be replaced by the relation

$$\frac{\nu_s}{\nu_o} = \frac{df}{ds_s}(s_s). \tag{4}$$

The function f in Eq. (3) must be found for each particular set-up separately. In particular, the process of eliminating p and z_A from Eqs. (2) may result in an equation that admits a multiplicity of solutions of the form (3). For instance, frequently one obtains two such solutions, determined by two functions f_+ and f_-,

$$s_{1o} = f_+(s_s); \qquad s_{2o} = f_-(s_s), \tag{5}$$

which to the same value of s_s assign two different values s_{1o} and s_{2o} of the parameter s_o. There may be various underlying causes of such a situation: a split of rays produced by the gravity field itself, by an optical apparatus introduced in the space between S and O, or by two different optical fibers which join S and O. Regardless of what is the cause, Eq. (5) describes two light rays which, after being sent out from the same point P_s on the line S, with the coordinates $\xi^\alpha(s_s)$, intersect the observer's world line O at two different points which correspond to two values, $s_{1o} = s_o$ and $s_{2o} = s_o + \Delta_o$, of the proper time parameter s_o along O, where

$$\Delta_o = f_-(s_s) - f_+(s_s). \tag{6}$$

The difference Δ_o of arrival times of the slower light signal, represented by f_-, and the faster one, described by f_+, defined above will be called here the Sagnac time. It can be easily expressed as a function of the proper time s_o along O.

The observer can measure Δ_o along his world line by performing interference experiments. Moreover, by shuttering off one of the light signals, represented respectively by f_+ or f_-, he can also measure the frequencies ν_- and ν_+ of the corresponding signals as functions of his time s_o. It can be shown [2] that the three, independently measurable quantities $\Delta_o(s_o)$, $\nu_-(s_o)$, and $\nu_+(s_o)$ must always satisfy the relation

$$\frac{d\Delta_o}{ds_o} + 1 = \frac{\nu_+(s_o)}{\nu_-(s_o + \Delta_o)}. \tag{7}$$

Its immediate consequence is the following observation.

Corollary 2 *The Sagnac difference of times $\Delta_o(s_o)$ remains constant along the observer's world line O if and only if the equality $\nu_+(s_o) = \nu_-(s_o + \Delta_o)$ is there taking place.*

EXAMPLES

The discrete case

The general procedure just described can be illustrated on examples. In this section we consider two examples in which a finite number of time like world lines serve as retransmitters of the light signal which is represented by a piecewise null geodesic.

The first is the Sagnac effect in Minkowski space-time. Here n $(n > 2)$ time like world lines O_k, $k = 1, 2, ..., n$, are given by the equations

$$\begin{aligned}
x_k^0 &= \gamma s_k, \\
x_k^1 &= r \cos\left[\tfrac{\omega\gamma}{c} s_k + \tfrac{2\pi}{n}(k-1)\right], \\
x_k^2 &= r \sin\left[\tfrac{\omega\gamma}{c} s_k + \tfrac{2\pi}{n}(k-1)\right], \\
x_k^3 &= 0,
\end{aligned} \qquad (8)$$

where r, ω, and $\gamma = (\sqrt{1 - \omega^2 r^2/c^2})^{-1}$ are constant, and s_k denotes the proper time taken along O_k. When, at a proper time $s_1 = s_s$, O_1 simultaneously sends out two light signals to his nearest neighbors on his right and left, each of the signals, after being immediately retransmitted by every next observer, returns to O_1, in general, at a different proper time. The signal which arrives to O_1 as the first one is in what follows called the *fast* one, and the second the *slow* one. From the equation of the Minkowski null cone, by making use of Eqs. (8), we find that in the case of the slow signal the proper times s_k and s_{k+1}, for $k = 1, 2, \ldots, n$, where $s_{n+1} = s_{2o}$ is the arrival time of the second (slow) signal to O_1, satisfy the transcedental equation

$$\frac{\gamma}{2r}(s_{k+1} - s_k) - \sin\left[\frac{\omega\gamma}{2c}(s_{k+1} - s_k) + \frac{\pi}{n}\right] = 0, \qquad (9)$$

which, after introducing a dimensionless variable $x = (s_{k+1} - s_k)\gamma/(2r)$ and putting $\beta = \omega r/c$, is equivalent to

$$x - \sin(\beta x + \frac{\pi}{n}) = 0. \qquad (10)$$

For $0 < \beta < 1$, in the interval $(0, \pi(n-1)/(\beta n))$, Eq. (10) has a single root $x = F_-(\beta, n)$ which is a function of the indicated parameters only. Therefore

$$s_{k+1} - s_k = \frac{2r}{\gamma} F_-(\beta, n),$$

and the consecutive proper times at which the slow light signal meets the time like world lines O_k, for $k = 1, 2, \ldots, n, 1$, form an arithmetic sequence. Hence, for the last term s_{2o} in this sequence, denoting s_1 by s_s, we obtain

$$s_{2o} = s_s + \frac{2rn}{\gamma} F_-(\beta, n). \tag{11}$$

The equation above is just an example of the second of Eqs. (5): $s_{2o} = f_-(s_s)$.
In the case of the fast light signal, Eq. (10) has to be replaced by

$$y + \sin(\beta y - \frac{\pi}{n}) = 0, \tag{12}$$

where $y = (s_k - s_{k+1})\gamma/(2r)$, and s_k, s_{k+1} are the proper times of the points at which the fast signal meets O_k and O_{k+1}, respectively.

For $0 < \beta < 1$, in the interval $(0, \pi/(\beta n))$, Eq. (12) also admits only a single root, which is denoted by $y = F_+(\beta, n)$. In terms of this root, one can express the proper time s_{1o} at which the fast light signal f_+ which was emitted by O_1 at the proper time instant s_s returns to the world line O_1:

$$s_{1o} = s_s + \frac{2rn}{\gamma} F_+(\beta, n). \tag{13}$$

This equation defines the first of the functions (5), $s_{1o} = f_+(s_s)$, for the present case.

From Eqs. (11), (13), and (4) it follows that there is no frequency shift during any of the transmissions of the light signals in the example considered now.

Since for $\beta \neq 0$ the proper times of arrivals of the two light signals, f_- and f_+, to O_1 are different from one another, in accordance with Eq. (6), a nonvanishing Sagnac difference Δ_o of the proper times will arise,

$$\Delta_o = \frac{2rn}{\gamma} [F_-(\beta, n) - F_+(\beta, n)]. \tag{14}$$

The solutions F_- and F_+ of Eqs. (11) and (13), correspondingly, can be found either by means of an approximation procedure or numerically.

The simplest approach to solving Eqs. (11) and (13) is to linearize them with respect to β. The solution of Eq. (11) is then

$$F_-(\beta, n) = \frac{\sin \frac{\pi}{n}}{1 - \beta \cos \frac{\pi}{n}},$$

and Eq. (12) leads similarly to

$$F_+(\beta, n) = \frac{\sin \frac{\pi}{n}}{1 + \beta \cos \frac{\pi}{n}}.$$

From here the Sagnac difference of times, due to Eqs. (14), (11), and (13), is

$$\Delta_o = \frac{2\omega r^2 n \sin \frac{2\pi}{n}}{c} \frac{\sqrt{1 - \frac{\omega^2 r^2}{c^2}}}{1 - \frac{\omega^2 r^2}{c^2} \cos^2 \frac{\pi}{n}}.$$

In the approximation considered now, one must, however, disregard factors which contain the term $(\omega^2 r^2)/c^2$. One obtains therefore the equation

$$\Delta_o = 2\beta r^2 n \sin(2\pi/n), \tag{15}$$

which is identical with the classical formula for the Sagnac effect derived already in the prerelativistic physics. Relativistic corrections to Eq. (15) can be obtained either by the power series expansion of the sin function in Eqs. (11) and (13) or by applying to them the Newton approximation method.

The second example is a set-up of n straight, time like world lines O_k, $k = 1, 2, ..., n$, in Minkowski space-time which are defined by the equations

$$\begin{aligned} x_k^0 &= \gamma s_k, \\ x_k^1 &= -\beta\gamma s_k \sin\phi_k + R\cos\phi_k, \\ x_k^2 &= \beta\gamma s_k \cos\phi_k + R\sin\phi_k, \\ x_k^3 &= 0, \end{aligned} \tag{16}$$

where $\beta = v/c$, $\gamma^{-2} = 1 - \beta^2$ are constant, s_k is the proper time along the k-th straight line O_k, and $\phi_k = 2\pi(k-1)/n$.

Here again the observer O_1, at an instant $s_1 = s_s$ of his proper time, sends out a light signal to his nearest two observers O_2 and O_n. The signal is then instantly retransmitted by every next observer from the family (16). As a result, the light returns to the first observer from two opposite directions at the instants of his proper time respectively equal to $s_{1o} = f_-(s_s)$ and $s_{2o} = f_+(s_s)$, where $s_{1o} \geq s_{2o}$.

Let us now take into account the segment of the null geodesic which corresponds to the piece of path of the slow signal, characterized by f_-, between two neighboring world lines O_k and O_{k+1}. To find the proper time s_{k+1} at which O_{k+1} receives the signal retransmitted by O_k at his proper time s_k, we must find the point of intersection of the light cone with the vertex at $O_k(s_k)$ with the line O_{k+1}. Solving this exercise leads us to the quadratic equation for $(s_{k+1} - s_k)$

$$(s_{k+1} - s_k)^2 - 4\beta\gamma\sin\frac{\pi}{n}\left[\beta\gamma\sin\frac{\pi}{n}s_k + R\cos\frac{\pi}{n}\right](s_{k+1} - s_k)$$
$$- 4\sin^2\frac{\pi}{n}\left[\beta^2\gamma^2 s_k^2 + R^2\right] = 0. \tag{17}$$

Denoting the positive root of Eq. (17) by $h_-(s_k)$, we see that $s_{k+1} = F_-(s_k) = s_k + h_-(s_k)$ is given by the expression

$$\begin{aligned} s_{k+1} &= F_-(s_k) \\ &= s_k + 2\sin\frac{\pi}{n}\left[\beta\gamma(\beta\gamma\sin\frac{\pi}{n}s_k + R\cos\frac{\pi}{n})\right. \\ &\quad \left. + \sqrt{\beta^2\gamma^2(\beta\gamma\sin\frac{\pi}{n} + R\cos\frac{\pi}{n})^2 + \beta^2\gamma^2 s_k^2 + R^2}\right]. \end{aligned} \tag{18}$$

In the case of the fast ray, characterized by f_+, the equation of the null cone with the vertex at $O_{k+1}(s_{k+1})$ leads to the equation

$$(s_k - s_{k+1})^2 - 4\beta\gamma \sin\frac{\pi}{n}\left[\beta\gamma \sin\frac{\pi}{n} s_{k+1} - R\cos\frac{\pi}{n}\right](s_k - s_{k+1})$$
$$-4\sin^2\frac{\pi}{n}\left[\beta^2\gamma^2 s_{k+1}^2 + R^2\right] = 0. \quad (19)$$

This equation determines the proper time s_k in terms of s_{k+1} for any two neighboring observers along f_+. Solving it for $(s_k - s_{k+1})$, and taking the positive root, one obtains

$$\begin{aligned} s_k &= F_+(s_{k+1}) \\ &= s_{k+1} + 2\sin\frac{\pi}{n}\left[\beta\gamma(\beta\gamma\sin\frac{\pi}{n} s_{k+1} - R\cos\frac{\pi}{n})\right. \\ &\left. + \sqrt{\beta^2\gamma^2(\beta\gamma\sin\frac{\pi}{n} - R\cos\frac{\pi}{n})^2 + \beta^2\gamma^2 s_{k+1}^2 + R^2}\right]. \end{aligned} \quad (20)$$

Since the corresponding roots of both Eqs. (17) and (19) are functions of the proper time, otherwise as in the case of the rotating family (8), the frequencies of the two signals, after their travel between O_k and O_{k+1}, are shifted. To compute the proper times s_{1o} and s_{2o} at which the two corresponding signals f_- and f_+ will return to the starting world line O_1, one must perform an n-fold composition of the respective functions F, that is

$$s_{1o} = f_-(s_s) = F_-(\ldots F_-(F_-(s_s))\ldots),$$

and

$$s_{2o} = f_+(s_s) = F_+(\ldots F_+(F_+(s_s))\ldots),$$

Making use of Eqs. (4), and (5), one could find from here the global frequency shifts of the two signals f_- and f_+. In accordance with (6), there is here, surprisingly enough, also a nonvanishing Sagnac's difference of times. It is equal to

$$\begin{aligned} \Delta_o &= f_-(s_s) - f_+(s_s) \\ &= F_-(\ldots F_-(F_-(s_s))\ldots) - F_+(\ldots F_+(F_+(s_s))\ldots). \end{aligned} \quad (21)$$

Exact application of formula (21), even for such expressions like (18) and (20), would be rather tedious. If one, however, expands (18) and (20) into power series with respect to β, keeping under control the linear terms only, then one can easily obtain the leading term for the "Sagnac effect" produced by a family (8) of nonrotating, inertial motions with constant velocities of equal magnitude:

$$\Delta_o = 2\beta\gamma Rn \sin\frac{2\pi}{n} + O(\beta^3), \quad (22)$$

where the neglected terms of higher order in β depend on s_1. This indicates, in accordance with Corollary 2, that the frequency shifts of the two signals considered here will be not the same.

Thus one cannot maintain the sometimes expressed opinion that the classical Sagnac effect generated within a family of rotating motions, like those described by Eqs. (8), is an optical counterpart of the Foucault pendulum, and produces a measure of how the motions in a system under consideration deviate from the inertial motion. It is interesting to observe that the leading terms in Eqs. (15) and (22) which are proportional to β are in the two cases of the same order of magnitude.

The continuous case

It is also possible to formulate a continuous version of the approach proposed here. To do it, let us consider a family of time like world lines in a given pseudo-Riemannian spacetime M_4. In a coordinate system, the family is defined with a help of some given functions ξ^α as

$$x^\alpha = \xi^\alpha(s, \rho), \qquad (23)$$

from where, for $R \supset [c,d] \ni \rho = \rho_o = \text{const}$, one obtains equations describing a single time like world line parametrized by its proper time $s \in (a,b) \subset R$.

Physically, the two dimensional surface Σ_2 defined by Eqs. (23) represents either a one parameter family of observers, in which every two neighbors are located close to each other in a space time, or a history of a deformable optical fiber.

The parameters s and ρ can be taken as coordinates in Σ_2. Any relation $s = h(\rho)$ describes a curve in Σ_2 whose image in M_4 is given by $x^\alpha = \xi^\alpha(h(\rho), \rho)$. Since

$$k^\alpha = \frac{d\xi^\alpha}{d\rho} = \frac{\partial \xi^\alpha}{\partial s}\frac{dh}{d\rho} + \frac{\partial \xi^\alpha}{\partial \rho}\bigg|_{s=h(\rho)} \qquad (24)$$

is a vector tangent to the world line $x^\alpha = \xi^\alpha(h(\rho), \rho)$, the requirement $g_{\alpha\beta} k^\alpha k^\beta = 0$ writes as

$$g_{\alpha\beta}(h(\rho), s) \left(\frac{\partial \xi^\alpha}{\partial s} h'(\rho) + \frac{\partial \xi^\alpha}{\partial s}\right) \left(\frac{\partial \xi^\beta}{\partial s} h'(\rho) + \frac{\partial \xi^\beta}{\partial \rho}\right)\bigg|_{s=h(\rho)} = 0. \qquad (25)$$

If $g_{\alpha\beta}$ and ξ^α are given as functions of their arguments, Eq. (25) turns into an ordinary differential equation for the function $h(\rho)$. Its solutions[2] $h(\rho, s_o)$ determined by an initial condition $s_o = h(\rho_o)$, for a fixed value of ρ_o taken from the interval $[c,d)$, form two one parameter families of null trajectories of the time like generators $\xi^\alpha(s, \rho_o)$ of Σ_2. Members of every of these two families are labeled by s_o, and are parametrized by ρ. Their images in M_4 are two families of null lines $\xi^\alpha(h(\rho, s_o), \rho_o)$ in M_4 which, in general, need not be null geodesics. Any of these lines, for a fixed value of s_o, is parametrized by values of ρ from $[c,d)$, and intersects the time like generator $\xi^\alpha(s, \rho_o)$ of Σ_2 at the point $\xi^\alpha(s_o, \rho_o)$.

From a physical point of view, such a null line may represent a history of a light ray guided by an optical fiber described by Eqs. (23). This is one of its

---

[2] As will be shown in the sequel, there are always exactly two such solutions.

possible interpretations. The null line may also be interpreted as a limit of a broken line formed piecewise by null geodesics which arises in the case when (23) is a continuous limit of a finite, but numerous family of time like observers $\xi^\alpha(s, \rho_i)$, for $i = 0, 1, \ldots, N$. In the case of the finite family, the broken line describes a history of a light signal sent out[3] by the observer labeled by ρ_o at a value s_o of his proper time to his nearest neighbor. The signal is then instantaneously retransmitted by every next observer at the very moment he receives it from his in turn nearest neighbor.

In order to disclose the geometric meaning hidden behind Eq. (25), let us chose s and ρ as coordinates in Σ_2, and denote them by x^A, $A = 1, 2$. Then the metric γ_{AB} induced in Σ_2 by the spacetime metric $g_{\alpha\beta}$ is

$$\gamma_{AB}(x^C) = g_{\alpha\beta}\left(\xi^\alpha(s,\rho)\right) \frac{\partial \xi^\alpha}{\partial x^A} \frac{\partial \xi^\beta}{\partial x^B}. \tag{26}$$

and Eq. (25) of null trajectories takes the form

$$\gamma_{00}(h')^2 + 2\gamma_{01}h' + \gamma_{11} = 0. \tag{27}$$

Since $\gamma = \det(\gamma_{AB}) < 0$, Eq. (27) admits exactly two different algebraic solutions

$$h'_\pm(\rho) = -\frac{\gamma_{01}}{\gamma_{00}} \pm \frac{\sqrt{-\gamma}}{\gamma_{00}} \tag{28}$$

for the first derivative $h'(\rho)$. Every of the two algebraic solutions (28) is a differential equation brought to its normal form $h'(\rho) = F(h(\rho), \rho)$ in which the r.h. side is a known function of its arguments. Note that the unknown function $h(\rho)$ appears, in general, as one of the arguments of the r.h. side of the equation. The solutions of the differential equations (28) determine thus in Σ_2 two linearly independent null vector fields $k^A_\pm = (h'_\pm(\rho), 1)$, where k_+ belongs to the future, and k_- to the past null cone of the induced metric γ_{AB}. The flows of these fields are the just discussed two families of null trajectories on Σ_2.

An interesting special case of the one parameter family (23) of observers arises when the functions $\xi^\alpha(s, \rho)$ are periodic with respect to the parameter ρ. In such a case, because the period after a suitable redefinition of ρ can always be set equal to 2π, we have

$$\xi^\alpha(s, \rho + 2\pi) = \xi^\alpha(s, \rho). \tag{29}$$

The equality above means that, for any s and for any arbitrarily chosen constant value of $\rho = \rho_o$ from their permissible ranges of values, the world line labeled by ρ_o is identical with that labeled by $\rho_o + 2\pi$. As a result, the family (23) forms a time like world tube Σ_2, and its generators, i.e. the world lines labeled by $\rho = $ const, may be considered as relativistic observers.

---

[3] It may also be an incoming signal.

Suppose now that at the moment s_o of his proper time an observer ρ_o, i.e. with the world line $\xi^\alpha(s,\rho_o)$, sends out in the direction k_+ a light signal that is retransmitted by all the other observers generating Σ_2. The history of this signal within Σ_2 is then described by the solution $h_+(\rho,s_o)$ of Eqs. (28) supplemented by the initial condition $h_+(\rho_o) = s_o$, and the signal returns to the observer ρ_o, after going around the tube Σ_2, at the value of his proper time equal to $h_+(2\pi, s_o)$. Similarly, the observer ρ_o can send out the light signal in the direction $-k_-$, which also belongs to his future light cone. The history of this second signal is given by the solution $h_-(-\rho, s_o)$ of Eqs. (28) together with the initial condition $h_-(\rho_o) = s_o$ of the same kind as before. The second signal returns to the observer ρ_o when his proper time is equal to $h_-(-2\pi, s_o)$. In general, the two times of return of signals forwarded simultaneously need not be equal to each other, and a Sagnac like difference Δ of the proper times arises,

$$\Delta = |h_-(-2\pi, s_o) - h_+(2\pi, s_o)|. \qquad (30)$$

For some particular set-ups of observers, especially when the spacetime M_4 is curved, the analytic approach to Sagnac like effects discussed in the present section appears to be easier to handle then the algebraic procedure used in the previous section devoted to discrete cases. However, if the number of observers who take part in the game is small, the algebraic approach is the only one possible. Research is currently being in progress on applications of the continuous approach to some metrics of general relativity, especially to those describing gravitational waves.

REFERENCES

1. Schrödinger, E., *Expanding Universes*, Cambridge: Cambridge University Press, 1956.
2. Bażański, S. L.,"The Split and Propagation of Light Rays in Relativity," in *On Einstein's Path*, ed. A. Harvey, New York: Springer, 1998.

On Some Class of Gravitational Lagrangians

Andrzej Borowiec

*Institute of Theoretical Physics, Wrocław University, pl. Maxa Borna 9,
50-204 Wrocław, Poland
E-mail: borow@ift.uni.wroc.pl*

Abstract. A class of (Ricci squared) Lagrangians which leads to the almost-complex and almost-product Einstein manifolds is described. In particular, a complex space-time appears as a solution of our variational problem.

INTRODUCTION

Almost-complex and almost-product structures are among the most fundamental geometric structures on a manifold [1]. Structures of this kind appear in a natural way from the first-order (Palatini) variational principle applied to general class of non-linear Lagrangians depending on the Ricci squared invariant constructed out of a metric and a symmetric connection [2]. Moreover, Einstein equations of motion and Komar energy-momentum complex are *universal* for this class of Lagrangians [3]. The non-linear gravitational Lagrangians which still genrate Einstein equations are particulary important since, at the classical level, they are equivalent to General Relativity. However, their quantum contents and divergences could be slightly improved. An important example of non-linear Lagrangian is given by Calabi variational principle.

This note is based on joint works with M. Ferraris, M. Francaviglia (Torino) and I. Volovich (Moscow).

NON-LINEAR RICCI SQUARED LAGRANGIANS

Einstein metrics are extremals of the Einstein-Hilbert purely metric variational problem. It is known that the non-linear Einstein-Hilbert type Lagrangians $f(R)\sqrt{g}$, where f is a function of one real variable and R is a scalar curvature of a metric g [1], lead to fourth order equations for g which are not equivalent to

---

[1] One simply writes $\sqrt{g}$ for $\sqrt{|det g|}$.

Einstein equations unless $f(R) = R - c$ (linear case), or to appearance of additional matter fields. It is also known that the linear "first order" Lagrangian $r\sqrt{g}$, where $r = r(g, \Gamma) = g^{\alpha\beta}r_{\alpha\beta}(\Gamma)$ is a scalar concomitant of the metric g and linear (symmetric) connection Γ, [2] leads to separate equations for g and Γ which turn out to be equivalent to Einstein equations for g.

In the sequel we shall use small letters $r^\alpha_{\beta\mu\nu}$ and $r_{\beta\nu} = r^\alpha_{\beta\alpha\nu}$ to denote the Riemann and Ricci tensor of an arbitrary (symmetric) connection Γ

$$r^\alpha_{\beta\mu\nu} = R^\alpha_{\beta\mu\nu}(\Gamma) = \partial_\mu \Gamma^\alpha_{\beta\nu} - \partial_\nu \Gamma^\alpha_{\beta\mu} + \Gamma^\alpha_{\sigma\mu}\Gamma^\sigma_{\beta\nu} - \Gamma^\alpha_{\sigma\nu}\Gamma^\sigma_{\beta\mu}$$
$$r_{\mu\nu} = r_{\mu\nu}(\Gamma) = r^\alpha_{\mu\alpha\nu} \qquad (1)$$

i.e. without assuming that Γ is the Levi-Civita connection of g.

Unlike in a purely metric case, an equivalence with General Relativity also holds for non-linear gravitational Lagrangians

$$L_f(g, \Gamma) = \sqrt{g}f(r) \qquad (2)$$

when they are considered within the first-order formalism [4] (see also [5]).

Our goal in the present note is to investigate the family of non-linear gravitational Lagrangians

$$\hat{L}_f(g, \Gamma) = \sqrt{g}f(s) \qquad (3)$$

parameterized by the real function f of one variable. Now, the scalar (*Ricci squared*) concomitant $s = s(g, \Gamma) = g^{\alpha\mu}g^{\beta\nu}s_{\alpha\beta}s_{\mu\nu}$, where $s_{\mu\nu} = r_{(\mu\nu)}(\Gamma)$ is the symmetric part of the Ricci tensor of Γ.[3]

Equations of Motion

We choose a metric g and a symmetric connection Γ on a space-time manifold M as independent dynamical variables (so-called Palatini method). Variation of $\hat{L}_f$ gives

$$\delta \hat{L}_f = \sqrt{g}(2f'(s)g^{\mu\nu}s_{\alpha\mu}s_{\beta\nu} - \frac{1}{2}f(s)g_{\alpha\beta})\delta g^{\alpha\beta} + 2\sqrt{g}f'(s)s^{\alpha\beta}\delta s_{\alpha\beta}$$

where for short $s^{\alpha\beta} = g^{\alpha\mu}g^{\beta\nu}s_{\mu\nu}$. Taking into account that from (1)

$$\delta s_{\alpha\beta} \equiv \delta r_{(\alpha\beta)} = \nabla_\mu \delta \Gamma^\mu_{\alpha\beta} - \nabla_{(\alpha}\delta \Gamma^\sigma_{\beta)\sigma}$$

with ∇_α being the covariant derivative with respect to Γ and performing the "covariant" Leibniz rule ("integrating by parts") one gets the variational decomposition formula which splits $\delta \hat{L}_f$ into the Euler-Lagrange part and the boundary term:

---

[2] Now, the scalar $r(g, \Gamma) = g^{\alpha\beta}r_{\alpha\beta}(\Gamma)$ is not longer the scalar curvature, since Γ is not longer the Levi-Civita connection of g.

[3] Thereafter () denotes a symmetryzation.

$$\delta \hat{L}_f = \sqrt{g}(2f'(s)g^{\mu\nu}s_{\alpha\mu}s_{\beta\nu} - \frac{1}{2}f(s)g_{\alpha\beta})\delta g^{\alpha\beta} - \nabla_\nu[2\sqrt{g}f'(s)(s^{\alpha\beta}\delta^\nu_\lambda$$
$$- s^{\nu\alpha}\delta^\beta_\lambda)]\delta\Gamma^\lambda_{\alpha\beta} + \partial_\mu[2\sqrt{g}f'(s)s^{\alpha\beta}(\delta\Gamma^\mu_{\alpha\beta} - \delta^\mu_\beta\delta\Gamma^\sigma_{\alpha\sigma})] \quad (4)$$

Since the boundary term in (4) transforms as a vector density of weight 1, one was allowed to replace the covariant derivative by the partial one [4]. The Euler-Lagrange field equations in this case are [2,3]:

$$f'(s)g^{\mu\nu}s_{\alpha\mu}s_{\beta\nu} - \frac{1}{4}f(s)g_{\alpha\beta} = 0 \quad (5)$$

$$\nabla_\lambda(\sqrt{g}f'(s)g^{\alpha\mu}g^{\beta\nu}s_{\mu\nu}) = 0 \quad (6)$$

In fact, variation of $\hat{L}_f$ with respect to Γ leads to the following equations (cf. (4)):

$$\nabla_\beta[2\sqrt{g}f'(s)(s^{\alpha\sigma}\delta^\beta_\lambda - s^{\beta(\alpha}\delta^{\sigma)}_\lambda)] = 0$$

which due to the symmetry of $s^{\mu\nu}$ reduce to (6).

Equations (5-6) must be considered together with the consistency condition obtained by contraction of (5) with $g^{\alpha\beta}$. It gives

$$f'(s)s - \frac{n}{4}f(s) = 0 \quad (7)$$

where n denotes a dimension of the space-time M. The last equation (except the case it is identically satisfied) forces s to take a set of constant values $s = c$, with c being solution of (7). In the *generic* case (simple roots, with $f'(c) \neq 0$, $n > 2$) equation (5) can be rewritten in the following (matrix) form [2,3]

$$(g^{-1}h)^2 = \frac{c}{|c|}I \quad (8)$$

where $c/|c| = \pm 1$ and h is defined by

$$h_{\alpha\beta} = \sqrt{\frac{n}{|c|}}\, s_{\alpha\beta}(\Gamma). \quad (9)$$

$h_{\alpha\beta}$ is a symmetric, twice-covariant and due to (8) non-degenerate tensor field on M i.e., it is simply a metric. By making use of the Ansätz (9), equations (6) can be converted into the form

$$\nabla_\lambda(\sqrt{h}h^{\alpha\beta}) = 0$$

with $h^{\alpha\beta}$ being the inverse of $h_{\alpha\beta}$. This, in turn, forces Γ to be the Levi-Civita connection of h. Replacing back into (9) we find

$$R_{\mu\nu}(h) = \Lambda(c)h_{\mu\nu}$$

---

[4] We should remember that $\delta\Gamma$ transforms as a tensor.

the Einstein equations for the metric h with the cosmological constant $\Lambda(c) = \sqrt{|c|}/n$. Therefore, the scalar curavatur R of the metric h takes a constant value $\sqrt{n|c|}$. It shows *universality* of Einstein equations for this class of Lagrangians. It means independence of the equations of motion on the choice of the Lagrangian (represented by the function f). This property holds true in any dimension $n > 2$.[5]

Symmetries and Superpotentials

We proceed now to analyse the Noether symmetries and the corresponding conservation laws for the Ricci squared Lagrangians. Our Lagrangians are *reparametrization invariant*, in the sense that diffeomorphisms of M transform $\hat{L}_f$ as a scalar density of weight 1. Therefore, as a symmetry transformation, consider a 1-parameter group of diffeomorphisms generated by the vectorfield $\xi = \xi^\alpha \partial_\alpha$ on M. This means that, at the infinitesimal level, variations of the field variables are represented by the Lie derivatives. (See e.g. [6] and [7] for a self-contained exposition of the Second Noether Theorem.)

The main contribution to the Noether current comes from the boundary term in (4) which when expressed in terms of a new metric (9) reads as follow

$$2f'(c)\Lambda(c)\sqrt{h}h^{\alpha\beta}(\delta\Gamma^\mu_{\alpha\beta} - \delta^\mu_\beta \delta\Gamma^\sigma_{\alpha\sigma})] \tag{10}$$

where now

$$\delta\Gamma^\alpha_{\alpha\rho} \equiv \mathcal{L}_\xi \Gamma^\beta_{\alpha\rho} = \xi^\sigma R^\beta_{\alpha\sigma\rho} + \nabla_\alpha \nabla_\rho \xi^\beta$$

and $\mathcal{L}_\xi$ stands for the Lie derivative along ξ. The expression (10) is proportional with a constant factor $2f'(c)\Lambda(c)$ to that which is known from the standard Einstein-Hilbert variational principle. As a consequence, one obtains the Komar expression

$$U^{\mu\nu}_f(\xi) = 2f'(c)\Lambda(c)|deth|^{\frac{1}{2}}(\nabla^\mu \xi^\nu - \nabla^\nu \xi^\mu) \tag{11}$$

for the corresponding superpotential [8,6,9] Therefore, energy-momentum flow as well as superpotential are proportional to already known from the standard Einstein-Hilbert formalism. This extends a notion of universality for the Ricci squared Lagrangians also to the energy-momentum complex [3].

Related Differential - Geometric Structures

The algebraic constraints (8) are of special interest by their own [2]. They provide on space-time some additional differential-geometric structure, namely a Riemannian almost-product structure and an almost-complex anti-Hermitian ($\equiv$ Norden) structure.

---

[5] See [3] for $n = 2$ case, where also non-generic cases have been described.

In the (psedo-)Riemannian almost-product case one equivalently deals with an almost-product structure given by the $(1,1)$ tensor field $P = g^{-1}h$ ($P^2 = I$) as well as with a compatible metric h [6] satisying the condition

$$h(PX, PY) = h(X, Y) \qquad (12)$$

which is also encoded in the simple algebraic relation (8). Here X, Y denote two arbitrary vecorfields on M.

There is a wide class of integrable almost-product structures, namely so called *warped product* structures [1,10], which are an intrinsic property of some well know exact solutions of Einstein equations: these include e.g. Schwarzschild, Robertson-Walker, Reissner-Nordström, de Sitter, etc. (but not Kerr!). Some other examples are provided by Kaluza-Klein type theories, 3+1 decompositions and more generally so called *split* structures [11]. The explicit form of the zeta function on product spaces and of the multiplicative anomaly has been derived recently in [12].

In the anti-Hermitian case one deals with $2m$ - dimensional manifold M, an almost complex structure $J = g^{-1}h$ ($J^2 = -I$) and an anti-Hermitian metric h: [7]

$$h(JX, JY) = -h(X, Y) \qquad (13)$$

This implies that the signature of h should be (m, m). In the Kählerian case ($\nabla J = 0$ for the Levi-Civita connection of h) the almost-complex structure is automaticly integrable. We have proved that in fact the metric h has to be a real part of certain holomorphic metric on a complex (space-time) manifold M [2].

It should be however remarked that a theory of complex manifolds with holomorphic metric (so called *complex Riemannian* manifods) has become one of the corner-stone of the twistor theory [13]. This includes a *non-linear graviton* [14], theory of *H-spaces* [15] and *ambitwistor* formalism [16].

CONCLUSIONS

We showed that the use of Palatini formalism leads to results essentially different from the metric formulation when one deals with non-linear Lagrangians: with the exception of special ("non-generic") cases we always obtain the Einstein equations as gravitational field equations and Komar complex as a Noether energy-momentum complex. In this sense non-linear theories are equivalent to General Relativity. They admit alternative Lagrangians for the Einstein equations with a cosmological constant.

Moreover, besides the initial metric g one gets the Einstein metric h. Both metrics are related by algebraic equation (8). A characterization of an anti-Kähler manifold as a complex manifold with a holomorphic metric has been obtained.

---

[6] In our case the metric h should be Einsteinian.
[7] Recall that for Hermitian metric $h(JX, JY) = h(X, Y)$.

Our results can be relevant for quantum gravity. In fact, in order to remove divergences one has to add counterterms to the Lagrangian which depend not only on the scalar curvature but also on the Ricci and Riemann tensor invariants. It follows from our results that in the first order formalism, such counterterms do not change the semiclassical limit, since genericly we still have the standard Einstein equation.

REFERENCES

1. Besse A., *Einstein Manifolds*, Berlin: Springer-Verlag, 1987.
2. Borowiec A, Ferraris M., Francaviglia M. and Volovich I., *Almost Complex and Almost Product Einstein Manifolds from a Variational Principle*, dg-ga/9612009 (Centro Vito Volterra, Universitá Degli Studi Di Roma "Tor Vergata", N.**292**, July 1997).
3. Borowiec A, Ferraris M., Francaviglia M. and Volovich I., *Class. Quantum Grav.* **15**, 43 (1998).
4. Ferraris M., Francaviglia M. and Volovich I., *Class. Quantum Grav.* **11**, 1505 (1994).
5. Jakubiec A. and Kijowski J., *J. Math. Phys.* **30**, 1073 (1989).
6. Ferraris M. and Francaviglia M., *Class. Quantum Grav.* **9**, S79 (Supplement 1992).
7. Borowiec A, Ferraris M. and Francaviglia M., *Lagrangian Symmetries of Cher-Simons Theories*, hep-th/9801126.
8. Kijowski J., *Gen. Rel. Grav.* **9**, 857 (1978).
9. Borowiec A, Ferraris M., Francaviglia M. and Volovich I., *Gen. Rel. Grav.* **26**, 637 (1994).
10. Carot J. and da Costa J., *Class. Quantum Grav.* **10**, 461 (1993).
11. Gladush V. D. and Konoplya R. A., *Split structures in general relativity and the Kaluza-Klein theories*, gr-qc/9804043.
12. Bytsenko A. A. and Williams F. L., *J. Math. Phys.* **39**, 1075 (1998).
13. Flathery E. J., *Gen. Rel. Grav.* **9**, 961 (1978).
14. Penrose R., *Gen. Rel. Grav.* **7**, 31 (1976).
15. Boyer C. P., Finley III J. D. and Plebański J. F., in *General Relativity and Gravitation*, Einstein memorial volume, ed. A. Held, New York: Plenum, 1980, pp. 241-281.
16. LeBrun C., *Trans. Amer. Math. Soc.* **278**, 209 (1983).

Remarks on Symplectic Connections

Michel Cahen

Département de Mathématique, Université Libre de Bruxelles
C. P. 218 boulevard du triomphe, B - 1050 Bruxelles

Abstract. We study a variational principle for symplectic connections and describe the moduli space of solutions of the field equations in the case of compact surfaces. Furthermore we show what happens in the Kähler situation and study some particular solutions in the purely symplectic context.

1. The initial motivation for this work comes from the deformation approach to quantization of classical systems introduced some 20 years ago by Bayen, Flato, Fronsdal, Lichnerowicz and Sternheimer [1]. Such a deformation quantization always exists on a given phase space (symplectic manifold) [2–4] but is by no-way unique. [5,6] In particular one can find in [3] an algorithmic construction of such a quantization using as basic data a symplectic connection; it is also known that a change of symplectic connection leads to "equivalent" quantization. The change of quantization inside an equivalence class corresponds, in the more traditional formulation of quantum mechanics, to a change of ordering of the operators. This says that equivalent quantizations leads, in general, to different spectral properties. It thus seems worthwhile to single out, among symplectic connections, some preferred ones, which would lead, in a given equivalence class, to preferred quantization.

We have introduced in [7] a variational principle of the Yang-Mills type, which led to nice moduli-space in the case of compact surfaces. This encouraged us to pursue the matter in more general situation. The Kähler situation also leads to a statisfactory framework. The description of the general symplectic situation is still in the early stages, but we can already make a few remarks.

The results presented here are joint work with Simone Gutt and John Rawnsley.

2. Let (M, ω) be a symplectic manifold. A linear connection ∇ on (M, ω) is said to be symplectic if it is torsion free and if ω is parallel. On a given symplectic manifold, there always exists a symplectic connection but this symplectic connection is not unique. In fact the space of smooth symplectic connections is isomorphic (in a non canonical way) to the space of completely symmetric covariant, smooth, 3-tensors on (M, ω). To a given symplectic connection ∇ and to any pair of vector

fields X, Y, one can associatate a curvature endomorphism R by

$$R(X,Y) = \nabla_X \nabla_Y - \nabla_Y \nabla_X - \nabla_{[X,Y]}$$

The curvature 4-tensor $\underline{R}$ is given by :

$$\underline{R}(X,Y,Z,T) = \omega(R(X,Y)Z, T)$$

It is antisymmetric in the 2 first arguments, symmetric in the 2 last arguments and satisfies a cyclic identity on the 3 first arguments :

$$\oint_{X,Y,Z} (\underline{R})(X,Y,Z,T) = 0$$

The Bianchi identities read :

$$\oint_{X,Y,Z} (\nabla_X \underline{R})(Y,Z,T,U) = 0$$

As in the riemannian context one can define a Ricci tensor, r, by

$$r(X,Y) = \text{tr}[Z \to R(X,Z)Y]$$

It is symmetric. There is no analog of the scalar curvature.

A variational principle for a symplectic connection can be defined using a functional J of the form

$$J = \int \mathcal{A} \frac{\omega^n}{n!}$$

if $\dim M = 2n$. Here $\mathcal{A}$ is an invariant polynomial in the curvature tensor $\underline{R}$ (and eventually some of its derivatives) and a rational function in ω. If one imposes (for simplicity reasons) that $\mathcal{A}$ be a polynomial of degree ≤ 2 in the curvature tensor then $\mathcal{A}$ has the form

$$\mathcal{A} = \lambda r_{ab} r^{ab} + \mu \underline{R}_{abcd} \underline{R}^{abcd}$$

where r_{ab} (resp. $\underline{R}_{abcd}$) are the components of the Ricci (resp. the curvature) tensor relative to a local basis and where the indices are raised by means of the inverse of the symplectic tensor; λ and μ are real numbers.

It was proved in [7] that the functionals corresponding to $r_{ab} r^{ab}$ and $\underline{R}_{abcd} \underline{R}^{abcd}$ lead to the same field equations which read

$$\oint_{X,Y,Z} (\nabla_X r)(Y,Z) = 0 \tag{1}$$

A symplectic connection solution of (1) is said to be preferred.

3. We briefly recall the description of the preferred connections in dimension 2 as this is the starting point of our study of higher dimensions.

In dimension 2, the field equation (1) are equivalent to the existence of a 1-form u such that

$$(\nabla_X r)(Y, Z) = \omega(X, Y)u(Z) + \omega(X, Z)u(Y) \tag{2}$$

From this one deduces that there exists a function β such that

$$(\nabla_X u)(Y) = \beta \omega(X, Y) \tag{3}$$

All the analysis of dimension 2 is based on the examination of the properties of this function β.

Let $\bar{u}$ denote the vector field such that

$$u = i(\bar{u})\omega$$

The 2 following identities involving the function β are particularly useful. There exist real numbers A, B such that

$$r(\bar{u}, \bar{u}) = \beta^2 + B \tag{4}$$

$$\frac{1}{4}r_{ab}r^{ab} = \beta + A \tag{5}$$

One then observes that the 1-forms u and $d\beta$ are linearly independent at all points where $r(\bar{u}, \bar{u}) \neq 0$. This leads to the following local solution of the field equations (1).

Proposition 1. *Assume $d\beta \neq 0$. Let $U = \{p \in M | r_p(\bar{u}, \bar{u}) \neq 0\}$. Then the preferred connection ∇ must be given on U by the formulae*

$$\begin{cases} \nabla_{\bar{u}}\bar{u} = -\beta \bar{u} \\ \nabla_{\bar{u}}X_\beta = -\beta X_\beta \\ \nabla_X X_\beta = (\beta^2 + 2A\beta - B)\bar{u} \end{cases} \tag{6}$$

where X_β is the Hamiltonian vector field associated to β.

Conversely given a 1-form u and a non constant function β such that $u(X_\beta) = \beta^2 + B$, the formulae (6) define a preferred connection on the open set $\beta^2 \neq -B$.

To complete this local result by a global one in the compact case one has the following propositions :

Proposition 2. *If (M, ω) is a compact symplectic surface and ∇ a preferred connection such that $d\beta = 0$, then $\beta = 0$ and ∇ is locally symmetric.*

To investigate the case $d\beta \neq 0$, one has to examine the nature of the critical points of β and the properties of the flow of the hamiltonian vector field X_β. This leads to

Theorem 1. *If (M, ω) is a compact symplectic manifold of dimension 2 and if ∇ is a preferred symplectic connection, then it is locally symmetric.*

The description of the moduli space of such locally symmetric connections can be achieved if one adds the assumption that ∇ is geodesically complete. We refer to [7] fot the detailed results.

To conlude this brief summary, let us mention, that in dimension 2, when M is not compact, proposition 1 can be completed by a similar statement when $\beta = \text{Cte} \neq 0$. One can thus construct all preferred symplectic connections on the plane, which are not locally symmetric and determine which are the geodesically complete ones.

4. In dimension $2n$ ($n > 1$), we first examine the case where (M, ω) admits a Kähler or pseudo-Kähler structure (g, J) and where the Levi Civita connection of g is preferred in the sense of (1).

Proposition 3. Let (M, ω, J) be a pseudo Kähler manifold. Then the Levi Civita connection ∇ is preferred if and only if the Ricci tensor is parallel.

Now, if (M, ω, J) is Kähler, simply connected and geodesically complete, then

$$(M, \omega, J) = (M_0, \omega_0, J_0) \times \prod_{i=1}^{p}(M_i, \omega_i, J_i)$$

where (M_0, ω_0, J_0) is a standard flat C^r manifold and where each of the factors (M_i, ω_i, J_i) is Kähler, simply connected, complete and has irreducible holomony. Furthermore the factors are uniquely determined up to permutations.

Corollory 1. If (M, ω, J) is Kähler and has preferred Levi Civita connection and an irreducible holomony, then it is either Ricci flat or Kähler Einstein.

In the compact case one knows [8] that the first Chern class is either 0 (Ricci flat) or has a definite sign. The Ricci flat case leads to tori and K_3 surfaces. The negative sign was solved by Calabi and Yau.

The Kählerian situation suggests to examine. in the purely symplectic context those preferred connections which have a parallel Ricci tensor, which is not identically zero.

Proposition 4. Let (M, ω) be a symplectic manifold and let ∇ be a symplectic connection whose Ricci tensor is parallel and has maximal rank. Then M admits an Einstein pseudo riemannian metric. If the manifold is complete and simply connected, the building blocks may be described with great detail

When the Ricci tensor is not of maximal rank the situation is more involved. The various types of Ricci tensor are classified by the adjoint orbits of the group $Sp(n, I\!R)$. In dimension 4, we proved that there are 11 types of distinct adjoint orbits, and in each case we describe the properties of the corresponding manifold.

5. In dimension 2 we showed that the field equations were equivalent to the existence of a 1-form u such that :

$$\nabla_X r(Y, Z) = \omega(X, Y)u(Z) + \omega(X, Z)u(Y)$$

This admits the following partial generalization.

Lemma 1. *Let V be a vector space of dimension n over $\mathbb{R}$; let $\alpha : S^2V^* \otimes V^* \to S^3V^*$ be defined by*

$$\alpha(t) = \frac{1}{3!} \sum_{\sigma \in S^3} \sigma t$$

where S^3 denotes the symmetric group on 3 elements. Let $\lambda : \Lambda^2V^* \otimes V^* \to S^2V^* \otimes V^*$ be defined by

$$\lambda(u \otimes v)(x,y,z) = u(x,z)v(y) + u(y,z)v(x)$$

Then the sequence

$$0 \to \Lambda^3V^* \xrightarrow{i} \Lambda^2V^* \otimes V^* \xrightarrow{\lambda} S^2V^* \otimes V^* \to S^3V^* \to 0$$

is exact.

Corollary 2. *Let (M,ω) be a symplectic manifold and let ∇ be a symplectic connection which is preferred. Let r denote the Ricci tensor of ∇. Then there exists a family of 2-forms $\overset{(i)}{\alpha}$ $(i \leq p)$ and a family of 1-forms $\overset{(i)}{\beta}$ such that :*

$$\nabla_X r(Y,Z) = \sum_i \overset{(i)}{\alpha}(X,Y) \overset{(i)}{\beta}(Z) + \overset{(i)}{\alpha}(X,Z) \overset{(i)}{\beta}(Y).$$

6. This note is dedicated to the memory of Richard Razcka, who has been a highly motivated and briliant physicist. The results described in section 3 have been presented at a meeting in the honour of Richard held in Lodz in April 1998.

REFERENCES

1. F. Bayeu, M. Flato, C. Fronsdal, A. Lichnerowicz, D. Sternheimer. *Deformation theory and quantization.* Ann. Phys. III (1978) pp. 61-110.
2. M. De Wilde, L. Lecomte. *Existence of * products and of formal deformations of the Poison Lie Algebra of arbitrary symplectic manifolds.* L. M. P. **7** (1983) pp. 487-496.
3. B. Fedosov. *A simple geometrical construction of deformation quantization.* J. diff. geom. **40** (1994), pp. 213-238.
4. H. Omari, Y. Maeda, A. Yoshioka. *Deformation quantizations of Poisson algebras.* Contemp. Math. AMS **179** (1994), pp. 213-240.
5. R. Nest, B. Tsygan. *Algebaic index theorem for families.* Adv. MAth. **113** (1995), pp. 154-205.
6. M. Bertelson, M. Cahen, S. Gutt. *Equivalence of * products.* Clas. Quant. Grav. **14** (1997), A93-A107.
7. F. Bourgeois, M. Cahen. *A variational principle for symplectic connections.* Warwick preprint, 21 May 1998.
8. A. Besse. Einstein manifolds. Springer Verlag 1987.
   see also
   A. Beauville. Surfaces $K3$. Séminaire Bourbaki 609.

Two-Dimensional Dilaton Gravity

Marco Cavaglià

Max-Planck-Institut für Gravitationsphysik, Albert-Einstein-Institut,
Schlaatzweg 1, D-14473 Potsdam, Germany [1]

Abstract. I briefly summarize recent results on classical and quantum dilaton gravity in 1+1 dimensions.

INTRODUCTION

In the last few years a great deal of activity has been devoted to the investigation of lower-dimensional gravity [1]. The interest on dimensionally reduced theories of gravity relies essentially on their relation to string theory, higher-dimensional gravity, black hole physics, and gravitational collapse. In this talk I will focus attention on the simplest, non-trivial, lower-dimensional theory of gravity: 1+1 (pure) dilaton gravity [2].

Dilaton gravity is described by the action

$$S[\phi, g_{\mu\nu}] = \int d^2x \sqrt{-g} [\phi R(g) + V(\phi)], \qquad (1)$$

where ϕ is the dilaton field, $V(\phi)$ is the dilatonic potential, and R is the two-dimensional Ricci scalar. In Eq. (1) we have used a Weyl-rescaling of the metric to eliminate the kinetic term of the dilaton field. Equation (1) describes a family of models whose elements are identified by the dilatonic potential. For instance, $V(\phi) = constant$ identifies the matterless sector of the Callan-Giddings-Harvey-Strominger model (CGHS) [3], $V(\phi) = \phi$ identifies the Jackiw-Teitelboim model, and $V(\phi) = 2/\sqrt{\phi}$ describes the two-dimensional sector of the four-dimensional spherically-symmetric Einstein gravity after having integrated on the two-sphere with area $4\pi\phi$.

Dilaton gravity is an interesting example of *Completely Integrable Model*, i.e. a model that can be expressed in terms of free fields by a canonical transformation. Completely integrable models play an important role from the point of view of the quantum theory because they can be quantized exactly (in the free-field representation). This property allows the discussion of quantization subtleties and

---

[1] E-mail: cavaglia@aei-potsdam.mpg.de, web page: http://www.aei-potsdam.mpg.de/~cavaglia

non-perturbative quantum effects. (For the CGHS model see for instance Refs. [4].) Since dilaton gravity can be used to describe black holes and/or gravitational collapse (in the case of coupling with matter), the quantization program is worth exploring.

A direct consequence of the complete integrability of dilaton gravity is that both the metric and the dilaton can be expressed in terms of a D'Alembert field and of a local integral of motion independent of the coordinates [5]. So, using the gauge in which the free field is one of the coordinates, one finds that all solutions depend on a single coordinate. This property constitutes a generalization of the classical Birkhoff Theorem. (For spherically-symmetric Einstein gravity, i.e. $V(\phi) = 2/\sqrt{\phi}$, the "local integral of motion independent of the coordinates" is just the Schwarzschild mass and the dependence of both the metric and the dilaton from a single D'Alembert field means that the four-dimensional line element can be written in a form depending on the radial coordinate only.)

So dilaton gravity can be quantized using two alternative, a priori non-equivalent, approaches. In the first one the theory is quantized by first reducing it to a 0+1 dynamical system, i.e. using first the classical Birkhoff theorem and then the quantization algorithm. Conversely, in the second approach the theory is quantized in the full 1+1 sector and the 0+1 dimensional nature of the system must be recovered a posteriori (*Quantum Birkhoff Theorem*) [6]:

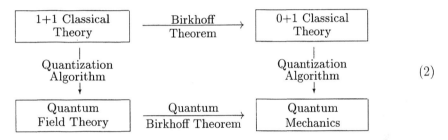

(2)

Furthermore, because of the gauge nature of the theory, the quantization of the system can be implemented according to two different procedures: the *Dirac method* – quantization of the constraints followed by gauge fixing – and the *reduced canonical method* – classical gauge fixing followed by quantization in the reduced space. Usually, the two methods do not lead to identical results.

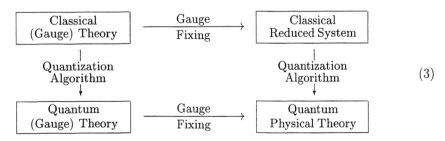

(3)

Here I will show that both diagrams close and the different approaches are equivalent. At the end of the talk I will briefly discuss why these conclusions fail in the case of dilaton gravity coupled to scalar matter.

0+1 QUANTIZATION

In the 0+1 approach the proof of the equivalence of Dirac and reduced methods is straightforward because we are able to pass, via a canonical transformation, to a maximal set of gauge-invariant canonical variables.

Using obvious notations the 0+1 action reads

$$S_{0+1} = \int dt [\dot{q}_i p_i - \mu H], \quad i = 1, 2, \tag{4}$$

where μ is a Lagrange multiplier enforcing the constraint $H = 0$. Thus in the 0+1 sector of the theory we can express the field equations as a canonical system in a finite, 2×2 dimensional, phase space.

Clearly, due to the complete integrability of the model, the equations of motion are analytically integrable and their solution coincides with the finite gauge transformation generated by the (single) constraint $H = 0$. So we can find a couple of gauge-invariant independent canonical quantities (M, P_M) and construct the maximal gauge-invariant canonical chart (M, P_M, H, T). Now T can be used to fix the gauge because its transformation properties for the gauge transformation imply that time defined by this variable covers once and only once the symplectic manifold, i.e. time defined by T is a global time. The quantization becomes trivial and both Dirac and reduced approaches lead to the same Hilbert space. The Hilbert space is spanned by the eigenvectors of the (gauge invariant) operator M corresponding to the "mass" of the system introduced in the previous section.

The quantization program illustrated above has been implemented in detail in Refs. [7] for the case of spherically-symmetric Einstein gravity but can be easily generalized to an arbitrary $V(\phi)$. In the case of Refs. [7] one can go further and prove that the Hermitian operator M in the gauge fixed, positive norm, Hilbert space is not self-adjoint, while its square is a self-adjoint operator with positive eigenvalues. This result is due to the fact that the conjugate variable to the "mass" M, P_M, has positive support, analogously to what happens for the radial momentum in ordinary quantum mechanics. It would be interesting to explore whether this conclusion holds for other choices of the dilatonic potential. In any case, what is important for the present discussion is that the mass M – or its square – is the only gauge-invariant observable of the system (apart from the conjugate variable, of course).

1+1 REDUCED QUANTIZATION

The reduced quantization of the full 1+1 theory can be implemented using "geometrodynamical-like" canonical variables similar to the canonical variables introduced by Kuchař for the canonical description of the Schwarzschild black hole [8]. The new variables are directly related to the spacetime geometry and the relation to the 0+1 formalism is straightforward.

Let us introduce the ADM parametrization of the metric

$$g_{\mu\nu} = \rho \begin{pmatrix} \alpha^2 - \beta^2 & \beta \\ \beta & -1 \end{pmatrix}, \tag{5}$$

where $\alpha(x_0, x_1)$ and $\beta(x_0, x_1)$ play the role of the lapse function and of the shift vector respectively, and $\rho(x_0, x_1)$ represents the dynamical gravitational degree of freedom. Using Eq. (5), the two-dimensional action in the Hamiltonian form reads

$$S[\phi, g_{\mu\nu}] = \int d^2x \sqrt{-g} [\dot\phi \pi_\phi + \dot\rho \pi_\rho - \alpha \mathcal{H}_0 - \beta \mathcal{H}_1]. \tag{6}$$

We can pass to a new canonical chart $(M, \pi_M, \bar\phi, \pi_{\bar\phi})$ using the canonical map

$$M = N(\phi) - \frac{\rho^2 \pi_\rho^2 - \phi'^2}{\rho}, \qquad \pi_M = \frac{\rho^2 \pi_\rho}{\rho^2 \pi_\rho^2 - \phi'^2}, \tag{7}$$

$$\bar\phi = \phi, \qquad \pi_{\bar\phi} = \pi_\phi - \frac{\rho^2 \pi_\rho}{\rho^2 \pi_\rho^2 - \phi'^2} \left[V(\phi) + 2\pi_\rho \left(\frac{\phi'}{\rho \pi_\rho} \right)' \right]. \tag{8}$$

The canonical quantity M corresponds to the local integral of motion mentioned in Sec. and can be identified with the mass of the system. In the 0+1 sector M reduces to the quantity defined in Sec. . In the new canonical chart the ADM super-Hamiltonian and super-momentum constraints read

$$\mathcal{H}_0 = [N(\phi) - M] \pi_{\bar\phi} \pi_M + [N(\bar\phi) - M]^{-1} \bar\phi' M', \qquad \mathcal{H}_1 = -\bar\phi' \pi_{\bar\phi} - M' \pi_M, \tag{9}$$

where $'$ means differentiation w.r.t. the spatial coordinate x_1.

The canonical action (6) must be complemented by a boundary term at the spatial infinities. This can be done along the lines of Refs. [8]. The resulting boundary term is of the form

$$S_\partial = -\int dx_0 (M_+ \alpha_+ + M_- \alpha_-), \tag{10}$$

where $M_\pm \equiv M(x_0, x_1 = \pm\infty)$ and $\alpha_\pm(x_0)$ parametrize the action at infinities.

Now we can solve the constraints and quantize the theory. It is easy to prove that the general solution of Eqs. (9) is given by

$$\pi_{\bar\phi} = 0, \qquad M' = 0. \tag{11}$$

(Note that M weakly commutes with the constraints, as expected for a local integral of motion.) Thus $M \equiv m(x_0)$ and the effective Hamiltonian is simply given by the boundary term (10). The reduced action reads

$$S[m] = \int d\tau \left[\frac{dm}{d\tau} p - m \right], \qquad (12)$$

where $p = \int_{\mathbb{R}} dx_1 \pi_M$ and $\tau(x_0) = \int^{x_0} dx_0' (\alpha_+ + \alpha_-)$. Now quantization can be carried on as usual. The Schrödinger equation is

$$i \frac{\partial}{\partial \tau} \psi(m;\tau) = H_{\text{eff}} \, \psi(m;\tau), \qquad H_{\text{eff}} \equiv m. \qquad (13)$$

The stationary states are the eigenfunctions of m and the Hilbert space coincides with the Hilbert space obtained in the 0+1 approach.

1+1 DIRAC QUANTIZATION

The equivalence between 0+1 and 1+1 Dirac methods can be easily proved using the canonical transformation illustrated in the previous section. However, the same result can be obtained through a completely different quantization scheme. Let me sketch the main points.

Since dilaton gravity is completely integrable [5], it seems reasonable to assume the existence of a canonical transformation mapping the original system to a system described by a pair of (constrained) free fields A_α ($\alpha = 0, 1$) in a flat two-dimensional Minkowski background. (In the CGHS case, i.e. constant dilatonic potential, this canonical transformation is explicitly known since long time [4]. A generalization to linear and exponential dilatonic potentials has been recently derived by Cruz and Navarro-Salas, see Ref. [9].) Using the free fields A_α the super-Hamiltonian and super-momentum constraints read

$$\mathcal{H}_0 = \frac{1}{2} \pi^\alpha \pi_\alpha + \frac{1}{2} A'^\alpha A'_\alpha = 0, \qquad (14)$$

$$\mathcal{H}_1 = -\pi^\alpha A'_\alpha = 0, \qquad (15)$$

where π^α are the conjugate momenta of A_α. Now the theory can be quantized in the free field representation. This has been done in detail in Ref. [6] for the CGHS model. (See also Refs. [4].) Since in this case the canonical transformation is known explicitly, the equivalence with the previous approaches can be proved.

In the CGHS case the functional M defined in Sec. and Sec. is [6]

$$M = M_0 + M_1(A_\alpha, \pi_\alpha), \qquad (16)$$

where M_0 is a constant (zero mode). On the field equations we have $M = M_0$. Due to positivity conditions that are present in the model the constraints (14-15) can be linearized and the quantization is carried out by use of the standard Gupta-Bleuler method [6].

The quantum reduction of the theory to a 0+1 dynamical system can be made clear by investigating the matrix elements of the operator M. Adopting a normal ordering the matrix elements of M between physical states are

$$< \Psi_2|M|\Psi_1 > = < \Psi_2|M_0|\Psi_1 > . \qquad (17)$$

Since M_0 is a zero mode, it commutes with all the creation and annihilation operators of A_α. So the vacuum must be labeled by the eigenvalue of M_0, i.e.

$$M_0|0;m> = m|0;m> . \qquad (18)$$

The existence of infinite vacua, differing by the eigenvalue of the mass, implies that the theory reduces to quantum mechanics. Again, the only gauge invariant operator is the mass (and its conjugate momentum) and the resulting Hilbert space is spanned by the eigenvectors of M.

COUPLING TO A MASSLESS SCALAR FIELD

Let me conclude this talk spending few words on dilaton gravity coupled to a massless scalar field. We have seen that the topological nature of dilaton gravity is a direct consequence of the existence of a functional of the canonical variables which is conserved under time and space translations (the mass M): the original fields can be expressed in terms of a free field and a local integral of motion instead of two free fields, as one might expect from the counting of the degrees of freedom.

When a scalar field is coupled to the system non-static solutions appear, the Birkhoff theorem is no longer valid, and the topological nature of dilaton gravity is destroyed. This has an important consequence from the canonical point of view. Indeed, we can immediately conclude that no local integrals of motion like M do exist. A provocative interpretation of this result is that the mass of a spherically symmetric black hole coupled to scalar matter cannot be defined at the canonical level! In my opinion this is quite worrying, especially from the quantum point of view. Finally, a related point is that 0+1 dimensional solutions of dilaton gravity coupled to scalar matter have no horizons – see Refs. [5] and, for the case $V(\phi) = 2/\sqrt{\phi}$, Ref. [10] – or, in other words, "black holes have no scalar hair".

The difficulty in the quantization of dilaton gravity coupled to a massless scalar field is evident in any of the procedures described above. For instance, even though the canonical transformation to free fields described in Sect. can be formally implemented, the linearization of the constraints at the basis of the Gupta-Bleuler quantization is no longer possible. Anomalies are present and a consistent quantization requires a modification of the theory [4].

ACKNOWLEDGEMENTS

I am grateful to the organizers of the conference *Particles, Fields & Gravitation '98* for hospitality and financial support. I am indebted to my friends and collaborators Vittorio de Alfaro and Alexandre T. Filippov for interesting discussions and useful suggestions on various questions connected to the subject of this paper. This work has been supported by a Human Capital and Mobility grant of the European Union, contract no. ERBFMRX-CT96-0012.

REFERENCES

1. An updated collection of papers on lower-dimensional gravity can be found at the web page http://www.aei-potsdam.mpg.de/~cavaglia/ldg.html.
2. Recently, dilaton gravity coupled to various matter fields has also been extensively investigated. See for instance: Jackiw, R., in: *Procs. of the Second Meeting on Constrained Dynamics and Quantum Gravity, Nucl. Phys. (Proc. Suppl.)* **57**, 162 (1997); Klösch, T., *ibid.* p. 326; Strobl, T., *ibid.* p. 330; Cavaglià, M., *Phys. Rev. D* **57**, 5295 (1998); *Phys. Lett. B* **413**, 287 (1997); Pelzer, H., and Strobl, T., *Generalized 2d Dilaton Gravity with Matter Fields*, e-Print Archive: gr-qc/9805059; Cavaglià, M., Fatibene, L., and Francaviglia, M., *Two-Dimensional Dilaton-Gravity Coupled to Massless Spinors*, e-Print Archive: hep-th/9801155, and references therein.
3. Callan, C., Giddings, S., Harvey, J., and Strominger, A., *Phys. Rev. D* **45**, 1005 (1992); Verlinde, H., in: *Sixth Marcel Grossmann Meeting on General Relativity*, M. Sato and T. Nakamura eds. (World Scientific, Singapore, 1992).
4. Benedict, E., Jackiw, R., and Lee, H.-J., *Phys. Rev. D* **54**, 6213 (1996); Cangemi, D., Jackiw, R., and Zwiebach, B., *Ann. Physics (N.Y.)* **245**, 408 (1995); Kuchař, K.V., Romano, J.D., and Varadarajan, M., *Phys. Rev. D* **55**, 795 (1997).
5. Filippov, A.T., in: *Problems in Theoretical Physics*, Dubna, JINR, June 1996, p. 113; *Mod. Phys. Lett. A* **11**, 1691 (1996); *Int. J. Mod. Phys. A* **12**, 13 (1997).
6. Cavaglià, M., de Alfaro, V., and Filippov, A.T., *Phys. Lett. B* **424**, 265 (1998), e-Print Archive: hep-th/9802158.
7. Cavaglià, M., de Alfaro, V., and Filippov, A.T., *Int. J. Mod. Phys. D* **4**, 661 (1995); *Int. J. Mod. Phys. D* **5**, 227 (1996).
8. Kuchař, K.V., *Phys. Rev. D* **50**, 3961 (1994); Varadarajan, M., *Phys. Rev. D* **52**, 7080 (1995).
9. Cruz, J., and Navarro-Salas, J., *Mod. Phys. Lett. A* **12**, 2345 (1997).
10. Cavaglià, M., and de Alfaro, V., *Int. J. Mod. Phys. D* **6**, 39 (1997).

Integrable Hierarchies in Twistor Theory

Maciej Dunajski

The Mathematical Institute
24-29 St Giles, Oxford OX1 3LB, UK

INTRODUCTION

One of the most remarkable achievements of twistor program is the link it provides between integrable differential equations and unconstrained holomorphic geometry. What lies at heart of the twistor approach to the integrability is the existence of the Lax pair which enables one to express a given nonlinear equation as the compatibility condition (usually in the form of the zero curvature representation) for a system of linear first order PDEs. The two most prominent systems of nonlinear equations which fit into the program are anti-self-dual vacuum Einstein equations (ASDVE) [10] and anti-self-dual Yang–Mills equations (ASDYM) [15]. Let us start off with a linear example

Let (w, z, x, y) be the coordinates on $\mathbb{C}^4$ which are null with respect to the metric $2dwdx + 2dzdy$. Long before twistor theory was introduced it was know [1] that solutions to the complex wave equation

$$\Theta_{xw} + \Theta_{yz} = 0 \qquad (1)$$

are given by contour integral formulae

$$\Theta(w, z, x, y) = \frac{1}{2\pi i} \oint_\Gamma f(w + \lambda y, z - \lambda x, \lambda) d\lambda. \qquad (2)$$

Here $\lambda \in \mathbb{CP}^1$ and the contour Γ separates poles of the integrand. Let us make a few remarks about the last formula.

- The function f is an arbitrary function of three variables. It is not constrained by any equations.

- As stated, the correspondence between the solutions to (1) and integrands (2) is certainly not one to one; We may change f by adding a function which is singular on one side of the contour Γ but is holomorphic on the other. We may also move the contour Γ without touching the poles of f. Both changes will not affect a corresponding solution to the wave equation. The precise relation between Θ and the pairs (f, Γ) is described in the twistor theory by using the sheaf cohomology [16].

- The geometric reasons for the appearance of $\lambda \in \mathbb{CP}^1$ are not clear from the formula (2). In the twistor approach to the integrable systems λ plays the role of a spectral parameter and parametrises certain null planes passing through each point of $\mathbb{C}^4$.

From the applied mathematics point of view, the formula (2) gives only an alternative to other methods of solving the wave equation. The usefulness of the twistor approach is better illustrated on examples of nonlinear equations.

Modify (1) by adding a nonlinear term of the Monge-Ampere type

$$\Theta_{xw} + \Theta_{yz} + \Theta_{xx}\Theta_{yy} - \Theta_{xy}^2 = 0. \qquad (3)$$

The motivation for studying the second heavenly equation (3) comes from the work of Plebański [11]. He showed that if Θ is a solution of (3) then

$$\mathrm{d}s^2 = 2\mathrm{d}w\mathrm{d}x + 2\mathrm{d}z\mathrm{d}y + 2\Theta_{xx}\mathrm{d}z^2 + 2\Theta_{yy}\mathrm{d}w^2 - 4\Theta_{xy}\mathrm{d}w\mathrm{d}z \qquad (4)$$

is the hyper-Kähler metric on an open ball in $\mathbb{C}^4$. Each hyper-Kähler metric on a complex four-manifold can locally be put to the form (4). In four (real or complex) dimensions, hyper-Kähler metrics are solutions to anti self-dual Einstein vacuum equations (ASDV). This makes (3) worth studying both from the geometry and the general relativity perspectives.

A natural question which arises is whether one can generalise the formula (2) to solve the equation (3). In general such an explicit description will not be possible, but nevertheless the twistor approach assures the integrability of (3). This follows from Penrose's Nonlinear Graviton construction [10], in which ASDV metrics locally correspond to certain three dimensional complex manifolds - twistor spaces. The manifold structure of a twistor space is given by a set of patching functions. The process of recovering an ASDV metric on $\mathbb{C}^4$ from the patching functions involves solving the non-linear Riemann Hilbert factorisation problem. In the case of the wave equation the analogous Riemann Hilbert problem is linear and a solution can be given explicitly.

The twistor methods often fail with providing explicit solutions to integrable equations. However they are effective tools in studying properties of integrable systems other than solutions. One of such properties is the existence of an infinite number of symmetries. If the symmetries of a given integrable equation (viewed as functionals on the phase-space) are in involution with respect to some Poisson structure, then each of them can act as a Hamiltonian and therefore generate a new PDE. The infinite sequence of all the equations obtained in this way is called a hierarchy. Each flow in the hierarchy defines a new Poisson structure on the phase space. The aim of this paper is to find hierarchical structure for the second heavenly equation (3). It summarises and extends my joint work with Lionel Mason [4-6] on this subject. We owe many ideas to discussions with George Sparling.

In the next Section we summarise the twistor correspondences for flat and curved spaces. In Section 3 we introduce the recursion operator R for the second heavenly equation and look at its twistor counterpart. We illustrate the method on the example of the Sparling-Tod solution and show how R can be used to construct an $\mathcal{O}(1) \oplus \mathcal{O}(1)$ rational curve in a twistor space. In Section 4 we study generalised twistor spaces corresponding to heavenly hierarchies. Lagrangian and Hamiltonian aspects of the constructions will be discussed elsewhere. For clarity of presentation the usage of a spinor notation has been reduced to minimum.

We end this introduction with some bibliographical remarks. For the case of ASDYM the twistorial treatment of hierarchies has been dealt with in [8] and more fully in [9]. Significant progress towards understanding the symmetry structure of the heavenly equations was achieved by Boyer and Plebański [2,3] who obtained an infinite number of conservation laws for the ASDVE equations and established some connections with the nonlinear graviton construction. Some of their results were later extended in papers of Strachan [13] and Takasaki [14].

OUTLINE OF THE TWISTOR CORRESPONDENCE

In this section we give a brief outline of the twistor correspondence. For more detailed expositions see [16] or [9]. We use the double null coordinates on $\mathbb{C}^4$ in which the metric and the volume element are

$$ds^2 = 2dwdx + 2dzdy, \qquad \nu = dw \wedge dz \wedge dx \wedge dy.$$

A two plane in $\mathbb{C}^4$ is null if $ds^2(X, Y) = 0$ for every pair (X, Y) of vectors tangent to it. The null planes can be self-dual (SD) or anti self-dual (ASD) depending on whether the tangent bivector $X \wedge Y$ is SD or ASD. The SD null planes are called α-planes. The α-planes passing through a point in $\mathbb{C}^4$ are parametrised by $\lambda \in \mathbb{CP}^1$. Tangents to α-planes are spanned by two vectors

$$L_0 = \partial_y - \lambda \partial_w, \quad L_1 = \partial_x + \lambda \partial_z \tag{5}$$

or (∂_z, ∂_w) if $\lambda = \infty$. The set of all α-planes is called a projective twistor space and denoted $\mathcal{PT}$. It is a three-dimensional complex manifold biholomorphic to $\mathbb{CP}^3 - \mathbb{CP}^1$.

We will make use of a double fibration picture

$$\mathcal{M} \xleftarrow{p} \mathcal{F} \xrightarrow{q} \mathcal{PT}. \tag{6}$$

The five complex dimensional correspondence space $\mathcal{F} = \mathbb{C}^4 \times \mathbb{CP}^1$ fibres over $\mathbb{C}^4$ by $p(w, z, x, y, \lambda) = (w, z, x, y)$. The functions on $\mathcal{F}$ which are constant on α-planes, or equivalently satisfy $L_A f = 0$; $A = 0, 1$, push down to $\mathcal{PT}$. They are called twistor functions. An example of a twistor function was used in the formula (2). The twistor space $\mathcal{PT}$ is a factor space of $\mathcal{F}$ by the two-dimensional distribution spanned by L_A. It can be covered by two coordinate patches U and $\widetilde{U}$, where U is a compliment of $\lambda = \infty$ and $\tilde{U}$ is a compliment of $\lambda = 0$. If (μ^0, μ^1, λ) are coordinates on U and $(\tilde{\mu}^0, \tilde{\mu}^1, \tilde{\lambda})$ are coordinates on $\tilde{U}$ then on the overlap

$$\tilde{\mu}^0 = \mu^0/\lambda, \quad \tilde{\mu}^1 = \mu^1/\lambda, \quad \tilde{\lambda} = 1/\lambda.$$

The local coordinates (μ^0, μ^1, λ) on $\mathcal{PT}$ pulled back to $\mathcal{F}$ are

$$\mu^0 = w + \lambda y, \quad \mu^1 = z - \lambda x, \quad \lambda. \tag{7}$$

One can rephrase it is the spinor language: The tangent space $\mathbb{T}$ at each point of $\mathbb{C}^4$ is isomorphic to a tensor product of two spin spaces

$$\mathbb{T} = S^A \otimes S^{A'}. \tag{8}$$

where $A, B, ..., A', B'$ are two-dimensional spinor indices. The spin spaces S^A and $S^{A'}$ are equipped with symplectic forms ε_{AB} and $\varepsilon_{A'B'}$ such that $\varepsilon_{01} = \varepsilon_{0'1'} = 1$. These anti-symmetric objects are used to raise and lower the spinor indices. A point in $\mathbb{C}^4$ is represented by its position vector (w, z, x, y) and the isomorphism (8) is realised by

$$x^{AA'} := \begin{pmatrix} y & w \\ -x & z \end{pmatrix}, \quad \text{so that} \quad ds^2 = \varepsilon_{AB}\varepsilon_{A'B'} dx^{AA'} dx^{BB'}.$$

The homogeneous coordinates on the twistor space are

$$(\omega^0, \omega^1, \pi_{0'}, \pi_{1'}) = (\omega^A, \pi_{A'}).$$

They are related to (μ^0, μ^1, λ) by

$$\omega^0/\pi_{1'} = \mu^0, \quad \omega^1/\pi_{1'} = \mu^1, \quad \pi_{0'}/\pi_{1'} = \lambda.$$

For $\lambda \neq \infty$ the twistor distribution may be rewritten as

$$L_A = (\pi_{1'})^{-1}\pi^{A'}\frac{\partial}{\partial x^{AA'}}.$$

The relations between various structures on $\mathbb{C}^4$ and $\mathcal{PT}$ can be read off from the equation

$$\omega^A = x^{AA'}\pi_{A'}. \tag{9}$$

Assume that $\pi_{A'} \neq 0$ and consider $(\omega^A, \pi_{A'})$ to be fixed. Then (9) has as its solution a complex two plane spanned by vectors of the form $\pi^{A'}v^A$ for some v^A. The other way of interpreting (9) is fixing $x^{AA'}$ and solving for $(\omega^A, \pi_{A'})$. The solution, when factored out by the relation $(\omega^A, \pi_{A'}) \sim (k\omega^A, k\pi_{A'})$, becomes a rational curve $\mathbb{CP}^1$ with a normal bundle $\mathcal{O}(1) \oplus \mathcal{O}(1)$. This condition guarantees that the family of rational curves in $\mathcal{PT}$ is four complex dimensional, and that the conformal structure ds^2 on $\mathbb{C}^4$ is quadratic. Here $\mathcal{O}(n)$ denotes the line bundle over $\mathbb{CP}^1$ with transition functions λ^{-n}. The points p and q are null separated in $\mathbb{C}^4$ iff the corresponding rational curves l_p and l_q intersect at one point in $\mathcal{PT}$.

The Nonlinear Graviton construction

Assume that ds^2 is a curved metric on some complex four-dimensional manifold $\mathcal{M}$. The notion of an α-plane must be replaced by an α-surface - a null two dimensional surface such that its tangent space at each point is an α plane. Let X and Y be two vectors tangent to an α-surface. The Frobenius integability condition yields
$$[X, Y] = aX + bY$$
for some a and b. The last formula implies that W_+ (the self-dual part of the Weyl tensor) vanishes. Thus given $W_+ = 0$ we can define a twistor space $\mathcal{PT}$ to be a three complex dimensional manifold of α-surfaces in $\mathcal{M}$. If ds^2 is also Ricci flat (vacuum) then $\mathcal{PT}$ has more structures which are listed in the following theorem [10].

Theorem 1 (Penrose 76) *There is one to one correspondence between ASD vacuum metrics and three dimensional complex manifolds $\mathcal{PT}$ such that*

- *There exists a holomorphic projection $\mu : \mathcal{PT} \longrightarrow \mathbb{CP}^1$*

- *$\mathcal{PT}$ is equipped with a four complex parameter family of sections of μ each with a normal bundle $\mathcal{O}(1) \oplus \mathcal{O}(1)$*

- *Each fibre of μ has a symplectic structure $\Sigma_\lambda \in \Gamma(\Lambda^2(\mu^{-1}(\lambda)) \otimes \mathcal{O}(2))$, where $\lambda \in \mathbb{CP}^1$.*

RECURSION RELATIONS FOR THE SECOND HEAVENLY EQUATIONS.

Let $\mathcal{PT}$ be a twistor space from the Theorem 1 corresponding to the curved ASDV metric. Let $(\omega^A, \pi_{A'})$ be the homogeneous coordinates on $\mathcal{PT}$ around $\lambda = \pi_{0'}/\pi_{1'} = 0$.

We pull back twistor coordinates to $\mathcal{F}$ and dehomogenise them by putting $\mu^A = \mu^*(\omega^A/\pi_{1'})$. Now parametrise the moduli space of rational curves by

$$\mu^A(\lambda = 0) = (w, z), \qquad \left.\frac{\partial \mu^A}{\partial \lambda}\right|_{\lambda=0} = (y, -x).$$

Fix $\lambda \in \mathbb{CP}^1$. The pull back of the symplectic form Σ_λ to $\mathcal{F}$ is (λ is not differentiated) $(\pi_{1'})^2 d\mu^0 \wedge d\mu^1$. It satisfies

$$\mu^*(\Sigma_\lambda) \wedge \mu^*(\Sigma_\lambda) = 0, \qquad d\mu^*(\Sigma_\lambda) = 0, \qquad (10)$$

where $d = dw \otimes \partial_w + dz \otimes \partial_z + dx \otimes \partial_x + dy \otimes \partial_y$. These two equations give rise [4] to a function $\Theta(w, z, x, y)$ such that the twistor functions μ^A, holomorphic around $\lambda = 0$, have the following expansions

$$\mu^0 = w + \lambda y - \lambda^2 \Theta_x + \lambda^3 \Theta_z + \dots,$$
$$\mu^1 = z - \lambda x - \lambda^2 \Theta_y - \lambda^3 \Theta_w + \dots. \tag{11}$$

As a consequence of the algebraic relation in (10) the function Θ satisfies the second heavenly equation (3)

The existence of the Lax pair for equation (3) can be deduced directly from the structure of $\mathcal{PT}$. Let l_p be the line in $\mathcal{PT}$ that corresponds to $p \in \mathcal{M}$. The normal bundle to l_p consists of vectors tangent to p (horizontally lifted to $T_{(p,\lambda)}\mathcal{F}$) modulo the twistor distribution. The corresponding short sequence of sheaves over $\mathbb{CP}^1$ is

$$0 \longrightarrow D \longrightarrow \mathbb{C}^4 \longrightarrow \mathcal{O}(1) \oplus \mathcal{O}(1) \longrightarrow 0.$$

The map $\mathbb{C}^4 \longrightarrow \mathcal{O}(1) \oplus \mathcal{O}(1)$ is given by $V^{AA'} \longrightarrow V^{AA'}\pi_{A'}$. Its kernel consists of vectors of the form $\pi^{A'}v^A$ with v^A varying. The twistor distribution is therefore $D = O(-1) \otimes \mathbb{C}^2$. The condition that L_A - the element of $\Gamma(D \otimes \mathcal{O}(1) \otimes (\mathbb{C}^2)^*)$ - annihilates twistor functions (11) yields the explicit form

$$L_0 = \partial_y - \lambda(\partial_w - \Theta_{xy}\partial_y + \Theta_{yy}\partial_x), \quad L_1 = \partial_x + \lambda(\partial_z + \Theta_{xx}\partial_y - \Theta_{xy}\partial_x). \tag{12}$$

Let $\Box_\Theta$ denote the wave operator on an ASDV curved background given by Θ, and let $\delta\Theta$ be the linearised solution to the second heavenly equation (i.e. $\Theta + \delta\Theta$ satisfies (3) up to the linear terms in $\delta\Theta$) and let $\mathcal{W}_\Theta$ be the kernel of $\Box_\Theta$. It is straightforward to check that $\delta\Theta \in \mathcal{W}_\Theta$.

Let $\phi \in \mathcal{W}_\Theta$. Define a recursion operator $R : \mathcal{W}_\Theta \longrightarrow \mathcal{W}_\Theta$ by

$$\partial_y(R\phi) = (\partial_w - \Theta_{xy}\partial_y + \Theta_{yy}\partial_x)\phi, \quad -\partial_x(R\phi) = (\partial_z + \Theta_{xx}\partial_y - \Theta_{xy}\partial_x)\phi. \tag{13}$$

¿From (13) and from (3) it follows that if ϕ belongs to $\mathcal{W}_\Theta$ then so does $R\phi$. Therefore we have

Proposition 1 ([4]) *Let $\mathcal{W}_\Theta$ be the space of solutions of wave equation on the curved ASD background given by Θ.*

(i) Elements of $\mathcal{W}_\Theta$ are infinitesimal symmetries of the second equation.

(ii) The map $R : \mathcal{W}_\Theta \longrightarrow \mathcal{W}_\Theta$ given by (13) generates new elements of $\mathcal{W}_\Theta$ from the old ones.

Let $f \in H^1(\mathcal{PT}, \mathcal{O}(2))$ be a patching function on $\mathcal{PT}$ which corresponds to an ASD vacuum metric given by Θ, and let δf be the linearised patching function corresponding to $\delta\Theta$. It can be shown [6] that

$$\delta\Theta = \frac{1}{2\pi i}\oint_\Gamma \delta f \lambda^{-4} d\lambda.$$

The recursion relations (13) have a straightforward form on $\mathcal{PT}$:

Proposition 2 ([6]) *Let R be the recursion operator defined by (13). Its twistor counterpart is the multiplication operator*

$$R\,\delta f = \lambda^{-1}\delta f. \tag{14}$$

We see that the twistor description of the recursion operator is simpler than the space-time one. It is also better defined since R acts on δf without ambiguity (alternatively, the ambiguity in the boundary conditions for the definition of R on space-time is absorbed into the choice of explicit representative for the cohomology class determined by δf).

Example

Now we shall illustrate the Propositions 1 and 14 on the example of the Sparling-Tod solution [12]. The calculations involved in this subsection where performed on MAPLE. I would like to thank David Liebowitz for introducing me to the use of computers in mathematics. Consider

$$\Theta = \frac{\sigma}{wx + zy}, \tag{15}$$

where $\sigma = const$. It satisfies both (1) and (3).
(a) First we shall treat (15), with $\sigma = 1$, as a solution ϕ_0 to the wave equation on the flat background (1). Recursion relations are

$$(R\phi_0)_x = \frac{y}{(wx+zy)^2}, \quad (R\phi_0)_y = \frac{-x}{(wx+zy)^2}.$$

They have a solution $\phi_1 := R\phi_0 = (-y/w)\phi_0$. More generally we find that

$$\phi_n := R^n \phi_0 = \left(-\frac{y}{w}\right)^n \frac{1}{wx+zy}. \tag{16}$$

The last formula can be also found using the twistor methods. The twistor function corresponding to ϕ_0 is $1/(\mu^0\mu^1)$, where μ^A are given by (7). By Proposition 14 the twistor function corresponding to ϕ_n is $\lambda^{-n}/(\mu^0\mu^1)$. This can be seen by applying the formula (2) and computing the residue at the pole $\lambda = -w/y$. It is interesting to ask whether any ϕ_n (apart from ϕ_0) is a solution to the heavenly equation. Inserting $\Theta = \phi_n$ to (3) yields $n = 0$ or $n = 2$. We parenthetically mention that ϕ_2 yields (by formula (4)) a metric of type D which is conformal to the Eguchi-Hanson solution.
(b) Now let Θ given by (15) determine the curved metric

$$ds^2 = 2dwdx + 2dzdy + 4\sigma(wx+zy)^{-3}(wdz - zdw)^2.$$

The recursion relations are

$$-\partial_x(R\psi) = (\partial_z + 2\sigma w(wx+zy)^{-3}(w\partial_x - z\partial_y))\psi,$$
$$\partial_y(R\psi) = (\partial_w + 2\sigma z(wx+zy)^{-3}(w\partial_x - z\partial_y))\psi,$$

where ψ satisfies

$$\Box_\Theta \psi = (\partial_x\partial_w + \partial_y\partial_z + 2\sigma(wx+zy)^{-3}(z^2\partial_x^2 + w^2\partial_y^2 - 2wz\partial_x\partial_y))\psi = 0. \quad (17)$$

One solution to the last equation is $\psi_1 = (wx+zy)^{-1}$. We apply the recursion relations to find the sequence of linearised solutions

$$\psi_2 = \left(-\frac{y}{w}\right)\frac{1}{wx+zy}, \quad \psi_3 = -\frac{2}{3}\frac{\sigma}{(wx+zy)^3} + \left(-\frac{y}{w}\right)^2 \frac{1}{wx+zy},$$

$$\psi_n = \sum_{k=0}^{n} A_{(n)}^k \left(-\frac{y}{w}\right)^k (wx+zy)^{k-n}.$$

To find $A_{(n)}^k$ note that the recursion relations imply

$$R\left(\left(-\frac{y}{w}\right)^k (wx+zy)^j\right) =$$

$$\left(\left(-\frac{y}{w}\right) - \sigma\left(-\frac{y}{w}\right)^{-1}(wx+zy)^{-2}\frac{k}{j+2}\right)\left(-\frac{y}{w}\right)^k (wx+zy)^j.$$

This yields a recursive formula

$$A_{(n+1)}^k = A_{(n)}^{k-1} - 2\sigma\frac{k+1}{n-k+1}A_{(n)}^{k+1}, \quad A_{(1)}^0 = 1, \quad A_{(1)}^1 = 0, \quad A_{(n)}^{-1} = 0, \quad k=0...n,$$
$$(18)$$

which determines the algebraic (as opposed to a differential) recursion relations between ψ_n and ψ_{n+1}. It can be checked that functions ψ_n indeed satisfy (17). Notice that if $\sigma = 0$ (flat background) then we recover (16). We can also find the inhomogeneous twistor coordinates pulled back to $\mathcal{F}$

$$\mu^0 = w + \lambda y + \sum_{n=0}^{\infty} \sigma\lambda^{n+2} \sum_{k=0}^{n} B_{(n)}^k w\left(-\frac{y}{w}\right)^k (wx+zy)^{k-n-1},$$

$$\mu^1 = z - \lambda x + \sum_{n=0}^{\infty} \sigma\lambda^{n+2} \sum_{k=0}^{n} B_{(n)}^k z\left(\frac{x}{z}\right)^k (wx+zy)^{k-n-1}.$$

where

$$B_{(n+1)}^k = B_{(n)}^{k-1} - 2\sigma\frac{k+1}{n-k+2}B_{(n)}^{k+1}, \quad B_{(1)}^0 = 1, \quad B_{(1)}^1 = 0, \quad B_{(n)}^{-1} = 0, \quad k=0...n.$$

The polynomials μ^A solve $L_A(\mu^B) = 0$, where now

$$L_0 = -\lambda\partial_w - 2\lambda\sigma z^2(wz+zy)^{-3}\partial_x + (1 + 2\lambda\sigma wz(wz+zy)^{-3})\partial_y,$$
$$L_1 = \lambda\partial_z + (1 - 2\lambda\sigma wz(wz+zy)^{-3})\partial_x + 2\lambda\sigma w^2(wz+zy)^{-3}\partial_y.$$

HIERARCHIES

Using the recursion operator, we can obtain higher flows by acting on the linearised perturbations corresponding to evolution along coordinate vector fields with the recursion operator. This embeds the the second heavenly equations into an infinite system of over-determined but consistent PDEs. We introduce the new coordinates x^{Ai}, where for $i = 0, 1, x^{Ai} = x^{AA'}$ are coordinates on $\mathcal{M}$ and for $1 < i \leq n, x^{Ai}$ are the parameters for the new flows (with $2n - 2$ dimensional parameter space $\mathbb{X}$). The propagation of Θ along these parameters are determined by the recursion relations

$$\partial_y(\partial_{Bi+1}\Theta) = (\partial_w - \Theta_{xy}\partial_y + \Theta_{yy}\partial_x)\partial_{Bi}\Theta,$$
$$-\partial_x(\partial_{Bi+1}\Theta) = (\partial_z + \Theta_{xx}\partial_y - \Theta_{xy}\partial_x)\partial_{Bi}\Theta. \tag{19}$$

Here n is an arbitrarily large integer. The next proposition shows that $\mathcal{N} = \mathcal{M} \times \mathbb{X}$, the moduli space of rational curves in $\mathcal{PT}$, is equipped with a function Θ satisfying the over-determined system of equations which contains (19) as a special case, and with the Lax distribution.

Proposition 3 ([5]) *Let $\mathcal{PT}$ be a 3 dimensional complex manifold with the following structures*

1) *a projection $\mu : \mathcal{PT} \longrightarrow \mathbb{CP}^1$,*

2) *a $2(n+1)$-dimensional family of sections with normal bundle $\mathcal{O}(n) \oplus \mathcal{O}(n)$,*

3) *a canonical isomorphism $K = \mathcal{O}(-2n - 2)$, where $K \cong \Lambda^3(\mathcal{PT})$ is the canonical line bundle, and $\mathcal{O}(-2n - 2) = \mu^*(\bigotimes^{n+1} T^*\mathbb{CP}^1)$.*

and let $\mathcal{N}$ be the moduli space of sections from (2). Then

a) *There exists a function $\Theta : \mathcal{N} \longrightarrow \mathbb{C}$ which (with the appropriate choice of the coordinates) satisfies the set of equations $(i, j = 1, ..., n)$.*

$$\partial_{Ai}\partial_{Bj-1}\Theta - \partial_{Bj}\partial_{Ai-1}\Theta + \partial_{Ai-1}\partial_{00}\Theta\partial_{Bj-1}\partial_{10}\Theta - \partial_{Ai-1}\partial_{10}\Theta\partial_{Bj-1}\partial_{00}\Theta = 0. \tag{20}$$

b) *The correspondence space $\mathcal{F} = \mathcal{N} \times \mathbb{CP}^1$ is equipped with the $2n$-dimensional distribution $D \subset T(\mathcal{N} \times \mathbb{CP}^1)$ which, as a bundle on $\mathcal{F}$ has an identification with $\mathcal{O}(-1) \otimes \mathbb{C}^{2n}$ so that the linear system can be written as*

$$L_{Ai} = \partial_{Ai-1} - \lambda(\partial_{Ai} + [\partial_{Ai-1}, \Theta_y\partial_x - \Theta_x\partial_y]). \tag{21}$$

Equations (20) are equivalent to $[L_{Ai}, L_{Bj}] = 0$.

Sketch of Proof. The canonical line bundle of $\mathcal{PT}$ is $K = \mathcal{O}(-2n-2)$. To obtain a global, line bundle valued three-form on $\mathcal{PT}$ one must tensor it with $\mathcal{O}(2n+2)$. We pick a global section $\xi \in \Gamma(K \otimes \mathcal{O}(2n+2))$ and restrict ξ to a line $\xi|_l = \Sigma_\lambda \wedge \tau$, where $\tau \in \Omega^1 \otimes \mathcal{O}(2)$ is a volume form on $\mathbb{CP}^1$. A two-form

$$\Sigma_\lambda \in \Gamma(\Lambda^2(\mu^{-1}(\lambda)) \otimes \mathcal{O}(2n)) \tag{22}$$

is defined on vectors vertical with respect to μ by $\Sigma_\lambda(U,V)\tau = \xi(U,V,...)$.

Let $(\omega^A, \pi_{A'}) \sim (k^n\omega^A, k\pi_{A'})$ be the homogeneous coordinates on $\mathcal{PT}$. We pull them back to $\mathcal{F}$ and define two twistor functions homogeneous of degree 0 by $\mu^A = \omega^A/(\pi_{1'})^n$. Choose $2n+2$ coordinates on $\mathcal{N}$ by

$$x^{Ai} := \frac{1}{(n-i)!} \frac{\partial^{n-i} \mu^A}{\partial \lambda^{n-i}}\bigg|_{\lambda=0}$$

and expand (here s^{Ai} are some functions on $\mathcal{N}$)

$$\mu^A = \sum_{i=0}^{n} x^{Ai} \lambda^{n-i} + \lambda^{n+1} \sum_{i=0}^{n-1} s_i^A \lambda^i + \ldots \; . \tag{23}$$

The pull-back of the symplectic form Σ_λ to the correspondence space is (for fixed λ) given by $\mu^*(\Sigma_\lambda) = (\pi_{1'})^{2n} d\mu^0 \wedge d\mu^1$. It satisfies

$$\mu^*(\Sigma_\lambda) \wedge \mu^*(\Sigma_\lambda) = 0, \quad d\mu^*(\Sigma(\lambda)) = 0 \tag{24}$$

where λ is not differentiated. The symplectic form $\mu^*(\Sigma_\lambda)$ has an expansion in powers of λ that truncates at order $2n+1$ by globality and homogeneity so that we have $s^{Ai} = \varepsilon^{AB} \frac{\partial \Theta}{\partial x^{Bi}}$. The closure condition in (24) is satisfied identically, whereas the algebraic relations give rise to hierarchy of flows (20), which proves (a).

A normal bundle to the line l_p consists of vectors tangent to the corresponding point $p \in \mathcal{N}$ modulo the twistor distribution D. Therefore we have a sequence of sheaves over $\mathbb{CP}^1$

$$0 \longrightarrow D \longrightarrow \mathbb{C}^{2n+2} \longrightarrow \mathcal{O}(n) \oplus \mathcal{O}(n) \longrightarrow 0.$$

The corresponding long exact sequence of cohomology groups yields (after some manipulations)

$$0 \longrightarrow \Gamma(D^*(-1)) \overset{\delta}{\longrightarrow} H^1(\mathcal{O}(-n-1) \oplus \mathcal{O}(-n-1)) \longrightarrow 0.$$

¿From here we conclude that, since D has rank $2n$, the connecting map δ is an isomorphism $\delta : \Gamma(D^*(-1)) \longrightarrow \mathbb{C}^{2n}$. Therefore $\delta \in \Gamma(D \otimes \mathcal{O}(1) \otimes \mathbb{C}^{2n})$. We claim that δ is a Lax system for the hierarchy (20). To derive coordinate form (21) of the twistor distribution we demand that $L_{Ai} \in \Gamma(D \otimes \mathcal{O}(1) \otimes \mathbb{C}^{2n})$ annihilate the twistor functions (23). Note that L_{A1} is just the Lax pair (12) for the second heavenly equation.

By examining relevant sheaf cohomology groups and using Kodaira deformation theory [7], we can show that flat $\mathcal{PT}$ (i.e. $\Theta = 0$) admtis complex deformations preserving conditions (1) − (3) of Proposition 3.

REFERENCES

1. Bateman, H., *Proc. Lon. Math. Soc.* **1**, 451 (1904).
2. Boyer, C.P., and Plebański, J.F., *J. Math. Phys.* **18**, 1022 (1977).
3. Boyer, C.P., and Plebański, J.F., *J. Math. Phys.* **26**, 229 (1985).
4. Dunajski, M., and Mason, L.J., *Twistor Newsletter* **41**, 26 (1996).
5. Dunajski, M., and Mason, L.J., *Twistor Newsletter* **42**, 16 (1997).
6. Dunajski, M., and Mason, L.J., *Twistor Newsletter* **43**, 24 (1997).
7. Kodaira, K. *Am. J. Math.* **85**, 79 (1963).
8. Mason, L.J., and Sparling, G.A.J., *J. Geom. Phys.* **8**, 243 (1992).
9. Mason, L.J., and Woodhouse, N.M.J., *Integrability, Self-Duality, and Twistor Theory*, L.M.S. Monographs New Series, **15**, OUP (1996).
10. Penrose, R., *Gen. Rel. Grav.* **7**, 31 (1976).
11. Plebański, J. F., *J. Math. Phys.* **16**, 2395 (1975).
12. Sparling, G. A. J., and Tod, K.P., *J. Math. Phys.* **22**, 331 (1981).
13. Strachan, I.A.B., *J. Math. Phys.* **36**, 3566 (1995).
14. Takasaki, K., *J. Math. Phys.* **30**, 1515 (1989).
15. Ward, R.S., *Phys. Lett.* **61A**, 81 (1977).
16. Ward, R.S., and Wells, R.O., *Twistor Geometry and Field Theory*, CUP, (1989).

Nonsingular Cosmological Black Hole

Irina Dymnikova and Bozena Soltysek

Institute of Mathematics and Physics, Pedagogical University of Olsztyn, Zolnierska 14, 10-561 Olsztyn, Poland; e-mail: irina@tufi.wsp.olsztyn.pl

Abstract. We present the analytic spherically symmetric solution of the Einstein equations, with de Sitter asymptotics for both $r \to 0$ and $r \to \infty$, which describes globally regular spherically symmetric space-time with two cosmological constants. At the range of mass parameter $M_{cr1} < M < M_{cr2}$, our solution has three horizons and describes a neutral nonsingular cosmological black hole - a black hole, whose singularity is replaced by a cosmological constant Λ of some fundamental (Planck or may be GUT) scale, at the background of small remnant λ. Global structure of spacetime contains an infinite sequence of black and white holes, de Sitter-like (with a fundamental Λ) future and past regular cores in place of former singularities, and asymptotically de Sitter universes corresponding to remnant λ. In the range of mass parameter $M < M_{cr1}$ we have one-horizon solution describing recovered particlelike structure at the background of small λ, and for $M > M_{cr2}$ - one-horizon solution describing space-time with variable cosmological constant - "de Sitter bag".

INTRODUCTION

The idea of replacing a black hole singularity by de Sitter-like core (by cosmological constant Λ) goes back to the 60's papers by Sakharov, who suggested that $p = -\varepsilon$ can be the equation of state at superhigh densities [19], by Gliner who classified $T_{\mu\nu} \sim \Lambda g_{\mu\nu}$ as vacuum stress tensor, and suggested that $p = -\varepsilon$ can be a final state in a gravitational collapse [12], and by Zel'dovich who interpreted a cosmological constant Λ as coming from gravitational interaction of virtual particles [22].

De Sitter-Schwarzschild metric constructed by direct matching de Sitter solution inside a joint layer (represented by thin massive shell) to the Schwarzschild solution outside [2,9,20,10], has a jump at the junction surface. In such an approach, a density profile is described by $\rho_{Pl} \times \Theta(r_{junct} - r)$. Since tangent pressure ($p_t = T^\vartheta_\vartheta = T^\varphi_\varphi$) depends on derivative of ρ as $p_t(r) = -\rho(r) - r\rho'(r)/2$, a jump is unavoidable.

The situation has been analyzed by Poisson and Israel who stated that de Sitter space-time cannot be joined directly to the exterior vacuum since the O'Brien-Synge junction condition $T^{\mu\nu} N_\nu = 0$, expressing continuity of a pressure at the boundary, is violated [18]. They proposed to introduce a transitional layer of "a noninflationary material", of uncertain depth, where geometry remains effectively classical and governed by the Einstein equations

$$G_{\mu\nu} = 8\pi G T_{\mu\nu}, \qquad (1)$$

with the source term representing vacuum polarization effects at the scale $r \sim r_g^{1/3}$ where the curvature r_g/r^3 grows to order of unity [18].

In Ref. [4] the nonsingular modification of the Schwarzschild solution was found by solving the Einstein equations with a source term representing, in semiclassical limit, vacuum polarization by gravitational field (all fields couple to gravity by (1) including gravity itself), i.e. with an expectation value of stress-energy tensor completely determined by geometry.

De Sitter-Schwarzschild metric [4] has the form

$$ds^2 = \left(1 - \frac{R_g(r)}{r}\right)dt^2 - \frac{dr^2}{1 - \frac{R_g(r)}{r}} - r^2(d\vartheta^2 + \sin^2\vartheta d\varphi^2), \qquad (2)$$

where

$$R_g(r) = r_g\left(1 - \exp\left(-\frac{\Lambda r^3}{3 r_g}\right)\right). \qquad (3)$$

For $M > M_{cr1} \simeq 0.3 M_{Pl}\sqrt{\Lambda_{Pl}/\Lambda}$ it describes a nonsingular neutral black hole. In the course of Hawking quantum evaporation of a nonsingular black hole a second-order phase transition occurs which can be evidence for a space-time symmetry restoration to the de Sitter group. For $M < M_{cr1}$ the metric (2) describes a recovered selfgravitating particlelike structure at the background of the Minkowski space [6]. The qualitative properties of de Sitter-Schwarzschild configurations do not depend essentially on the particular form for energy density profile $T_t^t(r)$ if it is a smooth function decreasing monotonically and quickly enough to guarantee the finiteness of a mass [5].

The solution (2) belongs to the class of solutions for which the condition $g_{00} = -g_{11}^{-1}$ is satisfied. In this case algebraic structure of stress-energy tensor

$$T_t^t = T_r^r, \qquad T_\theta^\theta = T_\phi^\phi \qquad (4)$$

corresponds to anisotropic spherically symmetric vacuum [4]. Such a tensor has an infinite set of comoving reference frames, since comoving frame is defined up to rotations into the plane of eigenvectors corresponding to degenerate eigenvalue. In the case $T_t^t = T_r^r$ it means invariance under boosts in the radial direction. The anisotropic vacuum (2) satisfies the general definition of a vacuum by absence of a preferred comoving reference frame for it.

In the Ref. [8] this solution has been extended to the case of nonzero background cosmological constant. Recent astronomical data compellingly testify in favour of nonzero cosmological constant today, comparable with average observed density in the Universe [14,17]. To describe a nonsingular black hole at the background of nonzero astronomical cosmological constant, we have found nonsingular modification, due to gravitational vacuum polarization, of the Kottler-Trefftz solution [13,21], frequently labelled in the literature as the Schwarzschild-de Sitter metric.

This talk will be organized as follows. In Section 2 we present the nonsingular modification of the Schwarzschild-de Sitter metric. In Section 3 we consider the global structure of the space-time. In Section 4 we analyze the horizon-mass diagram, show the existence of the lower and upper limits for a mass of a nonsingular cosmological black hole, and specify the configurations existing beyond these limits.

TWO-LAMBDA SPHERICALLY SYMMETRIC SOLUTION

In the spherically symmetric static case a line element has the form:

$$ds^2 = e^\nu dt^2 - e^\mu dr^2 - r^2(d\vartheta^2 + \sin^2\vartheta d\varphi^2), \tag{5}$$

and the Einstein equations reduce to the system:

$$8\pi G T_t^t = -e^{-\mu}\left(\frac{1}{r^2} - \frac{\mu'}{r}\right) + \frac{1}{r^2}, \tag{6}$$

$$8\pi G T_r^r = -e^{-\mu}\left(\frac{\nu'}{r} + \frac{1}{r^2}\right) + \frac{1}{r^2}, \tag{7}$$

$$8\pi G T_\vartheta^\vartheta = 8\pi G T_\varphi^\varphi = -\frac{e^{-\mu}}{2}\left(\nu'' + \frac{\nu'^2}{2} + \frac{(\nu'-\mu')}{r} - \frac{\nu'\mu'}{2}\right). \tag{8}$$

The boundary condition at $r \to \infty$ is the Kottler-Trefftz [13,21] solution:

$$ds^2 = \left(1 - \frac{r_g}{r} - \frac{\lambda r^2}{3}\right)dt^2 - \frac{dr^2}{(1 - \frac{r_g}{r} - \frac{\lambda r^2}{3})} - r^2(d\vartheta^2 + \sin^2\vartheta d\varphi^2), \tag{9a}$$

$$T_{ik} = (8\pi G)^{-1}\lambda g_{ik}. \tag{9b}$$

Here $r_g = 2GMc^{-2}$, M is mass parameter, and small λ is for cosmological background.

This solution describes a black hole in asymptotically de Sitter space. It has two horizons: black hole horizon r_+ and cosmological horizon r_{++}. Its global structure is presented in Fig.1 [11].

It contains an infinite sequence of black and white holes II_H, III_H, event horizons $r = r_+$, singularities $r = 0$, asymptotically de Sitter universes I, IV, and cosmological cores II_C, III_C (regions between cosmological horizons r_{++} and spacelike infinities $r = \infty$).

At $r \to 0$ the boundary condition is de Sitter metric

$$ds^2 = \left(1 - \frac{(\Lambda + \lambda)r^2}{3}\right)dt^2 - \frac{dr^2}{\left(1 - \frac{(\Lambda+\lambda)r^2}{3}\right)} - r^2(d\vartheta^2 + \sin^2\vartheta d\varphi^2), \qquad (10a)$$

$$T_{ik} = (8\pi G)^{-1}(\Lambda + \lambda)g_{ik}, \qquad (10b)$$

with big Λ for the scale of symmetry restoration and small λ for cosmological background.

We consider the solution satisfying the condition $T_t^t = T_r^r$. In this case equations (6) and (7) becomes identical, and the metric is given by

$$g_{00}(r) = 1 - \frac{8\pi G}{r}\int_0^r T_t^t(r)r^2 dr. \qquad (11)$$

To obtain the metric we need a density profile $T_t^t(r)$, which connects in smooth way two de Sitter vacuum states with different expectation values of vacuum energy density. Generalizing the model [4,6] to the case of nonzero background cosmological constant λ, we make use of the energy density profile:

$$T_t^t(r) = T_r^r(r) = \varepsilon_\Lambda \exp\left(-\frac{\Lambda r^3}{3r_g}\right) + (8\pi G)^{-1}\lambda, \qquad (12)$$

where $\varepsilon_\Lambda = (8\pi G)^{-1}\Lambda$. Integrating the equation (11) we obtain the metric

$$ds^2 = \left(1 - \frac{R_g(r)}{r} - \frac{\lambda r^2}{3}\right)dt^2 - \frac{dr^2}{1 - \frac{R_g(r)}{r} - \frac{\lambda r^2}{3}} - r^2(d\vartheta^2 + \sin^2\vartheta d\varphi^2), \qquad (13)$$

where

$$R_g(r) = r_g\left(1 - \exp\left(-\frac{\Lambda r^3}{3r_g}\right)\right). \qquad (14)$$

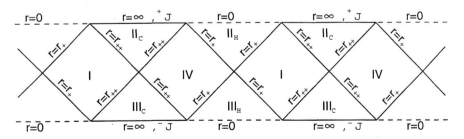

FIGURE 1. The Penrose-Carter diagram for the Schwarzschild-de Sitter space-time.

From the Equation (8), we calculate the angular components of the stress-energy tensor

$$T^\vartheta_\vartheta = T^\varphi_\varphi = \varepsilon_\Lambda \exp\left(-\frac{\Lambda r^3}{3r_g}\right)\left(1 - \frac{\Lambda r^3}{2r_g}\right) + (8\pi G)^{-1}\lambda. \tag{15}$$

The quadratic invariant of the Riemann curvature tensor $\mathcal{R}^2 = R_{iklm}R^{iklm}$ has the form

$$\mathcal{R}^2 = 4\frac{R_g^{\,2}(r)}{r^6} + 4\left(\Lambda e^{-r^3\Lambda/3r_g} - \frac{R_g^{\,2}(r)}{r^3}\right)^2 + \left(\frac{2R_g(r)}{r^3} - \frac{\Lambda^2}{r_g}r^3 e^{-r^3\Lambda/3r_g}\right)^2$$

$$+ \frac{8\lambda^2}{3} + \frac{16\lambda\Lambda}{3}e^{-\Lambda r^3/3r_g} - \frac{4\lambda\Lambda^2}{3r_g}, \tag{16}$$

and tends, for $r \to 0$, to the value $\mathcal{R}^2 = 8(\Lambda + \lambda)^2/3$, and, for $r \to \infty$, to the value $\mathcal{R}^2 = 8\lambda^2/3$. The other invariants of the Riemann curvature tensor are also finite.

GLOBAL STRUCTURE OF SPACE-TIME

Our solution is regular everywhere (see Fig.2).

In general case it has three horizons: cosmological horizon r_{++}, black hole horizon r_+, and internal (Cauchy [18,4,6]) horizon r_-. They are obtained by solving equation

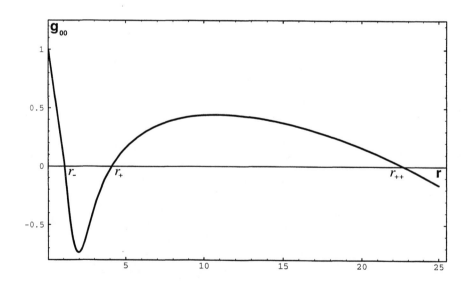

FIGURE 2. Two-lambda spherically symmetric solution of the Einstein equations.

$$1 - \frac{R_g(r)}{r} - \frac{\lambda r^2}{3} = 0. \tag{17}$$

Following Chandrasekhar approach [3], we divide the space-time into four regions: $\mathcal{A}: 0 < r < r_-$, $\mathcal{B}: r_- < r < r_+$, $\mathcal{C}: r_+ < r < r_{++}$, $\mathcal{D}: r > r_{++}$, and introduce the tortoise coordinate r_* defined by the equation $dr/g_{00}(r) = dr_*$. In regions $\mathcal{A}$ and $\mathcal{C}$, $g_{00}(r) > 0$ and $r_*(r)$ is monotonically increasing function of $r \in (0, r_-)$ and $r \in (r_+, r_{++})$, such that

$$\lim_{r \to r_-} r_* = \infty, \quad \lim_{r \to r_+} r_* = -\infty, \quad \lim_{r \to r_{++}} r_* = \infty. \tag{18}$$

In regions $\mathcal{B}$ and $\mathcal{D}$, $g_{00} < 0$ and $r_*(r)$ is monotonically decreasing function of $r \in (r_-, r_+)$ and $r \in (r_{++}, \infty)$, such that

$$\lim_{r \to r_-} r_* = \infty, \quad \lim_{r \to r_+} r_* = -\infty, \quad \lim_{r \to r_{++}} r_* = \infty. \tag{19}$$

To introduce the isotropic coordinates, we put in the regions $\mathcal{A}$ and $\mathcal{C}$

$$dt = 1/2(du + dv), \quad dr_* = 1/2(dv - du), \tag{20}$$

then:
$$u = t - r_*, \quad v = t + r_*. \tag{21}$$

In the regions $\mathcal{B}$ and $\mathcal{D}$ we put:

$$dt = 1/2(du - dv), \quad dr_* = 1/2(du + dv), \tag{22}$$

then:
$$u = t + r_*, \quad v = -t + r_*. \tag{23}$$

As a result the metric takes the Kruskal form

$$ds^2 = g_{00} du dv - r^2(d\vartheta^2 + \sin^2 \vartheta d\varphi^2) \tag{24}$$

in the regions $\mathcal{A}$ and $\mathcal{C}$, and

$$ds^2 = |g_{00}| du dv - r^2(d\vartheta^2 + \sin^2 \vartheta d\varphi^2) \tag{25}$$

in the regions $\mathcal{B}$ and $\mathcal{D}$.

In addition to the regions $\mathcal{A}$, $\mathcal{B}$, $\mathcal{C}$, $\mathcal{D}$, there exist also regions $\mathcal{A}'$, $\mathcal{B}'$, $\mathcal{C}'$ and $\mathcal{D}'$ which are copies of regions $\mathcal{A}$, $\mathcal{B}$, $\mathcal{C}$ and $\mathcal{D}$, respectively, obtained from (21) and (23) by replacing (u,v) with (-u,-v).

To specify these regions, we introduce the invariant quantity [16,1]

$$\Delta = g^{\mu\nu} r_{,\mu} r_{,\nu}.$$

Dependently on the sign of Δ, space-time is divided into R and T regions (see [16,1]): In the R regions the normal vector to the surface $r = const$, $N_\mu = r_{,\mu}$ is spacelike. Therefore in the R region an observer on the surface $r = const$ can send two radial signals: one directed inside and the other outside of this surface. In the T regions the normal vector N_μ is timelike. The surface $r = const$ is spacelike, and both signals propagate on the same side of this surface. In the T regions any observer can cross the surface $r = const$ only once and only in the same direction.

For the space-time considered here, $\Delta < 0$ in the regions $\mathcal{A}$, $\mathcal{A}'$, $\mathcal{C}$, $\mathcal{C}'$, and they are R regions. In the regions $\mathcal{B}$, $\mathcal{B}'$, $\mathcal{D}$, $\mathcal{D}'$, $\Delta > 0$, and they are T regions. The R and T regions are separated by horizons, where $\Delta = 0$.

For the metric in the Kruskal form $\Delta = (1/2)g_{00}^{-1}r_{,u}r_{,v}$ [1]. Since in the T regions $\Delta > 0$, the vector $r_{,u}$ cannot be zero, and the conditions $r_{,u} > 0$ and $r_{,u} < 0$ are invariant. When $r_{,u} < 0$, we have T_- region or the region of contraction (a black hole). When $r_{,u} > 0$, we have an expanding T_+ region (a white hole). Near horizons the sign of T region tells us that the vector $r_{,u}$ enters or goes out of the corresponding R region.

To obtain the Penrose-Carter diagram we introduce coordinates U and V by

$$\begin{cases} tgU = -e^{-u}, \\ tgV = +e^{v} \end{cases} \tag{26}$$

in the regions $\mathcal{A}$ and $\mathcal{C}$, and by

$$\begin{cases} tgU = e^{u}, \\ tgV = e^{v} \end{cases} \tag{27}$$

and in the regions $\mathcal{B}$ and $\mathcal{D}$.

The metric takes the form

$$ds^2 = -4|g_{00}(r)|cosec2U\,cosec2V\,dUdV - r^2(d\vartheta^2 + \sin^2\vartheta d\varphi^2), \tag{28}$$

and the global structure of the space-time is shown in Fig.3.

It contains an infinite set of the following structures: $\mathcal{B}$, $\mathcal{B}'$ regions which are T_- regions of contraction - black holes $\mathcal{BH}$, and T_+ regions of expansion - white holes $\mathcal{WH}$. Their former singularities are replaced with the regions $\mathcal{A}$, $\mathcal{A}'$ which are R_+, R_- regions: the future and past regular cores $\mathcal{RC}_1$, $\mathcal{RC}_2$, asymptotically de Sitter with fundamental Λ in the limit of $r \to 0$. Black and white holes are separated from their regular cores by horizons $r = r_-$. Beyond the horizons $r = r_+$ are located $\mathcal{C}$, $\mathcal{C}'$ regions which are R_+, R_- regions: asymptotically de Sitter (with small background λ) universes $\mathcal{U}_1$, $\mathcal{U}_2$. They in turn are separated by horizons $r = r_{++}$ from $\mathcal{D}'$, $\mathcal{D}$ regions which are T_+, T_- regions: cosmological cores $\mathcal{CC}_1$, $\mathcal{CC}_2$ corresponding to small background λ.

LOWER AND UPPER LIMITS FOR A BLACK HOLE MASS

Two-lambda spherically symmetric spacetime has three scale of length $r_g = 2GM$, $r_\Lambda = \sqrt{3/\Lambda}$, and $r_\lambda = \sqrt{3/\lambda}$. Normalizing to r_Λ, we obtain two dimensionless parameters: M (mass normalized to $c^2\sqrt{3}/G\sqrt{\Lambda}$) and $q = \sqrt{\Lambda/\lambda}$.

Horizon-mass diagram obtained by numerical solving the equation (17) is presented in the Fig.4.

We see that three horizons exist only at the range of mass parameter $M_{cr1} < M < M_{cr2}$. Within this range of masses, the metric (13) describes nonsingular cosmological black hole - a black hole, whose singularity is replaced by the regular de Sitter-like core with big Λ at the background of cosmological λ.

The value of critical mass M_{cr1} corresponding to the first extreme $(r_- = r_+)$ black hole state puts the lower limit for a black hole mass, given by

$$M_{cr1} \simeq 0.3 M_{Pl}\sqrt{\rho_{Pl}/\rho_\Lambda}, \qquad (29)$$

and does not depend practically on the parameter $q = \sqrt{\Lambda/\lambda}$. We see this extreme black hole at the Fig.5, where its mass is normalized to the quantity $c^2\sqrt{3}/G\sqrt{\Lambda}$.

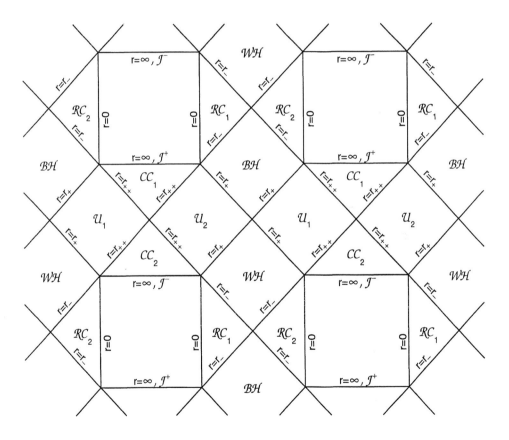

FIGURE 3. The Penrose-Carter diagram for two-lambda spherically symmetric solution

In these units $M_{cr1} \simeq 3.5$.

In the range of mass parameter $M < M_{cr1}$ two-lambda solution describes one-horizon selfgravitating particlelike structure at the background of cosmological λ. We can estimate its characteristic size in the following way.

The strong energy condition of singularities theorems,

$$\left(T_{\mu\nu} - \frac{1}{2}g_{\mu\nu}T\right)u^{\mu}u^{\nu} \geq 0,$$

where u^{ν} is any timelike vector, is violated inside of any configurations with a de Sitter-like core replacing a singularity by a big cosmological constant Λ. There exists, therefore, the characteristic surface $r = r_v$ beyond which the gravitational acceleration changes sign. For the case $q \gg 1$ it is located at [6]

$$r = r_v = (4GM/c^2\Lambda)^{1/3} = \frac{l_{Pl}}{(2\pi)^{1/3}}\left(\frac{M}{M_{Pl}}\right)^{1/3}\left(\frac{\rho_{Pl}}{\rho_{\Lambda}}\right)^{1/3}. \tag{30}$$

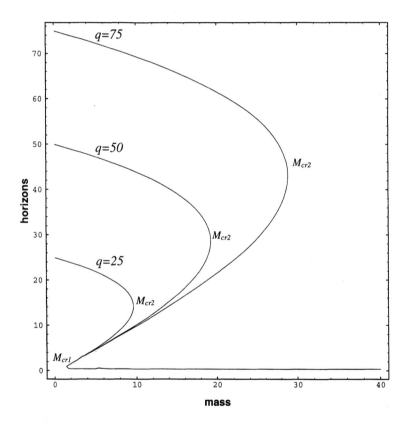

FIGURE 4. Horizon-mass diagrams for two-lambda spherically symmetric solution.

Outside it there exists one more characteristic surface, $r = r_c$, at which the scalar curvature $R = (8\pi G/c^4)T$ changes sign. For $q \gg 1$ it is located at

$$r = r_c = (8GM/c^2\Lambda)^{1/3} = \frac{l_{Pl}}{\pi^{1/3}}\left(\frac{M}{M_{Pl}}\right)^{1/3}\left(\frac{\rho_{Pl}}{\rho_\Lambda}\right)^{1/3} \qquad (31)$$

which can be a suitable estimate for a characteristic size of a particlelike structure. Setting it equal to the Compton wavelength of a particle with the mass $M = \sqrt{2}gM_0$, where M_0 is the vacuum expectation value of a relevant non-Abelian field with self-interaction g^2, we can roughly estimate a coupling constant g at the scale of symmetry restoration. Taking into account that $(M_0/M_{Pl})^4 = (\rho_\Lambda/\rho_{Pl})$, we obtain $g = (\pi/4)^{1/4}$. Corresponding fine structure constant $\alpha = g^2/4\pi$ is $\alpha = 1/(8\sqrt{\pi}) \simeq 1/15$ [7].

The second extreme black hole state exists at the critical value of the mass parameter $M = M_{cr2}$ which puts the upper limit for a black hole mass. As we see from the Fig.4, the magnitude of M_{cr2} essentially depends on the parameter q. The second extreme black hole state we see at the Fig.6.

In the units normalized to $c^2\sqrt{3}/G\sqrt{\Lambda}$, the mass corresponding to the second extreme black hole state in the case of $q = 10$ is $M_{cr1} \simeq 7.7$.

For $M > M_{cr2}$ we have another type of one-horizon solution describing spacetime with variable cosmological constant - "de Sitter bag". Unlike the case $M < M_{cr1}$, which is configuration with cosmological horizon, here we have internal (Cauchy) horizon.

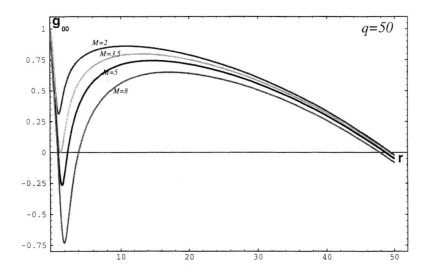

FIGURE 5. Two-lambda metric $g_{00}(r)$ for the case $q = 50$.

The upper limit M_{cr2} corresponds to nonsingular modification of the degenerate member of the Schwarzschild-de Sitter family known as the Nariai solution [15].

The existence of the lower limit for a black hole mass [6,8], does not have analogy in the singular family of black hole solutions. This is the direct consequence of replacing a black hole singularity by the de Sitter-like core - by the cosmological constant Λ of a fundamental scale.

ACKNOWLEDGEMENT

This work was supported by the Polish Committee for Scientific Research through the Grant Nr 2.PO3D.017.11.

REFERENCES

1. Berezin V. A., Kuzmin V. A., Tkachev I. I., ιSov. Phys. JETP **93**, 1159 (1987).
2. Bernstein M. R., *Bull. Amer. Phys. Soc.* **16**, 1016 (1984).
3. Chandrasekhar S., *The Mathematical Theory of Black Holes*, Clarendon Press Oxford, 1983.
4. Dymnikova I. G., *Gen. Rel. Grav.* **24**, 235 (1992).
5. Dymnikova I. G., in *14th International Conference on General Relativity and Gravitation*, Ed. M. Francaviglia, L. Lusanna, 1995, p. A24.

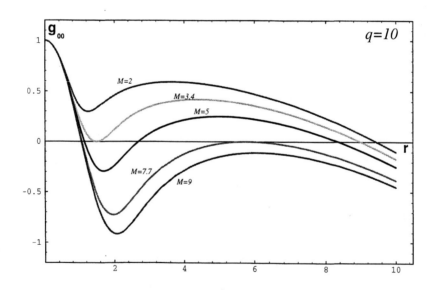

FIGURE 6. Two-lambda metric $g_{00}(r)$ for the case $q = 10$.

6. Dymnikova I. G., *Int. Journ. Mod. Phys.* **D5**, 529 (1996).
7. Dymnikova I., in *Internal Structure of Black Holes and Spacetime Singularities*, Ed. L.M.Burko and A.Ori, Institute of Physics Publishing, Bristol and Philadelphia, and The Israel Physical Society, Jerusalem, 1997, p.422.
8. Dymnikova I., Soltysek B., in *The Proceedings of the VIII Marcel Grossmann Meeting on General Relativity*, Ed. Tsvi Piran, World Scientific, 1997.
9. Fahri E., and Guth A., *Phys. Lett.* **B183**, 149 (1997).
10. Frolov V. P., Markov M. A., Mukhanov V. F., *Phys. Rev.* **D41**, 3831 (1990).
11. Gibbons G. W. and Hawking S. W., *Phys. Rev.* **D15**, 2738 (1977).
12. Gliner E. B., *Sov. Phys. JETP* **22**, 378 (1966).
13. Kottler F., *Encykl. Math. Wiss.* **22a**, 231 (1922).
14. Krauss L. and Turner M., *astro-ph/9504003* (1995).
15. Nariai H., *Sci. Rep. Tohoku Univ. Ser. I* **35**, 62 (1951).
16. Novikov I.D., Frolov V. P., *Physics of Black Holes*, Kluwer Acad. Publ., 1989.
17. Ostriker J. P., Steinhardt P. J., *Nature* **377**, 600 (1995).
18. Poisson E., and Israel W., *Class. Quant. Grav.* **5**, L201 (1988).
19. Sakharov A. D., *Sov. Phys. JETP* **22**, 241 (1966).
20. Shen W., and Zhu S., *Phys. Lett.* **A126**, 229 (1988).
21. Trefftz E., *Math. Ann.* **86**, 317 (1922).
22. Zel'dovich Ya. B., *Sov. Phys. Letters* **6**, 883 (1967).

Relativistic Particle in Singular Gravitational Field[1]

George Jorjadze[†] and Włodzimierz Piechocki[‡]

[†]*Razmadze Mathematical Institute, Tbilisi, Georgia*
[‡]*Sołtan Institute for Nuclear Studies, Warsaw, Poland*

Abstract. Classical and quantum dynamical ambiguities for the system of relativistic particle moving in 2-dimensional singular gravitational field is investigated.

INTRODUCTION

Classical general relativity permits the occurrence of gravitational singularities. This can be treated as a warning that the theory is incomplete. The classical dynamics of a particle in a singular field usually contains some ambiguities. We find it intriguing to examine what is the impact of gravitational singularities on the quantization procedure and investigate this problem for 2-dimensional theory.

DYNAMICS OF A PARTICLE IN SINGULAR SPACETIME

Let us consider a relativistic particle of mass m_0 moving in a gravitational field $g_{\mu\nu}(x^0, x^1)$; $\mu, \nu = 0, 1$. Action describing such a system is proportional to the length of a particle world-line

$$S = -m_0 \int d\tau \sqrt{g_{\mu\nu}\left(x^0(\tau), x^1(\tau)\right) \dot{x}^\mu(\tau) \dot{x}^\nu(\tau)}, \tag{1}$$

where τ is an evolution parameter along the trajectory $x^\mu(\tau)$ and $\dot{x}^\mu(\tau) := dx^\mu(\tau)/d\tau$.

In the case of 2d Minkowskian manifold one can always choose local coordinates in such a way that

---

[1)] Dedicated to the memory of Professor Ryszard Rączka.

$$g_{\mu\nu}(x^0, x^1) = \exp \varphi(x^0, x^1) \begin{pmatrix} 1 & 0 \\ 0 & -1 \end{pmatrix}, \qquad (2)$$

where φ is a field.

Since the scalar curvature $R(x^0, x^1)$ for (2) is

$$R(x^0, x^1) = \exp\left(-\varphi(x^0, x^1)\right) (\partial_1^2 - \partial_0^2)\varphi(x^0, x^1), \qquad (3)$$

the Einstein-Hilbert Lagrangian reads

$$\mathcal{L} = -\sqrt{|g|} \, R = (\partial_0^2 - \partial_1^2)\varphi(x^0, x^1), \qquad (4)$$

and it does not lead to dynamical equation for the field φ. We can specify the 2d general relativity model by requiring that φ satisfies a certain dynamical equation.

Suppose φ is a solution to the Liouville equation [1]

$$(\partial_0^2 - \partial_1^2)\varphi(x^0, x^1) + R_0 \exp \varphi(x^0, x^1) = 0, \qquad (5)$$

where $R_0 < 0$ is a constant.

Eq.(5) is usually used as a model of 2d gravity [2,3]. This equation is conformally invariant and according to (3) describes a spacetime manifold with a constant curvature ($R(x^0, x^1) = R_0$).

Let us consider the case when the spacetime manifold is R^2 and the solution to (5) is singular

$$\varphi(x^0, x^1) = \ln \frac{1}{(mx^0)^2}, \qquad (6)$$

where $m := \sqrt{-R_0/2}$.

The solution (6) gives singular spacetime metric at $x^0 = 0$

$$g_{\mu\nu}(x^0, x^1) = \frac{1}{(mx^0)^2} \begin{pmatrix} 1 & 0 \\ 0 & -1 \end{pmatrix}, \qquad (7)$$

but the scalar curvature is finite and constant

$$R(x^0, x^1) = R_0 < 0, \quad \text{for} \quad (x^0, x^1) \in R^2 \quad \text{and} \quad x^0 \neq 0. \qquad (8)$$

Proper time of a particle

$$\int_{x_0'}^{x_0''} dt \sqrt{g_{00}(t, x^1(t))} = \int_{x_0'}^{x_0''} \frac{dt}{m|t|} \qquad (9)$$

diverges at $x^0 = 0$. Particle needs infinite proper time to reach the singularity. Therefore, dynamics of a particle can be considered for $x^0 < 0$ and $x^0 > 0$ separately. In such interpretation we deal with two independent systems each without singularity. Let us call them $\mathcal{S}_-$ for $x^0 < 0$ and $\mathcal{S}_+$ for $x^0 > 0$.

Lagrangian of (1) for the singular solution (6) reads

$$L = -2c\sqrt{\frac{\dot{x}^+\dot{x}^-}{(x^+ + x^-)^2}}, \qquad (10)$$

where $c := m_0/m$, $x^\pm := x^0 \pm x^1$.

This Lagrangian formally is invariant under the fractional-linear transformations

$$x^+ \longrightarrow \frac{ax^+ + b}{cx^+ + d}, \quad x^- \longrightarrow \frac{ax^- - b}{-cx^- + d}, \quad ad - bc = 1 \qquad (11)$$

The infinitesimal transformations for (11) are

$$x^\pm \longrightarrow x^\pm \pm \alpha_0, \quad x^\pm \longrightarrow x^\pm + \alpha_1 x^\pm, \quad x^\pm \longrightarrow x^\pm \pm \alpha_2 (x^\pm)^2, \qquad (12)$$

and the dynamical integrals corresponding to (12) read

$$P := p_+ - p_-, \quad K := p_+ x^+ + p_- x^-, \quad M := -p_+(x^+)^2 - p_-(x^-)^2, \qquad (13)$$

where $p_\pm := \partial L/\partial \dot{x}^\pm$.

The integrals P, K and M are not independent, since (1) is invariant under reparametrization $\tau \to f(\tau)$. This gauge invariance leads to constraint dynamics [4]. The constraint is

$$\Phi := m^2(x^+ - x^-)^2 p_+ p_- - m_0^2 = 0 \qquad (14)$$

and it leads to the relation

$$K^2 - PM = c^2. \qquad (15)$$

Eqs.(13) and (15) define the trajectories of a particle

$$x^1(x^0) = \begin{cases} -(M\epsilon)/(2c), & \text{for } P = 0 \\ (K + \epsilon\sqrt{(x^0 P)^2 + c^2})/P, & \text{for } P \neq 0 \end{cases} \qquad (16)$$

where $\epsilon := x^0/|x^0|$.

It is clear that points (P, K, M) of the hyperboloid (15) specify particle trajectories. However, from (16) we see that the trajectories with

$$P = 0, \quad K = -c \quad \text{for} \quad S_- \qquad (17)$$

and

$$P = 0, \quad K = +c \quad \text{for} \quad S_+ \qquad (18)$$

do not exist.

Since P, K and M are constant along the trajectories (16), and since they are gauge invariant on the constraint surface

$$\Gamma_c := \{(x^+, x^-, p_+, p_-) \in \Gamma \mid \Phi = 0\}, \tag{19}$$

(where $\Gamma := \{(x^+, x^-, p_+, p_-)\} \subset R^4$ is an extended phase space) it is natural to choose them as the observables of our classical system. Hovever, according to (15), only two of them are funcionally independent on Γ_c.

Defining the Poisson bracket on Γ, we get

$$\{P, K\} = P, \quad \{K, M\} = M, \quad \{P, M\} = 2K. \tag{20}$$

Eq.(20) defines a Lie algebra, which is isomorphic to the $sl(2, R)$ algebra.

QUANTIZATION

Let us consider canonical quantization of reduced Hamiltonian systems. It turns out that the canonical variables for our systems are (see (20))

$$P \quad \text{and} \quad Q_\epsilon := (K + \epsilon c)/P. \tag{21}$$

Making use of these variables we get

$$K = PQ_\epsilon - \epsilon c, \quad M = PQ_\epsilon^2 + 2\epsilon c Q_\epsilon. \tag{22}$$

One can check (see [5,6]) that in the case observables are of the form

$$f(P, Q) = PA(Q) + B(Q), \tag{23}$$

where $A(\cdot)$ and $B(\cdot)$ are functions of Q only, the prescription

$$f \longrightarrow \hat{f} := \frac{1}{2}[A(Q)\hat{P} + \hat{P}A(Q)] + B(Q), \tag{24}$$

where $\hat{P} := -i\partial/\partial Q$, defines the homomorphism

$$[\hat{f}, \hat{h}] = -i\widehat{\{f, h\}} \tag{25}$$

of classical and quantum algebras.

Applying (24) to (22) we get

$$\hat{P} = -i\frac{\partial}{\partial Q_\epsilon}, \quad \hat{K} = -iQ_\epsilon \frac{\partial}{\partial Q_\epsilon} - \epsilon c - \frac{i}{2}, \quad \hat{M} = -iQ_\epsilon^2 \frac{\partial}{\partial Q_\epsilon} - (i + 2\epsilon c)Q_\epsilon. \tag{26}$$

The operators $\hat{P}, \hat{K}$ and $\hat{M}$ are Hermitian (symmetric) on the Hilbert space $L^2(R)$ and give representation of $sl(2, R)$ algebra. However, quantum observables should be represented by self-adjoint operators. It turns out [6,7] that $\hat{M}$ has non-unique self-adjoint extensions. In fact there are infinitely many unitarily inequivalent quantum systems corresponding to our single classical system $\mathcal{S}_+$ or $\mathcal{S}_-$. This ambiguity is parametrized by $\alpha \in S^1$. The non-uniqueness we are dealing with is connected with the fact that the points of the hyperboloid (15) defined by (17) and (18) are not available for the dynamics. For this reason $SL(2,R)/Z_2$ cannot be the group of symmetry of $\mathcal{S}_+$ or $\mathcal{S}_-$.

Recently, we have shown [7] that there exists a unique quantum theory, but it corresponds to the new classical system $\mathcal{S}$ which 'consists of' the system $\mathcal{S}_-$, the system $\mathcal{S}_+$, the singularity and having $SL(2,R)/Z_2$ as the group of symmetry. The system $\mathcal{S}$ is defined by specification of the particle trajectories. It turns out that continuity of trajectories and conservation of P, K and M integrals at $x_0 = 0$ are incompatible. The trajectories with constant P, K and M are defined in the following way:

1. For $P \neq 0$ and arbitrary K trajectories are defined by (16) and have discontinuity $2c/P$ at $x^0 = 0$, i.e., particle is 'annihilated' at $(x^0, x^1) = (0, (K-c)/P)$ and then it is 'created' at $(x^0, x^1) = (0, (K+c)/P)$.

2. For $P = 0, K = c$ and any M trajectories are defined by $x^1 = M/2c$ for $x^0 < 0$, Eq.(16), and there are no trajectories for $x^0 > 0$, i.e., particle is annihilated at $(x^0, x^1) = (0, M/2c)$ and it cannot appear for $x^0 > 0$.

3. For $P = 0, K = -c$ and any M there are no trajectories for $x^0 < 0$; for $x^0 > 0$ trajectories are defined by $x^1 = -M/2c$, Eq.(16), i.e., particle is created by singularity at $(x^0, x^1) = (0, -M/2c)$.

To quantize the system we use the following parametrization of the hyperboloid (15):

$$P = J(1 - \cos\beta) - c\sin\beta, \quad M = J(1 + \cos\beta) + c\sin\beta, \quad K = -J\sin\beta + c\cos\beta, \tag{27}$$

where $J \in R$, $\beta \in S^1$.

According to (20) the new coordinates J and β are canonically conjugated variables. Applying the quantization rules (23–24) with $\hat{J} = -i\partial/\partial\beta$ we get the following operators corresponding to (27):

$$\hat{P} = -i(1 - \cos\beta)\partial/\partial\beta - (c + i/2)\sin\beta, \tag{28}$$

$$\hat{M} = -i(1 + \cos\beta)\partial/\partial\beta + (c + i/2)\sin\beta, \tag{29}$$

$$\hat{K} = i\sin\beta\, \partial/\partial\beta + (c + i/2)\cos\beta. \tag{30}$$

The operators $\hat{P}, \hat{M}$ and $\hat{K}$ are self-adjoint on $L^2(S^1)$ and define the unitary irreducible representation of the group $SL(2,R)/Z_2$.

CONCLUSION

Making the assumption that spacetime has the topology of R^2 and the group $SL(2,R)/Z_2$ is the symmetry group of both classical and quantum systems, we have found a unique quantum system corresponding to the classical system $\mathcal{S}$. The system $\mathcal{S}$ has gravitational singularity with the properties:

(i) singularity annihilates and/or creates a particle,

(ii) there is the violation of causality due to the discontinuity of trajectories at the singularity.

Taking into account the gravitational singularity made possible unique quantization of the classical system.

Acknowledgments

One of us (WP) would like to thank the organizers of the Conference 'Particles, Fields and Gravitation' for a stimulating atmosphere and for financial support. WP is also grateful to J.Plebański for inspiration.

REFERENCES

1. Liouville J., *J. Math. Pures Appl.* **18**, 71 (1853).
2. Jackiv R., *Quantum Theory of Gravity*, ed. by Christensen S.M., Bristol: Adam Hilger Ltd., 1984, pp. 404–420.
3. Seiberg N., *Rundom Surfaces and Quantum Gravity*, ed. by Alvarez et al., New York: Plenum Press, 1991, pp. 363–395.
4. Faddeev L. D., *Theor. Math. Phys.* **1**, 3 (1969).
5. Jorjadze G., *Mem. Differential Equations Math. Phys.* **13**, 1 (1998).
6. Jorjadze G. and Piechocki W., *Preprint hep-th/9709059*.
7. Jorjadze G. and Piechocki W., *Class. Quantum Grav.* **15**, L41 (1998).

Black Holes with Yang-Mills Hair

B. Kleihaus, J. Kunz, A. Sood and M. Wirschins[1]

Fachbereich Physik, Universität Oldenburg, Postfach 2503
D-26111 Oldenburg, Germany

Abstract. In Einstein-Maxwell theory black holes are uniquely determined by their mass, their charge and their angular momentum. This is no longer true in Einstein-Yang-Mills theory. We discuss sequences of neutral and charged $SU(N)$ Einstein-Yang-Mills black holes, which are static spherically symmetric and asymptotically flat, and which carry Yang-Mills hair. Furthermore, in Einstein-Maxwell theory static black holes are spherically symmetric. We demonstrate that, in contrast, $SU(2)$ Einstein-Yang-Mills theory possesses a sequence of black holes, which are static and only axially symmetric.

INTRODUCTION

The "no hair" conjecture for black holes states, that black holes are uniquely characterized by their mass, their charge and their angular momentum. This conjecture presents a generalization of rigorous results obtained for scalar fields coupled to gravity as well as for Einstein-Maxwell (EM) theory. Notably, the static black hole solutions in EM theory are spherically symmetric, and the stationary black holes are axially symmetric.

In recent years, counterexamples to the "no hair" conjecture were established in various theories with non-abelian fields, including Einstein-Yang-Mills (EYM) theory, Einstein-Yang-Mills-dilaton (EYMD) theory, Einstein-Yang-Mills-Higgs (EYMH) theory, and Einstein-Skyrme (ES) theory. These non-abelian black holes are asymptotically flat and possess a regular event horizon with non-trivial matter fields outside the event horizon.

In $SU(2)$ EYM theory, there exists a sequence of neutral static spherically symmetric non-abelian black hole solutions, labelled by the node number n of the gauge field function [1]. Existing for arbitrary horizon radius x_H, these black hole solutions with Yang-Mills hair approach globally regular solutions [2] in the limit $x_H \to 0$. In contrast, all static spherically symmetric $SU(2)$ EYM black hole solutions with non-zero charge are embedded Reissner-Nordstrøm (RN) solutions [3]. However, this "non-abelian baldness theorem" no longer holds for $SU(3)$ EYM theory [4].

---

[1] supported by DFG

Non-abelian static spherically symmetric solutions of $SU(N)$ EYM theory are obtained by embedding the N-dimensional representation of $su(2)$ in $su(N)$. The gauge field ansatz then involves $N-1$ gauge field functions, and the solutions are labelled by the corresponding node numbers $(n_1, \ldots, n_{N-1})$ [5-8]. When all gauge field functions are non-trivial, neutral globally regular and black hole solutions are obtained. But when one or more of these functions are identically zero, magnetically charged black hole solutions emerge, whose charge resides in the $su(N)$ Cartan subalgebra (CSA). All these solutions possess EYMD counterparts [9,6-8].

We here first consider static spherically symmetric $SU(N)$ EYM and EYMD solutions, presenting examples of neutral globally regular $SU(4)$ solutions and charged $SU(5)$ black hole solutions, and consider also the interior of the black hole solutions.

Then we discuss globally regular and black hole solutions of $SU(2)$ EYM and EYMD theory, which are asymptotically flat, static and only axially symmetric but not spherically symmetric [10]. These solutions are characterized by two integers, the winding number $k > 1$ and the node number n of the gauge field functions. The black hole solutions possess a regular event horizon and a constant surface gravity. We argue, that non-abelian theories even possess static black hole solutions with only discrete symmetries.

$SU(N)$ EYMD ACTION

We here consider the action of $SU(N)$ EYMD theory

$$S = \int \left(\frac{R}{16\pi G} + L_M \right) \sqrt{-g} d^4 x \tag{1}$$

with matter Lagrangian

$$L_M = -\frac{1}{2} \partial_\mu \Phi \partial^\mu \Phi - e^{2\kappa\Phi} \frac{1}{2} \text{Tr}(F_{\mu\nu} F^{\mu\nu}) , \tag{2}$$

field strength tensor $F_{\mu\nu} = \partial_\mu A_\nu - \partial_\nu A_\mu + ie[A_\mu, A_\nu]$, gauge field $A_\mu = \frac{1}{2}\lambda^a A_\mu^a$, dilaton field Φ, and e and κ are the Yang-Mills and dilaton coupling constants, respectively.

STATIC SPHERICAL BLACK HOLES

As shown by Bartnik and McKinnon [2], $SU(2)$ EYM theory possesses a sequence of neutral globally regular solutions. These have black hole counterparts with a regular horizon [1]. $SU(N)$ EYM theory possesses a with N increasing number of sequences of neutral globally regular and black hole solutions [6,7] as well as of charged black hole solutions [4,8], which all have EYMD counterparts.

Ansätze

To construct static spherically symmetric $SU(N)$ EYM and EYMD solutions we employ Schwarzschild-like coordinates and adopt the spherically symmetric metric

$$ds^2 = g_{\mu\nu}dx^\mu dx^\nu = -\mathcal{A}^2 \mathcal{N} dt^2 + \mathcal{N}^{-1} dr^2 + r^2(d\theta^2 + \sin^2\theta d\phi^2) , \qquad (3)$$

with the metric functions $\mathcal{A}(r)$ and $\mathcal{N}(r) = 1 - (2m(r)/r)$.

The static spherically symmetric ansätze for the gauge field of $SU(N)$ EYM theory are based on the $su(2)$ subalgebras of $su(N)$. Considering the N-dimensional representation of $su(2)$, the ansatz is [5]

$$A_\mu^{(N)} dx^\mu = \frac{1}{2e} \begin{pmatrix} (N-1)\cos\theta d\phi & \omega_1 \Theta & 0 & \cdots & 0 \\ \omega_1 \bar{\Theta} & (N-3)\cos\theta d\phi & \omega_2 \Theta & \cdots & 0 \\ \vdots & & \ddots & & \vdots \\ 0 & \cdots & 0 & \omega_{N-1}\bar{\Theta} & (1-N)\cos\theta d\phi \end{pmatrix} \qquad (4)$$

with $\Theta = id\theta + \sin\theta d\phi$, and $A_0 = A_r = 0$. The ansatz contains $N-1$ matter field functions $\omega_j(r)$. The field strength tensor component $F_{\theta\phi}$ is diagonal,

$$F_{\theta\phi} = (1/2e)\text{diag}(f_1, ..., f_N)\sin\theta \qquad (5)$$

with $f_j = \omega_j^2 - \omega_{j-1}^2 + \delta_j$, $\delta_j = 2j - N - 1$ ($\omega_0 = \omega_N = 0$). In EYMD theory this is supplemented with the ansatz for the dilaton field, $\Phi = \Phi(r)$.

We employ the dimensionless coordinate $x = er/\sqrt{4\pi G}$, the dimensionless mass function $\mu = em/\sqrt{4\pi G}$, and the scaled matter field functions [5]

$$u_j = \frac{\omega_j}{\sqrt{\gamma_j}}, \quad \gamma_j = j(N-j) , \qquad (6)$$

and, in EYMD theory, the dimensionless dilaton function $\varphi = \sqrt{4\pi G}\Phi$ and the dimensionless dilaton coupling constant $\gamma = \kappa/\sqrt{4\pi G}$. ($\gamma = 1$ corresponds to string theory, $\gamma = 0$ to EYM theory.)

Neutral solutions

When all $N-1$ gauge field functions are non-trivial, neutral solutions are obtained. The boundary conditions are chosen to have asymptotically flat solutions with a regular origin for the globally regular solutions and a regular horizon for the black hole solutions.

As an example, we consider static spherically symmetric globally regular solutions of $SU(4)$ EYM theory [7]. The $SU(4)$ EYM solutions are labelled by the node numbers (n_1, n_2, n_3) of the three gauge field functions u_1, u_2 and u_3, and can be

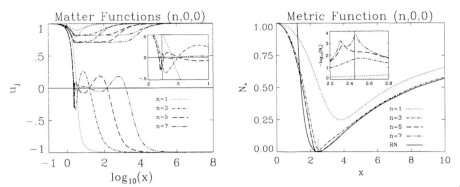

FIGURE 1. The matter functions $u_1(x)$, $u_2(x)$ and $u_3(x)$ (left) and the metric function $\mathcal{N}(x)$ (right) for the globally regular $SU(4)$ EYM solutions with node structure $(n,0,0)$.

classified into sequences. In Fig. 1 we present the globally regular solutions of the sequence with node structure $(n,0,0)$, $n = 1-7$, with odd n. With increasing n, the solutions approach a limiting solution.

To all regular solutions black hole counterparts exist. When all node numbers differ, these black hole solutions exist for arbitrary horizon radius. When two or more node numbers are the same, bifurcation phenomena occur at critical horizon radii [6,7].

Charged solutions

When one or more of the $N-1$ gauge field functions are identically zero, magnetically charged solutions are obtained.

Decomposition of the ansatz

When one gauge field function is identically zero ($\omega_{j_1} \equiv 0$), the ansatz for the gauge field splits into two parts

$$A_\mu^{(N)} dx^\mu = \left(\begin{array}{cc} \boxed{A_\mu^{(j_1)} dx^\mu} & \\ & \boxed{A_\mu^{(N-j_1)} dx^\mu} \end{array} \right) + \mathcal{H}_{j_1} \qquad (7)$$

with $\mathcal{H}_{j_1} = \frac{\cos\theta d\phi}{2e} h_{j_1}$ and

$$h_{j_1} = \left(\begin{array}{cc} \boxed{(N-j_1)\mathbf{1}_{(j_1)}} & \\ & \boxed{-j_1 \mathbf{1}_{(N-j_1)}} \end{array} \right) . \qquad (8)$$

$A_\mu^{(j_1)}$ and $A_\mu^{(N-j_1)}$ denote the non-abelian spherically symmetric ansätze for the $su(j_1)$ and $su(N-j_1)$ subalgebras of $su(N)$ (based on the j_1 and $(N-j_1)$-dimensional embeddings of $su(2)$, respectively), referred to by $su(\bar{N})$ in the following. $\mathcal{H}_{j_1}$ represents the ansatz for the element h_{j_1} of the CSA of $su(N)$. The field strength tensor splits accordingly with

$$F^{(\mathcal{H}_{j_1})} = -\frac{\sin\theta}{2e} d\theta \wedge d\phi\, h_{j_1} . \tag{9}$$

Considering the charge of the solution, the $su(\bar{N})$ parts of the solutions are neutral, because their field strength tensor decays at least like $O(r^{-1})$. In contrast, $F^{(\mathcal{H}_{j_1})}$ does not depend on r. A solution based on the element h_{j_1} of the CSA then carries charge of norm P,

$$P^2 = \frac{1}{2}\text{Tr}\, h_{j_1}^2 . \tag{10}$$

By applying these considerations again to the subalgebras $su(\bar{N})$ of eq. (7), one obtains the general case for $SU(N)$ EYM theory [8].

RN black hole solutions exist only for horizon radius $x_\text{H} \geq P$, and the extremal RN solution has $x_\text{H} = P$. As first observed for $SU(3)$ EYM theory [4], the same is true for charged EYM black holes. In contrast, charged EYMD black holes exist for arbitrary $x_\text{H} > 0$, like their Einstein-Maxwell-dilaton (EMD) counterparts.

Extremal black hole solutions

We now apply the above general analysis to $SU(5)$ EYM theory, presenting numerical examples for the case $u_4 \equiv 0$ [8]. In that case the non-abelian part of the gauge field corresponds to an $su(4)$ part, and the solutions carry charge $P = \sqrt{10}$. In Fig. 2 and Fig. 3 (left) we present the black hole solutions of the sequence with node structure $(n,0,0)$, $n = 1-7$, with odd n, and extremal horizon $x_\text{H} = \sqrt{10}$, also in the interior of the black hole (for $x < x_\text{H}$). Beside the matter function $\bar{u}_1$ we exhibit the charge function $P(x)$, $P^2(x) = 2x\,(\mu(\infty) - \mu(x))$ [2,4], and the metric function $\mathcal{N}(x)$.

Non-extremal black hole solutions

As seen from Fig. 2 and Fig. 3 (left), extremal EYM black hole solutions vary smoothly inside the horizon. In contrast, non-extremal EYM black hole solutions exhibit the phenomenon of mass inflation inside the horizon, analogously to neutral EYM black hole solutions [11]. In Fig. 3 (right) we demonstrate this phenomenon for the metric function $\mathcal{N}$ for the same sequence with node structure $(n,0,0)$, $n = 1-7$, n odd, and non-extremal horizon $x_\text{H} = 4.5$. In contrast, for EYMD solutions with $\gamma = 1$ no mass inflation occurs.

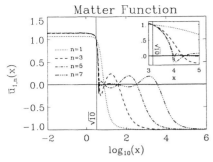
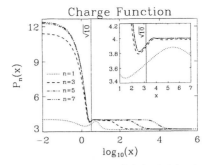

FIGURE 2. The matter function $\bar{u}_1(x)$ (left) and the charge function $P(x)$ (right) for the $SU(4)$ EYM black hole solutions with node structure $(n,0,0)$ and extremal event horizon $x_H = \sqrt{10}$.

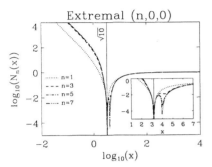
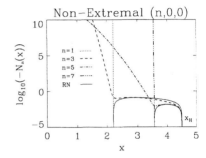

FIGURE 3. The metric function $\mathcal{N}(x)$ for the $SU(4)$ EYM black hole solutions with node structure $(n,0,0)$ and extremal event horizon $x_H = \sqrt{10}$ (left) and non-extremal event horizon $x_H = 4.5$ (right). Also shown (right) are the corresponding functions of the RN solutions with charges $P = \sqrt{10}$ and $P = 4$.

STATIC AXIAL BLACK HOLES

Ansätze

To obtain static axially symmetric solutions, we now employ isotropic coordinates for the metric [10,12]

$$ds^2 = -f dt^2 + \frac{m}{f} dr^2 + \frac{mr^2}{f} d\theta^2 + \frac{lr^2 \sin^2\theta}{f} d\phi^2 , \qquad (11)$$

where f, m and l are only functions of r and θ. We parametrize the purely magnetic gauge field ($A_0 = 0$) by [10,12,13]

$$A_\mu dx^\mu = \frac{1}{2er}\left[\tau_\phi^k \left(H_1 dr + (1-H_2)r d\theta\right) - k\left(\tau_r^k H_3 + \tau_\theta^k (1-H_4)\right) r \sin\theta d\phi\right] , \qquad (12)$$

with the Pauli matrices $\vec{\tau} = (\tau_x, \tau_y, \tau_z)$ and $\tau_r^k = \vec{\tau} \cdot (\sin\theta \cos k\phi, \sin\theta \sin k\phi, \cos\theta)$, $\tau_\theta^k = \vec{\tau} \cdot (\cos\theta \cos k\phi, \cos\theta \sin k\phi, -\sin\theta)$, $\tau_\phi^k = \vec{\tau} \cdot (-\sin k\phi, \cos k\phi, 0)$. We refer to k as the winding number of the solutions. Again, the four gauge field functions H_i and the dilaton function Φ depend only on r and θ. For $k = 1$ the spherically symmetric ansatz of ref. [9] is recovered with $H_1 = H_3 = 0$, $H_2 = H_4 = w(r)$ and $\Phi = \Phi(r)$.

Denoting the stress-energy tensor of the matter fields by T_μ^ν, with this ansatz the energy density $\epsilon = -T_0^0 = -L_M$ becomes

$$
\begin{aligned}
-T_0^0 = \quad & \frac{f}{2m}\left[(\partial_r\Phi)^2 + \frac{1}{r^2}(\partial_\theta\Phi)^2\right] + e^{2\kappa\Phi}\frac{f^2}{2e^2 r^4 m}\left\{\frac{1}{m}\left(r\partial_r H_2 + \partial_\theta H_1\right)^2\right. \\
& + \frac{k^2}{l}\left[(r\partial_r H_3 - H_1 H_4)^2 + (r\partial_r H_4 + H_1(H_3 + \mathrm{ctg}\theta))^2\right. \\
& \left.\left. + (\partial_\theta H_3 - 1 + \mathrm{ctg}\theta H_3 + H_2 H_4)^2 + (\partial_\theta H_4 + \mathrm{ctg}\theta(H_4 - H_2) - H_2 H_3)^2\right]\right\}.
\end{aligned}
\tag{13}
$$

With respect to $U = e^{i\Gamma(r,\theta)\tau_\phi^k}$ the system possesses a residual abelian gauge invariance. To fix the gauge we choose the gauge condition $r\partial_r H_1 - \partial_\theta H_2 = 0$.

Regular solutions

We first briefly consider static axially symmetric solutions of the field equations, which have a finite mass and are globally regular and asymptotically flat.

We change to dimensionless quantities again. The dimensionless mass $\mu = (e/\sqrt{4\pi G})GM$ is determined by the derivative of the metric function f at infinity, $\mu = \frac{1}{2}x^2 \partial_x f|_\infty$. Similarly, the derivative of the dilaton function at infinity determines the dilaton charge $D = x^2 \partial_x \varphi|_\infty$.

As for the spherically symmetric solutions, the following relations between the metric and the dilaton field hold [6]

$$\varphi(x) = \frac{1}{2}\gamma \ln(-g_{tt}),\tag{14}$$

$$D = \gamma\mu.\tag{15}$$

For the globally regular EYM solution with $k = 2$ and $n = 1$ the energy density of the matter fields is shown in Fig. 4, together with a surface of constant energy density, which is toruslike.

Black hole solutions

Now we consider static axially symmetric black hole solutions with a regular event horizon. The event horizon of static black hole solutions is characterized by $g_{tt} = -f = 0$. We impose that the regular horizon resides at a surface of constant r.

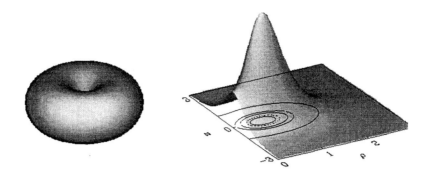

FIGURE 4. A surface of constant energy density (left) and the energy density (right) of the globally regular EYM solution with $k = 2$ and $n = 1$.

According to the zeroth law of black hole physics, the surface gravity κ_{sg} [14,15],

$$\kappa_{sg}^2 = -(1/4)g^{tt}g^{ij}(\partial_i g_{tt})(\partial_j g_{tt}) , \tag{16}$$

is constant at the horizon of the black hole solutions. It is proportional to the temperature $T = \kappa_{sg}/(2\pi)$. The dimensionless area A of the event horizon of the black hole solutions is proportional to the entropy $S = A/4$ [14]. The black hole solutions satisfy the general mass relation

$$\mu = \mu_o + 2TS , \tag{17}$$

with $\mu_o = -(e/\sqrt{4\pi G})G \int_0^{2\pi} \int_0^{\pi} \int_{r_H}^{\infty} d\phi d\theta dr \sqrt{-g} \left(2T_0{}^0 - T_\mu{}^\mu\right)$ [14], and the relation (compare (15))

$$D = \gamma(\mu - 2TS) . \tag{18}$$

In Fig. 5 we exhibit surfaces of constant energy density and in Fig. 6 the energy density for the EYM black hole solution with $k = 2$ and $n = 1$. Also the event horizon of the black hole solutions is not spherically symmetric, but only axially symmetric [10].

CONCLUSIONS

In EYM and EYMD theory black hole solutions are not uniquely specified by their mass, charge and angular momentum. Here we have considered static EYM and EYMD black hole solutions with non-abelian hair. The spherically symmetric solutions are unstable [16], and there is all reason to believe, that the axially symmetric black hole solutions are unstable as well. However, we expect analogous solutions in EYMH theory [15] and in ES theory, and for $n = 2$ these axially symmetric black hole solutions should be stable [15]. In contrast, the stable black hole

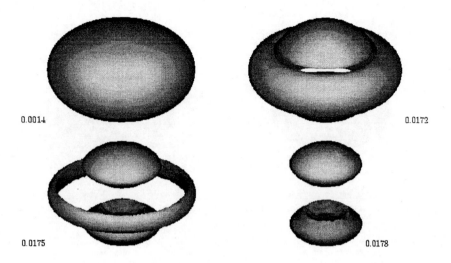

FIGURE 5. Surfaces of constant energy density for four values of the dimensionless energy density of the EYM black hole solution with $k = 2$ and $n = 1$.

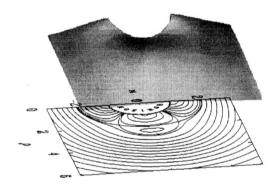

FIGURE 6. The energy density of the EYM black hole solution with $k = 2$ and $n = 1$.

solutions with higher magnetic charges (EYMH) or higher baryon numbers (ES) should not correspond to such axially symmetric solutions with $n > 2$. Instead these stable solutions should have much more complex shapes, exhibiting discrete crystal-like symmetries [15]. Analogous but unstable black hole solutions of this kind should also exist in EYM and EYMD theory.

REFERENCES

1. Volkov, M.S., and Gal'tsov, D.V., *Sov. J. Nucl. Phys.* **51** 747 (1990); Bizon, P., *Phys. Rev. Lett.* **64** 2844 (1990); Künzle, H.P., and Masoud-ul-Alam, A.K.M., *J. Math. Phys.* **31** 928 (1990).
2. Bartnik, R., and McKinnon, J., *Phys. Rev. Lett.* **61** 141 (1988).
3. Gal'tsov, D.V., and Ershov, A.A., *Phys. Lett.* **A138** 160 (1989); Bizon, P., and Popp, O.T., *Class. Quantum Grav.* **9** 193 (1992).
4. Gal'tsov, D.V., and Volkov, M.S., *Phys. Lett.* **B274** 173 (1992).
5. Künzle, H.P., *Comm. Math. Phys.* **162** 371 (1994).
6. Kleihaus, B., Kunz, J., and Sood, A., *Phys. Lett.* **B354** 240 (1995); *Phys. Lett.* **B374** 289 (1996); *Phys. Rev.* **D54** 5070 (1996).
7. Kleihaus, B., Kunz, J., Sood, A., and Wirschins, M., *Phys. Rev.* **D** in press.
8. Kleihaus, B., Kunz, J., and Sood, A., *Phys. Lett.* **B418** 284 (1998).
9. Donets, E.E., and Gal'tsov, D.V., *Phys. Lett.* **B302** 411 (1993); Lavrelashvili, G., and Maison, D., *Nucl. Phys.* **B410** 407 (1993).
10. Kleihaus, B., and Kunz, J., *Phys. Rev. Lett.* **79** 1595 (1997); *Phys. Rev.* **D57** 6138 (1998).
11. Gal'tsov, D.V., Donets, E.E., and Zotov, M.Yu., *Phys. Rev.* **D56** 3459 (1997); *JETP Lett.* **65** 895 (1997); Breitenlohner, P., Lavrelashvili, G., and Maison, D., *Nucl. Phys.* **B524** 427 (1998); *gr-qc/9708036*; Sarbach, O., Straumann, N., and Volkov, M.S., *gr-qc/9709081*.
12. Kleihaus, B., and Kunz, J., *Phys. Rev. Lett.* **78** 2527 (1997); *Phys. Rev.* **D57** 6138 (1998).
13. Rebbi, C., and Rossi, P., *Phys. Rev.* **D22** 2010 (1980).
14. Wald, R.M., *General Relativity*, Chicago: University of Chicago Press, 1984.
15. Ridgway, S.A., and Weinberg, E.J., *Phys. Rev.* **D52** 3440 (1995).
16. Straumann, N., and Zhou, Z.H., *Phys. Lett.* **B243** 33 (1990).

Gravitational Instantons with Source

Kishore B. Marathe
City University of New York, Brooklyn College

Abstract. We discuss various geometric formulations of the equations of gravitational instantons with sources. We show that these equations include as a special case the classical Einstein equations with or without the cosmological constant. We discuss the notion of an Einstein pair to clarify the distinction between the Palatini and the metric variational equations. Our approach naturally leads to the introduction of a cosmological function and a generalized conservation law for the energy-momentum tensor. We also discuss the physical significance of the new conservation condition.

PACS 04.20 - Classical general relativity.

INTRODUCTION

It is well known that the geometry of the space-time manifold plays a fundamental role in Einstein's theory of gravity. The physical implications of the theory depend essentially on the Lorentzian nature of the space-time manifold. In trying to incorporate electromagentic field together with gravitation a higher dimensional version of Einsteins equations was considered by Kaluza and Klein. However, this theory had a rather limited success and the problem of considering Einstein's equations (with or without source) on manifolds other than four dimensional Lorentz manifolds had not been considered until recently. The situation was similar for other classical field theories. Solution of the Euclidean Yang-Mills instanton equations and their possible use in the path integral quantization scheme has now led to a study of the corresponding problem for solutions of Einstein's vacuum field equations in the Euclidean setting. Finite action solutions of the vacuum Einstein equations with or without the cosmological constant are called gravitational instantons by analogy with the Yang-Mills instantons. One possible use of gravitational instantons is in application of Feynman's path integral method to quantization of the gravitational field. This method should take into account the geometry and topology of the space-time manifolds which admit gravitational instantons. At present there is no precise definition of Feynman's path integral over all gravitational instantons (or more precisely over the moduli space of the equivalence classes of gravitational instantons). It is well known that the standard Hilbert-Einstein action is not necessarily positive but the action corresponding to gravitational instantons is always non-negative. All gravitational instantons fall in a class of spaces that are well known in differential geometry as the Einstein spaces or Einstein manifolds. A detailed discussion of the structure of Einstein spaces and their moduli spaces may be found in Besse [2]. It is well known that there are topological obstructions to a manifold being an Einstein manifold. An early result in this direction is the following theorem of Hitchin [6].

Theorem .1 *Let M be a compact, four-dimensional Einstein manifold with signature τ and Euler characteristic χ. Then*

$$|\tau| \leq \frac{2}{3}\chi.$$

If, in addition, M has non-positive (or non-negative) sectional curvature then

$$|\tau| \leq (\frac{2}{3})^{2/3}\chi.$$

An application of this theorem shows that the manifolds $S^1 \times S^3$, $2T^4$ (the connected sum of two 4-tori), and $n\mathbf{CP}^2$, $n > 3$ (the connected sum of n copies of the complex projective space $\mathbf{CP}^2$) do not admit any Einstein structure. There are also examples of manifolds which admit both Einstein and non-Einstein metrics. An example of a non-Einstein metric for $\mathbf{CP}^2$ is well known. These manifolds are possible candidates for the base spaces of generalized gravitational instantons with a non-zero source term.

In studying the geometry and topology of spaces admitting gravitational fields, a characterization of Einstein's field equations of gravitation was obtained by Marathe in [8]. This result was further extended and applied to the study of gravitational instantons and their couplings in [9–12]. In this paper we use the results developed in these papers to give a generalization of the gravitational instanton equations. The conservation condition expressed by the divergence-free character of the energy momentum tensor is also generalized. The new equations can be expressed in a very simple form using a linear transformation of the space of second order differential forms (pointwise) on the usual pseudo-Riemannian, four-dimensional manifold. As partial differential equations, Einstein's field equations are highly non-linear and this has prevented a development of an effective method of quantization of gravitational field. The new equations expressed in terms of linear transformations seem to have a form suitable for quantization. In the following section we discuss the Palatini principle and define an Einstein pair and the generalized vacuum gravitational instantons. In the next section we consider a gravitational instanton coupled to classical fields. The resulting field equations include the usual gravitational equations with source term with or without the cosmological constant. For background material on geometry and topology with applications to gauge theory see, for example, Marathe and Martucci [14].

PALATINI PRINCIPLE AND EINSTEIN PAIRS

A large number of Lagrangians satisfying the naturality and regularity requirements are used in the physics literature. They are broadly classified into Lagrangians where gauge and other force fields are coupled to fields of bosonic matter which have integral spin and to fields of fermionic matter which have half-integral spin. In the standard derivation of the Einstein equations from the Hilbert-Einstein action it is the metric that is the fundamental variable and one uses the Levi-Civita connection uniquely determined by the metric to construct the Lagrangian of the theory. Considering the metric and connection as independent variables is a natural generalization that usually goes under the name of the Palatini principle. Note that the connection is not on an arbitrary principal bundle but on the bundle of frames $L(M)$ of (M,g). In [4] the Palatini principle has been used to study the field equations obtained from the variation of the action

$$S(\Gamma, g) := \int_M f(\rho)\sqrt{g}\, d^n x \tag{1}$$

where (M, g) is an n-dimensional pseudo-Riemannian manifold with metric g and Γ is a torsion-free connection on $L(M)$. The function f is taken as an analytic function of one variable ρ which is the scalar curvature determined by the pair (Γ, g). They obtain the the following Euler-Lagrange equations.

$$f'(\rho)\rho_{(\mu\nu)} - \frac{1}{2}f(\rho)g_{\mu\nu} = 0 \tag{2}$$

$$\nabla_\Gamma(f'(\rho)\sqrt{g}g^{\mu\nu}) = 0 \tag{3}$$

where $\rho_{(\mu\nu)}$ is the symmetric part of $(\rho)_{\mu\nu}$. These equations must be considered together with the consistency condition obtained by taking the trace of equation (2), namely

$$f'(\rho)\rho - \frac{n}{2}f(\rho) = 0 \tag{4}$$

If we require that the Lagrangian function f be not a constant function then it follows from equations (2) and (4) that the pair (Γ, g) satisfy the equations

$$\rho_{(\mu\nu)} - \frac{1}{n}\rho g_{\mu\nu} = 0.$$

These equations contain as a very special case the vacuum Einstein equations but are much more general. We use them to make the following definition [13].

Definition .1 *(Einstein pair) Let M be a pseudo-Riemannian manifold with metric g and let Γ be a torsion-free connection on the bundle of frames $L(M)$. Let $\rho_{\mu\nu}$ be the Ricci tensor obtained by contracting the curvature tensor of Γ and let ρ*

be its scalar curvature. We call the pair (Γ, g) an Einstein pair if it satisfies the equations (5) given below.

$$\rho_{(\mu\nu)} - \frac{1}{n}\rho g_{\mu\nu} = 0 \tag{5}$$

Definition .2 *(Generalized Gravitational Instanton) Let M be a pseudo-Riemannian manifold with metric g and let Γ be a torsion-free connection on the bundle of frames $L(M)$. Let $\rho_{\mu\nu}$ be the Ricci tensor obtained by contracting the curvature tensor of Γ and let ρ be its scalar curvature. We call the pair (Γ, g) a generalized gravitational instanton if it satisfies the following equations*

$$\rho_{(\mu\nu)} = \phi g_{\mu\nu} \tag{6}$$

where ϕ is a real scalar field on M.

The equations (6) generalize the gravitational instanton equations which in turn generalize the vacuum Einstein equations. Thus the universal equations obtained from the non-constant Lagrangian $f(\rho)$ by the Palatini principle are not the Einstein equations but the generalized gravitational instanton equations. It is important to note that the tensor appearing in the equations (5) is not the usual Einstein tensor but is the trace-free part of the Ricci tensor ρ_{ij}. When the symmetric connection Γ is a Weyl connection on a Weyl manifold [5] then the corresponding Einstein pair satisfies the Einstein-Weyl condition. We note that the definition of an Einstein pair can be applied to the case of complex manifolds and Hermitian vector bundles. It then corresponds to the weak Einstein-Hermitian condition as defined by Kobayashi in [7]. We shall consider only real manifolds in this paper. The symmetric connection Γ can always be written as the sum of the Levi-Civita connection λ and a $(1, 2)$ tensor θ. If we regard the metric (and hence the Levi-Civita connection and its curvature) as corresponding to pure gravity, then we can take the tensor θ as a deformation of λ leading to an energy-momentum tensor of an interacting field. On a four-dimensional space-time manifold the equations (5) can then be written as

$$R_{\mu\nu} - \frac{1}{4}R g_{\mu\nu} = T_{\mu\nu} - \frac{1}{4}T g_{\mu\nu} \tag{7}$$

where $R_{\mu\nu}$ is the standard Ricci tensor of λ and R is its scalar curvature. The right hand side is a function of θ, λ and g. When the tensor T is zero the equations (7) become the equations of gravitational instanton. Thus we can regard the equations (7) as equations of generalized gravitational instantons or gravitational instantons with source T.

GENERALIZED GRAVITATIONAL INSTANTONS

By the pseudo-Riemannian structure Γ_g on M corresponding to the symmetric fundamental tensor g, we shall understand the unique torsion-free Levi-Civita connection such that g is parallel with respect to Γ_g. We use the same letter g to denote the fundamental tensor of type $(0,2)$ or its dual of type $(2,0)$ and a similar usage is followed for other tensors. In what follows we assume M to be a pseudo-Riemannian manifold with fundamental tensor g of signature $(-,-,-,+)$. The space $\Lambda_x^2(M)$ is a real six-dimensional vector space. The Hodge star operator J on $\Lambda_x^2(M)$ defines a complex structure on it (i.e. $J^2 = -I$ is the identity transformation of $\Lambda_x^2(M)$)). Using the complex structure $J, \Lambda_x^2(M)$ can be made into a complex three-dimensional vector space. The space of complex linear transformations of this complex vector space can be identified with the subspace of real transformations S of the real vector space defined by $\{S \mid SJ = JS\}$. We begin by defining a tensor of curvature type.

Definition .3 *Let C be a tensor of type $(4,0)$ on M. We can regard C as a quadrilinear mapping (pointwise) so that for each $x \in M$, C_x can be identified with a multilinear map*

$$C_x : T_x^*(M) \times T_x^*(M) \times T_x^*(M) \times T_x^*(M) \to \mathbf{R}.$$

where $\mathbf{R}$ is the real number field. We say that the tensor C is of curvature type if C_x satisfies the algebraic properties of the Riemann-Christoffel curvature tensor.

We observe that if C is of curvature type then C_x can be identified with a self-adjoint linear transformation of $\Lambda_x^2(M)$ for each $x \in M$, where $\Lambda_x^2(M)$ is the space of second order differential forms at x.

Example .1 *The Riemann-Christoffel curvature tensor is of curvature type. Indeed, the definition of tensors of curvature type is modelled after this fundamental example. Another important example of a tensor of curvature type is the tensor G defined by*

$$G_x(\alpha, \beta, \gamma, \delta) = g_x(\alpha, \gamma)g_x(\beta, \delta) - g_x(\alpha, \delta)g_x(\beta, \gamma), \quad \forall x \in M$$

where g is the fundamental or metric tensor of M.

We now define the curvature product of two symmetric tensors of type $(2,0)$ on M. It was introduced in [8] and used in [10] to obtain a geometric formulation of Einstein's equations.

Definition .4 *Let g and T be two symmetric tensors of type $(2,0)$ on M. The* **curvature product** *of g and T, denoted by $g \times_c T$, is a tensor of type $(4,0)$ defined by*

$$\begin{aligned}(g \times_c T)_x(\alpha, \beta, \gamma, \delta) := \ &\tfrac{1}{2}[g(\alpha,\gamma)T(\beta,\delta) + g(\beta,\delta)T(\alpha,\gamma) \\ &- g(\alpha,\delta)T(\beta,\gamma) - g(\beta,\gamma)T(\alpha,\delta)],\end{aligned}$$

for all $x \in M$ and $\alpha, \beta, \gamma, \delta \in T_x^*(M)$.

When g is the fundamental tensor of M and T is the energy-momentum tensor, we call $g \times_c T (= T \times_c g)$ the interaction tensor between the field and the source (energy-momentum tensor). It plays an essential role in definition of the new field equations given here. We now define the gravitational tensor W_g, of curvature type, which includes the source term.

Definition .5 *Let M be a space-time manifold with fundamental tensor g and let T be a symmetric tensor of type $(2,0)$ on M. Then the* **gravitational tensor** W_g *is defined by*

$$W_g := R + g \times_c T, \tag{8}$$

where R is the Riemann-Christoffel curvature tensor of type $(4,0)$.

We are now in a position to give a geometric formulation of the field equations of gravitational instantons with source.

Theorem .2 *Let W_g denote the gravitational tensor defined by (8) with source tensor T. We also denote by W_g the linear transformation of $\Lambda_x^2(M)$ induced by W_g. Then the following are equivalent:*

1. *g satisfies the generalized field equations of gravitation*

$$R_{\mu\nu} - \frac{1}{4} R g_{\mu\nu} = T_{\mu\nu} - \frac{1}{4} T g_{\mu\nu} \tag{9}$$

2. *W_g commutes with J, i.e.*

$$[W_g, J] = 0 \tag{10}$$

3. *W_g induces a complex linear transformation of the complex vector space $\Lambda_x^2(M)$ for each $x \in M$.*

4. *Let G denote the inner product on $\Lambda_x^2(M)$ induced by g. Define the submanifolds D_+ and D_- of $\Lambda_x^2(M)$ by*

$$D_{\pm} = \{v \in \Lambda_x^2(M) | G(v,v) = \pm 1 \text{ and } v \wedge v = 0\}. \tag{11}$$

Define a real valued function f on $D_+ \cup D_-$, called the **gravitational sectional curvature**, *by*

$$f(v) = s(v) \times G(W_g v, v), \tag{12}$$

where $s(v) = +1$, if $v \in D_+$ and $s(v) = -1$, if $v \in D_-$. Then

$$f = f \circ J. \tag{13}$$

We shall call the triple (M, g, T) a **gravitational instanton with source** T if any one of the conditions of Theorem .2 is satisfied. In the special case when the energy momentum tensor T is zero the results of the above theorem were obtained in [17].

We note that the above theorem can be used to discuss the Petrov classification of gravitational fields (see Petrov [15]). We observe that these equations contain Einstein's equations with or without the cosmological constant as special cases. Thus the equations (9) can be regarded as a generalization of the gravitational field equations. Solutions of these generalized gravitational field equations which are not solutions of Einstein's equations have been discussed in [3].

We note that, over a compact, 4-dimensional, Riemannian manifold (M, g), the gravitational instantons that are not solutions of the vacuum Einstein equations are critical points of the quadratic, Riemannian functional or action $\mathcal{A}_2(g)$ defined by

$$\mathcal{A}_2(g) = \int_M R^2 dv_g.$$

Furthermore, the standard **Hilbert-Einstein action**

$$\mathcal{A}_1(g) = \int_M R \, dv_g$$

also leads to the generalized field equations when the variation of the action is restricted to metrics of volume 1.

There are several differences between the Riemannian functionals used in theories of gravitation and the Yang-Mills functional used to study gauge field theories. The most important difference is that the Riemannian functionals are dependent on the bundle of frames of M or its reductions, while the Yang-Mills functional can be defined on any principal bundle over M. However, we have the following interesting theorem [1].

Theorem .3 *Let (M, g) be a compact, 4-dimensional, Riemannian manifold. Let $\Lambda_+^2(M)$ denote the bundle of self-dual 2-forms on M with induced metric G_+. Then the Levi-Civita connection λ_g on M satisfies the gravitational instanton equations if and only if the Levi-Civita connection λ_{G_+} on $\Lambda_+^2(M)$ satisfies the Yang-Mills instanton equations.*

Recall that Einstein's field equations may be written as

$$R^{ij} - \frac{1}{2} R g^{ij} = -T^{ij}, \tag{14}$$

where the coupling constant accompanying T^{ij} is taken to be unity. If one wishes to retain the coupling constant, say k, in Einstein's equations one need only introduce it in the definition of W as follows:

$$W = R + k g \times_c T. \tag{15}$$

The equations (9) are then replaced by

$$R^{ij} - \frac{1}{4}Rg^{ij} = -k(T^{ij} - \frac{1}{4}Tg^{ij}). \tag{16}$$

In particular by taking $R = kT$, in the equations (16) we obtain

$$R^{ij} - \frac{1}{2}Rg^{ij} = -kT^{ij} \tag{17}$$

which are Einstein's equations with the coupling constant k.

We observe that the equations (16) does not lead to any relation between the scalar curvature and the trace of the source tensor, since both sides of the equations (16) are trace-free.

With each symmetric tensor S of type $(2,0)$ we associate the trace-free tensor $\hat{S}$ by the definition

$$\hat{S}^{ij} = S^{ij} - \frac{1}{4}Sg^{ij}. \tag{18}$$

Then equations (16) may be written as

$$\hat{R}^{ij} = -k\hat{T}^{ij}. \tag{19}$$

Taking divergence of both sides of the equations (16) and using the Bianchi identities we obtain the generalized conservation condition

$$\nabla_i T^{ij} - \frac{1}{k}g^{ij}\Phi_i = 0, \tag{20}$$

where ∇_i is the covariant derivative with respect to the vector $\frac{\partial}{\partial x^i}$,

$$\Phi = \frac{1}{4}(kT - R) \tag{21}$$

and $\Phi_i = \frac{\partial}{\partial x^i}\Phi$. Using this function Φ the generalized field equations can be written as

$$R^{ij} - \frac{1}{2}Rg^{ij} - \Phi g^{ij} = -kT^{ij}. \tag{22}$$

In this form the new field equations appear as Einstein's field equations with the cosmological constant replaced by the function Φ, which we may call the cosmological function. The cosmological function is intimately connected with the classical conservation condition expressing the divergence-free nature of the energy-momentum tensor as is shown by the following proposition.

Proposition .4 *The energy-momentum tensor satisfies the classical conservation condition*

$$\nabla_i T^{ij} = 0 \qquad (23)$$

if and only if the cosmological function Φ is a constant. Moreover, in this case the generalized field equations reduce to Einstein's field equations with cosmological constant.

PROOF. The first part of the proposition follows directly from the equations (20). If $\Phi(x)$ has the constant value λ for each $x \in M$, then using the form (22) of the generalized field equations, we obtain

$$R^{ij} - \frac{1}{2} R g^{ij} - \lambda g^{ij} = -k T^{ij}. \qquad (24)$$

In particular, if the cosmological function Φ is zero, one obtains the usual field equations.

Equation (23) is called the differential or local law of conservation of energy and momentum. To obtain global conservation laws one needs to integrate equations (23). However such integration is possible only if the space-time manifold admits Killing vectors. In general, a manifold does not admit any Killing vectors. Thus equations (23) (or the equivalent condition that Φ be constant) have, in general, no clear physical meaning. An interesting discussion of this point is given in Sachs and Wu [16]. We note that, if the energy momentum tensor is non-zero but is localized in the sense that it is negligible away from a given region, then the scalar curvature acts as a measure of the cosmological constant. The cosmological function Φ can interpolate between regions where the equations (20 and (23 are satisfied. In particular, we may have regions corresponding to solutions of the field equations with different cosmological constants connected by the cosmological function Φ.

In conclusion we note that the generalized equations arise naturally as a deformation of the gravitational instanton equations. The new equations lead to a generalized conservation condition for the energy-momentum tensor. Relation of the cosmological function and the classical conservation condition is discussed. In particular it is shown that Einstein's field equations with and without the cosmological constant are special cases of the new field equations. The new field equations can be expressed in several different forms and some of these are found to lead to solution of the problem of classification of gravitational fields. Equation (10) is suggestive of quantum theoretical formalism. The new field equations can also be expressed in terms of the electric and magnetic parts of the Riemann curvature tensor using a duality tranformation between the active and passive electric parts. This aspect of the field equations and some special solutions will be considered elsewhere.

ACKNOWLEDGEMENTS

This work was supported in part by the Istituto Nazionale di Fisica Nucleare, sezione Firenze and was carried out during the Fall 97 Fellowship leave from Brooklyn College of the City University of New York. The author would like to thank Prof. Martucci (Florence University) and Prof. Dadhich (Inter University Center for Astronomy and Astrophysics, Pune) for useful discussions.

REFERENCES

1. M. F. Atiyah, N. J. Hitchin, and I. M. Singer. Self-duality in Four Dimensional Riemannian Geometry. *Proc. Roy. Soc. Lond.*, A362:425–461, 1978.
2. A. Besse. *Einstein Manifolds*. Springer-Verlag, Berlin, 1986.
3. D. Canarutto. Marathe's Generalized Gravitational Fields and Singularities. *Il Nuovo Cimento*, 75B:134–144, 1983.
4. M. Ferraris, M. Francaviglia, and I. Volovich. The universality of vacuum Einstein equations with cosmological constant. *Class. Quantum Grav.*, 11:1505–1517, 1994.
5. G. B. Folland. Weyl Manifolds. *J. Diff. Geom.*, 4:145–153, 1970.
6. N. J. Hitchin. On compact four-dimensional Einstein manifolds. *J. Diff. Geo.*, 9:435–442, 1974.
7. S. Kobayashi. *Differential Geometry of Complex Vector Bundles*. Princeton University Press, Princeton, 1987.
8. K. B. Marathe. *Structure of Relativistic Spaces*. PhD thesis, University of Rochester, 1971.
9. K. B. Marathe. Generalized Field Equations of Gravitation. *Rend. Mat. (Roma)*, 6:439–446, 1972.
10. K. B. Marathe. Spaces admitting gravitational fields. *J. Math. Phys.*, 14:228–233, 1973.
11. K. B. Marathe. The Mean Curvature of Gravitational Fields. *Physica*, 114A:143–145, 1982.
12. K. B. Marathe. Generalized Gravitational Instantons. In *Proc. Coll. on Diff. Geom., Debrecen (Hungary) 1984*, pages 763–775, Hungary, 1987. Colloquia Math Soc. J. Bolyai.
13. K. B. Marathe and G. Martucci. Geometry of Gravitational Instantons. *Il Nuovo Cimento*, to appear, 1998.
14. K. B. Marathe and G. Martucci. *The Mathematical Foundations of Gauge Theories*. Studies in Mathematical Physics, vol. 5. North-Holland, Amsterdam, 1992.
15. A. Z. Petrov. *Einstein Spaces*. Pergamon Press, New York, 1969.
16. R. K. Sachs and H. Wu. *General Relativity for Mathematicians*. SpringerVerlag, Berlin, 1977.
17. I. M. Singer and J. Thorpe. The Curvature of 4-dimensional Einstein Spaces. In *Global Analysis, Papers in honor of K. Kodaira*, pages 355–365, Princeton, 1969. Princeton Uni. Press.

High Current in General Relativity

Boris E. Meierovich

P.L.Kapitza Institute of Physics Problems
2 Kosygina Str., Moscow 117334, Russia
E-mail: TRC@meyer.msk.su

Abstract. Physical nature of equilibrium of current-carrying filaments is studied on the basis of Einstein equations of General Relativity. Considering a conducting filament as an element of the structure of Universe, one has to take into account both electromagnetic and gravitational interactions of charges. The energy of the proper electromagnetic field, stored in the current channel, contributes to the gravitation field of the system together with the energies of individual charges. Depending on the relation between the parameters of a filament there are two different physical situations of weak and strong current. In both limiting cases of weak and strong current we derive analytically energy balance condition (a generalization of famous Bennet relation), valid for arbitrary equation of state of matter within a filament. Even if the current is small the relative motion of electrons with respect to ions, connected with the current, results in additional gravitation attraction of these two subsystems. This relativistic effect does not exist within Newtonian approximation. In the limit of high current energy balance relation displays common physical nature of gravitational collapse and electromagnetic one. Boundaries of equilibrium domain determine limiting parameters of a filament: maximum energy per unit length stored in the magnetic field is $0.927 \cdot 10^{48}$ erg/cm, limiting current is $0.94 \cdot 10^{25}$ A, and maximum gravitational mass per unit length is $0.81 \cdot 10^{28}$ g/cm.

INTRODUCTION

Intergalactic currents are playing a very important role in modern Plasma Astrophysics [1]. Understanding that the Universe is largely a Plasma Universe came from the fact that electromagnetic forces exceed gravitational forces by a factor of 10^{36}, and even in a neutral as a whole system a relatively small electromagnetic fluctuation can lead to non-uniform distribution of matter. Volumewise 99.999% of matter in the Universe is in the plasma state – in the state of free electrons and protons. Relative motion of the charges results in formation of current-conducting filaments. The filaments can undergo radial contraction to extremely dense states (pinch-effect), resulting in various phenomena, such as acceleration of particles to very high energies, for instance. Extrapolation of plasma properties from laboratory phenomena to the scale of Universe dictates the necessity to take gravitation into account.

The basis for theoretical understanding of processes in Plasma Universe is General Relativity with account of electromagnetism. We consider a problem of equilibrium structure of a current-conducting channels. Understanding physical nature of equilibrium of a filament is a necessary step for further investigations of stability and evolution of matter in the scale of the Universe.

Curvature of spacetime, caused by intergalactic currents, is small. Condition of equilibrium of current channels with account of gravitation in the Newtonian approximation had been derived in [2]. However Newtonian approximation is applicable when two conditions are satisfied: the motion must be non-relativistic, and gravitation field must be weak ([3], paragraphs 87,99). For this reason the condition of equilibrium, derived in Newtonian approximation, can be used for non-relativistic drift velocities only. In order to get the condition of equilibrium for a high current channel with relativistic drift velocity it is necessary to use Einstein equations even if the gravitation field is week.

We derive condition of equilibrium (of Bennet type) with account for both electromagnetic and gravitation forces, and applicable for arbitrary drift velocity. We show that the energy of gravitation compression consists of two terms. One is the ordinary Newtonian energy of gravitational attraction. The other is the energy of gravitational attraction of the two counterstreaming subsystems (electrons and ions), caused by the relative motion. This relativistic effect that does not exist in Newtonian approximation.

Another topic that requires General Relativity is the old Alfven's [4] problem of a limiting current. If we ignore the effects of General Relativity, then self-consistent theory [5] does not impose any limitations on the current values of equilibrium relativistic beams. In General Relativity the matter curves the space-time, and this results in gravitational self-attraction of matter. If total energy (or mass) of matter exceeds some limit, the forces of contraction can not be balanced by the pressure. In this case equilibrium is not possible, and the matter undergoes infinite contraction, which is called gravitational collapse. In the case of a current-conducting filament total energy of the matter is the sum of two parts – energy of the particles, and energy of the electromagnetic field. The energy in equilibrium can not exceed some limiting value, no matter how the energy is distributed between particles and electromagnetic field. The limiting value is a boundary of gravitational collapse. As far as there is a limiting value of total energy, clearly electromagnetic energy is also restricted. Magnetic energy is a regular function of current, and we come to conclusion that there must be a limiting value of current in General Relativity.

We show that the current of an equilibrium filament can not exceed $I_{max} = 0.94 \cdot 10^{25} A$. Currents $I \sim 10^{20} A$ in the Galactic and Intergalactic Medium are discussed by Peratt [6]. I didn't find in the literature any mentioning of currents higher than $10^{20} A$. Nevertheless solution of the problem of limiting current in General Relativity is interesting in principle, especially taking into account filamentary structure of the Universe. Our analysis realizes common physical nature of gravitational and electromagnetic collapse, and displays peculiarities of space distribution of matter and gravitational field near the collapse boundary.

PHYSICAL APPROACH

We consider a current-conducting filament as a cylindrically symmetric system of counterstreaming electrons and protons. Electric current of the filament is caused by the relative motion of electrons with respect to ions with drift velocity W. Distribution of particles and fields depends only on one coordinate – the distance from the axis. Our approach is based on the fact, that in limiting cases of high and small drift velocities, we can consider the current channel as a superposition of two equilibrium subsystems, interacting with each other only via electromagnetic and gravitational forces of collective interaction. This is true in limiting cases of high and small drift velocity. This approach is a traditional one. A new step forward is the account of gravitation on the basis of Einstein equations of General Relativity.

Distribution of the particles and fields depends only on one curved coordinate x^1 – the "distance" from the axis. If electric current is connected with drift velocity, the subsystems are in the state of relative motion in any frame of reference. In this case non-diagonal components of the metric tensor $g_{0\alpha}$ are not zeroes. In the case $W/c \sim 1$ $g_{0\alpha}$, g_{00}, and g_{33} give contribution of the same order to the equilibrium energy balance. Gravitation field of a high current channel as stationary, but not static ([3], paragraph 88).

We are using standard notations [3]. Latin indexes are used for 4-vectors and 4-tensors. Greek ones - for 3-vectors and 3-tensors. $h = g_{00}$ is a scalar function, $g_\alpha = -g_{0\alpha}/h$ is a 3-vector. $\gamma_{\alpha\beta}$ (three dimensional metric tensor) is used for raising and lowering Greek indexes,

$$\gamma_{\alpha\beta} = -g_{\alpha\beta} + \frac{g_{0\alpha}g_{0\beta}}{g_{00}}. \tag{1}$$

In the state of thermal equilibrium distribution function is an integral of motion and a relativistic invariant. In any reference frame it can depend only on an invariant combination of additive integrals of motion (projections of generalized 4-momentum P_i on cyclic coordinates) and 4-velocity U^i of a subsystem, moving as a whole. In our case $P_i U^i$ is such a combination,

$$U^0 = \frac{dx^0}{\sqrt{g_{ik}dx^i dx^k}} = \frac{1}{\sqrt{g_{00} + 2g_{0\alpha}\frac{W^\alpha}{c} + g_{\alpha\beta}\frac{W^\alpha W^\beta}{c^2}}}$$

$$U^\alpha = \frac{dx^\alpha}{\sqrt{g_{ik}dx^i dx^k}} = \frac{\frac{W^\alpha}{c}}{\sqrt{g_{00} + 2g_{0\alpha}\frac{W^\alpha}{c} + g_{\alpha\beta}\frac{W^\alpha W^\beta}{c^2}}}$$

U^i is 4-velocity of a subsystem in the laboratory frame, $W^\alpha = dx^\alpha/dx^0$. $W = W_i^3 - W_e^3$ is the drift velocity. In the presence of electromagnetic field generalized 4-momentum P_i is connected with kinetic 4-momentum p_i and 4-potential of electromagnetic field A_i by the relation $p_i = P_i - (e/c)A_i$.

Dependence of distribution function on $P_i U^i$ is determined by statistics of the particles. In case of fermions (electrons and protons) it has the form

$$F(X) = \frac{1}{1 + \exp\left(B + \frac{X}{T}\right)} \qquad X = cP_i U^i, \qquad (2)$$

B and T are scalars. As far as any distribution function is an integral of motion, its argument $B + X/T$ does not depend on coordinates. B is simply a constant. Projections of 4-momentum P_i on the direction of current and on the time coordinate are integrals of motion. Hence, U_i/T, i=0,3, do not depend on coordinates. For this reason $\frac{W^3}{c} = U^3/U^0 = const$, and we come to a conclusion, that W^3 does not depend on the coordinate x^1. Coordinate dependence of the temperature is determined by

$$T\sqrt{g_{00} + 2g_{0\alpha}\frac{W}{c} + g_{\alpha\beta}\frac{W^\alpha W^\beta}{c^2}} = \frac{T}{U_0} = const. \qquad (3)$$

The covariant formula for the stress-energy tensor T^{ik} as an integral of distribution function

$$T^{ik} = \frac{2cg}{(2\pi\hbar)^3} \int p^i p^k F(X) \delta\left(p_s p^s - m^2 c^2\right) \sqrt{-g} d\Omega_p \qquad (4)$$

gives a well known expression:

$$T^{ik} = (\mathcal{E} + \mathcal{P})\mathcal{U}^i \mathcal{U}^{\|} - \mathcal{P}\}^{\|}. \qquad (5)$$

However the energy $\mathcal{E}$ and the pressure $\mathcal{P}$ are now expressed via potentials of electromagnetic field A_i:

$$\mathcal{E} = \frac{\triangle \pi g}{(\in \pi \{)^{\ni}} \int_{\updownarrow]\in}^{\infty} \frac{\mathcal{E}^\in \sqrt{\mathcal{E}^\in - \Updownarrow^\in]^\triangle} \lceil \mathcal{E}}{\infty + \exp\left[\frac{\varepsilon - \mu}{\mathcal{T}}\right]} \qquad (6)$$

$$\mathcal{P} = \frac{\triangle \pi g}{\ni(\in\pi\{)^\ni} \int_{\updownarrow]\in}^{\infty} \frac{\left(\mathcal{E}^\in - \Updownarrow^\in]^\triangle\right)^{\ni/\in} \lceil \mathcal{E}}{\infty + \exp\left[\frac{\varepsilon-\mu}{\mathcal{T}}\right]} \qquad (7)$$

$$-\mu = BT + eA_i U^i \qquad (8)$$

g is g-factor, μ is chemical potential. Fermi-gas equation of state is already included in (3), (6), (7), and (8). The stress-energy tensor is expressed by these formulae via potentials of electromagnetic field and 4-velocity of the subsystem moving as a whole. 4-vector of flux n^i is the following integral of distribution function:

$$n^i = nU^i = \frac{2g}{c(2\pi\hbar)^3} \int p^i F(X)\delta\left(p_s p^s - m^2 c^2\right) \sqrt{-g}d\Omega_p \qquad (9)$$

Here

$$n = \frac{4\pi g}{(2\pi\hbar)^3} \int_{mc^2}^{\infty} \frac{E\sqrt{E^2 - m^2 c^4}dE}{1 + \exp\left[\frac{E-\mu}{T}\right]} \qquad (10)$$

n is density of charges in rest frame of reference.

Stress-energy tensor of all the charges is the sum over the sorts of the charges:

$$T_{P\;i}^{\;k} = \sum_a [(\mathcal{E}_a + \mathcal{P}_a)U_{a\,i} U_a^{\;k} - \mathcal{P}_a \delta_i^k] \qquad (11)$$

Index of summation $a = i, e$ stands for ions and electrons respectively. 4-vector of electric current density is equal

$$j^i = c\sum_a e_a n_a^{\;i} = c\sum_a e_a n_a U_a^{\;i} \qquad (12)$$

Total current I and charge Q per unit length of a filament can be found from (12) by invariant integration over the cross-section:

$$\begin{aligned}\frac{I}{c} &= 2\pi \int_{-\infty}^{\infty} dx^1 \sqrt{-g} \sum_a e_a n_a U_a^{\;3} \\ Q &= 2\pi \int_{-\infty}^{\infty} dx^1 \sqrt{-g} \sum_a e_a n_a U_a^{\;0}\end{aligned} \qquad (13)$$

Stress-energy tensor of electromagnetic field is

$$T_{F\;i}^{\;k} = \frac{1}{4\pi}\left(-F_{il}F^{kl} + \frac{1}{4}\delta_i^k F_{lm}F^{lm}\right)$$

Without azimuth current the only non-zero components of 4-potential of electromagnetic field are A_0 and A_3. In the electromagnetic field tensor

$$F_{ik} = A_{k;\,i} - A_{i;\,k} = \frac{\partial A_k}{\partial x^i} - \frac{\partial A_i}{\partial x^k} \qquad (14)$$

only $F_{10} = -F_{01} = \frac{\partial A_0}{\partial x^1}$ components of electric field and $F_{13} = -F_{31} = \frac{\partial A_3}{\partial x^1}$ components of magnetic field are not zeroes. It is convenient to express the stress-energy tensor via mixed component of electric field $F_0^{\;1} = g_{00}F^{01} + g_{03}F^{31}$ and contravariant component of magnetic field F^{31}:

$$T_{00} = \frac{\gamma_{11}}{8\pi}\left[\gamma_{33}h(F^{13})^2 + (F^1_{\;0})^2\right] \qquad (15)$$

$$T_0^{\;3} = \frac{\gamma_{11}}{4\pi}F_0^{\;1}F^{31} \qquad (16)$$

$$T^{11} = \frac{1}{8\pi}\left[\gamma_{33}(F^{13})^2 - \frac{(F^1_0)^2}{h}\right] \tag{17}$$

$$T^{22} = \frac{\gamma^{22}\gamma_{11}}{8\pi}\left[-\gamma_{33}(F^{13})^2 + \frac{(F^1_0)^2}{h}\right] \tag{18}$$

$$T^{33} = \frac{\gamma^{33}\gamma_{11}}{8\pi}\left[\gamma_{33}(F^{13})^2 + \frac{(F^1_0)^2}{h}\right] \tag{19}$$

Total stress-energy tensor of matter (to be used in Einstein equations) is the sum of corresponding expressions for particles and electromagnetic field:

$$T_i{}^k = T_P{}_i{}^k + T_F{}_i{}^k.$$

We shall write down complete set of Einstein equations

$$R_{ik} = \frac{8\pi G}{c^4}\left(T_{ik} - \frac{1}{2}g_{ik}T\right), \tag{20}$$

describing stationary current-conducting filament with arbitrary drift velocity, with no restrictions on the strength of gravitation field.

Components of Ricci tensor R_{ik} for a stationary gravitation field we find with the aid of problem after paragraphs 95 and 92 in [3]. Introducing the notations: $g_{00} = h = e^{2F_0}$,

$$\gamma_{\alpha\beta} = \begin{vmatrix} e^{2F_1} & 0 & 0 \\ 0 & e^{2F_2} & 0 \\ 0 & 0 & e^{2F_3} \end{vmatrix},$$

and specifying the coordinate x^1 by imposing the following connection between functions F_i [12]:

$$F_1 = F_0 + F_2 + F_3, \tag{21}$$

we come to the following expressions for non zero components of Ricci tensor:

$$R_{00} = e^{2(F_0-F_1)}F_0'' + \frac{1}{2}e^{4F_0-2F_1-2F_3}g_3'^2 \tag{22}$$

$$R_0{}^3 = -\frac{1}{2}e^{-2F_1}\left(e^{2(F_0-F_3)}g_3'\right)' \tag{23}$$

$$R^{11} = e^{-4F_1}\{[2(F_2'F_3' + F_3'F_0' + F_0'F_2') - F_1''] + \frac{1}{2}e^{2F_0-2F_3}g_3'^2\} \tag{24}$$

$$R^{22} = -e^{-2F_2-2F_1}F_2'' \tag{25}$$

$$R^{33} = -e^{-2F_3-2F_1}F_3'' + \frac{1}{2}e^{2F_0-2F_1-4F_3}g_3'^2 \tag{26}$$

Prime stands for differentiation over x^1.

It follows from Einstein equations (20) that divergence of stress-energy tensor

$$T_i{}^k = \sum_a (W_\dashv U_\dashv)_i U_\dashv^{\|} - \mathcal{P}_\dashv \delta_{\rangle}^{\|}) + \frac{\infty}{\Delta\pi}(-\mathcal{F}_{\rangle\ddagger}\mathcal{F}^{\|\ddagger} + \frac{\infty}{\Delta}\delta_{\rangle}^{\|}\mathcal{F}_{\ddagger\ddagger}\mathcal{F}^{\ddagger\ddagger})$$

is zero:
$$T_i{}^k{}_{;k} = 0$$

If one calculates directly $T_i{}^k{}_{;k}$ with $W_\dashv = \mathcal{E}_\dashv + \mathcal{P}_\dashv$; $\mathcal{E}_\dashv$ and $\mathcal{P}_\dashv$ from (6) and (7), he gets finally

$$T_i{}^k{}_{;k} = \frac{1}{4\pi}F_{im}\left(F^{mk}{}_{;k} + \frac{4\pi}{c}j^m\right)$$

So from Einstein equations it follows that

$$F_{im}\left(F^{mk}{}_{;k} + \frac{4\pi}{c}j^m\right) = 0. \tag{27}$$

The conclusion that Maxwell equations

$$F^{mk}{}_{;k} + \frac{4\pi}{c}j^m = 0 \tag{28}$$

are contained in Einstein equations can be made only if the determinant of matrix F_{im} is not zero: $det F_{im} \neq 0$. However in our case electric and magnetic vectors are mutually perpendicular. One (of the two) field invariant is zero. It is exactly that separate case, when Maxwell equations are not the direct consequence of Einstein equations ([13], paragraph 20). In our case $det F_{ik} = 0$, and hence there is a non trivial solution of (27). Generally speaking, expressions $F^{km}{}_{;m} + \frac{4\pi}{c}j^k$ are not zeroes. Out of four equations (27) only one is effective,

$$F_{10}\left(F^{0m}{}_{;m} + \frac{4\pi}{c}j^0\right) + F_{13}\left(F^{3m}{}_{;m} + \frac{4\pi}{c}j^3\right) = 0, \tag{29}$$

because $F_{2k} = 0$, $j^1 = 0$, $F^{1m}{}_{;m} = 0$, and rank of F_{ik} matrix is unity. In the case of relative electron-ion motion $F_{13} \neq 0$ in any frame of reference. So one can conclude from (29) that Einstein equations are fulfilled for arbitrary value of $F^{0m}{}_{;m} + \frac{4\pi}{c}j^0$, if

$$F^{3m}{}_{;m} + \frac{4\pi}{c}j^3 = -\frac{F_{10}}{F_{13}}\left(F^{0m}{}_{;m} + \frac{4\pi}{c}j^0\right)$$

Maxwell equations reflect mathematically the fact that electromagnetic field A_i is created by the charges of our system only. The field could have been created by some external charges as well. Therefore one of the two Maxwell equations $F^{ik}{}_{;k} = -\frac{4\pi}{c}j^i$, $i = 0, 3$, is independent. The second one is the consequence of the first one together with Einstein equations (20).

COMPLETE SET OF EQUATIONS

After substitution of (11), (15)-(19), and (22)-(26) into (20) one comes to the following equations of gravitation field:

$$e^{2F_1-2F_0}R_{00} = F_0'' + \tfrac{1}{2}e^{2(F_0-F_3)}g_3'^{\,2} =$$
$$= \tfrac{8\pi G}{c^4}e^{2F_1}\left\{\sum_a\left[\frac{(1-g_3\frac{W_a}{c})^2(\mathcal{E}_\dashv+\mathcal{P}_\dashv)}{(1-g_3\frac{W_a}{c})^2-(\frac{W_a}{c})^2 e^{(2F_3-2F_0)}} - \frac{\mathcal{E}_\dashv-\mathcal{P}_\dashv}{2}\right] + \frac{e^{2F_1}}{8\pi}[e^{2F_3}(F^{13})^2 + e^{-2F_0}(F^1{}_0)^2]\right\}$$
(30)

$$e^{2F_1}R_0{}^3 = -\tfrac{1}{2}\left[e^{2(F_0-F_3)}g_3'\right]' =$$
$$= \tfrac{8\pi G}{c^4}e^{2F_1}\left\{\sum_a\frac{\frac{W_a}{c}(1-g_3\frac{W_a}{c})(\mathcal{E}_\dashv+\mathcal{P}_\dashv)}{(1-g_3\frac{W_a}{c})^2-(\frac{W_a}{c})^2 e^{(2F_3-2F_0)}} + \frac{e^{2F_1}}{8\pi}F^1{}_0 F^{13}\right\}$$
(31)

$$e^{4F_1}R^{11} = [2(F_2'F_3' + F_3'F_0' + F_0'F_2') - F_1''] + \tfrac{1}{2}e^{2F_0-2F_3}g_3'^{\,2} =$$
$$= \tfrac{8\pi G}{c^4}e^{2F_1}\left\{\sum_a\frac{\mathcal{E}_\dashv-\mathcal{P}_\dashv}{2} + \frac{e^{2F_1}}{8\pi}\left[e^{2F_3}(F^{13})^2 - e^{-2F_0}(F^1{}_0)^2\right]\right\}$$
(32)

$$e^{2F_1+2F_2}R^{22} = -F_2'' = \tfrac{8\pi G}{c^4}e^{2F_1}\left\{\sum_a\frac{\mathcal{E}_\dashv-\mathcal{P}_\dashv}{2} - \frac{e^{2F_1}}{8\pi}\left[e^{2F_3}(F^{13})^2 - e^{-2F_0}(F^1{}_0)^2\right]\right\}$$
(33)

$$e^{2F_1+2F_3}R^{33} = -F_3'' + \tfrac{1}{2}e^{2(F_0-F_3)}g_3'^{\,2} =$$
$$= \tfrac{8\pi G}{c^4}e^{2F_1}\left\{\sum_a\left[\frac{(\frac{W_a}{c})^2 e^{(2F_3-2F_0)}(\mathcal{E}_\dashv+\mathcal{P}_\dashv)}{(1-g_3\frac{W_a}{c})^2-(\frac{W_a}{c})^2 e^{(2F_3-2F_0)}} + \frac{\mathcal{E}_\dashv-\mathcal{P}_\dashv}{2}\right] + \frac{e^{2F_1}}{8\pi}[e^{2F_3}(F^{13})^2 + e^{-2F_0}(F^1{}_0)^2]\right\}$$
(34)

We have 5 equations (30)-(34) for 5 functions $(F_0, F_1, F_2, F_3, g_3)$, describing gravitation field, and 2 functions (A_0 and A_3), describing electromagnetic field. Seven functions altogether. The system becomes completed with the aid of relation (21) and one Maxwell equation

$$e^{-2F_1}(e^{2F_1}F^{01})' = -4\pi\sum_a\frac{e_a n_a e^{-F_0}}{\sqrt{(1-g_3\frac{W_a}{c})^2 - (\frac{W_a}{c})^2 e^{(2F_3-2F_0)}}}$$
(35)

As we have already demonstrated, the second Maxwell equation is a consequence of the set of the above equations.

Two useful relations

$$F_0'' + F_3'' = \frac{16\pi G}{c^4}e^{2F_1}\sum_a\mathcal{P}_\dashv$$
(36)

$$F_2'F_3' + F_3'F_0' + F_0'F_2' + \tfrac{1}{4}e^{2F_0-2F_3}g_3'^2 =$$
$$= \tfrac{8\pi G}{c^4}e^{2F_1}\left\{\sum_a \mathcal{P}_{\dashv} + \tfrac{]\in\mathcal{F}\infty}{\forall\pi}\left[]\in\mathcal{F}_3(\mathcal{F}^{\infty 3})\in -]^{-\in\mathcal{F}_i}(\mathcal{F}_i^\infty)\in\right]\right\} \quad (37)$$

are the consequences of the set (30)-(34). There are no second derivatives in (37), so it is a first integral of the system.

In order to find out the connection of the coordinate x^1 with real distance from the axis, let us for a moment "switch off" electromagnetic field ($F_0^1 = F^{13} = 0$) and "stop" the charges ($W_a = 0$). Then we have from (33) and (34):

$$F_3'' = F_2'', \quad (F_0^1 = F^{13} = 0, \; W_a = 0).$$

If one puts the integration constant equal unity: $F_2' = F_3' + 1$, then the coordinates x^2 and x^3 will coincide with cylindrical coordinates φ and z, provided that the coordinate x^1 is connected with the radius by the relation $x^1 = \ln r$. This corresponds to Galilean metric. $x^1 \to \infty$ is the direction outside the channel, and $x^1 = -\infty$ corresponds to the axis of the filament.

Values of current and charge per unit length are finite, if the integrals (13) are convergent. In this case $\mathcal{E}_{\dashv}$, $\mathcal{P}_{\dashv}$, n_a, as functions of x^1, quickly tend to zero far from the axis:

$$\mathcal{E}_{\dashv} = {\prime}, \quad \mathcal{P}_{\dashv} = {\prime}, \quad \backslash_{\dashv} = {\prime}, \quad \S^\infty \to \infty. \quad (38)$$

Then one finds from Maxwell equations:

$$e^{2F_1}F^{01} = -2Q, \quad e^{2F_1}F^{31} = -\frac{2I}{c}, \quad x^1 \to \infty. \quad (39)$$

One can find total mass of a filament using Tolman's formula ([3], p. 425):

$$M = \frac{c^2}{4\pi G}\int dV\, e^{2F_1} R_0^{\,0}$$

Total mass is proportional to the length L of a filament. Mass per unit length is

$$\mathrm{M} = \frac{M}{L} = \frac{c^2}{2G}\left[F_0'(\infty) + \int_{-\infty}^{\infty} dx^1 e^{2(F_0-F_3)}g_3'^{\,2}\right] \quad (40)$$

For an infinitely long system, however, the metric is not Galilean at distances large compared with transverse dimensions. Functions F_i, satisfying equations (30)-(34), are not zeroes at $x^1 \to \infty$. It is because cylindrical symmetry with dependence on only one coordinate x^1 is valid for small distances, compared to the length L of the filament. For very large distances $r \gg L$ metric would be Galilean. However the equations (30)-(34) do not describe any real system at that long distances, because x^3 is not a cyclic variable there.

Let r_0 be characteristic radius of the region occupied by the charges. Physical situation is different depending on the relation between two parameters: $\ln(L/r_0)$

and $\mathbf{i}^{-1} = (GI^2/c^6)^{-1/2}$. The first one indicates the distance from the axis (in curved coordinate x^1) where the properties of the filament are no more depending on only one coordinate. The second parameter, $\mathbf{i}^{-1}$, is the size of the region, occupied by electromagnetic field, and giving main electromagnetic contribution to the curvature of space-time. From the point of view of General Relativity current I is small, if

$$\mathbf{i} \ll \ln^{-1}(L/r_0), \quad \text{or} \quad \ln(L/r_0) \ll (GI^2/c^6)^{-1/2}. \tag{41}$$

For intergalactic currents currently discussed in literature [6] parameter $\mathbf{i}$ is about 10^{-5} or less. In the limit of small current (41) curvature of space-time is practically negligible in the whole range of applicability of equations (30)-(34). In most cases Newtonian approximation is sufficient for taking gravitation into account. And only in case of relativistic drift velocity one still has to use Einstein equations to consider gravitation.

The opposite limit of strong current

$$\mathbf{i} \gg \ln^{-1}\left(\frac{L}{r_0}\right), \quad \text{or} \quad \ln\left(\frac{L}{r_0}\right) \gg \sqrt{\frac{c^6}{GI^2}} \tag{42}$$

is most interesting from the point of view of limiting parameters. As far as $L \gg r_0$, parameter $\mathbf{i}$ still can be arbitrary as compared to unity. In case (42) gravitation forces are strong enough to confine the matter (both particles and electromagnetic field) within much smaller distance from the axis than the length of the filament.

ENERGY BALANCE RELATION

In General Relativity one can get the balance relation from the first integral (37), just tending x^1 to infinity (outside the region, occupied by the charges). In the limit of small current (41) in the region $1 \ll x^1 \ll \ln(L/r_0)$ with account of (38), (39), and $F_0{}^1 = e^{2F_0}(F^{01} - g_3 F^{31})$, we have:

$$\frac{1}{4}e^{2F_0 - 2F_3}g_3'{}^2 + F_3'F_0' + F_2'(F_3' + F_0') = \frac{4GI^2}{c^6}\left[e^{2F_3} - e^{2F_0}\left(\frac{cQ}{I} - g_3\right)^2\right]. \tag{43}$$

Integrating the relation (36), one finds

$$F_0' + F_3' = \frac{8G}{c^4}\sum_a P_a \qquad P_a = 2\pi \int_{-\infty}^{\infty} dx^1 \, e^{2F_1} \mathcal{P}_{\dashv} \tag{44}$$

P_a is pressure of type "a" charges, integrated over the cross-section of the channel.

Two first terms in (43) are proportional to G^2. In order to neglect gravitation at all one should first cut common factor G in all the terms, and after that put $G = 0$, and substitute for F_i their Galilean values. This way we get the condition of equilibrium of a filament, valid for Fermi-gases of electrons and ions with arbitrary degree of degeneracy (and not only for Boltzmann ideal gases):

$$\sum_a P_a = \frac{I^2}{2c^2} - \frac{Q^2}{2}.$$

In the limit of small current (41) one has to take into account gravitation in general balance of energies when electromagnetic contribution is so small, and total number of particles is so large, that $\frac{I^2}{c^2} - Q^2$ and $G(\sum_a N_a m_a)^2$ are of the same order of magnitude. This situation can be realized in intergalactic currents either if both – current and charge – are not too large, or if magnetic compression is compensated by electrostatic repulsion of the space charge to a very high degree of accuracy. In the latter case the drift velocity can be relativistic, so that Newtonian approximation is not valid.

In the limit of small current (41) $GI^2/c^6 \ll (\ln\frac{L}{r_0})^{-2} \ll 1$ gravitation field is week: $GN_a m_a/c^2 \ll 1$. In a week gravitation field in the region $(I/c)^2 - Q^2 \sim G(m_a N_a)^2$ (where magnetic and gravitation attractions are of the same order) curvature of space-time, caused by electromagnetic field, is small compared to the curvature, caused by the charges:

$$\frac{G}{c^4}\left(\frac{I^2}{c^2} - Q^2\right) \sim \frac{(Gm_a N_a)^2}{c^4} \ll \frac{GN_a m_a}{c^2}$$

In expansion in terms of $1/c^2$ curvature of space-time by the charges is of the order of $1/c^2$, while curvature by electromagnetic field is of the order of $1/c^4$. One can substitute Galilean values (zeroes) for F_0, F_3, and g_3 into (43) and put $F_2' = 1$. We have:

$$\frac{g_3'^2}{4} + F_0'F_3' + \frac{8G}{c^4}\sum_a P_a = \frac{4G}{c^4}\left[\frac{I^2}{c^2} - Q^2\right] \quad (45)$$

It is sufficient to find both factors in the product $F_0'F_3'$, as well as g_3', with accuracy to the first order in $1/c^2$. With this degree of accuracy we find from (30), (31), and (34):

$$F_0' = -F_3' = \frac{2G}{c^4}\sum_a E_a \frac{1 + \frac{W_a^2}{c^2}}{1 - \frac{W_a^2}{c^2}}, \quad g_3' = \frac{8G}{c^4}\sum_a E_a \frac{\frac{W_a}{c}}{1 - \frac{W_a^2}{c^2}}, \quad (46)$$

$$E_a = 2\pi \int_{-\infty}^{\infty} dx^1\, e^{2F_1} \mathcal{E}_+, \quad (47)$$

E_a is the energy of type "a" subsystem per unit length of the channel. Substituting (46) and (47) into (45), we get:

$$\frac{G}{2}\left(\sum_a \frac{E_a}{c^2}\right)^2 + \frac{2GE_eE_i}{c^6}\frac{(W_e-W_i)^2}{\left(1-\frac{W_e^2}{c^2}\right)\left(1-\frac{W_i^2}{c^2}\right)} + \frac{I^2}{2c^2} = \frac{Q^2}{2} + \sum_a P_a \qquad (48)$$

Two first terms in (48) are proportional to G. They are both positive and represent the energy of gravitational compression. The first one is the ordinary Newtonian energy of compression per unit length. The second term in left hand side of (48) describes additional gravitational attraction of the subsystems, caused by their relative motion. There is no such effect in Newtonian approximation. It is negligible when $W/c \ll 1$, however it can dominate in case of ultra-relativistic drift velocity. (48) is applicable for arbitrary degree of degeneracy of Fermi-gases of electrons and/or ions.

Gravitational attraction of the subsystems, connected with their relative motion, in high current channels is additionally small because of $m_e \ll m_i$. Relative simplicity of Einstein equations in cylindrical geometry allowed us to achieve this result analytically. However this effect might be displayed much better in super-fluid stars, where the masses of normal and super-fluid components can be of the same order. There is no friction between normal and super-fluid subsystems, and their relative velocity of rotation is not small.

In case of non-relativistic Boltzmann gases ($P_a = N_a T_a, E_a = N_a m_a c^2$, N_a is the number of type "a" particles per unit length) formula (48) gives:

$$\frac{G}{2}\left(\sum_a N_a m_a\right)^2 + \frac{2GN_eN_im_em_i}{c^2}\frac{(W_e-W_i)^2}{\left(1-\frac{W_e^2}{c^2}\right)\left(1-\frac{W_i^2}{c^2}\right)} + \frac{I^2}{2c^2} = \frac{Q^2}{2} + \sum_a N_a T_a$$

(49)

In the limit of non-relativistic drift velocity (49) reduces to:

$$\frac{G}{2}\left(\sum_a N_a m_a\right)^2 + \frac{I^2}{2c^2} = \frac{Q^2}{2} + \sum_a N_a T_a$$

This formula was derived earlier [2] in Newtonian approximation under the assumption that electrons and ions are Boltzmann ideal gases.

Electromagnetic forces exceed gravitational forces by a factor of 10^{36}. For this reason magnetic compression of the current channel can compress the charges to ultrarelativistic state, while gravitation forces still remain small. In equilibrium $\mathcal{E}_\dashv$ is of the order of the energy of collective interaction per one charge $e^2 N \beta^2$ [9]:

$$\mathcal{E}_\dashv \sim]^\epsilon \mathcal{N} \beta^\epsilon$$

Here $-e$ is electron charge, N is number of particles (electrons or ions) per unit length, $\beta = W/c, W$ is drift velocity, c is the speed of light. Subsystem of type "a" charges is ultrarelativistic if the energy $\mathcal{E}_\dashv$ is much larger than the rest energy $m_a c^2$:

$$e^2 N \beta^2 \gg m_a c^2.$$

So N must be relatively large:
$$N \gg \frac{m_a c^2}{e^2 \beta^2} \tag{50}$$

Curvature of space-time by the matter (gravitation) is small if
$$\frac{G}{c^4} N \mathcal{E}_\dashv = \frac{\mathcal{G}}{\mathrm{J}^\triangle} \mathrm{l}^\in \mathcal{N}^\in \beta^\in \ll \infty,$$

so N must be relatively small:
$$N \ll \frac{c^2}{\sqrt{G} e \beta} \tag{51}$$

If number of particles per unit length N satisfies both conditions (50) and (51):
$$\frac{c^2}{\sqrt{G} e \beta} \gg N \gg \frac{m_a c^2}{e^2 \beta^2},$$

then the equation of state of subsystem "a" is ultrarelativistic even in a weak gravitation field. For ions the conditions (50) and (51) are compatible if
$$\beta \gg \frac{\sqrt{G} m_i}{e} \sim 10^{-18}.$$

It is clear now that magnetic self-contraction is able to compress the matter to ultrarelativistic state even in case of non-relativistic drift velocity $\beta \ll 1$. Considering below the equilibrium in the limit of strong current we assume
$$\frac{\sqrt{G} m_i}{e} \ll \beta \ll 1.$$

Assumption $\beta \ll 1$ allows to simplify the equations essentially.

The set of Einstein equations (30)-(34) is written down without any limitations on β and g_{ik}. Gravitation field can be arbitrary strong. For $\beta \ll 1$ only diagonal components of metric tensor g_{ik} are zeros, and gravitation field is static. For non-relativistic drift velocity in equilibrium electric energy of the space charge is much smaller than magnetic energy of the current:
$$Q^2 < \frac{I^2}{c^2} \beta^2 \ll \frac{I^2}{c^2}, \quad \beta \ll 1.$$

It is a consequence of a confinement condition $N_i > N_e(1-\beta^2)$ [9] (without account of gravitation, however). We consider equilibrium in the limit of strong current for an electrically neutral filament, $Q = 0$. The system of Einstein equations (30)-(34) reduces to:

$$F_0'' = \frac{8\pi G e^{2F_1}}{c^4}\left[\frac{1}{2}\sum_a (\mathcal{E}_\dashv + \ni\mathcal{P}_\dashv) + \frac{\infty}{\sqrt\pi}\right]^{\in\mathcal{F}_\infty \,+\, \in\mathcal{F}_\ni(\mathcal{F}^{\infty\ni})\in}$$

$$2(F_2'F_3' + F_3'F_0' + F_0'F_2') - F_1'' = \frac{8\pi G e^{2F_1}}{c^4}\left[\frac{1}{2}\sum_a (\mathcal{E}_\dashv - \mathcal{P}_\dashv) + \frac{\infty}{\sqrt\pi}\right]^{\in\mathcal{F}_\infty \,+\, \in\mathcal{F}_\ni(\mathcal{F}^{\infty\ni})\in}$$

$$-F_2'' = \frac{8\pi G e^{2F_1}}{c^4}\left[\frac{1}{2}\sum_a (\mathcal{E}_\dashv - \mathcal{P}_\dashv) - \frac{\infty}{\sqrt\pi}\right]^{\in\mathcal{F}_\infty \,+\, \in\mathcal{F}_\ni(\mathcal{F}^{\infty\ni})\in} \quad (52)$$

$$-F_3'' = \frac{8\pi G e^{2F_1}}{c^4}\left[\frac{1}{2}\sum_a (\mathcal{E}_\dashv - \mathcal{P}_\dashv) + \frac{\infty}{\sqrt\pi}\right]^{\in\mathcal{F}_\infty \,+\, \in\mathcal{F}_\ni(\mathcal{F}^{\infty\ni})\in}$$

Maxwell equation

$$e^{-2F_1}[e^{2F_1}F^{31}]' = -4\pi e^{-F_0}\sum_a e_a n_a \frac{W_a}{c}$$

in case of electrically neutral filament is a consequence of Einstein equations (52).

In General Relativity one can exclude gravitation field in any arbitrary chosen point. Let space-time be flat at the axis of the filament, $x^1 \to -\infty$. Corresponding boundary conditions are

$$F_0 = 0, \quad F_0' = 0, \quad F_2' = 1, \quad F_3 = 0, \quad F_3' = 0, \quad x^1 = -\infty. \quad (53)$$

In case (42) analysis of physical nature of equilibrium is still based on the first integral (37). For electrically neutral filament

$$F_2'F_3' + F_3'F_0' + F_0'F_2' = \frac{8\pi G e^{2F_1}}{c^4}\left[\sum_a \mathcal{P}_\dashv + \frac{\infty}{\sqrt\pi}\right]^{\in\mathcal{F}_\infty \,+\, \in\mathcal{F}_\ni(\mathcal{F}^{\infty\ni})\in} \quad (54)$$

Let $x^1 \to \infty$. In the case of strong current pressure and magnetic energy tend to zero within the range of applicability of equations (52). Right-hand side in (54) tends to zero, and we have

$$F_2'F_3' + F_3'F_0' + F_0'F_2' = 0, \quad x^1 \to \infty. \quad (55)$$

Using equations (52) and boundary conditions (53), we express derivatives F_0', F_2', and F_3' via dimensionless pressure $\mathbf{P} = \frac{G}{c^4}\sum_a P_a$, dimensionless energy of particles per unit length $\mathbf{E} = \frac{G}{c^4}\sum_a E_a$, and dimensionless magnetic energy $\mathbf{M}$ also per unit length:

$$F_0' = 2(\mathbf{E} + 3\mathbf{P} + 2\mathbf{M}), \quad F_2' = 1 - 2(\mathbf{E} - \mathbf{P} - 2\mathbf{M}), \quad F_3' = -2(\mathbf{E} - \mathbf{P} + 2\mathbf{M}), \quad (56)$$

$$M = \frac{G}{4c^4} \int_{-\infty}^{\infty} dx^1 e^{4F_1 + 2F_3}(F^{13})^2,$$

P_a and E_a (pressure and energy of type "a" particles, integrated over cross-section) are defined in (44) and (47), respectively. In accordance with relation (36)

$$F_3' + F_0' = 8P.$$

In equilibrium parameters $\mathbf{E}, \mathbf{M}$, and $\mathbf{P}$ are not independent. We find the following condition of equilibrium from (55):

$$2P[1 - 2(\mathbf{E} - \mathbf{P} - 2\mathbf{M})] - (\mathbf{E} - \mathbf{P} + 2\mathbf{M})(\mathbf{E} + 3\mathbf{P} + 2\mathbf{M}) = 0, \quad \ln^{-2}(L/r_0) \ll GI^2/c^6 \tag{57}$$

Formula (57) represents the energy balance relation in the region of strong current – an analog of Bennet relation.

If curvature of space-time, created by the charges and magnetic field, is small, condition of equilibrium (57) reduces to

$$2P = (\mathbf{E} + 2\mathbf{M})^2, \quad \mathbf{E} \ll 1, \quad \mathbf{M} \ll 1. \tag{58}$$

For $\mathbf{E} \ll \mathbf{M} \ll 1$ we can express magnetic energy $\mathbf{M}$ via current I. In this case magnetic field is spread at much longer distances from the axis than the charges. Consider the last equation (52) in the region occupied by magnetic field, where there are no more charges. Taking into account (38), we get the following equation for F_3:

$$F_3{}'' = -4i^2 e^{2F_3}, \quad i^2 = \frac{GI^2}{c^6}$$

Condition $\mathbf{E} \ll \mathbf{M}$ allows to neglect the contribution of particles, and we have

$$F_3 = -\ln\left\{\cosh\left[2i\left(x^1 - x_0\right)\right]\right\}, \quad \mathbf{E} \ll \mathbf{M} \ll 1 \tag{59}$$

At $x^1 \to \infty$ we find from (59) $F_3' = -2i$. At the same time according to (56) $F_3' = -4\mathbf{M}$. We see that in the limit of strong current and small gravitation field magnetic energy per unit length is a linear function of current:

$$\mathbf{M} = \frac{i}{2} = \frac{\sqrt{G}I}{2c^3}, \quad \ln^{-1}\left(\frac{L}{r_0}\right) \ll i \ll 1, \quad \mathbf{E} \ll \mathbf{M}. \tag{60}$$

Condition of equilibrium (58) takes the form of a regular Bennet relation [11], if the energy of magnetic field $\mathbf{M}$ exceeds the energy of the particles $\mathbf{E}$:

$$\sum_a P_a = \frac{I^2}{2c^2}, \quad \mathbf{E} \ll \mathbf{M} \ll 1. \tag{61}$$

For a gravitating electrically neutral filament without current ($\mathbf{M} = 0$) condition (57) reduces to

$$2\mathbf{P}[1 - 2(\mathbf{E} - \mathbf{P})] - (\mathbf{E} - \mathbf{P})(\mathbf{E} + 3\mathbf{P}) = 0, \qquad \mathbf{M} \ll \mathbf{E}.$$

In the limit $\mathbf{E} \ll 1$ it gives the balance of pressure, integrated over the cross-section, and energy of gravitational compression per unit length:

$$\sum_a P_a = \frac{G}{2c^4}\left(\sum_a E_a\right)^2, \qquad \mathbf{E} \ll 1.$$

For ultrarelativistic pressure $\mathbf{P} = \mathbf{E}/3$ we would have $\mathbf{E}(\mathbf{E} - 0.3) = 0$. $\mathbf{E_m} = 0.3$ is the boundary of gravitational collapse for cylindrical geometry.

We derived condition of equilibrium (57) without any assumptions on equation of state. It is applicable for arbitrary degree of degeneracy of electrons and ions. For Boltzmann gases $P_a = N_a T_a$, T_a is temperature of type "a" charges.

Energy balance relation (57) is a quadratic equation with respect to $\mathbf{M}$. Its positive solution is

$$\mathbf{M} = \frac{1}{2}\left[\sqrt{2\mathbf{P}(1 + 4\mathbf{P} - 4\mathbf{E})} - \mathbf{E} + \mathbf{P}\right] \qquad (62)$$

In equilibrium magnetic energy $\mathbf{M}$ is a growing function of pressure $\mathbf{P}$ for fixed energy $\mathbf{E}$. In order $\mathbf{M}$ to reach its maximum value pressure $\mathbf{P}$ must be as big as possible. Maximum possible pressure is $\mathcal{P} = \mathcal{E}/\mathsf{э}$ (after integration over cross-section $\mathbf{P} = \mathbf{E}/3$), it corresponds to the ultrarelativistic limit.

Gravitational interaction of electrically neutral particles does not compress the matter to ultrarelativistic state. For example, Oppenheimer and Volkoff [8] showed that for neutron stars boundary of gravitational collapse corresponds to relativistic, but not ultrarelativistic state of matter.

In case of current-conducting filaments magnetic self-attraction of charges can be much stronger than their gravitational contraction. It is because electromagnetic interaction is 36 orders of magnitude stronger (for protons), than gravitational interaction. If drift velocity is not extremely small,

$$\beta \gg \frac{\sqrt{G}m_i}{e} \sim 10^{-18},$$

and condition (50) is fulfilled, proper magnetic field of the current can compress the charges to ultrarelativistic state, no matter how weak or strong the gravitation field is. In order to determine limiting parameters we are interested in maximum possible magnetic energy. For this reason we consider ultrarelativistic equation of state.

For $\mathbf{P} = \mathbf{E}/3$ expression (62) reduces to

$$\mathbf{M} = \sqrt{\frac{\mathbf{E}}{6} - \frac{4\mathbf{E}^2}{9}} - \frac{\mathbf{E}}{3} \qquad (63)$$

Magnetic energy (63) reaches its maximum value $\mathbf{M_{max}} = 0.07725$ at $\mathbf{E_0} = 0.103647$. Magnetic energy per unit length, stored in an equilibrium current-conducting filament, can not exceed $\frac{c^4}{G}\mathbf{M_{max}} = 0.927 \cdot 10^{48}$ erg/cm. For $\mathbf{E} > \mathbf{E_0}$ magnetic energy decreases with rising $\mathbf{E}$, and finally $\mathbf{M}$ becomes zero at $\mathbf{E_m} = 0.3$. Equilibrium of cylindrically symmetric distribution of matter is impossible beyond this point. If $\mathbf{E} > \mathbf{E_m}$ even ultrarelativistic pressure is not enough to balance the energy of gravitational compression. $\mathbf{E} = \mathbf{E_m}$ is the boundary of gravitational collapse for cylindrical geometry.

Energy balance condition (57) together with ultrarelativistic equation of state $\mathbf{P} = \mathbf{E}/3$ determines the boundary of equilibrium domain. It shows a fundamental connection between electromagnetic and gravitational collapse and displays their common physical nature. If (even in ultrarelativistic case) pressure is not enough to compensate total energy of self-compression (whatever the relation between $\mathbf{E}$ and $\mathbf{M}$ is), equilibrium can not exist, and a filament will undergo infinite contraction. If $\mathbf{E} \gg \mathbf{M}$ it is a gravitational collapse, and in the opposite limit $\mathbf{M} \gg \mathbf{E}$ it is an electromagnetic one.

LIMITING CURRENT

Magnetic energy $\mathbf{M}$ is a regular function of current I. As far as magnetic energy is limited by $\mathbf{M_{max}} = 0.07725$, we make a conclusion, that the current also can not exceed some limiting value I_{max}. We derived the value $\mathbf{M_{max}}$ from most general properties of Einstein equations in cylindrical one-dimensional case. Actually we used the first integral (37) of the system (30)-(34) without specification of particular charges and their equations of state. Electrical neutrality of the filament as a whole was the only important assumption. However finding I_{max} requires integration of the complete set of Einstein equations for particular composition of the charges. Relation (60) between $\mathbf{M}$ and I is valid only for $I \ll I_{max}$ and $\mathbf{E} \ll \mathbf{E_0}$.

$\mathbf{M_{max}}$ corresponds to ultrarelativistic state of matter. At present time insufficient knowledge of strong nuclear interactions does not allow us to make any definite conclusions about the state of matter compressed to extremely high densities, exceeding nuclear ones. If the energy of a proton and/or electron in a filament is much higher than the rest mass of a proton, neutrons and other particles are likely to appear as a result of weak and strong interactions. These additional particles can also contribute to the sum $\sum_a \mathcal{E}_\dashv$ in Einstein equations. It seems likely that increasing portion of the energy of additional particles would result in decreasing magnetic contribution to the equilibrium balance. In the sum $\sum_a \mathcal{E}_\dashv$ we take into account only two types of ultrarelativistic charges of opposite sign. Neglecting the contribution of neutrons and other particles, we can only estimate the upper limit for the current. Real limiting current of a particular filament can be still smaller.

In the ultrarelativistic limit the difference in masses of electrons and ions does not play any role. The only difference is the sign of a charge: $e_i = -e_e = e$. For electrically neutral filament as a whole in the center-of-mass frame the velocities of the two counterstreaming subsystems also differ only in sign: $W_i = -W_e = W/2$. Drift velocity is relative velocity of electron and ion subsystems $W = W_i - W_e$. The system is symmetric with respect to mutual exchange of electrons and protons. If temperatures $T_a \ll E_F$, E_F is Fermi energy, chemical potentials (8) for both subsystems are identical:

$$\mu = -\frac{e}{2c} W A_3 e^{-F_0} \tag{64}$$

With the aid of (64) we express the magnetic field via chemical potential:

$$F^{31} = -e^{-2F_1 - 2F_3} \frac{\partial A_3}{\partial x^1} = \frac{2c}{eW} e^{-2F_1 - 2F_3} (e^{F_0} \mu)'$$

Magnetic term in Einstein equations has the form:

$$\frac{1}{8\pi} e^{4F_1} e^{2F_3} (F^{13})^2 = \frac{1}{2\pi} \left(\frac{c}{eW}\right)^2 e^{-2F_3} \left[(e^{F_0}\mu)'\right]^2 \tag{65}$$

From (6) and (10) we find the sums of the energies and densities:

$$\sum_a \mathcal{E}_\dashv = \begin{cases} \frac{\mu^4}{2\pi^2(c\hbar)^3}, & \mu \geq 0 \\ 0, & \mu < 0 \end{cases} \qquad \sum_\dashv \backslash_\dashv = \begin{cases} \frac{\mu^3}{3\pi^2(c\hbar)^3}, & \mu \geq 0 \\ 0, & \mu < 0 \end{cases} \tag{66}$$

Contributions of particles (66) and magnetic field (65) to Einstein equations are now expressed in terms of the same chemical potential μ. This reflects the fact that we consider a filament, consisting of equal number of counterstreaming electrons and ions only. Particles occupy all energy levels in the interval $0 < \mu \leq E_F$, E_F is Fermi energy. Chemical potential has maximum value $\mu = E_F$ at the axis, $x^1 \to -\infty$, and we have boundary conditions:

$$\mu(-\infty) = E_F, \quad \mu'(-\infty) = 0.$$

For numerical integration we take the following set of four independent equations:

$$F_0'' = \frac{16\pi G}{3c^4} e^{2F_1} \sum_a \mathcal{E}_\dashv + \mathcal{F}_\epsilon' \mathcal{F}_\ni' + \mathcal{F}_\ni' \mathcal{F}_\iota' + \mathcal{F}_\iota' \mathcal{F}_\epsilon'$$

$$-F_2'' = \frac{16\pi G}{3c^4} e^{2F_1} \sum_a \mathcal{E}_\dashv - (\mathcal{F}_\epsilon' \mathcal{F}_\ni' + \mathcal{F}_\ni' \mathcal{F}_\iota' + \mathcal{F}_\iota' \mathcal{F}_\epsilon') \tag{67}$$

$$-F_3'' = F_2' F_3' + F_3' F_0' + F_0' F_2'$$

$$e^{-2F_1} [e^{-2F_3} (e^{F_0} \mu)']' = -\pi \left(\frac{eW}{c}\right)^2 e^{-F_0} \sum_a n_a$$

Equations (67) do not contain the coordinate x^1. As a result the solution depends on x^1 in combination $x^1 - x_0$, x_0 is a constant of integration. We can add to (53) one more boundary condition $F_2 = x^1 - x_0$, $x^1 \to -\infty$. Specifying x_0, we choose the scale of the radius. To simplify the equations we assume

$$x_0 = \frac{1}{2} \ln \left[\frac{E_F^2}{3\pi (c\hbar)^3} \left(\frac{eW}{c} \right)^2 \right].$$

Introducing dimensionless functions we reduce the set of equations to a form, convenient for numerical integration. It depends on only one dimensionless parameter ζ:

$$\zeta = 8G \left(\frac{E_F}{ecW} \right)^2 = \frac{8GNE_F}{c^4} \frac{NE_F}{\left(\frac{I}{c} \right)^2}.$$

$\zeta \to 0$ when $NE_F \to 0$, because in this limit $\frac{NE_F}{\left(\frac{I}{c}\right)^2}$ is of the order of unity. Near the collapse boundary $\zeta \to \infty$, because in this limit the current tends to zero. ζ changes from zero to infinity and overlaps the whole domain of possible equilibrium states.

Dependence of dimensionless current **i** on dimensionless mass $\mathbf{m} = \frac{G}{2c^2}M$ per unit length, found numerically, has maximum. We have found numerically that **current in an equilibrium filament can not exceed $\mathbf{I_{max}} = 0.94 \cdot 10^{25}$ A.**

In the vicinity of the collapse we can solve the equations (67) analytically. For $\zeta \to \infty$ the system (67) reduces to

$$F_0'' = -3F_3''', \quad F_2'' = F_3''', \quad F_3'' + F_2'F_3' + F_3'F_0' + F_0'F_2' = 0$$

For F_3' we get the equation:

$$F_3'' - 2F_3' - 5F_3'^2 = 0.$$

Solution, vanishing at $x^1 = -\infty$, has the form

$$F_3' = -\frac{1}{\frac{5}{2} + e^{-2(x^1 - x_0)}}.$$

Mass per unit length, located to the left from x^1, is equal to

$$\mathbf{M}(\mathbf{x}^1) = \frac{c^2}{2G} \frac{3}{\frac{5}{2} + e^{-2(x^1 - x_0)}} = \frac{M_{max}}{1 + \frac{2}{5}e^{-2(x^1 - x_0)}}. \tag{68}$$

$M_{max} = 3c^2/5G = 0.81 \cdot 10^{28}$ g/cm is total mass per unit length on the boundary of gravitational collapse.

In the close vicinity of the origin ($\zeta \to 0$), as well as very near the collapse boundary ($\zeta \to \infty$), the current tends to zero. For zero current ultrarelativistic equation of state is not applicable. Without magnetic contraction gravitational forces alone are too weak to compress the matter to ultrarelativistic state even at the collapse boundary.

CONCLUSION

Our analysis clarifies common physical nature of equilibrium of current-conducting channels in a very broad range from laboratory pinches to filamentary structures in the scale of Universe. Energy balance condition, derived analytically from Einstein equations, clarifies fundamental connection between gravitational collapse and electromagnetic one. Boundaries of equilibrium domain determine limiting parameters of a filament. This presentation demonstrates a possibility to achieve analytical results from Einstein equations with account of electromagnetism. Understanding physical properties of equilibrium filaments provides a reliable basis for treating the problems of stability and evolution of complicated cellular structure of Plasma Universe.

ACKNOWLEDGEMENTS

I am grateful to Col. William P. Schneider, President of Technical Research Institute, for his support of Russian fundamental science during the difficult transition period in the life of my country. I am thankful to Prof. Anthony L. Peratt for discussion.

This work is supported by Russian Foundation for Basic Research (RFBR). Grant # 98-02-16268.

REFERENCES

1. A.L. Peratt, Astrophysics and Space Science **227**, 97 (1995).
2. B.E. Meierovich and A.L. Peratt, IEEE Transaction on Plasma Science **20**, 891 (1992).
3. L.D. Landau and E.M. Lifshitz, Field Theory ("Nauka", Moscow, 1973).
4. H. Alfven, Proc. Royal Swedish-Academy of Sciences (Kungliga Svenska Vetenskapakademiens Handlingar **18**, 139, 1939)
5. B.E. Meierovich, S.T. Sukhorukov, Sov. Phys. JETP **41**, 895 (1975).
6. A.L. Peratt, Physics of the Plasma Universe (Springer-Verlag New York, Inc).
7. L.D.Landau, Phys. Zs. Sowjet. **1**, 285 (1932).
8. J.R. Oppenheimer, G.M. Volkoff, Phys. Rev. **55**, 374 (1939).
9. B.E. Meierovich, Phys. Repts. **92**, 83 (1982).
10. J. Ehlers, Kinetic theory of gases in general relativity. In: Relativity and gravitation. (Gordon and Breach science publishers, 1971).
11. W.H. Bennet, Phys. Rev. **45**, 890 (1934).
12. K.A. Bronnikov, J.Phys.A: Math.Gen., **12**, 201 (1979).
13. C.W. Misner, K.S. Thorne, J.A. Wheeler, Gravitation. (W.H.Freeman and Co., San Francisco, 1973)

Reflections and Spinors on Manifolds[1]

Andrzej Trautman

Instytut Fizyki Teoretycznej, Uniwersytet Warszawski
Hoża 69, 00-681 Warszawa, Poland
e-mail: amt@fuw.edu.pl

Abstract. This paper reviews some recent work on (s)pin structures and the Dirac operator on hypersurfaces (in particular, on spheres), on real projective spaces and quadrics. Two approaches to spinor fields on manifolds are compared. The action of reflections on spinors is discussed, also for two-component (chiral) spinors.

INTRODUCTION

This paper contains a brief review of the work, done mostly in collaboration with Ludwik Dąbrowski [8], Michel Cahen, Simone Gutt [5] and Thomas Friedrich [11], on pin structures and the Dirac operator on higher-dimensional Riemannian manifolds; see also [18] and the references given there. In physics, there is now interest in higher dimensions motivated by research on unified theories, on supersymmetries, strings and their generalizations. There is also an intrinsic motivation: the Dirac operator is a fundamental object of an importance comparable to that of the Laplace operator and of the Maxwell, Yang-Mills and Einstein equations.

Reflections in a quadratic space generate the orthogonal group of automorphisms of that space; according to the Cartan-Dieudonné theorem, every orthogonal transformation in an m-dimensional quadratic space can be written as the product of a sequence of no more than m reflections. Reflections are of considerable interest in physics: invariance of electromagnetism under space reflections leads to selection rules; their violation is a striking feature of weak interactions. The PCT theorem describes a fundamental property of relativistic quantum field theories.

There are several "spinorial" extensions of orthogonal groups; each of them can be used to define a "(s)pin structure" that is required to describe spinor fields on a curved manifold. For these structures to exist, the manifold should satisfy certain topological conditions; see [10] and the literature listed there. In this paper, I recall two definitions of spinor fields on manifolds and give several simple examples of spin

---

[1] Research supported in part by the Polish Committee for Scientific Research (KBN) under grant no. 2 P03B 017 12 and by the Foundation for Polish-German Cooperation with funds provided by the Federal Republic of Germany.

and pin structures. The next section summarizes the definitions and terminology of Clifford algebras and their representations.

CLIFFORD ALGEBRAS AND SPINORS

There are important differences—and similarities—between spinors associated with vector spaces of even and odd dimensions.

Let h be a quadratic form on a real vector space V of dimension m. The pair (V, h) is said to be a quadratic space. The Clifford algebra $\mathsf{Cliff}(h)$ associated with (V, h) is generated by elements of V subject to relations of the form $u^2 = h(u)$; see [1,7,10] and Ch. IX of [3]. Let α be the involutive automorphism of $\mathsf{Cliff}(h)$ such that $\alpha(1) = 1$ and $\alpha(v) = -v$ for every $v \in V$. This (main) automorphism defines a $\mathbb{Z}_2$-grading of the Clifford *algebra*, $\mathsf{Cliff}(h) = \mathsf{Cliff}^{\mathrm{even}}(h) \oplus \mathsf{Cliff}^{\mathrm{odd}}(h)$ and V is a vector subspace of the odd part. Let $(e_i), i = 1, \ldots, m$ be an orthonormal frame in V. As a *vector* space, the algebra $\mathsf{Cliff}(h)$ is $\mathbb{Z}$-graded, $\mathsf{Cliff}(h) = \oplus_{p=0}^{m} \mathsf{Cliff}^p(h)$, where $\mathsf{Cliff}^p(h)$ is the vector space spanned by all elements of the form $e_{i_1} \ldots e_{i_p}$ such that $1 \leqslant i_1 < \cdots < i_p \leqslant m$. In particular, $\mathsf{Cliff}^m(h)$ is spanned by the *volume element* $\eta = e_1 \ldots e_m$; its square is either 1 or -1, depending on the signature of h. If $u \in V$ is not null, $u^2 \neq 0$, then the linear map $V \to V$, $v \mapsto -uvu^{-1}$, is a reflection in the hyperplane orthogonal to u. Since $\eta v = (-1)^{m+1} v \eta$ for every $v \in V$, if m is even, then one can write $-uvu^{-1} = (u\eta) v (u\eta)^{-1}$. One says that $u \in V$ is a unit vector if $h(u) = 1$ or -1. The group $\mathsf{Pin}(h)$ is defined as the set of products $u_1 u_2 \ldots u_r$ of all sequences of unit vectors, with a composition induced by Clifford multiplication and $\mathsf{Spin}(h) = \mathsf{Pin}(h) \cap \mathsf{Cliff}^{\mathrm{even}}(h)$. The *adjoint* representation Ad of $\mathsf{Pin}(h)$ in V is defined by $\mathrm{Ad}(a)v = ava^{-1}$, where $a \in \mathsf{Pin}(h)$ and $v \in V$. For m even, the map Ad is a homomorphism onto $\mathsf{O}(h)$ with kernel $\mathbb{Z}_2 = \{1, -1\}$; for m odd, Ad is a homomorphism onto $\mathsf{SO}(h)$. In both cases, to obtain a double cover of $\mathsf{O}(h)$ that coincides with Ad when restricted to $\mathsf{Spin}(h)$, one can use the *twisted adjoint* representation $\widetilde{\mathrm{Ad}}$ of $\mathsf{Pin}(h)$, defined by $\widetilde{\mathrm{Ad}}(a)v = \alpha(a)va^{-1}$. If h is of signature (k, l), $k + l = m$, then one writes $\mathsf{Cliff}_{k,l}$, $\mathsf{Pin}_{k,l}$, and $\mathsf{Spin}_{k,l}$ instead of $\mathsf{Cliff}(h)$, $\mathsf{Pin}(h)$ and $\mathsf{Spin}(h)$, respectively.

For $m = 2n$, the algebra $\mathsf{Cliff}(h)$ is central simple and has one, up to equivalence, irreducible and faithful representation γ in a complex, 2^n-dimensional space S of *Dirac* spinors. Given such a representation, one identifies $\mathsf{Cliff}(h)$ with its image in End S. Upon restriction to $\mathsf{Cliff}^{\mathrm{even}}(h)$, this representation decomposes into the direct sum of two irreducible and complex-inequivalent representations in spaces of Weyl (chiral, reduced or half) spinors so that $S = S_+ \oplus S_-$. The Dirac operator changes the chirality of spinors. Introducing the $2^n \times 2^n$ Dirac matrices γ^i and writing $\gamma_\pm^i = \gamma^i \mid S_\pm$, one obtains the well-known decomposition of the Dirac operator, $\gamma^i \partial_i = \begin{pmatrix} 0 & \gamma_-^i \partial_i \\ \gamma_+^i \partial_i & 0 \end{pmatrix}$.

For $m = 2n-1$, the algebra $\mathsf{Cliff}^{\mathrm{even}}(h)$ is central simple and has one, up to equivalence, representation in a complex, 2^{n-1}-dimensional space of *Pauli* spinors. The full algebra has two complex-inequivalent, in general not faithful, representations in spaces of Pauli spinors. The direct sum of these representations is a decomposable, but faithful, representation of $\mathsf{Cliff}(h)$ in the 2^n-dimensional space of *Cartan* spinors. Similarly as in this case of an even-dimensional space, the Dirac operator interchanges here the spinors belonging to the two Pauli representations. Namely, let σ^i, where $i = 1, \ldots, 2n-1$, be the $2^{n-1} \times 2^{n-1}$ Pauli matrices. The (modified) Dirac operator, acting on Cartan spinor fields, can be written as $\begin{pmatrix} 0 & \sigma^i \partial_i \\ \sigma^i \partial_i & 0 \end{pmatrix}$. The Cartan representation is essential when one considers the Dirac operator on non-orientable, odd-dimensional manifolds [4,17].

SPINORS ON MANIFOLDS

There are (at least) two approaches to spinors on manifolds; both of them can be traced to early work by mathematicians and physicists; see [14] and the references to the period 1928-1931 given there.

The classical approach

The first approach to be summarized here, initiated by Wigner, Weyl and Fock, consists in referring spinors to tetrads ("Vierbeine"); its modern formulation uses the notion of a (s)pin structure involving a "prolongation" of the bundle P of orthonormal frames to the principal (s)pin bundle Q. More precisely, given a Riemannian manifold M with a metric tensor of signature (k, l), a $\mathsf{Pin}_{k,l}$-structure on M is given by the maps

$$\begin{array}{ccc} \mathsf{Pin}_{k,l} & \longrightarrow & Q \\ \widetilde{\mathrm{Ad}} \downarrow & & \downarrow \chi \\ \mathsf{O}_{k,l} & \longrightarrow & P \xrightarrow{\pi} M, \end{array} \qquad (1)$$

such that $\chi(qa) = \chi(q)\widetilde{\mathrm{Ad}}(a)$, $(q, a) \mapsto qa$ denotes the action map of $\mathsf{Pin}_{k,l}$ in the principal spin bundle Q, etc. If $\mathsf{Pin}_{k,l}$ in (1) is replaced by $\mathsf{Pin}_{l,k}$, then one obtains the definition of a $\mathsf{Pin}_{l,k}$-structure. They are both referred to as pin structures; if the manifold is orientable and has a pin structure, then it has a spin structure. The diagram describing a spin structure is shortened to

$$\mathsf{Spin}_{k,l} \to Q \to P \to M. \qquad (2)$$

The spinor connection is obtained as the lift of the Levi-Civita connection from P to Q. This approach, standard in mathematics [10], is sometimes criticized by physicists who say that they have no use for *principal bundles* and are willing to consider only spinor fields.

One can present this approach in a language familiar to physicists, by referring everything to local sections of the bundle $Q \to M$ and using the terminology of *gauge fields.* For simplicity, consider an even-dimensional manifold, $k+l = 2n$, put $G = \mathsf{Pin}_{k,l}$ and let a representation of $\mathsf{Cliff}_{k,l}$ in S be given by the Dirac matrices $\gamma_i \in \mathsf{End}\, S$. A spinor field is now a map $\psi : M \to S$; given a function $U : M \to G$, one defines the gauge-transformed spinor field as $\psi' = U^{-1}\psi$, $\psi'(x) = U(x)^{-1}\psi(x)$ for $x \in M$. A spinor connection ("gauge potential") is a 1-form ω on M with values in the Lie algebra of G, i.e. in $\mathsf{Cliff}_{k,l}^2 \subset \mathsf{End}\, S$; therefore, it can be written as $\omega = \frac{1}{4}\gamma^i\gamma^j\omega_{ij}$, where $\omega_{ij} = -\omega_{ji}$ are 1-forms. The covariant ("gauge") derivative of ψ is

$$D\psi = \mathrm{d}\psi + \omega\psi. \tag{3}$$

The gauge transformation U induces a change of the connection, $\omega \mapsto \omega' = U^{-1}\omega U + U^{-1}\mathrm{d}U$ so that $(D\psi)' = U^{-1}D\psi$. Since the dimension of M is even, the adjoint representation is onto $\mathsf{O}_{k,l}$ and one can define, for every $a \in \mathsf{Pin}_{k,l} \subset \mathsf{GL}(S)$, the (orthogonal) matrix $(\rho^i{}_j(a))$ by $a^{-1}\gamma^i a = \rho^i{}_j(a)\gamma^j$, so that $a^{-1}\gamma_i a = \gamma_j \rho^j{}_i(a^{-1})$. From the Lemma

if $a \in \mathsf{Cliff}_{k,l}^p$, then $g^{ij}\gamma_i a \gamma_j = (-1)^p(n - 2p)a$,

taking into account that $U^{-1}\mathrm{d}U$ is in the Lie algebra of G—therefore of degree $p = 2$—one obtains

$$g^{ij} U^{-1}\gamma_i U \mathrm{d}(U^{-1}\gamma_j U) = 4 U^{-1}\mathrm{d}U$$

so that

$$\omega'^i{}_j = \rho^i{}_k(U^{-1})\omega^k{}_l \rho^l{}_j(U) + \rho^i{}_k(U^{-1})\mathrm{d}\rho^k{}_j(U).$$

Let (e_i) be a field of orthonormal frames on M and let (e^i) denote the dual field of coframes. Since, by definition, $\omega_{ij} + \omega_{ji} = 0$, the 1-forms (ω_{ij}) define a *metric* linear connection. Its *torsion* $\mathrm{d}e^i + \omega^i{}_j \wedge e^j$ need not be zero.

The action of the Dirac operator $\mathcal{D}$ on a spinor field is $\mathcal{D}\psi = \gamma^i e_i \lrcorner D\psi$, where $\lrcorner$ denotes contraction.

Spinor fields according to Schrödinger and Karrer

The second approach can be traced back to work by Tetrode; it has been clearly formulated by Schrödinger [14][2] and Karrer [12]; it sometimes appears in texts written by physicists; see, e.g., [2]. Consider a Riemannian manifold (M, g) and let g_x denotes the quadratic form induced by g in the vector space $T_x M$ tangent to M at x. One assumes now the existence of a representation γ of the Clifford bundle $\mathsf{Cliff}(g) = \bigcup_{x \in M} \mathsf{Cliff}(g_x)$ in a *vector bundle* $\Sigma \to M$ of spinors so that $\gamma(u)^2 =$

---

[2] I thank Engelbert Schücking for having drawn my attention to this remarkable paper. It contains a derivation of the formula for the square of the Dirac operator on Riemannian manifolds.

$g(u,u)\mathrm{id}_{T_xM}$ for every $u \in T_xM$. One then introduces a spinor covariant derivative on Σ, compatible with a metric covariant derivative on TM. Such a structure is weaker (more general) than a classical spin structure. For example, it exists on every almost Hermitean manifold even though some of these manifolds—such as the even-dimensional complex projective spaces—do not admit a spin structure. The precise relation between those two approaches, in the case of even-dimensional orientable manifolds is described in [11]: the second method is equivalent to the introduction of a spin^c-structure on M.

In the physicist's local approach one introduces—following Schrödinger—a spinor field as a smooth map $\psi : M \to S$. The set of all such fields is a module $\mathcal{S}$ over the ring $\mathcal{C}$ of smooth functions on the Riemannian manifold M assumed here to be of dimension $m = 2n$. Let ∇ be a covariant derivative in the module $\mathcal{V}$ of vector fields: if $u, v \in \mathcal{V}$, then $\nabla_u v \in \mathcal{V}$ is the covariant derivative of v in the direction of u. The representation γ mentioned above associates with a vector field u and a spinor field ψ another spinor field $\gamma(u)\psi$; the map $\mathcal{V} \times \mathcal{S} \to \mathcal{S}$, $(u, \psi) \mapsto \gamma(u)\psi$, is bilinear and $\gamma(u)f\psi = f\gamma(u)\psi$ for every $f \in \mathcal{C}$; moreover, it has the Clifford property:

$$\gamma(u)^2 \psi = g(u,u)\psi. \tag{4}$$

One postulates now the existence of a *spinor covariant derivative* $\nabla^s_u : \mathcal{S} \to \mathcal{S}$ compatible with ∇ in the sense that

$$\nabla^s_u(\gamma(v)\psi) = \gamma(\nabla_u v)\psi + \gamma(u)\nabla^s_u \psi \quad \text{for} \quad u, v \in \mathcal{V} \text{ and } \psi \in \mathcal{S}. \tag{5}$$

Let (e_μ), where $\mu = 1, \ldots, m$, be a field of frames and let (e^μ) be the dual field of coframes; they need not be orthonormal; e.g., given local coordinates (x^μ), one can take $e^\mu = \mathrm{d}x^\mu$. The field $\gamma_\mu = \gamma(e_\mu) : M \to \mathsf{End}\, S$ satisfies

$$\gamma_\mu \gamma_\nu + \gamma_\nu \gamma_\mu = 2g_{\mu\nu}, \quad \text{where} \quad g_{\mu\nu} = g(e_\mu, e_\nu).$$

The coefficients of the linear connection defined by ∇ can be read off from

$$\nabla_{e_\nu} e_\mu = e_\rho \Gamma^\rho_{\mu\nu}.$$

The spinor covariant derivative of ψ in the direction of u can be written as the contraction

$$\nabla^s_u \psi = u \lrcorner D\psi,$$

where $D\psi$ has the form (3). The compatibility condition (5) is equivalent to

$$\mathrm{d}\gamma_\mu + [\omega, \gamma_\mu] - \Gamma^\rho_{\mu\sigma}\gamma_\rho e^\sigma = 0. \tag{6}$$

The metricity of ∇ can be justified by covariant-differentiating both sides of (4) and using (5). If ω is a solution of (6) and A is a complex-valued 1-form on M, then $\omega + \mathrm{i}A\,\mathrm{id}_S$ is another solution. Therefore, the connection ω can be interpreted as including an interaction with the electromagnetic field (of potential equal to the real part of A). Such an interpretation has been clearly formulated by Fock [9] and Schrödinger [14]: they may be considered as precursors of the idea of spin^c-structures.

EXAMPLES

Hypersurfaces in $\mathbb{R}^{m+1}$

Every hypersurface M in the Euclidean space $\mathbb{R}^{m+1}$, defined by an isometric immersion $f : M \to \mathbb{R}^{m+1}$, has a $\text{Pin}_{0,m}$-structure, canonically defined by f. Moreover, the *associated bundle* Σ of spinors on M is *trivial*: it is isomorphic to the Cartesian product $M \times S$ and a spinor field can be (globally!) described by a funtion $\psi : M \to S$. A Dirac (resp., Cartan) spinor field on an even (resp., odd) dimensional hypersurface is the restriction of a Pauli (resp., Dirac) field on the surrounding space. In terms of the trivialization of Σ, the Dirac operator assumes a rather simple form [17]. Let ν_i, where $i = 1,\ldots,m+1$, be the Cartesian components of a unit, normal vector field on M. Let $m = 2n$ or $2n - 1$; introduce the Dirac $2^n \times 2^n$ matrices γ_i satisfying

$$\gamma_i\gamma_j + \gamma_j\gamma_i = -2\delta_{ij}, \quad i,j = 1,\ldots,m+1.$$

Then

$$\mathcal{D} = \tfrac{1}{2}(\gamma^k\nu_k)(\gamma^i\gamma^j(\nu_j\partial_i - \nu_i\partial_j) - \operatorname{div}\nu), \tag{7}$$

where $\operatorname{div}\nu$ is the intrinsic divergence of ν, $\operatorname{div}\nu = \sum_{i,j}(\delta_{ij} - \nu_i\nu_j)\partial_i\nu_j$. If M is the hyperplane of equation $x^{m+1} = 0$, then $\nu_i = \delta_i^{m+1}$ and $\mathcal{D} = \sum_{\mu=1}^{m}\gamma^\mu\partial_\mu$. Formula (7) has been used to find, in a simple manner, the spectrum and the eigenfunctions of the Dirac operator on spheres [18].

Spheres, projective spaces, and quadrics

1. The spin structures on *spheres* are well-known: for every $m > 1$, the m-dimensional sphere $\mathbb{S}_m$ has a unique spin structure; in the style of (2) it is given by the sequence of maps

$$\text{Spin}_m \to \text{Spin}_{m+1} \to \text{SO}_{m+1} \to \mathbb{S}_m.$$

The spectrum of $\mathcal{D}$ on $\mathbb{S}_m$ is of the form:

Fig. 1. The spectrum of the Dirac operator on the m-sphere.

The eigenfunctions $\psi : \mathbb{S}_m \to S$ are either symmetric or antisymmetric; these two types of symmetries are indicated here by bullets and crosses; which of the eigenfunctions (bullets or crosses) are even depends on the trivialization of the bundle of spinors; only the relative parity matters.

2. The real projective spaces $\mathbb{RP}_m = \mathbb{S}_m/\mathbb{Z}_2$ are orientable iff m is odd; for $m > 1$ there are either two inequivalent (s)pin structures or none [8]:

$$
\begin{array}{lccccc}
m \equiv & 0 & 1 & 2 & 3 & \text{mod } 4 \\
\text{structure:} & \text{Pin}_{m,0} & \text{no} & \text{Pin}_{0,m} & \text{Spin}_m &
\end{array}
$$

Since $\mathbb{RP}_m$ is locally isometric to $\mathbb{S}_m$, the spectrum of $\mathcal{D}$ on such a space can be obtained from that of the sphere: the symmetric eigenfunctions descend to one (s)pin structure on $\mathbb{RP}_m$ and the antisymmetric functions—to the other; these spectra are thus asymmetric.

3. The real projective quadrics are defined as conformal compactifications of pseudo-Euclidean spaces; they generalize the Penrose construction of compactified Minkowski space-time. The quadric $\mathbb{S}_{k,l} = (\mathbb{S}_k \times \mathbb{S}_l)/\mathbb{Z}_2$ admits two natural metrics, descending from the spheres: a proper Riemannian metric and a pseudo-Riemannian one, of signature (k, l). The quadrics $\mathbb{S}_{k,0}$ and $\mathbb{S}_{0,k}$ can be identified with $\mathbb{S}_k$; a quadric is said to be *proper* if $kl \neq 0$. A proper quadric is orientable iff its dimension is even. If $kl > 1$, then $\mathrm{H}^1(\mathbb{S}_{k,l}, \mathbb{Z}_2) = \mathbb{Z}_2$; therefore, for $kl > 1$, the quadric has either 2 inequivalent (s)pin structures or none. The following table, based on [4], summarizes the results on the existence of (s)pin structures on $\mathbb{S}_{k,l}$ for $kl > 1$:

$k+l = 2n$	n	proper Riem.	pseudo-Riem.
either k or $l = 1$	any	yes	yes
k and l even	even	no	yes
k and l even	odd	yes	no
k and l odd	even	no	no
k and l odd	odd	yes	yes
k even and l odd			
$k+l \equiv 1 \bmod 4$		$\text{Pin}_{0,k+l}$	$\text{Pin}_{l,k}$
$k+l \equiv 3 \bmod 4$		$\text{Pin}_{k+l,0}$	$\text{Pin}_{k,l}$

For example, the quadric $\mathbb{S}_{3,5}$ has no spin structure for either of the metric tensors.

The spectrum of the Dirac operator on the projective quadrics can be obtained from that of the spheres; this is facilitated by the following Lemma: if λ_i is an eigenvalue of the Dirac operator on a Riemannian spin manifold M_i, $i = 1, 2$, then the numbers $\sqrt{\lambda_1^2 + \lambda_2^2}$ and $-\sqrt{\lambda_1^2 + \lambda_2^2}$ are eigenvalues of the Dirac operator on $M_1 \times M_2$ [5].

ACTION OF SPACE AND TIME REFLECTIONS ON SPINORS

Charge conjugation

Space and time reflections seem to be closely related to charge conjugation. Consider the Dirac equation

$$(\gamma^\mu(\partial_\mu - iqA_\mu) - m)\psi = 0$$

for the Dirac wave function $\psi : \mathbb{R}^{2n} \to S$ of a particle of mass m and charge q moving in a $2n$-dimensional flat space-time with a metric tensor of signature $(2n-1, 1)$ and an electromagnetic field with potential A_μ. Let $C : S \to \bar{S}$ be the isomorphism such that $\overline{\gamma_\mu} = C\gamma_\mu C^{-1}$, where bar denotes complex conjugation; the *charge conjugate* spinor field $\mathsf{C}\psi = \overline{C\psi}$ satisfies the equation

$$(\gamma^\mu(\partial_\mu + iqA_\mu) - m)\mathsf{C}\psi = 0.$$

If $\psi \sim \exp(-iEt)$, then $\mathsf{C}\psi \sim \exp(+iEt)$: charge conjugation is said to transform particles into antiparticles.

Wigner's time inversion

Let $\psi : \mathbb{R}^3 \times \mathbb{R} \to \mathbb{C}$ be a wave function in non-relativistic quantum theory. The time-reversed wave function $\mathsf{T}_W\psi$ is defined by [20]

$$(\mathsf{T}_W\psi)(\mathbf{r}, t) = \overline{\psi(\mathbf{r}, -t)}.$$

If ψ is a solution of the Schrödinger equation with a time-independent, real potential, then so is $\mathsf{T}_W\psi$. If $\psi \sim \exp(-iEt)$, then also $\mathsf{T}_W\psi \sim \exp(-iEt)$.

Time inversion of spinor fields: Feynman *versus* Wigner

In the relativistic theory, there are two ways of defining time inversion [13]. Recall first the general statement about the relativistic invariance of the free Dirac equation in Minkowski space. Let S denote, as before, the complex vector space of Dirac spinors. A representation of the Clifford algebra $\mathsf{Cliff}_{3,1}$ in S being given in terms of the Dirac matrices γ_μ, one can identifiy the group $\mathsf{Pin}_{3,1}$ with a subgroup of $\mathsf{GL}(S)$ and embed $\mathbb{R}^4$ in $\mathsf{End}\, S$ by $x \mapsto x^\mu \gamma_\mu$, as usual. There is the exact sequence

$$1 \to \mathbb{Z}_2 \to \mathsf{Pin}_{3,1} \xrightarrow{\mathrm{Ad}} \mathsf{O}_{3,1} \to 1,$$

where $\mathrm{Ad}(U)x = UxU^{-1}$. There is a similar, but inequivalent, extension of $\mathsf{O}_{3,1}$ by $\mathbb{Z}_2$ corresponding to $\mathsf{Pin}_{1,3}$, as well as several other extensions described in [7]; see also [6] and the references given there. Every $U \in \mathsf{Pin}_{3,1}$ acts on spinor fields by sending a solution $\psi : \mathbb{R}^4 \to S$ of the free Dirac equation to another solution $\mathsf{U}\psi$,

$$(\mathsf{U}\psi)(x) = U(\psi(\mathrm{Ad}(U^{-1})x)).$$

In particular, with γ_4 and $\gamma_1\gamma_2\gamma_3 \in \mathsf{Pin}_{3,1}$, there are associated the space and time inversion operators P and T, respectively. The operator T is the *geometrical time inversion*; if $\psi \sim \exp(-\mathrm{i}Et)$, then $\mathsf{T}\psi \sim \exp(+\mathrm{i}Et)$. One can justify the interpretation of T as the time inversion operator by the Feynman idea of viewing antiparticles as particles travelling backwards in time. Physicists favour nowadays the *Wigner time inversion*

$$\mathsf{T}_W = \mathsf{T} \circ \mathsf{C}.$$

Since charge conjugation is involutory, $\mathsf{C}^2 = \mathrm{id}$, the product of operators considered in the PCT theorem is equivalent to $\mathsf{P} \circ \mathsf{T}$. This is the space-time reflection R corresponding to γ_5. The space-time reflection is in the connected component of the identity of the "complex" Lorentz group ($=\mathsf{SO}_4(\mathbb{C})$). The idea underlying the PCT theorem is that, in a quantum field theory invariant only with respect to the connected component of the Poincaré group, holomorphic functions such as the vacuum expectation values of field operators, are also invariant with respect to the "complex rotation" R [16].

In non-relativistic quantum mechanics, there is no place for T because that theory is obtained as a limit of the relativistic theory implying, for a free particle, the removal of all negative energy states.

Space and time inversion of Weyl spinors

Complex conjugation appears also in the realization of space and time reflections in the space of Weyl (chiral) spinors proposed by Staruszkiewicz [15]; see also [6].

Recall that the connected component of the group $\mathsf{Spin}_{3,1}$ is isomorphic to $\mathsf{SL}_2(\mathbb{C})$. It has two inequivalent representations in the spaces of 2-component (Weyl) dotted and undotted spinors.[3] These representations are complex conjugate one to another; their direct sum defines the representation in the space of Dirac spinors,

$$a \mapsto \begin{pmatrix} a & 0 \\ 0 & \bar{a} \end{pmatrix}, \quad a \in \mathsf{SL}_2(\mathbb{C}).$$

This decomposable representation is a restriction to $\mathsf{SL}_2(\mathbb{C})$ of the representation of $\mathsf{Pin}_{3,1}$ determined by the Dirac matrices

$$\gamma_1 = \begin{pmatrix} 0 & \mathrm{i}\sigma_1 \\ -\mathrm{i}\sigma_1 & 0 \end{pmatrix}, \quad \gamma_2 = \begin{pmatrix} 0 & I \\ I & 0 \end{pmatrix}, \quad \gamma_3 = \begin{pmatrix} 0 & \mathrm{i}\sigma_3 \\ -\mathrm{i}\sigma_3 & 0 \end{pmatrix}, \quad \gamma_4 = \begin{pmatrix} 0 & \sigma_2 \\ -\sigma_2 & 0 \end{pmatrix}.$$

---

[3] This terminology is due to van der Waerden; many people follow now Penrose and use the names: primed and unprimed "reduced" spinors; see [19] for references and further comments.

In this representation one has $C = \gamma_2$ and a Dirac spinor of the form

$$\psi = \begin{pmatrix} u \\ \bar{u} \end{pmatrix}, \quad \text{where} \quad u \in \mathbb{C}^2, \tag{8}$$

is real in the sense that it satisfies the Majorana condition, $C\psi = \psi$. Space and time inversions, as defined in the previous section, induce the following transformations of the Weyl part u of the Majorana spinor (8),

$$\mathsf{P}: u \mapsto \sigma_2 \bar{u} \quad \text{and} \quad \mathsf{T}: u \mapsto \mathrm{i}\sigma_2 \bar{u},$$

respectively.

REFERENCES

1. Barut, A. and Rączka, R., *Theory of group representations and applications*, Warszawa: PWN, 1977.
2. Benn, I. M., and Tucker, R. W., *An introduction to spinors and geometry with applications in physics*, Bristol: Hilger, 1988.
3. Bourbaki, N., *Algèbre*, Paris: Hermann, Masson, and Diffusion C.C.I.S. 1959–80.
4. Cahen, M., Gutt, S. and Trautman, A., *J. Geom. Phys.*, **10**, 127–154 (1993) and **17**, 283–297 (1995).
5. Cahen, M., Gutt, S. and Trautman, A., article in: *Clifford algebras and their applications in mathematical physics*, V. Dietrich et al. (eds.), Dordrecht: Kluwer, 1998, pp. 391–399.
6. Chamblin, A. and Gibbons, G. W., *Class. Quantum Grav.* **12**, 2243–2248 (1995).
7. Dąbrowski, L., *Group actions on spinors*, Naples: Bibliopolis, 1988, pp. 8–13.
8. Dąbrowski, L. and Trautman, A., *J. Math. Phys.* **27**, 2022–28 (1986).
9. Fock, V., *Z. Physik.*, **57**, 261 (1929).
10. Friedrich, T., *Dirac-Operatoren in der Riemannschen Geometrie*, Wiesbaden: Vieweg, 1997.
11. Friedrich, T. and Trautman, A., *Clifford structures and spinor bundles*, Sfb 288 Preprint No. 251, Inst. für Reine Mathematik, Humboldt University, Berlin 1997.
12. Karrer, G., *Ann. Acad. Fennicae*, Ser. A, Math. 336/5, 1–16 (1963).
13. Pauli, W., article in: *Niels Bohr and the development of physics*, W. Pauli and L. Rosenfeld (eds.), London and New York: Pergamon Press, 1955, pp. 30–51.
14. Schrödinger, E., *Sitzungsber. preuss. Akad. Wissen.*, Phys.-Math. Kl. **XI**, 105–128 (1932).
15. Staruszkiewicz, A., *Acta Physica Polonica* **B7**, 557–565 (1976).
16. Streater, R. F. and Wightman, A. S., *PCT, spin & statistics, and all that*, New York: Benjamin, 1964.
17. Trautman, A., *J. Math. Phys.* **33**, 4011–4019 (1992).
18. Trautman, A., *Acta Physica Polonica* **B26**, 1283–1310 (1995).
19. Trautman, A., *Contemporary Mathematics* **203**, 3–24 (1997).
20. Wigner, E. P., *Göttinger Nachrichten*, Math. Phys. Kl., pp. 546–559 (1932).

Exact Solutions in Locally Anisotropic Gravity and Strings

Sergiu I. Vacaru

Institute of Applied Physics, Academy of Sciences, 5 Academy str., Chişinău 2028, Moldova[1]
& Institute for Basic Research, P. O. Box 1577, Palm Harbor, FL 34682, U. S. A.[2]

Abstract. In this Report we outline some basic results on generalized Finsler–Kaluza–Klein gravity and locally anisotropic strings. There are investigated exact solutions for locally anisotropic Friedmann–Robertson–Walker universes and three dimensional and string black holes with generic anisotropy.

Devoted to the memory of Professor Ryszard Raczka (1931–1996)

INTRODUCTION

The theory of locally anisotropic field interactions and (super)strings is recently descussed [6–9] in the context of development of unified approaches to generalized Finsler like [3] and Kaluza–Klein gravity [4]. A number of present day cosmological models are constructed as higher–dimensional extensions of general relativity with a general anisotropic distribution of matter and in correlation with low–energy limits of string perturbation theory. In non–explicit form it is assumed the **postulate: the matter always (even being anisotropic) gives rise to a locally isotropic geometry,** which is contained in the structure of Einstein equations for metric $g_{ij}(x^k)$ on (pseudo)Riemannian spaces:

$$G_{ij}(x^k) \simeq T_{ij}(x^k, y^a)$$
Einstein tensor — Energy–momentum tensor
(for a locally isotropic curved space) — (in general anisotropic)

where $x^i, i = 0, 1, ..., n-1$ are coordinates on space–time M and $y^a, a = 1, 2, ..., m$ are parameters (coordinates) of anisotropies.

---

[1] e-mail: vacaru@lises.as.md
[2] ibr@gte.net, http://home1.gte.net/ibr

Anisotropic cosmological and locally anisotropic self–gravitating models are widely used in order to interpret the observable anisotropic structure of the Universe and of background radiation. Our basic idea to be developed in this paper is that cosmological anisotropies are not only consequences of some anisotropic distributions of matter but they reflect a generic space–time anisotropy induced after reductions from higher to lower dimensions and by primordial quantum field fluctuations. If usual

Kaluza–Klein theories routinely require compactification mechanisms, we suggest a more general scenarios of possible decompositions of higher dimensional (super)space into lower dimensional ones being modelled by a specific "splitting field" defined geometrically as a nonlinear connection.

A geometry of manifolds provided with a metric more general than the usual Riemann one, $g_{ij}(x^k) \Longrightarrow g_{ij}(x^k, \lambda^s y^n)$, where $y^n \simeq \frac{dx^n}{dt}$ and λ^s is a parameter of homogeneity of order s, was proposed in 1854 by B. Riemann and it was studied for the first time in P. Finsler (1918) and E. Cartan (1934) (see historical overviews, basic results and references in [3,8,9]). At first sight there are very substantial objections of physical character to generalized Finsler like theories: One was considered that a local anisotropy crucially frustrates the local Lorentz invariance. Not having even local (pseudo)rotations and translations it is an unsurmountable problem to define conservation laws and values of energy–momentum type, to apply the concept of fundamental particles fields (for example, without local rotations we can not define local groups and algebras and their representations). A difficulty with Finsler like gravity was also the problem of its inclusion into the framework of modern approaches based on (super)strings, Kaluza–Klein and gauge theories.

The main purpose of a series of our works (see [6–9] and references) is the development of a general approach to locally anisotropic gravity imbedding both type of Kaluza–Klein and Finsler–like theories. It should be emphasized that a subclass of such models can be constructed as to have a local space–time Lorentz invariance. We proved that the general higher order anisotropic gravity can be treated as alternative low energy limits of (super)string theories with a dynamical reduction given by the nonlinear connection field and that there are natural extensions of the Einstein gravity to locally anisotropic theories constructed on generic nonholonomic vector bundles provided with nonlinear connection structure.

The field equations of locally anisotropic gravity are of type

$$\boxed{G_{\alpha\beta}(x^\alpha, y^\beta) \simeq T_{\alpha\beta}(x^\alpha, y^\beta)}$$

where the Einstein tensor is defined on a bundle (generalized Finsler) space, x^α are usual coordinate on the base manifold and y^β are coordinates on the fibers (parameters of anisotropy), in general $\dim\{x^\alpha\} \neq \dim\{y^\beta\}$.

This paper is organized as follows. In Sec. II we briefly review the geometric background of locally anisotropic gravity. Models of locally anisotropic Friedmann–Robertson–Walker universe are considered in Sec. III. In Sec. IV we analyze anisotropic black hole solutions in three dimensional space–times and extend such solutions to the string theory. Conclusions are drawn in Sec. V.

GENERALIZED FINSLER–KALUZA–KLEIN GRAVITY

In Einstein gravity and its locally isotropic modifications of Kaluza–Klein, lower dimensional, or of Einstein–Cartan–Weyl types, the fundamental
space–time is considered as a real $4+d$–dimensional ($d = -2, -1, 0, 1, ..., n$) manifold of necessary smoothly class and signature, provided with independent metric (equivalently, tetrad) and linear connection (in general nonsymmetric). In order to model spaces with generic local anisotropy instead of manifolds one considers vector, or tangent/cotangent, bundles (with possible higher order generalizations) enabled with nonlinear connection and distinguished (by the nonlinear connection) linear connection and metric structures. The coordinates in fibers are treated as parameters of possible anisotropy and/or as higher dimension coordinates which in general are not compactified.

In this section we outline the basic results from the so–called locally anisotropic (la) gravity [3,6–9] (in brief we shall use la–gravity, la–space and so on).

Let $\mathcal{E} = (E, \pi, F, Gr, M)$ be a locally trivial vector bundle (v–bundle) over a base M of dimension n, where $F = \mathcal{R}^m$ is the typical real vector space of dimension m, the structural group is taken to be the group of linear transforms of $\mathcal{R}^m$, i. e. $Gr = GL(m, \mathcal{R})$. We locally parametrize $\mathcal{E}$ by coordinates $u^\alpha = (x^i, y^a)$, where $i, j, k, l, m, ..., = 0, 1, ..., n-1$ and $a, b, c, d, ... = 1, 2, ..., m$. Coordinate transforms $(x^k, y^a) \to (x^{k'}, y^{a'})$ on $\mathcal{E}$, considered as a differentiable manifold, are given by formulas $x^{k'} = x^{k'}(x^k), y^{a'} = M_a^{a'}(x)y^a$, where $rank(\frac{\partial x^{k'}}{\partial x^k}) = n$ and $M_a^{a'}(x) \in Gr$.

One of the fundamental objects in the geometry of la–spaces is the **nonlinear connection**, in brief **N–connection**. The N–connection can be defined as a global decomposition of v–bundle $\mathcal{E}$ into horizontal, $\mathcal{HE}$, and vertical, $\mathcal{VE}$, subbundles of the tangent bundle $\mathcal{TE}, \mathcal{TE} = \mathcal{HE} \oplus \mathcal{VE}$. With respect to a N–connection in $\mathcal{E}$ one defines a covariant derivation operator $\nabla_Y A = Y^i \left\{ \frac{\partial A^a}{\partial x^i} + N_i^a(x, A) \right\} s_a$, where s_a are local linearly independent sections of $\mathcal{E}, \Lambda = \Lambda^a s_a$ and $Y = Y^i s_i$ is the decomposition of a vector field Y with respect to a local basis s_i on M. Differentiable functions $N_i^a(x, y)$ are called the coefficients of the N–connection. One holds these transformation laws for components N_i^a under coordinate transforms: $N_{i'}^{a'} \frac{\partial x^{i'}}{\partial x^i} = M_a^{a'} N_i^a + \frac{\partial M_a^{a'}}{\partial x^i} y^a$. The N–connection is also characterized by its **curvature**

$$\Omega_{ij}^a = \frac{\partial N_j^a}{\partial x^i} - \frac{\partial N_i^a}{\partial x^j} + N_j^b \frac{\partial N_i^a}{\partial y^b} - N_i^b \frac{\partial N_j^a}{\partial y^b},$$

and by its linearization which is defined as $\Gamma_{.bi}^a(x, y) = \frac{\partial N_i^a(x,y)}{\partial y^b}$. The usual linear connections $\omega_b^a = K_{.bi}^a(x)dx^i$ in a v–bundle $\mathcal{E}$ form a particular class of N–connections with coefficients parametrized as $N_i^a(x, y) = K_{.bi}^a(x)y^b$.

Having introduced in a v–bundle $\mathcal{E}$ a N–connection structure we must modify the operation of partial derivation and introduce a locally adapted (to the N–connection) basis (frame)

$$\frac{\delta}{\delta u^\alpha} = (\frac{\delta}{\delta x^i} = \partial_i - N_i^a(x,y)\frac{\partial}{\partial y^a}, \frac{\delta}{\delta y^a} = \frac{\partial}{\partial y^a}), \qquad (2.1)$$

instead of the local coordinate basis $\frac{\partial}{\partial u^\alpha} = (\frac{\partial}{\partial x^i}, \frac{\partial}{\partial y^a})$. The basis dual to $\frac{\delta}{\delta u^\alpha}$ is written as

$$\delta u^\alpha = (\delta x^i = dx^i, \delta y^a = dy^a + N_i^a(x,y)dx^i). \qquad (2.2)$$

We note that a v–bundle provided with a N–connection structure is a generic nonholonomic manifold because in general the nonholonomy coefficients $w^\gamma_{.\alpha\beta}$, defined by relations $[\frac{\delta}{\delta u^\alpha}, \frac{\delta}{\delta u^\beta}] = \frac{\delta}{\delta u^\alpha}\frac{\delta}{\delta u^\beta} - \frac{\delta}{\delta u^\beta}\frac{\delta}{\delta u^\alpha} = w^\gamma_{.\alpha\beta}\frac{\delta}{\delta u^\gamma}$, do not vanish.

By using bases (2.1) and (2.2) we can introduce the algebra of tensor distinguished fields (d–fields, d–tensors) on $\mathcal{E}$, $\mathcal{C} = \mathcal{C}^{pr}_{qs}$, which is equivalent to the tensor algebra of the v–bundle $\mathcal{E}_d$ defined as $\pi_d : \mathcal{HE} \oplus \mathcal{VE} \to \mathcal{TE}$ [3]. An element $t \in \mathcal{C}^{pr}_{qs}$, of d–tensor of type $\begin{pmatrix} p & r \\ q & s \end{pmatrix}$, are written in local form as

$$t = t^{i_1...i_p a_1...a_r}_{j_1...j_q b_1...b_s}(u) \frac{\delta}{\delta x^{i_1}} \otimes ... \otimes \frac{\delta}{\delta x^{i_r}} \otimes dx^{j_1} \otimes ... \otimes dx^{j_p} \otimes$$

$$\frac{\partial}{\partial y^{a_1}} \otimes ... \otimes \frac{\partial}{\partial y^{a_r}} \otimes \delta y^{b_1} \otimes ... \otimes \delta y^{b_s}.$$

In addition to d–tensors we can consider different types of d-objects with group and coordinate transforms adapted to a global splitting of v-bundle by a N–connection.

A **distinguished linear connection**, in brief **a d–connection**, is defined as a linear connection D in $\mathcal{E}$ conserving as a parallelism the Whitney sum $\mathcal{HE} \oplus \mathcal{VE}$ associated to a fixed N-connection structure in $\mathcal{E}$. Components $\Gamma^\alpha_{.\beta\gamma}$ of a d–connection D are introduced by relations $D_\gamma(\frac{\delta}{\delta u^\beta}) = D_{(\frac{\delta}{\delta u^\gamma})}(\frac{\delta}{\delta u^\beta}) = \Gamma^\alpha_{.\beta\gamma}(\frac{\delta}{\delta u^\alpha})$.

We can compute in a standard manner but with respect to a locally adapted frame (2.1), the components of **torsion and curvature of a d–connection** D:

$$T^\alpha_{.\beta\gamma} = \Gamma^\alpha_{.\beta\gamma} - \Gamma^\alpha_{.\gamma\beta} + w^\alpha_{.\beta\gamma} \qquad (2.3)$$

and

$$R^{.\alpha}_{\beta.\gamma\delta} = \frac{\delta\Gamma^\alpha_{.\beta\gamma}}{\delta u^\delta} - \frac{\delta\Gamma^\alpha_{.\beta\delta}}{\delta u^\gamma} + \Gamma^\varphi_{.\beta\gamma}\Gamma^\alpha_{.\varphi\delta} - \Gamma^\varphi_{.\beta\delta}\Gamma^\alpha_{.\varphi\gamma} + \Gamma^\alpha_{.\beta\varphi}w^\varphi_{.\gamma\delta}. \qquad (2.4)$$

The global decomposition by a N-connection induces a corresponding invariant splitting into horizontal $D^h_X = D_{hX}$ (h–derivation) and vertical $D^v_X = D_{vX}$ (v–derivation) parts of the operator of covariant derivation $D, D_X = D^h_X + D^v_X$, where $hX = X^i\frac{\delta}{\delta u^i}$ and $vX = X^a\frac{\partial}{\partial y^a}$ are, respectively, the horizontal and vertical components of the vector field $X = hX + vX$ on $\mathcal{E}$.

Local coefficients $\left(L^i_{.jk}(x,y), L^a_{.bk}(x,y)\right)$ of covariant h-derivation D^h are introduced as $D^h_{\left(\frac{\delta}{\delta x^k}\right)}\left(\frac{\delta}{\delta x^j}\right) = L^i_{.jk}(x,y)\frac{\delta}{\delta x^i}$, $D^h_{\left(\frac{\delta}{\delta x^k}\right)}\left(\frac{\partial}{\partial y^b}\right) = L^a_{.bk}(x,y)\frac{\partial}{\partial y^a}$ and $D^h_{\left(\frac{\delta}{\delta x^k}\right)}f = \frac{\delta f}{\delta x^k} = \frac{\partial f}{\partial x^k} - N^a_k(x,y)\frac{\partial f}{\partial y^a}$, where $f(x,y)$ is a scalar function on $\mathcal{E}$.

Local coefficients $\left(C^i_{.jk}(x,y), C^a_{.bk}(x,y)\right)$ of v-derivation D^v are introduced as $D^v_{\left(\frac{\partial}{\partial y^c}\right)}\left(\frac{\delta}{\delta x^j}\right) = C^i_{.jk}(x,y)\frac{\delta}{\delta x^i}$, $D^v_{\left(\frac{\partial}{\partial y^c}\right)}\left(\frac{\partial}{\partial y^b}\right) = C^a_{.bc}(x,y)$ and $D^v_{\left(\frac{\partial}{\partial y^c}\right)}f = \frac{\partial f}{\partial y^c}$.

By straightforward calculations we can express respectively the coefficients of torsion (2.3) and curvature (2.4) [3] via h- and v-components parametrized as $T^\alpha_{\beta\gamma} = \{T^i_{.jk}, T^i_{ja}, T^i_{aj}, T^i_{.ja}, T^a_{.bc}\}$ and $R^\alpha_{.\beta.\gamma\delta} = \{R^i_{h.jk}, R^a_{b.jk}, P^i_{j.ka}, P^c_{b.ka}, S^i_{j.bc}, S^a_{b.cd}\}$.

The components of the Ricci d-tensor $R_{\alpha\beta} = R^\tau_{\alpha.\beta\tau}$ with respect to locally adapted frame (2.2) are as follows: $R_{ij} = R^k_{i.jk}$, $R_{ia} = -{}^2P_{ia} = -P^k_{i.ka}$, $R_{ai} = {}^1P_{ai} = P^{.b}_{a.ib}$, $R_{ab} = S^{.c}_{a.bc}$. We point out that because, in general, ${}^1P_{ai} \neq {}^2P_{ia}$ the Ricci d-tensor is nonsymmetric.

Now, we shall analyze the compatibility conditions of N- and d–connections and metric structures on the v-bundle $\mathcal{E}$. A metric field on $\mathcal{E}$, $G(u) = G_{\alpha\beta}(u)du^\alpha du^\beta$, is associated to a map $G(X,Y): \mathcal{T}_u\mathcal{E} \times \mathcal{T}_u\mathcal{E} \to R$, parametrized by a non degenerate symmetric $(n+m) \times (n+m)$-matrix with components $\hat{G}_{ij} = G\left(\frac{\partial}{\partial x^i}, \frac{\partial}{\partial x^j}\right)$, $\hat{G}_{ia} = G\left(\frac{\partial}{\partial x^i}, \frac{\partial}{\partial y^a}\right)$ and $\hat{G}_{ab} = G\left(\frac{\partial}{\partial y^a}, \frac{\partial}{\partial y^b}\right)$. One chooses a concordance between N-connection and G-metric structures by imposing conditions $G\left(\frac{\delta}{\delta x^i}, \frac{\partial}{\partial y^a}\right) = 0$, equivalently, $N^a_i(x,u) = \hat{G}_{ib}(x,y)\hat{G}^{ba}(x,y)$, where $\hat{G}^{ba}(x,y)$ are found to be components of the matrix $\hat{G}^{\alpha\beta}$ which is the inverse to $\hat{G}_{\alpha\beta}$. In this case the metric G on $\mathcal{E}$ is defined by two independent d-tensors, $g_{ij}(x,y)$ and $h_{ab}(x,y)$, and written as

$$G(u) = G_{\alpha\beta}(u)\delta u^\alpha \delta u^\beta = g_{ij}(x,y)dx^i \otimes dx^j + h_{ab}(x,y)\delta y^a \otimes \delta y^b. \quad (2.5)$$

The d-connection $\Gamma^\alpha_{.\beta\gamma}$ is compatible with the d-metric structure $G(u)$ on $\mathcal{E}$ if one holds equalities $D_\alpha G_{\beta\gamma} = 0$.

Having defined the d-metric (2.5) in $\mathcal{E}$ we can introduce the scalar curvature of d-connection $\overleftarrow{R} = G^{\alpha\beta}R_{\alpha\beta} = R + S$, where $R = g^{ij}R_{ij}$ and $S = h^{ab}S_{ab}$.

Now we can write the Einstein equations for la–gravity

$$R_{\alpha\beta} - \frac{1}{2}G_{\alpha\beta}\overleftarrow{R} + \lambda G_{\alpha\beta} = \kappa_1 \mathcal{T}_{\alpha\beta}, \quad (2.6)$$

where $\mathcal{T}_{\alpha\beta}$ is the energy–momentum d-tensor on la–space, κ_1 is the interaction constant and λ is the cosmological constant. We emphasize that in general the d-torsion does not vanish even for symmetric d–connections (because of nonholonomy coefficients $w^\alpha_{\beta\gamma}$). So the d-torsion interactions plays a fundamental role on la–spaces. A gauge like version of la–gravity with dynamical torsion was proposed in [10]. We can also restrict our considerations only with algebraic equations for d–torsion in the framework of an Einstein–Cartan type model of la–gravity.

Finally, we note that all presented in this section geometric constructions contain as particular cases those elaborated for generalized Lagrange and Finsler spaces [3], for which a tangent bundle TM is considered instead of a v–bundle $\mathcal{E}$. We also note that the Lagrange (Finsler) geometry is characterized by a metric of type (2.5) with components parametrized as $g_{ij} = \frac{1}{2}\frac{\partial^2 \mathcal{L}}{\partial y^i \partial y^j}$ $\left(g_{ij} = \frac{1}{2}\frac{\partial^2 \Lambda^2}{\partial y^i \partial y^j}\right)$ and $h_{ij} = g_{ij}$, where $\mathcal{L} = \mathcal{L}(x, y)$ is a Lagrangian (($\Lambda = \Lambda(x, y)$) is a Finsler metric) on TM, see details in [3,6–10]. The usual Kaluza–Klein geometry could be obtained for corresponding parametrizations of N–connection and metric structures on the background v–bundle.

LOCALLY ANISOTROPIC FRIEDMANN–ROBERTSON–WALKER UNIVERSES

In this section we shall construct solutions of Einstein equations (2.6) generalizing the class of Friedmann–Robertson–Walker (in brief FRW) metrics to the case of ($n = 4, m = 1$) dimensional locally anisotropic space. In order to simplify our considerations we shall consider a prescribed N-connection structure of type $N_0^1 = n(t, \theta), N_1^1 = 0, N_2^1 = 0, N_3^1 = 0$, where the local coordinates on the base M are taken as spherical coordinates for the Robertson-Walker model, $x^0 = t, x^1 = r, x^2 = \theta, x^3 = \varphi$, and the anisotropic coordinate is denoted $y^1 \equiv y$.

The la–metric (2.5) is parametrized by the anzats

$$\delta s^2 = ds_{RW}^2 + h_{11}(t, r, \theta, \varphi, y)\delta y^2 \qquad (3.1)$$

where the Robertson-Walker like metric ds_{RW}^2 is written as

$$ds_{RW}^2 = -dt^2 + a^2(t, \theta)\left[\frac{dr^2}{1 - kr^2} + r^2\left(d\theta^2 + \sin^2\theta \cdot d\varphi^2\right)\right],$$

$k = -1, 0$ and 1, respectively, for open, flat and closed universes, $H(t, \theta) = \partial a(t, \theta)/\partial t$ is the anisotropic on angle θ (for our model) Hubble parameter, the containing the N-connection coefficients value δy, see (2.2), is of type $\delta y = dy + n(t, \theta)dt$ and coefficients $n(t, \theta)$ and $h_{11}(t, r, \theta, \varphi, y)$ are considered as arbitrary functions, which are prescribed on la–spaces defined as nonholonomic manifolds (in self–consistent dynamical field models one must find solutions of a closed system of equations for N- and d–connection and d–metric structure).

Considering an anisotropic fluctuation of matter distribution of type $\mathcal{T}_{\alpha\beta} = T_{\alpha\beta}^{(a)} + T_{\alpha\beta}^{(i)}$, with nonvanishing anisotropic components $T_{10}^{(a)}(t, r, \theta) \neq 0$ and $T_{20}^{(a)}(t, \theta) \neq 0$ and diagonal isotropic energy-momentum tensor $T_{\alpha\beta}^{(i)} = diag(-\rho, p, p, p, p_{(y)})$, where ρ is the matter density, p and $p_{(y)}$ are respectively pressures in 3 dimensional space and extended space, we obtain from the Einstein equations (2.6) this generalized system of Friedmann equations:

$$\left(\frac{1}{a}\frac{\partial a}{\partial t}\right)^2 = \frac{8\pi G_{(gr)}}{3}\rho - \frac{k}{a^2}, \qquad (3.2)$$

$$\frac{1}{a}\frac{\partial^2 a}{\partial t^2} - n(t,\theta)\frac{1}{a}\frac{\partial a}{\partial t} = -\frac{4\pi G_{(gr)}}{3}(\rho + 3p) \qquad (3.3)$$

with anisotropic additional relations between nonsymmetric, for la–spaces, Ricci and energy–momentum d–tensors:

$$R_{10} = -n(t,\theta)\left(\frac{kr}{1-kr^2} + \frac{2}{r}\right) \simeq T_{10}^{(a)}(t,r,\theta),\ R_{20} = -n(t,\theta)\cdot ctg\theta \simeq T_{20}^{(a)}(t,\theta)$$

when $R_{01} = 0$ and $R_{02} = 0$. The $G_{(gr)}$ from (3.2) and (3.3) is the usual gravitational constant from the Einstein theory.

For the locally isotropic FRW model, when $\rho = -p$, the equations (3.2) and (3.3) have an exponential solution of type $a_{FRW}^{(exp)} = a_0 \cdot e^{\omega_\rho \cdot t}$, where $a_0 = const$ and $\omega_\rho = \sqrt{\frac{8\pi G_{(gr)}}{3}\rho}$. This fact is widely applied in modern cosmology.

Substituting (3.2) into (3.3) we obtain the equation

$$\frac{\partial^2 a}{\partial t^2} - n(t,\theta)\frac{\partial a}{\partial t} - \omega_\rho^2 a = 0 \qquad (3.4)$$

where the function $a(t,\theta)$ depends on coordinates t and (as on a parameter) θ. Introducing a new variable $u = a \cdot \exp\left[-\frac{1}{2}\int n(t,\theta)dt\right]$ we can rewrite the (3.4) as a parametric equation

$$\frac{d^2 u(t,\theta)}{dt^2} - \widetilde{\omega}(t,\theta)u(t,\theta) = 0$$

for $\widetilde{\omega}(t,\theta) = \omega_\rho^2 + \left(\frac{n}{2}\right)^2 + \frac{1}{2}\frac{\partial n}{\partial t}$ which admits expressions of the general solution as series (see [2]).

It is easy to construct exact solutions and understand the physical properties of the equations of type (3.4) if the nonlinear connection structure does not depend on time variable, i.e. $n = n(\theta)$. By introducing the new variable $\tau = \omega_\rho t$ and function $a = v \cdot \exp\left(-D_0(\theta)\tau\right)$, where $D_0(\theta) = -n(\theta)/2\omega_\rho$, we transform (3.4) into the equation

$$\frac{d^2 v}{d\tau^2} + \left(1 - D_0^2(\theta)\right)v = 0$$

which can be solved in explicit form:

$$v = \begin{cases} C \cdot e^{-D_0(\theta)\tau} \cdot \cos(\varsigma\tau - \tau_0), & \varsigma^2 = 1 - D_0^2(\theta),\ D_0(\theta) < 1; \\ C \cdot ch\left(\varepsilon\tau + \tau_0\right), & \varepsilon^2 = D_0^2(\theta) - 1,\ D_0(\theta) > 1; \\ e^{-\tau}\left[v_0(1+\tau) + v_1\tau\right], & D_0(\theta) \to 1, \end{cases} \qquad (3.5)$$

where C, τ_0, v_0 and v_1 are integration constants.

It is clear from the solutions (3.5) that a generic local anisotropy of space–time (possibly induced from higher dimensions) could play a crucial role in Cosmology. For some prescribed values of nonlinear connection components we can obtain exponential anisotropic acceleration, or damping for corresponding conditions, of the inflational scenarios of universes, for another ones there are possible oscillations.

THREE DIMENSIONAL LOCALLY ANISOTROPIC BLACK HOLES AND STRINGS

Three dimensional la–solutions

We first consider the simplest possible case when (2+1)–dimensional space–time admits a prescribed N–connection structure. The anzats for la–metric (2.5) is chosen in the form

$$\delta s^2 = -N_*^2(r)dt^2 + S_*^2(r)dr^2 + P_*^2(r)\delta y^2, \tag{4.1}$$

where $\delta y = d\varphi + n(r)dr$. The metric (4.1) is written for a la–space with local coordinates $x^0 = t, x^1 = r$ and fiber coordinate $y^1 = \varphi$ and has components: $g_{00} = -N_*^2(r), g_{11} = S_*^2(r)$ and $h_{11} = P_*^2(r)$. The prescription for N–connection from (2.2) is taken $N_0^1 = 0$ and $N_1^1 = n(r)$.

The Einstein equations (2.6) are satisfied if one holds the condition $n = \frac{\ddot{N}_*}{N_*} - \frac{\dot{S}_*}{S_*}$, where, for instance, $\dot{S}_* = \frac{dS_*}{dt}$. So on a (2+1)–dimensional space–time with prescribed generic N–connection there are possible nonsingular la–metrics.

Nevertheless (2+1)–like black hole solutions with singular anisotropies can be constructed, for instance, by choosing the parametrizations

$$P_*^2(r) = P^2(r) = \rho^2(r), \quad N_*^2(r) = N^2(r) = \left(\frac{r}{\rho}\right)\cdot\left(\frac{r^2-r_+^2}{l}\right), \tag{4.2}$$

$$S_*^2 = S^2 = \left(\frac{r}{\rho N}\right)^2, \quad n(r) = N^\varphi(r) = -\frac{J}{2\rho^2}$$

where

$$\rho^2 = r^2 + \frac{1}{2}\left(Ml^2 - r_+^2\right), \quad r_+^2 = Ml^2\sqrt{1-\left(\frac{J}{Ml}\right)^2}$$

and J, M, l are constants characterizing some values of rotational momentum, mass and fundamental length type. In this case the la–metric (4.1) transforms in the well known BTZ–solution for three dimensional black holes [1].

We can also parametrize solutions for la–gravity of type (4.1) as to be equivalent to a locally isotropic anti–de Sitter space with cosmological constant $\Lambda = -\frac{1}{l^2}$ when coefficients (4.2) are modified by the relations $N(r) = N^\perp = f = \left(-M + \frac{r^2}{l^2} + \frac{J^2}{4r^2}\right)^{1/2}, N^\varphi(r) = -\frac{J}{2r^2}$, where $M > 0$ and $|J| \leq Ml$ and the solution has an outer event horizon at $r = r_+$ and inner horizon at $r = r_-$, $r_\pm^2 = \frac{Ml^2}{2}\{1 \pm \sqrt{1-\left(\frac{J}{Ml}\right)^2}\}$. We conclude that the N–connection could model both singular and nonsingular anisotropies of (2+1)–dimensional space–times.

Three dimensional la–solutions and strings

We proceed to study the possibility of imbedding of 3–dimensional solutions of la–gravity into the low energy dynamics of la–strings [7–9].

A la–metric

$$\delta s^2 = -K^{-1}(r)f(r)dt^2 + f^{-1}(r)dr^2 + K(r)\delta y^2, \qquad (4.3)$$

where $\delta y = dx_1 + n(r)dt$, i.e. $N_0^1 = n(r)$ and $N_0^i = 0$, solves the Einstein la–equations (2.6) if

$$n(r) = \frac{3}{4}\varsigma(r) + \frac{\dot\varsigma(r)}{\varsigma} \qquad (4.4)$$

with $\varsigma(r) = \dot f/f - \dot K/K$, where, for instance $\dot f = df/dt$.

Metrics of type (4.3) are considered [5] in an isotropic manner in connection to solutions of type IIA supergravity that describes a non–extremal intersection of a solitonic 5–brane, a fundamental string and a wave along one of common directions.

We have an anisotropic plane wave solution in $D+1$ dimensions if

$$\delta s^2 = -K^{-1}(r)f(r)dt^2 + f^{-1}(r)dr^2 + r^2d\Omega_{D-2}^2 + K(r)\delta y_{(\alpha)}^2 \qquad (4.5)$$

where $\delta y_{(\alpha)} = dx_1 + [1/K'(r) - 1 + \tan\alpha]dt$, $K(r) = 1 + \mu^{D-3}\sinh^2\alpha/r^{D-3}$, $(K'(r))^{-1} = 1 - \mu^{D-3}\sinh\alpha\cosh\alpha/(r^{D-3}K)$, $f(r) = 1 - \mu^{D-3}/r^{D-3}$ for isotropic solutions but $f(r)$ is a function defined by the prescribed component of N–connection (4.4) for la–spaces, $r^2 = x_2^2 + ...x_D^2$, and the parameter α define shift translations.

A la–string [7–9] solution is constructed by including (4.5) into a 10–dimensional la–metric with trivial shift $\delta y_{(\alpha=0)}$

$$\delta s_{(10)}^2 = H_f^{-1}\left[-\frac{f(r)}{K(r)}dt^2 + K(r)\delta y_{(0)}^2\right] + dx_2^2 + ... + dx_5^2 + H_{S5}\left[f^{-1}(r)dr^2 + r^2d\Omega_3^2\right]$$

where the la–string dilaton fields and antisymmetric tensor are defined [5] by the relations $e^{-2\phi} = H_{S5}^{-1}H_f$, $B_{01} = H_f^{-1} + \tanh\alpha_f$, $r^2 = x_6^2 + ...x_8^2$, $H_{ijk} = \frac{1}{2}\epsilon_{ijkl}\delta_l H_{S5}$ (in general, one considers la–derivations of type (2.2)) and $i,j,d,l = 6, ..., 9$.

Considering dimensional reductions in variables x_1, x_2, x_3, x_4, x_5 one can construct non–extremal, under singular anisotropies, 5–dimensional la–black hole solutions

$$\delta s_{(5)}^2 = -\lambda^{-2/3}f(r)dt^2 + \lambda^{1/3}\left[f^{-1}(r)dr^2 + r^2d\Omega_3^2\right]$$

where $\lambda = H_{S5}H_fK = \left(1 + \frac{Q_{S5}}{r^2}\right)\left(1 + \frac{Q_f}{r^2}\right)\left(1 + \frac{Q_K}{r^2}\right)$; Q_{S5}, Q_f and Q_K are constants. Finally we note that for la–backgrounds the function $f(r)$ is connected with the components of N–connection via relation (4.4), i.e. the N–connection structure could model both type of singular (like black hole) and nonsingular locally anisotropic string solutions.

CONCLUSIONS

The scenario of modelling of physical theories with generic locally anisotropic interactions on nonholonomic bundles provided with nonlinear connection structure has taught us a number of interesting things about a new class of anisotropic cosmological models, black hole solutions and low energy limits of string theories. Generic anisotropy of space–time could be a consequence of reduction from higher to lower dimensions and of quantum filed and space–time structure fluctuations in pre–inflationary period. This way an unification of logical aspects, geometrical background and physical ideas from the generalized Finsler and Kaluza–Klein theories was achieved.

The focus of this paper was to present some exact solutions with prescribed nonlinear connection for the locally anisotropic gravity and string theory. We have shown that a generic anisotropy of Friedmann–Robertson–Walker metrics could result in drastic modifications of cosmological models. It was our task here to point the conditions when the nonlinear connection will model singular, or nonsingular, anisotropies with three dimensional black hole solutions and to investigate the possibility of generalization of such type constructions to string theories.

Acknowledgments

The author would like to thank the Organizing Committee of International Conference "Particle, Fields and Gravitation", Lodz, April 15–19, 1998, for support of his participation. He is very grateful to Profs. A. Trautman and S. Bazanski for hospitality during his visit to Warsaw University.

REFERENCES

1. M. Banados, C. Teitelboim and J. Zanelli, Phys. Rev. Lett. **69**, 1849-1855 (1992); M. Henneaux, C. Teitelboim and J. Zanelli, Phys. Rev. **D48** 1506–1541 (1993).
2. E. Kamke, *Differentialgleichungen, Losungsmethoden und Losungen, I. Gewohnliche Differntialgleichungen* (Leipzig, 1959).
3. R. Miron and M. Anastasiei, *The Geometry of Lagrange Spaces: Theory and Applications* (Kluwer Academic Publishers, Dordrecht, Boston, London, 1994).
4. J. M. Overduin and P. S. Wesson, Phys. Rep. **283**, 303–378 (1997).
5. K. Sfetsos and K. Slenderis, hep–th/9711138.
6. S. Vacaru, J. Math. Phys. **37**, 508–523 (1997); gr-qc/9604015.
7. S. Vacaru, Annals of Physics (NY) **256**, 39-61 (1997); gr-qc/9604013.
8. S. Vacaru, Nucl. Phys. **B434**, 590–654 (1997); hep–th/9611034.
9. S. Vacaru, *Interactions, Strings and Isotopies in Higher Order Anisotropic Superspaces* (Hadronic Press, Palm Harbor, USA, 1998); summary in: physics/9706038.
10. S. Vacaru and Yu. Goncharenko, Int. J. Theor. Phys. **34**, 1955-1980 (1995); gr-qc/9604013.

Statistical Mechanics of Black Holes in Induced Gravity

Andrei Zelnikov[*][†][1]

[*]*Theoretical Physics Institute, University of Alberta, Edmonton, AB, T6G 2J1, Canada*
[†]*P.N.Lebedev Physics Institute, Leninskii Prospect 53, Moscow 117924, Russia*

Abstract. The density matrix of a black hole is constructed. On an example of induced gravity it is shown that the black hole entropy has a statistical mechanical meaning of the entropy of all dynamical degrees of freedom of matter fields that remain independent after resolution of effective constraint equations. This statement is independent of the particular variant of underlying microscopic field theory since it is based only on the low-energy asymptotic of the corresponding effective action.

INTRODUCTION

The problem of statistical-mechanical explanation of the black hole entropy has a long history. Inspite of an enormous variety of different approaches to the problem [1,20], till now there is no agreement about the answer on the fundamental question: "What are the dynamical degrees of freedom of the black hole, responsible for the origin of it's entropy ?".

In this talk of freedom of matter fields and the black hole entropy. On an example of induced gravity I demonstrate explicitly that the statistical-mechanical entropy $S = -\mathbf{Tr}[\hat{\rho}\ln\hat{\rho}]$ of all the matter degrees of freedom that remain independent after the resolution of the constraint equations is exactly the entropy of a black hole. It is important that the background metric and other background fields are assumed to satisfy the effective equations of motion, that include the back reaction effect. Taking into account the constraints on quantum matter fields drastically reduces the number of independent dynamical degrees of freedom. It resolves the problem of over-counting of independent oscillators, that appears in some other approaches [1,17,19].

The representation of the density matrix in terms of the Euclidean functional integral is quite general and applicable not only to black holes but to other static objects, e.g. stars etc.. It provides a possibility to describe in the same way entropies of the star before collapse and after it collapsed into a black hole. It can be shown that the entropy produced during the collapse of matter formed the black

---
[1] E-mail: zelnikov@phys.ualberta.ca

hole is exactly the Bekenstein-Hawking entropy $S_{BH} = \mathcal{A}/4l_{Pl}^2$, the Plank length l_{Pl} being determined in terms of induced gravitational constant.

PHYSICAL DEGREES OF FREEDOM

Let $\hat{\phi} \equiv \hat{\phi}^A = (\hat{g}_{\mu\nu}(x), \hat{A}_\mu(x), \hat{\Psi}(x), \ldots)$ be an arbitrary set of quantum fields described by classical action

$$I[\phi] = \int_\Sigma d^D x \, L(\phi) + \int_{\partial\Sigma} d^{D-1}y \, L_b(\phi) \tag{1}$$

All these fields are supposed to be dynamical, while some of them can be classically non-propagating. In other words it means that all the fields are allowed to evolve preserving the boundary conditions. The generating functional for connected Green's functions $W[J]$ is given by

$$e^{iW[J]} = \int D\phi \, e^{i(I[\phi]+J\phi)} \tag{2}$$

The effective action $\Gamma[\bar\phi]$ is the Legendre transform $\Gamma[\bar\phi] = W[J] - J\bar\phi$ where $\bar\phi = \frac{\delta W[J]}{\delta J}$ and, hence, when the background fields $\bar\phi$ satisfy the effective equations, $\frac{\delta \Gamma[\bar\phi]}{\delta \bar\phi} = -J$,

Then Eq.(2) can be rewritten in the form

$$e^{i\Gamma[\bar\phi]} = \int D\phi \, e^{i\left[I[\phi] - \frac{\delta\Gamma[\bar\phi]}{\delta\bar\phi}(\phi-\bar\phi)\right]} \tag{3}$$

In the absence of external currents, the system is on shell and the equation for the effective action simplifies to the problem

$$e^{i\Gamma[\bar\phi]} = \int D\phi \, e^{i \, I[\bar\phi+\phi]} \, , \qquad \frac{\delta \Gamma[\bar\phi]}{\delta \bar\phi} = 0 \, . \tag{4}$$

All the fields ϕ are divided onto propagating fields η and non-propagating λ, which do not have a kinetic term in the action and play the role similar to that of Lagrange multipliers.

DENSITY MATRIX OF A BLACK HOLE

To construct a density matrix we follow the ideology of the paper [21]. For any static background one can define the no boundary wave function, which is a functional integral over the half-instanton with fixed values of quantum fields $\phi(y) = \phi|_\Sigma$ on the boundaries (at infinity, and at $\tau = (0, \beta/2)$). We define the black hole density matrix by identification of the part of the boundary Σ at $\tau = \beta/2$ of the half instanton corresponding to the wave function and that of the corresponding to the conjugate wave function. Let Σ be separated by a $(D-2)$-surface σ on two parts, inside and outside of the surface, Σ_{in} and Σ_{out} correspondingly. If we identify the values of the quantum fields on Σ_{out} and then trace over all possible values, we get the **density matrix** of the system of fields located inside the surface B. Then the density matrix $\hat\rho = \exp\left[\Gamma - \beta\hat H\right]$ can be rewritten as a functional integral over the **complete instanton** $0 \leq \tau \leq \beta$, the arguments of the density matrix being the values of the quantum fields on different branches of the cut of the instanton

$$\rho(\phi';\phi) = <\phi'\mid\hat\rho\mid\phi> = e^{\Gamma[\varphi]} \int_{\phi(\Sigma_{in})=\phi;\ \phi(\Sigma'_{in})=\phi'} D\phi\mu(\phi)\ e^{-I^E[\phi]} \tag{5}$$

where the Euclidean effective action $\Gamma[\varphi]$ is determined from the condition $\mathrm{Tr}\ \hat\rho = 1$

$$\exp(-\Gamma[\varphi]) = \int D\phi\mu(\phi)\ \exp(-I^E[\phi]) = \int \frac{d\pi\, d\phi}{2\pi}\ e^{-\int_0^\beta d\tau\{\pi\dot\phi - H(\pi,\phi)\}} \tag{6}$$

and the mean field $\varphi = <\phi>$ is the solution of the effective equation $\frac{\delta\Gamma[\varphi]}{\delta\varphi(x)} = 0$. Note that the density matrix is defined in terms of the **canonical** Hamiltonian $\hat H$. The Liouville measure $d\pi d\phi$ in the phase space path integral guarantees unitarity of evolution in Minkowskian signature while manifestly violates general covariance. Integration over the momenta of physical fields in the phase space functional integral gives in general case a non-covariant local measure $\mu(\phi)$ for the path integral in configuration space [22]. This fact could create some problems but in induced gravity models, we are interested in, the number of degrees of freedom of bosons is equal to that of fermions and, as a consequence, the non-covariant terms cancel in the total measure in configuration space.

INDUCED GRAVITY MODEL

Consider a model of induced gravity, where Einstein gravity arises effectively only after quantization of matter fields. On classical level spacetime excitations do not propagate and, therefore, there is no way to ascribe entropy to the gravitational field. On the other hand, there is a black hole solution of the effective gravitational equations and, hence, the corresponding Bekenstein-Hawking entropy. The question arises: "Is there any relation of the entropy of excitations of constituent matter fields and Bekenstein-Hawking entropy?".

To answer this question consider a particular induced gravity model in four dimensions where matter fields are only minimal scalars, vector fields, and Dirac spinors. All the fields are assumed to be massive.

Here we restrict ourselves with minimal scalars. In the opposite case there would be $R\phi^2$ terms in classical Lagrangian which would make metric perturbations propagating even if there is no pure Einstein R term in Lagrangian, and the gravity is not induced in a strict sense. If there were terms in the classical Lagrangian explicitly containing scalar curvature, one could always redefine fields and metric in such a way, that only minimally coupled fields and pure Einstein action will remain in the classical Lagrangian. On the other hand the functional integral representation for the density matrix and the dynamics of the whole system is invariant under the field redefinitions, Thus for such systems the Einstein gravity is not induced and exists at the very beginning.

In this paper we consider only minimally coupled to geometry fields what allows us to relate the entropy of microscopic constituents to that of the self-consistent solutions of **effective** gravitational equations.

The total classical action of the system in question reads

$$I = \sum_{i=1}^{n_\phi} I_\phi^i + \sum_{i=1}^{n_\psi} I_\Psi^i + \sum_{i=1}^{n_A} I_A^i \tag{7}$$

where classical actions I^i are

$$I_\phi = -\frac{1}{2} \int \sqrt{g} \, [\, (\nabla \phi)^2 + m^2 \phi^2 \,], \tag{8}$$

$$I_\psi = -\int \sqrt{g} \, [\, \bar{\psi} \gamma^\mu \nabla_\mu \psi + \mu^2 \bar{\psi} \psi \,], \tag{9}$$

$$I_A = -\frac{1}{4} \int \sqrt{g} \, [\, F_{\mu\nu} F^{\mu\nu} + M^2 A_\mu A^\mu \,]. \tag{10}$$

One should impose the condition of cancellation of leading ultraviolet divergences to obtain the effective gravity with finite cosmological and gravitational constants.

Then we obtain 6 algebraic constraint equations on the number and masses of the fields

$$c_1 = n_\phi - 4n_\psi + 3n_A = 0,$$

$$c_2 = n_\phi + 8n_\psi - 9n_A = 0,$$

$$c_3 = \sum_{i=1}^{n_\phi} m_i^2 - 4\sum_{i=1}^{n_\psi} \mu_i^2 + 3\sum_{i=1}^{n_A} M_i^2 = 0,$$

$$c_4 = \sum_{i=1}^{n_\phi} m_i^2 + 8\sum_{i=1}^{n_\psi} \mu_i^2 - 9\sum_{i=1}^{n_A} M_i^2 = 0,$$

$$c_5 = \sum_{i=1}^{n_\phi} m_i^4 - 4\sum_{i=1}^{n_\psi} \mu_i^4 + 3\sum_{i=1}^{n_A} M_i^4 = 0,$$

$$c_6 = \sum_{i=1}^{n_\phi} m_i^4 \ln m_i^2 - 4\sum_{i=1}^{n_\psi} \mu_i^4 \ln \mu_i^2 + 3\sum_{i=1}^{n_A} M_i^4 \ln M_i^2 = const. \tag{11}$$

The last constraint means that induced cosmological constant Λ_{ind} is finite. For simplicity we put here $\Lambda_{ind} = 0$. The induced gravitational constant is finite and is a function of masses of fields

$$\frac{1}{G_{ind}} = \frac{1}{12\pi} \left[\sum_{i=1}^{n_\phi} m_i^2 \ln m_i^2 + 8\sum_{i=1}^{n_\psi} \mu_i^2 \ln \mu_i^2 - 9\sum_{i=1}^{n_A} M_i^2 \ln M_i^2 \right] \tag{12}$$

These constraints Eq.(11) have a number of solutions if $n_\phi = n_\psi = n_A \geq 2$. We do not consider divergences quadratic in curvatures, since they are irrelevant to the problem of the entropy of black holes in Einstein gravity.

CANONICAL HAMILTONIAN IN INDUCED GRAVITY

To describe the time evolution of the system one needs to know the canonical Hamiltonian

$$H = \int_\Sigma d^{D-1}x \, [\pi_A \dot\phi^A - L(\phi)] - \int_{\partial\Sigma} d^{D-2}y \, L_b(\phi) \tag{13}$$

where $\pi_A = \delta I/\delta\dot\phi^A$ is the canonically conjugated momentum. As usual, one should substitute the operators of field $\hat\phi$ and momentum $\hat\pi$ instead of ϕ and π to get the operator of Hamiltonian $\hat H$. Statistical mechanics of the system is also described by the canonical Hamiltonian. In order to construct the density matrix of the system, one needs to know what are it's degrees of freedom.

The Euclidean functional integral representation for the density matrix for the set of fields $\{g_{\mu\nu}, A, \psi, \phi, A\}$ reads

$$\rho(g', A', \psi', \phi'; g, A, \psi, \phi) = e^\Gamma \int_{g,a,\psi,\phi}^{g',a',\psi',\phi'} Dg \, DA \, D\psi \, D\phi \, e^{-I^E[g,A,\psi,\phi]} \tag{14}$$

In induced gravity models the number of bosonic dynamical degrees of freedom is equal to the number of fermionic ones $(c_1 = 0)$ and, hence, the functional integral measure $\mu = [g^{1/4}(g^{00})^{1/2}]^{c_1} = 1$. Therefore, after we integrate over the momenta in the phase space functional integral only the usual covariant measure appears in the total resulting configuration space functional integral.

The canonical Hamiltonian $\hat H$ of the whole system is the sum of canonical Hamiltonians of matter fields only, since metric is non-propagating classically. Of course, after the quantization of matter fields effective gravity becomes propagating. Now return to the gravitational effective action $\Gamma^E_g[g, \bar A, \bar\psi, \bar\phi]$

$$e^{-\Gamma} = \int Dg \, DA \, D\psi \, D\phi \, e^{-I^E[g,A,\psi,\phi]} = \int Dg \, e^{-\Gamma^E_g[g,\bar A,\bar\psi,\bar\phi]} \tag{15}$$

It is a covariant functional of the metric and its expansion in powers of curvatures begins from local terms corresponding to the Einstein action with the proper boundary term Ref. [23]

$$\Gamma^E_g[g, \bar A, \bar\psi, \bar\phi] = -\frac{1}{8\pi G_{ind}} \left[\int_\mathcal{M} R + 2\int_{\partial\mathcal{M}} K \right] + O(R^2) \tag{16}$$

Terms, that are quadratic in curvatures, are generally nonlocal and depend on boundary conditions imposed on quantum fields at the boundary. It should be emphasized that there are boundary terms only on the external boundary of Hartle-Hawking instanton. No boundary terms appear on the horizon, which is a regular point on the instanton. This observation is not quite trivial, since the effective action has been derived from the canonical formalism, which is singular on the Euclidean horizon. The remarkable property of induced gravity models is that one can easily resolve the complications connected with the canonical quantization of the set of fields and get manifestly covariant representation for the resulting functional integral in configuration space.

STATISTICAL-MECHANICAL ENTROPY AND CONSTRAINTS

The statistical-mechanical entropy of the system is defined by the formula

$$S^{SM} = -\text{Tr}\,[\hat\rho\,\ln\hat\rho] = -\text{Tr}\,[\hat\rho\,\{\Gamma - \beta\hat H\}] = \beta <\hat H> -\Gamma = \beta E - \Gamma \qquad (17)$$

This is exactly the integral form of the second thermodynamical law and, hence, the statistical-mechanical and thermodynamical entropies coincide.

Thus, to calculate the entropy we have to calculate the average Hamiltonian and the effective action on shell. Let us begin with the Hamiltonian. For any generally covariant field theory the canonical Hamiltonian can always be represented in the form of a sum

$$\hat H = \hat{\mathcal H} + \hat H_s \qquad (18)$$

where $\mathcal H$ is a linear combinations of effective gravitational constraints and the second term $\hat H_s$ is the "surface" Hamiltonian - spatial integral from a total derivative. The Hamiltonian constraint comes from the variation of action over the metric and reads

$$\mathcal H = \int_M N\frac{\delta I}{\delta N} + N^i\frac{\delta I}{\delta N^i} = \int_M \frac{1}{8\pi G_\infty}[R_0^0 - \frac{1}{2}R + \Lambda_\infty] - \int_M T_0^0 \qquad (19)$$

where T_ν^μ is the energy-momentum tensor of all matter fields. The quantum average of the canonical Hamiltonian which enters the entropy is the sum of the average of bulk Hamiltonian, which is proportional to constraints, and the average of H_s. For minimal scalar fields and Dirac fermions canonical Hamiltonian coincides with $\mathcal H$, while for massive vector fields $H_s \neq 0$. Thus in our induced gravity model we have

$$H_s = -\int_M \nabla_\mu(A_0 F^{0\mu}) \qquad (20)$$

All the quantities we are dealing with in this approach are to be taken on shell of effective equations. So the gravitational constraints are automatically satisfied and $<\hat{\mathcal H}> = 0$. Then the entropy Eq.(17) reads

$$S = \beta <\hat{H}_s> -\Gamma \tag{21}$$

We are interested in the case of a black hole solution and. Black holes do not have massive vector hairs and, hence, background vector fields are zero. So in this case quantum average of the surface Hamiltonian is zero and the canonical Hamiltonian (on shell) is zero too. Finally we obtain

$$S = -\Gamma \tag{22}$$

Note that in this derivation we obtained the entropy of the whole system of matter fields inside the external boundary at infinity, which includes the entropy of the quantum atmosphere of the black hole. It may also contain the entropy related to the external boundary. The total Γ differs from the classical Einstein effective action on boundary terms. Now the problem is to separate the entropy of the black hole itself from that of Hawking radiation and boundary contributions. To do this, in the next section we compare the black hole entropy and the entropy of ordinary matter distribution of the same total mass.

There was a lot of attempts to explain the whole entropy of the black holes as the entropy of the atmosphere. One can show that Hawking quanta can provide only small quantum corrections $O(\hbar)$ to the Bekenstein-Hawking entropy, while the main effect comes from the collective excitations made of zero-point modes that remain independent after resolution of the gravitational constraint equations. At first sight it sounds strange that zero-point fluctuations can contribute to the entropy, but it's not surprising if one remembers that the effective gravitational constraint equations entangle vacuum fluctuations and thermal fluctuations of the renormalized matter. The importance of zero-point oscillators has been also noticed in Ref. [24].

ENTROPY OF A BLACK HOLE VS ENTROPY OF A STAR

The same consideration remains valid in application to the ordinary star. The construction for the density matrix is exactly the same. The only difference is the topology of Euclidean solution. For the Gibbons-Hawking instanton it is $Disk \times S^2$ while for the star it is $Cylinder \times S^2$. The horizon is an internal point of the Gibbons-Hawking instanton, and there is only an external boundary. The Euclidean star also has only an external boundary. At infinity the geometries of the black hole and the star of the same mass asymptotically coincide. So we can compare the entropies of these two systems to determine how much entropy that has been produced during the collapse of the star. In the difference the the dependence of the entropy on boundary terms and the entropies of thermal atmosphere of the black hole and the star of the same mass and temperature cancel each other. Of course the star has its own entropy, but it turns out to be negligibely small compare to black hole entropy. What is not quite surprising because the Planck constant comes as $1/\hbar$ in the expression Bekenstein-Hawking entropy, while the entropy of any ordinary (renormalized) matter is proportional to $\hbar^0$.

To calculate the difference of entropies it is convenient to use the effective equations. The effective action consists from a vacuum part, which is local and gives Einstein action with induced gravitational constant and renormalized (nonlocal) effective action of matter fields, describing matter excitations.

$$\Gamma = I_{grav\ ind} + \Gamma_{ind} = -\frac{1}{16\pi G_{ind}}\left[\int_\mathcal{M} dxg^{1/2}R + 2\int_{\partial \mathcal{M}} dxh^{1/2}K\right] + \Gamma_{ind} \quad (23)$$

Thus, the amount of entropy that has been produced during the collapse of the star reads

$$\Delta S = -\bar{\Gamma}_{bh} + \bar{\Gamma}_\odot = \frac{1}{16\pi G_{ren}}\int_{\mathcal{M}_{bh}} dxg^{1/2}[R - 2\Lambda_{ren}] - \Gamma_{bh\ ren}$$
$$-\frac{1}{16\pi G_{ren}}\int_{\mathcal{M}_\odot} dxg^{1/2}[R - 2\Lambda_{ren}] + \Gamma_{\odot\ ren} \quad (24)$$

Here the boundary terms $2\int K$ for the star and black hole are equal and canceled each other.

On shell (on the self-consistent solution) the answer can be rewritten as

$$\Delta S = \int_{\mathcal{M}_{bh}} dxg^{1/2}\left[\frac{1}{8\pi G_{ren}}R_0^0 - <T_0^0>_{ren}\right] - \Gamma_{bh\ ren}$$
$$-\int_{\mathcal{M}_\odot} dxg^{1/2}\left[\frac{1}{8\pi G_{ren}}R_0^0 - <T_{\odot 0}^{\ 0}>_{ren}\right] + \Gamma_{\odot\ ren} \quad (25)$$

where $\Gamma_{ind} = -\int \mathcal{L}$ and $<T_0^0>_{ren} = \bar{N}^\mu \delta\Gamma_{ind}/\delta\bar{N}^\mu$.

The contribution to the entropy that comes from $<T_\mu^\nu>_{ren}$ and Γ_{ind} is finite on any regular manifolds. It includes the entropy of the thermal atmosphere of a black hole and adds some amount to the total entropy of the system. Note that though the local temperature of thermal atmosphere goes to infinity at the horizon $<T_\mu^\nu>_{ren}$ and Γ_{ind} contain also vacuum polarization parts that exactly cancel the thermal divergences. This property takes place for any fields on the on shell background solutions.

The integrals $\int dxg^{1/2}R_0^0$ calculated for the star and the pure Schwarzschild manifold of the same total mass differ on $O(\hbar)$. Therefore, one gets

$$-\frac{1}{8\pi}\int_{\mathcal{M}_{bh}} dxg^{1/2}R_0^0 = O(\hbar) \quad (26)$$

On the other hand, the spatial integral from R_0^0 gives ADM mass of the matter inside this volume for any equilibrium distribution of matter. For the star we have

$$-\frac{1}{8\pi}\int_{\mathcal{M}_\odot} dxg^{1/2}R_0^0 = -\frac{1}{2}\beta_\infty M + O(\hbar) \quad (27)$$

where M is the total mass of the star (equal to the mass of the black hole) and $\beta_\infty = 8\pi M$ is the inverse temperature of the black hole at infinity (equal to that of the star). Substituting formulas (26,27) into (25) we finally obtain the entropy produced during the formation of the black hole

$$\Delta S = \frac{1}{2\hbar}\pi\beta_\infty M + O(\hbar^0) = \frac{\mathcal{A}_H}{4\hbar G_{ren}} + O(\hbar^0) \qquad (28)$$

where we expressed the inverse temperature and the area of the event horizon in terms of the black hole mass. This result coincides with the Bekenstein-Hawking entropy, ascribed to the black hole. Quantum corrections are always much smaller. Even if the entropy of thermal atmosphere surrounding the black hole is large quantitavily, because it is proportional to the space volume inside the boundary, the difference of the entropy and that of the thermal atmosphere of the star is small, since their asymptotics at infinity coincide.

ACKNOWLEDGMENTS

I am most grateful to my colleagues A.Barvinskii, V.Frolov, and D.Fursaev for inspiring discussions of this paper. This work was supported by Killam Trust.

REFERENCES

1. t'Hooft G., Nucl.Phys., **B256**, 727 (1985).
2. Zurek W.H., Thorne K.S., Phys.Rev.Lett., **54**, 2171 (1985).
3. Bombelli L., Koul R., Lee J., and Sorkin R., Phys.Rev., **D34**, 373 (1986).
4. Brown J.D., Martinez E.A., and York J.W., Phys.Rev.Lett., **D66**, 2281 (1991).
5. Coleman S., Krauss L.M., Preskill J., Wilczek F., Gen.Rel.Grav., **24**, 9 (1992).
6. Srednicki M., Phys.Rev., **71**, 666 (1993).
7. Frolov V.P. and Novikov I.D., Phys.Rev., **D48**, 4545 (1993).
8. Maggiore M., Nucl.Phys. **B429**, 205 (1994).
9. Carlip S., Phys.Rev., **D51**, 632 (1995).
10. Frolov V.P., Fursaev D.V., and Zelnikov A.I., Phys.Rev., **D54**, 2711 (1996).
11. Bekenstein J.D., Mukhanov V.F., Phys.Lett., **B360**, 7 (1995).
12. Strominger A., Vafa C., Phys.Lett., **B379**, 99 (1995).
13. Callan C.G., Maldacena J.M., Nucl.Phys., **B472**, 591 (1996).
14. Horowitz G.T., Strominger A., Phys.Rev.Lett., **77**, 2368 (1996).
15. Larsen F., Wilczek F., Nucl.Phys., **B475**, 627 (1996).
16. Susskind L., Uglum J., Nucl.Phys. **B** *(Proc.Suppl)*, 115 (1996).
17. Frolov V.P., Fursaev D.V., and Zelnikov A.I., Nucl.Phys., **B486**, 339 (1997).
18. Balachandran A.P., Chandar L., and Momen A., *Int.J.Mod.Phys.*, **A12**, 62 (1997).
19. Frolov V.P., Fursaev D.V., Phys.Rev., **D56**, 2212 (1997).
20. Brotz T., Phys.Rev. **D57** 2349 (1998).
21. Barvinskii A.O., Frolov V.P., and Zelnikov A.I., Phys.Rev., **D51**, 1741 (1995).
22. Fradkin E.S., Vilkoviskii G.A., Phys.Rev., **D8**, 4241 (1973).
23. Barvinskii A.D., Solodukhin S.N., Nucl.Phys., **B479**, 305 (1996).

24. Belgiorno F., Liberati S., Gen.Rel.Grav., **29**,1181 (1997).

List of Participants

Ali Riza Akcay
aakcay@bus.netas.com.tr
NETAS
Alemdag Cad. 151, Umraniye
Istanbul 81244
Turkey

Henryk Arodź
afarodz@thp1.if.uj.edu.pl
Institute of Physics
Jagiellonian University
ul. Reymonta 4
30-059 Cracow
Poland

Dumitru Baleanu
baleanu@venus.ifa.ro
Institute for Space Sciences
P.O. Box MG-36
Magurele–Buchurest 76900
Romania

Stanisław L. Bażański
Stanislaw.Bazanski@fuw.edu.pl
Institute of Theoretical Physics
University of Warsaw
ul. Hoża 69
00-681 Warszawa
Poland

Carl M. Bender
cmb@howdy.wustl.edu
Department of Physics
Washington University
St. Louis, MO 63130
U.S.A.

Iwo Białynicki-Birula
birula@cft.edu.pl
Polish Academy of Sciences
al. Lotników 32/46
02-668 Warszawa
Poland

Andrzej Borowiec
borow@ift.uni.wroc.pl
Institute of Theoretical Physics
University of Wrocław
pl. M. Borna 9
50-204 Wrocław
Poland

Marek Bożejko
bozejko@math.uni.wroc.pl
Institute of Mathematics
University of Wrocław
pl. Grunwaldzki 2/4
50-384 Wrocław
Poland

Yves Brihaye
Yves.Brihaye@umh.ac.be
Dept. of Mathematical Physics
University of Mons
Av. Maistriau
B-7000 Mons
Belgium

Bogusław Broda
bobroda@mvii.uni.lodz.pl
Dept. of Mathematical Physics
University of Łódź
ul. Pomorska 149/153
90-236 Łódź
Poland

Tomasz Brzeziński
tb10@york.ac.uk
Department of Mathematics
University of York
Heslington
York YO1 5DD
England

Paolo Budinich
Interdisciplinary Lab. for
Advanced Sciences
SISSA
via Beirut 4
I-34014 Trieste
Italy

Paweł Caban
caban@mvii.uni.lodz.pl
Dept. of Mathematical Physics
University of Łódź
ul. Pomorska 149/153
90-236 Łódź
Poland

Michel Cahen
mcahen@ulb.ac.be
Departement de Mathematiques
Université Libre de Bruxelles
Campus de la Plaine
Boulevard du Triomphe
1050 Bruxelles
Belgium

Marco Cavaglià
cavaglia@aej-potsdam.mpg.de
Max-Planck-Inst. für
Gravitationsphysik
Schlaatzweg 1
D-14476 Potsdam
Germany

Jacek Ciborowski
cib@hozavx.fuw.edu.pl
Institute of Experimantal Physics
University of Warsaw
ul. Hoża 69
00-681 Warszawa
Poland

Robert Coquereaux
Robert.Coquereaux@cpt.univ-mrs.fr
Groupe « Interactions
Fondamentales »
Centre de Physique Theorique
Centre National de la Recherche
Scientifique
Case 907 – Campus de Luminy
13288 Marseille Cedex 9
France

Vladimir Dobrev
dobrev@bgearn.acad.bg
Institute of Nuclear Research and
Nuclear Energy
Bulgarian Academy of Sciences
72 Tsarigradsko Chaussee
1784 Sophia
Bulgaria

Heinz-Dietrich Doebner
asi@pt.tu-clasthal.de
Arnold Sommerfeld Inst. für
Mathematische Physik
Technische Universität Clausthal
Leibnizstrasse 10
D-38678 Clausthal-Zellerfeld
Germany

Maciej Dunajski
dunajski@math.ox.ac.uk
Mathematical Institute
University of Oxford
24-29 St. Giles'
Oxford OX1 3LB
U.K.

Irina Dymnikova
irina@tufi.wsp.olsztyn.pl
Inst. of Mathematics and Physics
Pedagogical University of Olsztyn
ul. Żołnierska 14
10-561 Olsztyn
Poland

Jan Fischer
fisher@fzu.cz
Institute of Physics
Acad. of Sciences of Czech Rep.
Na Slovance 2
CZ-18040 Praha
Czech Republic

Moshé Flato
flato@u-bourgogne.fr
Department de Mathématique
Univesité de Bourgogne
BP 130
F-21004 Dijon
France

Sebastian Formański
sforman@ck-sg.p.lodz.pl
Institute of Physics
Technical University of Łódź
ul. Wólczańska 219
93-005 Łódź
Poland

Klaus Fredenhagen
fredenha@x4u2.desy.de
II. Inst. für Theoretische Physik
Universität Hamburg
Luruper Chaussee 149
D-22761 Hamburg
Germany

Christian Fronsdal
fronsdal@physics.ucla.edu
Physics Department
University of California
Los Angeles, CA 90095-1547
U.S.A.

Andrzej Frydryszak
amfry@ift.uni.wroc.pl
Institute of Theoretical Physics
University of Wrocław
pl. M. Borna 9
50-204 Wrocław
Poland

Stefan Giller
sgiller@mvii.uni.lodz.pl
Dept. of Theoretical Physics II
University of Łódź
ul. Pomorska 149/153
90-236 Łódź
Poland

Gerard Goldin
gagoldin@dimacs.rutgers.edu
Dept. of Mathematics and Physics
Rutgers University
New Brunswick, NJ 08903
U.S.A.

Cezary Gonera
cgonera@mvii.uni.lodz.pl
Dept. of Theoretical Physics II
University of Łódź
ul. Pomorska 149/153
90-236 Łódź
Poland

Eduard Gorbar
gorbar@ap3.gluk.apc.org
13-a Sholom-Aleykhema str., ap. 34
Kiev 253156
Ukraine

Michael Heller
mheller@alumn.wsd.tarnow.pl
Vatican Obervatory
V-12000 Vatican City State

Andrzej Horzela
horzela@solaris.ifj.edu.pl
H. Niewodniczański Inst. of Nuclear Physics
ul. Radzikowskiego 152
31-342 Cracow
Poland

Georg Junker
junker@faupt100.physik.uni-erlangen.de
Inst. für Theoretische Physik I
Universität Erlangen-Nürnberg
Staudstraße 7
D-91058 Erlangen
Germany

Igor Kanatchikov
ikonat@ippt.gov.pl
Tallin Technical University
Ehitajate tee 5
EE0026 Tallin
Estonia

Edward Kapuścik
sfkapusc@cyf-kr.edu.pl
Inst. of Physics and Computer Sci.
Pedagogical University of Cracow
ul. Podchorążych 2
30-084 Cracow
Poland

Louis Kauffman
kauffman@uie.edu
Dept. of Mathematics, Statistics
and Computer Sci.
University of Illinois at Chicago
Chicago, IL 60607-7045
U.S.A.

Richard Kerner
rk@ccr.jussieu.fr
Laboratiore de Gravitation
et Cosmologie Relaivistes
Université Pierre et Marie Curie –
CNRS URA 769
4, Place Jussieu
F-75005 Paris
France

George Khimshiashvili
khimsh@rmi.acnet.ge
Dept. of Geometry and Topology
Tbilisi Institute of Mathematics
M. Aleksidze str.1
380093 Tbilisi
Georgia

Maurice Kibler
kibler@lyoaxp.in2p3.fr
Institute de Physique Nucléaire
Université Claude Bernard Lyon I
43, Bd. du 11 Novembre 1918
F-69622 Villeurbanne Cedex
France

Jerzy Kijowski
kijowski@theta1.cft.edu.pl
Dept. of Mathematical Methods
of Physics
University of Warsaw
ul. Hoża 74
00-682 Warszawa
Poland

Jan Kłosiński
janklos@mvii.uni.lodz.pl
Dept. of Theoretical Physics
University of Łódź
ul. Pomorska 149/153
90-236 Łódź
Poland

Mikhail Knyazev
knyazev@iaph.bas.net.by
National Acad. of Sciences of
Belarus
Institute of Applied Physics
16 F. Scarina str.
220072 Minsk
Belarus

Piotr Kosiński
pkosink@mvii.uni.lodz.pl
Dept. of Theoretical Physics II
University of Łódź
ul. Pomorska 149/153
90-236 Łódź
Poland

Marcin Kościelecki
kosciej@fuw.edu.pl
Dept. of Mathematical Methods
of Physics
University of Warsaw
ul. Hoża 74
00-682 Warszawa
Poland

Krzysztof Kowalski
kowalski@krysia.uni.lodz.pl
Dept. of Mathematical Physics
University of Łódź
ul. Pomorska 149/153
90-236 Łódź
Poland

Sebastian Kubis
kubis@solaris.ifj.edu.pl
Institute of Nuclear Physics
im. H. Niewodniczańskiego
ul. Radzikowskiego 152
31-342 Cracow
Poland

Petr Kulish   kulish@pdmi.ras.ru
St. Petersburg Department of
Steklov Mathematical Institute
Fontanka 27
191 011 St. Petersburg
Russia

Jutta Kunz
kunz@merlin.physik.uni-oldenburg.de
Appollostraat 3
CN Sassenheim
NL 2172
The Netherlands

Andrzej K. Kwaśniewski
Institute of Physics
University of Białystok
ul. M. Curie-Skłodowskiej 14
Białystok
Poland

Julian Ławrynowicz
Department of Solid State Physics
University of Łódź
ul. Pomorska 149/153
90-236 Łódź
Poland

Jerzy Lewandowski
Jerzy.Lewandowski@fuw.edu.pl
Institute of Theoretical Physics
University of Warsaw
ul. Hoża 69
00-681 Warszawa
Poland

Jan Łopuszański
lopus@proton.ift.uni.wroc.pl
Institute of Theoretical Physics
University of Wrocław
pl. M. Borna 9
50-204 Wrocław
Poland

Jerzy Lukierski
lukier@proton.ift.uni.wroc.pl
Institute of Theoretical Physics
University of Wrocław
pl. M. Borna 9
50-204 Wrocław
Poland

John Madore
madore@qcd.th.u-psud.fr
Laboratoire de Physique
Theoretique et Hautes Energies
Université de Paris-Sud
91405 Orsay
France

Michał Majewski
mimajew@mvii.uni.lodz.pl
Dept. of Theoretical Physics
University of Łódź
ul. Pomorska 149/153
90-236 Łódź
Poland

Władysław A. Majewski
fizwam@univ.gda.pl
Dept. of Mathematical Methods
of Physics
University of Gdańsk
ul. Wita Stwosza 57
80-952 Gdańsk-Oliwa
Poland

Hans-Jürgen Mann
pthjm@pt.tu-clasthal.de
Institute for Theoretical Physics
Technical University Clausthal
Arnold-Sommerfeld-Str. 10
D-38678 Calusthal-Zellerfeld
Germany

Kishore Marathe
kmb@sci.brooklyn.cuny.edu
Brooklyn College
City University of New York
2900 Bedford Avenue
City Brooklyn, NY 11210-2889
USA

Władysław Marcinek
wmar@ift.uni.wroc.pl
Institute of Theoretical Physics
University of Wrocław
pl. M. Borna 9
50-204 Wrocław
Poland

Paweł Maślanka
pmaslan@krysia.uni.lodz.pl
Dept. of Theoretical Physics II
University of Łódź
ul. Pomorska 149/153
90-236 Łódź
Poland

Sergey Mayburov
maybur@x4u.lpi.ruhep.ru
Russian Academy of Sciences
Lebediev Institute of Physics
Leninsky pr. 53
117924 Moscow
Russia

Boris Meierovich
trc@meyer.msk.su
Kapitza Institute of Physical
Problems
2 Kosygina Str.
117334 Moscow
Russia

Pierre Minnaert
minnaert@bortibm4.in2p3.fr
Laboratoire de Physique
Théorique
Université Bordeaux I
33175 Gradignan
France

Marek Mozrzymas Institute of
Theoretical Physics
University of Wrocław
pl. M. Borna 9
50-204 Wrocław
Poland

Jiří Niederle
Institute of Physics
Acad. of Sciences of Czech Rep.
Na Slovance 2
18040 Praha 8
Czech Republic

Bogdan Nowak
bnowak@krysia.uni.lodz.pl
Dept. of Theoretical Physics
University of Łódź ul. Pomorska
149/153
90-236 Łódź
Poland

Jean Nuyts
Jean.Nuyts@umh.ace.be
Physique Théoretique et
Mathématique
Université de Mons-Hainaut
20 Place du Parc
7000 Mons
Belgium

Anatol Odzijewicz
aodzijew@cksr.ac.bialystok.pl
Institute of Physics
University of Białystok
ul. M. Curie-Skłodowskiej 14
Białystok
Poland

Victor Olkhov
olkhov@dpc.asc.rssi.ru
Lebedev Physical Institute
Profsouznaia 84/32
117810 Moscow
Russia

Marek Pawłowski
pawlowsk@fuw.edu.pl
A. Sołtan Inst. of Nuclear
Problems
ul. Hoża 69
00-681 Warsaw
Poland

Włodzimierz Piechocki
piech@fuw.edu.pl
A. Sołtan Inst. of Nuclear
Problems
ul. Hoża 69
00-681 Warsaw
Poland

Jerzy Plebański
Jerzy.Plebanski@fis.cinvestav.mx
Departamento de Física
Centro de Investigación y
de Estudos Avanzados del I.P.N.
Mexico D.F. 07000
Mexico

Piotr Podleś
podles@katedra.fuw.edu.pl
Dept. of Mathematical Methods
of Physics
University of Warsaw
ul. Hoża 74
00-682 Warsaw
Poland

Ziemowit Popowicz
ziemek@proton.ift.uni.wroc.pl
Institute of Theoretical Physics
University of Wrocław
pl. M. Borna 9
50-204 Wrocław
Poland

Maciej Przanowski
mprzan@ck-sg.p.lodz.pl
Institute of Physics
Technical University of Łódź
ul. Wólczańska 219
93-005 Łódź
Poland

Piotr Rączka
Piotr.Raczka@fuw.edu.pl
Institute of Theoretical Physics
University of Warsaw
ul. Hoża 69
00-681 Warsaw
Poland

Jakub Rembieliński
jaremb@mvii.uni.lodz.pl
Dept. of Mathematical Physics
University of Łódź
ul. Pomorska 149/153
90-236 Łódź
Poland

Olena Roman
asor@pt.tu-clausthal.de
Institute for Theoretical Physics
Technical University Clausthal
Arnold-Sommerfeld-Str. 10
D-38678 Calusthal-Zellerfeld
Germany

Leszek Roszkowski
Leszek.Roszkowski@cern.ch
Department of Physics
Lancaster University
Lancaster LA1 4YB
England

Gerd Rudolph
rudolph@rz.uni-leipzig.de
Institut für Theoretische Physik
Univerität Leipzig
Augustplatz 10/11
D-04109 Leipzig
Germany

Serge Salata
sca@blizzard.sabbo.kiev.ua
Astronomical Observatory
Kiev University
Zhukova 45 apt. 40
253166 Kiev
Ukraine

Yurii Shil'nov
visit2@ieec.fcr.es
Espacials de Catalunya
Edifici Nexus 104
Gran Capita 2-4
E-08034 Barcelona
Spain

Dmitry Shirkov
shirkovd@thsun1.jinr.dubna.su
JINR
141980 Dubna, Moscow reg.
Russia

Kordian A. Smoliński
xmolin@mxii.uni.lodz.pl
Dept. of Mathematical Physics
University of Łódź
ul. Pomorska 149/153
90-236 Łódź
Poland

Jan Sobczyk
jsobczyk@proton.ift.uni.wroc.pl
Institute of Theoretical Physics
University of Wrocław
pl. M. Borna 9
50-204 Wrocław
Poland

Dmytro Sorokin
sorokin@gft2.physik.hu-berlin.de
Institut für Physik
Humboldt-Universität zu Berlin
Invalidenstrasse 110
D-10115 Berlin
Germany

George Sparling
sparling@math.pitt.edu
Department of Mathematics
301 Thackeray Hall
University of Pittsburgh
Pittsburgh, PA 15260
USA

Daniel Sternheimer
sternheimer@u-bourgogne.fr
Department de Mathématique
Univesité de Bourgogne
BP 130
F-21004 Dijon
France

Aleksander Strasburger
aleksta@fuw.edu.pl
Dept. of Mathematical Methods
of Physics
University of Warsaw
ul. Hoża 74
00-682 Warsaw
Poland

Osamu Suzuki
Department of Mathematics
College of Humanities and
Sciences
Nihon University
Sakurajoshi 3-25-40, Setagaya-Ku
Tokyo 156
JAPAN

Barbara Szczerbińska
H. Niewodniczański Inst. of
Nuclear Physics
ul. Radzikowskiego 152
31-342 Cracow
Poland

Valeri Tolstoy
tolstoy@auua19.npi.msu.su
Institute for Nuclear Physics
Moscow State University
119899 Moscow
Russia

Nabila Touhami
touhami@elbahia.cerist.dz
Laboratoire de Physique
Theorique
Universite d'Es-Senia
Cite Emir Abdelkader 5 rue 8
Oran
Algeria

Andrzej Trautman
Andrzej.Trautman@fuw.edu.pl
Institute of Theoretical Physics
University of Warsaw
ul. Hoża 69
00-681 Warsaw
Poland

Włodzimierz Tulczyjew
Dipartimento di Matematica
Universita di Camerino
Piazza Cavour
I-62032 Camerino
Italy

Francisco Turrubiates
Departamento de Física
Centro de Investigación y
de Estudos Avanzados del I.P.N.
Mexico D.F. 07000
Mexico

Reidun Twarock
ptrt@rz.tu-clasthal.de
Arnold Sommerfeld Inst. für
Mathematische Physik
Technische Universität Clausthal
Leibnizstrasse 10
D-38678 Clausthal-Zellerfeld
Germany

Wacław Tybor
Dept. of Theoretical Physics
University of Łódź
ul. Pomorska 149/153
90-236 Łódź
Poland

Alexander Ushveridze
alexush@mvii.uni.lodz.pl
Dept. of Theoretical Physics
University of Łódź
ul. Pomorska 149/153
90-236 Łódź
Poland

Sergiu Vacaru
lises@cc.acad.md
Institute of Applied Physics
Academy of Siences of Moldova
Academy Str. 5
2028 Chisinanu
Moldova

Józef Werle
Institute of Theoretical Physics
University of Warsaw
ul. Hoża 69
00-681 Warsaw
Poland

Julius Wess
Julius.Wess@physik.uni-muenchen.de
Section Physik
Universität München
Theresienstraße 37
D-80333 Munich
Germany

Stanisław L. Woronowicz
Dept. of Mathematical Methods
of Physics
University of Warsaw
ul. Hoża 74
00-682 Warsaw
Poland

Stanisław Zakrzewski
Dept. of Mathematical Methods
of Physics
University of Warsaw
ul. Hoża 74
00-682 Warsaw
Poland

Andrei Zelnikov
zelnikov@phys.ualberta.ca
Deptartment of Physics
University of Alberta
Edmonton, T6G 2J1
Alberta, Canada

Author Index

A

Arodź, H., 293

B

Bakalarska, M., 303, 312
Baleanu, D., 413
Bażański, S. L., 421
Bender, C. M., 167
Borowiec, A., 431
Brihaye, Y., 177
Broda, B., 303
Brzeziński, T., 3
Budinich, P., 186

C

Caban, P., 199
Cahen, M., 437
Cannata, F., 209
Cavaglià, M., 442
Cisło, J., 219
Coquereaux, R., 9

D

Devchand, C., 317
Dobrev, V. K., 24
Dunajski, M., 449
Dütsch, M., 324
Dymnikova, I., 460

E

Elizade, E., 334

F

Fiore, G., 39
Fischer, J., 344
Flato, M., 49

Fredenhagen, K., 324
Frydryszak, A., 53

G

Giller, S., 226
Gorbar, E. V., 349

H

Heller, M., 234
Horzela, A., 242

J

Jorjadze, G., 472
Junker, G., 209

K

Kanatchikov, I. V., 356
Kapuścik, E., 242
Kauffman, L. H., 368
Kijowski, J., 382
Kleinhaus, B., 478
Kosiński, P., 67, 177
Kostadinov, B. S., 24
Kowalski, K., 249
Kulish, P. P., 75
Kunz, J., 478

L

Lukierski, J., 53

M

Madore, J., 39
Majid, S., 3
Marathe, K. B., 488

Marcinek, W., 86
Maślanka, P., 67
Mayer, D., 257
Meierovich, B. E., 498
Milczarski, P., 226
Minnaert, P., 53
Mozrzymas, M., 53

Sołtysek, B., 460
Sood, A., 478
Sorokin, D., 404
Sternheimer, D., 107
Stichel, P. C., 219

N

Niederle, J., 268
Nuyts, J., 317

T

Trautman, A., 518
Trost, J., 209
Twarock, R., 146
Tybor, W., 312

O

Olkhov, V., 276

U

Ushveridze, A., 257

P

Papaloucas, L. C., 249
Pawłowski, M., 394
Petrov, S. T., 24
Piechocki, W., 472
Podleś, P., 97
Popowicz, Z., 285

V

Vacaru, S. I., 528
Vrkoč, I., 344

R

Rembieliński, J., 199, 249
Rudolph, G., 382
Rudolph, M., 382

W

Walczak, Z., 257
Werle, J., 0
Wess, J., 155
Wirschins, M., 478

S

Sasin, W., 234
Schieber, G. E., 9
Shil'nov, Y. I., 334

Z

Zelnikov, A., 538